T0327721

**Design of Three-Phase AC Power
Electronics Converters**

Design of Three-Phase AC Power Electronics Converters

Fei "Fred" Wang
University of Tennessee, Knoxville, TN, USA

Zheyu Zhang
Rensselaer Polytechnic Institute, Troy, NY, USA

Ruirui Chen
University of Tennessee, Knoxville, TN, USA

IEEE PRESS

WILEY

Published by John Wiley & Sons, Inc., Hoboken, New Jersey.
Published simultaneously in Canada.

For general information on our other products and services or for technical support, please contact our Customer Care Department within the United States at (800) 762-2974, outside the United States at (317) 572-3993 or fax (317) 572-4002.

Wiley also publishes its books in a variety of electronic formats. Some content that appears in print may not be available in electronic formats. For more information about Wiley products, visit our web site at www.wiley.com.

Library of Congress Cataloging-in-Publication Data applied for

[ISBN: 9781119794233,ePDF: 9781119794240, epub: 9781119794257, oBook: 9781119794264]

Cover Image: Wiley
Cover Design: Courtesy of Fei Wang, Zheyu Zhang, Ruirui Chen

Set in 9.5/12.5pt STIXTwoText by Straive, Pondicherry, India

Contents

About the Authors *xiii*
Preface *xv*
Acknowledgments *xvii*

1   Introduction *1*
   1.1   Basics of Three-Phase AC Converters *1*
      1.1.1   Basic Applications *2*
      1.1.2   Basic Topologies *10*
      1.1.3   Composition of Three-Phase AC Converters *16*
   1.2   Basics of Three-Phase AC Converter Design *20*
      1.2.1   Essence of the Design and Design Tasks *20*
      1.2.2   Design Procedure, Strategy, and Philosophy *23*
   1.3   Goal and Organization of This Book *26*
      References *29*

Part I   Components *31*

2   Power Semiconductor Devices *33*
   2.1   Introduction *33*
   2.2   Static Characteristics *35*
      2.2.1   Output Characteristics *35*
      2.2.2   On-state Characteristics *38*
      2.2.3   Transfer Characteristics of Active Power Switch *41*
      2.2.4   Leakage Current and Breakdown Voltage *43*
      2.2.5   Junction Capacitance *46*
      2.2.6   Gate Charge *49*
   2.3   Switching Characteristics *50*
      2.3.1   Model *50*
      2.3.2   Method *54*
   2.4   Thermal Characteristics *57*
      2.4.1   Model *58*
      2.4.2   Method *58*
   2.5   Other Attributes *60*
      2.5.1   SOA *60*
      2.5.2   Reliability Characteristics *60*

2.5.3   Mechanical Characteristics   *63*
2.5.4   Nonstandard Characteristics   *66*
2.6   Scalability (Parallel/Series)   *68*
2.7   Relevance to Converter Design   *70*
2.8   Summary   *72*
References   *73*

3   Capacitors   *75*
3.1   Introduction   *75*
3.2   Capacitor Types and Technologies   *75*
3.2.1   Ceramic Capacitors   *76*
3.2.2   Paper   *76*
3.2.3   Mica   *76*
3.2.4   Poly-Film   *77*
3.2.5   Aluminum Electrolytic Capacitors (AECs)   *77*
3.2.6   Tantalum Electrolytic Capacitors (TECs)   *78*
3.2.7   Capacitor Technologies Comparison   *78*
3.2.8   Emerging Capacitor Technologies   *78*
3.3   Capacitor Selection in a Converter Design   *82*
3.4   Capacitor Characteristics and Models   *84*
3.4.1   Capacitor Equivalent Circuit Model and Capacitance   *84*
3.4.2   Voltage and Current Capability Models   *88*
3.4.3   Loss and Thermal Models   *93*
3.4.4   Lifetime Model   *96*
3.5   Capacitor Bank (Parallel/Series)   *98*
3.5.1   Capacitor Bank Configuration and Voltage Balancing   *98*
3.5.2   Capacitor Bank Layout for Parasitic Inductance Reduction   *99*
3.6   Relevance to Converter Design   *100*
3.6.1   Capacitor Scaling   *101*
3.6.2   AC Capacitor Classification   *101*
3.7   Summary   *102*
References   *102*

4   Magnetics   *105*
4.1   Introduction   *105*
4.2   Magnetic Core Materials and Construction   *105*
4.2.1   Soft Magnetic Alloy-Based Laminated, Tape Wound and Cut Cores   *105*
4.2.2   Powder Cores   *107*
4.2.3   Ferrite Cores   *108*
4.3   Inductor Design in a Converter   *108*
4.4   Inductor Characteristics and Models   *110*
4.4.1   Inductance and Permeability   *110*
4.4.2   Flux Density and Core Saturation   *113*
4.4.3   Fill Factor   *113*
4.4.4   Current Density and Core Window Area Product A_p *113*
4.4.5   Core Loss   *114*
4.4.6   Winding Loss   *120*
4.4.7   Temperature Rise   *121*

4.4.8 Leakage Inductance *123*
4.4.9 Fringing Effect of Gapped Cores *124*
4.5 Relevance to Converter Design *127*
4.5.1 Capacitor Winding Capacitance *127*
4.6 Summary *128*
References *129*

Part II Subsystems Design *131*

5 Passive Rectifiers *133*
5.1 Introduction *133*
5.2 Passive Rectifier Design Problem Formulation *135*
5.2.1 Passive Rectifier Design Variables *136*
5.2.2 Passive Rectifier Design Constraints *136*
5.2.3 Passive Rectifier Design Conditions *138*
5.2.4 Passive Rectifier Design Objectives and Design Problem Formulation *139*
5.3 Passive Rectifier Models *140*
5.3.1 AC Input Harmonic Current *141*
5.3.2 Minimum and Maximum DC Voltages Under Normal Operating Conditions *148*
5.3.3 Ride-Through or Holdup Time Without Input Power *149*
5.3.4 DC-Link Stability *149*
5.3.5 Device-Related Constraints – Inrush *149*
5.3.6 Inductor-Related Constraints and Design *154*
5.3.7 Capacitor-Related Constraints and Selection *157*
5.4 Passive Rectifier Design Optimization *157*
5.5 Interface to Other Subsystem Designs *161*
5.5.1 General Classifications *161*
5.5.2 Discussion *162*
5.6 Summary *164*
References *165*

6 Load-side Inverters *167*
6.1 Introduction *167*
6.2 Load-side Inverter Design Problem Formulation *167*
6.2.1 Load-side Inverter Design Variables *168*
6.2.2 Load-side Inverter Design Constraints *169*
6.2.3 Load-side Inverter Design Conditions *170*
6.2.4 Load-side Inverter Design Objectives and Design Problem Formulation *172*
6.3 Load-side Inverter Models *173*
6.3.1 AC Load Harmonic Current *174*
6.3.2 Inverter Power Loss *188*
6.3.3 Control Performance *198*
6.3.4 Device Maximum Junction Temperature – Maximum Thermal Impedance Requirement *200*
6.3.5 Device Switching Overvoltage *204*
6.3.6 Decoupling Capacitor *218*
6.3.7 Decoupling Inductor *222*

6.4   Load-side Inverter Design Optimization *230*

6.5   Load-side Inverter Interfaces to Other Subsystem Designs *236*

    6.5.1   General Classifications *236*

    6.5.2   Discussion *238*

6.6   Summary *241*

    References *241*

7   Active Rectifiers and Source-side Inverters *245*

7.1   Introduction *245*

7.2   Active Rectifier and Source-side Inverter Design Problem Formulation *245*

    7.2.1   Active Rectifier and Source-side Inverter Design Variables *246*

    7.2.2   Active Rectifier or Source-side Inverter Design Constraints *247*

    7.2.3   Active Rectifier and Load-side Inverter Design Conditions *248*

    7.2.4   Active Rectifier and Source-side Inverter Design Objectives and Design Problem Formulation *249*

7.3   Active Rectifier and Source-side Inverter Models *251*

    7.3.1   AC Source Harmonic Current *252*

    7.3.2   Control Performance *264*

    7.3.3   DC-Link Stability *266*

    7.3.4   Reliability *267*

7.4   Active Rectifier and Source-side Inverter Design Optimization *280*

7.5   Impact of Topology *280*

    7.5.1   Circuit Modeling for Different Topologies *280*

    7.5.2   Topology Impact on Device Models *286*

7.6   Active Rectifier and Source-side Inverter Interfaces to Other Subsystem Designs *300*

7.7   Summary *300*

    References *302*

8   EMI Filters *305*

8.1   Introduction *305*

8.2   EMI Filter Design Basics *306*

    8.2.1   EMI/EMC Standards *306*

    8.2.2   Definition of CM and DM Noise *306*

    8.2.3   EMI Noise Measurement *312*

    8.2.4   Basic EMI Filter Design Method *316*

    8.2.5   EMI Filter Topology *322*

8.3   EMI Filter Design Problem Formulation *324*

    8.3.1   EMI Filter Design Variables *324*

    8.3.2   EMI Filter Design Constraints *325*

    8.3.3   EMI Filter Design Conditions *325*

    8.3.4   EMI Filter Design Objectives and Design Problem Formulation *326*

8.4   EMI Filter Models *326*

    8.4.1   EMI Noise Source Model *327*

    8.4.2   EMI Propagation Path Impedance Model *332*

    8.4.3   EMI Filter Corner Frequency vs. Switching Frequency *342*

8.5   EMI Filter Design Optimization and Some Practical Considerations *343*

    8.5.1   Grounding Effect *344*

    8.5.2   EMI Filter Coupling *346*

8.5.3   Mixed-Mode Noise   *350*

8.5.4   EMI Noise Mode Transformation Due to Propagation Path Unbalance   *353*

8.6   EMI Noise and Filter Reduction Techniques   *353*

8.6.1   Switching Frequency   *354*

8.6.2   Modulation Scheme   *354*

8.6.3   EMI Filter Topology   *354*

8.6.4   Active/Hybrid Filter   *358*

8.6.5   Paralleled Converters Interleaving Angle Optimization   *363*

8.6.6   EMI Filter Integration   *365*

8.7   Interface to Other Subsystem Designs   *365*

8.7.1   Voltage Distribution   *365*

8.7.2   Current Distribution   *365*

8.7.3   Input/Output Terminals   *366*

8.7.4   Load-side *dv/dt*   *366*

8.8   Summary   *366*

References   *367*

9   Thermal Management System   *369*

9.1   Introduction   *369*

9.2   Cooling Technology Overview   *369*

9.2.1   Basic Conventional Cooling Methods for Power Electronics   *370*

9.2.2   Advanced Cooling Techniques   *372*

9.2.3   Comparison of Cooling Technologies   *375*

9.2.4   Heatsinks and Other Components   *377*

9.3   Thermal Management System Design Problem Formulation   *384*

9.3.1   Thermal Management System Design Variables   *385*

9.3.2   Thermal Management System Design Constraints   *385*

9.3.3   Thermal Management System Design Conditions   *386*

9.3.4   Thermal Management System Design Objectives and Design Problem Formulation   *386*

9.4   Thermal Management System Models   *388*

9.4.1   Thermal Impedance   *388*

9.4.2   Heatsink Dimensions   *393*

9.5   Thermal Management System Design Optimization   *393*

9.5.1   Design Optimization Example   *394*

9.5.2   Design Verification   *397*

9.6   Thermal Management System Interface to Other Subsystems   *397*

9.6.1   General Classification   *397*

9.6.2   Discussion   *399*

9.7   Other Cooling Considerations   *400*

9.7.1   Force-Liquid Convection Cooling   *400*

9.7.2   Cooling for Passives   *403*

9.8   Summary   *407*

References   *408*

10   Control and Auxiliaries   *411*

10.1   Introduction   *411*

10.2   Control Architecture   *411*

10.2.1   System Control Layer   *411*

10.2.2   Application Control Layer   *412*

10.2.3   Converter Control Layer   *413*

10.2.4   Switching Control Layer   *413*

10.2.5   Hardware Control Layer   *414*

10.3   Control Hardware Selection and Design   *414*

10.4   Isolation   *415*

10.4.1   Signal Isolator   *415*

10.4.2   Isolated Power Supply   *417*

10.4.3   Discussion on Isolation Strategies for Low-Power Converter Design   *418*

10.5   Gate Driver   *418*

10.5.1   Gate Driver Fundamentals   *419*

10.5.2   Gate Driver-Related Key Device Characteristics   *420*

10.5.3   Gate Driver Design   *422*

10.5.4   Bootstrap Gate Driver   *429*

10.6   Sensors and Measurements   *430*

10.6.1   Voltage Sensors   *431*

10.6.2   Current Sensors   *431*

10.6.3   Temperature Sensors   *432*

10.6.4   High-Voltage Sensors   *432*

10.6.5   Sensing Circuit Design Considerations for High-Frequency WBG
Converters   *434*

10.7   Protection   *445*

10.7.1   Device-Level Protection   *445*

10.7.2   Converter-Level Protection   *450*

10.8   Printed Circuit Boards   *452*

10.9   Deadtime Setting and Compensation   *455*

10.9.1   Deadtime Setting   *456*

10.9.2   Deadtime Compensation   *461*

10.10   Interface to Other Subsystems   *466*

10.11   Summary   *467*

References   *468*

11   Mechanical System   *471*

11.1   Introduction   *471*

11.2   Mechanical System Design Problem Formulation   *475*

11.2.1   Mechanical System Design Variables   *475*

11.2.2   Mechanical System Design Constraints   *476*

11.2.3   Mechanical System Design Conditions   *477*

11.2.4   Mechanical System Design Objectives and Design Problem Formulation   *478*

11.3   Busbar Design   *482*

11.3.1   Busbar Design Problem Formulation   *482*

11.3.2   Busbar Design Procedures and Considerations   *483*

11.3.3   Busbar Layout Design Example for a Three-Level ANPC Converter   *491*

11.4   Mechanical System Interface to Other Subsystems   *502*

11.4.1   General Classifications   *502*

11.4.2   Discussion   *503*

11.5   Summary   *504*

References   *505*

12 Application Considerations *507*

12.1 Introduction *507*

12.2 Motor Drive Applications *507*

   12.2.1 Harmonics in Motors *509*

   12.2.2 CM Voltage in Motors *513*

   12.2.3 Switching *dv/dt* Impact on Motors *523*

   12.2.4 Motor Drive Grounding *544*

12.3 Grid Applications *547*

   12.3.1 Baseline Design *548*

   12.3.2 Impact of High- and Low-Voltage Ride-Through *556*

   12.3.3 Impact of Grid Faults *559*

   12.3.4 Impact of Frequency Ride-Through and Grid Voltage Angle Change *562*

   12.3.5 Impact of Lightning Surge *563*

   12.3.6 Impact of Grid Requirements on Converter Hardware Design *566*

   12.3.7 Other Factors in Grid Applications *571*

12.4 Summary *577*

   References *578*

Part III Design Optimization *581*

13 Design Optimization *583*

13.1 Introduction *583*

13.2 Design Optimization Concept and Procedure *583*

   13.2.1 Concept and Mathematical Formulation *583*

   13.2.2 Optimization-Based Design Procedure *584*

   13.2.3 Multi-objective Optimization and Pareto Front *585*

   13.2.4 Mathematical Properties of Power Converter Design Optimization Problems *586*

13.3 Optimization Algorithms *587*

   13.3.1 Optimization Algorithm Classification *587*

   13.3.2 Optimization Algorithm Selection *589*

13.4 Partitioned Optimizers vs. Single Optimizer for Converter Design *590*

   13.4.1 Partitioned Optimizers *590*

   13.4.2 Single Optimizer *594*

13.5 Design Tool Development *596*

   13.5.1 Design Tool and Its Desired Features *596*

   13.5.2 A GA-Based Optimization Tool for General-Purpose Industrial Motor Drive *597*

   13.5.3 A Design Tool for High-Density Inverters *616*

   13.5.4 Partition-Based Design vs. Whole Converter Design *635*

13.6 Virtual Prototyping *638*

13.7 Summary *642*

   References *643*

Index *647*

About the Authors

Fei "Fred" Wang has been a Professor and Condra Chair of Excellence in Power Electronics at the Min. H. Kao Department of Electrical Engineering and Computer Science, University of Tennessee (UTK) since 2009. He is a Co-founder and the Technical Director of the US NSF-DOE Engineering Research Center for Ultra-wide-area Resilient Electric Energy Transmission Networks (CURENT) at UTK. He holds a joint appointment with Oak Ridge National Laboratory. Dr. Wang received his BS degree from Xi'an Jiaotong University, Xi'an, China, and his MS and PhD degrees from the University of Southern California, Los Angeles, in 1982, 1985, and 1990, respectively, all in electrical engineering. Dr. Wang was a Research Scientist at Electric Power Lab, University of Southern California, from 1990 to 1992. He joined GE Power Systems Engineering Department, Schenectady, NY, as an Application Engineer in 1992. From 1994 to 2000, he was a Senior Product Development Engineer with GE Drive Systems, Salem, VA. During 2000–2001, he was Manager of Electronic and Photonic Systems Technology Lab, GE Global Research Center, Schenectady, NY, and Shanghai, China. In 2001, he joined the Center for Power Electronics Systems (CPES) at Virginia Tech, Blacksburg, VA, as a Research Associate Professor and became an Associate Professor in 2004. From 2003 to 2009, he also served as CPES Technical Director. Dr. Wang has published over 600 journal and conference papers, authored one book and seven book chapters, and holds 20 US patents. His achievements have resulted in the 2018 IEEE IAS Gerald Kliman Innovation Award, 12 IEEE prize paper awards, Dushman Award – GE's highest award for technical team contributions, and four University of Tennessee Engineering Faculty Research Achievements Awards. He is a fellow of IEEE and a fellow of the US National Academy of Inventors. His research mainly focuses on wide bandgap device-based power electronics and power electronics applications in transportation, motor drives, renewable energy systems, and electric power grids.

Zheyu Zhang received the BS and MS degrees from Huazhong University of Science and Technology, Wuhan, China, and the PhD degree from The University of Tennessee, Knoxville, TN, in 2008, 2011, and 2015, respectively, all in electrical engineering.

Dr. Zheyu Zhang is an Assistant Professor at Rensselaer Polytechnic Institute. He was the Warren H. Owen-Duke Energy Assistant Professor of Engineering at Clemson University from 2019 to 2023. He was a Research Assistant Professor in the Department of Electrical Engineering and Computer Science at the University of Tennessee, Knoxville, from 2015 to 2018. Afterward, he joined General Electric Research as the Lead Power Electronics Engineer at Niskayuna, NY, USA, from 2018 to 2019. He has published 100+ papers in the most prestigious journals and conference proceedings, filed 10+ patent applications, authored one book and one book chapter, and presented 10 IEEE tutorial seminars and webinars. His research interests include wide bandgap-based power electronics characterization and applications. Dr. Zhang is currently the Standard Vice-Chair of IEEE IAS Power Electronics Devices and Components Committee and Associate Editor for *IEEE*

Transactions on Power Electronics and *IEEE Transactions on Industry Applications*. He was the recipient of three prize paper awards from the IEEE Industry Applications Society and IEEE Power Electronics Society, 2021 IEEE IAS Andrew W. Smith Outstanding Young Member Achievement Award, and 2022 NASA Early Career Faculty Award. He is a senior member of IEEE.

Ruirui Chen received a BS degree from Huazhong University of Science and Technology, Wuhan, China; an MS degree from Zhejiang University, Hangzhou, China; and a PhD degree from the University of Tennessee, Knoxville, USA, in 2010, 2013, and 2020, respectively, all in electrical engineering.

Dr. Ruirui Chen is a Research Assistant Professor with the Department of Electrical Engineering and Computer Science, the University of Tennessee, Knoxville, USA. From 2013 to 2015, he was an electrical engineer at FSP-Powerland Technology Inc., China. He has published over 50 journal and conference papers, authored one book chapter, and received one IEEE prize paper award. His research interests include wide bandgap devices and applications, medium-voltage power electronics, cryogenic power electronics, EMI, and power electronics for electrified transportation and grid applications.

Preface

My first involvement in three-phase AC converter design was on the development team for medium-voltage, MW-class motor drives in GE Drive Systems during 1990s. The products I helped to design included thyristor-based cycloconverters, IGBT-based three-level neutral-point-clamped (NPC) pulse-width modulation (PWM) inverters with diode front-end rectifiers, and IGCT-based three-level NPC back-to-back AC/DC/AC converters. The development of each of these products was a multi-year effort with dozens to hundreds of world-class GE engineers through many design, simulation, and testing iterations.

When I joined Virginia Tech in the early 2000s, one of the first projects I helped lead was the design and cost optimization of general-purpose industrial motor drives sponsored by Schneider-Toshiba Inverter Europe. With the objective to minimize the motor drive cost, as well as to increase the designer's efficiency and reduce time, an automated design tool was built based on analytical models for the front-end passive rectifier, the inverter power stage and thermal management system, and the electromagnetic interference (EMI) filter. Later, the design and optimization method for three-phase AC converters was used and further developed in several projects, especially, in projects sponsored by Boeing on SiC-based high-density inverters and rectifiers. In the meantime, I felt the need to train students systematically on three-phase AC converter design and started a graduate-level course ECE5984 "Application and Design of Multi-phase PWM Converters." After I moved to the University of Tennessee, Knoxville (UTK) in 2009, I have continued to teach a similar course, ECE683 "Drive System Control and Converter Design," while continuing to conduct research on high-power, high-density three-phase AC converters for electrified transportation, grid, and industry applications. As a result of the research and course work at UTK, another design automation tool was developed focusing on optimizing wide bandgap (WBG) device-based high-density three-phase AC converters.

My co-author, Prof. Zheyu Zhang, also used similar materials in the ECE4930/6930 "Fundamentals of Power Electronics" and ECE8930 "Advanced Power Electronics" courses at Clemson University. The materials we use are largely based on our research project reports and papers, as well as published papers by other researchers. Three-phase AC converters are not new, and there are many books on their circuit topologies, operating principles, and control. However, in our teaching and research, we see a clear value and need for a book focusing on three-phase AC converter design, with sufficient coverage of analyses, models, and methods related to design. Given that the converter design is essentially an optimization process, this book presents the design of three-phase AC converters and their subsystems as clearly formulated optimization problems. The book has also incorporated recent developments in three-phase AC converter design, especially WBG device enabled technologies and applications, and related new requirements. The intent is that the book can be used by professors, graduate students, and practicing engineers working in the area of three-phase AC converters, as a classroom textbook or as a reference.

In addition to Chapter 1 "Introduction," this book consists of three parts. Part I, from Chapters 2 to 4, is on components, including power semiconductor devices, capacitors, and magnetics. Part II, from Chapters 5 to 12, is on the design of subsystems, including passive and active rectifiers, inverters, EMI filters, thermal management systems, control and auxiliaries, mechanical systems, and application considerations. Part III comprises Chapter 13 "Design Optimization." If used as a textbook, it is probably challenging to cover the whole book in a one-semester course. In my own ECE683 course at UTK, I have focused on Chapter 5 "Passive Rectifiers," Chapter 6 "Load-side Inverter," Chapter 8 "EMI Filters," Chapter 9 "Thermal Management System," Chapter 10 "Control and Auxiliaries," and part of Chapter 12 "Application Considerations" on motor drives, and part of Chapter 13. Other chapters can be used as reference and further study materials. Similar contents were covered by Prof. Zhang's graduate-level ECE8930 course at Clemson. In addition, Professor Zhang also delivered part of Chapter 2 "Power Semiconductor Devices" and Chapter 4 "Magnetics" in his undergraduate and graduate combined 4930/6930 course at Clemson.

The AC three-phase converter design is a complex multidisciplinary task. In this book, we have tried to cover the subject comprehensively but surely have missed some important topics due to our knowledge limitations. Still we hope the readers will find the book useful. We sincerely welcome your feedback.

Fei "Fred" Wang
Knoxville, Tennessee, USA
August 2023

Acknowledgments

In preparing this book, the authors received generous help from many former and current colleagues and students: Dr. Puqi Ning for help and discussion on thermal management system design, Dr. Qian Liu on decoupling capacitor design, Dr. Ren Ren on design optimization and design tool, Dr. Zhou Dong on harmonics analysis and design tool, Dr. Haiguo Li on mechanical system design and grid applications, Dr. Bo Liu on topology impact and sensing circuit, Dr. Wen Zhang on switching transients and motor filters, Dr. Shimul Dam on design optimization examples, Mr. Dingrui Li on grid applications, Dr. Le Kong and Mr. Liang Qiao on stability, and Mr. Bill Giewont and Mr. Bob Martin on mechanical system design. Mr. Jiaohao Niu and Dr. Venkata Itte prepared many figures, and their help is greatly appreciated.

This book is largely a result of the authors' research, development, and teaching activities related to three-phase AC power electronics converters over the last 30 years. The authors have been very fortunate to have the opportunity to work with and learn from many sponsors, mentors, collaborators, colleagues, and students on these activities. In addition to individuals mentioned earlier, the authors would like to acknowledge the following former and current colleagues and students from GE, Virginia Tech, and University of Tennessee, Knoxville: Dr. Jim Lyons, Mr. Paul Espelage, Mr. Lee Tupper, Dr. Dushan Boroyevich, Dr. Fred Lee, Dr. Daan van Wyk, Dr. Rolando Burgos, Dr. Kai Ngo, Dr. G.Q. Lu, Dr. Zhenxian Liang, Dr. Wei Shen, Dr. Rixin Lai, Dr. Di Zhang, Dr. Shuo Wang, Dr. Ruxi Wang, Dr. Fang Luo, Dr. Gang Chen, Dr. Zheng Chen, Dr. Xuning Zhang, Mr. Yoann Millet, Dr. Leon Tolbert, Dr. Daniel Costinett, Dr. Ben Blalock, Dr. Kevin Bai, Dr. Dong Jiang, Dr. Ben Guo, Dr. Jing Xue, Dr. Fan Xu, Dr. Zhuxian Xu, Dr. Weimin Zhang, Dr. Shuoting Zhang, Dr. Yalong Li, Dr. Wenchao Cao, Dr. Yiwei Ma, Dr. Handong Gui, Dr. Zhe Yang, Dr. Xingxuan Huang, Dr. Shiqi Ji, Dr. Cheng Nie, Dr. Li Zhang, Mr. Craig Timms, Mr. Jacob Dyer, Mr. Jimmy Palmer, Dr. Edward Jones, Ms. Paige Williford, and Mr. Zihan Gao.

The authors would like to thank the following sponsors: Boeing, Schneider-Toshiba Inverter Europe, SAFRAN, Thales, Rolls-Royce, ABB, GE, II-VI Foundation, Keysight, Volkswagen, NSF, DOE, ARPA-E, Oak Ridge National Laboratory, NASA, PowerAmerica, Office of Naval Research, and Army Research Lab. This book made use of Engineering Research Center Shared Facilities supported by the Engineering Research Center Program of the National Science Foundation and DOE under NSF Award Number EEC-1041877 and the Center for Ultra-wide-area Resilient Electric Energy Transmission Networks (CURENT) Industry Partnership Program.

The authors would like to thank Wiley staff, especially, Brett Kurzman, Kimberly Monroe-Hill, and Infanta Ravikumar. Without their support, this book would not have been possible.

1

Introduction

This book focuses on the design of power electronics converters for three-phase AC applications. Out of many different types of power electronics converters, three-phase AC converters are among the most important and most popular, especially for medium- and high-power (e.g. kilowatts and higher) applications.

While three-phase AC converters are not new, they have gone through significant new developments in recent years as a result of new applications, new advances in power electronics technologies, and corresponding new requirements. Applications, such as electrified transportation and renewable energy systems, demand converters with ever-increasing power density and efficiency, at the same time achieving low cost and high reliability. Technologies like wide bandgap (WBG) power semiconductor devices offer tremendous performance improvement opportunities but also pose many new challenges for the converter design. More than ever, there is a need for converter design that can accurately resemble the actual converter, with only the necessary and scientifically determined design margins. The design should also be able to consider the interaction and integration of various functions and subsystems in a converter. In addition, the design method should be very efficient in finding desired solutions for a given application and its associated requirements.

Considerable works have been conducted in recent years by power electronics researchers and engineers, including the authors, on advanced design methodology, optimization, and automation for three-phase AC converters. These works are generally scattered in research papers, technical reports, and application notes. This book intends to provide a systematic treatment of the topic, covering both new and conventional materials.

1.1 Basics of Three-Phase AC Converters

Corresponding to three basic types of AC power converters, there are three basic types of three-phase AC power converters as illustrated in Figure 1.1. They are three-phase AC-to-DC (AC/DC) rectifiers, DC-to-AC (DC/AC) three-phase inverters, and three-phase AC to three-phase AC (AC/AC) converters.

- Three-phase rectifiers: three-phase rectifiers convert three-phase AC to DC. Depending on device types, they can be passive (diode-based), phase-controlled (thyristor-based), or active (active switch-based).
- Three-phase inverters: three-phase inverters convert DC to three-phase AC. Inverters generally require active switching devices.

Design of Three-phase AC Power Electronics Converters, First Edition. Fei "Fred" Wang, Zheyu Zhang, and Ruirui Chen.
© 2024 The Institute of Electrical and Electronics Engineers, Inc. Published 2024 by John Wiley & Sons, Inc.

Figure 1.1 Basic types of three-phase AC converters: (a) three-phase rectifier, (b) three-phase inverter, and (c) three-phase AC/AC converter. *Source:* General Electric.

- Three-phase AC/AC converters: three-phase AC/AC converters directly convert one three-phase AC to another, either only the voltage magnitude or both magnitude and frequency. The former can also be called AC switch, and the latter can be called frequency changer.

Each of the converter types can be further clarified. For example, a converter can have unidirectional power flow or bidirectional power flow. Diode-based rectifiers should be unidirectional as a result of the diode's single-quadrant characteristics. Note that although there exist direct AC/AC conversion topologies like cycloconverters or matrix converters, most three-phase AC/AC converters today are realized by cascading AC/DC and DC/AC converters.

1.1.1 Basic Applications

1.1.1.1 Motor Drives

Three-phase AC converters are widely used. The best-known applications are motor drives for three-phase AC induction or synchronous machines (including permanent magnet or PM synchronous machines). Figure 1.2 shows a typical configuration for a three-phase AC-fed motor drive. Note that some motor drives can be DC-fed, where only a three-phase inverter will be needed. Also note that in general an input filter is required for the motor drive for power quality enhancement (e.g. harmonics attenuation) and/or electromagnetic interference (EMI) mitigation. Motor drives can be found in almost all industries. Some examples are listed below [1].

- pulp and paper industry: paper machines, dryer fans, boiler fans and pumps, chippers, refiners, and conveyors;
- metal industry: rolling mill stands, reels, and winders;
- material handling industry: cranes and conveyors;
- mining industry: excavators, conveyors, and grinding mills;
- cement industry: kiln drives and fans and conveyors;
- oil and chemical industry: pipeline compressors and pumps, oil well drilling equipment (draw works, top drives, mud pumps, and cement pumps), water and waste water pumps, and rubber and plastics equipment (extruders, inlet pumps, pelletizers, and mixers);
- transportation industry: locomotive traction, ship propulsion, aircraft generators, off-highway vehicles, elevators, and escalators;
- automotive industry: electric vehicles, dynamometers, and wind tunnels;

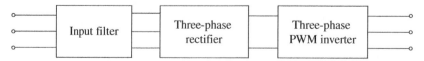

Figure 1.2 A typical motor drive configuration (PWM stands for pulse-width modulation or pulse-width modulated).

- appliance industry: washing machines, heating, ventilation, and air conditioning (HVAC) and robotics;
- electric utility: turbine starters, boiler and cooling tower fans and pumps, wind turbines, photo-voltaic (PV) stations, energy storage systems, and micro-turbines;
- IT industry: cooling

Figure 1.3 shows two traditional applications for motor drives: paper and steel mills. These applications generally involve several motor drives working in a coordinated manner to make paper or roll steel, and therefore they are system drives and required to have high performance in terms of control. Figure 1.4 shows pictures of different types of motor drive products. The integrated motor drive refers to combining the drive converter and motor within the same assembly.

1.1.1.2 High Power Supply

In addition to motor drives, three-phase AC converters are also used in many other applications. Figure 1.5 shows a typical medium- or high-power (kilowatts and higher) power supply configuration using a three-phase AC converter as a rectifier driving a DC load, in this case, a DC/DC converter. This configuration corresponds to a three-phase power factor correction (PFC) converter in many power supplies, including battery chargers, data center power supplies, and large DC current sources for industrial processing (e.g. metal plating). Note that the filter between the rectifier and

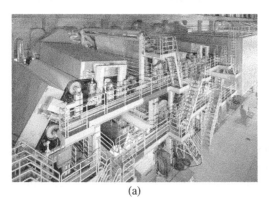 

(a)   (b)

Figure 1.3 Application examples for traditional high-performance motor drives: (a) paper mill and (b) steel mill. *Source:* Jose Luis Stephens/Adobe Stock; rukhmalev/Adobe Stock.

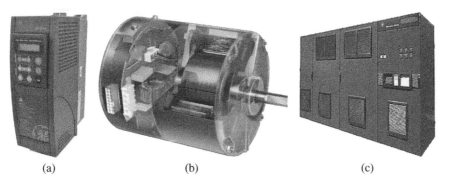

(a)   (b)   (c)

Figure 1.4 Example motor drives: (a) low power (kW level) drive. *Source:* General Electric. (b) Integrated motor drive. *Source:* Regal Rexnord Corporation. (c) high power (MW level) drive. *Source:* General Electric.

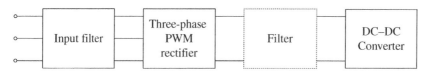

Figure 1.5 A typical medium- or high-power power supply configuration.

DC/DC converter is often for EMI attenuation, and the dashed box indicates that the filter is optional.

Although most power supplies are for DC loads, there are power supplies for AC loads, e.g. uninterruptible power supplies (UPS). In the case of a three-phase UPS, a DC to three-phase AC inverter is needed.

1.1.1.3 Emerging Applications

There have been tremendous growths for three-phase AC converters in recent years. Some are driven by more economical and higher performance power electronics converters, and by demand for better energy efficiency in conventional applications, such as in the case of HVAC, where more motors are driven by motor drives for improved efficiency. More growths are driven by new and emerging applications, including renewable energy systems, electrified transportation (electric vehicles (EVs), electrified trains, all-electric ships, more-electric aircraft (MEA), and electrified aircraft propulsion (EAP)), EV charging stations and other energy storage systems (e.g. UPS), and ever-growing data centers.

No matter what these new applications are, due to their nature of either the DC/AC or the AC/DC conversion, their operating principles are in essence similar to those of the basic applications of motor drives, DC power supplies, or AC power supplies. However, the converter design for each application can be different due to different application requirements, as will be discussed later in this book.

1.1.1.3.1 Electrified Transportation Figure 1.6 shows a typical powertrain electric power system architecture for EV. The three AC converters in the figure: the front-end AC/DC rectifier for the battery charger and the DC/AC inverters for the traction electric machine and air-conditioning compressor motor are power supply and motor drives, respectively. Similar observations can be

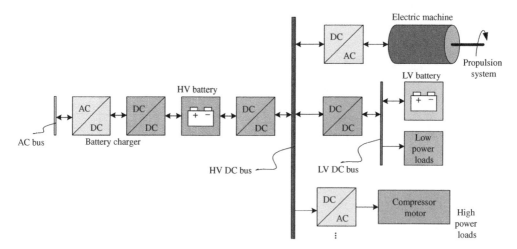

Figure 1.6 EV powertrain power system architecture. *Source:* Adapted from [2].

made for other electrified transportation systems, such as MEA, electrified trains, and all-electric ships, although the larger systems tend to involve more complex applications. Figure 1.7 shows an example of a state-of-the-art shipboard electric power system based on zonal distribution with pod-ded propulsors. The various buses of the system are all of AC three phases. Three-phase AC power converters are an essential part of the system, converting the three-phase AC power produced by generators at fixed voltage and frequency (e.g. 13.8 kV and 60 Hz) to three-phase AC power at various voltage levels and frequencies, and DC power at various voltage levels. These various AC and DC powers are needed for supplying and controlling propulsion motors, ship service loads, mission loads, and distribution buses.

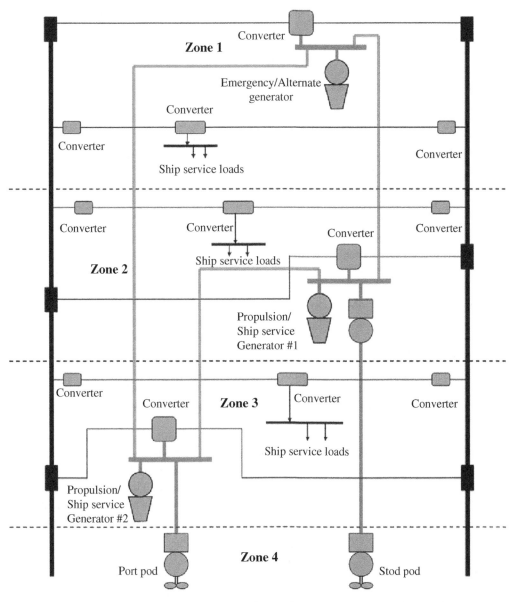

Figure 1.7 Shipboard electric power system example: zonal distribution with podded propulsors. *Source:* Adapted from [2, 3].

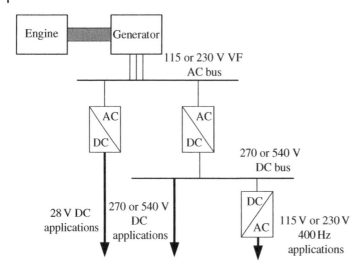

Figure 1.8 A conceptual MEA electrical system. *Source:* From [2]/with permission of John Wiley & Sons.

MEA powers almost all loads in the aircraft with electric power, except the propulsion. On the other hand, aircraft with electrified propulsion are getting increased interests due to economic and environmental considerations. Figure 1.8 shows the architecture of a conceptual commercial MEA electrical system, which covers the environmental control system (EMS), entertainment system, actuation, etc. Three-phase power conversion equipment includes the AC/DC rectifier from 115 or 230 V variable frequency AC bus to 270 or 540 V DC bus, the AC/DC rectifier from AC bus to 28 V DC bus, the DC/AC inverter from DC bus to AC 400 Hz bus, and the DC/AC inverter from 270 or 540 DC bus to drive motors. The power ratings of the three-phase converters in a MEA are typically around 100 kW or less. Figure 1.9 shows several example electrical architectures for future EAP electrical systems, which will require three-phase AC/DC rectifiers and DC/AC motor drives ranging from hundreds of kilowatts to tens of megawatts.

Figure 1.10 is a typical configuration of electrified railway traction system. In this configuration, the single-phase AC catenary is connected to primary winding of the low-frequency transformer (LFT) and grounded through the rail. The secondary winding of LFT is connected to a single-phase AC/DC rectifier to provide the DC voltage for the motor drive system, which is comprised of a three-phase inverter and a three-phase motor. Note that a two-level three-phase voltage source inverter is illustrated in Figure 1.10 as the motor drive, although other three-phase topologies have also been used in practice.

With the rapid growth of EV, one important related power electronics application is the charging station. The battery DC voltage for EV normally ranges in the hundreds of volts. The input AC voltage can be single-phase or two-phase residential voltage, or three-phase low voltage (<1 kV) or medium voltage (up to tens of kV) used in commercial fast chargers. For medium-power Level 3 (60 kW) or higher power off-board fast charger, a three-phase rectifier plus a DC/DC converter is often used. With many fast chargers within a future charging station, it can be expected that even MW-level, medium-voltage three-phase rectifiers will be needed.

1.1.1.3.2 Renewable Energy Systems Renewable energy sources, especially wind and PV solar energy, are increasingly used for electricity generation and expected to largely displace conventional fossil fuel-based power plants in the future. Modern wind and PV solar energy sources rely

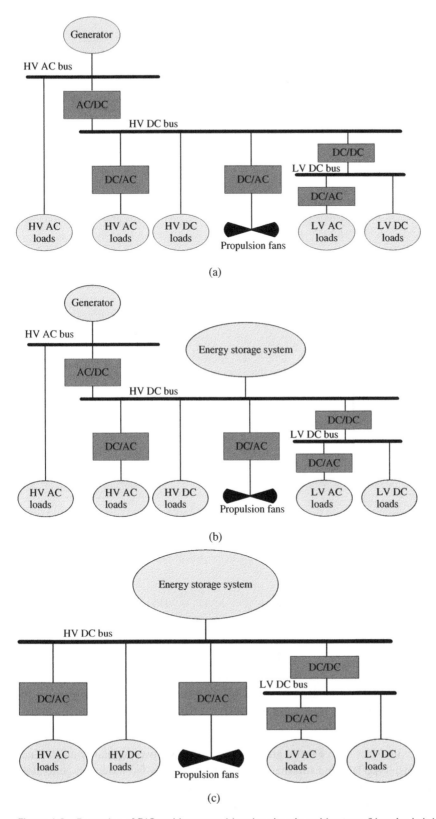

Figure 1.9 Examples of EAP architectures: (a) turbo-electric architecture, (b) series hybrid-electric architecture, and (c) all-electric architecture.

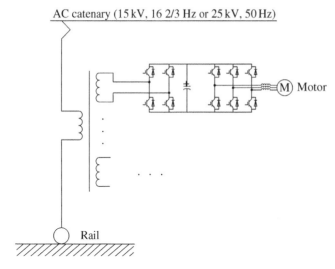

Figure 1.10 Configurations of a typical electrified rail traction system. *Source:* From [2]/with permission of John Wiley & Sons.

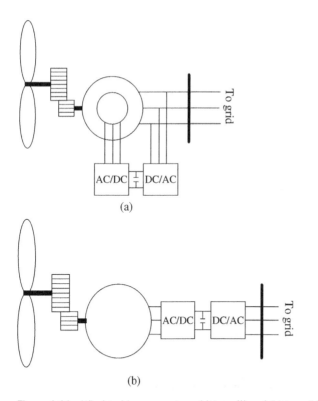

Figure 1.11 Wind turbine generators: (a) type III and (b) type IV.

on power electronics to interface with electric grid. Figure 1.11 shows two popular wind turbine generator configurations: Type III wind turbine generator using doubly-fed induction generator (DFIG) and Type IV wind turbine generator using PM generator. In both configurations, three-phase AC rectifiers and inverters are used. For Type III configuration, the rectifier is interfaced to the DFIG rotor windings; for Type IV configuration, it is interfaced to the PM generator stator windings. In both configurations, the inverter is interfaced to the grid.

PV panels generate DC at a voltage level of tens of volts and normally require a DC/DC converter to boost the voltage to the required level, e.g. hundreds of volts, before interfacing to AC grid through a DC/AC inverter. For small residential rooftop PV, the inverter is often single phase; however, for large utility-scale PV generation with several hundreds of kilowatts to megawatts, three-phase inverters are used. Figure 1.12 shows a typical configuration of utility-scale PV farm using a three-phase central inverter.

1.1.1.3.3 Data Center Our economy and daily lives are increasingly going digital and data-driven. Data centers are becoming an important type of load. Figure 1.13a shows a widely used AC power architecture in data centers. The power supply train of this architecture contains several parts, including UPS, power distribution unit (PDU), power supply unit (PSU), and voltage regulator (VR). It can be seen clearly that there are many power electronics converters: three-phase 480 V AC to DC, DC to single-phase 120 V AC, single-phase AC to 400 V DC, 400–12 V DC/DC, and 12–1 V VR. Figure 1.13b shows a more recent 400 V (or 380 V) DC power architecture [2], with fewer power conversion stages, but having a three-phase 480 V AC to 400 V DC converter.

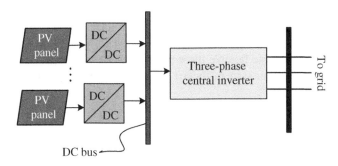

Figure 1.12 Utility-scale PV system.

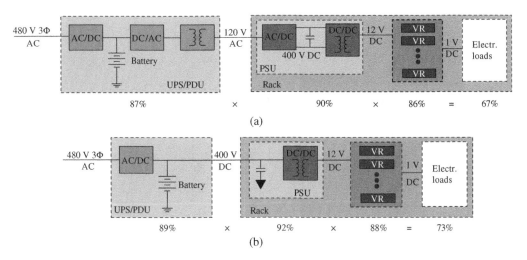

Figure 1.13 Power distribution architecture of data center power supply. (a) 120-V AC power architecture and (b) 400-V DC power architecture. *Source:* From [2]/with permission of John Wiley & Sons.

With reduced conversion stages, the efficiency of the DC power architecture is improved over that of the AC architecture, e.g. 73% vs. 67% in Figure 1.13. Other benefits include space-saving, reliability improvement, and simplified wiring.

Other notable emerging applications involving three-phase converters include those related to increased use of energy storage systems (e.g. battery chargers, traditional UPS, and utility-scale energy storage interface converters), and various power control and conditioning equipment for AC power transmission and distribution systems.

1.1.2 Basic Topologies

There are a number of converter topologies for each type of three-phase converter. The most commonly used basic topologies for three-phase rectifiers and inverters are shown in Figures 1.14 and 1.15, respectively. Note that in these figures, insulated gate bipolar transistors (IGBTs) and diodes have been assumed to be the semiconductor devices. In practice, any switching devices can be used, including metal-oxide-semiconductor field-effect transistors (MOSFETs). These basic topologies can be expanded through paralleling or series of devices and/or converters to achieve higher current and voltage ratings. It should be pointed out that in some applications, a three-phase converter can be formed by combining three single-phase converters, which are not considered as basic three-phase converter topologies here.

Note that nowadays current source-based converters are less popular than voltage source-based converters due to limited availability of bidirectional voltage blocking switching devices; however, historically CSCs have played an important role as a feasible three-phase converter topology. For most of the three-phase applications today, the basic state-of-the-art topology is the VSC shown in

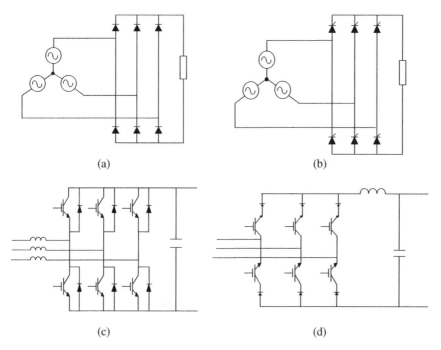

(a)   (b)

(c)   (d)

Figure 1.14 Commonly used three-phase rectifier topologies. (a) Diode-based rectifier, (b) thyristor-based phase-controlled rectifier, (c) voltage source converter (VSC) based or boost rectifier, and (d) current source converter (CSC)-based or buck rectifier.

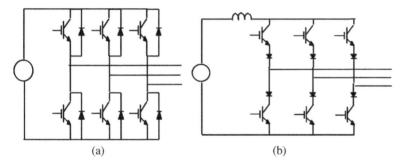

Figure 1.15 Commonly used three-phase inverter topologies. (a) Voltage source inverter (VSI) or buck inverter and (b) current source inverter (CSI) or boost inverter.

Figures 1.14c and 1.15a and their variations. The VSC based on Si IGBT (to a lesser extent Si power MOSFET) becomes the topology of choice in three-phase AC/DC or DC/AC applications due to its circuit and control simplicity and overall good performance and reliability.

In order to increase power rating and improve efficiency, power density, and reliability, many other topologies beyond basic VSCs have been developed for three-phase inverters and rectifiers. These topologies can be classified as multilevel converters, interleaved converters, and soft-switched converters.

1.1.2.1 Multilevel Converters

Multilevel converters refer to a family of converters that synthesize AC voltages using more than two DC voltage levels as in the case of the basic VSC in Figures 1.14c and 1.15a. The benefits of multilevel converters include lower harmonic ripple, lower voltage stress on semiconductor devices and loads (e.g. motors), and higher equivalent switching frequency. Many types of multilevel converters have been proposed. The commercially adopted converters include the neutral-point clamped (NPC) multilevel converter, the flying capacitor multilevel converter, the cascaded multilevel converter, the modular multilevel converter (MMC), and the Vienna-type multilevel converter. These topologies are shown in Figure 1.16. The three-level NPC converter is a suitable topology for medium voltage when using Si IGBT or integrated gate-commutated thyristor (IGCT), or emerging SiC devices, considering the power and voltage levels of these converters (MWs and kVs) and availability of medium-voltage Si IGBT, Si IGCT, or upcoming SiC MOSFET (kVs and kAs). The NPC converter has the advantage of high power density due to its less need on passive components. Therefore, NPC and its variants (including active neutral-point clamped (ANPC), T-type, and Vienna-type) are often preferred for applications where high power density is important, e.g. shipboard or aircraft electrical systems. Cascaded converters and MMC are modular and fault-tolerant, can switch slower, use lower voltage devices, and have high efficiency and good harmonic characteristics. On the other hand, they require a large number of capacitors and have inferior power density.

1.1.2.2 Interleaved Converters

Multilevel converters fundamentally employ the series converter strategies, with each converter switching phase-shifted with respect to others to achieve better ripple reduction while attaining higher voltage capability or switch slower to achieve higher equivalent switching frequency. Interleaving refers to the strategy of paralleling identical converters, with each converter switching phase-shifted to achieve better ripple reduction while attaining higher current capability.

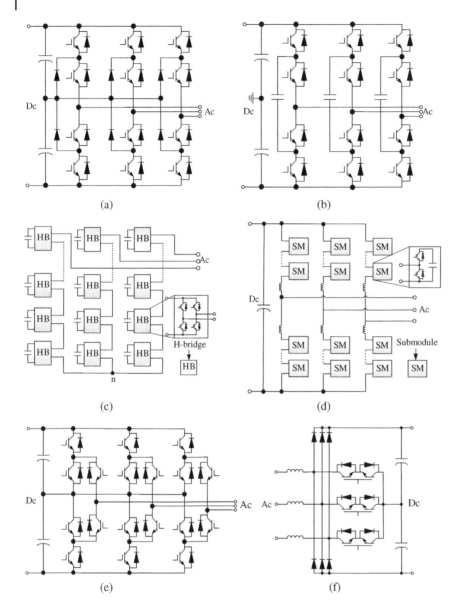

Figure 1.16 Multilevel converters: (a) three-level NPC; (b) three-level flying capacitor (FC) converter; (c) cascaded H-bridge (CHB) converter; (d) MMC; (e) three-level active NPC (ANPC); and (f) a Vienna-type rectifier.

Figure 1.17 illustrates the principle of interleaving. Similar to multilevel converters, interleaved converters have lower harmonic ripple, lower current stress on semiconductor devices, and higher equivalent switching frequency. Note that interleaving can cause circulating current problems for paralleled converters. The circulating current can be both in common mode (CM) and differential mode (DM). As a result, interphase inductors (also called coupled inductors) may be necessary for interleaved converters, which will increase design and control complexity and add more components. For motor drives, the interphase inductor function could be integrated as part of the motor windings.

Interleaving and multilevel technologies can be combined for further performance and power density gains.

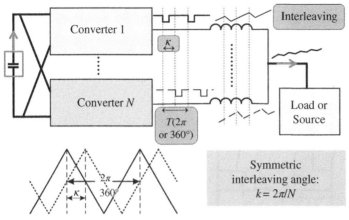

Synchronized switching with shifted phase angle of PWM

Figure 1.17 *N* paralleled interleaved voltage source inverters. *Source:* From [2]/with permission of John Wiley & Sons.

1.1.2.3 Multiphase Converters

Most converters in high-power applications such as renewable energy system and grid have three-phase AC input and/or output. For loads, such as motors, multiphase motors with phase number greater than 3 may be beneficial in terms of cost, performance, power density, and/or fault tolerance. Correspondingly, the converters can have more than three phases, which can be realized with multiple three-phase converters or with dedicated multiphase converters. In addition to modularity and redundancy, multiphase converters may generate lower CM voltages, which will be beneficial to loads (e.g. motors) and sources (e.g. grids), and lead to less filter requirement. The impact of CM voltage on motors and the corresponding mitigation strategies will be further discussed in Chapter 12.

1.1.2.4 Three-Phase Four-Wire Converters

The topologies in Figures 1.14, 1.15, and 1.16 all have three wires on the three-phase AC side. As a result, these converter topologies are suitable for the cases with three-phase sources or loads that are balanced or do not require zero-sequence or CM current path. Otherwise, a fourth wire, which should be connected to the neutral point of the three-phase source or load, will be needed for the three-phase converter to accommodate the unbalance or to provide zero-sequence current path. The commonly used schemes to realize three-phase four-wire converter topology include adding a fourth phase leg, creating a DC-link midpoint with split DC-link capacitors, and adding a neutral-forming network (the most popular one is a transformer). With increasing applications in electric power systems (grid-tied or autonomous), the three-phase four-wire converters are becoming more popular.

1.1.2.5 Soft-Switched Converters

DC–DC converters often employ resonant topologies to realize zero-voltage switching (ZVS) or zero-current switching (ZCS) to reduce switching loss and enable high-frequency switching. Many soft-switching topologies have been proposed for three-phase AC converters by adding auxiliary circuits to the original hard-switching topologies, including two-level VSCs, three-level NPC, and ANPC converters [4]. As an example, the zero-current transition (ZCT) converter potentially suitable for high-power applications is shown in Figure 1.18. The ZCT techniques can almost

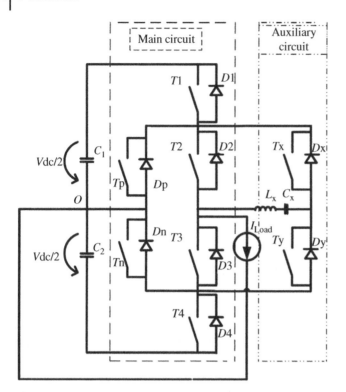

Figure 1.18 A phase-leg circuit of the 3L-ANPC ZCT converter. *Source:* From [4] with permission of IEEE.

eliminate turn-off losses and overvoltage spikes during the turn-off transient and greatly reduce turn on losses and alleviate reverse-recovery issue during turn-on. In addition, the low-power rating auxiliary switches have no switching losses. However, the soft-switched converters have not been much adopted in practical applications due to their added complexity and cost, and limited benefits in motor drive type of applications, where extra high switching frequencies are usually not required.

1.1.2.6 Other Topologies
Although less popular than VSCs, current source converters (CSCs) are also used in three-phase applications. Similar to VSCs, CSCs can have topology variations including multilevel CSCs, interleaved CSCs, and soft-switched CSCs. As its name indicates, a CSC and its variations are based on current source. For example, multilevel CSCs are for multilevel currents, with direct benefits of reduced device current ratings and improved current ripple. As mentioned earlier, one of the reasons that CSCs are less used than VSCs is lack of bidirectional voltage devices with good switching capabilities. To achieve bidirectional voltage blocking, two unidirectional devices, e.g. IGBTs or MOSFETs, have to be connected in series, resulting in increased conduction loss. Figure 1.19 shows a delta-connected three-phase CSC to relieve the conduction loss issue. In conventional CSCs, often current only flows through one phase, while in delta-connected CSCs, the current will be shared by two phases, resulting in a reduced conduction loss of up to 20% [5]. Although the number of diodes needed will increase, the total current ratings of these diodes will remain the same.

Some three-phase applications require AC/AC power conversion, as in the case of AC-fed motor drives and wind turbine generators. As mentioned earlier, in most cases, back-to-back (i.e. cascaded) AC/DC rectifier and DC/AC inverter are used; however, direct AC/AC converter topologies

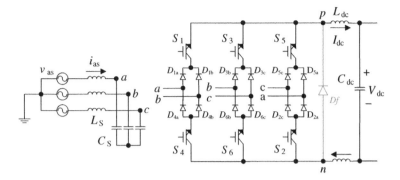

Figure 1.19 Delta-connected current source rectifier. *Source:* From [5] with permission of IEEE.

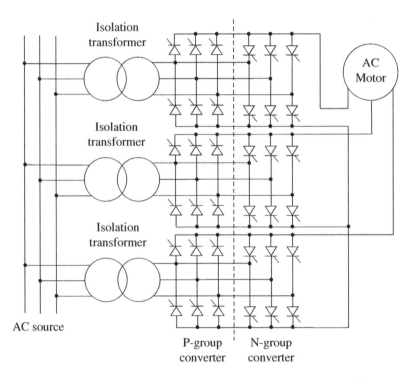

Figure 1.20 Cycloconverter using phase-controlled thyristors for motor drive.

including cycloconverter and matrix converter can be used. Figure 1.20 shows the cycloconverter topology, which essentially uses six phase-controlled rectifiers with each responsible for one output phase current in one polarity. Although the cycloconverter topology is simple and easy to control, its drawbacks include poor device utilization, poor power quality and power factor, and limited output frequency range (less than 1/3 or 1/2 of the input frequency).

Figure 1.21 shows a basic three-phase matrix converter for direct AC/AC conversion, which directly maps the three-phase input AC voltage to the desired three-phase AC output voltage like a matrix through proper switching of the nine switches. Clearly, it requires these switches to be four-quadrant, which in reality will require four unidirectional switches to realize one switch in Figure 1.21. To overcome the difficulty, several matrix converter topology variations have been

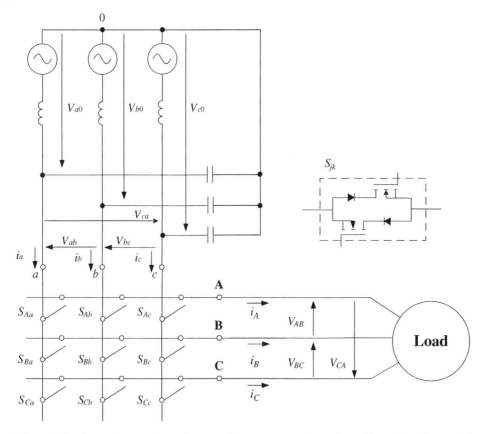

Figure 1.21 Three-phase to three-phase matrix converter topology. *Source:* From [6] with permission of IEEE.

developed such that reduced number of regular unidirectional IGBTs or MOSFETs can be used. One other often-cited drawback of the matrix converters is their limited output voltage range (maximum output voltage $<\sqrt{3}/2$ of input voltage) [6].

Despite issues with cycloconverters and matrix converters, they can be advantageous in certain applications and have been applied.

1.1.3 Composition of Three-Phase AC Converters

The real functional physical three-phase AC converters, whose design is the focus of this book, are more than just the converter circuits as those illustrated in the topology section above, where only the essential electrical circuit components are shown. A fully functional power converter also requires other electrical elements, such as controllers. In addition, power converters need to meet performance requirements, such as power quality and electromagnetic compatibility (EMC); as a result, filters are generally needed. Furthermore, even though the basic function of a power converter is electrical, nonelectrical functions are also necessary, including thermal management and mechanical structure. Given that this book is for the design of three-phase AC converters, it is important to understand the composition of these converters. Without loss of generality, a typical AC-fed motor drive system in Figure 1.22 is used to illustrate the composition of a functional physical three-phase AC converter.

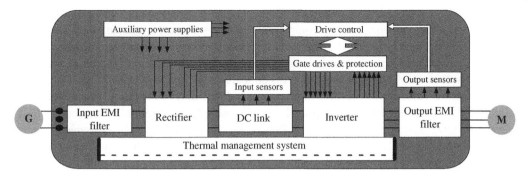

Figure 1.22 Composition of a typical AC-fed motor drive. *Source:* From [7] with permission of IEEE.

1.1.3.1 Switching Devices

At the heart of any power electronics converters, three-phase AC converters included, are power semiconductor switching devices. The switching networks or converter circuits, e.g. rectifier and inverter circuits in Figure 1.22, are made up of switching devices. The advancement of power electronics has been largely driven by the advancement of power semiconductor devices. Since the advent of power diodes and thyristors in the late 1950s, power semiconductor devices have gone through many different types and generations, including gate turn-off (GTO), power bipolar junction transistor (BJT), power MOSFET, IGBT, and IGCT. Currently, IGBT dominates the three-phase AC converter markets, except at low-voltage (e.g. < hundreds of volts), low-power (e.g. <1 kW) level where MOSFET may be used, and in some high-power cases where thyristors (including GTO and IGCT) are used.

The above-mentioned devices are all Si-based. In recent years, the emergence of WBG semiconductor devices including silicon carbide (SiC) and gallium nitride (GaN) devices has promised to revolutionize power electronics converters. Compared with the mature Si devices, WBG devices feature higher breakdown electric field, lower specific on-resistance, faster switching speed, and higher junction temperature capability. All these attributes can benefit power electronics converters for higher efficiency, higher power density, higher reliability, and/or lower cost. WBG devices have been employed in some three-phase AC converters in commercial products and R&D prototypes. More applications are expected in the near future. As a result, this book will cover converters based on both Si and WBG devices.

1.1.3.2 Energy Storage Passives

Clearly from different circuits in Figures 1.14, 1.15 and other figures in the topology section, all converters have inductors and/or capacitors, whose role is mainly for energy storage. In fact, energy storage elements are essential for any power electronics converters. The essence of the power conversion is to convert electricity from one form to another. Power electronics converters realize the voltage or current form conversion by using the energy storage passives to temporarily make up the power/energy difference between the source and load, with the help of the switching network. Even the AC/AC direct conversion will need energy storage elements to help with the output voltage and current wave forming. In Figure 1.22, the energy storage passives include DC-link and rectifier AC line passives (capacitors and inductors).

1.1.3.3 Filters

Power electronics converters rely on switching devices, which will inevitably generate noise, ripples, and harmonics. The switching transients can also lead to voltage and current spikes.

As a result, filters are needed for protection, EMC, and power quality. The fast-switching WBG devices may result in more demanding requirements for filters. Since the filters are mostly realized with passive components, they can be important contributors to converter size and weight, as well as cost. The main filter types include harmonic filters to meet power quality requirements, EMI filters for EMC, and *dv/dt* filters to relieve stress on loads or sources, especially motor loads with limited insulation capability. Note that different types of filters may be realized with the same passives, such that the filter can be an integrated one. For example, the harmonic filter and the EMI filter can be integrated in some applications. Depending on applications and requirements, some filters are optional. Note also that the passives in a converter can often play dual roles, acting as both energy storage and filtering elements.

1.1.3.4 Thermal Management System

Thermal management is essential for power electronics converters. All electrical components, including power semiconductor devices and passives, are nonideal and will incur power losses during operation. In addition, components all have limited operating temperature range. Appropriate thermal management (mostly cooling) is necessary in order for power semiconductor devices and other components to be maintained within their respective safe operating temperature ranges. Thermal management system can critically impact the performance of a converter, such as its power loss, efficiency, and reliability. Additionally, the thermal management system usually takes up a significant portion of the total converter size, weight, and cost. A sample design shows that for a GaN FET-based rooftop PV inverter, its heatsink is the highest cost item, even more than the GaN devices used in the design. It is important to identify and choose efficient and effective thermal management strategies.

1.1.3.5 Controller and Auxiliary Circuits

All power electronics converters, three-phase AC converters included, function through controlled switching actions of power semiconductor devices. Control is an essential part of the power electronics converters. Modern power converter control is generally hybrid in nature with both analog and digital controllers, especially for more complex converters like three-phase AC converters. The control includes both controller hardware and software.

The three-phase AC converter control usually has a layered controller architecture, as commonly the case for high-power applications [2]. As will be described in more detail in Chapter 10, in general, the control can be divided into five layers: (i) hardware control layer; (ii) switching control layer; (iii) converter control layer; (iv) application control layer; and (v) system control layer. Some of the control layers will be implemented with hardware, some with software, and others with combined hardware and software. They can also be analog, digital, or mixed. Note that not all converters need all control function layers above. In general, the converter should have the first three control layers, which will be covered in the converter design discussion in this book. The impact of the application and system level needs will also be considered.

To realize the control functions, several important auxiliary components are needed, including sensors and auxiliary power supplies. Sensors are needed to provide converter operation information, e.g. voltages, currents, and temperature, to the controllers. For motor drives, often speed or position sensors are needed. Auxiliary power supplies, which are usually AC/DC or DC/DC converters themselves, are needed to provide power for controller hardware and circuit boards. Note that sensors and power supplies, together with hardware control layer circuits (e.g. gate drive), need to interface both with the power electronics converter main circuit, which is often at higher converter voltage level, and with digital control circuit, which is at lower signal voltage level. Therefore, isolation is important and is often achieved with optical means such as an opto-coupler in lower voltage, lower power (<1 kV, hundreds of kW) case or optical fiber in higher voltage, higher power case. The isolation for the gate drive power supply is often achieved through magnetic transformer. With higher voltage applications like grid applications and faster switching devices like WBG devices, isolation in power electronics converters becomes an increasingly challenging issue.

1.1.3.6 Mechanical Assembly

Mechanical assembly in power electronics converters plays multiple roles, not only having structural function but also having electrical and thermal functions. In addition, mechanical assembly contributes significantly to the overall converter size, weight, cost, and reliability. It is therefore also important to consider the mechanical assembly in the overall design.

In high-power converters as is the case for most three-phase AC converters, one particularly important piece of the mechanical assembly is the busbar. Figure 1.23 shows a busbar structure for a basic two-level VSC. Note that a busbar contains not only the metal conductors but also the electrical insulation between conductors. The busbar provides connections for DC and AC bus terminals, power semiconductor device modules, and passive components including DC-link capacitors and filters. Mechanically, the busbar provides structural support for the components connected to it. Electrically, the busbar conducts the current through the buses and bears the voltage between buses. Thermally, the busbar contributes to power loss due to its electrical resistance and also can impact thermal performance as thermal conductors.

In addition to DC and low-frequency AC components, the busbar currents and voltages also have high-frequency and transient components as a result of power device switching. These high-frequency and transient components often pose more stringent requirements for busbar design. For example, high di/dt during fast switching can cause overvoltage issues for the switching device as a result of parasitic inductance of the switching loop. Busbar is a central piece of the power loop, and its parasitic inductance should be minimized. In fact, one of the reasons to adopt busbar structure in high-power converters is due to its intrinsic property of low parasitic inductance. The capability of the busbar insulation to withstand switching pulse voltage and switching transient overvoltage is another important requirement. The eddy current, proximity effect, and impact on EMI should also be considered in busbar design. Note that WBG devices can switch faster than Si counterparts, which helps reduce switching loss and enable higher switching frequency.

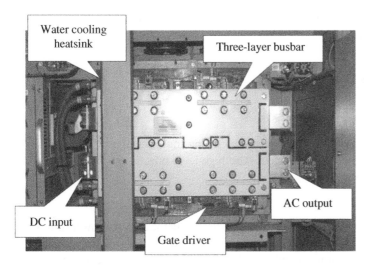

Figure 1.23 Example of a laminated busbar. *Source:* From [8]/with permission of IEEE.

However, parasitics impact becomes stronger as the *dv/dt* and *di/dt* during a switching transition significantly increase. Therefore, a proper design of the busbar with low parasitics like low loop inductance is even more critical for fully utilizing the switching capability of WBG power devices.

Enclosure, another essential part of the power converter, can also contribute significantly to the size, weight, and cost of the converter. The importance of the enclosure to the structural integrity of the converter is obvious. Electrically, the enclosure can affect grounding, insulation coordination, and EMI; thermally, it provides the interface to the ambient.

It should be pointed out that the three-phase AC converter composition is discussed here to highlight the key components, technologies, and function blocks. There has been more systematic partitioning of converter functions and technologies. In [9], Van Wyk identified constituent technologies of a modern converter, independent of the converter type, power level, or application. These constituent technologies include:

a) power switch technology (covering device technology, driving, snubbing, and protection technology);
b) Power switching network technology (i.e. topology or what is classically termed converter technology, covering the switching technologies such as hard switching, soft switching, resonant transition switching, and all the topological arrangements)
c) Passive component technology (covering magnetic, capacitive, and conductive components)
d) Packaging technology (covering materials technology, interconnection technology, layout technology, and mechanical construction technology)
e) Electromagnetic environmental impact technology (covering harmonics and network distortion, EMI, and EMC)
f) Physical environmental impact technology (covering acoustic interaction, physical material interaction, i.e. recycling, pollution).
g) Cooling technology (cooling fluids, circulation, heat extraction and conduction, and heat exchanger construction)
h) Manufacturing technology
i) Converter sensing and control technology

All these constituent technologies are interactively related to one another, contributing to the complex nature of designing, building, and operating power electronics converters. All of them will be considered in this book, in most cases, directly, and in some cases (physical environmental impact and manufacturing technologies), through their impact.

1.2 Basics of Three-Phase AC Converter Design

1.2.1 Essence of the Design and Design Tasks

In principle, design of a power electronics converter, including the three-phase AC power converter, is like design of any equipment or system. The high-level process can be illustrated in Figure 1.24 and will be discussed here.

1.2.1.1 System Conditions
System conditions refer to known conditions or input to the design. In the case of a power converter, they refer to specified operating conditions of the converter, including the source, load, and ambient conditions. More generally, the system conditions for a converter can be regarded as the interface conditions of the converter to its external environment, which includes electrical, thermal, and

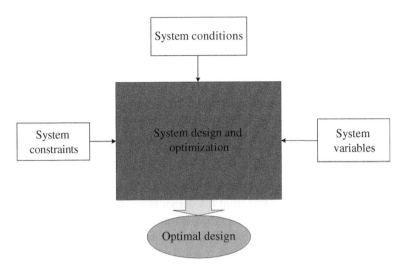

Figure 1.24 Design procedure illustration. *Source:* Adapted from [7].

mechanical environment. For the AC-fed motor drive in Figure 1.22, the system conditions include: (i) AC source voltage conditions – amplitude, frequency, and quality (voltage range, harmonics, equivalent impedance, etc.); (ii) motor load conditions – speed/torque and corresponding voltage, current, and power, and motor characteristics including motor electrical and mechanical models; and (iii) ambient conditions – ambient temperature range and to a lesser extent air pressure. Note that the conditions may include transient and abnormal conditions, such as overload for a motor drive.

1.2.1.2 System Constraints

System constraints can be regarded as the "output" of the designed system, which are often related to the performance requirements and physical constraints. For the AC-fed motor drive, the performance requirements may include general requirements like reliability, efficiency, power quality (e.g. source-side AC current harmonics), power factor, EMI, as well as specific requirements like motor speed and torque dynamic response, and motor-side current quality (e.g. current harmonics, DC offset, and negative sequence components). The physical constraints mainly refer to limits associated with the components. Examples of physical constraints are junction temperature limits for power semiconductor devices, temperature limits for inductor and capacitor materials, and saturation flux density limits for magnetic materials. Mechanical size constraints such as magnetic core window area also belong to physical constraints.

1.2.1.3 System Variables

System variables refer to the variables in the design process. They can be selected/designed such that the system constraints (design output) will be satisfied under the system conditions (design input). In other words, the design is simply the process of finding suitable set(s) of system variables that will satisfy system constraints under given system conditions. For the AC-fed motor drive in Figure 1.22, the system variables include design variables for all function blocks or subsystems (rectifier, inverter, DC link, input/out filters, thermal management system, controller, etc.) and their corresponding components. One important aspect of a successful design is to understand the design space, or available selections of system variables. Specific to power converters, the design space refers to available components and technology. Examples for the three-phase AC converter include available semiconductor devices, magnetic cores, windings, capacitors, heatsinks, three-phase converter topologies, and control methods.

1.2.1.4 Objective Function

Objective function refers to design objectives. For power converters, the objective function often involves a single design objective, such as minimum cost, minimum size or weight, or even maximum efficiency. In some cases, the objective function can have multiobjectives, meaning that some form of combination of cost, size, and efficiency will be used as the design objective.

1.2.1.5 System Design and Optimization

System design represents the process of calculating the output values associated with system constraints, using the selected system variables available from the design space and under the given system conditions (i.e. input conditions). If the calculated values satisfy all the system constraints, then the design is a feasible one, otherwise it is not. For example, a feasible design of the AC-fed motor drives will require the designed efficiency, harmonics, and all other performance indices to meet the specifications (i.e. performance constraints) for the given source, load, and other environmental conditions. The design should also satisfy physical constraints, e.g. not exceeding component temperature limits. On the other hand, there can be many feasible designs for a given converter. Optimization is the process of finding the best design among all feasible designs based on the objective function. Assume that the cost is the objective function for the design of a particular converter. For each feasible design, there should be a corresponding cost. The design with the lowest cost is the optimal design. Clearly, the system design optimization is an iterative process.

In summary, the converter design should be an optimization process, involving identifying the optimal design from all feasible designs. The feasible designs are found from available design space under given system conditions that meet system constraints. The high-level converter design tasks can be categorized as follows:

1) Understand system conditions and constraints, as well as the objective function. In the case of the converter design, this means to understand the design specifications and requirements for the converter, as well as the design target (e.g. cost or size).
2) Understand the system variables and available design space. Gather information and data needed for the design. For converter design, this means finding available candidate technologies and components. The attributes associated with the objective function are also needed for the candidate technologies and components. For example, if the design objective is minimum cost, then the cost information for candidate components should be collected.
3) Perform the design process to find feasible design solutions. For converter design, this means understanding/establishing relationships and models among system conditions, constraints, objective function, and variables; and based on these relationships and models, determining whether a given design will satisfy performance requirements and specifications.
4) Find the optimal design among all available solutions based on the predetermined objective function.

All four categories of tasks are necessary for the converter design. However, for power electronics design engineers, Tasks 2 and 3 are core tasks, especially when dealing with new types of converters, using new technologies, and/or for new applications. Task 1 is relatively straightforward and serves mainly as a preparation step for other design tasks. Task 4 deals with optimization, which can employ many techniques commonly used in other fields. On the other hand, Task 2 requires the designer to have the knowledge to properly select the candidate technologies and components unique to power electronics for the design. Task 3 requires the designer to have the knowledge on the relationships among the converter design output (constraints), input (conditions), and variables, such that the feasibility of a design can be determined. In other words, the designer needs models depicting such relationships. For a new type of converters, using new technologies and components, and/or for new applications, new models will likely be needed.

Power electronics converters are complex equipment, while the three-phase AC converters are more complex among converters. Hence, the converter design is a complex process. There are many parameters and variables associated with the converter system conditions, constraints, and design variables. There are also many relationships among these parameters and variables. Not all relationships are of the same physical nature, as we already know that a converter has electrical, thermal, and mechanical functions. Behaviors and performance of a converter can cover very different timescales (e.g. switching transients vs. motor speed control events for a motor drive) and frequency ranges (e.g. AC line or motor frequency vs. EMI noise frequency). As a result, these numerous relationships and relationship types will be best depicted by many different models and different types of models.

Therefore, in principle, a converter design only involves four categories of tasks described earlier, and the whole converter can be designed together as one design and optimization problem. In practice, due to the complexity of the converter and the converter design, it is often more appropriate to break the design into several subdesign problems. As a result, appropriate partition and design procedure will be needed, which will be discussed in Section 1.2.2. The partition is also needed for design methodology discussion, which is the focus of this book.

It should be mentioned that there have been noticeable efforts in converter design automation. Conceivably, with all models available, the design and optimization can be performed by computers automatically. The design problem partition and the design procedure will not matter much. However, the models still need to be established based on the relationships per the physical nature between various parameters and variables. In other words, automated design is still based on the fundamental understanding and modeling of relationships between various design parameters and variables. The modeling is often easier for partitioned function blocks or subsystems. For example, it is easier and more natural to model the electrical and thermal functions of a converter separately.

1.2.2 Design Procedure, Strategy, and Philosophy

1.2.2.1 Design Problem Partition
As concluded from the design task discussion in Section 1.2.1, it is a useful practice to break a converter design problem into several subdesign problems. This will be particularly helpful when discussing and formulating the converter design problem. The principle to follow in partitioning the converter design problem is that the subdesign problems can be relatively decoupled, and each subdesign problem is relatively simple.

The logical approach for partitioning is by converter functions or function blocks that generally correspond to a physical part or subsystem. Figure 1.22 shows the function blocks for a motor drive. Section 1.1.3 listed key components and functions composing a modern power electronics converter and also associated constituent technologies summarized in [9]. In general, the function blocks or subsystems suitable for subdesign problems are:

1) Converter main circuit: The converter main circuit is also referred to as power stage. It is the core of the converter and responsible for the main power conversion function. The design of the power stage includes the selection of topology, devices, energy storage passives, and control and operation strategies to meet required power conversion function between the input source and output load.

2) Controller and auxiliaries: Controllers and auxiliaries (e.g. sensors, power supplies, and communications) are to be designed to accommodate the control and operation requirements of the main circuit. Controllers include hardware and software. The gate drives for devices are often included as part of the controller design, although they can also be included as part of the power stage.

3) Interface to source and load: This includes the design of various filters needed for harmonics, EMI, and/or *dv/dt* management to meet the associated performance requirements. This also includes protection devices like overvoltage and overcurrent suppression devices. Note that the interface design can be closely related to the power stage design.

4) Thermal management system: Thermal management system design is to select and design the cooling system to keep the operating temperatures of the converter components within their safe operating range under the given ambient conditions. The popular power converter cooling types include air cooling and liquid cooling. The air cooling design includes heatsink, fan, and airflow design; and the liquid cooling design can include cold plate, liquid coolant type, liquid loop, pump, and heat exchanger to the ambient. The thermal management system design is closely related to power stage design, through its impact on converter loss, efficiency, and reliability.

5) Mechanical system: Section 1.1.3 discussed busbars, the enclosure, and their impact on converters. Other components in the mechanical system include connectors, fasteners, boards, supporters/standoffs, cables, and wires. The mechanical system design is to design all these mechanical structural support subsystems and components, considering the requirement of and their impact on the converter from the structural, electrical, and thermal standpoints.

It should be noted that each function block listed above can be further partitioned during the design. The main circuit may include more than one basic converter, like in the case of the AC-fed motor drive, which includes both a rectifier and an inverter. The components in the controller categories are often designed separately. Different filters perform different functions and are normally designed separately, although integrated filter design is possible.

Moreover, the design also should include functional and physical design steps, for each function block and component. Take the main circuit or power stage design as example. For the functional design, the designer needs to select a circuit that can perform the required power conversion function. In addition, the design should cover the selection of circuit parameters, including current and voltage ratings of the components and inductance and capacitance values. The control or operation strategy is also part of the power stage functional design.

After the functional design, the next step is physical design, i.e. selection and/or design of physical components to achieve electrical functions. The real power semiconductor devices need to be selected based on voltage and current ratings; inductors and capacitors need to be selected or designed according to required inductance or capacitance values, together with voltage and current ratings.

Thermal and mechanical design also has functional and physical design steps. Take a forced-air cooling design for power semiconductor devices as an example. The functional design for the thermal management system will determine the cooling needed for the devices, based on temperature limits of the devices, ambient temperatures, thermal characteristics of the devices, and the device loss. The functional design results are often in terms of the heatsink thermal conductance. The physical design will involve the selection and/or design of the actual physical heatsink and cooling fan to achieve the required thermal conductance.

Like the function block partition, where different function blocks are interrelated, the physical design and functional design are intertwined. Although it makes sense to start with functional design and then conduct physical design, iterations are often necessary.

1.2.2.2 Design Procedure

The design problem partition, and functional and physical design discussions entail the discussion of design procedure. Section 1.2.1 defines the design problem. In principle, as long as the relationships (i.e. models) among design conditions, constraints, and variables are established, the sequence of the design does not matter. On the other hand, for power electronics designers, there are normally some logical procedures with a design. In order to illustrate the procedure clearly, we will use the topology study for a three-phase AC-fed motor drive as an example. The basic converter architecture and main weight contributors are shown in Figure 1.25 [10]. The objective of the design is to identify a circuit topology that will yield the minimum weight for the corresponding motor drive.

Figure 1.26 illustrates the design procedure of the problem, which is discussed here briefly in steps:

1) Device type and topology: Device type and topology are both design variables, which are selected in this step. It starts with a candidate topology and also the device type. Obviously, a database for different types of devices and different candidate topologies is needed.
2) Device rating: Device rating for a given device type is determined from power, voltage, and current requirements. Here, database of physical devices for the selected device type with various ratings is needed.
3) Switching frequency: Assuming that the three-phase converter switching is PWM-controlled, switching frequency f_{sw} is an important design variable. It will impact switching loss, and harmonic and EMI noise.

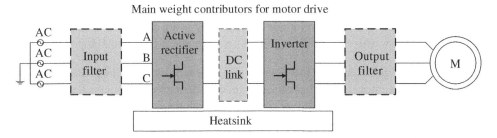

Figure 1.25 Key components and weight contributors of an AC-fed motor drive. *Source:* From [10] with permission of IEEE.

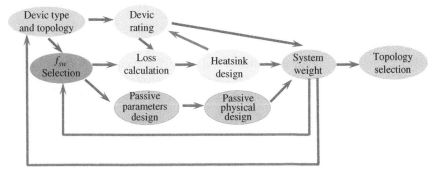

Figure 1.26 A design procedure illustration.

4) Loss calculation: With selected device, operating conditions, and control, the device loss including conduction and switching loss can be calculated. Here, the loss characteristics of the device and models for loss are needed. Note that the loss calculation requires an assumption or selection of the device junction temperature.

5) Heatsink design: Based on device loss and junction temperature requirements in step (4), the required heatsink can be selected. Note that in this example, the airflow is given, so only heatsink is needed. The functional and physical designs are both covered here. A heatsink database with heatsink thermal models is needed. The device thermal models are also needed as part of the device database. Note that a feasible heatsink may not be physically possible to achieve from the database. In this case, a device with different ratings can be selected to lower the loss.

6) Passive parameter design: In this example, this step includes both the energy storage passives and filter passives. From the candidate topology, control, and f_{sw}, the passive parameters, mainly the capacitance and inductance values, can be determined to meet various performance requirements and constraints, such as total harmonic distortion (THD) limit and EMC standard. This step requires the models that link inductance and capacitance parameters with harmonics and EMI for the candidate topology and control.

7) Passive physical design: Given the designed parameters in step (6), this step selects capacitors and designs inductors with available magnetic cores, wires, and other materials. Databases for capacitors and materials for inductors are needed.

8) Converter weight calculation: In this example, the minimum weight is the design objective. With devices, passive components, and heatsink selected, their total weight can be calculated. Then, the whole process can be iterated from step (1). Note that a full converter should also include other parts, like mechanical structure, controllers, and auxiliaries. Since this example is for weight comparison of different topologies, and those items can be considered more or less the same for different topologies, their impact is excluded.

9) Optimal topology selection: Once all minimum weight designs are completed for all candidate topologies, the topology with the lowest weight will correspond to the optimal topology.

Some observations should be made about this example design procedure. First, the procedure is not unique. For example, when iterating the heatsink design, the alternative to changing device rating can be changing switching frequency. In the end, different procedures should be equivalent, as long as the whole design space can be covered. Second, physical devices generally have nonideal characteristics. A capacitor has parasitic resistance and inductance; and an inductor, on the other hand, has parasitic resistance and capacitance. These parameters are often frequency-dependent, and/or subject to saturation. The nonidealities must be properly accounted for and modeled in the design.

1.3 Goal and Organization of This Book

This book is intended for professors, graduate students, and practicing engineers working in the area of three-phase AC converters, as a classroom textbook or as a reference. The focus is on the design of three-phase AC converters, specifically the design methodology. Given that, several points should be made about this book.

- This book is for design of practical three-phase AC converters. However, it is not intended to be a design handbook to list and cover all aspects and steps needed in a converter design; rather, it intends to focus more on modeling, theory, and methodology of the design.

- Even though applications are important for converters, this book will focus on the design of the converters themselves and will not discuss any applications in detail. The impact of applications on the design of three-phase AC converters will be considered. Some of the important application-related technologies (e.g. filters for motor drives and lightning protection for grid-tied converters) will also be covered.
- Clear from the earlier discussion, components and function blocks or subsystems are integral parts of the converters, and their selection and design will be important parts of this book. Here, the focus will be on modeling and design of the components and subsystems, not their working principles. For example, we will not spend much effort on the physics and structure of semiconductor devices; rather, we will focus on their terminal models needed for converter design.
- Topology selection is an important task for converter designs. In fact, often the converter design goal is to determine the optimal topology for a given application. Practically, the optimal topology for a three-phase AC converter design will be selected from several candidate topologies. Since the commonly used topologies for three-phase AC converters are well known and well covered in other publications, they will not be discussed in detail in this book.
- Although this book targets the three-phase AC converters, many approaches should apply to other converters. Generality of a given method will be discussed, where appropriate.

The book organization, chapter contents, and their logics are explained below. The main content of the book consists of three parts. Part 1, from Chapters 2–4, is on components. Part 2, from Chapters 5–12, is on subsystems or function blocks. Part 3, Chapter 13, is on design optimization.

Chapter 2 focuses on characteristics important to the converter design for devices most commonly used in three-phase converters, including Si IGBTs, diodes and thyristors, SiC MOSFETs and Schottky diodes, and GaN HEMTs. The characteristics include static and switching electrical characteristics, thermal characteristics, and mechanical characteristics. Most characteristics can be directly obtained from the datasheet. Some important ones to be obtained through testing or simulation will also be discussed. Device reliability, scalability (parallel/series), and application tips that are important for converter design and performance will be included.

Chapter 3 covers various capacitors used in three-phase converters and their corresponding characteristics. The focused capacitor types are film capacitors, aluminum electrolytic capacitors, and ceramic capacitors. Electrochemical capacitors (i.e. super capacitors) and some other less commonly used types will also be discussed. The characteristics include capacitance values, voltage/current ratings and capabilities, loss, temperature rating. Size, weight, reliability, and lifetime will also be explained. The nonidealities associated with capacitors such as parasitics will be included.

Chapter 4 covers the characteristics and design of magnetics, with focus on inductors. The characteristics are mainly for magnetic core materials including ferromagnetic materials (silicon steel, amorphous, nanocrystalline, and powder) and ferrite. The characteristics include permeability, saturation flux density, loss, and temperature rating. Various core loss models will be described. Winding loss model considering eddy current and proximity effect will be presented. The impact of insulation, cooling, and mechanical structure will be discussed. The inductor and transformer design methods pertaining to three-phase AC converters will be discussed.

Chapter 5 is on the design of the passive three-phase rectifiers using diodes or thyristors. Selection of devices, AC line inductors, and DC bus passives considering both steady-state and transient conditions is discussed. The focus is on the relationships between circuit parameters (e.g. AC line inductance, DC bus capacitance, and DC inductance) and the rectifier performance (e.g. AC line current THD and inrush current).

Chapter 6 is on the design of the three-phase PWM inverters connected to loads. Switching device selection, PWM scheme, decoupling capacitors and inductors, and the requirements for thermal management are the main tasks of the load-side inverter design. The focus will be on: PWM voltage harmonic spectrum; device loss and voltage overshoot as functions of device characteristics, control (switching and gate drive), temperature, and parasitics; and decoupling capacitor and inductor design.

Chapter 7 is on the design of the three-phase PWM rectifiers and inverters connected to sources, especially the three-phase AC grid. Many of the design tasks are similar to the PWM load-side inverters and passive rectifiers, but there are also unique tasks related to the source-side requirements such as filters. The focus will be on source-side *LCL* filter design and the reliability consideration. The impact of the topology on the three-phase PWM converter design will be discussed.

Chapter 8 is on the design of EMI filters for three-phase AC converters, including filters (mainly the passive EMI filters) for input and output sides which can be AC or DC. Various EMI/EMC standards, filter topologies, and design methods will be presented. The focus will be on relationships between filter circuit parameters and filter performance (EMI noise level) as functions of component characteristics and operating conditions. The nonideal characteristics of filter components (capacitors and inductors) will be considered in the design. EMI reduction techniques including circuit, control, and filter will be discussed. Some active EMI filter techniques will be discussed.

Chapter 9 covers the design of thermal management systems commonly used for three-phase converters. The focus is on forced-air cooling and liquid cooling system, while natural cooling and phase-change cooling will also be discussed. Thermal models of the cooling system will be presented. Selection of heatsink, fan, cold plate, and thermal interface materials will be described. Cooling of active and passive devices will be covered.

Chapter 10 includes the design of controller architecture, design and selection of digital controller hardware, isolation strategy, and isolation power supply. It will also include gate drive design and sensor selection, including voltage considerations for medium-voltage design and conditioning circuit design under the high-frequency high-noise environment. The protection will focus on device-level protection, such as short -circuit and crosstalk, as well as converter-level protection, including overcurrent, over and under voltage, and over temperature. Printed circuit boards (PCBs) will also be discussed. The deadtime setting and compensation, particularly considering the unique characteristics of WBG power semiconductors and WBG-based converters, will be covered.

Chapter 11 discusses the mechanical system design for three-phase converters from the electrical designer's point of view, with the focus on mechanical parts selection and design to complement and support the electrical and thermal components and subsystems. The mechanical parts for the converters include busbars, enclosure (cabinets), PCB and other supporting boards, connectors, wires and cables, and other supports. The busbar design optimization will be presented.

Chapter 12 considers the impact of applications on three-phase converter design. The focus will be on motor drive and grid applications. For motor drives, the design of the motor-side filters and other techniques to limit the impact of high *dv/dt*, CM voltage, and long cables will be discussed in relation with the three-phase PWM inverter. For grid applications, the impact of abnormal voltage and frequency ride through, grid faults, lighting surge, and unbalance will be considered. The converter grounding requirements and design will be discussed.

Chapter 13 discusses methodology to achieve optimal design results efficiently for three-phase AC converters through design automation. Design optimization concept, general procedure, and desired features for a design automation tool will be introduced. Mathematical formulation and classification for optimization problems will be introduced. Optimization algorithms will be discussed with a focus on nonlinear, hybrid (with both continuous and discrete variables), and constrained problems, as generally all converter design problems fall into these categories. The partitioned and centralized optimizer structures for converter design will be discussed. The development of example three-phase converter design tools will be presented. Example designs for using such a tool for whole three-phase AC converters or their subsystems will be demonstrated.

References

1 B. K. Bose, "*Power Electronics and Motor Drives: Advances and Trends,*" Elservier, 2006.

2 B. K. Bose and F. Wang, "Energy, environment, power electronics, renewable energy systems, and smart grid," in *Power Electronics in Renewable Energy Systems and Smart Grid: Technology and Applications*, B. K. Bose, ed., IEEE Press, John Wiley & Sons, 2019.

3 F. Wang, Z. Zhang, T. Ericsen, R. Raju, R. Burgos and D. Boroyevich, "Advances in power conversion and drives for shipboard systems," *Proceedings of the IEEE*, vol. 103, no. 12, pp. 2285–2311, 2015.

4 J. Li, J. Liu, D. Boroyevich, P. Mattavelli and Y. Xue, "Three-level active neutral-point-clamped zero-current-transition converter for sustainable energy systems," *IEEE Transactions on Power Electronics*, vol. 26, no. 12, pp. 3680–3693, 2011.

5 B. Guo, F. Wang and E. Aeloiza, "A novel three-phase current source rectifier with delta-type input connection to reduce device conduction loss," *IEEE Transactions on Power Electronics*, vol. 31, no. 2, pp. 1074–1084, 2016.

6 L. Huber and D. Borojevic, "Space vector modulated three-phase to three-phase matrix converter with input power factor correction," *IEEE Transactions on Industry Application*, vol. 31, no. 6, pp. 1234–1246, 1995.

7 F. Wang, W. Shen, D. Boroyevich, S. Ragon, V. Stefanovic and M. Arpilliere, "Voltage source inverter – development of a design optimization tool," *IEEE Industry Applications Magazine*, vol. 15, no. 2, pp. 24–33, March–April 2009.

8 C. Chen, X. Pei, Y. Chen and Y. Kang, "Investigation, evaluation, and optimization of stray inductance in laminated busbar," *IEEE Transactions on Power Electronics*, vol. 29, no. 7, pp. 3679–3693, 2014.

9 J. V. Wyk, "Power electronics quo vadis?," in *15th International Power Electronics and Motion Control Conference (EPE/PEMC)*, Novi Sad, Serbia, 2012.

10 R. Lai, F. Wang, R. Burgos, Y. Pei, D. Boroyevich, T. A. Lipo, B. Wang, V. Immanuel and K. Karimi, "A systematic topology evaluation methodology for high-density PWM three-phase AC-AC converters," *IEEE Transactions on Power Electronics*, vol. 23, no. 6, pp. 2665–2680, November 2008.

Part I

Components

2

Power Semiconductor Devices

2.1 Introduction

This chapter is on the characteristics of power semiconductor devices for the design of power electronics converters. Modern power converters, three-phase AC converters included, are based on switching mode circuits to realize various electric power conversion needs from one voltage and frequency form to another. At the heart of these converters are power semiconductors devices. The desired device characteristics are low on-state voltage drop and low off-state leakage current for low steady-state loss, and fast turn-on and turn-off speed for low switching loss. For needed power conversion capabilities, these switches should also possess sufficient blocking voltage ratings during off-state and conducting current ratings during on-state.

Silicon (Si) based semiconductor switching devices have dominated power electronics applications. Since the advent of Si thyristors in 1957, many Si-based switching devices have been developed to meet different application and performance needs. Today, for three-phase AC applications, Si-insulated gate bipolar transistors (IGBTs) are used predominately, especially when the voltages are above several hundred volts and power above several kilowatts, due to their overall superior characteristics in these ranges. Si power metal–oxide field-effect transistors (MOSFETs), which have better switching characteristics but worse conduction loss and lower voltage and power capabilities than IGBTs, can be found mainly in low-power applications. In some AC systems, integrated gate-commutated thyristors (IGCTs) are used in place of IGBTs, especially for MW-level medium or high-voltage applications, such as medium-voltage motor drives and high-voltage grid converters. Conventional thyristors including gate turn-off (GTO) thyristors can also be found in applications that use bridge rectifier, cycloconverter, and load-commutated inverter (LCI) topologies.

Si power semiconductor devices have gone through many generations of development in the past 60 years and are approaching their material theoretical limits in terms of conduction and switching performance characteristics. Although the Si-based power converters today are already highly efficient, generally with efficiencies greater than 95% or even as high as 98%, they are not as efficient as the traditional electromagnetic converters, i.e. magnetic transformers. The relatively high loss and limited switching speed of the Si devices also limit further improvement on power density, mainly due to the cooling system need for removing loss and passive components need for smoothing and filtering switching ripples. For example, the typical power density and specific power for MW medium voltage drive are about 1.1 MW/m^3 and 1.3 kW/kg [1]. Further significant improvements with Si devices will be difficult.

Design of Three-phase AC Power Electronics Converters, First Edition. Fei "Fred" Wang, Zheyu Zhang, and Ruirui Chen.
© 2024 The Institute of Electrical and Electronics Engineers, Inc. Published 2024 by John Wiley & Sons, Inc.

The emergence of WBG semiconductor devices promises to revolutionize next-generation power electronics converters. Compared with Si devices, WBG devices feature high breakdown electric field, low specific on-resistance, fast switching speed, and high junction temperature capability. The WBG devices under rapid development and commercialization include SiC and GaN devices, with SiC mainly targeting high-voltage high-power (650 V, kilowatts or above) applications and GaN for low-voltage low-power (650 V, kilowatts or below) applications. With high breakdown electric field, WBG devices have higher voltage capability and smaller packages; with low loss and high temperature capability, the cooling requirements can be reduced; with fast switching speed and low switching loss, high switching frequency can be utilized, which will lead to better dynamics and reduced filter and passive components needs in general.

In this chapter, six types of power devices that are commonly used in three-phase AC applications will be discussed, including widely applied Si IGBTs, Si PiN diodes, Si thyristors, and emerging SiC MOSFETs, SiC Schottky diodes, and GaN high-electron-mobility transistors (HEMTs). From the perspective of converter design, this chapter aims at illustrating device characteristics that could significantly affect the converter performances, such as efficiency, power density, specific power, reliability, and cost. Specifically, the critical device characteristics for the converter design include static characteristics (e.g. on-state characteristics, transfer characteristics, leakage, and breakdown characteristics), switching characteristics (e.g. switching time, switching energy loss), and thermal characteristics (e.g. junction-to-case thermal resistance, transient thermal impedance). Other device attributes, such as reliability, mechanical, and nonstandard characteristics (e.g. dynamic on-state resistance of GaN HEMTs), play a crucial role on the converter design and will also be discussed. Note that this chapter will not focus on the power semiconductor physics; instead, we attempt to model the relationship between device characteristics and converter design variables and conditions based on the understanding of power semiconductor physics. Taking SiC MOSFETs as an example, as seen in Figure 2.1, different techniques, such as planar structure and trench structure, lead to different behaviors; however, a general model can be used for the converter design while the specific parameters associated with the model could be very different. In this case, instead of describing different device technologies separately, we will only provide a single set of models. Readers could apply these models with the specific parameters in their own applications.

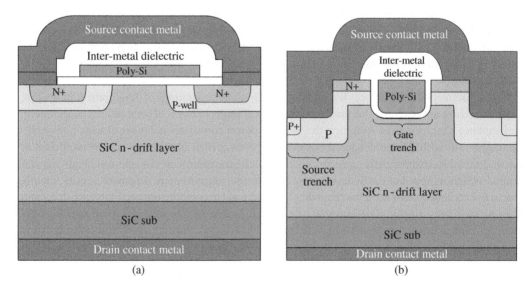

Figure 2.1 Structure of SiC MOSFETs: (a) planar DMOS structure and (b) trench structure.

Additionally, we will identify the design constraints originated from power devices and provide methodology to obtain accurate model considering best engineering practices. Finally, a relevance between the derived models at the device level and converter design will be illustrated and a package of delivery from this chapter to support the converter design will be summarized.

2.2 Static Characteristics

Static characteristics primarily include output characteristics, on-state characteristics, transfer characteristics, and leakage/breakdown characteristics. Other important device parameters based on device static characterization, such as junction capacitance and gate charge, are also discussed.

2.2.1 Output Characteristics

Output characteristics, i.e. V–I curve, is the most critical behavior for the device static characteristics, of which a single chart illustrates the V–I relationship depending on gate control, if applicable, in the wide operating ranges. The on-state characteristics and transfer characteristics, which are highly relevant to the converter design, can be derived based on the output characteristics.

Table 2.1 displays the typical output characteristics of diode, thyristor, IGBT, MOSFET, and HEMT.

Depending on the voltage polarity and current direction defined in Table 2.1, the passive diode, including PiN diode and Schottky diode, blocks positive voltage and conducts reverse current; thyristor is able to block bipolar voltages and conduct forward current; IGBT can block positive voltage and conduct forward current; while MOSFET and HEMT can block positive voltage and conductor bidirectional currents.

Thyristor, also known as silicon-controlled rectifier (SCR), starts to conduct current triggered by the gate current pulse and will be latched in current conduction mode until its current naturally goes to zero, the voltage across it is reversed biased, or until the voltage is removed. It is also observed that the thyristor can handle positive and negative voltages when it is off.

For IGBT, MOSFET, and HEMT, the device on/off status and corresponding performance are a function of controlled gate voltages, as indicated by multiple lines in the V–I curve. Unlike IGBT, which only conducts the forward current, both MOSFET and HEMT are able to handle reverse current through either the device channel or body diode of MOSFET/diode-like-behavior (DLB) of HEMT. The gate voltage regulates the device channel resistance and then determines reverse conduction performance. Therefore, a set of lines can also be observed in the reverse conduction region where the leftmost one (the most lossy curve) is purely the diode behavior when the gate turns off the device channel.

In addition to on/off states, the V–I curve of the active devices also illustrates the behavior during the active regions, i.e. the region with relatively large current and high voltage through/across the active device simultaneously. It indicates the saturation current and affects the switching behavior, which are critical to the converter design.

It is worth noting that the terminologies to describe different regions for bipolar transistors (e.g. IGBT) versus unipolar transistors (e.g. MOSFET) are quite different: saturation region for IGBT and ohmic region for MOSFET refer to devices in the on state while active region for IGBT and saturation region for MOSFET indicate devices with relatively large current and high voltage through/across the active device simultaneously.

Lastly, V–I curve of all devices listed in Table 2.1 are dependent on temperature, which is an irreplaceable variable in the device models.

Table 2.1 Typical output characteristics dependent on device types.

Device type	Symbol	Output characteristics
Diode		
Thyristor		
IGBT		
MOSFET		
HEMT		

2.2.1.1 Model

Two important device models could be obtained based on the device V–I curve, including on-state characteristics and saturation characteristics (e.g. transfer characteristics) for the active devices, which will be discussed in Sections 2.2.2 and 2.2.3.

2.2.1.2 Method

Typically, device manufacturer datasheet provides the V–I curve as a function of temperature. A SiC MOSFET based on the design example in Chapter 6 is displayed in Figure 2.2. Also note, multiple temperature-dependent curves are provided in most device datasheet.

For emerging power devices, like SiC and GaN engineering samples, sometimes, the datasheet is not available. Even with datasheet, verification is often desired. Therefore, a dedicated test needs to be performed to obtain the V–I curve. Pulsed V–I testing is widely applied.

The basic concept of the pulsed V–I testing is to pulse the device in a temporary on-state, allowing a precise control of the operating conditions without significant increase in junction temperature. Unlike the dynamic characterization in Section 2.3, the transients at the beginning of the pulse are only used to turn on the device without loads. The area of interest is the middle of each pulse, neglecting any dynamic effects caused by the turn-on or turn-off transients. It is therefore desirable to avoid hard switching, which injects resonant energy into the circuit and causes ringing.

Figure 2.3 displays a circuit representation of a typical V–I testing setup with MOSFET's terminology (gate, drain, and source) as an example. Two controlled pulse voltage sources are introduced across gate-source terminal and drain-source terminal in series with two current limiting resistors. Two ammeters and one voltmeter are applied to measure drain current, gate current, and drain-source voltage, respectively. During the test, (i) gate-source voltage is established by one controllable voltage source; (ii) after gate-source voltage is in steady state, the other controllable voltage source provides the drain-source voltage, and therefore the drain current; (iii) after certain delay when drain-source voltage and drain current are in steady state, gate-source voltage, drain-source voltage, and drain current are measured using a four-wire Kelvin connection; (iv) then one V–I point is obtained; and (v) repeat these steps to obtain entire V–I curve.

The timing of the gate-source and drain-source voltage pulses must be precisely tuned to accurately capture steady-state behavior in a pulsed V–I test. The drain-source voltage pulse must begin

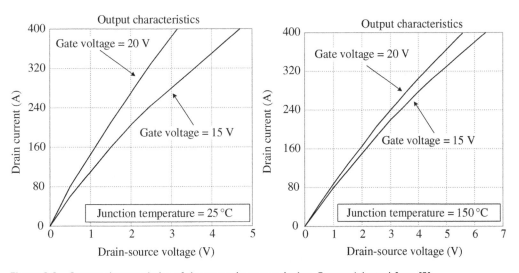

Figure 2.2 Output characteristics of the example power device. *Source:* Adapted from [2].

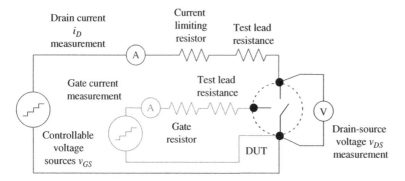

Figure 2.3 Circuit representation of a typical pulsed V–I setup.

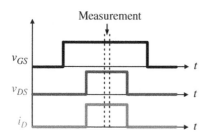

Figure 2.4 Waveform timing settings for pulsed I–V testing.

after the gate-source voltage pulse, and only after the gate has had sufficient time to turn fully on, as illustrated in Figure 2.4.

Typically, power device analyzer (i.e. curve tracer) is used for the pulsed V–I testing to obtain the V–I curve [3]. As emphasized earlier, the settings are critical for the measurement accuracy, which is well-documented in [4] and will not be repeated here.

2.2.2 On-state Characteristics

Device on-state characteristics significantly affect the conduction loss calculation, which is critical for the converter design. Different types of devices have different on-state characteristics, as displayed in Figure 2.5. Generally, on-state voltage is used to indicate the on-state characteristics, which is contributed by on-state resistance along with the knee voltage, if applicable. As can be observed in Figure 2.5, knee voltage exists for minority-carrier devices PiN diode, thyristor, and IGBT, while there is no knee voltage for majority-carrier devices MOSFET and HEMT. It is noted that Schottky diode, a majority-carrier device, also has knee voltage though the value is lower than that of PiN diode. The channel of MOSFET and HEMT can conduct reverse current when device is on, and the reverse conduction on-state resistance is a function of gate voltage determined by the turn-on driving voltage. If the channel turns off, the body diode of MOSFET and DLB of HEMT are also able to conduct the reverse current but less efficiently. Also note, the knee voltage of the body diode of MOSFET and DLB of HEMT is a function of gate voltage controlled by the turn-off driving voltage.

2.2.2.1 Model

For different types of devices, the on-state voltages as a function of current directions are summarized in Table 2.2. V_{on} is the on-state voltage; V_{knee} is the knee voltage, the point on a device's V–I curve where conductivity increases rapidly, which exists for diode, IGBT, and thyristor. For IGBT, collector-to-emitter saturation voltage $V_{CE(SAT)}$ is typically applied, including the impact of V_{knee} and on-state resistance R_{on}. For MOSFET and HEMT, R_{dson} is usually used for R_{on}. $i_L(t)$ indicates the instantaneous current flowing through the device; V_{CC} and V_{EE} refer to the on-state and off-state gate voltages, respectively; T_j is the device junction temperature. Note that, V_{knee} and R_{on} are device parameters and highly dependent on converter operating conditions (T_j) and converter design variables (V_{CC} and V_{EE}). Therefore, instead of a single dataset, a curve should be provided from the

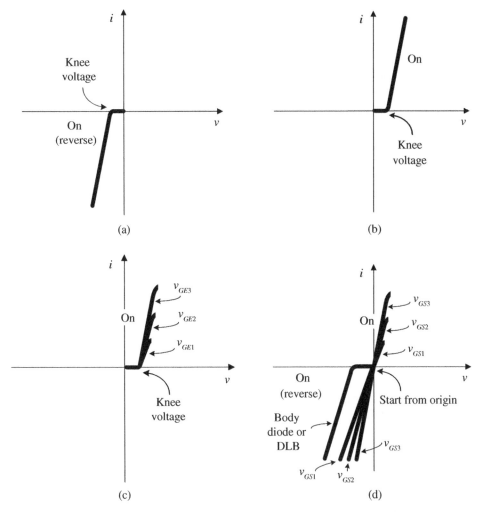

Figure 2.5 On-state characteristics of power devices. (a) Diode, (b) thyristor, (c) IGBT, and (d) MOSFET including body diode or HEMT with DLB.

device model to illustrate V_{knee}/R_{on} as a function of $T_j/V_{CC}/V_{EE}$. It is also worth noting that the on-state resistance R_{on} follows a first-order approximation of the V–I curve when the device is on, of which the accuracy can be further improved with higher order approximation. Therefore, it is found that, sometimes, the designer may consider R_{on} as a function of current i_L to model such relationship more accurately. Additionally, for MOSFET and HEMT when channel in on, device channel and its body diode (MOSFET) or DLB (HEMT) branch are in parallel. Typically, device channel is more conductive and takes most of the current if not all.

2.2.2.2 Method
Device on-state characteristics can be extracted from the V–I curve provided by the datasheet. In the meantime, device manufacturer usually provides separate curves to illustrate the relationship between on-state voltage (e.g. $V_{CE(SAT)}$ for IGBTs, V_f for diodes) and on-state resistance (e.g. R_{dson} for MOSFTs and HEMTs) as a function of device junction temperature and load current at the recommended driving voltage.

Table 2.2 On-state models of different types of power devices.

Current direction	Device type	On-state voltage model
Forward conduction	IGBT, thyristor	$$V_{on} = V_{knee}(T_j) + R_{on}(V_{CC}, T_j) \times i_L(t) \qquad (2.1)$$
	MOSFET, HEMT	$$V_{on} = R_{on}(V_{CC}, T_j) \times i_L(t) \qquad (2.2)$$
Reverse conduction	Diode (PiN and Schottky)	$$V_{on} = -V_{knee}(T_j) - R_{on}(T_j) \times i_L(t) \qquad (2.3)$$
	MOSFET, HEMT when channel is off	$$V_{on} = -V_{knee}(V_{EE}, T_j) - R_{on}(T_j) \times i_L(t) \quad (2.4)$$
	MOSFET, HEMT when channel is on	$$V_{on} \approx -R_{on}(V_{CC}, T_j) \times i_L(t) \qquad (2.5)$$

The on-state characteristics of the same device example displayed in Figure 2.2 are illustrated in Figure 2.6, including SiC MOSFET and SiC diode.

In case the datasheet of emerging power devices has limited data or is not available, the on-state characteristics could be simply obtained based on the tested output characteristics using the pulsed V–I setup in Figure 2.3. It is noted that for MOSFET and HEMT where the on-state resistance is usually adopted, its value under given temperature and load current is determined by the ratio of absolute value of V and I (i.e. the slope from V–I point to origin) rather than the slope determined

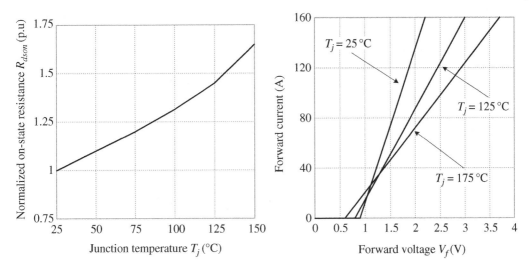

Figure 2.6 On-state characteristics of the example power device. *Source:* From [2] with permission of Microsemi.

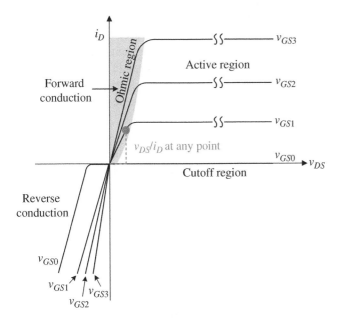

Figure 2.7 On-state resistance derived based on V–I curve.

by small-signal dv/di, as illustrated in Figure 2.7. Based on the series of V–I curve dependence on temperature, the model of on-state resistance as a function of temperature and load current can be derived.

2.2.3 Transfer Characteristics of Active Power Switch

Although active power switch is not designed to operate at high voltage and large current simultaneously (i.e. active region for switches), from the converter design perspective, it is essential to understand its characteristics because (i) it affects the switching trajectory and the corresponding switching behavior and (ii) it dominates the switch behavior at the overload condition and shoot-through event, such as the saturation current. Typically, transfer characteristics provide the relationship for active power switch at high voltage and large current simultaneously. Additionally, users could obtain threshold voltage and transconductance information based on transfer characteristics, which are critical data for the modeling of device dynamic behavior (e.g. voltage overshoot, etc.), presented in Chapter 6.

2.2.3.1 Model

The primary model derived from transfer characteristics is the relationship between device channel current and gate voltage when device operates in the active region. Like other models for device, this relationship is also a function of device temperature. Considering the "Early effect" [5], it also varies as the device voltage changes in the active region (i.e. collector-emitter voltage for IGBT and drain-source voltage for MOSFET and HEMT), yielding

$$i_{ch} = f\left(v_{GE} \text{ or } v_{GS}, v_{CE} \text{ or } v_{DS}, T_j\right) \tag{2.6}$$

where i_{ch} is the device channel current; v_{GE} and v_{GS} are the gate-emitter voltage for IGBT and gate-source voltage for MOSFET/HMET, respectively; v_{CE} and v_{DS} are the voltage across power terminals of IGBT and MOSFET/HMET, respectively. Please notice that channel current i_{ch} in (2.6) is not identical to the current flowing through the collector terminal of IGBT i_C or drain terminal of MOSFET/HEMT i_D with the consideration of displacement current induced by junction capacitance during the dv/dt transient. However, transfer characteristics represent the device static behavior, therefore, sometimes, it is observed that i_C and i_D are used instead of i_{ch}. Once transfer characteristics are adopted for the device switching performance analysis, i_{ch} has to be applied.

Ignoring the impact of T_j and v_{CE}/v_{DS}, a simple analytical formula can be derived as

$$i_{ch} = g_{fs}\left(v_{GE} - v_{th}\right) \text{ or } i_{ch} = g_{fs}(v_{GS} - v_{th}) \tag{2.7}$$

where g_{fs} is the transconductance and v_{th} is the threshold voltage. Apparently, this is a simple first-order approximation. In fact, both g_{fs} and v_{th} are nonconstant and are functions of T_j at least. v_{CE}/v_{DS} and v_{GE}/v_{GS} also affect g_{fs}

$$v_{th} = f\left(T_j\right) \tag{2.8}$$

$$g_{fs} = f\left(v_{GE} \text{ or } v_{GS}, v_{CE} \text{ or } v_{DS}, T_j\right) \tag{2.9}$$

2.2.3.2 Method

Similar to on-state characteristics, transfer characteristics could be extracted from the V–I curve provided by the datasheet. In the meantime, device manufacturer usually provides separate curves as a function of temperature with the given v_{CE}/v_{DS}. One SiC MOSFET example is listed in Figure 2.8.

In the meantime, transfer characteristics could also be derived by device V–I curve discussed in Section 2.2.1. As illustrated in Figure 2.9a taking a MOSFET as an example, at the device active region with a fixed v_{DS}, the relationship of i_D (actually channel current) and v_{GS} could be established. Furthermore, v_{th} and g_{fs} can be derived based on the transfer characteristics, as displayed in Figure 2.9b. It is noted that, practically, v_{th} under the given temperature is determined by a predefined current. Different device manufacturers may use different currents. Typically, a range of mA is adopted. Additionally, g_{fs} is the slope of transfer characteristics. Based on the series of V–I curve dependence on temperature, transfer characteristics dependence on temperature can be derived.

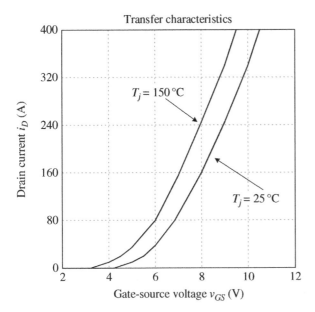

Figure 2.8 Transfer characteristics of the example power device. *Source:* From [2] with permission of Microsemi.

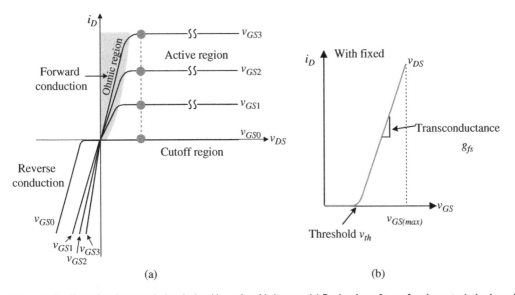

Figure 2.9 Transfer characteristics derived based on V–I curve. (a) Derivation of transfer characteristics based on V–I curve and (b) derived transfer characteristics.

2.2.4 Leakage Current and Breakdown Voltage

In this section, power terminal leakage current and breakdown voltage are presented. More importantly, from the converter design perspective, the determination of maximum operating voltage is discussed.

Regarding the devices with the gate terminal, the gate oxide breakdown is also crucial. From the converter designer's point of view, to avoid gate oxide breakdown, the best practice is to follow the recommended turn-on and turn-off gate voltages provided by the device manufacturers. These ratings are usually available, even for the emerging WBG devices. Generally, the selection of gate voltages is a tradeoff between efficient conduction, reliable gate voltage stress, and sufficient safety margin to tolerant any disturbances, like the crosstalk [6]. This can be achieved in the collaboration with the gate driver design, as will be discussed in Chapter 10.

Unlike gate voltage ratings, the determination of power terminal maximum operating voltage is not that straightforward, and in many cases, is highly related to applications, which will be discussed next.

2.2.4.1 Model

Device breakdown voltage is a derived value based on the device V–I curve at a predefined current (i.e. leakage current) when the device is off (i.e. reverse biased for power diode and gate voltage below threshold voltage for active switches), as expressed by

$$V_{BD} = f(T_j) @ \text{ given } I_{lkg} \tag{2.10}$$

where V_{BD} is the device breakdown voltage and I_{lkg} is the corresponding leakage current. The converter designer would not use this value directly as the maximum operating voltage; instead, practically, a voltage margin based on engineering insights is implemented. This voltage margin is typically determined based on the application and fundamentally driven by the reliability consideration. Considering the device could suffer increased dynamic voltage spike as compared to the voltage in the steady state, it is suggested to introduce two different voltage margins for the steady state and dynamic transition, respectively. Obviously, static voltage margin should be larger than the dynamic one as the former should cover the latter. For example, 50% (based on V_{BD}) is often applied as the steady-state voltage margin considering the failure modes, such as cosmic ray, but close to 0% could be adopted for the dynamic margin for applications like industrial motor drive with Si IGBT. Therefore, two associated models are introduced here and their values are determined based on applications.

$$V_{design\_steady} = (1 - VM_{steady}(application)) \times V_{BD} \tag{2.11}$$

$$V_{design\_dynamic} = (1 - VM_{dynamic}(application)) \times V_{BD} \tag{2.12}$$

where V_{design_steady} and $V_{design\_dynamic}$ are steady-state and dynamic voltage design constraints, respectively. VM_{steady} and $VM_{dynamic}$ are steady and dynamic voltage margins in %, respectively.

2.2.4.2 Method

Typically, the breakdown voltage information is provided as the first parameter in the electrical characteristics table of the device datasheet, including the drain current (i.e. leakage current I_{lkg}) data under the given breakdown voltage V_{BD} with zero gate voltage. Table 2.3 displays a datasheet example of SiC MOSFET. The predefined leakage current is manufacturer-dependent and selected in the range of hundreds of micro-amperes. It is also noted that only a single data point at the room temperature is provided, which, usually, is sufficient to support most of the converter design. For converter design under the harsh environment, such as cryogenic temperature or extremely high

Table 2.3 Breakdown voltage of the example power device.

Symbol	Description	Test conditions	Min.	Typ.	Max.	Unit
I_{DSS}	Zero gate voltage drain current	$V_{GS} = 0\,\text{V}, V_{DS} = 1200\,\text{V}$			400	μA

Source: Adapted from [2].

temperature (e.g. up to 200 °C), breakdown voltage in a wide range of the temperature is desired. As a result, a dedicated test needs to be performed.

Theoretically, V–I curve discussed in Section 2.2.1 can provide the information to determine the breakdown voltage based on Eq. (2.10), but due to the measurement accuracy issue (it is challenging, based on the same test setup, to measure the current at the device rated value, in the meantime, at the micro-amperes scale), a separate testing is designed to characterize the device leakage/breakdown.

The basic testing circuit and operation principle are highlighted in Figures 2.10 and 2.11 with MOSFET's terminology (gate, drain, and source) as an example. Gate-source voltage is held below threshold to ensure device in the cutoff region (terminals may directly be shorted with zero-volt gate-source voltage). A high-voltage controllable source is connected across drain-source terminals and then controlled gradually with an increased v_{DS} with a current limiting resistor in series. In the meantime, leakage current i_D and v_{DS} are measured. Once i_D approaches to the preset leakage current threshold, high-voltage controllable source stops increasing, the corresponding v_{DS} may be identified as the breakdown voltage. The test is repeated under different temperatures to obtain the breakdown voltage information dependence on temperature.

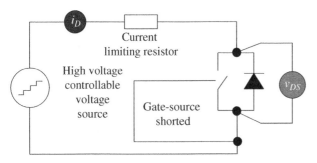

Figure 2.10 Circuit representation of a typical breakdown/leakage testing.

Figure 2.11 Waveform illustration for breakdown/leakage testing.

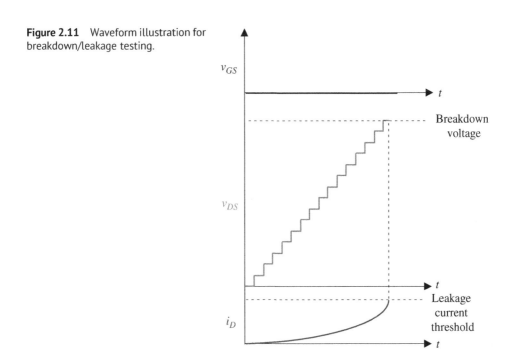

2.2.5 Junction Capacitance

Parasitic capacitances are formed within the power device between any two terminals at the die level and in the package. At the die level, these capacitances partly represent the charge separation that must occur to create the depletion region that blocks voltage and also the recombination of these charges that restores the conductive channel. The capacitance representing this charge separation is highly nonlinear with respect to voltage. The nonlinearity depends on the geometry and material properties of the device. There are also additional sources of capacitance between contact metallizations, field plates, and the epitaxial layers of the die. Metal in the package may add capacitance as well. While it is not necessary to understand the physical origin of these capacitances to characterize them, it is helpful to remember that the physical structure of the device has a tremendous impact on these characteristics. For example, in a Super Junction MOSFET, the capacitance between drain and source may change by several orders of magnitude over the rated voltage range, while in others it may only change moderately. Similarly, the capacitance from gate-source terminals typically changes more dramatically for GaN HEMTs over the rated gate voltage range than in typical MOSFETs.

From the perspective of converter design, junction capacitances significantly affect the switching behavior for both hard switching and soft switching applications. Their influence becomes even more critical for WBG devices with high switching speed capability. Therefore, junction capacitances are critical data for modeling of device dynamic behavior (e.g. switching time, switching energy loss, voltage overshoot, etc.).

2.2.5.1 Model

Based on the symbolic illustration in Figure 2.12, depending on the types of devices, the notations are different. Nevertheless, similar nonlinear characteristics as a function of power terminal voltages remain.

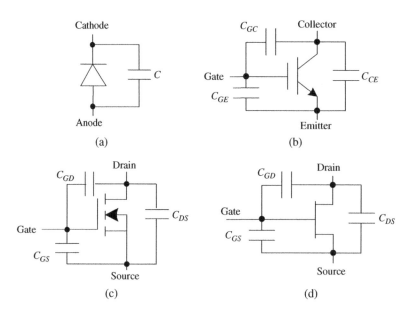

(a)   (b)

(c)   (d)

Figure 2.12 Junction capacitances in different power devices. (a) Diode (PiN diode symbol as an example, Schottky diode is similar), (b) IGBT, (c) MOSFET, and (d) HMET.

For diodes in Figure 2.12a where only two terminals exist, there is one junction capacitance across anode and cathode terminals, yielding

$$C = f(v_{AK}) \tag{2.13}$$

where C is the junction capacitance of the diode, and v_{AK} is the diode voltage across anode and cathode terminals.

For active devices with three terminals, there are three junction capacitances.

For IGBTs in Figure 2.12b, they are capacitances between the collector and emitter terminals C_{CE}, between the gate and emitter terminals C_{GE}, and between the gate and collector terminals C_{GC}. In practice (like the parameters displayed in the datasheet), they are more often described using the following definitions.

Output capacitance, C_{oes}, represents the total capacitance seen at the collector terminal with the emitter and gate shorted together.

$$C_{oes} = C_{CE} + C_{GC} = f(v_{CE}) \tag{2.14}$$

Input capacitance, C_{ies}, represents the total capacitance seen at the gate terminal with the collector and emitter shorted together.

$$C_{ies} = C_{GE} + C_{GC} = f(v_{CE}) \tag{2.15}$$

Reverse transfer capacitance, C_{res}, represents the capacitance seen between the gate and collector terminal with the emitter terminal floating. The reverse transfer capacitance, often referred to as the Miller capacitance, is one of the major parameters affecting voltage rise and fall times during switching.

$$C_{res} = C_{GC} = f(v_{CE}) \tag{2.16}$$

MOSFETs and HEMTs have three capacitance but with different notations as compared to IGBTs, as illustrated in Figure 2.12c and d: capacitances between the drain and source terminals C_{DS}, between the gate and source terminals C_{GS}, and between the gate and drain terminals C_{GD}. Similar to IGBTs, three lumped parameters are usually observed in the device datasheet, as expressed by

$$C_{oss} = C_{DS} + C_{GS} = f(v_{DS}) \tag{2.17}$$

$$C_{iss} = C_{GS} + C_{GD} = f(v_{DS}) \tag{2.18}$$

$$C_{rss} = C_{GD} = f(v_{DS}) \tag{2.19}$$

where C_{oss} is the output capacitance, C_{iss} is the input capacitance, and C_{rss} is the reverse transfer capacitance, i.e. Miller capacitance.

2.2.5.2 Method

Usually, comprehensive junction capacitance curve(s) are provided in the device datasheet as a function of power terminal voltage. Figure 2.13 displays the datasheet example. For emerging devices without available junction capacitance data, C–V testing is used based on a precision impedance analyzer. A test setup with MOSFET's terminology (gate, drain, and source) as an example is provided in Figure 2.14. Detailed operation principles and test procedures are described in [4] and will not be repeated here.

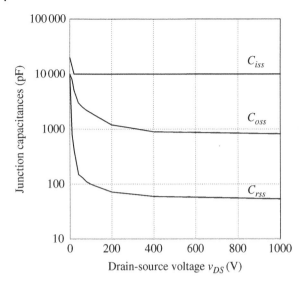

Figure 2.13 Junction capacitance of the example power device. *Source:* From [2] with permission of Microsemi.

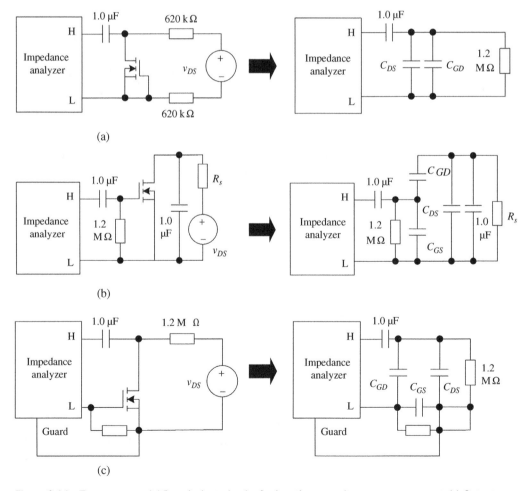

Figure 2.14 Test setups and AC equivalent circuits for junction capacitances measurement. (a) Output capacitance C_{oss}, (b) input capacitance C_{iss}, and (c) reverse transfer capacitance C_{rss}.

2.2.6 Gate Charge

The gate charge characteristic shows the relationship between gate charge and gate voltage for a particular operating condition, including operating voltage, load current, and junction temperature. Therefore, it is a series of curves rather than a single data point. Many informative data could be derived based on the gate charge curve, such as total gate charge supplied by the gate drive, charge needed to bring gate voltage to its threshold voltage, gate charge needed to bring gate voltage up to the Miller voltage, and Miller charge.

From the perspective of converter design, gate charge determines the gate driver design to be discussed in Chapter 10.

2.2.6.1 Model

In general, gate charge characteristics can be derived based on junction capacitance curve in Section 2.2.5. The critical information for the converter design, such as device switching analysis and gate driver design, can be obtained based on following models. Please note the following Eqs. (2.20)–(2.24) depend on the MOSFETs' terminology. Other types of devices could follow the similar formulas to obtain the corresponding information.

$$Q_{GS,th} = \int_{V_{EE}}^{V_{th}} C_{GS}(v_{GS})dv_{GS} \tag{2.20}$$

where $Q_{GS,th}$ is the gate charge needed to bring gate voltage to its threshold voltage, indicating the switching delay time; v_{th} is the threshold voltage, which can be obtained based on transfer characteristics in Section 2.2.3; V_{EE} is the turn-off gate voltage, which is a design variable associated with the gate driver.

$$Q_{GS,Miller} = \int_{V_{EE}}^{V_{Miller}} C_{GS}(v_{GS})dv_{GS} \tag{2.21}$$

where $Q_{GS,Miller}$ is the gate charge needed to bring gate voltage up to the Miller voltage, indicating the switching delay time plus the *di/dt* transition; V_{Miller} is the Miller plateau voltage to be discussed in Section 2.3 and can be obtained from transfer characteristics in Section 2.2.3 as a function of junction temperature and load current.

$$Q_{GD} = \int_{0}^{V_{DC}} C_{rss}(v_{DS})dv_{DS} \tag{2.22}$$

where Q_{GD} is the gate charge during the Miller plateau, indicating the switching *dv/dt* transition. V_{DC} is the operating voltage.

$$Q_{G,total} = Q_{GS,Miller} + Q_{GD} + \int_{V_{Miller}}^{V_{CC}} (C_{GS}(v_{GS}) + C_{rss}(v_{DS} = 0V))dv_{GS} \tag{2.23}$$

where $Q_{G,total}$ is the total gate charge supplied by the gate driver, and it is important for the gate driver design; V_{CC} is the turn-on gate voltage which is a design variable associated with the gate driver.

Overall, the entire gate charge curve can be interpolated using the piecewise function as

$$Q_G = \begin{cases} \int_{V_{EE}}^{v_{GS}} C_{GS}(v_{GS})dv_{GS}, & \text{for} \quad v_{GS} < V_{Miller} \\ Q_{GS,Miller} + Q_{GD} + \int_{V_{Miller}}^{v_{GS}} [C_{GS}(v_{GS}) + C_{rss}(v_{DS} = 0V)]dv_{GS}, & \text{for} \quad v_{GS} \geq V_{Miller} \end{cases} \tag{2.24}$$

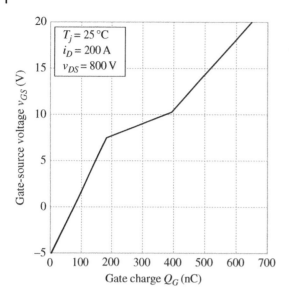

$T_j = 25\,^{\circ}C$
$i_D = 200\,A$
$v_{DS} = 800\,V$

Figure 2.15 Gate charge curve of the example power device. *Source:* From [2] with permission of Microsemi.

2.2.6.2 Method

Gate charge curve is provided in the device datasheet as a function of operating voltage, load current, and temperature. Figure 2.15 displays a datasheet example. For different operating conditions, Eq. (2.24) could be used to calculate the corresponding gate charge.

2.3 Switching Characteristics

Device switching characteristics determine the converter performance and is one of the most critical data for the converter design. Typically, it includes two sets of information: switching time and switching energy loss, which dominate the switching loss calculation, switching frequency selection, deadtime setting, and so on. Additionally, dynamic spikes, such as spurious gate voltage due to crosstalk [7] and overvoltage because of the parasitic ringing [8], also impact the device selection and converter optimization. In this section, basic model on the switching transient analysis and practical considerations on the switching time and switching energy loss acquisition are the focus; models considering device nonlinear characteristics (e.g. transconductance, junction capacitances) and circuit parasitics, together with the more precise high-order analytical expressions, will be discussed in Chapter 6.

2.3.1 Model

Switching characteristics is not only dependent on device parameters but also affected by gate driver, operating condition, and converter circuit parasitics. Figure 2.16 illustrates an equivalent circuit and device channel model in different states for switching analysis.

It is noted that this is a simplified model with a first-order approximation. The purpose here is to display the key waveforms and corresponding relationships during the switching transition and establish a basic version of the switching model to support the later discussion with more holistic considerations. Therefore, readers will not be overwhelmed at the first step by high-order differential equations. In the meantime, it is expected the analysis below is oversimplified with limited accuracy.

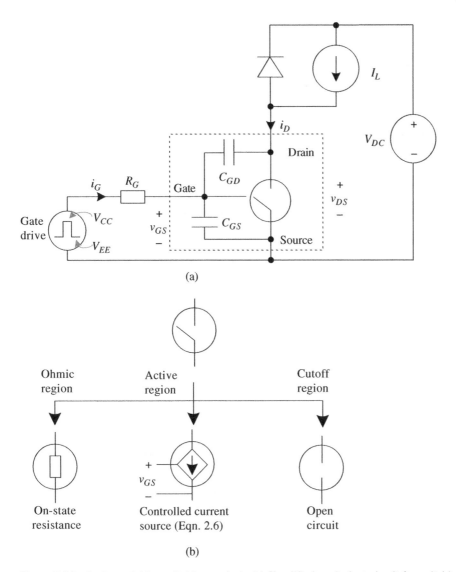

Figure 2.16 Basic model for switching analysis. (a) Simplified equivalent circuit for switching analysis; (b) device channel model in different states.

During the turn-on transient in Figure 2.17a, there are four subintervals: subinterval 1 – turn-on delay, subinterval 2 – current rise, subinterval 3 – voltage fall, and subinterval 4 – gate voltage rise. Table 2.4 summarizes the turn-on switching performance under subintervals 1–4 considering device states, gate loop behavior, and power loop behavior.

Similarly, during the turn-off transient in Figure 2.17b, there are four subintervals: subinterval 5 – turn-off delay, subinterval 6 – voltage rise, subinterval 7 – current fall, and subinterval 8 – gate voltage fall. Table 2.5 summarizes the turn-off switching performance under subintervals 5–8.

Based on the earlier analysis, Figure 2.18 displays the switching trajectory, which is another interesting view to better understand the switching transition. It also emphasizes the importance of the active region (i.e. transfer characteristics) on the switching behavior since during most of the time (subintervals 2, 3, 6, and 7), device V–I operating points are located in the active region.

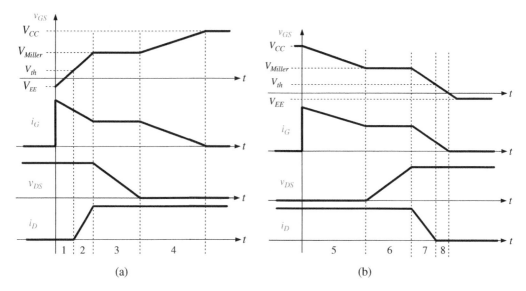

Figure 2.17 Typical waveforms for switching analysis. (a) Turn-on and (b) turn-off.

Table 2.4 Turn-on switching analysis.

Sub-interval	Device state	Gate loop	Power loop
1	Cutoff region	C_{GS} charged from V_{EE} to v_{th}	$i_D = 0$, $v_{DS} = V_{DC}$
2	Active region	C_{GS} charged from V_{th} to V_{Miller}	$i_D = g_{fs}[v_{GS}(t) - V_{th}]$, $v_{DS} = V_{DC}$
3	Active region	$v_{GS} = V_{Miller}$, C_{GD} charged	$i_D = I_L$, $v_{DS} = V_{DC} - \dfrac{V_{CC} - V_{Miller}}{C_{gd}R_g}t$
4	Ohmic region	C_{GS} charged from V_{Miller} to V_{CC}	$i_D = I_L$, v_{DS} further decreased

Table 2.5 Turn-off switching analysis.

Sub-interval	Device state	Gate loop	Power loop
5	Ohmic region	C_{GS} discharged from V_{CC} to V_{Miller}	$i_D = I_L$, $v_{DS} \approx 0$
6	Active region	$V_{GS} = V_{Miller}$, C_{GD} charged	$i_D = I_L$, $v_{DS} = \dfrac{V_{Miller}}{C_{GD}R_g}t$
7	Active region	C_{GS} discharged from V_{Miller} to V_{th}	$i_D = g_{fs}[v_{GS}(t) - V_{th}]$, $v_{DS} = V_{DC}$
8	Cutoff region	C_{GS} discharged from V_{th} to V_{EE}	$i_D = 0$, $v_{DS} = V_{DC}$

Apparently, the simplified models derived in Tables 2.4 and 2.5 ignore some important features during an actual switching transition, such as voltage spikes, current spikes, reverse recovery for non-Schottky diode, tail current for IGBT, and parasitic ringing. Therefore, unlike "ideal" switching waveform in Figure 2.17, an actual waveform based on a test example [9] displayed in Figure 2.19 is quite different. Extensive studies have been performed to derive more accurate models and some of them will be discussed in Chapters 6 and 7.

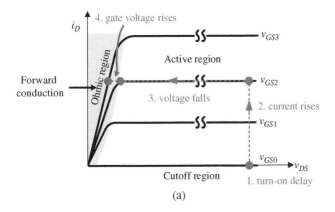

(a)

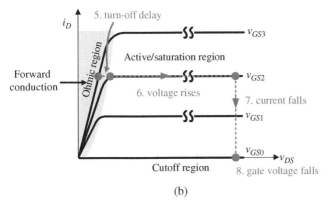

(b)

Figure 2.18 Switching trajectory. (a) Turn-on and (b) turn-off.

From converter design perspective, instead of deriving high-order mathematical models to represent an accurate device dynamic behavior, practically, a set of testing data is used. Therefore, a data-driven model is derived, as expressed

$$t_{on} = \int_{subinterval\ 1}^{subinterval\ 4} dt = f\left(V_{DC}, I_L, T_j, R_G, V_{CC}, V_{EE}, Z\right) \tag{2.25}$$

$$t_{off} = \int_{subinterval\ 5}^{subinterval\ 8} dt = f\left(V_{DC}, I_L, T_j, R_G, V_{CC}, V_{EE}, Z\right) \tag{2.26}$$

$$E_{on} = \int_{subinterval\ 1}^{subinterval\ 4} v_{DS}(t) i_D(t) dt = f\left(V_{DC}, I_L, T_j, R_G, V_{CC}, V_{EE}, Z\right) \tag{2.27}$$

$$E_{off} = \int_{subinterval\ 5}^{subinterval\ 8} v_{DS}(t) i_D(t) dt = f\left(V_{DC}, I_L, T_j, R_G, V_{CC}, V_{EE}, Z\right) \tag{2.28}$$

where t_{on} is the turn-on time, including subintervals 1–4 in Figure 2.17a; t_{off} is the turn-off time, including subintervals 5–8 in Figure 2.17b; E_{on} is the turn-on switching energy loss, the integral of device voltage and current product during subintervals 1–4; E_{off} is the turn-off switching energy loss, the integral of device voltage and current product during subintervals 5–8; V_{DC}, I_L, and T_j refer to the operating conditions; R_G is the gate resistance; and Z represents circuit parasitic impedance.

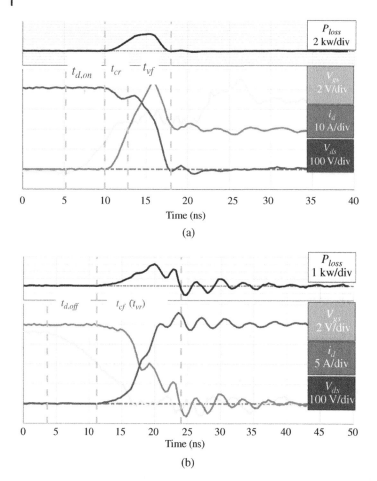

Figure 2.19 Actual switching waveform based on test using GaN Systems GS66508P. (a) Turn-on (b) turn-off.

In practice, R_G could be set differently for turn-on and turn-off, respectively, if so, separate R_{G_on} and R_{G_off} should be used.

It should also not be ignored that if the nonoperating device, e.g. the upper diode in Figure 2.16a has reverse recovery current when it turns off, that is, when the operating device, e.g. lower switch in Figure 2.16a turns on, there will be reverse recovery energy loss

$$E_{rr} = f\left(V_{DC}, I_L, T_j, R_G, V_{CC}, Z\right) \tag{2.29}$$

2.3.2 Method

Switching behavior data are usually provided by the device datasheet in a relatively comprehensive way. Figure 2.20 displays a datasheet example. Sometimes, customized test is needed due to (i) relevant data is limited or not available, such as [10]; (ii) device switching behavior is highly sensitive to the physical circuit design and implementation. Many factors, such as gate driver design, parasitic management, and diode characteristics of nonoperating switch, may affect the results. This is more important for WBG devices, as evidenced by Figure 2.21. Therefore, beyond the device datasheet, it may be preferred to perform an in-house test to obtain more accurate data.

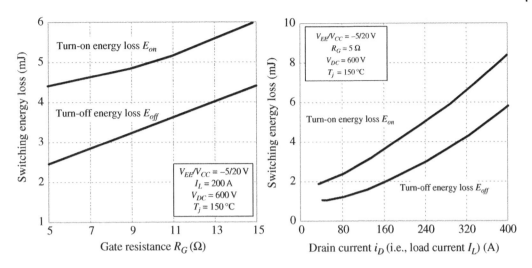

Figure 2.20 Switching characteristics of the example power device. *Source:* From [2] with permission of Microsemi.

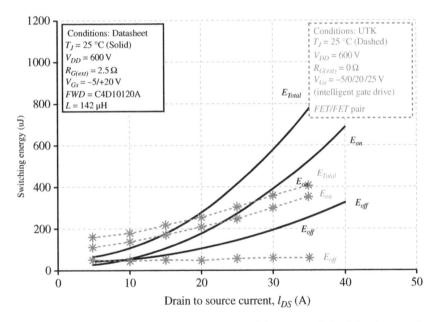

Figure 2.21 Datasheet value versus test data considering actual circuit implementation.

Double pulse test (DPT) is typically used to assess the dynamic performance of power devices [11]. Two pulses are sent to the lower switch (i.e. device under test (DUT)) in a clamped inductive load circuit, as shown in Figure 2.22b. During the first pulse, DUT is on, DC bus voltage applies across the load inductor, and load current linearly increases assuming the load inductor works in its unsaturated region with a constant inductance. When the current reaches the desired value I_0, the first pulse ends, the DUT is turned off and its switching performance is obtained. After the first pulse and the DUT is off, the inductor current flows through the upper freewheeling diode and remains almost constant during the short interval between the first and second pulses. At the

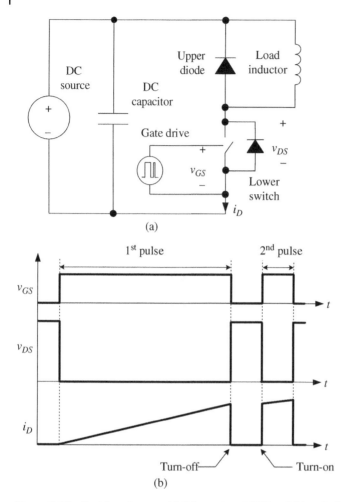

Figure 2.22   Double pulse test. (a) Schematics of DPT and (b) typical waveforms of DPT.

beginning of the second pulse, the DUT is turned on, and the turn-on switching behavior is assessed under essentially identical operating condition (e.g. voltage and current) under which the turn-off switching behavior is evaluated. After the second pulse, DUT turns off. The inductor current flows through the upper diode and naturally decays to zero due to the loss dissipation of the load inductor and diode. DPT completes.

By adjusting the DC bus voltage and the first pulse duration, the DUT's switching transients can be captured under different desired voltage and current conditions. Since the DUT switches only twice for each test and also the pulses are short, the device junction temperature rise due to the switching and conduction losses should be negligibly small even without deliberate cooling.

Note that the DPT circuit in Figure 2.22a features a diode as the sole upper device. In a real voltage source converter, the device can be a switch with freewheeling diode, or simply a switch. Thus, as shown in Figure 2.23, there are two typical DPT configurations: switch/diode pair and switch/switch pair. Device manufacturers generally rely on the switch/diode pair for the dynamic characterization.

In addition to active devices, it is also worthwhile to evaluate the dynamic characteristics (like reverse recovery) of the diode, including external diode (e.g. Si PiN diode, SiC Schottky diode), body

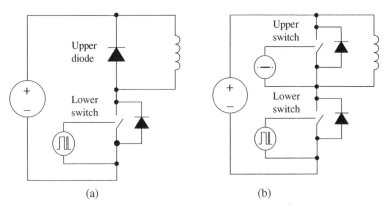

Figure 2.23 DPT configuration. (a) Switch/diode pair and (b) switch/switch pair.

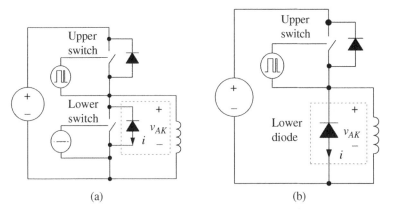

Figure 2.24 Dynamic characterization of WBG diode based on DPT. (a) Diode characterization and (b) body diode or DLB (e.g. GaN HEMTs) characterization.

diode (e.g. SiC MOSFET), and DLB (e.g. GaN HEMTs). Based on the similar DPT circuit discussed earlier, dynamic characterization configuration for diode and device body diode or DLB is illustrated in Figure 2.24. As can be observed, diode under test is located on the lower side of the phase-leg due to the stable measurement ground. In this case, upper switch becomes the active switch and is responsible for generating the switching action by double-pulse control. The load inductor is placed in parallel with the lower diode under test. Detailed DPT implementation is well-documented in [4] and will not be repeated here.

2.4 Thermal Characteristics

This section only focuses on the power device. More general discussions on the thermal characteristics and models are presented in Chapter 9. Device thermal characteristics are an important part of the entire thermal path for the converter thermal management design. More importantly, device maximum junction temperature, as another critical thermal parameter, poses a device-level constraint for the converter design.

2.4.1 Model

Depending on the converter design purpose, two sets of models are usually applied: one is for steady-state condition and the other is for thermal transient.

For steady-state model, a device junction-to-case thermal resistance R_{th_jc} is needed. It is also noted that if a device and diode are co-packaged, there might be two different thermal resistances associated with device die and diode die, respectively.

$$R_{th\_jc} = f(T) \tag{2.30}$$

For transient case, a thermal impedance model considering the thermal capacitance is needed. Typically, two approaches are applied: one is the partial-fraction circuit, also known as the Foster model or Π-model; the other is the continued-fraction circuit, also known as Cauer model, T-model or ladder network, as illustrated in Figure 2.25.

Analytically, thermal impedance model as a function of time is derived, as

$$Z_{th\_jc} = f(t \text{ or } f, T) \tag{2.31}$$

Also, the thermal impedance can be expressed in the frequency domain based on the Fourier analysis [12]. Therefore, (2.31) can also be expressed as a function of frequency f.

2.4.2 Method

Thermal resistance is provided in the device datasheet thermal characteristics table while thermal impedance is usually provided as a curve. The *RC* values in Figure 2.25 can be estimated based on the thermal impedance curve according to the curve fitting method (sometimes they are given in datasheet). Table 2.6 and Figure 2.26 display a datasheet example.

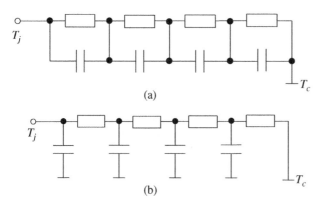

(a)

(b)

Figure 2.25 Thermal impedance model. (a) Partial-fraction circuit, also known as the Foster model or Π model (b) continued-fraction circuit, also known as Cauer model, T-model or ladder network.

Table 2.6 Thermal resistance of the example power device.

Symbol	Description	Min.	Typ.	Max.	Unit
$R_{th,jc}$	Junction to case thermal resistance			0.1	°C/W

Source: Adapted from [2].

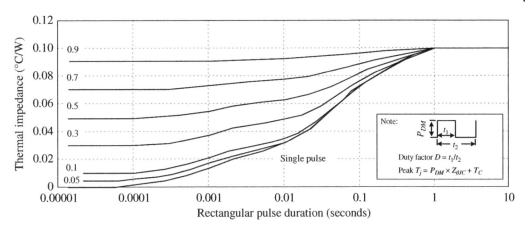

Figure 2.26 Thermal impedance of example power devices. *Source:* From [2] with permission of Microsemi.

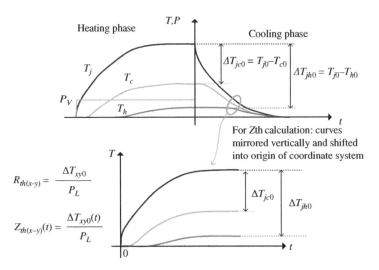

Figure 2.27 Thermal impedance measurement.

Table 2.7 Operating junction temperature of the example power device.

Symbol	Description	Min.	Max.	Unit
T_j	Operating junction temperature range	−40	150	°C

Source: Adapted from [2].

If needed, thermal characteristics could also be tested. The principle of the thermal impedance measurement includes: (i) a constant power is fed into the power device by a current flow, so that a steady junction temperature T_j is reached after a transient period; and (ii) after turning off the power, the cooling down of the module is recorded to calculate the thermal resistance and impedance. Details are documented in [13] and illustrated in Figure 2.27.

Furthermore, device operating junction temperature range is provided in the thermal characteristics table of the device datasheet, as illustrated in Table 2.7.

2.5 Other Attributes

In addition to the models discussed in Sections 2.2 to 2.4, there are other critical attributes associated with the devices, which play a significant role in the converter design. These include safe operating area (SOA), reliability characteristics, mechanical characteristics, and other nonstandard characteristics, like dynamic on-state resistance for GaN HEMTs.

2.5.1 SOA

SOA describes the maximum time a device can be exposed to a specific voltage and current condition below which the device can be expected to operate without self-damage. SOA is usually presented in device datasheets as a graph with voltage on the abscissa and current on the ordinate; the safe "area" referring to the area under the series of curves dependence on duration. Typically, the SOA curves consider three limits, taking MOSFET as an example illustrated in Figure 2.28: maximum current, maximum voltage, and maximum power dissipation and the resultant thermal limits.

As illustrated in Figure 2.28 based on an example datasheet associated with a SiC MOSFET [14], there are three major sections consisting of the SOA curves.

When the drain-source voltage is low, i.e. the device is in the on-state on the left-hand-side of the coordinate, the curve is limited by the on-state resistance R_{dson}.

As the drain-source voltage increases, the device transitions from the ohmic region to the active region where the relatively large drain current and drain-source voltage occur simultaneously, then the curve is determined by the power dissipation P_{loss} generated by $i_D \times v_{DS}$, which fundamentally is limited by the junction temperature, following the expression

$$T_j(t) = T_C(t) + P_{loss}(t) \times Z_{th\_jc}(t) \tag{2.32}$$

where T_j is the device junction temperature, T_C is the case temperature, and Z_{th_jc} is the device thermal impedance discussed in Section 2.4, which is a function of time. Therefore, multiple curves dependence on the duration are generated: a shorter duration indicating a smaller thermal impedance and a larger tolerable power dissipation. This provides a qualitative explanation that the shorter duration results in larger current with a given voltage. It is also noted that once the duration approaches sub-μs scale, the maximum current is somehow clamped which is corresponding to the saturation current discussed in transfer characteristics in Section 2.2.3.

Once the drain-source voltage approaches the breakdown voltage, a hard limit with respect to the SOA occurs, i.e. the corresponding current drops to zero immediately, indicating the device transitions to the cutoff region, any voltage beyond this sharply declining curve causes device breakdown.

Clearly, the SOA curve is highly dependent on the operating conditions. For example, if the case temperature T_C changes, the thermal calculation in (2.32) would lead to different junction temperature T_j estimates, therefore, reshaping the SOA curve accordingly. In summary, the SOA specification combines the various limitations of the device – maximum voltage, maximum current, maximum power, and maximum junction temperature – into one curve. In the meantime, users cannot oversimplify the usage of the SOA curve as the conditions defined in the device datasheet may not be the same as the actual operating conditions.

2.5.2 Reliability Characteristics

Typical failure modes for power devices are either due to (i) thermomechanical degradation of the package or (ii) electrical degradation of the components. Table 2.8 summarizes the failure mode and effect analysis (FMEA) for power devices.

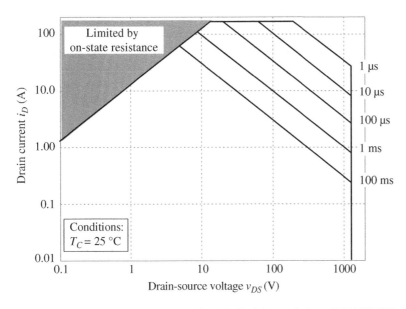

Figure 2.28 Example of SOA. *Source:* From [14] with permission of WOLFSPEED, INC.

Table 2.8 FMEA of power devices.

Failure modes	Failure mechanisms	Critical stressors
Increased on-state voltage and open circuit	Bond wire lift-off driven by coefficient of thermal expansion (CTE) mismatch primarily between bond wire and device die and bond wire heel cracking due to thermo-mechanical effects – bond wire expanding and contracting under temperature cycling	Thermal cycling and power cycling
Increased junction-case thermal resistance and deterioration of device electrical performance (longer turn-off time due to increased "transit time" of minority carriers through the n–body of the device)	Device die attachment to substrate fatigue due to CTE mismatch, leading to void formation and figure-induced cracking	Power cycling
Increased case-to-ambient thermal resistance	Substrate attachment to cold plate due to long-term CTE mismatch	Thermal cycling
Gate breakdown and open/short circuit	Time-dependent gate oxide breakdown leading to gate oxide short	Temperature and gate voltage
	Gate oxide breakdown due to strong electric field (e.g. electrostatic discharge)	Gate voltage beyond rating
Open/short circuit	Single-event burnout due to neutron and heavy ion radiation especially in case of high altitude and cold installation in high voltage devices (>200 V)	External radiation (cosmic rays)

(Continued)

Table 2.8 (Continued)

Failure modes	Failure mechanisms	Critical stressors
	Leakage current due to hot carrier	High current density
	Termination ring structure failure to alleviate effects of high electric field, cracks in passivation (oxide, nitride), surface leakage (can result from manufacturing flaws)	High electric fields and manufacturing flaws
Conducting pathway through encapsulant and ultimately power module failure	Degradation of power module encapsulant (e.g. silicone gel) due to "electrical treeing" and "water treeing," leading to increased local electric field	High electric field, and humidity

2.5.2.1 Reliability Model

Reliability model of power devices is usually illustrated with respect to dedicated failure mode and mechanism. Although extensive studies have been performed, it is still an ongoing research topic and evolving rapidly. Two example models regarding two typical failure mechanisms, bond wire lift-off and solder joint fatigue, are highlighted later.

Bond wire lift-off usually occurs at the bond wire terminations bonded onto power device dies due to large temperature excursions. The number of thermal cycles to failure, N_f, can be modeled using a simple bimetallic approach to approximate the thermo-mechanical stresses arising at the interface between a bond wire joint and die under a temperature excursion of ΔT, and can be expressed by the Manson–Coffin relation as

$$N_f = a(\Delta T)^{-n} \tag{2.33}$$

where coefficients a and n can be experimentally measured based on thermal and/or power cycling of power devices. This simple model has been shown to accurately predict the bond wire fatigue for temperatures less than 120 °C. For higher temperature ranges, alternative models that consider the deviation from the power law are necessary [15].

The die attachment to the substrate can also suffer from deterioration due to CTE mismatch and eventual void formation and fatigue-induced cracking from shear strain due to repeated power cycling. While fast power cycling is responsible for degradation of the bond wire or die attach, the substrate attach layer is also affected by slower thermal cycling. This slow temperature variation can gradually increase case-to-ambient thermal resistance, which can raise case temperature, leading to failures after around 1×10^3 cycles. This compares to power cycling, which typically leads to failures at $<1 \times 10^6$ cycles. The number of cycles to the failure of large solder joints due to thermo mechanical fatigue is given by the modified Manson–Coffin relationship in the form of

$$N_f = \frac{1}{2} \left(\frac{\Delta a \Delta TL}{\gamma x} \right)^{c-1} \tag{2.34}$$

where L is the lateral size of the solder joint, Δa is the CTE mismatch between the upper and the lower joint materials, ΔT is the temperature swing, c is the fatigue exponent, x is the thickness of the solder joint, and γ is the ductility factor of the solder.

2.5.2.2 Method

As illustrated in the earlier two examples, reliability model of power devices is highly dependent on reliability testing, which can be classified as environmental testing and endurance testing. These tests are usually subjected to international standards such as JEDEC and MIL-STD-883.

Environmental testing includes: (i) thermal shock (MIL-STD-883 Method 1011.9, IEC68-2-14), (ii) temperature cycling (MIL-STD-883 Method 1010 Condition C, IEC68-2-14, JEDS22-A-104D), (iii) vibration (IEC68-2-6), (iv) lead integrity (IEC-2-21), (v) solderability (IEC-2-20), (vi) autoclave (JESD22-A110AB), and (vii) mounting torque.

Endurance testing includes (i) power cycling (JEDS22-A-122), (ii) high-temperature storage (IEC68-2-2), (iii) low-temperature storage (IEC68-2-1), (iv) moisture resistance (IEC-2-3), (v) high-temperature reverse bias, (vi) high-temperature gate bias, and (vii) highly accelerated stress test.

Sometimes, different test methods are utilized to quantify the power device reliability as a function of different failure mechanisms, as summarized in Figure 2.29. It includes (i) high-temperature gate bias (HTGB) to apply DC gate stress at high junction temperature, (ii) high-temperature gate switching (HTGS) to apply AC gate stress at high temperature, (iii) chopper mode bias (CMB) to apply pulse load current to the body diode; (iv) thermal cycling to switch between ovens with different temperatures; (v) DC power cycling to generate junction temperature swing through its conduction loss; and (vi) AC power cycling to generate junction temperature swing through its conduction and switching loss.

2.5.3 Mechanical Characteristics

Device mechanical parameters also affect the converter design primarily because of the different packages and the associated electrical and thermal interfaces.

Generally speaking, the power device package can be categorized into two groups: discrete package and power module. Also note emerging WBG packaging technology is currently an active research topic, and there are some other advanced packaging techniques under development, which are out of the scope of this discussion.

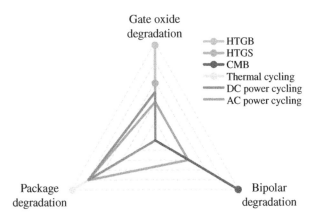

Figure 2.29 Different stress test for power device reliability assessment.

Table 2.9 Summary of discrete packages.

	Non-Kelvin connection	Kelvin connection
Through-hole	TO-247-3 Gate Drain Source	TO-247-4 Drain Source Kelvin source Gate
Surface mount	TO-268-2 Gate Drain (TAB) Source	TO-263-7 Gate Kelvin source Drain (TAB) Source

Specifically, TO-series package is popular for discrete devices. As illustrated in Table 2.9, the discrete packages can be grouped from two aspects: (i) through-hole versus surface mount and (ii) non-Kelvin connection versus Kelvin connection.

The primary difference between through-hole and surface mount packages is the tradeoff between cooling and parasitics. The device with through-hole-based package has better cooling capability since its case can directly attach with the thermal management system via the thermal interface material, while the surface mount device usually has to utilize the thermal via through PCB for heat dissipation, leading to large thermal resistance. However, surface mount device has short lead and less parasitics than that of the through-hole device. Also, the optimized layout design method can be utilized for the surface mount device as compared to the through-hole device since more flexibility is provided due to the single-sided PCB capability of the surface mount device. As a result, switching loop parasitics for surface mount device-based design are smaller.

In comparison with the non-Kelvin connection, the key improvement for Kelvin connection is the separated source terminals for gate and power loops and its resultant less common-source inductance. This allows the device to have better switching performance and less switching loss.

Power module is popular for high current high-power applications. Instead of single device, usually a basic circuit cell (e.g. a half-bridge phase-leg configuration) with multiple dies is integrated into the power module. Also, Kelvin connection is provided for common-source inductance minimization and switching performance improvement. Furthermore, a clear separation between thermal and electrical interfaces makes the power module easy to design and optimize in a converter. Several examples from the device manufacturers are summarized in Figure 2.30.

Regarding the device package for emerging WBG devices, typically, commercially available SiC parts follow the traditional Si-based design, including discrete package and power module as illustrated in Table 2.9 and Figure 2.30. For GaN transistor, lead-free surface mount customized package is popular due to the high-speed switching capability and ultralow parasitic requirement, while TO-series package also exists. Also, Kelvin connection is usually offered. Several examples are listed in Figure 2.31.

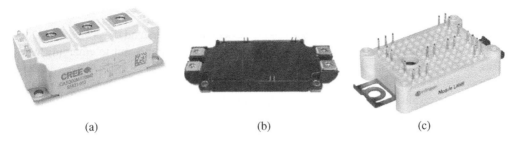

Figure 2.30 Examples of the package for power modules. (a) 62 mm half bridge power module. *Source:* Wolfspeed, Inc. (b) Low profile half bridge power module. (c) Lightweight EASY package. *Source:* Infineon Technologies AG.

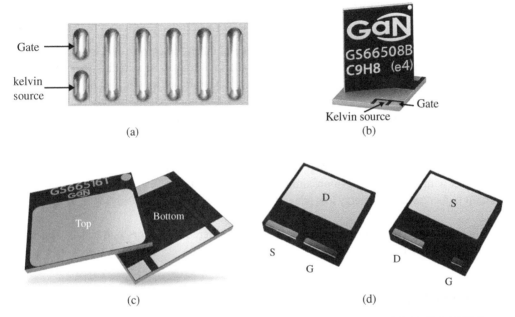

Figure 2.31 Examples of the package for GaN transistors. (a) EPC2010C from EPC, (b) GS 66508B from GaN Systems, (c) top-side cooled. *Source:* From GaN systems, and (d) common drain and common. *Source:* From Transphorm.

In addition to package, the information of device mechanical dimension and weight are also important to the converter design.

2.5.3.1 Model
Regarding the mechanical characteristics, the package information and the parasitics associated with the package are crucial. Specifically, they include a CAD drawing of the device and parasitic inductances associated with device terminals. In the meantime, the mechanical dimension information (e.g. height, length, and depth) and weight are needed.

2.5.3.2 Method
Typically, device datasheet would provide the data discussed. Some parasitics may need to be charatcerized.

2.5.4 Nonstandard Characteristics

To support the converter design, sometimes, it is also essential to understand the device characteristics that do not exist in a standard datasheet. This is particularly important to emerging WBG devices, such as dynamic on-state resistance R_{dson} for GaN HEMTs.

It is reported that, for GaN HEMTs, R_{dson} after a switching transient becomes larger as compared to its steady-state value provided by the manufacturer datasheet, which significantly affects the GaN-based converter design.

From the perspective of power semiconductor physics, increase of dynamic R_{dson} originates primarily from electron trapping by surface traps in the gate-to-drain access region and buffer traps beneath the two-dimensional electron gas (2DEG) channel induced by gate injection in the off state or hot electrons injection from channel during the switching transients, as illustrated in Figure 2.32 [16]. The impact factors on dynamic R_{dson} include blocking time, blocking voltage, load current, junction temperature, switching frequency, duty cycle, switching speed, switching scheme, and gate technology [17]. Specifically,

- It increases as blocking time increases. This is because of the accumulation of electrons from the off-state leakage current trapped at the gate drain access region under a high electric field.
- It increases as blocking voltage and load current increase. Drain-source voltage and load current generate hot electrons during hard-switching transient. The hot electrons can be trapped in the buffer layer or the gate passivation region, making current collapse more severe.
- It increases as junction temperature elevates. This is because of the elevation of the kinetic energy in the electrons, which makes them more easily become hot electrons to get trapped.
- It increases with higher switching frequency and longer duty cycles.
- It increases with both larger turn-on and turn-off gate resistances, i.e. slower switching. The resultant higher switching energy loss generates more hot electrons and leads to larger dynamic R_{dson}.
- For different gate technologies, dynamic R_{dson} exhibit different behaviors under hard and soft switching conditions.

2.5.4.1 Dynamic R_{dson} Model
To support GaN-based converter design, a behavior model for dynamic R_{dson} is sufficient. One way is to adjust the gate voltage and then modify the equivalent R_{dson}, as illustrated in the example [18] in Figure 2.33 with Eqs. (2.35)–(2.37).

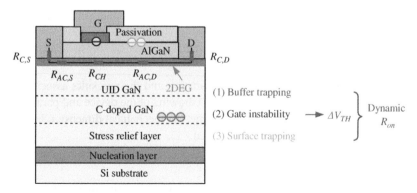

Figure 2.32 Schematic cross section of the GaN-on-Si power transistor using the p-GaN HEMT as an example.

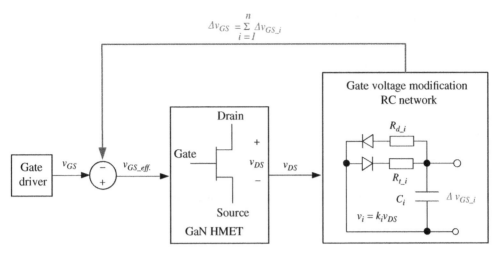

Figure 2.33 Block diagram of the dynamic R_{dson} model based on the RC circuit network.

$$V_i = \begin{cases} k_i v_{ds} & \text{(off state)} \\ 0 & \text{(on state)} \end{cases} \tag{2.35}$$

$$\Delta V_{GS_i}(t_1) = V_i \left(1 - e^{-\frac{t_1}{R_{L\_i} C_i}}\right) \tag{2.36}$$

$$\Delta V_{GS_i}(t_2) = \Delta V_{GS_i}(t_1) \times e^{-\frac{t_1}{R_{L\_i} C_i}} \tag{2.37}$$

2.5.4.2 Method

To build a behavior model, such as the earlier example, a set of measurement data as a function of impact factors is needed. Dynamic R_{dson} detection is challenging since the GaN transistor is off right before the dynamic R_{dson} measurement. As a result, a clamping circuit is needed to reliably block the high DC voltage before the measurement, promptly capture the on-state voltage (i.e. dynamic R_{dson}), and accurately measure its low amplitude in the submicrosecond scale. Figure 2.34 summarizes the basic idea. The details can be found in [19].

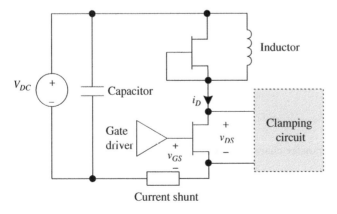

Figure 2.34 Measurement technique for dynamic R_{dson}.

2.6 Scalability (Parallel/Series)

In practical implementation, devices can be connected in parallel and/or in series to increase the current and/or voltage capability, or to reduce stress/loss. Therefore, numbers of devices in parallel connection and in series connection are design variables. As a result, the equivalent device characteristics and the corresponding models need to be derived.

Table 2.10 summarizes the equivalent device static characteristics and models considering number of devices in parallel and in series, where N_P is the number of devices in parallel and N_S is the number of devices in series [20].

Table 2.10   Equivalent device static characteristics and models with the consideration of number of devices in parallel (N_P) and in series (N_S).

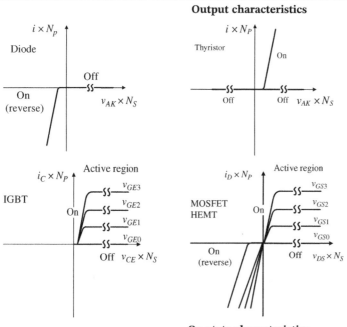

Output characteristics

On-state characteristics

Forward, IGBT/thyristor $\qquad V_{on} = V_{knee}(T_j) \times N_S + R_{on}(V_{CC}, T_j) \times i_L(t) \times N_S/N_P$ $\qquad$ (2.38)

Forward, MOSFET/HEMT $\qquad V_{on} = R_{dson}(V_{CC}, T_j) \times i_L(t) \times N_S/N_P$ $\qquad$ (2.39)

Reverse, diode $\qquad V_{on} = -V_{knee}(T_j) \times N_S - R_{on}(T_j) \times i_L(t) \times N_S/N_P$ $\qquad$ (2.40)

Reverse, MOSFET/HEMT when channel is off $\qquad V_{on} = -V_{knee}(V_{EE}, T_j) \times N_S - R_{on}(T_j) \times i_L(t) \times N_S/N_P$ $\qquad$ (2.41)

Reverse, MOSFET/HEMT when channel is on $\qquad V_{on} \approx -R_{dson}(V_{CC}, T_j) \times i_L(t) \times N_S/N_P$ $\qquad$ (2.42)

Transfer characteristics

$$i_{chan} = g_{fs} \times (v_{GE} - v_{th}) \times N_P \text{ or } i_{chan} = g_{fs} \times (v_{GS} - v_{th}) \times N_P \qquad (2.43)$$

$$v_{th} = f(T_j) \qquad (2.44)$$

$$g_{fs} = f(v_{GE} \text{ or } v_{GS}, v_{CE} \text{ or } v_{DS}, T_j) \times N_P \qquad (2.45)$$

Table 2.10 (Continued)

Leakage and breakdown

$$V_{BD} = f(T_j) \times N_S \text{ @given } I_{lkg} \times N_P \tag{2.46}$$

Junction capacitances

Diode
$$C = f(v_{AK}) \times N_P/N_S \tag{2.47}$$

IGBT
$$C_{oes} = C_{CE} + C_{GC} = f(v_{CE}) \times N_P/N_S \tag{2.48}$$

$$C_{ies} = C_{GE} + C_{GC} = f(v_{CE}) \times N_P/N_S \tag{2.49}$$

$$C_{res} = C_{GC} = f(v_{CE}) \times N_P/N_S \tag{2.50}$$

MOSFET/HEMT
$$C_{oss} = C_{DS} + C_{GS} = f(v_{DS}) \times N_P/N_S \tag{2.51}$$

$$C_{iss} = C_{GS} + C_{GD} = f(v_{DS}) \times N_P \times N_S \tag{2.52}$$

$$C_{rss} = C_{GD} = f(v_{DS}) \times N_P/N_S \tag{2.53}$$

Gate charge (MOSFET/HMET as an example, similar relationship can be derived for IGBT with its terminology)

$$Q_G = \begin{cases} \int_{V_{EE}}^{v_{GS}} C_{GS}(v_{GS}) dv_{GS} \times N_P \times N_S, & \text{for } v_{GS}V_{Miller} \\ \left(Q_{GS,Miller} + Q_{GD} + \int_{V_{Miller}}^{v_{GS}} (C_{GS}(v_{GS}) + C_{rss}(0V)) dv_{GS} \right) \times N_P \times N_S, & \text{for } v_{GS} \geq V_{Miller} \end{cases} \tag{2.54}$$

where $Q_{GS,Miller}$ and Q_{GD} can be referred to (2.21) and (2.22)

Unlike device static characteristics and corresponding models, equivalent switching characteristics of multiple devices in parallel/series practically cannot be linearly scalable. Specifically, the switching time for parallel/series connected devices may be longer than that of a single device, the switching energy loss could be larger than $N_P \times N_S \times E_{SW}$, where E_{SW} is the switching energy loss of a single device under the same equivalent operating conditions, including turn-on switching energy loss E_{on}, turn-off switching energy loss E_{off}, and reverse recovery energy loss E_{rr}, if applicable. Practically, a derating factor k_{DF} (k_{DF} is greater than one) is applied to estimate the equivalent switching behaviors.

$$t_{on\_equiv} = t_{on} \times k_{DF} \tag{2.55}$$

$$t_{off\_equiv} = t_{off} \times k_{DF} \tag{2.56}$$

$$E_{on\_equiv} = E_{on} \times N_P \times N_S \times k_{DF} \tag{2.57}$$

$$E_{off\_equiv} = E_{off} \times N_P \times N_S \times k_{DF} \tag{2.58}$$

$$E_{rr\_equiv} = E_{rr} \times N_P \times N_S \times k_{DF} \tag{2.59}$$

where t_{on}, t_{off}, E_{on}, E_{off}, and E_{rr} are switching parameters associated with a single device determined in (2.25)–(2.29).

Equivalent thermal characteristics of multidevices in parallel/series depend on the physical implementation of thermal management system. Assuming individual devices are cooled the same (i.e. thermally decoupled), the equivalent thermal resistance/impedance could be scaled linearly, as expressed by

$$R_{th\_jc\_equiv} = R_{th\_jc}/N_P/N_S \tag{2.60}$$

$$Z_{th\_jc\_equiv} = Z_{th\_jc}(t)/N_P/N_S \tag{2.61}$$

where R_{th_jc} and Z_{th_jc} are thermal models associated with the single device in (2.30) and (2.31)

In practical implementation, thermal coupling leads to complex thermal network. In this case, finite element analysis may need to be considered, which is beyond the scope of this chapter and will not be discussed here.

For other attributes listed in Section 2.5, the way to scale the corresponding models is highly dependent on the specific design and some of them are ongoing research topics, which will not be covered here. In general, except for the equivalent weight, a linear scalability may no longer exist.

2.7 Relevance to Converter Design

Figure 2.35 overviews the focus of Chapter 2 and its relevance to converter design. Following the same design procedure diagram in Figure 1.24, three inputs (system conditions, system constraints, and system variables) associated with Chapter 2 are highlighted.

Specifically, system conditions include the operating voltage and current in Chapters 5–7, thermal-related parameters/models (e.g. thermal network, operating temperatures) in Chapter 9,

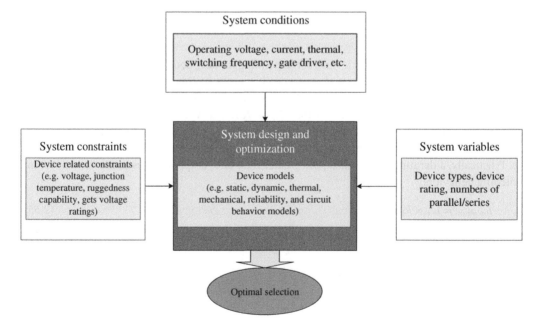

Figure 2.35 Overview of Chapter 2 and its relevance to converter design.

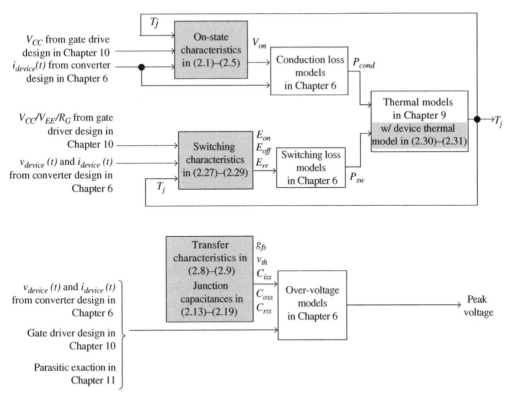

Figure 2.36 Integration of device models in Chapter 2 with other models for converter design: loss, thermal, and overvoltage models as an example.

and gate driver design in Chapter 10. System variables associated with power devices in Chapter 2 include device types, ratings, number of parallel/series, etc. System constraints posed by devices discussed in Chapter 2 include voltage considering steady state and dynamic margins separately (including gate voltage ratings), maximum junction temperature, SOA, and reliability.

Figure 2.36 illustrates the method of integrating device models in Chapter 2 with other models for the converter design. For instance, on-state characteristics models in (2.1)–(2.5) and switching characteristics models in (2.27)–(2.29) will be used for the device loss calculations. Device thermal models in (2.30) and (2.31) will be adopted as part of the converter thermal model for junction temperature determination. Also, transfer characteristics in (2.8) and (2.9) and junction capacitance in (2.13)–(2.19) will be essential device data inputs for the overvoltage model.

A package of delivery will be provided from Chapter 2 to support the later converter design. It includes:

- Models (2.1)–(2.31) and inherent relationship with converter design, such as examples highlighted in Figure 2.36.
- The Figure 2.37 device behavior model circuit (taking MOSFET as an example) for converter-level analysis with device behavior emphasized in Chapter 6.
- Other important attributes, such as weight, size, footprint, and cost.
- Lastly, design constraints related to device discussed in this Section 2.7.

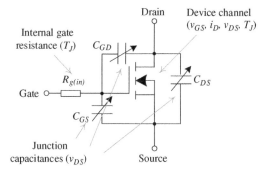

Figure 2.37 Device behavior model.

2.8 Summary

This chapter overviews six types of power devices that are commonly used in three-phase AC applications. It includes Si IGBTs, Si PiN diodes, Si thyristors, emerging SiC MOSFETs, SiC Schottky diodes, and GaN HEMTs. From the perspective of converter design, this chapter illustrates device characteristics that could significantly affect the converter performances, derives the models for the later subsystem and converter-level design optimization, and summarizes the method(s) and best practice to obtain the models. The main points and takeaways are summarized as follows.

1) Critical device characteristics for the converter design include static characteristics (e.g. on-state characteristics, transfer characteristics, leakage and breakdown characteristics), switching characteristics (e.g. switching time, switching energy loss), and thermal characteristics (e.g. junction-to-case thermal resistance, transient thermal impedance)
   a) Device output characteristics (i.e. V–I curves) can be used to derive two important device models: on-state characteristics and transfer characteristics, in (2.1)–(2.5) and (2.6)–(2.9), respectively. Beyond the device datasheet, pulsed V–I testing in Figure 2.3 is the widely applied method for the model derivation.
   b) Leakage current and breakdown voltage can be modeled in (2.10). The converter designer would not use this value directly as the maximum operating voltage, instead, practically, a voltage margin based on engineering insights is implemented. This voltage margin is typically determined based on the application, and fundamentally driven by the reliability consideration, as expressed in (2.11) and (2.12). Testing circuit in Figure 2.10 is the widely applied method for the model derivation.
   c) Switching characteristics, including turn-on/-off times, turn-on/-off switching energy losses, and reserves recovery energy loss, if applicable, can be modeled in (2.25)–(2.29). Double pulse test in Figure 2.22 is typically used to assess the switching performance of power devices.
   d) Two sets of thermal models are usually applied: one is for steady-state analysis, and the other is for thermal transient, as expressed in (2.30) and (2.31), respectively. Figure 2.27 illustrates a thermal impedance measurement method.
2) Other device attributes, such as SOA, reliability, mechanical, and nonstandard characteristics (e.g. dynamic on-state resistance of GaN HEMTs), play a crucial role on the converter design.
   a) SOA describes the maximum time a device can be exposed to a specific voltage and current condition below which the device can be expected to operate without self-damage. Typically, the SOA curves consider three limits: maximum current, maximum voltage, and maximum power dissipation and the resultant thermal limits.

b) Typical failure modes for power devices are either due to (i) thermomechanical degradation of the package or (ii) electrical degradation of the components. Two example models regarding bond wire lift-off and solder joint fatigue are expressed by (2.33) and (2.34) based on Manson–Coffin relationship. It is also noted that reliability model of power devices is highly dependent on reliability testing, which can be classified as environmental testing and endurance testing. These tests are usually subjected to international standards such as JEDEC and MIL-STD-883.

c) Device mechanical parameters affect the converter design primarily because of the different packages and the associated electrical and thermal interfaces. A CAD drawing of the device and parasitics associated with device terminals are required. In the meantime, the mechanical dimension information (e.g. height, length, and depth) and weight are needed. Typically, device datasheet would provide the data except parasitics.

d) To support the converter design, sometimes, it is also essential to understand the device characteristics that do not exist in a standard datasheet. This is particularly important to emerging WBG devices, such as dynamic on-state resistance R_{dson} for GaN HEMTs. The corresponding models are presented in (2.35)–(2.37), and a testing method is illustrated in Figure 2.34.

3) In practical implementation, devices can be connected in parallel and/or in series to increase the current and/or voltage capabilities and to reduce stress/loss. Therefore, numbers of devices in parallel connection and in series connection are design variables. The equivalent device characteristics and the corresponding models are provided in (2.38)–(2.61).

4) Finally, the relevance between Chapter 2 and the converter design is overviewed in Section 2.7. Also, a package of delivery provided from Chapter 2 to converter design chapters is summarized.

References

1 GE Power Conversion, "MV7000 Reliable, high performance medium voltage drive, 2013." [Online]. Available: https://www.gepowerconversion.com/sites/default/files/2022-01/ GEA30737_MV7_BCH_MV7000%20Medium%20Voltage%20Drive_EN_20180821.pdf

2 Microsemi, "APTMC120AM09CT3AG Datasheet, Rev 1," October 2014. [Online]. Available: https://www.microsemi.com/document-portal/doc_download/134001-aptmc120am09ct3ag-rev1-datasheet (accessed February 2023).

3 Keysight, "B1505a power device analyzer/curve tracer." [Online]. Available: http://www.keysight.com/en/pd-1480796-pn-B1505A/ (accessed December 2017).

4 F. Wang, Z. Zhang and E. Jones, *"Characterization of Wide Bandgap Power Semiconductor Devices,"* Institution of Engineering and Technology, September 2018.

5 B. Jayant Baliga, *"Fundamentals of Power Semiconductor Devices,"* New York, NY, USA: Springer-Verlag, 2008.

6 Z. Zhang, F. Wang, L. M. Tolbert and B. J. Blalock, "Active gate driver for crosstalk suppression of SiC devices in a phase-leg configuration," *IEEE Transactions on Power Electronics*, vol. 29, no. 4, pp. 1986–1997, April 2014.

7 Z. Zhang, W. Zhang, F. Wang, L. M. Tolbert and B. J. Blalock, "Analysis of the switching speed limitation of wide band-gap devices in a phase-leg configuration," in *Proc. IEEE Energy Conversion Congress and Exposition*, September 2012, pp. 3950–3955.

8 W. Zhang, "Modeling, Measurement and Mitigation of Fast Switching Issues in Voltage Source Inverters." PhD diss., University of Tennessee, 2021.

9 E. Jones, F. Wang, D. Costinett, Z. Zhang, B. Guo, B. Liu and R. Ren, "Characterization of an enhancement-mode 650-V GaN HFET," in *Proc. IEEE Energy Conversion Congress and Exposition*, September 2015, pp. 400–407.

10 GaNSystems, "GS-065-060-5-B-A Datasheet, Rev 2 210429." [Online]. Available: https://gansystems.com/wp-content/uploads/2021/05/GS-065-060-5-B-A-DS-Rev-210429-1.pdf (accessed February 2023).

11 Z. Zhang, B. Guo, F. Wang, E. Jones, L. M. Tolbert and B. J. Blalock, "Methodology for wide band-gap device dynamic characterization," in *IEEE Transactions on Power Electronics*, vol. 32, no. 12, pp. 9307–9318, December 2017.

12 F. Wang, W. Shen, D. Boroyevich, S. Ragon, V. Stefanovic and M. Arpilliere, "Voltage source inverter – development of a design optimization tool," *IEEE Industry Applications Magazine*, vol. 15, no. 2, pp. 24–33, March–April 2009.

13 Infineon, "Transient thermal measurements and thermal equivalent circuit models, 2015." [Online]. Available: https://www.infineon.com/dgdl/Infineon-Thermal_equivalent_circuit_models-ApplicationNotes-v01_02-EN.pdf?fileId=db3a30431a5c32f2011aa65358394dd2 (accessed February 2023).

14 CREE/Wolfspeed, "C3M0065090J Datasheet, Rev D," June 2019. [Online]. Available: https://assets.wolfspeed.com/uploads/2020/12/C3M0065090J.pdf (accessed February 2023).

15 H. Chung, H. Wang, F. Blaabjerg and M. Pecht, *"Reliability of Power Electronic Converter Systems,"* Institution of Engineering and Technology, December 2015.

16 S. Yang, S. Han, K. Sheng and K. J. Chen, "Dynamic on-resistance in GaN power devices: mechanisms, characterizations, and modeling," in *IEEE Journal of Emerging and Selected Topics in Power Electronics*, vol. 7, no. 3, pp. 1425–1439, September 2019.

17 R. Li, X. Wu, S. Yang and K. Sheng, "Dynamic on-state resistance test and evaluation of GaN power devices under hard- and soft-switching conditions by double and multiple pulses," in *IEEE Transactions on Power Electronics*, vol. 34, no. 2, pp. 1044–1053, February 2019.

18 K. Li, P. L. Evans and C. M. Johnson, "Characterization and modeling of gallium nitride power semiconductor devices dynamic on-state resistance," in *IEEE Transactions on Power Electronics*, vol. 33, no. 6, pp. 5262–5273, June 2018.

19 J. Gareau, R. Hou and A. Emadi, "Review of loss distribution, analysis, and measurement techniques for GaN HEMTs," in *IEEE Transactions on Power Electronics*, vol. 35, no. 7, pp. 7405–7418, July 2020.

20 H. Wang, A. Q. Huang and F. Wang, "Development of a scalable power semiconductor switch (SPSS)," in *IEEE Transactions on Power Electronics*, vol. 22, no. 2, pp. 364–373, March 2007.

3

Capacitors

3.1 Introduction

Capacitors are a basic type of passive components used for energy storage and harmonics/noises filtering in power electronics. A three-phase converter often utilizes the following capacitors: (i) DC-link capacitors; (ii) AC harmonic filter capacitors; (iii) electromagnetic interference (EMI) filter capacitors; (iv) *dv/dt* filter capacitors for motor loads; (v) DC decoupling capacitors; and (vi) capacitors in auxiliary circuits such as control circuitry for harmonics/noises filtering. Capacitors in a converter can often be multifunctional, acting as both energy storage and filter elements. In order to choose an appropriate capacitor for desired application, we must know the required voltage and current ratings, environment conditions, and most important, the characteristics and models of the capacitors.

This chapter will introduce various capacitors used in three-phase AC converters, the capacitor selection problem relevant to converter and converter subsystem design, and the capacitor characteristics and models needed for the capacitor selection. Note that the electrical design of different capacitors in a converter (for example, how to determine the required capacitance for DC-link capacitors, or how to determine the required capacitance and noise attenuation performance for EMI filter capacitors) will be discussed in converter subsystem design chapters in Part 2 and will not be covered here. Instead, the models which will be used in capacitor physical design will be provided.

3.2 Capacitor Types and Technologies

This section covers the types of capacitors that are widely available today, describing the materials used, highlighting their range of applications, and noting their failure modes. A capacitor is generally constructed by placing an insulating medium (dielectric) between two conducting plates or foils [1]. When a voltage is applied across the plates, charge accumulates on the plates, and electrical energy is stored in the dielectric. Different capacitor technologies arise as different dielectric and foil materials are used in the construction of a capacitor. Commercially available capacitor technologies fall under the following four broad categories based on dielectric materials [2]: ceramic, film, electrolytic, and electrochemical.

Design of Three-phase AC Power Electronics Converters, First Edition. Fei "Fred" Wang, Zheyu Zhang, and Ruirui Chen.
© 2024 The Institute of Electrical and Electronics Engineers, Inc. Published 2024 by John Wiley & Sons, Inc.

Each category may be further differentiated. Film capacitors can be of natural (paper or mica) or synthetic (polystyrene, polypropylene, etc.) type. Electrolytic capacitors are distinguished by the type of impregnant used, either liquid or dry (aluminum and electrolytic, respectively) [2, 3].

3.2.1 Ceramic Capacitors

Ceramic capacitors use ceramic dielectric with a very high relative permittivity or dielectric constant. These capacitors can be constructed using a single-layer structure for small capacitance or by stacking multiple layers to achieve high capacitance. The latter is commonly known as multilayer ceramic capacitor (MLCC). Ceramic capacitors are available in a wide range of values, from less than 1 pF to more than 10 μF. Low-voltage ceramic capacitors (up to 500 V DC) are generally used in control circuits as well as for decoupling, filtering, and bypassing [2]. High-voltage ceramic capacitors are constructed using a molded epoxy resin and show excellent stability and electrical characteristics. Because of these benefits, high-voltage ceramic capacitors are widely used in snubber circuits, resonant circuits, and filters in power electronic applications. Generally, high-voltage ceramic capacitors are considered unreliable for true high-voltage applications because their dielectric breakdown strength is low. The ceramic capacitor construction is susceptible to both dielectric breakdown failure and delamination. Excessive voltage applied to the capacitor may cause dielectric breakdown, which leads to a short circuit. Temperature cycling and thermal stress cause delamination, where the epoxy and ceramic layers separate due to differing thermal expansion coefficients. This type of failure may result in an open circuit or a capacitance value that is out of the specified tolerance range [4].

3.2.2 Paper

As mentioned earlier, paper capacitors are a subset of film capacitor technology, where a paper is used as the insulating dielectric medium. In general, paper capacitors are more reliable than ceramic capacitors and are available in the range from 100 pF to 100 mF. Paper capacitors may be designed to withstand up to 200 kV. This technology presents the highest possible DC voltage rating. While paper capacitors are moderately stable across their operating temperature range, they also present the highest dissipation factor of all the film capacitor technologies. This high dissipation factor is a major drawback for their applications in power electronics.

3.2.3 Mica

Mica capacitors also belong to film capacitors. They employ mica, an extremely inert and stable material, as the insulating dielectric. The intrinsic inertness of mica leads to mica capacitors being extremely stable and reliable. These capacitors also provide a very low capacitance variation from −55 to +125 °C. Second only to paper capacitors, mica capacitors provide a high-voltage capability of up to 100 kV. Additionally, parasitic series inductance can be extremely low (in the range of 10 nH). Mica capacitors are also stable at high frequencies and have a low dissipation factor. Despite all of the positive characteristics, drawbacks of mica capacitor technology include small capacitance range (up to 10 μF), as well as the lowest capacitance and power density of all capacitor technologies.

3.2.4 Poly-Film

Poly-film capacitors are generally constructed with either wound layers of foil and film, or in metallized capacitor construction, where metal deposited directly onto the film prior to winding forms the electrode. A major benefit of metallized construction is its self-healing ability with only a slight decrease in total capacitance. Metallized construction is, however, limited by a low current capability. A hybrid foil/metallized construction that gives higher current capability while retaining some self-healing characteristics is also possible [5, 6]. Poly-film capacitors exhibit excellent aging characteristics, as the capacitance decreases slightly with time. Even after the capacitance begins to drop below the specified tolerance, it will not exhibit catastrophic failure [3]. Several popular films include polyester, polycarbonate, and polypropylene. Others include polystyrene, polyphenylene sulfide, Kapton, and Teflon.

3.2.4.1 Polyester Film
Polyester has the highest dielectric constant and lowest minimum thickness of the three widely used polymer films, which results in the smallest size for a given capacitance value and voltage rating. Despite these advantages, polyester is rarely used in power electronics applications because of its high dissipation factor and poor stability. Polyester film capacitors are capable of a 5% tolerance rating, with 10% typical [5].

3.2.4.2 Polycarbonate Film
Polycarbonate provides lower dissipation factor than polyester over the given temperature and frequency ranges. Polycarbonate also provides the capability of 1% tolerance (although 10% is typical). Major drawbacks are the availability as metallized winding construction only (limits dv/dt and high-current capability), as well as the material being the most expensive of all poly-films [5].

3.2.4.3 Polypropylene Film
Compared to polyester and polycarbonate, polypropylene film offers the lowest dissipation factor over both temperature and frequency ranges. Tolerances identical to polycarbonate film are easily achieved. The minor drawback is that polypropylene film capacitors can only operate up to a temperature of 105 °C. Today, polypropylene capacitors are the preferred choice in power electronics [5].

3.2.4.4 Other Films
Polystyrene is limited by its maximum operating temperature of 85 °C, as well as its availability in foil winding construction only. For surface-mount applications, polyphenylene sulfide is sometimes used because of its high-temperature rating (150 °C); however, it is less reliable, more expensive, and has larger dissipation factor than polypropylene. Kapton and Teflon are both expensive films and are usually limited to foil winding construction [5].

3.2.5 Aluminum Electrolytic Capacitors (AECs)

In an AEC, the anode foil is coated with a thin oxide layer. The layer between anode and cathode foils is then impregnated with a liquid electrolyte. The oxide layer on the anode foil serves as the dielectric, while the electrolyte is a conductor. Because the foil layers are etched, increasing their surface area by 100 times, the electrolytic capacitor achieves the largest capacitance and energy density of the standard capacitor technologies (i.e. excluding some emerging technologies in

Section 3.2.8) [7]. AECs can provide energy density up to 1 kJ/kg. Disadvantages for this technology are the high dissipation factor, and its polarity dependence. Also, the voltage limit is about 700 V DC [3]. AECs are often used in high ripple current applications; however, their equivalent series resistance (ESR) can more than double over the rated service life (usually 10 000 hours). Recent advances in increased stability electrolytes have led to the possibility of extending service life to 100 000 hours. These new electrolytes are still highly temperature dependent, as increased temperature causes increased electrolyte chemical reactions. If kept below 85 °C, the ESR can remain virtually constant over the component lifetime; however, increasing the temperature to 125 °C can cause the ESR to double over the lifetime [7].

3.2.6 Tantalum Electrolytic Capacitors (TECs)

TECs are similar in construction to AECs, except that a dry impregnant is used [2]. While their dissipation factor is slightly lower than that of AECs, TECs have a smaller capacitance range (up to several mF). They also feature similar energy density as AECs. Because a dry impregnant is used, TECs are less susceptible to temperature variations than AECs. A major drawback is its failure mode. TECs often fail short, which can cause ignition of the dry impregnant. Obviously, this could pose a hazard depending on the surrounding environment.

3.2.7 Capacitor Technologies Comparison

The general advantages, disadvantages, failure modes, and applications for commonly used capacitor technologies are summarized in Table 3.1. The electrical characteristics of commonly used capacitors are summarized in Table 3.2. The energy and power densities of different type of capacitors versus their typical capacitance range are summarized in Figure 3.1 [8].

In general, electrolytic capacitors have the highest energy density and lowest cost; however, their short lifespan, limited current conduction capability, and low-frequency operation are concerns for industry applications.

In contrast, film capacitors show high current conduction capability, high-frequency operation, and lower ESR compared to electrolytic capacitors. Moreover, they show self-healing characteristics, which increases reliability and makes them a popular choice for industry applications. Also, film capacitors use plastic/polymers as the dielectric and have very low temperature dependence. However, the dielectric constants of these dielectrics are low (e.g. 2–3), and thus film capacitors are bulkier in comparison to electrolytic capacitors.

MLCCs can have a higher RMS current rating, higher temperature capability, and higher energy density than film capacitors [9]. MLCCs also have several shortcomings. The most common dielectric used in MLCCs is barium titanate ($BaTiO_3$), which is a Class II ferroelectric dielectric material and whose parameters are highly temperature dependent. The capacitance of the Class II ceramic capacitors decreases rapidly with the DC bias voltage. Also, there are reliability issues associated with ceramic capacitors as the ceramic dielectric material is rigid and can crack due to mechanical and thermal stress. Thus, ceramic capacitors are not suitable for reliability-critical applications.

3.2.8 Emerging Capacitor Technologies

3.2.8.1 PLZT-Based Ceramic Capacitors

To overcome the issues of the conventional MLCC capacitor, an emerging ceramic capacitor technology is the PLZT (lead lanthanum zirconium titanate) based ceramic capacitor. PLZT is an antiferroelectric dielectric material that can withstand higher currents and temperature. Unlike ferroelectric materials used in MLCCs, antiferroelectric material exhibits an incremental

Table 3.1 Summary of commonly used capacitor technologies.

Technology	Advantages	Disadvantages	Failure Modes	Applications
Ceramic (general purpose)	High capacitance-to-volume ratio; low cost	Poor temperature stability; susceptible to shock and vibrations; poor reliability	Mechanical cracking; surge currents; dielectric breakdown	High-frequency blocking, bypassing, filtering, and coupling
Ceramic (temperature compensating)	Temperature and time stable	Low capacitance values only; susceptible to shock and vibration; low capacitance-to-volume ratio	Mechanical cracking, surge currents, dielectric breakdown	Compensation for variation in temperature
Paper	Reliable; average stability; low cost	Average capacitance-to-volume ratio; high resistance	Dielectric breakdown	Low-frequency filtering and coupling; power-factor correction; contact protection; motor start and run
Mica	Low dielectric losses; good temperature, frequency, and aging characteristic; low AC loss; reliable; low cost	Low capacitance-to-volume ratio	Dielectric breakdown	High-frequency blocking, bypassing, filtering, and coupling; resonant circuits; high voltage
Poly-Film	Wide range of capacitance and voltage ratings; low dissipation factor; stable; high temperature stability	Relatively high cost	Dielectric breakdown; open circuit	Mid-frequency blocking, bypassing, filtering, and coupling; timing
Aluminum electrolytic	Highest electrolytic capacitance-to-volume ratio; highest electrolytic voltage rating; high ripple capacity	High leakage current; requires reforming after long storage periods; high cost for high reliability; susceptible to dynamic environments; damaged by chlorinated hydrocarbons	Dielectric breakdown; open circuit; loss of capacitance; increased ESR; and leakage current	Low-frequency blocking, bypassing, filtering, and coupling
Chemical double layer (super cap)	Highest capacitance; highest energy density	Low operating temperature; low-frequency capability; relatively high cost	Similar to AEC	Energy storage

capacitance with respect to the bias voltage thus PLZT capacitors have increased capacitance with DC bias voltage. Furthermore, the commercially available PLZT capacitors are more reliable than the MLCCs. Their reliability is achieved by utilizing the series connection of two MLCC geometries in one component, such that the capacitor will be operational in the event of a crack in the dielectric. The copper electrodes on these capacitors improve the electrical and thermal performance and can withstand a higher operation temperature (150 °C) [9]. One commercial PLZT capacitor is the

Table 3.2 Electrical characteristics of commonly used capacitors.

Technology	Typical dielectric constant (Air = 1.0)	Capacitance range (typical)	DC voltage range	Volume efficiency ($\mu F \cdot V/in^3$)	Dissipation factor (25 °C)	Capacitance change from −55 to 125 °C (%Δ°C from nominal)
Ceramic	12–400 000	0.1 pF to 12 μF	25 V to 30 kV	273–55 000	0.006–2.5 (1 kHz) 0.006–7 (1 MHz)	0 (temp. comp.) −10 to +2
Paper	4.0	100 pF to 200 μF	50 V to 200 kV	130–300	0.2–0.5 (120 Hz) 0.2–1 (1 kHz)	−10 to +8
Mica	7.5	1 pF to 10 μF	0 V to 100 kV	11–600	0.04–0.06 (120 Hz) 0.05–0.5 (1 kHz) 0.05–0.9 (1 MHz)	−2 to +1
Poly-film	3.0	20 pF to 400 μF	30 V to 10 kV	34–5 000	0.007–0.9 (1 kHz) 0.02–0.45 (1 MHz)	+1 to −1 (Polystyrene)
Aluminum electrolytic	10	0.5 μF to 2 F	3–700 V	16 000–82 000	3–80 (120 Hz)	−30 to +25
Tantalum electrolytic (solid)	26	0.1 μF to 8 mF	3–125 V	5 800–87 500	1–6 (120 Hz)	−8 to +10

CeraLink capacitor from TDK. The CeraLink capacitor has higher capacitance and energy densities compared with the common ceramic and film capacitors as shown in Figure 3.2 [10].

3.2.8.2 Nanostructure Multilayer Capacitors

Nanostructure multilayer capacitors can be made of multilayer polymer or multilayer ceramic. These capacitors use inorganic coatings built up in an interleaving dielectric-electrode structure and can have as many as 10 000 layers, with each layer as thin as 0.1 μm [3]. An example is the NanoLam™ capacitor from PolyCharge [11]. NanoLam capacitors utilize superthin polymer dielectric layers that are formed in line with the metal electrode deposition process, eliminating the need for extruded films, film metallization, and capacitor winding. The submicron, self-healing, and polymer dielectrics result in capacitors with much high-energy density compared to polypropylene film capacitors. Nano-layer dielectrics also reduce weight significantly compared to multilayer ceramic capacitors. Thermoset dielectrics are formulated to withstand temperature from −196 to 200 °C and thus cooling requirements can be reduced or eliminated. According to the vendor, NanoLam capacitors also exhibit stable capacitance, breakdown strength, and equivalent series resistance across a broad temperature and frequency with lifetimes measured in hundreds of thousands of hours.

3.2.8.3 Supercapacitors

Supercapacitor is also known as electrochemical capacitor. Types of supercapacitors include electrochemical double-layer capacitor, pseudo capacitor, and hybrid capacitor. Supercapacitors are designed for use below the impregnant's electrolysis point, which yields the capability for very high

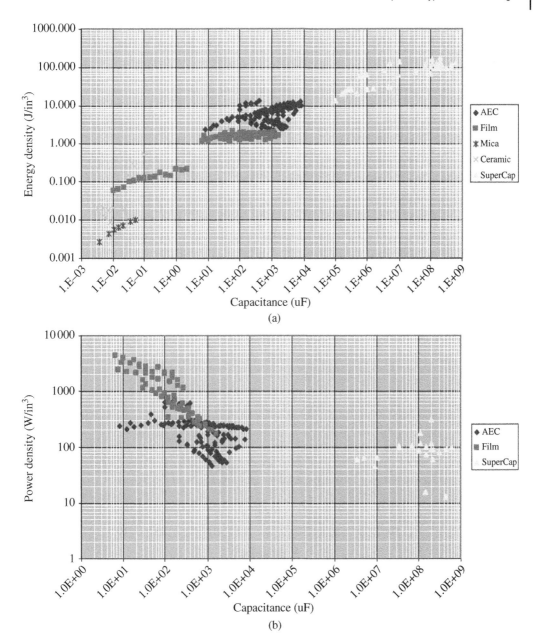

Figure 3.1 Capacitor technologies: (a) energy density vs. capacitance and (b) power density vs. capacitance. *Source:* Form [8] with permission of IEEE.

capacitances. Supercapacitors have instantaneous power densities more significant than batteries and energy densities much larger than dielectric capacitors. They can be charged/discharged more than 100 000 times without significant energy loss. These capacitors provide lower ESR than traditional electrolytic capacitors and are capable of bipolar operation [3]. Supercapacitors are attractive for a wide variety of power applications such as power quality conditioning equipment, new energy power generation system, and electric vehicles.

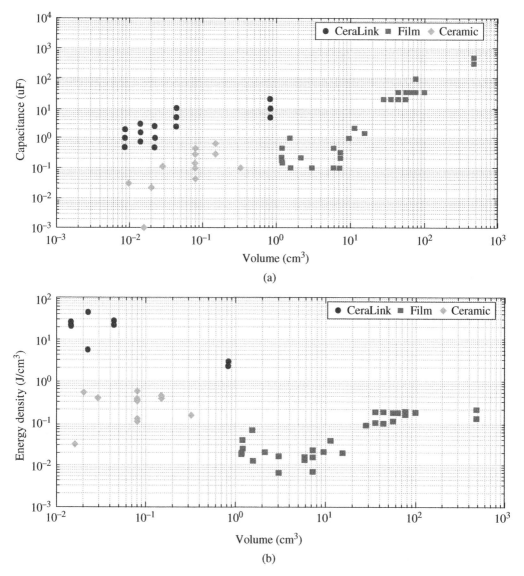

Figure 3.2 CeraLink, ceramic, and film capacitor technologies: (a) capacitance density and (b) energy density. *Source:* From [10] with permission of IEEE.

3.3 Capacitor Selection in a Converter Design

In a converter design, capacitors are usually selected from commercial products, rather than custom designed, similar to semiconductor devices. When selecting capacitors, physical and performance constraints need to be considered. The capacitor physical constraints include

1) Temperature limit: The capacitor operating temperature should not exceed the maximum permissible temperature specified by manufacturers in datasheets. For example, the maximum permissible case temperature for film capacitors is usually 105 °C. Some datasheets specify the maximum permissible hotspot temperature, usually below 125 °C. Note that the hotspot

temperature should be the true limit for a capacitor. When a maximum permissible case temperature is specified, it implies that the capacitor should be used within its specified current limits such that the corresponding power loss, and temperature difference between the case and hotspot will stay within the permissible range.

2) Voltage limits: These include both continuous and transient voltage limits.
3) Current limits: i.e. ripple current limits since capacitors can only carry AC currents. This can be seen as an equivalent specification for hotspot temperature limit. Therefore, current limits are functions of ambient or case temperature.
4) Pulse handling capability: Voltage pulse with rapid voltage changes either during normal or abnormal conditions (e.g. fault) will lead to strong peak currents in a capacitor. These currents will generate heat in the contact regions between the sprayed metal and the metallization. Thus, there is a limit to the voltage pulse or peak current that will cause high dissipated heat to damage contact regions and capacitors. High heat can also cause capacitor bulging. The *dv/dt* limit and peak current limit are usually provided in datasheet.

The capacitor performance constraints include (i) the capacitance C, which usually has a nominal value C_{nom} as the lower bound, although too high a C could cause performance issues such as low power factor, higher loss, and undesirable (i.e. too low) resonance frequencies and leakage current (especially in EMI filter cases) and (ii) reliability and lifetime requirements, which are usually functions of temperature and voltage stresses.

Following the terminology and approach used in this book, the capacitor selection problem can be illustrated by Figure 3.3. Since the commercial capacitors are used, the selection variable is really

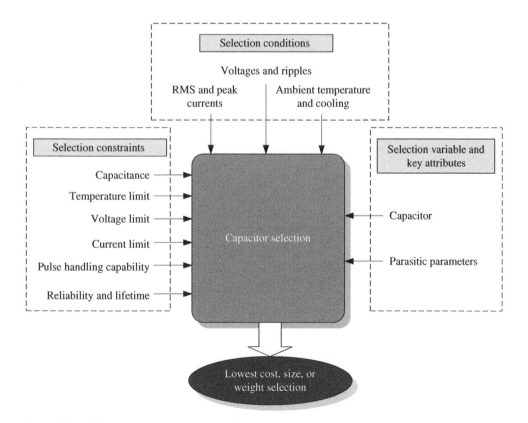

Figure 3.3 DC-link capacitor selection problem.

only the physical capacitor itself. However, some capacitor characteristics, especially its parasitic resistance and inductance, will impact the capacitor performance, and therefore are important selection attributes. In addition to the variable/attributes and constraints, the input (selection) conditions are needed for the capacitor selections. For example, in a DC-link capacitor case, they include RMS and peak currents, normal and abnormal DC-link voltages, DC voltage ripples, and ambient temperature and cooling condition (e.g. air flow).

3.4 Capacitor Characteristics and Models

This section will focus on the capacitor characteristics and models that are needed for the selection constraints, including: (i) the equivalent impedance to check the capacitance limit and determine parasitic parameters – key attributes, (ii) the voltage, and current capability characteristics needed to check voltage and current limits, (iii) loss and thermal models needed to check temperature limit, and (iv) reliability and lifetime models needed to check lifetime limit.

3.4.1 Capacitor Equivalent Circuit Model and Capacitance

Figure 3.4 shows a nonideal capacitor equivalent circuit including parasitics. The nominal capacitance C is shunted by a resistance R_P, which represents insulation resistance or leakage. Resistance R_s appears in series with the capacitance C, which represents resistance of capacitor leads and plates. Inductance L_s is also in series with the capacitance C, which represents the inductance of the leads and plates. Resistance R_d represents the dielectric loss due to dielectric absorption and molecular polarization, and C_d is the equivalent capacitance to account for the inherent dielectric absorption [12]. In reality, different capacitor phenomena are not easy to separate out. The matching of phenomena and models is for convenience in explanation.

The equivalent circuit in Figure 3.4 can usually be simplified to a series connection of the capacitance C, an ESR, and an equivalent series inductance (ESL) L_s, as shown in Figure 3.5. An example of the capacitor impedance curve is shown in Figure 3.6. In low-frequency range, the impedance is mainly determined by its capacitance value C. At the resonant frequency point, the negative impedance of capacitance and the positive impedance of the ESL are equal, and the impedance is equal to the ESR. Above the resonant frequency, the capacitor becomes inductive as the ESL dominates the impedance.

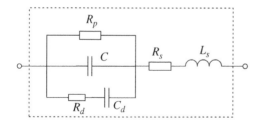

Figure 3.4 A nonideal capacitor equivalent circuit model.

Figure 3.5 Simplified capacitor equivalent circuit model.

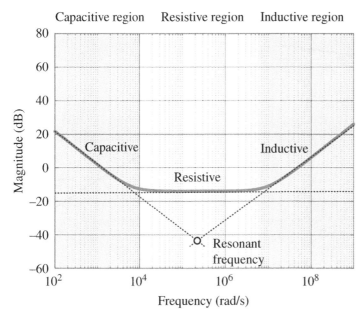

Capacitive region   Resistive region   Inductive region

Figure 3.6 Impedance characteristics of a capacitor.

3.4.1.1 Capacitance

Capacitance represents the amount of electric charge a capacitor can store per unit voltage. Capacitors contain at least two electrical conductors often in the form of metallic plates or surfaces separated by a dielectric medium. The capacitance of such a parallel plate structure is proportional to the area.

In general, capacitance is frequency, temperature, and voltage dependent. For example, temperature has considerable effect on the capacitance of an electrolytic capacitor. With decreasing temperature, the viscosity of the electrolyte increases, thus reducing its conductivity [13]. The resulting typical behavior is shown in Figure 3.7a. The AC capacitance also depends on the measuring frequency. Figure 3.7b shows the typical behavior. Typical values of the effective capacitance can be derived from the impedance curve in datasheet as $C = 1/(2\pi f Z)$, as long as the impedance Z is still in the range where the capacitive component is dominant.

Note that electrolytic capacitor presents significant capacitance reduction in the high-frequency range (e.g. >10 kHz) because the tunnel-shaped pit structure of anode foil is unsuitable for the response to high-frequency switching. An example of the curve-fitted frequency-dependent capacitance in high-frequency range is shown in Figure 3.8 [14]. In high-frequency applications, electrolytic capacitors may need to have their rated capacitances at 100 Hz several times higher than the required value at high frequencies to account for the frequency-dependent capacitance drop. Capacitor power density worsens compared to low-frequency applications. Compared with electrolytic capacitors, the capacitance of film and ceramic capacitors are much more stable with frequency. As mentioned earlier, ceramic capacitors exhibit significant capacitance variation with the DC bias voltage. An example is shown in Figure 3.9, where the capacitance of the Class II ferroelectric material based MLCC decreases rapidly with DC bias voltage, while the antiferroelectric material-based PLZT capacitor shows an opposite trend [10].

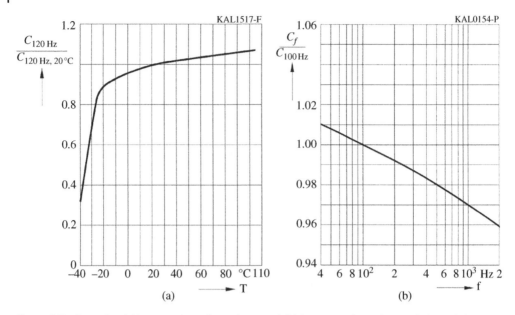

Figure 3.7 Example of (a) temperature dependence and (b) frequency dependence of electrolytic capacitor. *Source:* From [12].

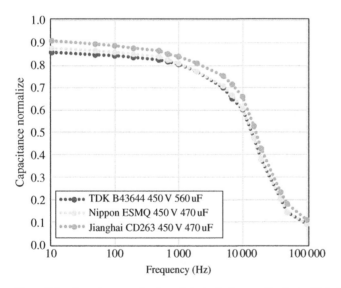

Figure 3.8 Capacitance reduction of electrolytic capacitor in the high-frequency range. *Source:* From [14] with permission of IEEE.

3.4.1.2 ESR

ESR represents the power loss of a capacitor, which includes the resistances of the dielectric material, the terminal leads, the connections to the dielectric, and the capacitor plate. ESR is frequency f and temperature T dependent. It can be generally modeled as

$$ESR\,(C, f, T) = ESR_{rated}(C) \times \alpha_{ESR}(f, T) \tag{3.1}$$

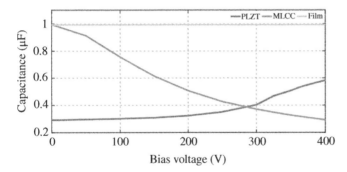

Figure 3.9 Capacitance variation of ceramic capacitors with respect to DC bias voltages. *Source:* From [10] with permission of IEEE.

$ESR_{rated}(C)$ is the ESR at specified frequency (e.g. 100 Hz for AECs and 10 kHz for film capacitors from datasheet), which is a function of capacitance C. $\alpha_{ESR}(f, T)$ is the coefficient to represent the ESR frequency and temperature dependency.

ESR is frequency dependent. At low frequencies, ESR is usually dominated by the dielectric loss and decreases as frequency increases. At medium to high frequencies, losses in the metallic conductors are dominant and ESR becomes relatively constant. At very high frequencies, ESR increases due to the skin effect. ESR is also temperature dependent. ESR usually decreases at higher temperature as it is usually dominated by the variation in dielectric viscosity. But the trend could be complex and varies case by case as it is essentially determined by the properties of the dielectric, the capacitor construction, and manufacturing parameters of the capacitor. Figure 3.10a, b show the typical ESR versus frequency and temperature curves for AECs and film capacitors, respectively. Note that the ESR variation of film capacitors with temperature is very small and is negligible [15].

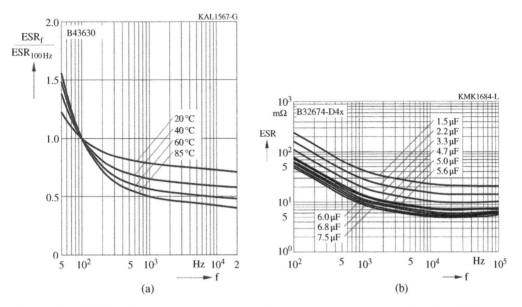

Figure 3.10 (a) ESR vs. frequency and temperature of electrolytic capacitor. *Source:* Data from TDK B43630 series datasheet. (b) ESR vs. frequency of film capacitor. *Source:* Data from TDK B32674 series datasheet.

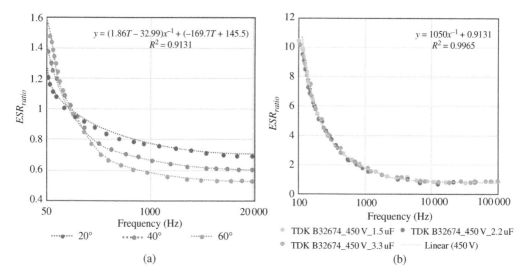

Figure 3.11 (a) Relationship among ESR ratio, frequency, and temperature of electrolytic capacitors. *Source:* Data from TDK B43630 series datasheet. (b) Relationship between ESR ratio and frequency of film capacitors. *Source:* Data from TDK B32674 series datasheet.

By measuring the ESR in frequency and temperature ranges (or from capacitor datasheets), $\alpha_{ESR}(f, T)$ can be obtained using curve fitting. Figure 3.11 shows $\alpha_{ESR}(f, T)$ (ESR ratio) of the electrolytic capacitors and film capacitors as an example [14].

3.4.1.3 ESL

ESL of capacitors is introduced by winding layers and connected power terminals. Typical ESL for electrolytic capacitors ranges from 20 to 40 nH for snap-in types. The value for film capacitors with snap-in type ranges from 5 to 30 nH. ESL of ceramic capacitors is usually several nH or smaller due to their smaller profile. Generally, especially for film capacitors, ESL increases with the terminal spacing, terminal-tab loop area, and winding itself as well as capacitance. The model for ESL can be described as a linear function of the capacitance C

$$ESL = k_{1,ESL}C + k_{2,ESL} \tag{3.2}$$

where $k_{1,ESL}$ is the coefficient to describe the winding inductance related to capacitance, and $k_{2,ESL}$ is the terminal inductance of the capacitor. $k_{1,ESL}$ and $k_{2,ESL}$ can be obtained through measurements or from datasheets.

3.4.2 Voltage and Current Capability Models

3.4.2.1 Overvoltage Capability

The rated capacitor voltage V_R is defined as the maximum operating peak voltage that can be applied continuously to the capacitor terminals at any temperature between the minimum temperature (T_{min}) and the rated temperature (T_R). The overvoltage capability is usually a function of voltage duration time specified by capacitor manufacturer. For example, Table 3.3 shows the allowed overvoltage versus duration for film capacitor TDK B32674 series.

Table 3.3 Overvoltage capability of film capacitor TDK B32674 series.

Overvoltage	Maximum duration
$1.1 \times V_R$	30% of on-load duration per day
$1.15 \times V_R$	30 minutes per day
$1.2 \times V_R$	5 minutes per day
$1.3 \times V_R$	1 minutes per day
$1.5 \times V_R$	30 ms for 1000 times during the life of the capacitor

Source: Data from TDK B32674 series datasheet.

3.4.2.2 Maximum Permissible Continuous DC Voltage vs. Temperature

The maximum permissible continuous operating DC voltage is impacted by temperature. When the temperature is between the rated temperature T_R and the upper bound temperature (T_{max}), derating should be applied. The derating factor is dielectric material dependent and specified by manufacturer. Still taking film capacitor TDK B32674 series as an example, for temperature between 85 and 105 °C, a derating factor 1.2%/°C of continuous operating voltages should be applied compared to that at 85 °C.

3.4.2.3 Maximum AC Voltage/Current vs. Frequency

The ability of a capacitor to withstand a continuous sinusoidal AC voltage V_{RMS} or AC current I_{RMS} is a function of the frequency. Use film capacitor as an example. The AC voltage/current capability is limited by three different factors as shown in Figure 3.12 [15].

Region (a): Limit to avoid corona discharge. Below a certain frequency limit f_1, the applied AC voltage V_{RMS} should not exceed the threshold voltage V_{CD} at which corona discharge would start to occur with some intensity in air pockets in the capacitor. It may degrade the film metallization and occasionally endanger its dielectric strength. For sinusoidal waveforms, the following relation must be taken into consideration for RMS voltage V_{RMS} and current I_{RMS}

$$V_{RMS} \leq V_{CD} \text{ or } I_{RMS} \leq V_{CD} \cdot 2\pi f \cdot C \tag{3.3}$$

For nonsinusoidal waveforms, peak-to-peak voltage

$$V_{pp} \leq 2\sqrt{2}V_{CD} \tag{3.4}$$

Region (b): Limit due to thermal power dissipation. Above a certain frequency limit f_1, the permissible AC voltage must be reduced with increasing frequency to keep the power loss generated in the capacitor body P_{loss} below the power that can be dissipated by the surface of the capacitor P_{diss}. The two powers are defined as

$$P_{loss} = I_{RMS}^2 \cdot ESR \approx \left(\frac{V_{RMS}}{2\pi f \cdot C} \right)^2 \cdot ESR = I_{RMS}^2 \cdot 2\pi f \cdot C \cdot \tan \delta \tag{3.5}$$

$$P_{diss} = \alpha \cdot A \cdot \Delta T \tag{3.6}$$

where $\tan \delta = \dfrac{ESR}{2\pi f \cdot C}$ is the dissipation factor, α represents the heat transfer coefficient, ΔT denotes the self-heating or steady-state overtemperature attained at the hottest part of the capacitor surface in relation to the surrounding atmosphere, and A is the surface area of the capacitor.

To prevent permanent damage to the capacitor, the self-heating ΔT must not exceed a certain value, depending on the dielectric material. For polyester (PET) and polyethylene naphthalate

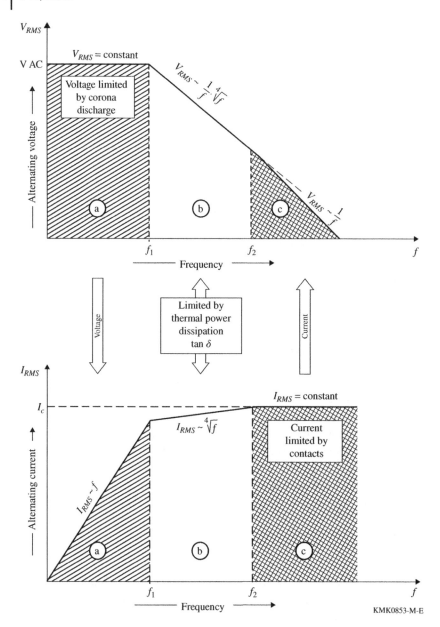

Figure 3.12 AC voltage and current load limits. *Source:* From [15].

(PEN), $\Delta T_{max} = 15\,°C$. For polypropylene (PP), $\Delta T_{max} = 10\,°C$. Since $P_{loss} \leq P_{diss}$, the conditions for the maximum permissible AC voltage and current in this region can be deduced as

$$V_{RMS} \leq \sqrt{\frac{\alpha \cdot A \cdot \Delta T}{2\pi f \cdot C \cdot \tan \delta}} \text{ or } I_{RMS} \leq \sqrt{\frac{2\pi f \cdot C \cdot \alpha \cdot A \cdot \Delta T}{\tan \delta}} \tag{3.7}$$

Their frequency dependency can be approximated by

$$V_{RMS(max)} \sim \frac{1}{f} \cdot \sqrt[4]{f} \text{ or } I_{RMS(max)} \sim \sqrt[4]{f} \tag{3.8}$$

Note the above equations assume tan δ is a constant, which is a reasonable assumption for the frequency range considered (also well below the natural resonance frequency).

The heat transfer coefficient α and the dissipation factor tan δ depend on technology, construction, material, and geometry of each capacitor. tan δ also depends on frequency and temperature. This complicates the use of these equations to calculate the maximum allowed V_{RMS} (or I_{RMS}) or the self-heating (ΔT) under practical operating conditions. The limit lines for the maximum V_{RMS} (or I_{RMS}) at moderate temperatures have consequently been obtained empirically and are illustrated in the datasheets relative to each series. For example, Figure 3.13 shows the permissible current I_{RMS} versus frequency f at 70 °C for film capacitor B32674D series. For intermediate values of capacitance (denoted as C') not included in the graphs, the permissible RMS voltage or current can be approximately calculated for the closest existing curves according to

$$V'_{RMS} = V_{RMS} \cdot \sqrt{C/C'} \text{ or } I'_{RMS} = I_{RMS} \cdot \sqrt{C/C'} \tag{3.9}$$

The above calculations correspond to pure sinusoidal waveforms. Usually, practical applications will not involve loads with pure sine waves. In some common applications, the loads can be estimated accurately enough by approximating them to sine waves. In other cases, the power loss generated must consider harmonic power. One approximate approach is to decompose voltage via Fourier transform and then find the total power loss as

$$P_{loss} = \sum_i V^2_{RMS,i} \cdot 2\pi f_i \cdot C \cdot \tan \delta(f_i) \tag{3.10}$$

The temperature rise ΔT can be determined using (3.6) by setting $P_{loss} = P_{diss}$.

The relationship described for the maximum AC voltage/current vs. frequency is valid for moderate temperatures. For high temperatures, derating needs to be applied. For example, Figure 3.14

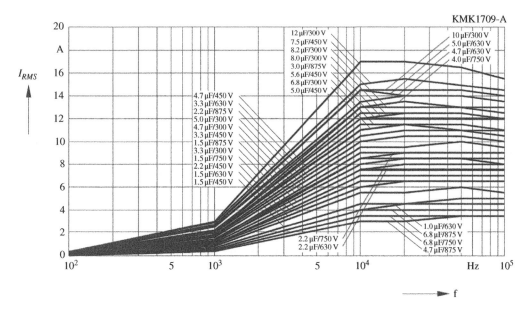

Figure 3.13 The permissible current I_{RMS} versus frequency f at 70 °C for TDK film capacitor B32674D series. *Source:* Data from TDK B32674 series datasheet.

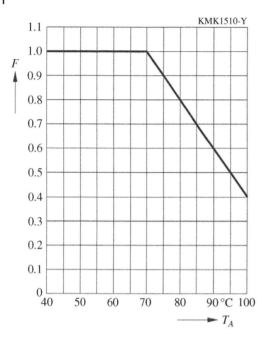

Figure 3.14 I_{RMS} derating versus temperature for TDK film capacitor B32674 series. *Source:* Data from TDK B32674 series datasheet.

shows derating factor F as a function of ambient temperature (i.e. case temperature) T_A, such that the maximum current I_{RMS} at T_A is

$$I_{RMS}(T_A) = F \cdot I_{RMS}(T_{rated}) \tag{3.11}$$

Region (c): Limit due to maximum current handing capability. Above another frequency limit f_2, the permissible AC voltage is limited by the maximum current I_C that can pass through the connection between sprayed metal and film metallization, without causing overheating due to associated resistive losses.

$$V_{RMS} \leq \frac{I_C}{2\pi f \cdot C} \text{ or } I_{RMS} \leq I_C \tag{3.12}$$

In practice, this region is only evident for smaller capacitors with short contact lengths.

3.4.2.4 Pulse Withstanding Capability

If the peak current i_p associated with a voltage pulse is considered, the maximum voltage pulse slope dv/dt of a pulse can be defined as in (3.13). Values of maximum dv/dt are given in capacitor datasheets.

$$\frac{dv}{dt} = \frac{i_p}{C} \tag{3.13}$$

The heat energy Q generated by a pulse in the capacitor can be calculated from

$$Q = \int_0^\tau i^2 R_i dt = \int_0^\tau C^2 \left(\frac{dv}{dt}\right)^2 R_i dt \tag{3.14}$$

where τ is the pulse width and R_i is the resistance of the contacts.

To relate the heat energy and the pulse slope, a characteristic factor k_0 depending on the waveform can be defined as

$$k_0 = 2\int_0^T \left(\frac{dv}{dt}\right)^2 dt \qquad (3.15)$$

The k_0 value of a particular waveform should not exceed the maximum permissible values given in the datasheets of each capacitor series. k_0 values are valid assuming complete dissipation of heat between consecutive pulses, which is the case for relatively long duty cycles (e.g. longer than 100 μs). For short duty cycles (e.g. for repetition frequencies higher than 10 kHz), heat dissipation from pulse to pulse may be incomplete, and a dangerous overheating can build up, which might destroy the capacitor. In these cases, a derating of the permissible k_0 value should be applied.

The dv/dt value and/or k_0 value are usually provided by datasheet. For example, Table 3.4 shows the dv/dt limits for film capacitor TDK B32674 series.

3.4.3 Loss and Thermal Models

Thermal stress through temperature rise is also one of the critical limiting factors for capacitors. Over times, high thermal stress can result in capacitance reduction and ESR increase due to wear out. The power loss and ambient temperature are the contributors to the internal thermal stress and temperature rise of the capacitor. Both factors lead to an increase of ESR, resulting in increased power loss and higher operating temperature inside the capacitor.

As ESR is provided in datasheet, which is the sum of resistances of the leads, the contacts, and the winding (that also includes the dielectric loss), power loss can be expressed as a function of ripple current and ESR. The power losses model is simplified as in (3.16), where N is the upper bound of ripple current harmonics order that needs to be considered, and other variables and parameters have been defined before. Note that (3.16) is equivalent to (3.10).

$$P_{loss} = \sum_{i=1}^{N} ESR(C, f_i, T_h) \times I_{RMS}^2(f_i) \qquad (3.16)$$

Depending on the converter application, two sets of models are usually used: one for steady-state analysis, and the other is for thermal transient analysis.

For the steady-state analysis, the capacitor hotspot to ambient thermal resistance is needed. Then, the hotspot temperature is expressed as

$$T_h = T_a + R_{th-ha} \cdot P_{loss} \qquad (3.17)$$

where R_{th-ha} is the equivalent thermal resistance from the hotspot to ambient, T_h is the hotspot temperature, and T_a is the ambient temperature.

Table 3.4 dv/dt capability of film capacitor TDK B32674 series.

Type	Film capacitor TDK B32674				
Lead spacing	27.5 mm				
V_R (V DC)	300	450	630	750	875
dv/dt (V/μs)	40	75	100	125	150

Source: Data from TDK B32674 series datasheet.

For thermal transient analysis, a thermal impedance considering the thermal capacitance is needed. The thermal impedance can be described as an *RC* network using the Foster model or Cauer model described in Chapter 2. In this case, Laplace transform can be used to determine the temperature rise, and the temperature rise in s domain is expressed as

$$\Delta T_h(s) = Z_{th-ha}(s) \cdot P_{loss}(s) \tag{3.18}$$

where $Z_{th-ha}(s)$ is the equivalent thermal impedance from the hotspot to ambient, and $\Delta T_h(s)$ is the capacitor temperature rise, i.e. the temperature difference between the hotspot and ambient.

The thermal resistance is usually provided in capacitor datasheet. Some datasheets provide the equivalent heat coefficient G (mW/ °C), which is a function of the capacitor dimension. Again, take film capacitor TDK B32674 series as an example. For a given dimension 11 mm×19 mm ×31.5 mm, the equivalent heat coefficient G is 25 mW/ °C as provided in the capacitor datasheet. Note that the equivalent heat coefficient G is given by measuring the temperature on the surface of the capacitor not the hotspot temperature in the datasheet.

Sometimes the thermal resistance information is not provided by datasheet. In this case, finite element analysis (FEA) based simulation or experiment can be adopted to obtain the thermal resistance. The methods for FEA simulation and experiment of film capacitors are briefly introduced here based on [16].

For FEA simulation, as the thickness of the film (typically on the order of micro-meters) and the metal layer (typically on the order of nano-meters) is too small to be computed, some assumptions are made to approximate realistically the metallized films without the need to have such small features. The wound element is modeled as a single solid body, without any micrometer or nanometer thick layers inside. The material for this body is the dielectric used in the capacitor since it constitutes almost all the mass of the roll due to the high plastic to metal thickness ratio. The heat is generated in the winding, in the contacts between the winding and the spray, and in the terminations. Power losses are imposed as volumetric heat generation on the interested elements. Through thermal conduction, the heat moves from the inside to the outside of the capacitor, where it is then dissipated through the external surfaces of the plastic case, the top resin layer, and the terminals. The main contributions to the external heat dissipation are convection and radiation heat transfer, which are given by

$$q_{conv} = h_{conv}A(T_s - T_a) \tag{3.19}$$

$$q_{rad} = \epsilon\sigma A\left(T_s^4 - T_a^4\right) \tag{3.20}$$

where q_{conv} and q_{rad} are transferred thermal power (in W) through convection and radiation, respectively; h_{conv} is the convection heat transfer coefficient of the material with a unit of W/(m$^2 \cdot$ K); A is the surface of heat body, T_s is the surface temperature; ϵ is the emissivity of the radiating surface; and σ is the Stefan–Boltzmann constant (5.67×10^{-8} W/m^2/K^4). The 3D simulation model and mesh of a film capacitor (R76MR382050H3J) are shown in Figure 3.15.

For experiment, the test circuit scheme as shown in Figure 3.16a can be used. An AC ripple voltage or current is applied to provide fixed power loss to the capacitor. The setup can be put in a temperature chamber to control the ambient temperature. A thermocouple is attached on the wall of the plastic box to record the temperature of external surface (T3) as shown in Figure 3.16b. If the capacitor can be opened, the internal temperature can be measured. Thermocouples can be attached on the spray head (T1) and on the wound element (T2) as shown in Figure 3.16c. The temperature should be recorded after reaching a stable point.

Thermal resistance can then be calculated as the temperature rise is measured and power loss is known. Note that in the experimental approach, it is usually not practical to measure the internal

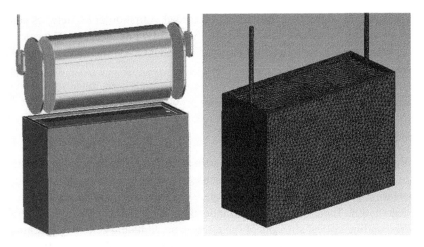

Figure 3.15 3D FEA simulation model of a film capacitor: left – the wound element modeled as a single solid body, together with the case; right – mesh.

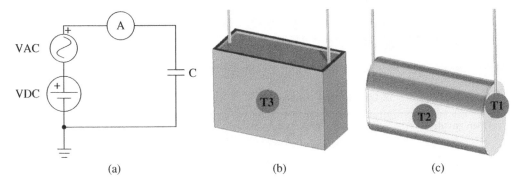

(a)   (b)   (c)

Figure 3.16 (a) Experiment test circuit: (b) thermocouple position on outside surface: (c) thermocouple position inside of the capacitor.

temperature of the capacitor, thus only the thermal resistance from capacitor case to ambient can be obtained. However, the thermal resistance from capacitor hotspot to capacitor case is usually small compared to the thermal resistance from the capacitor case to ambient. Most of the temperature rise is between ambient and the capacitor case. As shown in [16], when 0.5 W power loss is applied to the test sample (R76MR382050H3J), the capacitor case temperature, wound element temperature, and expected hotspot temperature are 84.5, 85, and 85.4 °C when the ambient temperature is 70.9 °C, while the capacitor case temperature, wound element temperature, and expected hotspot temperature are 112, 112.4, and 114.4 °C when the ambient temperature is 99.9 °C. Therefore, we can use the capacitor case temperature to approximate its hotspot temperature without much error when the internal structure of the capacitor is not known. And the thermal resistance from case to ambient is mainly determined by capacitor surface area as indicated by (3.19) and (3.20) instead of internal structure of the capacitor. On the other hand, sometimes it is prudent to leave some margins between the surface temperature and hotspot temperature, e.g. 3–5% margin according to [16].

FEA simulation and experiment can also be used to obtain thermal impedance for thermal transient analysis. The procedure and setup are similar to the earlier discussion. But in this case, instead of recording the steady-state temperature, the temperature response curves of the capacitor during the current charging and natural cooling periods need to be recorded. A first-order *RC* network can be obtained to fit the temperature response data, and then the thermal impedance can be calculated. More details on thermal impedance modeling of a capacitor using FEA simulation and experiment test methods can be found in [17].

The above thermal models are for natural cooling. If forced convection is used, models similar to those for inductors can be used. Detailed discussions will be provided in Chapters 4 and 9.

3.4.4 Lifetime Model

Capacitors could fail due to intrinsic and external factors, such as design defect, material wear out, operating temperature, voltage, current, moisture, and mechanical stress. Generally, the capacitor failures can be divided into catastrophic failure due to single-event overstress and wear out failure due to the long-term degradation. Table 3.5 gives a systematic summary of the failure modes, failure mechanisms, and corresponding critical stressors of the electrolytic, film, and ceramic capacitors [18]. Table 3.6 shows the comparison of failure and self-healing capability of the three types of capacitors [18].

Electrolyte vaporization is the major wear out mechanism of small size Al-capacitors (e.g. snap-in type) due to their relatively high ESR and limited heat dissipation surface. For AECs,

Table 3.5 Overview of failure modes, critical failure mechanisms, and critical stressors of the three main types of capacitors.

Capacitor type	Failure modes	Critical failure mechanisms	Critical stressors
AECs	Open circuit	Self-healing dielectric breakdown	V_C, T_a, i_C
		Disconnection of terminals	Vibration
	Short circuit	Dielectric breakdown of oxide layer	V_C, T_a, i_C
	Wear out: electrical parameter drift	Electrolyte vaporization	T_a, i_C
		Electrochemical reaction (e.g. degradation of oxide layer, anode foil capacitance drop)	V_C
MPP-capacitors	Open circuit (typical)	Self-healing dielectric breakdown	V_C, T_a, dV_C/dt
		Connection instability by heat contraction of a dielectric film	T_a, i_C
		Reduction in electrode area caused by oxidation of evaporated metal due to moisture absorption	Humidity
	Short circuit (with resistance)	Dielectric film breakdown	V_C, dV_C/dt
		Self-healing due to overcurrent	T_a, i_C
		Moisture absorption by film	Humidity
	Wear out: electrical parameter drift	Dielectric loss	V_C, T_a, i_C, humidity
MLCCs	Short circuit (typical)	Dielectric breakdown	V_C, T_a, i_C
		Cracking, damage to capacitor body	Vibration
	Wear out: electrical parameter drift	Oxide vacancy migration; dielectric puncture; insulation degradation; micro-crack within ceramic	V_C, T_a, i_C, vibration

V_C, capacitor voltage stress; i_C, capacitor ripple current stress; i_{LC}, leakage current; T_a, ambient temperature.
Source: Adapted from [18].

Table 3.6 Comparison of failure and self-healing capability of the three types of capacitors.

	AECs	MPP capacitors	MLCCs
Dominant failure modes	Wear out		
	Open circuit	Open circuit	Short circuit
Dominant failure mechanisms	Electrolyte vaporization; electrochemical reaction	Moisture corrosion; dielectric loss	Insulation degradation; flex cracking
Most critical stressors	V_C, T_a, i_C	V_C, T_a, humidity	V_C, T_a, vibration
Self-healing capability	Moderate	Good	No

Source: Adapted from [18].

the wear out lifetime is dominantly determined by the increase of leakage current, which is relevant with the electrochemical reaction of oxide layer. The most important reliability feature of film (metallized polypropylene or MPP) capacitors is their self-healing capability. Initial dielectric breakdowns (e.g. due to overvoltage) at local weak points of a film capacitor will be cleared, and the capacitor regains its full capability except for a negligible capacitance reduction. With the increase of these isolated weak points, the capacitance is gradually reduced to reach the end of life. The metallized layer in MPP-capacitors is typically less than 100 nm, which are susceptible to corrosion due to the ingress of atmospheric moisture. Severe corrosion occurs at the outer layers resulting in the separation of metal film from edge and therefore the reduction of capacitance. The corrosion in the inner layers is less advanced as it is less open to the ingress of moisture. Unlike the dielectric materials of AECs and MPP-capacitors, the dielectric materials of ceramic capacitors are expected to last for thousands of years at use level conditions without showing significant degradation. Therefore, wear out of ceramic capacitors is typically not an issue. However, a ceramic capacitor could degrade much more quickly due to the "amplifying" effect from the large number of dielectric layers. It has been shown that a modern MLCC could wear out within 10 years due to increasing miniaturization through the increase of the number of layers. Moreover, the failure of MLCCs may induce severe consequences to power converters due to the short-circuit failure mode. The dominant failure causes of MLCCs are insulation degradation and flex cracking. Insulation degradation due to the decrease of the dielectric layer thickness results in increased leakage currents.

A widely used empirical lifetime model for capacitors, which describes the influence of temperature and voltage stress, is

$$LT = LT_0 \left(\frac{V}{V_0} \right)^{-n} \exp\left[\left(\frac{E_a}{K_B} \right) \left(\frac{1}{T} - \frac{1}{T_0} \right) \right] \tag{3.21}$$

where LT and LT_0 are the lifetimes for the operating and test conditions, respectively; V and V_0 are corresponding voltages, respectively; T and T_0 are corresponding temperatures (in K), respectively; E_a is the activation energy (in eV); K_B is the Boltzmann constant (8.62×10^{-5} eV/K); and n is the voltage stress exponent. Therefore, the values of E_a and n are the key parameters for the lifetime model. The ranges of E_a and n for MLCC are 1.3–1.5 and 1.5–7, respectively.

For AECs and film capacitors, a simplified model is popularly applied

$$LT = L_0 \left(\frac{V}{V_0} \right)^{-n} 2^{\frac{T_0 - T}{10}} \tag{3.22}$$

This model is corresponding to a specific case when $E_a = 0.94$ eV, and T_0 and T are substituted by 398 K. For MPP capacitors, the exponent n is from around 7 to 9.4 used by leading capacitor manufacturers. For AECs, the value of n typically varies from 3 to 5.

3.5 Capacitor Bank (Parallel/Series)

In many cases, instead of using a single capacitor, multiple capacitors connected in parallel and/or in series are used to achieve higher capacitance, lower ESR, lower ESL, higher RMS current capability (parallel connection), or higher voltage capability (series connection). For example, in motor drives with DC-link voltages over 500 V, AECs are connected in series and then in parallel to withstand the voltages and to filter harmonics. Capacitor paralleling or series may also achieve smaller size and weight than a single capacitor.

3.5.1 Capacitor Bank Configuration and Voltage Balancing

If both parallel and series connections are necessary, there are two possible structures for capacitor banks, which are shown in Figure 3.17. The two structures have different properties, especially with respect to balancing the capacitor voltage. The series–parallel connection requires a separate balancing device (for example, a resistor) for each capacitor, while in the parallel–series arrangement the whole group of capacitors connected can share the same device. The results of failure of one capacitor in the bank are also different. Consider the worst-case scenario when a capacitor fails due to a short circuit. In the case of series–parallel structure, such failure first affects the remaining capacitors in the same branch and subjects them to an overvoltage, which can then also cause them to fail. This can result in either an open circuit or a short circuit for the affected branch, but without spreading to the remaining branches. In the case of the parallel–series connection, the remaining capacitors in the same group first discharge to the failed part. The other groups are then exposed to the full DC-link voltage and can also fail due to the applied voltage. As a result, the whole bank can be damaged.

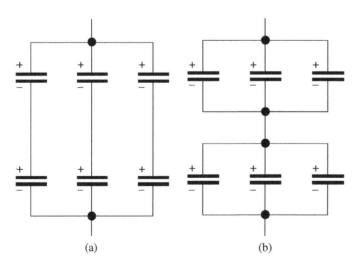

(a)   (b)

Figure 3.17 Possible capacitor bank structure: (a) series–parallel connection, and (b) parallel–series connection.

When a capacitor bank is being built, it is recommended that all individual capacitors are subject to as equal stresses (voltage, current, temperature, and cooling conditions) as possible. This will ensure a similar aging rate for all capacitors and minimize the risk of failures. Voltage stress equalization in series-connected capacitors merits particular concern. The cause of potential voltage asymmetry is the spread of leakage current versus voltage characteristics. This spread is unavoidable and can be extensive. The effective voltage unbalance is reduced by the nonlinearity of the leakage current characteristic near the rated voltage (the current rises quickly with a small increase of voltage). Nevertheless, it is widely accepted that some additional balancing measures are needed. The most popular voltage balancing method is the use of a series resistor voltage divider as shown in Figure 3.18. In order to ensure the necessary voltage symmetry, the resistance value must be relatively low (compared with the capacitor shunt resistance corresponding to the leakage current) and have a tight tolerance (as this directly affects the precision

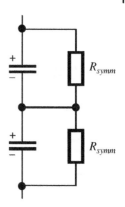

Figure 3.18 Passive volatge balancing method using a resistor divider.

of the voltage equalization). The typical recommendation is that the current flowing through the resistor should be approximately 20 times higher than the expected leakage current.

The balancing measure described above may be omitted in cases where the total capacitor bank voltage V_{bank} to be applied is substantially lower than the sum of the rated voltage of the series connected capacitors. Experience has shown that this is possible if the total V_{bank} does not exceed $0.8N_sV_R$, where N_s is the number of capacitors in series. Although some voltage asymmetry may be observed in such cases, it will not lead to capacitor failure.

3.5.2 Capacitor Bank Layout for Parasitic Inductance Reduction

It is important to minimize the capacitor bank parasitic inductance for power electronics converters, especially with fast-switching WBG devices. Use multiple smaller capacitors connected in parallel may offer a lower ESL than one large capacitor with equal total capacitance. In this case, attention should be paid to the capacitor bank layout design to maximize the ESL reduction.

The concept of magnetic field cancelling can be used to minimize parasitic inductance for both series and parallel of capacitors [19]. For the case of series capacitors, the schematic is shown in Figure 3.19a. A three-layer printed circuit board (PCB) or busbar can be used to connect the capacitors, and the layout is shown in Figure 3.19b. The series capacitor MID connection is on the outer two pins, and the DC+ and DC− connections are on the inner two pins. For each capacitor, the currents flow through the capacitor body and the PCB/busbar copper below the capacitor are in opposite directions, which enhances the magnetic cancellation and thus reduces the loop inductance.

For the parallel capacitor's case, the schematic is shown in Figure 3.20a and the PCB/busbar layout is shown in Figure 3.20b. The adjacent two paralleled capacitors are placed opposite. Figure 3.20b shows the layout of DC+ layer, while the DC− layer is similar and therefore not repeated. With this configuration, for the two adjacent capacitors, the currents flow through the two PCB/busbar layers are in opposite directions and the currents flow through the two capacitor bodies are also in opposite directions, which enhances the magnetic cancellation and thus reduces the loop inductance.

The low parasitic inductance busbar design utilizing magnetic field cancellation will be further discussed in Chapter 11.

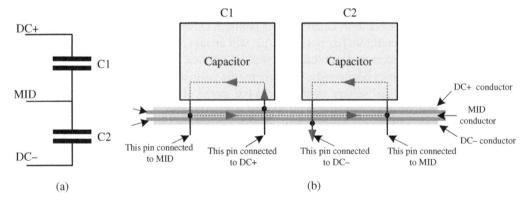

Figure 3.19 Series capacitors layout: (a) schematic; (b) layout for reduced inductance.

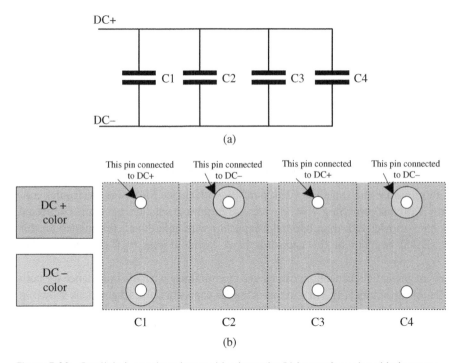

Figure 3.20 Paralleled capacitors layout: (a) schematic; (b) layout for reduced inductance.

3.6 Relevance to Converter Design

The relevance of the capacitor selection to the converter subsystem design has already been discussed in Section 3.3 with capacitor selection input (conditions) and output (variables and constraints) clearly defined. In addition, the capacitor selection will generate associated attributes. Some of these attributes have been discussed above, including parasitic parameters ESR and ESL. Here, we will further discuss capacitance scaling and AC capacitor classification.

3.6.1 Capacitor Scaling

The capacitor attributes important to converter design include their size, weight, and cost. For commercial capacitors, their size and weight are provided in datasheet. In general, electrolytic capacitors have higher energy density and lower cost compared with film and ceramic capacitors. For the same type of capacitors, size, weight, and cost of capacitors are linear with the energy storage capability in general, and thus they are linear with the capacitance for capacitors of the same voltage rating in the same series.

3.6.2 AC Capacitor Classification

In AC applications, there are two special classes of capacitors, Class-X and Class-Y, which must be connected directly to AC lines (line-to-line or line-to-ground) in order to perform EMI filtering functions. They are commonly referred to as "safety capacitors."

Class-X capacitors are commonly referred to as "line-to-line" or "across the line" capacitors, which are placed across the line and neutral connections to mitigate negative effects that may be caused by conducted EMI, overvoltage surges, and voltage transients. Class-X capacitors are subject to all of the AC line variations, which can create a hazardous situation if the voltage or power threshold of the capacitor's capabilities are exceeded. In an overstress situation, Class-X capacitors are designed to fail short circuit, in order to trigger the circuit breaker or fuse to break the supply circuit. Class-Y capacitors are commonly referred to as "line-to-ground" or "line bypass" capacitors, which are placed between the AC line and ground, to handle EMI noise. Class-Y capacitors are also subject to AC line variations via conducted EMI, overvoltage surges, and voltage transients, which can also lead to hazardous situations if the threshold of the capacitor's capabilities are exceeded and the capacitor fails. Different from Class-X capacitors, Class-Y capacitors are designed to fail open circuit; otherwise, it could lead to a fatal electric shock due to the loss of the ground connection if it fails short. As with many safety-critical devices, varying standards and respective classifications are used to indicate the capabilities and threshold of safety capacitors. One commonly used standard IEC 60384-14 defines the safety classification of Class-X and Class-Y per various levels of peak voltage pulse before failure. Per the IEC 60384-14, the Class-X capacitor subclasses are ranked as follows: Subclass X3 – peak voltage pulse of less than or equal to 1.2 kV; Subclass X2 – peak voltage pulse of less than or equal to 2.5 kV; Subclass X1 – between 2.5 kV and less than or equal to 4.0 kV. The Class-Y subclasses are ranked as follows: Subclass Y4 – under 150 V AC; Subclass Y3 – less than or equal to 150 V AC up to 250 V AC; Subclass Y2 – less than or equal to 150 V AC up to 300 V AC; and Subclass Y1 – less than or equal to 500 V AC.

Ceramic capacitors and film capacitors can both be used for Class-X or Class-Y applications. Film capacitors are better choice when higher capacitance values are needed. Film capacitors offer self-healing, that is, they can recover from a dielectric breakdown. Their capacitance and dissipation factor are stable over a wide temperature range. Ceramic capacitors come with different formats and are suitable for space-sensitive applications. Three-terminal ceramic capacitors are available with improved high-frequency characteristics compared with two-terminal devices. One lead in a three-terminal capacitor has two connections, which can lead to reduced ESL and is beneficial when used for EMI attenuation.

3.7 Summary

This chapter covers the capacitors used in three-phase converter design with focus on their characteristics and models.

1) Different types of capacitors are introduced and compared. Electrolytic capacitors have polarized structure, high capacitance density, relatively low-voltage capability, and relatively high loss, and are suitable as DC-link capacitors for low-voltage converters. Ceramic capacitors have low capacitance values and low parasitic inductances and are suitable for EMI filters and decoupling capacitors. Film capacitors with their overall more balanced characteristics are widely used for DC, AC, as well as EMI filters at different voltage levels.
2) Capacitor selection as part of the converter design problem is discussed with physical constraints of voltage, current, and temperature limits, and performance constraints of minimum capacitance and reliability/lifetime.
3) Capacitor characteristics and models needed for the converter design are covered. The equivalent circuit model needs to include capacitance, ESR, and ESL, and their dependence on frequency and temperature. Voltage capability depends on applied voltage duration, and voltage and current capabilities are also temperature and frequency dependent. The loss can be characterized through ESR. Capacitor lifetime is strongly influenced by temperature and voltage stresses.

References

1 W. J. Sarjeant, I. W. Clelland and R. A. Price, "Capacitive components for power electronics," *Proceedings of the IEEE*, vol. 89, pp. 846–855, 2001.
2 W. J. Sarjeant and D. T. Staffiere, "A report on capacitors," in *Proc. APEC*, 1996.
3 W. Sarjeant, J. Zirnheld and F. W. MacDougall, "Capacitors," *IEEE Transactions on Plasma Science*, vol. 26, pp. 1368–1392, 1998.
4 J. W. Kim, S. H. Shin, D. S. Ryu and S. W. Chang, "Reliability evaluation and failure analysis for high volatge ceramic capacitor," in *Int'l Symposium on Elect. Materials and Packaging*, 2001.
5 R. Anderson, "Select the right plastic film capacitor for your power electronic applications," in *Proc. Ind. Applicat. Conf.*, 1996.
6 M. H. El-Husseini, P. Venet, G. Rojat and M. Fathallah, "Effect of the geometry on the aging of metalized polypropylene film capacitors," in *Proc. Power Elect. Specialists Conf.*, 2001.
7 L. L. Macomber, "Reduce capacitor count 50% in bus-capacitor arrays using new electrolyte," in *Proc. APEC*, 2003.
8 F. Wang, Z. Zhang, T. Ericsen, R. Raju, R. Burgos and D. Boroyevich, "Advances in power conversion and drives for shipboard systems," *Proceedings of IEEE*, vol. 103, no. 12, pp. 2285–2311, 2015.
9 S. Chowdhury, E. Gurpinar and B. Ozpineci, "Capacitor technologies: characterization, selection, and packaging for next-generation power electronics applications," *IEEE Transactions on Transportation Electrification*, vol. 8, no. 2, pp. 2710–2720, 2022.
10 S. Chowdhury, E. Gurpinar and B. Ozpineci, "High-energy density capacitors for electric vehicle traction inverters," in *IEEE Transportation Electrification Conference & Expo (ITEC)*, 2020.
11 A. Yializis, "A disruptive DC-link capacitor technology for use in electric drive inverters," in *E-mobility Reinvented*, Santa Clara, CA, 2018.

12 B. W. Williams, "*Principles and Elements of Power Electronics: Devices, Drivers, Applications, Passive Components,*" Glasgow: Univ. Strathclyde, 2006.

13 TDK, "*Aluminum Electrolytic Capacitors General Technical Information,*" Tokyo, Japan: TDK, 2018.

14 H. Wang, C. Li, G. Zhu, Y. Liu and H. Wang, "Model-based design and optimization of hybrid DC-link capacitor banks," *IEEE Transactions on Power Electronics*, vol. 35, no. 9, pp. 8910–8925, 2020.

15 TDK, "*Film Capacitor General Technical Information,*" Tokyo, Japan: TDK, 2018.

16 D. Zuffi, E. Boni, W. Bruno, H. Nieves and M. Totaro, "K-TEM: KEMET thermal expectancy model," in *IEEE APEC*, Houston, TX, 2022.

17 Z. Yin, Y. Yang and H. Wang, "Thermal modeling of an electrolytic capacitor bank," in *IEEE 21st Workshop on Control and Modeling for Power Electronics (COMPEL)*, Aalborg, Denmark, 2020.

18 H. Wang and F. Blaabjerg, "Reliability of capacitors for DC-link applications in power electronics converters – an overview," *IEEE Transactions on Industry Applications*, vol. 50, no. 5, pp. 3569–3578, 2014.

19 CREE, "*Design Considerations for Designing with Cree SiC Modules Part 2: Techniques for Mimimizing Parasitic Inductance,*" CREE, 2014.

4

Magnetics

4.1 Introduction

Magnetics are also the basic passive components needed for power electronics converters. Magnetics include transformers and inductors, with the former mainly for AC voltage conversion and isolation, and the latter for energy storage and harmonics/noise filtering. Depending on topology and performance requirements, there can be many different types of magnetics in high-power three-phase converters, including: (i) DC-link inductors; (ii) AC line inductors; (iii) EMI filter inductors on AC and/or DC sides; (iv) *dv/dt* filter inductors for motor loads; (v) AC decoupling inductors; (vi) transformers on AC source side and for isolated converter topologies; and (vii) inductors in auxiliary circuits such as control circuitry. Magnetics in a converter can be multifunctional. For example, the harmonics filter and the EMI filter can be integrated in some applications. Transformers on the AC source side can provide the needed AC line inductance with their leakage inductance.

Transformers and inductors have much in common in their constructions and design. In this book, we will focus on inductors. This chapter will introduce magnetic core materials and types, the inductor design problem relevant to the converter or converter subsystem design, and the characteristics and models needed for the inductor design. Note that the electrical design of different inductors in a converter (e.g. how to determine the required inductance for rectifier AC line inductors, or how to determine the required inductance and noise attenuation performance for EMI filter inductors) will be discussed in converter subsystem design chapters in Part 2 and will not be covered here. Instead, the models which will be used in inductor physical design will be provided.

4.2 Magnetic Core Materials and Construction

Magnetic cores can be made with a wide variety of materials and manufacturing processes. Magnetics devices are usually constructed using a standard core shape and size. Popular core shapes include E, I, C, U, and toroid.

4.2.1 Soft Magnetic Alloy-Based Laminated, Tape Wound and Cut Cores

Laminated cores are made of soft magnetic alloys of thin sheets, which are oxidized or varnished to provide a relatively high resistance between them. Tape wound cores are made using a thin strip of soft magnetic alloy and then wound. The resulting core is toroidal in shape. The alloy strips range in

Design of Three-phase AC Power Electronics Converters, First Edition. Fei "Fred" Wang, Zheyu Zhang, and Ruirui Chen.
© 2024 The Institute of Electrical and Electronics Engineers, Inc. Published 2024 by John Wiley & Sons, Inc.

thickness from 0.5 to 14 mils and can be 1/8 of an inch to several inches wide [1]. For high-frequency applications, a thinner strip is used. Among the beneficial characteristics of tape wound cores are low core loss, high permeability, and high flux density [2]. Similar to tape wound cores are cut cores, which are made by the same winding process, and then cut in half, providing an air gap. Cut cores come in a variety of shapes, including E, C, and U.

4.2.1.1 Si Steel (e.g. 3% Si and 97% Fe)

Known by various trade names, including Silectron, Magnesil, Microsil, and Supersil, Si steel is used extensively in low-frequency (e.g. 60 Hz) applications. It has a high flux density and is relatively inexpensive [1]. The raw material has a saturation flux density of 2.0 Tesla (T). In tape wound or cut core form, the effective saturated flux density can range from 1.4 to 1.8 T, depending on the tape thickness. Higher-frequency applications would require thinner tapes [3].

4.2.1.2 Supermendur (e.g. 49% Co, 49% Fe, and 2% V)

Supermendur is often used in high temperature as well as miniaturized applications [1]. At 2.3 T, the raw material has the highest saturation flux density of all the soft magnetic alloys. When processed as a tape wound or cut core, the maximum flux density is 2.0–2.1 T. The high flux density lends this material to be extensively used in 400-Hz aircraft transformer applications [3].

4.2.1.3 Orthonol (e.g. 50% Ni and 50% Fe)

This material has also been known as Deltamax. Although its maximum flux density is considerably lower than silicon steel, it provides lower losses (by about one-half) [2] and therefore can be used in higher frequency for equivalent tape thickness [1]. Raw material flux density is 1.6 T, which reduces to 1.5 T when processed [3]. Orthonol is also popularly used in 400-Hz transformer applications.

4.2.1.4 Permalloy (e.g. 79% Ni, 17% Fe, and 4% Moly)

Variations of Permalloy include Supermalloy (78% Ni, 17% Iron, and 5% Moly) and Mo-Permalloy (79% Ni, 16% Fe, and 5% Moly). Saturation flux densities range from 0.65 T (Supermalloy) to 0.75 T (Permalloy) after processing. These materials see extensive use in applications at low signal levels or high frequencies and are the ideal choice for audio transformers [3].

4.2.1.5 Amorphous Alloys

Amorphous alloys are a general category of alloys commonly used at a 1 mil thickness. Popular ones are Alloy 2605 (iron-based, also known as Namglass 1) and Alloy 2714A (cobalt-based). Used in high-frequency applications, they exhibit low core losses. Saturation flux density ranges from 0.58 T for Alloy 2714A to 1.6 T for Alloy 2605 [1].

4.2.1.6 Nanocrystalline Alloys

An example of this category of materials is Namglass 4 (82% iron, qual parts silicon, niobium, boron, copper, carbon, nickel, and molybdenum). These materials are made by first quenching the elements into an amorphous state and then recrystallizing it into a controlled mix of amorphous and nanocrystalline states. Used in high-frequency applications, nanocrystalline alloys have a saturation flux density of 1 T after processing [3].

Nanocomposites, a new class of nanocrystalline materials, are being developed with better performance. They are composed of ferromagnetic nanocrystals embedded in an amorphous matrix, where nanocrystals contribute to desired magnetic properties and amorphous phase enables small grain size and large resistivity. The nanocomposites offer several better properties as compared to

Table 4.1 Fundamental characteristics of materials used in laminated, and/or tape wound and cut cores.

Material	Saturation flux density^a (T)	Electrical resistivity (μΩ-cm)	Upper frequency limit^b (kHz)	Material density (g/cm^3)	Loss density^c (kW/m^3)
Silicon steel	1.8	47	0.1–2	7.67	5800–8400
Supermendur	2.1	26	0.75–1.5	8.15	
Orthonol	1.5	45	1.5–8	8.25	
Permalloy	0.65–0.75	55–60	4–40	8.72–8.77	1200–3400
Alloy 2605	1.5	130	20–500	7.19	2200
Nanocrystalline	~1.0	120	500	7.3	300

^{a} Maximum achievable after processing.
^{b} Dependent on tape thickness; based on operation near the saturation flux density.
^{c} At 100 kHz, 0.2 T flux density, and 25 °C temperature.

typical commercially available nanocrystalline materials (e.g. FINEMET), including high operating temperature ($\geq$770 °C) and high saturation flux density (1.5–2.1 T). Low hysteresis and eddy current losses, compact size and light weight, improved mechanical properties, and enhanced corrosion resistance can be achieved by nanocomposite materials as well.

Table 4.1 summarizes the soft magnetic alloy materials characteristics used in laminated, and/or tape wound and cut cores.

4.2.2 Powder Cores

Powder cores are made from an alloy that has been ground to a fine powder. Next, an insulating material is mixed with the powder in order to increase the resistivity and create a distributed air gap. After pressing the powder into shape, an insulating layer of paint is applied [1, 2]. The only standard core shape available in this method of construction is the toroid core. These cores are used mostly in high-frequency applications, e.g. EMI filters. Following is a brief overview of common powder core materials.

4.2.2.1 Molypermalloy Powder
Molypermalloy powder (MPP) cores are made from an alloy similar to the Permalloy used in tape wound cores. These cores are good choices for high-frequency chokes and power inductors due to their low losses. Saturation flux density of MPP cores is about 0.7 T [1].

4.2.2.2 Nickel/Iron Powder (High Flux Cores)
These cores are made from a 50% Ni and 50% Fe material (the same raw material as Orthonol). With saturation flux densities on the order of 1.5 T, they are capable of smaller and lighter cores than MPP for the same characteristics. The one drawback is increased losses compared to MPP cores.

4.2.2.3 Kool Mμ
Made from a ferrous alloy, these cores have losses greater than MPP cores, but less than high flux cores. Their flux density, 1 T, is also between those offered by the other powder cores.

4.2.3 Ferrite Cores

Ferrite cores are ceramics, made of oxides of metals such as iron, zinc, and manganese. Because of their intrinsic insulating property, ferrite cores have high resistivity, which leads to low losses. As such, ferrite cores can be used for extremely high frequencies. The oxide is fired, ground, pressed into shape, and refired to create a core. This process allows the construction of many core shapes, including standard E, I, U, etc., as well as toroid and pot cores. Additionally, any custom shape may be machined to meet user-specified requirements. Ferrite cores are good choices for high-frequency applications, including EMI filters [1, 2]. Saturation flux densities for ferrite cores are on the order of 0.4–0.5 T.

4.3 Inductor Design in a Converter

The inductors for power converters are usually custom-designed. The physical design of an inductor includes selecting the magnetic cores and wire types, determining number of winding turns and air gap length, and choosing insulation and cooling if extra insulation and heat transfer capabilities are needed. Other auxiliaries such bobbins, connectors, and mechanical support/frame should also be included when determining the layout, size, weight, and cost.

The inductor physical constraints include:

1) Flux density limit B_{max}: B_{max} should be less than the magnetic core saturation flux density B_{sat}, although during abnormal conditions (e.g. inrush conditions to be discussed in Chapter 5), saturation may be allowed. B_{sat} of commonly used magnetic materials are introduced in last section. Some margin needs to be applied when selecting B_{max}. This is because the flux density may not be evenly distributed in the core and local saturation could occur if B_{max} is very close to B_{sat}. Moreover, the permeability and inductance can decrease significantly when the flux density is close B_{sat}, which is generally undesirable. In high-power applications, B_{max} is usually selected as 50~90% of B_{sat}.

2) Temperature limits: The maximum permissible temperature of an inductor is usually determined by the wire insulation material as the magnetic core Curie temperature is generally higher than the wire insulation material temperature limit. One exception is ferrite cores, whose Curie temperature is around 100 °C. The temperature limit of commonly used insulation materials for enameled wire and Litz wire is usually in the 105–200 °C range.

3) Mechanical limit: The windings must fit through core window. The maximum fill factor exists.

Note that the inductor design should also include insulation design to withstand the required voltage, which is omitted here and will be separately considered in Chapter 12 when discussing medium-voltage inductors for grid applications.

The inductor design performance constraints include (i) the inductance L, which usually has a nominal value L_{nom} as the lower bound, although too high an L could cause performance issues such as instability or excessive voltage drop and (ii) reliability and lifetime.

Following the terminology and approach used for design in this book, the inductor design problem can be illustrated by Figure 4.2 for a gapped inductor with core structure in Figure 4.1. In addition to the variables and constraints, the design conditions for the inductor include winding RMS and peak currents I_{RMS} and I_{pk}, voltage V, ambient temperature T_a, and cooling condition (e.g. air flow).

As mentioned earlier, in this book, the inductor design is not treated as a separate design problem but part of the converter subsystem design, e.g. the design of the rectifier power stage or the EMI filter. Here, we will identify the inductor characteristics and models needed for converter subsystem design. From Figure 4.2, the models should include (i) inductance, (ii) flux density, (iii) fill factor, (iv) temperature rise, and (v) reliability.

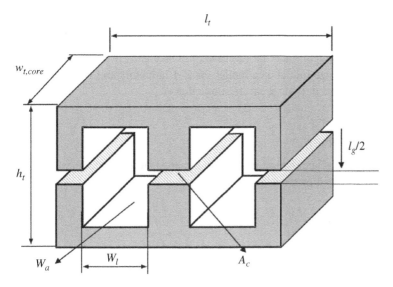

Figure 4.1 EE cores for a gapped DC or single-phase AC inductor.

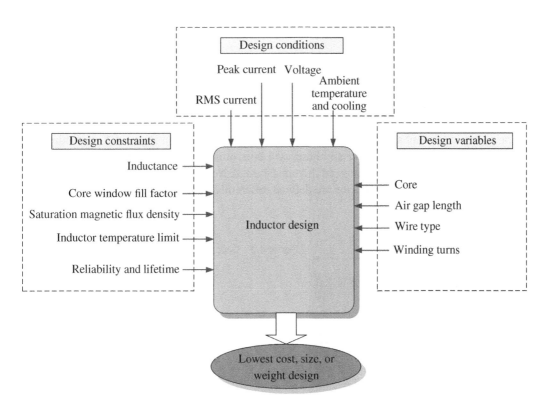

Figure 4.2 Inductor design problem for a gapped inductor.

4.4 Inductor Characteristics and Models

This section focuses on the inductor characteristics and models needed for the design constraints. Since the reliability model for inductors will be discussed in Chapter 7, it will be omitted here. Given their importance, the leakage inductance model for CM inductors and fringing effect for gapped core magnetics are covered in their separate subsections.

4.4.1 Inductance and Permeability

Without considering the fringing effect, the inductance for a gapped DC or single-phase AC inductor based on the Figure 4.1 core can be determined as:

$$L = \frac{\mu_0 N_t^2 A_C}{\left(l_g + \frac{l_C}{\mu_C}\right)} \tag{4.1}$$

where μ_0 is permeability of free space, μ_C is the relative permeability of the core material, N_t is the winding turns, and l_C is the equivalent core magnetic path length. When there is no saturation and $l_g \gg \frac{l_C}{\mu_C}$, (4.1) reduces to

$$L \approx \frac{\mu_0 N_t^2 A_C}{l_g} \tag{4.2}$$

When $l_g = 0$ (i.e. gapless inductor), it reduces to

$$L = \frac{\mu_C \mu_0 N_t^2 A_C}{l_C} \tag{4.3}$$

For a gapped three-phase inductor based on an E-I core structure in Figure 4.3, neglecting the core impact, the magnetic field for all phases will be symmetric. The equivalent inductance per phase will be equal and can be expressed the same as (4.2). Note that l_g's for the three-phase inductor and the DC/single-phase inductor are defined differently. Also, the three-phase inductance is

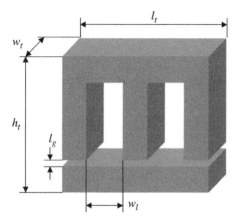

Figure 4.3 Gapped three-phase inductor core.

the equivalent positive or negative sequence inductance and has already included the mutual inductances from the other two phases. $L = \dfrac{3}{2}L_s$ (L_s is the phase winding self-inductance).

Permeability is a critical property of magnetic materials. There are different permeability definitions: initial permeability μ_i, effective permeability μ_e, incremental permeability μ_Δ, reversible permeability μ_{rev}, and complex permeability $\bar{\mu}$.

μ_C in (4.1) corresponds to μ_i, which is the relative permeability of the magnetic material near zero magnetic field strength H:

$$\mu_i = \frac{1}{\mu_0}\frac{\Delta B}{\Delta H}, \quad H \to 0 \tag{4.4}$$

If there is an air gap in magnetic core, the equivalent or the effective permeability μ_e is much lower than μ_i of the same core without an air gap. μ_e depends on μ_i and the dimensions of the core and the air gap. For Figure 4.1 cores with small air gaps, the effective permeability is given by:

$$\mu_e = \frac{\mu_i}{1 + \dfrac{l_g \mu_i}{l_c}} \tag{4.5}$$

which agrees with (4.1). If the air gap is long, some flux passes outside the air gap and this additional flux results in an increased value of the effective permeability. Therefore, (4.5) is valid only when fringing effect is neglected.

Incremental permeability μ_Δ is defined when an AC magnetic field strength H_{AC} is superimposed on a DC magnetic field strength H_{DC}. The B-H loop follows a minor loop path. The incremental permeability is

$$\mu_\Delta = \left(\frac{1}{\mu_0}\frac{\Delta B}{\Delta H}\right)_{H_{DC}} \tag{4.6}$$

The limiting value of the incremental permeability, when the amplitude of H_{AC} is very small, is termed reversible permeability μ_{rev}:

$$\mu_{rev} = \frac{1}{\mu_0}\frac{\Delta B}{\Delta H}, \quad \Delta H \to 0 \tag{4.7}$$

In practice, under sinusoidal excitation, there is a phase shift between the fundamental components of B and H. By using a complex quantity of the relative permeability $\bar{\mu}$, consisting of a real part and an imaginary part, these effects are represented. The imaginary part of $\bar{\mu}$ is associated with the small signal losses of the material.

4.4.1.1 Frequency-Dependent Permeability

For commonly used magnetic materials, both the real part and imaginary part of the permeability are frequency-dependent. Figure 4.4 shows an example of the permeability versus frequency for ferrite and nanocrystalline materials. Generally, nanocrystalline cores have a much higher initial permeability than ferrite cores. However, unlike the flat curve of the ferrite permeability, the permeability of nanocrystalline quickly drops from around 20 to 30 kHz. For ferrite cores, the real permeability dominates the magnitude of whole permeability below 1 MHz, while the real permeability only dominates the magnitude of whole permeability below 40 kHz for nanocrystalline cores.

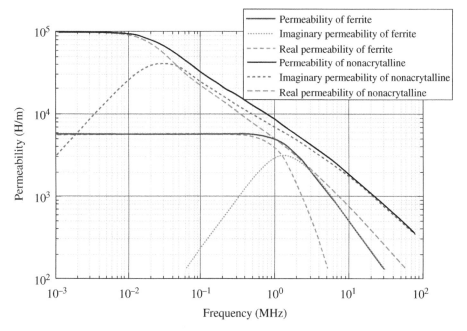

Figure 4.4 Absolute permeability curves for ferrite and nanocrystalline cores.

4.4.1.2 DC Current Bias-Dependent Permeability

Powder and amorphous cores with distributed air gaps are popular in high-power electronic applications. As shown in Figure 4.5, powder core with distributed gap structure can withstand much higher maximum DC current bias or magnetic flux density compared to ferrite. However, its permeability is more sensitive to the current bias. Figure 4.5 shows the permeability percentage curve with DC bias for powder and amorphous cores. In general, the permeability decreases when the bias current or magnetic field increases. Different materials have different current bias-dependent permeability characteristics, which should be considered and modeled in inductor design.

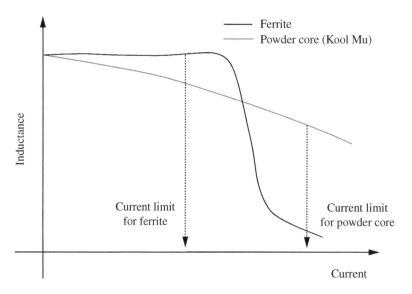

Figure 4.5 Inductance curve with current bias for ferrite and powder cores.

4.4.2 Flux Density and Core Saturation

Once the magnetic core material is selected, the relative permeability μ_C and saturation flux density B_{sat} are given. The saturation flux density of some commonly used magnetic materials can be found in Table 4.1.

For Figure 4.1 cores, the instantaneous flux density can be found as:

$$B(t) = \frac{\mu_0 N_t i(t)}{\left(l_g + \dfrac{l_C}{\mu_C}\right)} \tag{4.8}$$

Neglecting the core impact, the following (somewhat conservative) relationship should be satisfied:

$$B_{max} \geq \frac{\mu_0 N_t I_{pk}}{l_g} \tag{4.9}$$

which also applies to the three-phase core case.

With EMI filter terminology, cores in Figures 4.1 and 4.2 are for DM inductors, and their saturation is caused by main flux in cores generated by AC or DC winding currents. For CM inductor used in EMI filters, the situation for core saturation constraint is different as the main fluxes in cores generated by AC or DC winding currents are canceled. There are two possible effects that could cause CM core saturation: saturation due to CM currents and saturation due to leakage flux generated by DM currents. The real CM noise current is an impulse-like waveform and is partially bypassed by the parasitic capacitances of the windings. Therefore, it will not flow through the windings to contribute to the flux in core. When operating (i.e. DM) currents and leakage inductance are large, the CM choke (CMC) is more prone to saturation. Often CM inductors use single-layer structure to keep the window fill factor low for smaller leakage. At the same time, the winding parasitic capacitances will also be reduced, which is usually desirable for EMI filter performance but could make CM current generated core flux higher. More details for CM inductor design will be presented in Chapter 8.

4.4.3 Fill Factor

The windings should be able to fill into the core windows, i.e.

$$K_u \cdot W_A > b \cdot N_t \cdot A_w \tag{4.10}$$

where K_u is the fill factor, A_w is the effective wire conductor cross-sectional area, $b = 1$ for single-phase and DC inductors, and $b = 2$ for three-phase inductors. K_u can be affected by wire types/shapes, bobbins, and insulation, and can be as high as 0.8 or as low as 0.05.

4.4.4 Current Density and Core Window Area Product A_p

As a simple alternative to the inductor temperature limit, which will require loss and thermal modeling, a maximum current density J_{max} has been used as a constraint. In this case, the wire conductor cross-sectional area A_W should be selected to satisfy

$$A_W \geq \frac{I_{RMS}}{J_{max}} \tag{4.11}$$

Ideally, J_{max} should be determined based on detailed thermal model and design. For simplicity, a constant value is often used. Generally, J_{MAX} is selected to be <400 A/cm^2 for natural convection and $<1000\ \text{A/cm}^2$ for forced air cooling. Combining constraints (4.2), (4.9), (4.10), and (4.11), the well-known core window area product A_p constraint can be obtained:

$$A_p = q \cdot A_C \cdot W_A \geq q \cdot b \cdot \frac{L \cdot I_{RMS} \cdot I_{pk}}{K_u \cdot J_{max} \cdot B_{max}} \tag{4.12}$$

where the coefficient $q = 1$ for DC and single-phase inductor; and $q = \dfrac{3}{2}$ for the three-phase inductor core per definition [4].

Using A_p, a magnetic core can be selected, and core materials and dimensions are given. From (4.2) and (4.9), N_t and l_g can be found as in (4.13) and (4.15), and wire gauge can be determined via (4.11).

$$N_t \geq \frac{LI_{pk}}{A_C B_{max}} \tag{4.13}$$

$$B_{ratd} = \frac{LI_{pk}}{A_C N_t} \leq B_{max} \tag{4.14}$$

$$l_g = \mu_0 \frac{N_t I_{pk}}{B_{rated}} \tag{4.15}$$

J_{max} can be determined based on A_p and desired temperature rise ΔT [4]:

$$J_{max} = \sqrt{\frac{\Delta T}{K_9}} \cdot A_p^{-0.125} = K_j \cdot A_p^{-0.125} \tag{4.16}$$

where K_9 is a constant related to core configuration. ΔT is in $°C$, A_p is in cm^4, and J_{max} is in A/cm^2. K_j can be found in Table 4.2 for some cores.

4.4.5 Core Loss

Current density limit is merely a simplified approach to consider the inductor temperature limit. In general, to satisfy the temperature limit constraint, inductor loss needs to be calculated, including core loss and winding loss. The physical origin of core loss is the energy consumed to damp the magnetic domain wall movement by eddy currents and spin relaxation. Knowledge about the core loss origin does not necessarily provide a practical means for loss calculation. In general, the rather chaotic time and space distribution of the magnetization change is unknown and cannot be described exactly. Furthermore, the manufacturer's datasheets only provide core loss for sinusoidal excitations, and most power electronics applications have nonsinusoidal waveforms. To deal with this lack of a microscopic physical remagnetization model, several macroscopic and empirical approaches have been formulated. Two common categories are the loss separation approach and empirical equations.

Table 4.2 Temperature constant K_j [4].

Core type	$\Delta T = 25\ °C$	$\Delta T = 50\ °C$
Laminations	366	534
Tape wound core	250	365
Powder core	403	590

4.4.5.1 Loss Separation Approach

The loss separation approach calculates the total core loss separately by static hysteresis loss P_h, and dynamic eddy current loss P_d, including classic eddy current loss P_c and excess eddy current loss P_e. Note that all power losses in (4.16) are for per unit volume.

$$P_{core} = P_h + P_d = P_h + P_c + P_e \tag{4.17}$$

The physical reason for such decomposition is that the hysteresis loss originates from the discontinuous characteristic of the magnetization process, whereas the eddy current losses are associated with the magnetic domain structure. In principle, the separation approach can provide the solution to the core loss calculation of arbitrary waveforms. Hysteresis loss is only determined by the operating flux density peak value B_{max} and frequency f, and is not a function of waveform shapes. The hysteresis loss density is equal to f times of static B-H loop area, as:

$$P_h = f \cdot \int H dB \tag{4.18}$$

The static B-H loop can be obtained from measurement or models [5, 6]. The classic eddy current loss can be calculated by introducing bulk resistivity of the magnetic material, as:

$$P_c = \frac{(\pi B_{max} f d)^2}{6\rho} \tag{4.19}$$

where d represents material thickness and ρ is resistivity. However, results of the sum of static hysteresis loss (P_h) and dynamic eddy current loss (P_c) usually do not match with measurement, so the excess eddy current loss is introduced to count for the difference. According to [7], for a wide variety of ferromagnetic materials, the excess eddy current loss can be estimated as:

$$P_e = C_1 (B_{max} f)^{3/2} \tag{4.20}$$

The major drawback of this method is that it requires some extra measurements and parameters that are not available in the datasheet. Efforts have been made to make this approach more practical [8].

4.4.5.2 Empirical Approach

Another major category of core loss calculation approach is empirical equations based on measurements. Easy-to-use is a main advantage of these empirical methods, which are usually applicable to particular material and operating conditions. Steinmetz equation is a widely used tool for core loss calculation:

$$P_{core} = k f^\alpha B_m^\beta \tag{4.21}$$

where k, α, and β are parameters of magnetic materials determined by the material characteristics and usually available from the manufacturer's datasheet, f is the frequency, and B_m is the peak flux density.

The Steinmetz equation is basically the "curve-fitting" of measured core loss density under sinusoidal magnetization excitation. Therefore, it can be extracted from data provided by manufacturers without knowing the detailed material characteristics. However, to cover a wide range of operating conditions, it is impossible to represent the complex relationship among loss, flux density, and frequency by one set of explicit exponential functions. Therefore, a set of Steinmetz equations have to be used to fit experimental results, each of which may only be valid for a part of whole frequency and/or flux density range. Another problem with the original Steinmetz equation (OSE) is the limitation on waveform. The OSE and the corresponding set of parameters are only valid for a sinusoidal excitation condition. Intuitively, people have tried to apply Fourier transform to arbitrary periodic waveforms to obtain a series of sine waves and then to apply OSE to each frequency component. However, using the summation of calculated losses of each frequency as the total loss will lead to large error, because the loss and peak flux density do not have a linear relationship as shown in Steinmetz equation, and superposition is not valid [9]. OSE also cannot consider DC bias.

A number of modifications to OSE have been developed to account for the nonsinusoidal waveforms in core loss calculation. They are overviewed below.

4.4.5.2.1 Modified Steinmetz Equation Modified Steinmetz equation (MSE) [9] extends the Steinmetz equation parameters by equating the weighted time derivative of B for arbitrary magnetizing currents with that of a sine wave (ΔB is peak-to-peak flux density):

$$\left\langle \frac{dB_w}{dt} \right\rangle = \int_0^T \frac{\frac{dB^2}{dt}}{\Delta B} dt \tag{4.22}$$

Next, equating the average with the sine wave gives an equivalent frequency f_{eq}:

$$\left\langle \frac{dB_w}{dt} \right\rangle = \left\langle \frac{dB_{w,sin}}{dt} \right\rangle \tag{4.23}$$

$$f_{eq} = \frac{2}{(\Delta B)^2 \pi^2} \int_0^T \left(\frac{dB}{dt} \right)^2 dt \tag{4.24}$$

The corresponding MSE equation is

$$P_{core(MSE)} = k f_{eq}^{\alpha-1} B_m^\beta f \tag{4.25}$$

One issue of MSE is implicitly enforcing loss proportional to f^2 while still assuming loss proportional to f^α. Thus, MSE is only accurate for $\alpha \approx 2$.

4.4.5.2.2 Generalized Steinmetz Equation (GSE) Generalized Steinmetz equation (GSE) [10] hypothesizes that instantaneous core loss can be given by the "physically plausible" equation:

$$p_{core}(t) = k_1 \left| \frac{dB}{dt} \right|^\alpha |B(t)|^{\beta-\alpha} \tag{4.26}$$

Thus, P_{core} can be given by

$$P_{core} = \frac{1}{T} \int_0^T k_1 \left| \frac{dB}{dt} \right|^\alpha |B(t)|^{\beta-\alpha} dt \tag{4.27}$$

If $B(t) = B_m \sin \omega t$, then

$$P_{core} = k_1 \omega^\alpha B_m^\beta \int_0^T \frac{1}{T} |\cos \omega t|^\alpha |\sin \omega t|^{\beta-\alpha} dt \tag{4.28}$$

This equation will match with the Steinmetz equation if

$$k_1 = \frac{k}{(2\pi)^{\alpha-1} \int_0^{2\pi} |\cos \theta|^\alpha |\sin \theta|^{\beta-\alpha} d\theta} \tag{4.29}$$

The validation of GSE actually requires a particular waveform of $B(t)$, such that derivation dB/dt can be expressed in the format of $C_{coefficient} \cdot B(t) \cdot f$, which is true for sinusoidal and linear waveforms only.

4.4.5.2.3 Improved Generalized Steinmetz Equation (iGSE)

The improved generalized Steinmetz equation (iGSE) recognizes that core loss depends not only on B and dB/dt but also on the time history of the flux density waveform $B(t)$. Without the history of $B(t)$, it is difficult to determine whether its segment is a part of the minor or major B-H loop. So, GSE modification is made by replacing the instantaneous $B(t)$ with peak-to-peak flux density ΔB. The expression for time average loss calculation becomes

$$P_{core} = \frac{1}{T} \int_0^T k_1 \left| \frac{dB}{dt} \right|^\alpha (\Delta B)^{\beta - \alpha} dt \tag{4.30}$$

$$k_1 = \frac{k}{(2\pi)^{\alpha - 1} \int_0^{2\pi} |\cos\theta|^\alpha 2^{\beta - \alpha} d\theta} \tag{4.31}$$

At each point on the waveform, ΔB is taken as the peak-to-peak amplitude of major or minor loop that contains that point. This formulation can then calculate loss appropriately in the presence of minor loops. In order to apply this equation to each major or minor loop of a waveform in an automated design tool, it is necessary to have an algorithm capable of splitting an arbitrary waveform into a major loop and one or more minor loops [11]. With a set of major and minor loops, the loss of each one can be calculated. The total loss is then found by a weighted average, weighing the contribution of each by its ratio of the total period, i.e.

$$P_{core} = \sum_j P_j \frac{T_j}{T} \tag{4.32}$$

where P_j is the loss density for major or minor loop j, T_j is the period of loop j, and T is the total period.

A particularly common type of flux density waveforms in power electronics is piecewise linear (PWL). For PWL waveforms, the time average core loss is

$$P_{core} = \frac{(k_1 \Delta B)^{\beta - \alpha}}{T} \sum_j \left| \frac{B_{j+1} - B_j}{t_{j+1} - t_j} \right|^\alpha (t_{j+1} - t_j) \tag{4.33}$$

iGSE with separation of minor loops overcomes problems with the previous methods and achieves good accuracy. Only Steinmetz parameters are needed without additional measurements or curve fitting. Thus, it becomes the state-of-art approach. As discussed, the Steinmetz parameters are still limited by the frequency range. An option to consider a wide range of frequency could be the summing of power terms, but this comes with additional complexity. The iGSE also ignores the impact of DC bias.

4.4.5.2.4 Natural Steinmetz Equation (NSE)

Natural Steinmetz equation (NSE) [12] is an independently developed equation matching iGSE and therefore is not considered a separate method:

$$P_{core} = \left(\frac{\Delta B}{2} \right)^{\beta - \alpha} \frac{k_N}{T} \int_0^T \left| \frac{dB}{dt} \right|^\alpha dt \tag{4.34}$$

$$k_N = \frac{k}{(2\pi)^{\alpha-1} \int_0^{2\pi} |\cos\theta|^\alpha d\theta} \tag{4.35}$$

4.4.5.2.5 Improved iGSE (iiGSE) The improved iGSE (iiGSE) has implicitly assumed that core loss will be zero when the voltage is zero. However, an increase of core loss is observed during the zero-voltage period. When the voltage becomes zero, the change in flux density dB/dt becomes zero. A sudden shift in magnetization tends to rearrange the magnetic domains until a new thermal equilibrium is achieved. During the transition, the motion of the magnetic domains will continue to cause loss. This phenomenon is known as the magnetic relaxation process. The residual loss occurs due to magnetic relaxation. Note the effect of magnetic relaxation becomes significant at higher frequencies. The iiGSE [13] adds an extra term to the iGSE:

$$P_{core} = \frac{1}{T} \int_0^T k_1 \left| \frac{dB}{dt} \right|^\alpha (\Delta B)^{\beta-\alpha} dt + \sum_{l=1}^n Q_{rl} P_{rl} \tag{4.36}$$

$$P_{rl} = \frac{1}{T} k_r \left| \frac{dB(t_-)}{dt} \right|^{\alpha_r} (\Delta B)^{\beta_r} \left(1 - e^{-\frac{t_1}{\tau}} \right)$$
$$Q_{rl} = e^{-q_r \left| \frac{dB(t_+)/dt}{dB(t_+)/dt} \right|} \tag{4.37}$$

where k_1, k_r, α, β, β_r, τ, and q_r are material parameters and need to be extracted from measurement results. Extracting these parameters are sometimes very difficult and complicated. P_{rl} is the time average power loss for each transition to zero voltage.

4.4.5.2.6 Waveform Coefficient Steinmetz Equation (WcSE) Waveform coefficient Steinmetz equation (WcSE) [14] intends to calculate the core loss accurately under sinusoidal transition square (STS) voltage waveforms, which are quite common for PWM hard switching and resonant converters. This method measures the ratio of a nonsinusoidal flux waveform and a sinusoidal waveform with the same period and peak flux density through the flux waveform coefficient (FWC):

$$FWC = \frac{\text{area of } B_{non-sinusoidal}(t)}{\text{area of } B_{sine}(t)} \tag{4.38}$$

The FWC_{sqr} for a square voltage waveform and FWC_{tri} of a triangular voltage waveform can be determined as:

$$FWC_{sqr} = \frac{\pi}{4} \text{ and } FWC_{tri} = \frac{\pi}{3} \tag{4.39}$$

The loss density for the square voltage waveform and triangular voltage waveform are respectively:

$$P_{core_{sqr}} = FWC_{sqr} k f^\alpha B^\beta \text{ and } P_{core_{sqr}} = FWC_{tri} k f^\alpha B^\beta \tag{4.40}$$

4.4.5.2.7 Composite Waveform Hypothesis (CWH) Model The composite waveform hypothesis (CWH) [15] model was developed based on a rectangular voltage waveform. As the rectangular waveform is widely used in power electronics applications, and this technique is developed to directly calculate core loss under various square voltage waveforms rather than extending the Steinmetz equation. The core loss for each pulse is a function of the voltage magnitude V and time t. The CWH model can be represented as:

$$P = \frac{1}{T}\left(P_{sqr}\left(\frac{V_1}{N}, t_1\right)t_1 + P_{sqr}\left(\frac{V_2}{N}, t_2\right)t_2\right) \tag{4.41}$$

where $P_{sqr}\left(\frac{V_1}{N}, t_1\right)$ and $P_{sqr}\left(\frac{V_2}{N}, t_2\right)$ are core loss for two pulse widths (positive and negative). A "Herbert curve" can be used to obtain the core loss on the desired voltage magnitude and time pulse. Herbert curve depicts the core loss as a function of pulse time t, parametrized by voltage per winding turn. The manufacturers will need to provide the measurement results in a graphical representation (Herbert curve).

4.4.5.2.8 Steinmetz Premagnetization Graph (SPG)
Steinmetz premagnetization graph (SPG) [16] shows the dependency of Steinmetz parameters (k, α, and β) on DC bias condition and will aid in calculation core loss under DC bias. The commonly used iGSE ignores the DC bias effect when calculating core loss. Experimental data show that the core loss follows the power function of f and ΔB, even with a DC bias. Therefore, the inclusion of the DC bias effect on iGSE is beneficial. For Steinmetz parameters, α is independent of DC bias H_{DC}, but β and k are H_{DC} dependent. A graphical representation of this modified Steinmetz parameters with respect to DC bias will be valuable to calculate core loss.

If the manufacturer provides the SPG graph of a material in datasheet, core loss can be calculated using iGSE, which can be expressed as:

$$P_{core} = \frac{1}{T}\int_0^T k'\left|\frac{dB}{dt}\right|^{\alpha}(\Delta B)^{\beta' - \alpha}dt \tag{4.42}$$

where k' and β' are improved Steinmetz parameters. The DC premagnetization (H_{DC}) of a core without an air gap can be obtained by:

$$H_{DC} = \frac{I_{DC}N_t}{l_e} \tag{4.43}$$

where I_{DC} is the DC current and l_e is the effective magnetic length. The DC premagnetization for a gapped core can be calculated by reluctance model. For a given H_{DC} value, the modified Steinmetz parameters (k', β') from the SPG for a fixed temperature can be obtained and then iGSE can be applied to obtain core loss.

4.4.5.2.9 Loss Map
The loss map methods [17, 18] refer to a loss map which stores core loss data at different operating points. The experimental results based on various aspects (temperature, DC bias, core shape, flux ripple, and frequency) are stored. These are basically data-driven methods and require extensive measurements, although they enable accurate core loss calculation.

Table 4.3 summarizes the different empirical core loss models. In general, the sate-of-art iGSE model can be used for core loss calculation with good accuracy in most cases. The SPG and loss map methods can be used if the required data for these models are available. WcSE and CWH can be used to calculate core loss for specific waveforms with reduced complexity.

Table 4.3 Comparison of different empirical core loss models.

	SE	MSE	GSE	iGSE/NSE	iiGSE	WcSE	CWH	SPG	Loss map
Waveform	S	S, T, SW	S, T, SW	All	All	STS	PWM	All	All
DC bias	No	No	No	No	No	No	No	Yes	Yes
Magnetic relaxation	No	No	No	No	Yes	No	No	Yes	Yes
Parameters	α,β,k	α,β,k	α,β,k	α,β,k	$\alpha,\beta,k,\alpha_r,\beta_r,$ k_r,τ,q_r	α,β,k	None	α,β',k'	All
Subloops	No	No	No	Yes	Yes	No	No	Yes	Yes
Error	45%	12%	12%	12%	<12%	28%	~12%	<iiGSE	<SPG

S, sinusoidal; T, triangular; SW, sawtooth; STS, sinusoidal transition square; error evaluated for a bipolar square voltage waveform with 1/3 duty cycle [19].

4.4.6 Winding Loss

The DC resistance of the inductor winding can be calculated as:

$$R_{DC} = \frac{N_t \cdot \rho \cdot MLT}{A_w} \tag{4.44}$$

where ρ is conductor resistivity, and N_t, MLT, A_w have been defined before. The corresponding winding loss is

$$P_{winding} = c \cdot I_{RMS}^2 R_{DC} \tag{4.45}$$

where $c = 3$ for the three-phase and three single-phase inductor cases, and $c = 1$ for the DC inductor case. However, for high-frequency power electronics converters, skin and proximity effects cannot be omitted. Both effects change the current distribution inside the conductor, so the resistance is increased compared with the DC value. For an operating frequency above 100 kHz, eddy current losses would be the dominant part of the winding loss.

The one-dimensional (1-D) calculation model [20, 21] with the orthogonal concept ensures the skin and proximity effects can be calculated separately. The AC resistance of a m-layer winding is given by:

$$R_{AC} = R_{DC} \frac{\xi}{2} \left[\frac{\sinh \xi + \sin \xi}{\cosh \xi - \cos \xi} + (2m - 1)^2 \frac{\sinh \xi - \sin \xi}{\cosh \xi + \cos \xi} \right] \tag{4.46}$$

where

$$\xi = \frac{\sqrt{\pi}}{2} \frac{d}{\delta} \tag{4.47}$$

$\delta = \sqrt{\dfrac{\rho}{\pi f \mu_r \mu_0}}$ is the skin depth, d is the wire diameter, and ρ and μ_r are the conductor resistivity and relative permeability, respectively.

Litz wires are usually adopted to reduce the high-frequency winding loss. A Litz wire is supposed to cancel the proximity effect by external magnetic field, so its AC resistance is only determined by the skin effect of each strand. However, there are still local proximity effects for each strand imposed by its surrounding strands. Reference [22] models the winding loss of round Litz wires in closed-form equation. With the strand internal field distribution simplified, it is shown that the eddy current effect would be the function of both strand number N_0 and winding layer number m. The AC resistance of an N-turn Litz wire winding is expressed as:

$$R_{AC} = \frac{\sqrt{2}N_t\rho}{\pi\delta N_0 d_0}\left[\psi_1(\zeta) - \frac{\pi^2 N_0\beta}{24}\left(16m^2 - 1 + \frac{24}{\pi^2}\right)\psi_2(\zeta)\right] \tag{4.48}$$

where β is the Litz wire packing factor (defined as the ratio of the total conductor cross-sectional area in the bundle to the area of the overall bundle), $\zeta = \dfrac{d_0}{\delta}$ with d_0 as the Litz wire strand diameter, and $\psi_1(\zeta)$ and $\psi_2(\zeta)$ can be analytically determined as in [22].

It is sometimes more convenient to use AC to DC resistance ratio representing the eddy current effect. The DC resistance of the Litz wire can be expressed as:

$$R_{DC} = \frac{4N \cdot \rho \cdot MLT}{N_0 \cdot \pi \cdot d_0^2} \tag{4.49}$$

Then the ratio is

$$K_{FR} = \frac{\zeta}{\sqrt{2}}\left[\psi_1(\zeta) - \frac{\pi^2 N_0\beta}{24}\left(16m^2 - 1 + \frac{24}{\pi^2}\right)\psi_2(\zeta)\right] \tag{4.50}$$

4.4.7 Temperature Rise

Once the inductor losses are determined, the inductor temperature rise needs to be determined. Temperature rise prediction is difficult without a thermal model. One reasonably accurate method for open core and winding construction is based on the assumption that core and winding losses may be added together as:

$$P_{total} = P_{core} + P_{winding} \tag{4.51}$$

Note P_{core} here refers to the total core loss, rather than loss density. Assume that the heat is dissipated uniformly throughout the surface area of the core and winding assembly. The effective surface area A_t can be estimated as a function of A_p as in (4.50), where K_s is a constant related to core configuration as in Table 4.4. A_t can also be directly calculated for a specific inductor configuration:

$$A_t = K_s \cdot A_p^{0.5}\ (\text{cm}^2) \tag{4.52}$$

The heat dissipation density γ can be determined as:

$$\gamma = \frac{P_{total}}{A_t}\ (\text{W/cm}^2) \tag{4.53}$$

Table 4.4 Surface area coefficient K_s [4].

Core type	K_s
Laminations	41.3
Tape wound core	50.9
Powder core	32.5

The temperature rise can be estimated as:

$$\Delta T = \xi \cdot \gamma^\eta = 450 \cdot \gamma^{0.826} \ (^\circ C) \tag{4.54}$$

Note that natural cooling is assumed for the above estimation, which considers radiation and natural convection (e.g. 55% radiation, 45% convection, and 0.95 emissivity). The coefficients ξ and η could be different in different publications based on different data and assumptions (e.g. fill factor assumptions), but their ranges are close.

For forced convection cooling, generally a thermal model represented by thermal resistance R_{th} can be used to predict temperature rise. The general expression for the convection-based R_{th} is given by:

$$R_{th} = \frac{1}{h_c A_t} \tag{4.55}$$

where h_c is the convection heat transfer coefficient. For the forced air convection cooling, h_c can be determined through

$$\frac{h_c d}{k_f} = C \left(\frac{v \cdot d}{\nu_f} \right)^n Pr_f^{1/3} \tag{4.56}$$

where d is the height of the magnetic component, C and n are constants related to component geometry (e.g. $C = 0.102$ and $n = 0.0675$ for a long component with a square cross section), v is the velocity of the air flow, ν_f is the kinematic viscosity evaluated at the film temperature, Pr_f is the Prandtl number evaluated at the film temperature, and k_f is the air thermal conductivity evaluated at the film temperature. The film temperature is the average of the component surface and air flow temperature [23]. At atmospheric pressure, the model can be simplified as:

$$h_c \approx \left(3.33 + 4.8 v^{0.8} \right) L^{-0.288} \tag{4.57}$$

where L is the total distance passed by the air cooling the component (e.g. the length of the boundary flow layer of the component). In general, L is "half of the length of the shortest path around a vertical midsection of the object" (Figure 4.6)

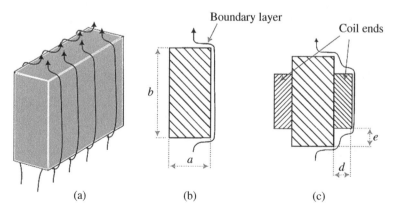

(a)   (b)   (c)

Figure 4.6 Parameter L calculation: (a) component and air flow; (b) $L = (a + b)$; (c) $L = (a + b) - 2e + 2\sqrt{d^2 + e^2}$ [23]. *Source:* From [23].

Note that (4.55) models both natural and forced convection processes. Once h_c is known, the average inductor temperature rise can be found as:

$$\Delta T = \overline{T_{inductor}} - T_a = P_{total} \cdot R_{th} = \frac{P_{total}}{h_c A_t} \tag{4.58}$$

In [23], more complex thermal models have been discussed to separately consider the core and winding losses and their temperature rises. In this case, a thermal resistance network will be needed to model winding and core thermal characteristics. During transient conditions such as overload conditions, thermal capacitances will need to be considered, which will be discussed in Chapter 5.

4.4.8 Leakage Inductance

Leakage inductance is a critical parameter for CM inductors in EMI filters. A three-phase toroidal CMC is used as an example to discuss leakage inductance model. One of the most commonly used CMC leakage inductance models is proposed by Nave [24]. It calculates the leakage inductance of CMC through an equivalent rod core inductor in Figure 4.7. The rod core inductance is calculated by the effective flux length l_{eff} and fitted effective permeability μ_{dm} as in (4.57), where A_e is the effective cross-sectional area of the CMC.

$$L_{leak} = \mu_{dm} L_{dm-air} = \mu_{dm} \frac{\mu_0 N_t^2 A_e}{l_{eff}} \tag{4.59}$$

l_{eff} is empirically derived as:

$$l_{eff,nave} = l_e \sqrt{\frac{\theta}{2\pi} + \frac{1}{\pi} \sin \frac{\theta}{2}} \tag{4.60}$$

where θ is the winding angle and l_e is the core effective path length as shown in Figure 4.7. μ_{dm} is given by an empirical formula defined by the core geometry:

$$\mu_{dm} = 2.5 \left(\sqrt{\frac{\pi}{A_e} \frac{l_e}{2}} \right)^{1.45} \tag{4.61}$$

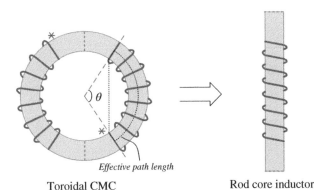

Effective path length

Toroidal CMC   Rod core inductor

Figure 4.7 Nave's method to predict leakage inductance [25].

An improved empririrical formula for l_{eff} is provided by Heldwin [26] as:

$$l_{eff,Heldwin} = \sqrt{\frac{OD^2}{\sqrt{2}}\left(\frac{\theta}{4} + 1 + \sin\frac{\theta}{2}\right)^2 + ID^2\left(\frac{\theta}{4} - 1 + \sin\frac{\theta}{2}\right)^2} \tag{4.62}$$

where ID and OD are inner and outter diameters of the toroid.

As shown in [25], the leakage inductance models (4.58) and (4.60) both have limitations on accuracy. More models are being developed including the curved rod core analogy method and data-driven method.

With the leakage inductance L_{leak} and the DM current (i.e. phase current) peak I_{dm-pk}, the CMC can be designed such that its core flux density due to the DM current generated leakage flux will be below B_{sat}:

$$\frac{L_{leak}I_{dm-pk}}{N_t A_e} < B_{sat} \tag{4.63}$$

4.4.9 Fringing Effect of Gapped Cores

For gapped cores, as the magnetic flux concentrated in the magnetic core approaches the air gap, it will not only pass through the air gap within the confinement of the core; rather, it tends to leak into the neighboring areas, resulting in fringing magnetic field. The fringing magnetic field at the air gap will interact with windings and induce eddy currents that will counteract the useful current. Fringing effect will lead to increased inductance and flux density as well as extra power loss in the inductor.

4.4.9.1 Fringing Effect on Inductance and Flux Density

Fringing effect increases the effective cross-sectional area of the air gap and hence decreases the magnetic reluctance of the air gap, which leads to increased inductance and flux density of the inductor. Take the DC or single-phase EE core in Figure 4.8 as an example, the inductance without considering fringing effect and saturation can be approximated as:

$$L \approx \frac{N_t^2}{R_{c,gap} + \dfrac{R_{o,gap}}{2}} = \frac{\mu_0 N_t^2 A_C}{l_g} \tag{4.64}$$

where $R_{c,gap}$ and $R_{o,gap}$ are the air gap reluctances of the center leg and outer legs, respectively, when neglecting the fringing effect.

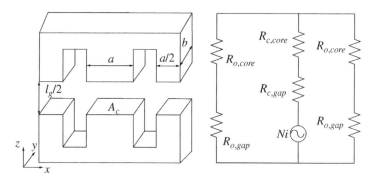

Figure 4.8 DC or single-phase EE core geometry and equivalent magnetic circuit.

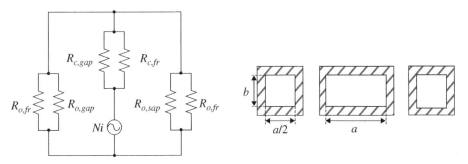

Figure 4.9 Fringing effects: magnetic circuit and enlarged cross-sectional area of center and outer legs.

The peak flux density B_{pk} should not exceed the saturation flux density or defined maximum flux density B_{max} to avoid saturation. At peak current I_{pk}, the peak flux density

$$B_{pk} = \frac{N_t I_{pk}}{A_c \left(R_{c,gap} + \dfrac{R_{o,gap}}{2} \right)} = \frac{\mu_0 N_t I_{pk}}{l_g} \tag{4.65}$$

To account for fringing effect, the magnetic circuit of the EE core becomes the one shown in Figure 4.9. The extra magnetic reluctance, $R_{o,ex}$ and $R_{c,ex}$, are to model the effectively enlarged cross-sectional area of the outer and center legs (shaded area in Figure 4.9 due to fringing flux). Therefore, the reluctance decreases and inductance increases. The inductance becomes

$$L = \frac{N_t^2}{R_{c,gap}//R_{c,ex} + \dfrac{R_{o,gap}//R_{o,ex}}{2}} \tag{4.66}$$

Models to calculate the enlarged cross-sectional area and reluctance of the air gap considering the fringing effect are illustrated here. One model is to increase each edge of the cross section by the same amount as the air gap length l_g. In this case, the effective cross-sectional area of the center leg and outer legs become $A_{ce,c} = (a + l_g)(b + l_g)$ and $A_{ce,o} = \left(\dfrac{a}{2} + l_g \right)(b + l_g)$ respectively. The lumped reluctances $R_{c,total}$ and $R_{o,total}$ in the center and outer legs become

$$R_{c,total} = R_{c,gap}//R_{c,ex} = \frac{\dfrac{l_g}{2}}{\mu_0 A_{ce,c}}$$

$$R_{o,total} = R_{o,gap}//R_{o,ex} = \frac{\dfrac{l_g}{2}}{\mu_0 A_{ce,o}} \tag{4.67}$$

The maximum flux density in the center leg and outer leg are equal and can be calculated as:

$$B_{o,pk} = B_{c,pk} = \frac{N_t I_{pk}}{A_c (R_{c,total} + R_{o,total}/2)} \tag{4.68}$$

Another method calculates the reluctance based on Schwarz–Christoffel Transformation. It first calculates the reluctance of a 2D basic geometry. Then, other complex geometries can be decomposed into paralleled and series combination of this basic geometry and each part can be calculated. A typical 2D basic geometry is shown in Figure 4.10, and the per unit length reluctance for this basic geometry is

$$R_b = \frac{1}{\mu_0 \left[\dfrac{w}{2l} + \dfrac{2}{\pi} \left(1 + \ln \dfrac{\pi h}{4l} \right) \right]} \tag{4.69}$$

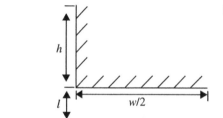

Figure 4.10 A 2D basic geometry.

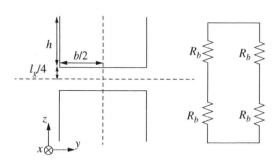

Figure 4.11 Decomposition of the center leg into four 2D basic geometry.

Each air gap in center and outer legs in the single-phase EE core can be decomposed into four basic geometries in both x- and y-directions. As an example, the center leg in x-direction is shown in Figure 4.11. The 2D fringing factors in these two directions (F_{c-x} and F_{c-y}), which are defined by the ratio between reluctances with and without considering the fringing effect, can be calculated by:

$$F_{c-x} = \frac{R_b}{\frac{\mu_0 l_g}{2a}} = \frac{\frac{2a}{l_g}}{\frac{2a}{l_g} + \frac{2}{\pi}\left(1 + \ln\frac{\pi h}{l_g}\right)}$$

$$F_{c-y} = \frac{R_b}{\frac{\mu_0 l_g}{2b}} = \frac{\frac{2b}{l_g}}{\frac{2b}{l_g} + \frac{2}{\pi}\left(1 + \ln\frac{\pi h}{l_g}\right)}$$

(4.70)

The 3D fringing factor is obtained by multiplying the 2D fringing factors in x- and y-directions. Therefore, for the single-phase EE core, 3D reluctance of the center leg and outer legs are

$$R_{c,total} = R_{c,gap}//R_{c,ex} = F_c \mu_0 \frac{l_g}{2ab}$$

$$R_{o,total} = R_{o,gap}//R_{o,ex} = F_o \mu_0 \frac{l_g}{2ab}$$

(4.71)

where $F_c = F_{c-x} \cdot F_{c-y}$ is the 3D fringing factor of the center leg, and $F_o = F_{o-x} \cdot F_{o-y}$ is the 3D fringing factor of the outer legs.

The effective cross-sectional areas of the center and outer legs are expressed as:

$$A_{ce,c} = \frac{\frac{l_g}{2}}{\mu_0 R_{c,total}}$$

$$A_{ce,o} = \frac{\frac{l_g}{2}}{\mu_0 R_{o,total}}$$

(4.72)

The peak flux density can also be calculated by (4.66).

4.4.9.2 Fringing Effect on Core Loss

Fringing flux causes eddy current in the inductor winding, increasing its AC resistance and thus resulting extra winding loss. For laminated or tape wound cores, the fringing field also causes eddy current and losses in the laminations. The gap losses (i.e. air gap fringing field caused loss on core) on laminated core is introduced here.

The prediction of gap losses typically uses an empirical formulation below that was orginally presented in [27] for steel laminated cores operating at power line frequencies:

$$P_g = Gl_g DfB_m^2 \tag{4.73}$$

where P_g is the gap loss, D is the lamination width, B_m is the peak flux density in the core with frequency f, and G is a constant.

Reference [28] investigated the sensitivity of the gap loss of nanocrystalline cores to the inductor design parameters using finite element analysis (FEA) simulations and concludes that the gap loss can be expressed as a power function of l_g, D, f, and B_m as:

$$P_g = K(l_g, D, f, B_m) = k_g l_g^{k_{l_g}} D^{k_D} f^{k_f} B_m^{k_{B_m}} \tag{4.74}$$

where k_g is a numerical constant, and k_{l_g}, k_D, k_f, and k_{B_m} are assumed to be constants independent of each other representing the sensitivity of the gap loss to the inductor parameters. The values of these constants can be determined by curve fitting the loss results from the sensitivity analysis. The equation to estimate nanocrystalline core gap loss is expressed as (units: l_g and D in mm, f in kHz, and B_m in T):

$$P_g = 1.68 \times 10^{-3} l_g D^{1.65} f^{1.72} B_m^2 \ (\text{W}) \tag{4.75}$$

4.5 Relevance to Converter Design

The relevance of the inductor design to the converter subsystem design has already been discussed in Section 4.3 with inductor design input (conditions) and output (variables and constraints) clearly defined. In addition, the inductor design will generate associated attributes that may not be explicitly considered in the design process. These attributes include parasitic parameters, such as ESR and parasitic capacitance. While ESR is somewhat considered through winding loss in the inductor design, the parasitic capacitance, which can impact loss and EMI performance, is not. Here, we will introduce the inductor parasitic capacitance model.

4.5.1 Capacitor Winding Capacitance

The per unit length capacitance for a single wire of circular cross section, uniformly charged, can be expressed as:

$$C_t = \varepsilon_0 \varepsilon_{eq} \frac{2\pi}{\ln\left(\dfrac{d_e}{d_i}\right)} \tag{4.76}$$

where ε_{eq} is equivalent dielectric constant, and d_i and d_e are the inner and outer diameters of the wire, respectively.

Following the analytical models derived in [29], according to Koch approach, the turn-to-turn capacitance is formulated as:

$$C_{tt,K} = 2\varepsilon_0 \left[m_L + \frac{2t_{eq}}{\varepsilon_{eq}} (d_e - t_{eq}) m_D \right] \tag{4.77}$$

where t_{eq} is the equivalent dielectric thickness, and the coefficients m_L and m_D are given respectively by:

$$m_L = \int_0^{\frac{\pi}{6}} \frac{\frac{1}{2} - \left[(\sin\theta)^2 + \cos\theta \sqrt{(\cos\theta)^2 - \frac{3}{4}} \right]}{\left[\cos\theta - \left(1 - \frac{2t_{eq}}{\varepsilon_{eq}d_e}\right)\left(\frac{1}{2} + \sqrt{(\cos\theta)^2 - \frac{3}{4}}\right) \right]^2} d\theta \tag{4.78}$$

$$m_D = \int_0^{\frac{\pi}{6}} \frac{(\sin\theta)^2 + \cos\theta \sqrt{(\cos\theta)^2 - \frac{3}{4}}}{\left[\cos\theta - \left(1 - \frac{2t_{eq}}{\varepsilon_{eq}d_e}\right)\left(\frac{1}{2} + \sqrt{(\cos\theta)^2 - \frac{3}{4}}\right) \right]^2} d\theta$$

According to Massarini approach, the turn-to-turn capacitance is formulated as:

$$C_{tt,M} = \varepsilon_{eq}\theta^* \ln^{-1}\left(\frac{d_e}{d_i}\right) + \varepsilon_0 \cot \frac{\theta^*}{2} - \varepsilon_0 \cot \frac{\pi}{12} \tag{4.79}$$

where

$$\theta^* = \cos^{-1}\left(1 - \frac{1}{\varepsilon_{eq}} \ln\left(\frac{d_e}{d_i}\right)\right) \tag{4.80}$$

Starting from the value of C_{tt}, one can calculate the value of the capacitance between two layers of n_t turns, then a section of the winding constitued by z layers, and finally of the whole winding made out of q sections.

The layer-to-layer capacitance is given by:

$$C_{ll} = \frac{n_t(n_t + 1)(2n_t + 1)}{6n_t^2} l C_{tt} \tag{4.81}$$

where l is the average length of a turn. Then the capacitance of a section is provide by:

$$C_s = C_{ll}(z-1)\left(\frac{2}{z}\right)^2 \tag{4.82}$$

For the overall winding, consitutued by q sections, the capacitance is

$$C_w = \frac{C_s}{q} \tag{4.83}$$

4.6 Summary

This chapter covers the magnetics used in three-phase converter design with focus on inductors and their characteristics and models.

1) Magnetic core materials and types are introduced. Laminated, tape wound, and cut cores based on ferromagnetic materials are more suitable for low-frequency and high flux density

applications, while powder cores and ferrites are more suitable for high-frequency EMI filter applications. Nanocrystalline and nanocomposite materials exhibit lower loss than traditional ferromagnetic materials and can be used for higher frequencies.

2) Inductor design as part of the converter design problem is discussed with physical constraints of saturation flux density, temperature limit and fill factor, and performance constraints of minimum inductance and reliability/life time.

3) Inductor characteristics and models needed for the converter design are covered. Inductance models for gapped and toroidal core inductors, core loss and winding loss models, and temperature rise models are presented. Core window area product A_p based core selection criteria is reviewed. Models for DC, single-phase, and three-phase inductor models are all covered. For core loss calculation, the empirical approaches based on datasheets, such as iGSE, are preferred considering accuracy and ease of use.

4) Leakage inductance models for toroidal CM inductors and fringing effects for gapped inductors are presented. The winding parasitic capacitance model is also discussed as an important inductor design attribute.

References

1 Magnetics Inc., "Magnetic cores for switching power supplies," [Online]. Available: https://www.mag-inc.com/Media/Magnetics/File-Library/Product%20Literature/General%20Information/ps-01.pdf?ext=.pdf. Accessed in 16 April 2023, 1999.

2 Magnetics Inc., "A single magnetic source for all high-tech applications," [Online]. Available: http://bitsavers.trailing-edge.com/components/magneticsInc/APB-2_Product_Line_Brochure_1993.pdf, 1993.

3 Magnetics Inc., "Tape wound cores," [Online]. Available: https://www.mag-inc.com/Media/Magnetics/File-Library/Product%20Literature/Strip%20Wound%20Core%20Literature/2016-Magnetics-Tape-Wound-Cores-Catalog.pdf, 2016.

4 C. W. T. McLyman, "*Transformer and Inductor Design Handbook*," Fourth Edition, CRC Press, 2011.

5 D. C. Jiles and D. L. Atherton, "Theory of ferromagnetic hysteresis," *Journal of Magnetism and Magnetic Materials*, vol. 61, pp. 48–60, 1986.

6 Y. Bernard, E. Mendes and F. Bouillault, "Dynamic hysteresis modeling based on Preisach model," *IEEE Transactions on Magnetics*, vol. 38, no. 2, pp. 885–888, 2002.

7 G. Bertotti, "General properties of power losses in soft ferromagnetic materials," *IEEE Transactions on Magnetics*, vol. 24, no. 1, pp. 621–630, 1988.

8 W. A. Roshen, "A practical, accurate and very general core loss model for nonsinusoidal waveforms," *IEEE Transactions on Power Electronics*, vol. 22, no. 1, pp. 30–40, 2007.

9 J. Reinert, A. Brockmeyer and R. W. A. A. D. Doncker, "Calculation of losses in ferro- and ferrimagnetic materials based on the modified Steinmetz equation," *IEEE Transactions on Industry Applications*, vol. 37, no. 4, pp. 1055–1061, 2001.

10 J. Li, T. Abdallah and C. R. Sullivan, "Improved calculation of core loss with nonsinusoidal waveforms," in *36th IEEE Industry Applications Conference*, 2001.

11 K. Venkatachalam, C. R. Sullivan, T. Abdallah and H. Tacca, "Accurate prediction of ferrite core loss with nonsinusoidal waveforms using only Steinmetz parameters," in *2002 IEEE Workshop on Computers in Power Electronics*, 2022.

12 A. V. D. Bossche, V. C. Valchev and G. B. Georgiev, "Measurement and loss model of ferrites with non-sinusoidal waveforms," in *35th Annual Power Electronics Specialists Conference*, 2004.

13 J. Muhlethaler, J. Biela, J. W. Kolar and A. Ecklebe, "Improved core-loss calculation for magnetic components employed in power electronic systems," *IEEE Transactions on Power Electronics*, vol. 27, no. 2, pp. 964–973, 2012.

14 W. Shen, F. Wang, D. Boroyevich and C. Tipton, "Loss characterization and calculation of nanocrystalline cores for high-frequency magnetics applications," *IEEE Transactions on Power Electronics*, vol. 23, no. 1, pp. 475–484, 2008.

15 C. R. Sullivan, J. H. Harris and E. Herbert, "Core loss predictions for general PWM waveforms from a simplified set of measured data," in *Twenty-Fifth IEEE Applied Power Electronics Conference and Exposition (APEC)*, 2010.

16 J. Muhlethaler, J. Biela, J. W. Kolar and A. Ecklebe, "Core losses under the DC bias condition based on steinmetz parameters," *IEEE Transactions on Power Electronics*, vol. 27, no. 2, pp. 953–963, 2012.

17 T. Shimizu and S. Iyasu, "A practical iron loss calculation for AC filter inductors used in PWM inverters," *IEEE Transactions on Industrial Electronics*, vol. 56, no. 7, pp. 2600–2609, 2009.

18 J. Wang, K. J. Dagan, X. Yuan, W. Wang and P. H. Mellor, "A practical approach for core loss estimation of a high-current gapped inductor in PWM converters with a user-friendly loss map," *IEEE Transactions on Power Electronics*, vol. 34, no. 6, pp. 5697–5710, 2019.

19 I. Villar, U. Viscarret, I. Etxeberria-Otadui and A. Rufer, "Global loss evaluation methods for nonsinusoidally fed medium-frequency power transformers," *IEEE Transactions on Industrial Electronics*, vol. 56, no. 10, pp. 4132–4140, 2009.

20 P. L. Dowell, "Effects of eddy currents in transformer windings," *Proceedings of IEE*, vol. 113, no. 8, pp. 1387–1394, 1966.

21 J. A. Ferreira, "Improved analytical modeling of conductive losses in magnetic components," *IEEE Transactions on Power Electronics*, vol. 9, no. 1, pp. 127–133, 1994.

22 F. Tourkhani and P. Viarouge, "Accurate analytical model of winding losses in round Litz wire windings," *IEEE Transactions on Magnetics*, vol. 37, no. 1, pp. 538–543, 2001.

23 V. C. Valchev and A. V. d. Bossche, "*Inductors and Transformers for Power Electronics*," CRC Press, 2005.

24 M. J. Nave, "On modeling the common mode inductor," in *IEEE International Symposium on Electromagnetic Compatibility*, 1991.

25 Z. Dong, R. Ren, B. Liu and F. Wang, "Data-driven leakage inductance modeling of common mode chokes," in *IEEE Energy Conversion Congress and Exposition (ECCE)*, Baltimore, MD, 2019.

26 M. L. Heldwein, L. Dalessandro and J. W. Kolar, "The three-phase common-mode inductor: modeling and design issues," *IEEE Transactions on Industrial Electronics*, vol. 58, no. 8, pp. 3264–3274, 2011.

27 R. Lee, "*Electronic Transformers & Circuits*," Wiley, 1947.

28 Y. Wang, G. Calderon-Lopez and A. J. Forsyth, "High-frequency gap losses in nanocrystalline cores," *IEEE Transactions on Power Electronics*, vol. 32, no. 6, pp. 4683–4690, 2017.

29 L. Dalessandro, F. da Silveira Cavalcante and J. W. Kolar, "Self-capacitance of high-voltage transformers," *IEEE Transactions on Power Electronics*, vol. 22, no. 5, pp. 2081–2092, 2007.

Part II

Subsystems Design

5

Passive Rectifiers

This is the first chapter of Part 2: the subsystem or function block design part. Each subsystem in a three-phase AC converter can be designed on its own, or together with one or more other subsystems. To facilitate the design problem formulation, we choose to discuss each subsystem design separately. The models and methods developed for a single subsystem can be used when designing several subsystems together, or designing the whole converter system. The relationships among the designs of different subsystems will be considered and discussed in each chapter of Part 2 (Chapters 5–12) and will be systematically presented in detail in Chapter 13.

5.1 Introduction

This chapter is on the design of three-phase passive rectifiers. "Passive" means no active switching devices, only diodes. Because diodes are uncontrolled, diode rectifiers are also often called uncontrolled rectifiers. Diode-based passive rectifiers are an important class of three-phase rectifies. They can be stand-alone converters, as in the case of some uncontrolled DC power supplies; or they can be part of the more complex converters, as in the case of serving as front-end rectifiers for controlled DC power supplies or inverters. For example, the Si IGBT-based two-level VSI with diode front-end rectifier (see Figure 6.1) has been the topology of choice for AC-fed general-purpose industrial motor drives.

Whether a passive rectifier works as a stand-alone converter or as the front-end for another converter, it performs a defined function – three-phase AC to DC rectification, and therefore, can be modeled and designed relatively independently. This chapter discusses the design of the power stage of the three-phase passive rectifier, as a function block or subsystem of the overall three-phase converter. Even for the rectifier working as a stand-alone converter, the power stage of the rectifier is still a subsystem of the whole converter, because there are still other subsystems for the whole rectifier, such as thermal management system, mechanical system, and EMI filter.

There are many different types of three-phase diode rectifiers. Figure 5.1 shows several topologies commonly used in medium- and high-power applications. Among them, the basic three-phase bridge rectifier in Figure 5.1a is the most popular. Multi-pulse topologies of Figure 5.1b and c are used mainly to reduce AC current harmonics.

In some cases, thyristors are used in place of diodes in passive rectifiers. In normal operation, thyristors are turned on fully with 0° firing angle such that they behave just like diodes. During the rectifier start-up, thyristor firing angles can be controlled to limit the inrush current.

Design of Three-phase AC Power Electronics Converters, First Edition. Fei "Fred" Wang, Zheyu Zhang, and Ruirui Chen.
© 2024 The Institute of Electrical and Electronics Engineers, Inc. Published 2024 by John Wiley & Sons, Inc.

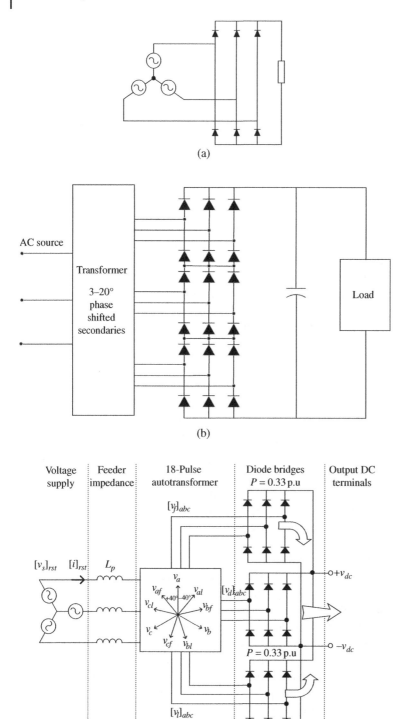

Figure 5.1 Three-phase diode rectifier topologies commonly used in medium- and high-power applications: (a) six-pulse bridge rectifier, (b) series 18-pulse bridge, and (c) shunt 18-pulse bridge.

The combination of diodes and thyristors, e.g. replacing the top or bottom three diodes in Figure 5.1a with thyristors, can also be used to achieve such a purpose.

By definition, with diodes, passive rectifiers allow only unidirectional power flow from three-phase AC sources to DC loads. However, the rectifiers with unidirectional power flow do not necessarily correspond to passive rectifiers. For example, the Vienna-type rectifier, as illustrated in Figure 1.16f, is also unidirectional but belongs to the active rectifier category because of the active switching devices used and corresponding control design needed. Three-phase active rectifiers will be discussed in Chapter 7.

5.2 Passive Rectifier Design Problem Formulation

Even though there are different topologies and applications for three-phase passive rectifiers, we will use the popular bridge rectifier topology employed in the three-phase AC-fed general-purpose industrial motor drive as example to facilitate the discussion and to formulate the design problem. The same principle and approaches can be applied to other topologies and applications. Figure 5.2 shows the circuit configuration of such a motor drive, with the front-end diode bridge rectifier and the inverter. For the power stage of the front-end rectifier, the components include devices (diodes),

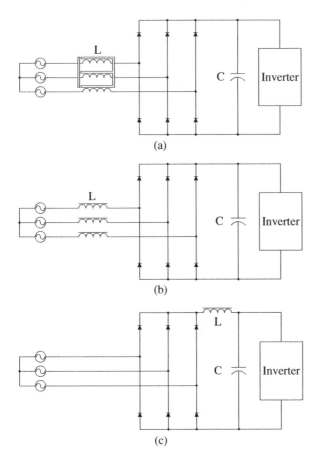

(a)

(b)

(c)

Figure 5.2 The motor drive front-end diode rectifier with three possible solutions for the input inductor: (a) with a three-phase inductor, (b) with three single-phase inductors, and (c) with a DC-link inductor.

inductor(s), and DC-link capacitor(s). It can be seen that even for the same diode bridge topology, there can be different configurations or solutions for the inductors.

It should be pointed out that the DC-link capacitor is needed for both the rectifier and inverter operation. As a result, its selection needs to consider the requirement of both sides. In this chapter, we will design and select the capacitor per the rectifier need, while also considering the inverter impact through load-side conditions and constraints as will be discussed in the following sections.

Note that thermal management system and mechanical system design will not be covered in the diode rectifier power stage design. Others that will not be considered include the EMI filter, which is often required for rectifiers, and control and auxiliary circuits, which are also required. The design of these subsystems will be considered separately.

5.2.1 Passive Rectifier Design Variables

Based on the earlier discussion, the design for the diode bridge power stage refers to the design and/or selection of diode or thyristor devices, AC or DC inductor(s), and DC-link capacitor(s). Therefore, the design variables include the following:

1) Devices: diodes or thyristors, including their types, ratings, and associated characteristics as described in Chapter 2.
2) DC capacitor(s): type, rating, and associated characteristics as described in Chapter 3. From the function standpoint, the capacitance value C is generally the most important variable for a capacitor selection. Although only one capacitor is needed on the DC link, in practice, multiple capacitors are often used in series and/or in parallel to meet the capacitance, voltage, and/or current needs.
3) Inductor(s): the variables associated with inductors include configuration and inductor design itself. As discussed in Chapter 4, the inductors for power converters are usually custom-designed. As a result, the design variables associated with an inductor are characteristics associated with magnetic cores, windings, insulation and heat transfer materials, and structures. Specifically, they include magnetic core type, size, wire types, number of winding turns, and air gap length. If extra insulation or heat transfer capabilities are needed, e.g. in the medium voltage and high power cases, additional insulation materials (e.g. potting materials) and/or heat transfer materials (e.g. thermal conductive fillers) may be needed. Note that inductance value L is not a final design variable, but it is an important intermediate variable.

5.2.2 Passive Rectifier Design Constraints

As discussed before, the constraints include performance constraints related to performance specifications and physical constraints related to component physical capabilities. There can be many different performance specifications for a diode rectifier depending on applications. For a motor drive application, the common performance specifications related to the rectifier include:

1) Power quality performance under normal operating conditions, specifically AC input current harmonics in terms of total demand distortion (TDD), which is the total harmonic distortion (THD) at the rated condition.
2) Minimum and maximum DC-link voltages for normal operating conditions. The normal operating conditions here refer to conditions when the converter (motor drive) is in operation and supplying loads, and should include short-term or overload conditions. This constraint is needed to support required inverter operation. For example, for a motor drive with a two-level VSI, the minimum DC-link voltage needs to be higher than the required peak line-to-line motor voltage under the peak load condition. Considering nonidealities like dead times, voltage drops on VSI

and other parts, often a margin (e.g. 5%) should be added to the minimum DC-link voltage requirement. The maximum DC-link voltage will be determined by device voltage rating. One of the worst-case DC-link voltages for motor drives is during the regenerative operation, such as when trying to stop a motor. Other conditions include inrush and AC line transient overvoltage, although these overvoltage conditions generally are too severe for the motor drive to operate normally. As a result, they are considered abnormal operating conditions.

3) Ride-through time or holdup time without input power. The definition for ride-through time varies with applications. At one extreme, it can refer to the maximum time that the converter (i.e. motor drive in this case) can continue to deliver the load power (up to rated or even overload power) before it has to trip. This normally is required for a critical load, such as a critical manufacturing process. Obviously, this extreme case requires the DC-link voltage not to drop at all during the ride-through period, and the topology in Figure 5.2 with only DC-link capacitor(s) will not be sufficient. At the other extreme, the ride-though time simply refers to the maximum time that the converter will not trip. In this case, it will also not need to provide any load power, but only to provide power loss in the converter. In fact, the load can even be used to help the ride-though (e.g. using motor inertia). In cases between these two extremes, ride-though can refer to various scenarios that require the converter to continue operation but not at full power or voltage. In these latter cases, given the topology of the drive in Figure 5.2, the ride-though energy can be drawn from DC-link capacitors, and the limit is the DC-link undervoltage fault trip level. Obviously, the ride-through or holdup time will impact DC-link capacitor selection.

4) DC-link stability, which is related to the dynamic interface requirement with the inverter. Both small-signal stability and large-signal dynamic stability need to be considered.

Efficiency is an important specification for any converters. For the diode rectifier subsystem, its power loss is generally a small portion of the motor drive, and its thermal management system design will be considered together with that of the inverter. Consequently, the efficiency or loss will not be considered in this chapter as an independent performance constraint; instead, it is treated as an attribute to the design variable "device," which will be an input to thermal management system of the overall converter system.

Motor drive generally should also comply with EMI/EMC standards. Since the EMI filter is treated as a separate subsystem, the EMI requirement will also not be included as a performance constraint.

Power factor is another important performance specification for rectifiers. For diode rectifiers, especially the six-pulse types in Figure 5.2, whose current harmonic contents are high, the displacement factor is more often used, which is defined as the power factor for fundamental frequency component. Generally, the displacement factor for diode rectifiers or thyristor rectifiers with 0° firing angle is high (>0.95) and therefore will not be a performance constraint for the design.

The physical constraints are those associated with components for the rectifier, including devices, DC-link capacitors, and DC or AC inductors. The physical constraints for diode or thyristor devices are parameters already described in Chapter 2, including: (i) device temperature limits, especially the maximum permissible junction temperature T_{jmax}; (ii) device current limits, especially the surge on-state current limit I_{TSM} or I_{FSM}; (iii) device voltage limits, especially the repetitive peak reverse voltage V_{RRM}.

The physical constraints for DC-link capacitors are some of the parameters already described in Chapter 3, including: (i) capacitor working temperature limits, especially the maximum permissible hotspot temperature $T_{cap(max)}$; (ii) capacitor voltage limits, both continuous and transient voltage limits; and (iii) capacitor ripple current limits. Note that both the rectifier and inverter will contribute to DC-link capacitor ripple current.

For the custom-designed inductors, the physical constraints are associated with components used in inductors. As explained in Chapter 4, they include saturation flux density, temperature limit set mainly due to wire insulation and core, and mechanical limits like maximum fill factor K_u for core windows. Sometimes, maximum winding current density J_{MAX} is used in addition to or instead of the temperature limit.

5.2.3 Passive Rectifier Design Conditions

Design conditions are boundary interface conditions, including source and load electrical conditions, and environmental conditions. For diode front-end rectifiers in motor drives, the source electrical conditions include:

1) Input AC voltage characteristics under nominal conditions, under which the rectifier and whole converter system (motor drive) need to operate normally and deliver required load power. The characteristics include: (i) nominal voltage, such as 380 V for 50 Hz system or 480 V for 60 Hz system; (ii) voltage range, typically $\pm 10\%$ of the nominal voltage; (iii) voltage unbalance, often measured by negative sequence voltage as percentage of the positive sequence voltage, typically below 2%; and (iv) voltage quality in terms of harmonics contents.
2) AC voltage characteristics under transient and abnormal conditions. These include over- and undervoltage conditions.
3) Source (e.g. grid) impedance, which is important during transient conditions like inrush.

The DC-side or load electrical conditions include:

1) Nominal load power, which includes motor load power plus the inverter loss. It should be noted that the nominal load power can sometimes be defined as function of input voltage. For example, even though the normal voltage range can be $\pm 10\%$ of the nominal voltage, the rectifier may only need to deliver rated power above certain input voltage level (e.g. 95% of the nominal voltage), below which a lower power is acceptable.
2) Overload, which should include both power level and time duration, for example, 150% overload for 60 seconds. Overload can be a power and time curve, depending on applications.
3) Peak power, which refers to the maximum transient load condition such as peak torque for the motor. In the case of general-purpose motor drive, this is related to and often determined by the peak current capability of the inverter.
4) Maximum regenerative energy, which among other things, can be tied to the requirement on how fast to stop a motor.
5) DC ripple current from the inverter side, contribution due to inverter switching.
6) DC-link under- and overvoltage trip levels, beyond which the converter system will be in under- or overvoltage fault state and will have to trip. These, together with other protection settings, should also be design variables. Since their design will be related to protection design, which will be discussed in Chapter 10, they will be treated as design input here.
7) DC harmonic impedance or frequency-domain impedance, which will impact the DC-link stability and capacitor design and selection.

The passive rectifier also interfaces with the thermal management system, the mechanical system, the AC source-side EMI filter when there is one, and to lesser extent the control system. In the case of the thermal management system and the mechanical system, the interface conditions are similar to those of inverters and active rectifiers, so we will discuss these conditions in Chapter 6. The thermal management system design will also not be considered as it will be the same as in the inverter case. As a result, the environmental conditions in this case mainly refer to the ambient temperature range for the converter.

There are no explicit interface conditions from the mechanical system, the EMI filter, and the control that will impact the passive rectifier design.

5.2.4 Passive Rectifier Design Objectives and Design Problem Formulation

The design objectives for each subsystem of an AC converter, including the front-end rectifier (passive or active), should follow those of the whole converter. If the design objective for the target motor drive is minimum cost, the rectifier design objective should also be associated with cost.

Based on the selected design objective(s), and the design variables, constraints, and conditions, the overall rectifier design problem can be formulated as the optimization problem in (5.1). Note that the peak inrush current and peak DC voltage during inrush are related to the diode component limits.

Minimize or maximize: $F(X)$ – objective function

Subject to inequality constraints:

$$G(X,U) = \begin{bmatrix} g_1(X,U) \\ g_2(X,U) \\ g_3(X,U) \\ g_4(X,U) \\ g_5(X,U) \\ g_6(X,U) \\ g_7(X,U) \\ g_8(X,U) \\ g_9(X,U) \\ g_{10}(X,U) \\ g_{11}(X,U) \\ g_{12}(X,U) \\ g_{13}(X,U) \end{bmatrix} = \begin{bmatrix} \text{AC input current TDD} \\ \text{Peak inrush current} \\ \text{Peak DC voltage during inrush} \\ -\text{Steady state } V_{DC(min)} \\ \text{Steady state } V_{DC(max)} \\ -\text{Ride through time} \\ -\text{DC link stability margin} \\ \text{Inductor core window fill factor} \\ \text{Inductor saturation flux density} \\ \text{Inductor temperature} \\ \text{Capacitor voltage} \\ \text{Capacitor ripple current} \\ \text{Capacitor temperature} \end{bmatrix} \leq \begin{bmatrix} \text{TDD limit} \\ I_{FSM} \\ V_{diode} \\ -V_{DC(min)}\,\text{limit} \\ V_{DC(max)}\,\text{limit} \\ -T_{r-t} \\ -\text{phase or gain margin} \\ K_{u(max)} \\ B_{sat} \\ T_{ind(max)} \\ V_{cap} \\ I_{cap} \\ T_{cap(max)} \end{bmatrix}$$

where design variables and design conditions:

$$X = \begin{bmatrix} x_1 \\ x_2 \\ x_3 \\ x_4 \\ x_5 \\ x_6 \end{bmatrix} = \begin{bmatrix} \text{Diode or thyristor} \\ \text{Capacitor} \\ \text{Inductor core} \\ \text{Inductor wire} \\ \text{Inductor air gap} \\ \text{Inductor winding turns} \end{bmatrix}$$

$$U = \begin{bmatrix} u_1 \\ u_2 \\ u_3 \\ u_4 \\ u_5 \\ u_6 \\ u_7 \\ u_8 \\ u_9 \\ u_{10} \\ u_{11} \end{bmatrix} = \begin{bmatrix} \text{Input AC voltage characteristics under nominal conditions} \\ \text{AC voltage characteristics under transient/abnormal conditions} \\ \text{Source impedance} \\ \text{Norminal load power} \\ \text{Overload and duration} \\ \text{Peak load} \\ \text{Maximum regenerative energy} \\ \text{DC link over and undervoltage trip levels} \\ \text{DC load ripple current} \\ \text{DC load impedance} \\ \text{Ambient temperature } T_a \end{bmatrix} \tag{5.1}$$

The optimization problem can be illustrated also as in Figure 5.3 [1, 2].

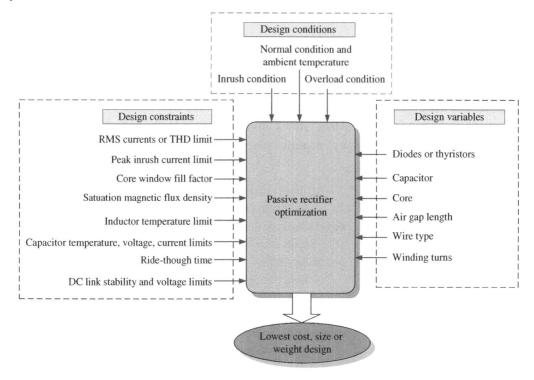

Figure 5.3 Passive rectifier design optimization problem.

In order to solve the optimization problem, relationships must be established among design variables, constraints, conditions, and objectives. Specifically, design constraints (output) as functions of design conditions (input) and variables need to be established. One set of the design variable values making up a design will be associated with one objective function value.

Note that in addition to the design variables, there are also attributes as a result of the design. Although they may not be part of the design problem, they will have impact on the design. This is especially true for the subsystem design. For example, the thermal characteristics (e.g. thermal impedances) of the diodes/thyristors, inductors, and capacitors are attributes of these components, not directly part of the design problem. However, they will impact the junction temperature or temperature rise, loss, and the thermal management system.

5.3 Passive Rectifier Models

Based on (5.1) and Figure 5.3, we need models for the following constraints as functions of design conditions and design variables:

1) AC input harmonic current TDD
2) Minimum and maximum DC-link voltage under normal operating conditions (normal condition includes regenerative overvoltage)
3) Ride-through or holdup time without input power
4) DC-link stability
5) Device selection-related constraints, including maximum junction temperature, peak surge current, and peak voltage

6) Capacitor-related constraints, including temperature, currents, and maximum voltage (both steady state and transient)
7) Inductor-related constraints, including temperature, maximum flux density to avoid saturation, and mechanical constraints (i.e. core window areas)

This section will focus on establishing these models. There are many types of models and many different ways to categorize them. For example, one way to categorize a model is whether it is physics-based or behavior-based; and another way to categorize is whether it is a closed-form analytical model or a numerical model based on differential equations or simulations. For models to be used in design, i.e. in optimization in our case, it is desirable for the models to be in analytical or algebraic forms, whether they are physics-based or behavioral.

5.3.1 AC Input Harmonic Current

The AC input line current harmonics for the three-phase rectifier in Figure 5.2 can be easily obtained through circuit simulation. However, as aforementioned, an analytical or algebraic model is desired for design optimization. This subsection will present an analytical model mainly based on work in [3, 4]. Instead of using time-domain circuit simulations, the modeling approach is to use switching functions to represent the converter under harmonic conditions as shown in Figure 5.4, where Z_n and Z_{on} are the nth harmonic impedance of the inverter and rectifier, respectively, viewed from the DC link. The DC-link voltage of the three-phase rectifier can be expressed by the sum of the average DC-link voltage and all the DC-link voltage harmonics (5.3). E_d is the average value of the DC-link voltage, and e_{dn} is the nth voltage harmonic, which is defined in (5.3) with ω as the fundamental angular frequency. e_d and its components E_d and e_{dn} are all equivalent open-circuit voltages, with the impact of impedance on DC-link voltage accounted for by Z_{on}.

$$e_d = E_d + \sum_{n=1}^{\infty} e_{dn} \tag{5.2}$$

$$e_{dn} = A_{dn} \cdot \cos(n\omega t) + B_{dn} \cdot \sin(n\omega t) \tag{5.3}$$

The derivation of E_d and e_{dn} are given in [3, 4] and highlighted here. Figure 5.5 illustrates the rectifier voltage waveform and the switching functions of S_a, S_b, and S_c for phases a, b, and c, respectively, for continuous conduction mode. Note that switching function S_a value 1 indicates phase a in noncommutation conducting period, 0 in off or nonconducting period, and 0.5 in commutation period. Positive values indicate upper diodes conducting, and negative values indicate lower diodes conducting. Using switching functions, e_d can be expressed as (5.4).

$$e_d = S_a e_a + S_b e_b + S_c e_c \tag{5.4}$$

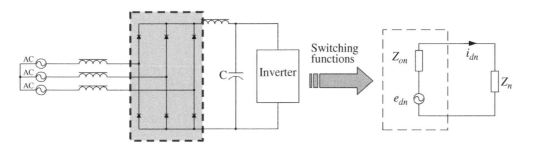

Figure 5.4 Equivalent circuit for obtaining harmonic components of DC current.

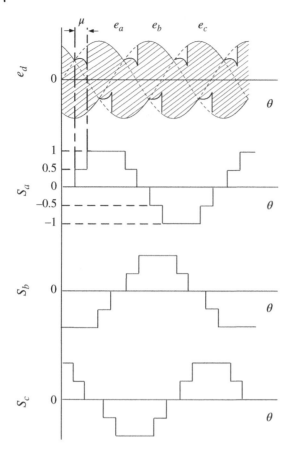

Figure 5.5 Rectifier AC terminal voltages and switching functions vs. angle θ. *Source:* From [3] with permission of IEEE.

where

$$
\begin{cases}
e_a = E_m \sin(\theta) \\
e_b = E_m \sin\left(\theta - \dfrac{2\pi}{3}\right) \\
e_c = E_m \sin\left(\theta - \dfrac{4\pi}{3}\right)
\end{cases}
\tag{5.5}
$$

In (5.5), E_m is the amplitude of the AC source phase voltages, and $\theta = \omega t$. The angle μ in Figure 5.5 is the commutation angle. Assuming that the DC current is nearly constant and equals to its average value I_d, and also the AC line impedance is dominated by reactance X_a, the commutation angle can be obtained using (5.6):

$$
\mu = \cos^{-1}\left(1 - \frac{2X_a I_d}{\sqrt{3}E_m}\right)
\tag{5.6}
$$

The switching functions can be obtained through Fourier series as in (5.7), where $j \in (a, b, c)$ phases; i is 0, 1, or 2, respectively, for a, b, c; and $l = 1, 2, 3, ..., k > 0$:

$$
S_j = \sum_{k = 6l \mp 1}^{\infty} \frac{\sqrt{3}}{\pi} \times \frac{(-1)^{l+1}}{k} \times \left\{ \sin k\mu \cos\left[k\left(\theta - i\frac{2\pi}{3}\right)\right] - (1 + \cos k\mu) \sin\left[k\left(\theta - i\frac{2\pi}{3}\right)\right] \right\}
\tag{5.7}
$$

With (5.7), the terms in Eqs. (5.2) and (5.3) can be obtained as in (5.8), (5.9), and (5.10), where $n = 6p$ $(p = 1, 2, 3,...)$:

$$E_d = \frac{3\sqrt{3}E_m}{2\pi}(1 + \cos\mu) = \frac{3\sqrt{3}E_m}{\pi} - \frac{3X_aI_d}{\pi} \tag{5.8}$$

$$A_{dn} = \frac{3\sqrt{3}E_m(-1)^p}{2\pi}\left\{\frac{1 + \cos(n + 1)\mu}{n + 1} - \frac{1 + \cos(n - 1)\mu}{n - 1}\right\} \tag{5.9}$$

$$B_{dn} = \frac{3\sqrt{3}E_m(-1)^p}{2\pi}\left\{\frac{\sin(n + 1)\mu}{n + 1} - \frac{\sin(n - 1)\mu}{n - 1}\right\} \tag{5.10}$$

Assuming that the inverter load on DC link can be represented by a resistance R_{INV}, Z_n for load can be calculated as in (5.11), where L_{DC} and C_{DC} are DC inductance and capacitance as shown in Figure 5.4. R_{INV} can be estimated as in (5.12), where P_{INV} is the inverter power. For motor drive inverter, its switching frequency, as well as the corresponding control bandwidth, is generally much higher than the AC source fundamental frequency and its important harmonics. Therefore, the approximation of (5.11) is acceptable. A more accurate impedance model can be found in Chapter 6 considering the inverter switching and control.

$$Z_n = jn\omega L_{DC} + \frac{R_{INV}}{1 + jn\omega C_{DC}R_{INV}} \tag{5.11}$$

$$R_{INV} = \frac{P_{INV}}{E_d} \tag{5.12}$$

The approximate rectifier harmonic impedance Z_{on} has been derived in [3] as in (5.13), where X_a has been defined in (5.6):

$$Z_{on} = j\left(2 - \frac{3\mu}{2\pi}\right)nX_a \tag{5.13}$$

The DC current i_d can be expressed as:

$$i_d = I_d + \sum_{n = 6,12,18, ...}^{\infty} i_{dn} = I_d + \sum_{n = 6,12,18, ...}^{\infty} \sqrt{2}I_{dn(RMS)}\cos(n\theta - \beta_n) \tag{5.14}$$

where with nth order DC harmonic current denoted as i_{dn}, the RMS value $I_{dn(RMS)}$ and phase angle β_n of i_{dn} can be found as:

$$I_{dn(RMS)} = \frac{E_{dn(RMS)}}{\|Z_{on} + Z_n\|} = \frac{\sqrt{(A_{dn}^2 + B_{dn}^2)/2}}{\|Z_{on} + Z_n\|} \tag{5.15}$$

$$\beta_n = \tan^{-1}\frac{B_{dn}}{A_{dn}} + \tan^{-1}\frac{Im(Z_{on} + Z_n)}{Re(Z_{on} + Z_n)} \tag{5.16}$$

The total DC RMS current is

$$I_{DC(RMS)} = \sqrt{I_d^2 + \sum_{n = 6, 12, ...}^{\infty} I_{dn(RMS)}^2} \tag{5.17}$$

The design constraint is on AC line current. Therefore, we are more interested in finding the RMS value of the AC line current. In some cases, the RMS value of the AC line current can be easily obtained from the DC RMS current. Figure 5.6 illustrates the three-phase diode rectifier AC-side phase voltage and current waveforms. Form the waveforms and the corresponding diode conduction patterns, the phase a current i_a can be found as in Table 5.1. In general, i_a is a function of the

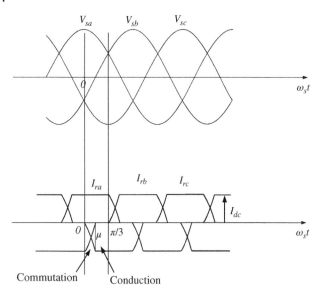

Figure 5.6 Three-phase diode rectifier AC phase voltage and current waveforms.

Table 5.1 Diode phase a current i_a as the function of the DC current i_d.

Region	Angle	General case	Special case 1: $\mu = 0$	Special case 2: $i_d = I_d$ and $\dfrac{di_a}{dt}$ constant during commutation
1	$0 < \theta \leq \dfrac{\pi}{3}$	$i_a = i_d$	$i_a = i_d$	$i_a = I_d$
2	$\dfrac{\pi}{3} < \theta \leq \dfrac{\pi}{3} + \mu$	$i_a = i_d - i_b$	$i_a = 0$	$i_a = I_d \dfrac{\mu - \theta - \dfrac{\pi}{3}}{\mu}$
3	$\dfrac{\pi}{3} + \mu < \theta \leq \dfrac{2\pi}{3}$	$i_a = 0$		$i_a = 0$
4	$\dfrac{2\pi}{3} < \theta \leq \dfrac{2\pi}{3} + \mu$	$i_a = -i_d - i_c$	$i_a = -i_d$	$i_a = -I_d \dfrac{\theta - \dfrac{2\pi}{3}}{\mu}$
5	$\dfrac{2\pi}{3} + \mu < \theta \leq \pi$	$i_a = -i_d$		$i_a = -I_d$
6	$\pi < \theta \leq \dfrac{4\pi}{3}$	$i_a = -i_d$	$i_a = -i_d$	$i_a = -I_d$
7	$\dfrac{4\pi}{3} < \theta \leq \dfrac{4\pi}{3} + \mu$	$i_a = -i_d - i_b$	$i_a = 0$	$i_a = -I_d \dfrac{\mu - \theta - \dfrac{4\pi}{3}}{\mu}$
8	$\dfrac{4\pi}{3} + \mu < \theta \leq \dfrac{5\pi}{3}$	$i_a = 0$		$i_a = 0$
9	$\dfrac{5\pi}{3} < \theta \leq \dfrac{5\pi}{3} + \mu$	$i_a = i_d - i_c$	$i_a = i_d$	$i_a = I_d \dfrac{\theta - \dfrac{5\pi}{3}}{\mu}$
10	$\dfrac{5\pi}{3} + \mu < \theta \leq 2\pi$	$i_a = i_d$		$i_a = I_d$

DC current i_d and also a function of the other phase currents i_b and i_c. In two special cases, i_a can be only the function of i_d. In these special cases, the RMS value of the AC current can be readily obtained from the DC RMS current.

Special case 1 corresponds to the commutation angle $\mu = 0$, which is a good approximation in many applications, when the AC-side inductance is relatively small or when there is no dedicated AC inductor. In this case, $I_{AC(RMS)}$ can be obtained as in (5.18). Note that the derivation has used the relation $I_{DC(RMS)} = \sqrt{\frac{3}{\pi} \int_0^{\pi/3} i_d^2 d\theta} = \sqrt{\frac{3}{\pi} \int_{\pi/3}^{2\pi/3} i_d^2 d\theta}$, given that i_d is periodic every $\pi/3$.

$$
I_{AC(RMS)} = I_{a(RMS)} = \sqrt{\frac{1}{T} \int_0^T i_a^2 dt} = \sqrt{\frac{1}{\pi} \int_0^\pi i_a^2 d\theta} = \sqrt{\frac{1}{\pi} \left[\int_0^{\pi/3} i_a^2 d\theta + \int_{2\pi/3}^\pi i_a^2 d\theta \right]}
$$

$$
= \sqrt{\frac{1}{\pi} \left[\int_0^{\pi/3} i_d^2 d\theta + \int_{2\pi/3}^\pi (-i_d)^2 d\theta \right]} = \sqrt{\frac{1}{\pi} \left[\frac{\pi}{3} I_{DC(RMS)}^2 + \frac{\pi}{3} I_{DC(RMS)}^2 \right]} = \sqrt{\frac{2}{3}} I_{DC(RMS)}
$$

$$
(5.18)
$$

Special case 2 corresponds to the case when i_d is constant and during the commutation period i_a changes linearly with time. The first condition is generally a reasonable assumption, since the DC current should only have a very small ripple with or without a large DC inductor. The second condition is approximate but is also reasonable, especially when the commutation angle μ is small. In this case, $I_{AC(RMS)}$ can be obtained as in (5.19). Clearly, when $\mu = 0$, Eqs. (5.18) and (5.19) are the same.

$$
I_{AC(RMS)} = I_{a(RMS)} = \sqrt{\frac{1}{T} \int_0^T i_a^2 dt} = \sqrt{\frac{1}{\pi} \int_0^\pi i_a^2 d\theta}
$$

$$
= \sqrt{\frac{1}{\pi} \left[\int_0^{\pi/3} i_a^2 d\theta + \int_{\pi/3}^{\pi/3+\mu} i_a^2 d\theta + \int_{2\pi/3}^{2\pi/3+\mu} i_a^2 d\theta + \int_{2\pi/3+\mu}^\pi i_a^2 d\theta \right]}
$$

$$
= \sqrt{\frac{1}{\pi} \left[\frac{\pi}{3} I_d^2 + \int_0^\mu \frac{I_d^2}{\mu^2} (\mu - \theta)^2 d\theta + \int_0^\mu \frac{I_d^2}{\mu^2} \theta^2 d\theta + \int_\mu^{\pi/3} I_d^2 d\theta \right]}
$$

$$
= I_d \sqrt{\left(\frac{2}{3} - \frac{\mu}{3\pi} \right)} = I_{DC(RMS)} \sqrt{\left(\frac{2}{3} - \frac{\mu}{3\pi} \right)}
$$

$$
(5.19)
$$

The TDD of AC current can be determined via (5.20), where $I_{1(RMS)}$ is the RMS value of fundamental frequency AC current at the rated condition:

$$
TDD = \frac{\sqrt{I_{AC(RMS)}^2 - I_{1(RMS)}^2}}{I_{1(RMS)}} \times 100\%
\tag{5.20}
$$

Note that harmonic current standards often specify individual harmonic limits. In this case, the expression for individual harmonic current can be found via (5.21) and (5.22):

$$
i_{AC} = i_a = \sum_{n=1,5,7,\dots}^\infty [A_n \cos(n\omega t) + B_n \sin(n\omega t)]
\tag{5.21}
$$

$$
I_{n(RMS)} = \sqrt{\frac{A_n^2 + B_n^2}{2}}
\tag{5.22}
$$

where

$$A_n = A_{0n} + \sum_{k = 6,12,18,\dots}^{\infty} \Delta A_{kn} \tag{5.23}$$

$$B_n = B_{0n} + \sum_{k = 6,12,18,\dots}^{\infty} \Delta B_{kn} \tag{5.24}$$

with

$$A_{0n} = \frac{\sqrt{3}I_d(-1)^{l+1}}{\pi}\left\{\frac{2\sin n\mu}{n} + \frac{1}{1-\cos\mu}\left[\frac{-2\sin n\mu}{n} + \frac{\sin(n+1)\mu}{n+1} + \frac{\sin(n-1)\mu}{n-1}\right]\right\} \tag{5.25}$$

$$B_{0n} = \frac{\sqrt{3}I_d(-1)^{l}}{\pi}\left\{\frac{2\cos n\mu}{n} + \frac{1}{1-\cos\mu}\left[\frac{2(1-\cos n\mu)}{n} - \frac{1-\cos(n+1)\mu}{n+1} - \frac{1-\cos(n-1)\mu}{n-1}\right]\right\} \tag{5.26}$$

$$\Delta A_{kn} = \frac{\sqrt{3}I_{dn}(-1)^{l+p}}{\pi}\left\{\frac{\sin\beta_k - \sin[(k+n)\mu-\beta_k]}{k+n} + \frac{\sin\beta_k - \sin[(k-n)\mu-\beta_k]}{k-n}\right\} \tag{5.27}$$

$$\Delta B_{kn} = \frac{\sqrt{3}I_{dn}(-1)^{l+p}}{\pi}\left\{\frac{\cos\beta_k + \cos[(k+n)\mu-\beta_k]}{k+n} - \frac{\cos\beta_k + \cos[(k-n)\mu-\beta_k]}{k-n}\right\}$$
$$n = 6l \mp 1 \ (l = 0,1,2,\dots, n > 0)$$
$$k = 6p \ (p = 1,2,3,\dots) \tag{5.28}$$

$I_{1(RMS)}$ in (5.20) can be obtained from (5.22) with $n = 1$. Approximately, $I_{1(RMS)}$ can also be estimated as in (5.29), where P_{REC} is the total power for the rectifier, which can be estimated from the inverter power, and E_m is the peak AC input voltage as defined in (5.5):

$$I_{1(RMS)} \approx \frac{P_{REC}}{3\frac{E_m}{\sqrt{2}}} \tag{5.29}$$

Table 5.2 shows some design conditions and constraints of a sample system. Figure 5.7 shows circuit simulation results of Figure 5.2a circuit for AC input RMS current (representing TDD) of the sample system. Clearly, L and C can be traded off for a given harmonics level. Table 5.3 shows the comparison of simulation and analytically calculated results of the RMS value of the AC currents using Eq. (5.19). It can be seen that the calculated and simulated results match sufficiently well for practical design purposes.

The peak value of the AC or DC current I_{pk} is a key parameter for AC or DC inductor design, respectively. It should also be noted that the AC and DC current peaks are the same, i.e.

$$I_{pk} = \text{Max}(i_{AC}) = \text{Max}(i_a) = \text{Max}(i_d) \tag{5.30}$$

The peak value can be obtained as the maximum value of i_d using (5.14) in a time range, e.g. in the range of $0 < \omega t < \pi$ of the fundamental cycle, which can be found through a search algorithm in the design optimization program. Note that the calculations assume that the AC line resistance and AC

Table 5.2 Design conditions and constraints of the sample system.

Design conditions and constraints		Values
Conditions	Maximum output power	50 kW
	Input frequency	50 Hz
	Maximum line-to-line voltage (RMS)	500 V
	Minimum line-to-line voltage (RMS)	380 V
	Initial DC-link voltage in inrush condition	400 V
	Equivalent source inductance	17 μH
	Overload condition	1.5 × 50 kW (60 seconds)
	Ambient temperature	60 °C
Constraints	Maximum RMS value of the input line current	98 A
	Maximum peak-to-peak DC voltage ripple	5% V_{DC}
	Maximum inrush current	1750A
	Saturation magnetic flux density	1.6 T (silicon steel)
	Wire current density	J_{MAX}
	Maximum core window fill factor	Ku
	Temperature rise	80 °C (normal load), 90 °C (overload)

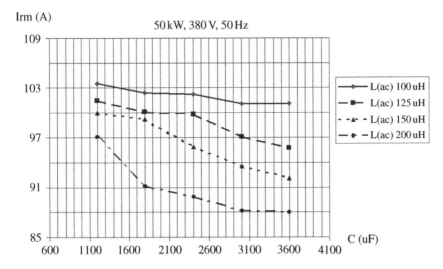

Figure 5.7 Simulated AC RMS current as function of *L* and *C* for the example system.

voltage distortion and unbalance can be ignored, and the rectifier is in the continuous conduction mode. The error caused by these assumptions in the design is acceptable. The impact of these assumptions can be considered in the detailed design if necessary, either through simulation or more elaborate analysis.

Table 5.3 Comparison of simulation and calculation of RMS current.

		$I_{(RMS)}$ **(A)**	
L (μH)	**C (μF)**	**Simulation**	**Analytical calculation**
150	2400	95.9	95.5
	3000	93.6	92.4
	3600	92.1	90.9
200	2400	89.9	87.2
	3000	88.2	86.2
	3600	88.0	85.7

5.3.2 Minimum and Maximum DC Voltages Under Normal Operating Conditions

From Eqs. (5.2), (5.3), (5.8), (5.9), and (5.10), the rectifier DC output voltage can be rewritten as in (5.31):

$$e_d = E_d + \sum_{n=6,12,18,\dots}^{\infty} e_{dn} = E_d + \sum_{n=6,12,18,\dots}^{\infty} \{A_{dn}\cos(n\omega t) + B_{dn}\sin(n\omega t)\} \tag{5.31}$$

To determine the DC-link voltage v_{DC}, the impact of the DC inductance L_{DC} needs to be considered, such that

$$v_{DC} = e_d - L_{DC}\frac{di_d}{dt} \tag{5.32}$$

It is quite complex to obtain an algebraic expression for Eq. (5.32). For the purpose of ripple estimation in design, the impact of L_{DC} can be neglected. In this case, the minimum and maximum DC voltages under normal operating conditions are in (5.33) and (5.34), respectively:

$$V_{DC(min)} = E_d + \text{Min}\left(\sum_{n=6,12,18,\dots}^{\infty} e_{dn}\right) = E_d + \text{Min}\left\{\sum_{n=6,12,18,\dots}^{\infty} \{A_{dn}\cos(n\omega t) + B_{dn}\sin(n\omega t)\}\right\},$$

$$(0 < \omega t < \pi) \tag{5.33}$$

$$V_{DC(max)} = E_d + \text{Max}\left(\sum_{n=6,12,18,\dots}^{\infty} e_{dn}\right) = E_d + \text{Max}\left\{\sum_{n=6,12,18,\dots}^{\infty} \{A_{dn}\cos(n\omega t) + B_{dn}\sin(n\omega t)\}\right\},$$

$$(0 < \omega t < \pi) \tag{5.34}$$

Although it is not a direct design constraint for the passive rectifier, the peak-to-peak DC-link voltage ripple $V_{DC(pp)}$ is often a useful piece of information. It can be obtained by (5.35) or algebraically in design through a search over $(0 < \omega t < \pi)$:

$$V_{DC(pp)} = V_{DC(max)} - V_{DC(min)} = \text{Max}\left(\sum_{n=6,12,18,\dots}^{\infty} e_{dn}\right) - \text{Min}\left(\sum_{n=6,12,18,\dots}^{\infty} e_{dn}\right)$$

$$= \text{Max}\left\{\sum_{n=6,12,18,\dots}^{\infty} \{A_{dn}\cos(n\omega t) + B_{dn}\sin(n\omega t)\}\right\} \tag{5.35}$$

$$- \text{Min}\left\{\sum_{n=6,12,18,\dots}^{\infty} \{A_{dn}\cos(n\omega t) + B_{dn}\sin(n\omega t)\}\right\} \quad (0 < \omega t < \pi)$$

5.3.3 Ride-Through or Holdup Time Without Input Power

Earlier we described several scenarios of ride-throughs, corresponding to different requirements on power and DC voltage levels during the period. Here, we consider the condition that the DC-link voltage is allowed to drop, while supplying the required output power P_{r-t}, which can range from the rated load power to only the loss (including loss of the DC-link bleeder resistor).

Assuming the required ride-through time is T_{r-t} and the minimum allowed DC-link voltage is $V_{DC(UV_{trip})}$, the relationship (5.36) should be satisfied, which will yield a constraint on DC-link capacitance in (5.37):

$$\frac{1}{2}C_{DC}\left(V_{DC(\min)}{}^2 - V_{DC(UV_{trip})}{}^2\right) \geq P_{r-t}T_{r-t} \tag{5.36}$$

$$C_{DC} \geq \frac{2P_{r-t}T_{r-t}}{\left(V_{DC(\min)}{}^2 - V_{DC(UV_{trip})}{}^2\right)} \tag{5.37}$$

5.3.4 DC-Link Stability

The rectifier and the inverter are two cascaded subsystems as shown in Figure 5.8a. In order to avoid instability, the output impedance Z_{out} of the rectifier should be lower than the input impedance Z_{in} of the inverter [5]. This is illustrated in Figure 5.8b showing the impedance relationship in a magnitude Bode plot. The inverter can be considered as a constant power load below the control bandwidth frequency f_{BW}. In contrast, Z_{out} can be assumed to be very low below the control bandwidth and is dominated by the DC-link capacitance impedance $1/\omega C$ beyond f_{BW}. Consequently, the constraint for the DC-link capacitance can be given by (5.38), where P_{out} is the rectifier output power and Z_m is the impedance margin (e.g. >6 dB). Then we can select the minimum capacitance accordingly.

$$20\log\left(\frac{V_{DC}{}^2}{P_{out}}\right) - 20\log\left(\frac{1}{2\pi f_{BW}C}\right) \geq Z_m \tag{5.38}$$

5.3.5 Device-Related Constraints – Inrush

There are three main constraints related to diode or thyristor devices, i.e. junction temperature, peak current, and peak voltage. Since the diodes or thyristors in passive rectifiers are line-frequency devices and their losses are relatively low, the junction temperature is usually not the main

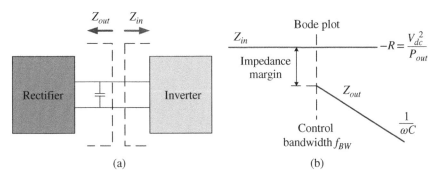

Figure 5.8 (a) DC-link impedance definition; (b) DC-link impedance Bode plot.

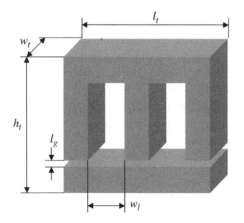

Figure 5.9 Three-phase inductor core.

constraint. Here, we will focus on the peak current and voltage. The loss will be discussed in Section 5.5 as input to thermal management system.

The peak current for a three-phase diode rectifier occurs during the inrush condition, when AC source voltage coming back from 0 to nominal voltage after a temporary loss of power or brown out, while the DC-link voltage is low as a result of voltage drop. It is conceivable that the worst-case scenario corresponds to when the DC-link voltage is close to the undervoltage trip level and AC voltage returns to the high-line condition (e.g. 110%).

During the inrush period, there are only two AC phases conducting most of the time. Considering the AC three-phase inductor with a core structure as shown in Figure 5.9, the equivalent circuit of the three-phase diode rectifier under a conduction condition can be simplified as shown in Figure 5.10a. Due to the high input current under the inrush condition, the line inductance (i.e. equivalent source inductance) L_{line} has been reckoned in.

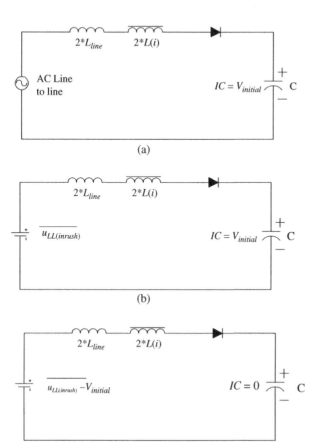

Figure 5.10 Equivalent circuits of the diode rectifier under the inrush condition: (a) original under diode conduction condition, (b) simplified for maximum inrush condition; and (c) further simplified with zero initial capacitor voltage.

The worst-case inrush for the front-end diode rectifier is when the input line-to-line voltage is near its peak value. The maximum average value of the input line-to-line voltage $u_{LL(inrush)}$ can be found as in (5.39), where T is the line period, T_{OSC} is the period corresponding to resonance between the equivalent inductance and capacitance in Figure 5.10, T_{inrush} is the duration of the inrush period, and $V_{ll(RMS)}$ is the line-to-line RMS voltage for rated or the worst condition (e.g. high-line condition):

$$\overline{u_{LL(inrush)}} = \frac{T}{2 \cdot \pi \cdot T_{OSC}} \int_{\frac{\pi}{2} - \frac{\pi \cdot T_{inrush}}{T}}^{\frac{\pi}{2} + \frac{\pi \cdot T_{inrush}}{T}} \sqrt{2} \cdot V_{ll(RMS)} \cdot \sin\theta \cdot d\theta = \frac{\sqrt{2} \cdot V_{ll(RMS)} \cdot T}{\pi \cdot T_{OSC}}$$
$$\cdot \sin\frac{\pi \cdot T_{OSC}}{T} \tag{5.39}$$

T_{inrush} can be approximated by Eq. (5.40), where L_{nom} is the nominal inductance value of the three-phase inductor under the normal operating condition. L_{nom} can be calculated as in (5.41), where N_t is the winding turns number per phase. Compared with the line period T, T_{inrush} is much smaller. L_{line} is neglected in (5.40), since it is usually much smaller than L_{nom} under normal operating condition.

$$T_{inrush} = \frac{1}{2} \cdot T_{OSC} \approx \pi \cdot \sqrt{2 \cdot L_{nom} \cdot C} \tag{5.40}$$

$$L_{nom} = L_{AC} = \frac{\mu_0 N_t^2 A_c}{l_g} \tag{5.41}$$

The maximum surge of the DC-link voltage occurs under the no-load condition. Its value is about

$$V_{surge} = 2 \cdot \overline{u_{LL(inrush)}} - V_{initial} \tag{5.42}$$

where $V_{initial}$ is the initial voltage on DC-link capacitor when inrush occurs and can be set at the undervoltage fault tripping level. Note that neither the DC-link capacitance nor the AC inductance impacts the surge voltage.

The peak inrush current appears when the voltage on the DC-link capacitor reaches $\overline{u_{LL(inrush)}}$ during the inrush period. For calculation of the inrush current, Figure 5.10c will have the same solution as Figure 5.10b. In the Figure 5.10c circuit, the maximum inductor current I_{inrush} occurs when the capacitor voltage reaches $\left(\overline{u_{LL(inrush)}} - V_{initial}\right)$, corresponding to $\frac{di}{dt} = 0$. At this time, the total stored magnetic energy is equal to the energy stored in the capacitor as (5.43), where $E_{L(i)}$ stands for the energy stored in the inductor:

$$L_{line} I_{inrush}^2 + E_{L(i)} = \frac{1}{2} C \cdot \left(\overline{u_{LL(inrush)}} - V_{initial}\right)^2 \tag{5.43}$$

If the inductor is designed to be linear and unsaturated during inrush, then $E_{L(i)} = L_{nom} I_{inrush}^2$, and the maximum inrush current can be determined as in (5.44), where L_{line} can be neglected with $L_{line} \ll L_{nom}$:

$$I_{inrush} = \sqrt{\frac{C}{2 \cdot (L_{line} + L_{nom})}} \cdot \left(\overline{u_{LL(inrush)}} - V_{initial}\right) \approx \sqrt{\frac{C}{2 \cdot L_{nom}}} \cdot \left(\overline{u_{LL(inrush)}} - V_{initial}\right) \tag{5.44}$$

If the inductor can become saturated during the inrush (often the case when a smaller inductor is desired for lower cost), the inductor nonlinear characteristics have to be considered. In this case, the energy stored in the three-phase inductor is

$$E_{L(i)} = \frac{1}{2} B(H_c) \cdot A_c \cdot \left(H_c \cdot l_c + 2 \cdot H_g \cdot l_g\right) \tag{5.45}$$

where A_c is cross-sectional area of the core, l_c is the core magnetic path length, l_g is the air gap length, H_g and H_c are the air gap and core magnetic field intensity, respectively, and $B(H_C)$ is the magnetic flux density in both the air gap and core. For a three-phase core (the same for a single-phase and DC core), l_c can be approximated as:

$$l_c \approx 2(h_t + w_l) \tag{5.46}$$

where h_t and w_l are the core height and window width as illustrated in Figure 5.9.

When the inductor is saturated, $B(H_C)$ nonlinearity should be considered, which can be obtained from the *B-H* curve of the core material. The air gap magnetic field intensity is

$$H_g = \frac{B(H_c)}{\mu_0} \tag{5.47}$$

where μ_0 is permeability of free space.

Applying Ampere's law to yield

$$I_{inrush} = \frac{1}{2 \cdot N_t} \cdot \left(H_c \cdot l_c + 2 \cdot H_g \cdot l_g \right) = \frac{1}{2 \cdot N_t} \cdot \left(H_c \cdot l_c + 2 \cdot \frac{B(H_c)}{\mu_0} \cdot l_g \right) \tag{5.48}$$

Combining (5.43), (5.45), and (5.48), we can obtain maximum H_C and the corresponding maximum inrush current I_{inrush}.

Figure 5.11 shows an example of simulated inrush currents with or without considering inductor core saturation (two cases with the same nominal inductance value). The peak current from the nonlinear inductance model is 879 A compared to 774 A for the linear model. The analytical calculation is 861 A.

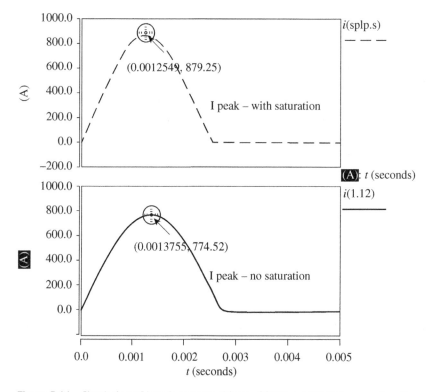

Figure 5.11 Simulation of inrush current with or without considering core saturation.

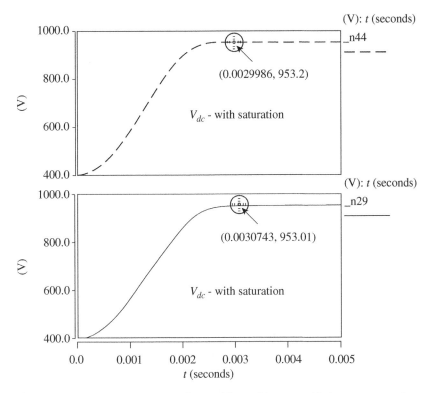

Figure 5.12 Simulation of surge voltage with or without considering core saturation.

Figure 5.12 shows the corresponding simulated results for surge voltage with or without considering inductor saturation. Clearly, the peak surge voltage is independent of saturation, which agrees with (5.42).

If a DC inductor or three single-phase AC inductors are used with a core structure in Figure 5.13, the inrush current calculation can follow the same approach as the three-phase inductor, but the expression will change.

For the unsaturated inductor case, Eq. (5.44) becomes (5.49), where $k = 1$ for DC inductor and $k = 2$ for single-phase AC inductors. L_{nom} can be calculated as in (5.50), which has the same format as (5.41), but l_g and A_c are defined for different core structures.

$$I_{inrush} = \sqrt{\frac{C}{(2L_{line} + k \cdot L_{nom})}} \cdot \left(\overline{u_{LL(inrush)}} - V_{initial}\right) \approx \sqrt{\frac{C}{k \cdot L_{nom}}} \cdot \left(\overline{u_{LL(inrush)}} - V_{initial}\right)$$

(5.49)

$$L_{nom} = L_{DC} = L_{AC} = \frac{\mu_0 N_t^2 A_c}{l_g}$$

(5.50)

For the saturated inductor case, Eqs. (5.45) and (5.48) become (5.51) and (5.52), respectively, which together with (5.43) can yield maximum H_C and the corresponding maximum I_{inrush}:

$$E_{L(i)} = k \cdot \frac{1}{2} B(H_c) \cdot A_c \cdot \left(H_c \cdot l_c + H_g \cdot l_g\right)$$

(5.51)

$$I_{inrush} = \frac{1}{N_t} \cdot \left[H_c \cdot l_c + H_g \cdot l_g\right] = \frac{1}{N_t} \cdot \left[H_c \cdot l_c + \frac{B(H_c)}{\mu_0} \cdot l_g\right]$$

(5.52)

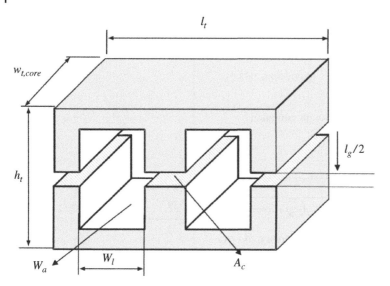

Figure 5.13 DC or single-phase AC core.

5.3.6 Inductor-Related Constraints and Design

The inductor-related physical constraints include: (i) saturation flux density; (ii) mechanical constraint, specifically, the core window size constraint; and (iii) temperature, which is related to materials, loss and cooling. Again, although the inductance value is not a design variable for the passive rectifier, it is an important intermediate parameter for the inductor design. The minimum inductance value has been determined earlier with the TDD constraint. Based on the value, the physical inductor design can be carried out following the design procedure presented in Chapter 4.

5.3.6.1 A_p-Based Design Method

In the inrush current calculation, we already assumed a core for a given type and value of an inductor. Here, we briefly review the inductor design and associated constraints. As discussed in Chapter 4, using the core widow area product A_p method, the relationship (5.53) should be satisfied, where L_{nom} has been determined for different inductor configurations earlier and the peak inductor current I_{pk} is the same for all inductor configurations in a diode rectifier. The calculation of I_{pk} has been discussed in the TDD subsection aforementioned. The winding RMS current $I_{winding(RMS)} = I_{AC(RMS)}$ for the three-phase and three single-phase inductor cases, and $I_{winding(RMS)} = I_{DC(RMS)} \approx \sqrt{\frac{3}{2}} I_{AC(RMS)}$ for the DC inductor case.

$$A_p = q \cdot A_C \cdot W_A \geq q \cdot \frac{L_{nom} \cdot I_{winding(RMS)} \cdot I_{pk}}{K_u \cdot J_{max} \cdot B_{max}} \tag{5.53}$$

In (5.53), the coefficient $q = \frac{3}{2}$ for the three-phase inductor case, and $q = 1$ for the single-phase and DC inductor cases. Maximum current density J_{max}, maximum flux density B_{max} corresponding to I_{pk}, and core window fill factor K_u have been discussed in Chapter 4. Using A_p, we can select core and obtain the associated parameters of the core such as A_C and W_A. Then the winding turns N_t and the air gap length l_g of the inductor can be determined using equations in Chapter 4. The selection of the wire conductor cross-sectional area A_W should satisfy J_{max} constraint, which can be determined based on cooling conditions and A_p as discussed in Chapter 4.

5.3.6.2 Power Loss and Temperature Rise

Methods to calculate inductor losses have been discussed in Chapter 4 and can be followed. Here, we will present one simplified approach for core loss suitable for passive rectifier design optimization.

The core loss for a three-phase inductor can be estimated by variation of Steinmetz's equation:

$$P_{core} = m \cdot k \cdot (n \cdot f)^{\alpha} \cdot \Delta B^{\beta} \cdot Vol_{core} \tag{5.54}$$

where Vol_{core} is the core volume; k, α, and β are the material Steinmetz parameters discussed in Chapter 4; m is the coefficient to account for the core configuration (DC, single-phase, or three-phase); and n is the coefficient to determine an equivalent frequency of the nonsinusoidal current in the inductor, with f being the fundamental frequency of the current.

In the case of passive rectifiers, the core loss calculation can be further simplified. Figure 5.14 shows experimental waveforms of AC phase voltage and current of a three-phase AC passive rectifier. It can be seen that the current waveform can be divided into three regions, a positive region of 120° where the current is positive with two 60° positive half-cycle currents, a negative region of 120° where the current is negative with two 60° negative half-cycle currents, and zero region of 120° where the current is zero with two 60° zero-current periods. As a result, from the standpoint of magnetic field variation, the current waveform can be approximately regarded as sinusoidal with a frequency of three times of the fundamental frequency f. With this approximation, the three-phase inductor core loss can be estimated as (5.55), which corresponds to $m = \dfrac{4}{9}$, $n = 3$, and $\Delta B = \dfrac{\mu_0 \cdot N_t \cdot I_{pk}}{l_g}$.

$$P_{core} = m \cdot k \cdot (n \cdot f)^{\alpha} \cdot \Delta B^{\beta} \cdot Vol_{core} = \frac{4}{9} k \cdot (3f)^{\alpha} \cdot \left(\frac{\mu_0 N_t I_{pk}}{l_g} \right)^{\beta} \cdot Vol_{core} \tag{5.55}$$

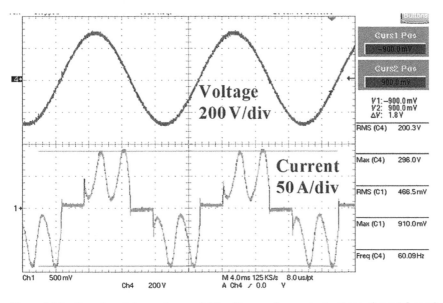

Figure 5.14 Experimental waveforms of AC voltage and current of a three-phase AC passive rectifier.

For the three single-phase AC inductor case, $m = 3$, $n = 3$, $\Delta B = \dfrac{\mu_0 \cdot N_t \cdot I_{pk}}{l_g}$, and the corresponding core loss is

$$P_{core} = 3k \cdot (3f)^{\alpha} \cdot \left(\frac{\mu_0 N_t I_{pk}}{l_g} \right)^{\beta} \cdot Vol_{core} \tag{5.56}$$

For the DC inductor case, $m = 1$, $n = 3$, $\Delta B = B_{AC} = \dfrac{\mu_0 \cdot N_t \cdot I_{pk}}{l_g}$, and the corresponding core loss is

$$P_{core} = k \cdot (3f)^{\alpha} \cdot \left[\frac{\mu_0 N_t (I_{pk} - I_d)}{l_g} \right]^{\beta} \cdot Vol_{core} \tag{5.57}$$

Considering only the DC/low-frequency resistance, the inductor winding loss is

$$P_{winding} = c \cdot I_{RMS}^2 R = c \cdot \frac{\rho \cdot N \cdot Mlt}{A_w} \cdot I_{winding(RMS)}^2 \tag{5.58}$$

where $c = 3$ for the three-phase and three single-phase inductor cases, and $c = 1$ for the DC inductor case; $I_{RMS}^2 R$ term is due to the inductor current and the DC resistance R of the winding; ρ is resistivity of winding material (e.g. copper), and Mlt is the mean length of the winding turns. The winding material resistivity is a strong function of temperature, which will have to be considered. For example, in the case of copper windings, if the maximum winding temperature is 130 °C, its corresponding resistivity $\rho \approx 2.54 \times 10^{-8}$ Ω · m should be used.

Once the power loss is determined, the temperature rise can be determined through thermal calculation. The methods in Chapter 4 and detailed finite element simulation can be adopted. Following the Chapter 4 method (also described in [6]), the equivalent surface area of the inductor A_t will be need. While A_t can be estimated from A_p, it can be also be estimated directly as (5.59), (5.60), and (5.61) for a three-phase inductor, three single-phase inductors, and a DC inductor, respectively.

$$A_t \approx 6 \cdot \left(w_t + w_l \cdot \frac{2 \cdot N_t \cdot A_w}{K_u \cdot W_A} \right) \cdot h_t + 3 \cdot \left(\frac{l_t - 2 \cdot w_l}{3} + w_l \cdot \frac{2 \cdot N_t \cdot A_w}{K_u \cdot W_A} \right) \cdot h_t \quad (\text{cm}^2) \tag{5.59}$$

$$A_t \approx 3 \cdot \left[2 \cdot \left(2 \cdot w_t + w_l \cdot \frac{2 \cdot N_t \cdot A_w}{K_u \cdot W_A} \right) \cdot h_t + \left((l_t - 2 \cdot w_l) + w_l \cdot \frac{2 \cdot N_t \cdot A_w}{K_u \cdot W_A} \right) \cdot h_t \right] \quad (\text{cm}^2) \tag{5.60}$$

$$A_t \approx 2 \cdot \left(2 \cdot w_t + w_l \cdot \frac{2 \cdot N_t \cdot A_w}{K_u \cdot W_A} \right) \cdot h_t + \left((l_t - 2 \cdot w_l) + w_l \cdot \frac{2 \cdot N_t \cdot A_w}{K_u \cdot W_A} \right) \cdot h_t \quad (\text{cm}^2) \tag{5.61}$$

The inductor temperature rise also needs to consider overload condition or other transient conditions, if there are any. First, the total loss during the overload condition, $P_{total(OL)}$, should be determined, which can follow the same procedure of the power loss calculation for the rated condition with the consideration of the overload RMS and peak current. Since the overload duration is short, transient thermal response should be taken into consideration. The temperature rise under the overload condition is

$$\Delta T_{transient} = \overline{T_{inductor(OL)}} - T_a = \Delta T_{rated} + R_{th} \cdot \left(P_{total(OL)} - P_{total} \right) \cdot \left(1 - e^{-\frac{t}{R_{th} \cdot C_{th}}} \right) \tag{5.62}$$

where ΔT_{rated} is the temperature rise at the rated condition. The equivalent thermal capacitance C_{th} can be obtained for different types of inductors, assuming the core is made of steel and winding of copper.

For a three-phase inductor,

$$C_{th} = c_{Fe} \cdot Wt_{core} + 3 \cdot c_{Cu} \cdot \rho_{Cu} \cdot N_t \cdot Mlt \cdot A_W \cdot 10^{-6} \tag{5.63}$$

For three single-phase inductors,

$$C_{th} = 3 \cdot c_{Fe} \cdot Wt_{core} + 3 \cdot c_{Cu} \cdot \rho_{Cu} \cdot N_t \cdot Mlt \cdot A_W \cdot 10^{-6} \tag{5.64}$$

For a DC inductor,

$$C_{th} = c_{Fe} \cdot Wt_{core} + c_{Cu} \cdot \rho_{Cu} \cdot N_t \cdot Mlt \cdot A_W \cdot 10^{-6} \tag{5.65}$$

In (5.63), (5.64), and (5.65), the specific heat capacity of the steel and copper is $c_{Fe} = 460 \, \text{J/(kg} \cdot {}^\circ\text{C)}$ and $c_{Cu} = 380 \, \text{J/(kg} \cdot {}^\circ\text{C)}$, respectively; and their density is $\rho_{Fe} = 7.833 \times 10^3 \, \text{kg/m}^3$ and $\rho_{Cu} = 8.954 \times 10^3 \, \text{kg/m}^3$, respectively. Note that the units of Mlt and A_W are cm and cm^2, respectively.

5.3.7 Capacitor-Related Constraints and Selection

As discussed in Chapter 3, the capacitor is defined by its capacitance value, voltage rating, current, and temperature capability. Capacitors are not custom-designed for general-purpose applications. Instead, commercially available standard capacitors from catalogs are selected. The capacitance values are therefore discrete and should be treated as such during the optimization procedure. As for the DC-link capacitor for the passive rectifier, the capacitance value is influenced by AC and DC current harmonics, stability, ride-through time, and the inrush current, which can be determined through the relationships established earlier in this section.

It should be pointed out that the capacitor voltage needs to consider steady state, inrush transient, and temporary conditions (e.g. during the inverter regenerative operation). The grid conditions, such as lightning and grid switching overvoltages, should also be considered as will be discussed in Chapter 12.

Additional constraints on capacitance can include the minimum capacitance value required by energy storage capability for control performance and AC unbalance conditions, which will be discussed in Chapters 7 and 12.

5.4 Passive Rectifier Design Optimization

Equation (5.1) and Figure 5.3 show design constraints, conditions, and variables for the passive rectifier. Since the diodes and thyristors are relatively low cost and also do not contribute much to the size or the weight of the rectifier, here we focus on the passive components (L and C) optimization for the lowest cost or size, as shown in Figure 5.15, which is a reduced optimization problem, as compared to Figure 5.3. The design constraints that include performance specifications and component physical limits are also a subset of the constraints in Figure 5.3. The design variables are the selection of the actual, commercially available components associated with capacitors and inductors. Note that the DC ripple current impact on the capacitor selection is not included. Instead, a minimum DC capacitance constraint is adopted, e.g. 50 μF/kW, to account for the current and dynamics need.

The diode rectifier is normally designed for the lowest cost or the smallest size. In the case of minimum size design, the volumes of components need to be determined. For the three-phase inductor in Figure 5.9, its volume can be calculated as (5.66), where D_w is the diameter of the wire, and N_{layer} is the number of layers of wires wrapped on the chosen core. The dimensions l_t, w_t, and h_t are indicated in Figure 5.9, which does not show the windings.

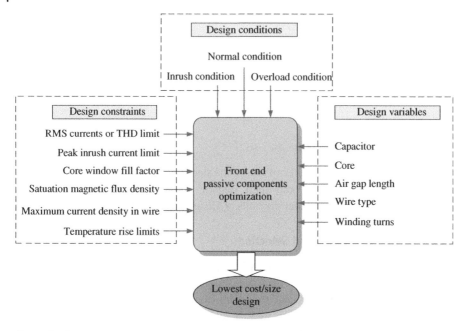

Figure 5.15 Front-end passive components design optimization.

$$Vol_L = \left(l_t + 2 \cdot N_{layer} \cdot D_w\right) \cdot \left(w_t + 2 \cdot N_{layer} \cdot D_w\right) \cdot h_t \tag{5.66}$$

The core window height w_h can be determined through (5.67). N_{layer} can be found through (5.68), where floor (x) returns the largest integer smaller than x. Note that (5.66) is a simplified expression and neglects the effect such as the bobbins and gaps between winding turns.

$$w_h = \frac{W_A}{w_l} \tag{5.67}$$

$$N_{layer} = \text{floor}\left[\frac{N_t}{\text{floor}\left(\frac{w_h}{D_w}\right)}\right] + 1 \tag{5.68}$$

For the DC inductor in Figure 5.13, its window height w_h and winding layer number N_{layer} can still be determined by (5.67) and (5.68), respectively. The inductor volume becomes (5.69).

$$Vol_L = l_t \cdot \left(w_t + 2 \cdot N_{layer} \cdot D_w\right) \cdot h_t \tag{5.69}$$

The single-phase inductor has the same structure as the DC inductor. Therefore, its volume can be determined the same way as the DC inductor. However, for the diode rectifier, three single-phase inductors will be needed, such that the total inductor volume will be

$$Vol_L = 3l_t \cdot \left(w_t + 2 \cdot N_{layer} \cdot D_w\right) \cdot h_t \tag{5.70}$$

The DC-link capacitor volume can be determined as in (5.71), where N_C is the number of capacitors, D_C is the diameter of the capacitor, and h_C is the height of the capacitor. The total volume of the passives is in (5.72). Note that the volume here refers to the total occupied space without considering the layout.

$$Vol_C = N_C \cdot D_C^2 \cdot h_C \tag{5.71}$$

$$Vol_{total} = Vol_L + Vol_C \tag{5.72}$$

An optimizer based on the genetic algorithm [2] provides for a systematic and efficient search of the databases of preferred components, which contain all (datasheet-based) component parameters necessary for evaluating the constraints and objective function. In this case, a genetic algorithm-based discrete variable optimizer called DARWIN [7] was selected to solve the optimization problem. This optimizer is capable of handling discrete variables (such as capacitors, cores, and wires), as well as a small number of continuous variables. Continuous variables are modeled directly by using a single real value, and specifying a lower and upper bound for the variable.

One advantage of using a genetic algorithm is that the result of the optimization process is not just one "optimal" design, but a population of designs. In fact, the optimization tool used in [2] presents the user with a list of the best 10 designs found during a given optimization run. Many of these designs may have objective functions very close to or identical to that of the optimal design. Users can use their engineering judgment to select the design that is most appropriate for a particular application. More discussions on genetic algorithms will be in Chapter 13.

Table 5.4 shows a comparison of a three-phase inductor design through manual and automated optimization design with the design conditions and constraints shown in Table 5.2. Design for the DC inductor case is also shown in the results. The objective function represents the total enclosed

Table 5.4 Design comparison of a three-phase line inductor with manual and automated approaches.

			Optimal design		
	Parameters	Manual design (three-phase)	Three-phase	DC 1	DC 2
Design variables and intermediate results	Capacitance (μF)	2400	2400	2400	2400
	Capacitor volume (cm^3)	1748	1748	1748	1748
	Core (silicon steel)	2-8-AC	EI-3P87/ 6814*4	EI-175H/ 0175_29	EI-175H/ 0175_29
	Winding turns	10	13	17	17
	Air gap l_g (mm)	1.5	1.8	2.1	2.4
	Wire type	AWG1	AWG2	AWG2	AWG2
	Inductance (μH)	155.7	233.1	341.7	299.0
	N_{layer}	3	2	2	2
Design constraints results	B_{rated} (T)	1.51	1.51	1.50	1.39
	I_{rms} (A)	92.4	88.9	84.6	86.7
	I_{pk} (A)	180.3	166.4	147.0	156.7
	I_{inrush} (A)	1643	1339	1665	1669
	J_{rated} (A/mm^2)	2.18	2.64	3.08	3.16
	ΔT (°C)	26.4	59.9	65.3	69.1
	$\Delta T_{transient}$ (°C)	27.0	61.0	66.3	70.1
	Window utilization factor	0.72	0.79	0.58	0.58
Volume of the inductor (cm^3)		4202	2037	1054	1054
Volume of the total passives (cm^3)		5950	3785	2802	2802

volume of the inductor, including the volume taken up by the windings, as calculated using (5.66) for three-phase inductors and (5.69) for DC inductors. Size reduction is achieved through the design optimization. Table 5.5 lists the parameters of the magnetic cores corresponding to the designs in Table 5.4. The optimization does not include the capacitor since the minimum capacitance is set at 2400 µF, which is the same in all design cases. The capacitor selected is the 450 V, 2400 µF aluminum electrolytic capacitor from TDK-EPCOS, with its parameters listed in Table 5.6. To satisfy the voltage and ripple current requirements, four capacitors are needed for series and parallel connections. In general, the optimization can also include capacitor as described earlier and the objective for optimization can also be cost.

Several observations can be made from the design optimization results in Table 5.4:

1) The optimized design results are clearly better than the manual design results, in terms of the inductor volume in this case;
2) All constraints in Table 5.2 are satisfied. The reason for better results achieved by the optimized designs is mainly due to better utilization of design margins for core and windings. For example,

Table 5.5 Magnetic core parameters.

Parameter	2-8-AC	EI-3P87/6814*4	EI-175H/0175_29
Core type	Three-phase AC	Three-phase AC	DC or single-phase
Weight density (g/cm^3)	7.63	7.63	7.63
B_{sat} (T)	1.6	1.6	1.6
Cross-sectional area A_C (cm^2)	18.58	19.76	19.76
Window area W_A (cm^2)	23.23	22.22	14.82
Mean-length-per-turn (MLT) (cm)	23.5	27.78	26.67
Core height h_t (cm)	13.7	11.11	11.11
Core length l_t (cm)	18.3	13.33	13.34
Core width w_t (cm)	9.1	8.88	4.5
Core loss coefficient k	5.05E-04	5.05E-04	5.05E-04
Core loss f coefficient m	1.67	1.67	1.67
Core loss B coefficient n	1.86	1.86	1.86
Material	Silicon steel	Silicon steel	Silicon steel

Table 5.6 DC-link capacitor parameters.

Voltage (V)	Capacitance (µF) (100 Hz, 20 °C)	Diameter (cm)	Height (cm)	ESR typical (mΩ) (100 Hz, 20 °C)
450	2400	6.43	10.57	40
ESR typical (mΩ) (30 Hz, 60 °C)	Z_{max} (mΩ) (10 kHz, 20 °C)	IAC, max (A) (300 Hz, 60 °C)	IAC, max (A) (100 Hz, 85 °C)	IAC, rated (A) (100 Hz, 105 °C)
11	60	31.8	17.9	11.6

the optimized design results all have significantly higher temperature rise and higher current density, as a result of using smaller cores and smaller wires (AWG#2 vs. AWG#1), but still within the capabilities of these components;

3) Using the developed optimization approach, we can trade off the AC inductor vs. the DC inductor. Table 5.4 lists one optimized three-phase AC inductor design and two DC inductor designs for the same application. In this case, the DC inductor solution results in a smaller inductor. Given that the capacitors are the same for both AC and DC cases, the overall passive component size favors the DC inductor solution;

4) Table 5.4 shows two different minimum volume DC inductor designs obtained using the automated optimization procedure. Both designs satisfy all constraints and have the same low volume. They have the same core, winding type, and winding turns. The difference is the air gap length or the equivalent inductance. The designer can choose the design based on other considerations, such as current and voltage behaviors.

It should be pointed out that in the design example results listed in Table 5.4, the winding fill factor was not considered in calculating the number of layers of the wires, N_{layer}, and only the bare copper conductor size was used. If the fill factor and bobbins are considered, the inductor volume will be larger. Another thing to note is that the calculated core window utilization factors (or fill factor) in Table 5.4 are quite large. Practically speaking, a fill factor of <0.5 will be more suitable. From this standpoint, the DC inductor is also preferred because of its more available window area. Each window of the DC indictor core will contain only one winding, while each window of the three-phase AC core will need to accommodate windings of two phases.

It should be mentioned that the results obtained are highly sensitive to the size of the databases. We used relatively small component databases to obtain these results (e.g. we only considered silicon steel cores). The software tool allows the user to add to and edit the component databases. More discussions will be in Chapter 13.

5.5 Interface to Other Subsystem Designs

5.5.1 General Classifications

The rectifier is often a subsystem of the whole converter, with its design and operation affected by other subsystems that directly interface with it. In the case of the motor drive in Figure 5.2, the directly interfaced subsystems with the passive rectifier include the inverter, the thermal management system, the AC-side EMI filter, and the mechanical system. The passive rectifier also has limited interface with the control system mainly for start-up, protection, and related sensing/measurement; and therefore, the control interface discussion will not be included here. The load (e.g. motor) and source (e.g. grid) are not directly connected to the rectifier.

In the earlier discussion, the rectifier is designed, or the models are established, under given interface conditions, which are treated as constants. In the overall converter system design, these conditions may change and the design may need to be iterated depending on the range of the variations for these conditions. In addition, the design also generates output that will be needed for the other subsystems interfaced with the rectifier. Here, we will examine the output needed one by one for the other subsystems.

5.5.1.1 Inverter

The output from the passive rectifier to the load-side inverter include (i) maximum and minimum DC voltages during normal operation and (ii) peak DC voltage during inrush, which will also determine the trip level for the overvoltage protection. They are both related to the design constraints as listed in (5.1). Sometimes, the inverter design also needs to use the DC-link capacitance value, which is an attribute of the design variable, capacitor.

5.5.1.2 EMI Filter

The output of the passive rectifier to the EMI filter include: (i) EMI bare noise, which can be represented by time-domain voltage waveforms, or frequency-domain characteristics. In principle, the EMI noise source characteristics should include both equivalent source and impedance, similar to the case of a Thevenin or Norton equivalent voltage or current source in a linear circuit. However, since the EMI filter design generally follows the impedance mismatch strategy, often, an ideal equivalent voltage source with zero series impedance or an ideal equivalent current source with zero paralleled admittance will be sufficient to represent the EMI noise source; (ii) AC line current information, which is needed for the design of the EMI filters. Although EMI filters are to attenuate high-frequency noise (e.g. 150 kHz or higher for conducted EMI), they have to take line-frequency current and voltage, as well as lower order harmonics. As a result, the design results on current from the rectifier will be needed.

5.5.1.3 Thermal Management System

As explained earlier, the main constraints for the diodes or thyristors in the passive rectifier are surge current and voltage. We have not discussed the device loss and device cooling in this chapter. Since the passive rectifier interface to the thermal management system will be similar to that of the active rectifier or the inverter to the thermal management system, Chapter 6 can be referred to. The device loss in passive rectifiers can follow the models in Chapter 2. In the earlier discussion for the passive components including the inductors and the DC-link capacitors, the temperature characteristics and constraints have been considered in their design and selection.

5.5.1.4 Mechanical System

As can be seen in (5.1), there is no input from the mechanical system to the passive rectifier. This is because there is no fast switching action for the passive rectifier, such that the mechanical system design will not impact the passive rectifier electrical design. In contrast, the passive rectifier will impact the mechanical system design. The output from the passive rectifier to the mechanical system include the dimensions of the components, their weight, as well as the voltage distribution and power loss. The voltage distribution is important as the mechanical system design will need to consider the insulation requirement. Since the mechanical system design is responsible for system layout, the temperature distribution is important and will need loss information to determine. The detailed discussion will be covered in Chapter 11.

5.5.2 Discussion

Table 5.7 summarizes all the output from the passive rectifier to its interfaced subsystems within the overall converter system (e.g. in this case, an AC-fed motor drive). As discussed earlier, some of the outputs are related to the design constraints, while others are attributes to the design variables. The value column indicates how to obtain the value for the particular output. "Design result" indicates that the particular value can be directly obtained as part of the design results. For example, in the case of the constraints, the peak DC voltage during inrush can be determined with (5.42); and

Table 5.7 Output from the passive rectifier to its interfaced subsystems.

Subsystem	Interface type	Item	Value
Inverter	Constraint	Max/min DC voltages during normal operation	Design result
		Peak DC voltage during inrush	Design result
	Attribute	DC-link capacitance	Design result
AC EMI filter	Constraint	AC line current	Design result
	Constraint	DC-link voltage	Design result
	Attribute	EMI bare noise	Discussion
Thermal management system	Attribute	Component loss (device and passives)	Discussion
		Component thermal impedance	Design result
Mechanical system	Constraint	Component dimensions	Design result
	Attribute	Component loss (device and passives)	Discussion
		Voltage distribution	Discussion

in the case of the attributes, the capacitance and the dimensions of the DC-link capacitors can be directly obtained from the capacitor datasheet.

"Discussion" in the value column of Table 5.7 indicates that further discussion is needed on how to obtain this particular output. Three items are further discussed in the following subsections.

5.5.2.1 EMI Bare Noise

Under normal operating conditions, the diodes in the passive rectifier are line-commutated. For a diode rectifier in Figure 5.2, each diode only turns on and off once through a line cycle. In addition, the line commutation takes place under soft switching condition, i.e. zero-voltage turn-on and zero-current turnoff. As a result, the diode rectifier main power circuit itself contributes very little EMI noise. In contrast, the conducted EMI noise will be generated by the load-side inverter in an AC-fed motor drive and passed through the diode rectifier to the AC input side. While the rectifier circuit and components including AC or DC inductors and DC-link capacitors will impact the AC input EMI, the impact is mainly as propagation path. The filter will be more impacted by the load-side inverter. As a result, the EMI bare noise characterization will be discussed in Chapter 8. An example of EMI bare noise and filter design for a motor drive with the front-end diode rectifier will be given in Chapter 13.

5.5.2.2 Component Loss

The main components include diodes or thyristors, inductors, and capacitors. The inductor losses have been discussed in Section 5.3.6. The capacitor loss can be determined from the ripple current and ESR as discussed in Chapter 3. And, the diode or thyristor loss can be determined using the currents calculated in Section 5.3.1 and diode/thyristor loss models. Note for the three-phase passive rectifier, the conduction periods for diodes or thyristors are 120°.

5.5.2.3 Voltage Distribution

The voltage potentials of various terminals of the components are needed for the mechanical system design. They depend on the system grounding. For example, if the DC-link midpoint is grounded, then the positive and negative DC bus potentials are $+V_{DC}/2$ and $-V_{DC}/2$, respectively.

Table 5.8 Voltage differences between buses and between a bus and the ground in a diode rectifier (independent of the reference ground point).

	Positive DC	Negative DC	Phase A	Phase B	Phase C	Ground
Positive DC	0	$+V_{DC}$	$\pm V_{DC}$	$\pm V_{DC}$	$\pm V_{DC}$	$+V_{DC}/2$
Negative DC	$-V_{DC}$	0	$\pm V_{DC}$	$\pm V_{DC}$	$\pm V_{DC}$	$-V_{DC}/2$
Phase A	$\pm V_{DC}$	$\pm V_{DC}$	0	$\pm V_{DC}$	$\pm V_{DC}$	$\pm V_{DC}/2$
Phase B	$\pm V_{DC}$	$\pm V_{DC}$	$\pm V_{DC}$	0	$\pm V_{DC}$	$\pm V_{DC}/2$
Phase C	$\pm V_{DC}$	$\pm V_{DC}$	$\pm V_{DC}$	$\pm V_{DC}$	0	$\pm V_{DC}/2$
Ground	$-V_{DC}/2$	$+V_{DC}/2$	$\pm V_{DC}/2$	$\pm V_{DC}/2$	$\pm V_{DC}/2$	0

Note that both the normal and transient conditions have to be considered, including maximum voltage like peak voltage during inrush. The AC terminals will be connected to the positive and negative DC buses during the rectifier operation, and therefore can be at either $+V_{DC}/2$ or $-V_{DC}/2$. As a result, the insulation between any two buses and between a bus and the ground has to consider all these possibilities. Table 5.8 lists the voltage differences between different buses in a rectifier, which are independent of the reference ground.

5.6 Summary

This chapter presents the power stage design of the passive three-phase AC rectifiers using diodes or thyristors, with AC and/or DC passives.

1) The passive rectifier design is formulated as an optimization problem, with its design variables, constraints, conditions, and objective(s) clearly defined. The design variables include selection and design of diodes or thyristors, AC or DC inductors, and DC-link capacitors, considering steady state as well as transient conditions both on the source and load sides. The design constraints include performance constraints for the rectifier (AC input current harmonics, DC-link voltage ripple, ride through, and DC-link stability) and physical constraints associated with the components (e.g. peak inrush current and voltage, inductor temperature rise, and inductor core window fill factor).

2) The focus is on establishing analytical or algebraic models for design constraints as functions of design conditions and variables. The AC input current harmonics and DC voltage ripples are obtained with the Fourier series based on rectifier switching functions considering the impact of the commutation angle. The peak inrush current and voltage are obtained considering the inductor saturation.

3) Minimum volume objective functions based on design variables and their attributes are derived for passive components in a diode rectifier. Design optimization examples using a genetic algorithm-based design tool have been presented. Comparison with manual design shows that the optimized designs can achieve lower volumes.

4) The output from the passive rectifier design as the interface parameters to other subsystem systems have been discussed, including interface parameters to the load-side inverter, the EMI filter, the thermal management system, and the mechanical system. Most of the interface parameters can be directly obtained from the design results. Others, including voltage distribution of different buses and EMI bare noise, need to be postprocessed.

References

1 F. Wang, W. Shen, D. Boroyevich, S. Ragon, V. Stefanovic and M. Arpilliere, "Voltage source inverter – development of a design optimization tool," *IEEE Industry Applications Magazine*, vol. 15, no. 2, pp. 24–33, March–April 2009.

2 F. Wang, G. Chen, D. Boroyevich, S. Ragon, V. Stefanovic and M. Arpilliere, "Analysis and design optimization of diode front-end rectifier passive components for voltage source inverters," *IEEE Transactions on Power Electronics*, vol. 24, no. 5, pp. 2278–2289, Septemper 2008.

3 M. Sakui, H. Fujita and M. Shioya, "A method for calculating harmonic currents of a three-phase bridge uncontrolled rectifier with DC filter," *IEEE Transactions on Industrial Electronics*, vol. 36, no. 3, pp. 434–440, August 1989.

4 M. Sakui and H. Fujita, "An analytical method for calculating harmonic currents of a three-phase diode-bridge rectifier with DC filter," *IEEE Transactions on Power Electronics*, vol. 9, no. 6, pp. 631–637, November 1991.

5 R. Lai, F. Wang, R. Burgos, Y. Pei, D. Boroyevich, T. A. Lipo, B. Wang, V. Immanuel and K. Karimi, "A systematic topology evaluation methodology for high-density PWM three-phase AC-AC converters," *IEEE Transactions on Power Electronics*, vol. 23, no. 6, pp. 2665–2680, November 2008.

6 C. W. T. McLyman, "*Transformer and Inductor Design Handbook*," Fourth Edition, CRC Press, 2011.

7 G. Soremekun and Z. Gürdal, "Stacking sequence blending of multiple composite laminates using genetic algorithms," in *42nd AIAA/ASME/ASCE/AHS/ASC Structures, Structural Dynamics, and Materials Conference*, Seattle, WA, 2001.

6

Load-side Inverters

6.1 Introduction

This chapter is on the design of three-phase load-side PWM DC/AC inverters. Inverters can be used to serve loads as in the case of motor drives, or to interface with the grid or other sources as in the case of renewable energy and other distributed energy resources (DERs). The design of the three-phase grid-side or source-side inverters will be covered together with that of active rectifiers in the next chapter, since their designs, especially the hardware designs, are more similar. Like three-phase rectifiers, inverters can be stand-alone converters, as in the case of DC-fed motor drives; or they can be part of the more complex converters, as in the case of AC-fed motor drives, connecting to front-end rectifiers.

Whether a three-phase inverter works as a stand-alone converter or works together with a front-end rectifier or other types of converters, it performs a defined function – DC to three-phase AC inversion, and therefore, it can be modeled and designed relatively independently. This chapter discusses the design of the power stage of the three-phase load-side inverter, as a function block or subsystem of the overall three-phase converter. Note that similar to the rectifier, even when the inverter works as a stand-alone converter, the power stage of the inverter is still a subsystem of the whole converter because there are still other subsystems for the whole inverter, such as thermal management system, mechanical system, control, and EMI filters.

From the topology standpoint, there are many different types of three-phase inverters. Many common topologies have been presented in Chapter 1 (see Figures 1.14 and 1.15). In principle, all of them can be and have been used in inverter applications. At present, the basic two-level VSI in Figure 1.14a is the most popular topology, and the motor drive application is the most popular application for load-side inverters. As a result, the two-level VSI motor drive will be mainly used in this chapter as an example to facilitate the design discussion. The same principle and approaches can be applied to other topologies and applications. The impact of topology and applications will be discussed in Chapter 7 and Chapter 12.

6.2 Load-side Inverter Design Problem Formulation

We will again use an AC-fed motor drive as the example to illustrate the load-side inverter design optimization problem. Figure 6.1 shows the circuit configuration of the example AC-fed motor drive, with the front-end rectifier and the two-level VSI. Clearly, in this case, the only main

Design of Three-phase AC Power Electronics Converters, First Edition. Fei "Fred" Wang, Zheyu Zhang, and Ruirui Chen.
© 2024 The Institute of Electrical and Electronics Engineers, Inc. Published 2024 by John Wiley & Sons, Inc.

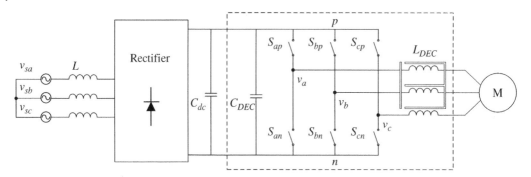

Figure 6.1 Circuit configuration of an AC-fed motor drive with a two-level VSI.

components for the inverter are semiconductor switches. If other topologies are used, there may be passive components; examples include capacitors in many multilevel VSCs in Figure 1.15. For soft-switching topologies, even more complex structures will be involved. Again, to facilitate the discussion, we will focus on the two-level VSI. The impact of the topology will be discussed separately.

As pointed out earlier, the DC-link capacitor is needed for both the rectifier and inverter operation, and its selection needs to consider the requirement of both sides. We have included the selection of the DC-link capacitor as a task in the rectifier design, with consideration of the inverter impact through load-side conditions and constraints.

Figure 6.1 includes a decoupling capacitor, which are often required for medium- and high-power converters to control switching overvoltage due to DC-bus parasitic inductances. Figure 6.1 also includes decoupling inductors, which are not yet popular in today's inverters. They play a counterpart role as the decoupling capacitors on DC link. Their role is to reduce switching overcurrent, therefore device switching loss due to AC cable and load parasitic capacitances. The selection of the decoupling capacitors and inductors is a task in the inverter design.

The switches in inverters are actively controlled, and therefore the inverter design is intimately tied with the control design. As a result, the inverter control design related to switching control will be part of the inverter design, while the controller hardware will be covered in Chapter 10.

Similar to the rectifier case, the thermal management system and the mechanical system design will not be covered in the inverter power stage design. EMI filters, which are sometimes required for load-side inverters, will also not be covered here. There are other application-specific auxiliaries such as the dv/dt filters for motor drives. The design of these functions or subsystems will be considered separately. On the other hand, their interactions with the inverter design are included here through design interfaces, as design input or output.

6.2.1 Load-side Inverter Design Variables

Based on the earlier discussion, the design for the load-side two-level VSI refers to the design and/or selection of switching devices and their control, decoupling capacitors, and decoupling inductors. Therefore, the design variables include the following:

1) Switching devices: their type, rating, and other associated characteristics as described in Chapter 2.
2) Switching scheme: i.e. PWM scheme. Different PWM schemes have different impacts on harmonics, switching and possibly conduction loss, and control performance. CM voltage or current, and EMI will also be impacted.

3) Switching frequency: like PWM scheme, switching frequency also directly impacts harmonics, CM voltage and current, EMI, switching loss, and control performance. Note that switching frequency can also be a design input or condition when the EMI filter is involved.

4) Gate control parameters: assuming a voltage-controlled gate drive, these parameters include gate resistance and gate voltage. Gate resistance in general determines the device switching speed, which in turn has a major impact on switching related behaviors, switching loss, and overvoltage. Gate voltage also impacts switching, conduction, as well as fault (e.g. short-circuit fault) behaviors.

5) Decoupling capacitors: type, capacitance value, voltage and current rating, and other associated characteristics as described in Chapter 3. Decoupling capacitors deal with switching transients and therefore their energy rating is important.

6) Decoupling inductors: like other inductors in power converters, decoupling inductors are usually custom designed. Unlike regular inductors, decoupling inductors deal with switching transients and need only small inductance value, and therefore can be of air-core type. As a result, the design variables are those characteristics associated with air-core inductors, including winding wire type and number of winding turns. As explained before, inductance value L is not a final design variable, but it is an important intermediate variable.

7) Thermal management system: in this case, it specifically refers to the thermal impedance of the cooling system. Since we treat the thermal management system as a separate subsystem, its detailed design and selection are not included in this chapter.

Mechanical system is also related to the load-side inverter power stage design, and it must be able to accommodate the mechanical and electrical needs of the inverter. Similar to the thermal management system, the mechanical system is also treated as a separate subsystem, and its detailed design will be covered separately. In addition, the impact of the inverter design on the mechanical system includes the device and component loss, as well as some other design attributes, which will be discussed later in Section 6.5.

6.2.2 Load-side Inverter Design Constraints

The load-side inverter design constraints include performance constraints related to performance specifications and physical constraints related to component physical capabilities. For a motor drive application, the common performance specifications related to the inverter include

1) Output power quality performance under normal operating conditions, specifically AC harmonic current performance in terms of THD. Note that the current THD is a function of the load, so a more reasonable performance requirement independent of loads should be voltage harmonics or voltage THD. However, design examples often use current THD at the rated condition, since the load is often specified for a load-side inverter design, e.g. in the case of an inverter for EV applications. In other cases, such as general-purpose industrial drives, motor loads are not specified. Even in this case, motor parameters are generally within a limited range on per unit basis and a representative motor load is often sufficient. In addition to harmonics, motor loads are also sensitive to negative sequence currents, which can cause undesirable torque ripple and loss in motors. DC offset also needs to be limited to avoid saturation, although saturation is relatively less a concern in motors than in transformers due to the existence of relatively large airgaps.

2) Power loss, mainly the conduction and switching losses of the semiconductor devices in the case of two-level VSI. This constraint often is expressed in terms of efficiency.

3) Control performance, including both the steady-state and dynamic performance. The steady-state performance includes regulation errors, e.g. speed regulation and torque (or current) regulation errors. For digital controllers, the steady-state errors are determined by digital controllers and sensors. The selection of digital controllers and sensors will not impact inverter power stage much and will be discussed in Chapter 10. The dynamic performance is often characterized through control bandwidth, which is directly related to switching frequency of the inverter. Another parameter that can impact load dynamics is the voltage margin. For example, sufficient voltage margin is needed to achieve a desired torque or current rate of change for a motor load at the rated voltage.

As discussed in Chapter 5, motor drives generally should comply with EMI/EMC standards. The requirements can be both for the source and load sides. Again, since the EMI filter is treated as a separate subsystem, the corresponding EMI requirement will not be included as a performance constraint for the inverter.

CM voltage or current is often a concern for inverter loads. For motor loads, CM voltage can cause shaft voltage and bearing current, and therefore may need to be limited. Another limiting factor for loads like motor is dv/dt, which can impact motor insulation. Both the CM voltage/current and dv/dt limiter or filter will be discussed in Chapter 12.

The physical constraints are those associated with components for the inverter, including switching devices, DC decoupling capacitor, and AC decoupling inductors. The physical constraints for switching devices are parameters already described in Chapter 2, including (i) device temperature limits, especially the maximum permissible junction temperature T_{jmax}; (ii) device voltage limits; and (iii) device current limits.

The physical constraints for DC decoupling capacitor are some of the parameters already described in Chapter 3, including (i) capacitor temperature limits, especially the maximum permissible hotspot temperature; (ii) capacitor voltage limits, both continuous and transient voltage limits; and (iii) capacitor ripple current limits.

Similar to other custom-designed AC line inductors, the physical constraints for AC decoupling inductors are associated with components used in the inductors. As explained in Chapters 4 and 5, they include saturation flux density (if magnetic cores are used), temperature limits set mainly due to winding insulation and core, and mechanical limits like core window area. As mentioned already, AC decoupling inductors generally only need low inductance and therefore can be realized with air-core inductors, which will be discussed in detail later in this chapter.

It should be noted that the reliability and lifetime should be important design constraints for any power electronics converters, including inverters. They will be discussed in Chapter 7 since the requirements can be treated the same for both inverters and rectifiers.

6.2.3 Load-side Inverter Design Conditions

A load-side inverter interfaces with the DC link as its source, the three-phase AC load such as a motor, the control, and the thermal and mechanical environment. Design conditions are boundary interface conditions, including source and load electrical conditions, and environmental conditions. For load-side inverters, the DC side electrical conditions include

1) Minimum DC voltage, including minimum voltages for both normal conditions and abnormal conditions. Worst-case normal conditions include rated and overload conditions. Abnormal conditions refer to low-voltage conditions as a result of fault or disturbance, e.g. DC voltage sag as a result of source disturbance for the AC-fed motor drive case.
2) Peak DC voltage, also including voltages under normal and abnormal conditions. Normal conditions include rated load conditions, as well as normal regenerative conditions, e.g. normal

motor braking conditions. Abnormal conditions refer to overvoltage as a result of grid-side fault or disturbance. In a rectifier-fed case like in Figure 6.1, grid-side conditions that can cause DC overvoltage include lightning, switching, etc. Lightning, switching, and other transient conditions can cause inrush, and corresponding overvoltage and/or overcurrent.

The AC load electrical conditions include

1) Load power, including nominal power, overload power, and peak power. As explained in Chapter 5, overload power should include both power level and time duration, e.g. 150% overload for 60 seconds. Peak power refers to the maximum transient load condition such as peak torque for motors. In practice, the peak power may not be a design condition and can be a design result. For example, in the case of a general-purpose motor drive, the peak power is often determined by the peak current capability of the inverter. However, in some other applications such as in EV applications, there can indeed be a peak torque or peak power requirement. Note also that the load power can also have a negative rating in addition to its positive rating, depending on applications. The negative rating is applicable to inverters with regenerative capability.
2) Maximum regenerative energy, which, as explained in Chapter 5, can be tied to the requirement on how fast to stop a motor in motor drive applications. This is more relevant in a nonregenerative inverter application, such as a motor drive with diode-front-end rectifier and a dynamic brake.
3) Load power factor, which is part of the load characteristics.
4) Load efficiency, which, together with the power factor, is important to determine the true kVA rating needed for the inverter. For example, the motor power generally refers to the mechanical output power. The true kVA rating S can be determined by output power P divided by the motor efficiency η and power factor PF, i.e. $S = P/(\eta \cdot PF)$.
5) Load voltage magnitude and frequency range.
6) Load impedance, including both the low frequency and high frequency impedances. The low frequency covers the range significantly below the switching frequency and will impact harmonics and control. The high frequency impedance represents the load characteristics around and above the switching frequency, and can impact converter switching behaviors and EMI. It should be pointed out that in many applications, including motor drives, connection wires or cables are used. The wire and cable impedance characteristics also need to be considered. Sometimes, the load, like a motor, may need to be modeled by more than impedances (e.g. back EMF), so this item becomes load model.

Since the thermal management system is designed as a separate subsystem, its interface condition is needed for the inverter power stage. The interface condition is the ambient temperature T_a, which is needed to determine the thermal impedance of the thermal management system as will be explained in Section 6.3.4.

The mechanical system design is also treated separately. Therefore, there needs to be an interface condition between the inverter power stage and the mechanical system. In this case, we focus on the item with significant impact on the electrical performance of the inverter, which is the busbar. The busbar parasitic parameters, mainly parasitic or stray inductances, will be the input conditions of the mechanical system. Parasitic capacitances from busbars and other part of the converters can also influence switching losses and EMI noise, and in general should also be considered.

EMI filters may be required for the load-side inverter on both the DC and AC side. The EMI filter design is closely related to the inverter design, especially the switching waveforms. Very often, as will be explained in Chapter 8, the switching frequency is better to be selected based on the EMI filter requirement. As a result, the switching frequency is considered as an input to the inverter design, when there is an EMI filter for the inverter on DC and/or AC side. If there is no EMI filter for the converter, then the switching frequency should be an inverter design variable as discussed earlier.

Different from the passive rectifier power stage in Chapter 5, the inverter is actively controlled and therefore strongly influenced by the control subsystem or controllers. From the inverter design standpoint, the controller components and their parameters should be designed and selected based on the inverter performance requirements. For example, the gate circuit parameters (i.e. gate resistance R_g and gate voltages V_{GH}/V_{GL}) are selected to meet requirements for device switching speed and on-state voltage drop, which determine the device switching loss, switching overvoltage, and conduction loss. Therefore, R_g and V_{GH}/V_{GL} are design variables for the inverter. On the other hand, the input of the gate driver to the inverter power stage design should be its actual characteristics, including drive circuit topology like regulated voltage source with gate voltages V_{GH}/V_{GL} and gate resistance R_g, as well as gate driver internal impedance and parasitic gate loop inductances. Similarly, sensors, isolated power supplies, protection, and controller hardware can all impact inverter performance and could be design variables. In three-phase medium- and high-power converters, since these control components are relatively minor contributors to the design objective functions (e.g. cost, size, weight, or efficiency), their design and selection are rarely traded off with other components. As a result, practically, the needed sensors, power supplies, protection, and controller hardware, and their parameters/settings can be determined after the design results of other inverter design variables to simplify the design optimization. For example, based on the topology and voltage/current range, we can determine sensor number, type, and range; based on the switching frequency, we can select power supply ratings and digital controller hardware capability. The impact of practical characteristics of control components on inverter performance will be ignored here and will be discussed in Chapter 10.

6.2.4 Load-side Inverter Design Objectives and Design Problem Formulation

The design objectives for each subsystem of an AC converter, including the load-side inverter, should follow those of the whole converter. If the design objective for the target motor drive is minimum cost, the inverter design objective should also be associated with cost.

Based on the selected design objective, design variables, design constraints, and design conditions, the overall load-side inverter design problem can be formulated as the optimization problem (6.1). Note that in (6.1), thermal management system is abbreviated as TMS. Note also V_{GH}/V_{GL} are used to represent positive and negative gate voltages here, which are the same as V_{CC}/V_{EE} in other chapters.

Minimize or maximize : $F(X)$ – objective function

Subject to constraints:

$$
G(X,U) = \begin{bmatrix} g_1(X,U) \\ g_2(X,U) \\ g_3(X,U) \\ g_4(X,U) \\ g_5(X,U) \\ g_6(X,U) \\ g_7(X,U) \\ g_8(X,U) \\ g_9(X,U) \\ g_{10}(X,U) \\ g_{11}(X,U) \end{bmatrix} = \begin{bmatrix} \text{Load current TDD} \\ \text{Power loss} \\ -\text{Control bandwidth} \\ -\text{Voltage margin} \\ \text{Device junction temperature} \\ \text{Device peak voltage} \\ \text{Device maximum current} \\ \text{Decoupling capacitor voltage} \\ \text{Decoupling capacitor temperature} \\ \text{Decoupling capacitor current} \\ \text{Decoupling inductor temperature} \end{bmatrix} \leq \begin{bmatrix} \text{TDD limit} \\ \text{Loss Limit} \\ -\text{minimum bandwidth} \\ -\text{minimum margin} \\ T_{jmax} \\ V_{limit} \\ I_{limit} \\ V_{CAP} \\ T_{CAP} \\ I_{CAP} \\ T_{inductor} \end{bmatrix}
$$

where design variables and design conditions:

$$
X = \begin{bmatrix} x_1 \\ x_2 \\ x_3 \\ x_4 \\ x_5 \\ x_6 \\ x_7 \\ x_8 \end{bmatrix} = \begin{bmatrix} \text{Devices} \\ \text{PWM scheme} \\ \text{Switching frequency } f_{SW} \text{ (if no EMI filter)} \\ \text{Gate resistance } R_g \\ \text{Gate voltage } V_{GH}/V_{GL} \\ \text{Decoupling capacitors} \\ \text{Decoupling inductors} \\ \text{Maximum TMS thermal impedance} \end{bmatrix}
$$

$$
U = \begin{bmatrix} u_1 \\ u_2 \\ u_3 \\ u_4 \\ u_5 \\ u_6 \\ u_7 \\ u_8 \\ u_9 \\ u_{10} \\ u_{11} \\ u_{12} \end{bmatrix} = \begin{bmatrix} \text{Mimimum DC voltage} \\ \text{Peak DC voltage} \\ \text{Load power} \\ \text{Maximum regenerative energy} \\ \text{Load power factor} \\ \text{Load efficiency} \\ \text{Load voltage magnitude and frequency range} \\ \text{Load model (impedance etc.)} \\ \text{Ambient temperature } T_a \\ \text{Busbar and connector parasitics} \\ \text{Switching frequency } f_{SW} \text{ (if with EMI filter)} \\ \text{Gate drive charatcristics} \end{bmatrix}
\tag{6.1}
$$

The inverter design optimization problem can be illustrated also as in Figure 6.2 [1]. Note that several of the design variables will be of discrete types, including devices, PWM scheme, and decoupling capacitors. The corresponding optimization problem will be a mixed-integer type, which will be discussed in more detail in Chapter 13.

6.3 Load-side Inverter Models

Based on (6.1) and Figure 6.2, we need the following models for the constraints as functions of design conditions and design variables:

1) AC load harmonic current THD
2) Power loss, mainly the device losses
3) Control performance, including control bandwidth and voltage margin
4) Device maximum junction temperature, which will lead to maximum thermal impedance requirement for the thermal management system
5) Device peak voltage
6) Decoupling capacitor parameters: capacitance, energy (current), and peak voltage
7) Decoupling inductor parameters: inductance and current

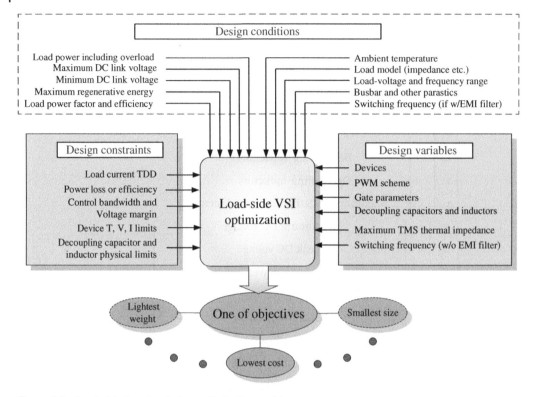

Figure 6.2 Load-side inverter design optimization problem.

This section will focus on how to establish these models. Again, we will focus on models in analytical or algebraic forms, whether they are physics-based or behavioral models. The simulation-based approach will be discussed as needed.

Note that the extreme overcurrent for devices that could hit current limit should only occur during fault conditions and therefore will not be modeled here. The overcurrent protection will be covered in Chapter 10.

6.3.1 AC Load Harmonic Current

The AC load can be a source of harmonics by itself. For example, the AC motor itself may produce harmonics as a result of winding distributions, winding slots, air gap, magnetic field saturation, and in the case of the synchronous motors, nonsinusoidal back-EMF. However, in general, a well-designed motor should have negligible self-generated harmonics. As a result, for the inverter design, we only focus on the current harmonics generated by the load-side inverter. In this case, the current harmonics of the AC load, e.g. a motor, are determined by the inverter AC voltage harmonics and the load (i.e. the motor) impedance. Since the load impedance is an input condition for the inverter design, the task to establish AC load harmonic current model is to determine the inverter AC voltage harmonics.

From the AC-fed motor drive case in Figure 6.1, it can be seen that if the input three-phase AC voltage and the output three-phase AC load are known, with the circuit components and control algorithms (i.e. PWM scheme) selected, the load-side AC voltage and current harmonics can be easily obtained through time-domain simulation. If the inverter is to be designed independently as a subsystem as in our case, then the DC voltage condition needs to be known.

Similar to the rectifier design, in the inverter design, it is desirable to avoid time-domain simulation based on detailed switching models. On the other hand, other modeling approaches such as the average model that is widely used for control design may not be suitable. For example, the average model neglects the switching harmonics, which are essential part of the AC harmonics. One alternative is the closed-form solution, and the other is the time-domain algebraic steady-state calculation, similar to the passive rectifier case in Chapter 5. Both of these approaches use ideal switching function to represent the actual switches and are more suitable for design optimization.

Figure 6.3 illustrates a switching network representation of the two-level VSI, where switching function S_{jk} is defined as in (6.2), with the constraint of (6.3).

$$S_{jk} = \begin{cases} 1, & \text{phase } j \text{ bus } k \text{ switch closed} \\ 0, & \text{phase } j \text{ bus } k \text{ switch open} \end{cases} \quad j \in \{a, b, c\} \text{ and } k \in \{p, n\} \tag{6.2}$$

$$S_{jp} + S_{jn} = 1 \tag{6.3}$$

Defining phase switching functions as in (6.4), the VSI AC line-to-line voltage can be expressed as in (6.5).

$$\begin{bmatrix} S_a \\ S_b \\ S_c \end{bmatrix} = \begin{bmatrix} S_{ap} - S_{an} \\ S_{bp} - S_{bn} \\ S_{cp} - S_{cn} \end{bmatrix} \tag{6.4}$$

$$\begin{bmatrix} v_{ab} \\ v_{bc} \\ v_{ca} \end{bmatrix} = \begin{bmatrix} v_a - v_b \\ v_b - v_c \\ v_c - v_a \end{bmatrix} = \begin{bmatrix} S_a - S_b \\ S_b - S_c \\ S_c - S_a \end{bmatrix} v_{DC} \tag{6.5}$$

Given that the DC-link voltage v_{DC} is an input for the VSI design, the AC line voltages and their harmonics can be determined if the switching functions are known. In general, v_{DC} in a VSI is the DC-link voltage from the front-end rectifier, and therefore will contain ripples. However, as discussed in Chapter 5 and will again be discussed in Chapter 7, the ripple contents of v_{DC} are generally small and their impact on AC side harmonics are small. Under the assumption that v_{DC} only has DC component V_{DC}, for given PMW algorithms, the AC side voltages and their harmonic contents can generally be determined analytically in closed form.

One well-known approach for closed-form harmonic solutions for PWM inverters are based on double Fourier series or double Fourier analysis (DFA) [2] as shown in (6.6), which describes the

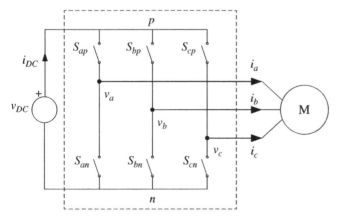

Figure 6.3 Switching function representation of the two-level VSI.

result of a reference variable (e.g. a sinewave voltage) with fundamental angular frequency ω_0 and initial angle θ_0 modulated by a PWM carrier with angular frequency ω_c and initial angle θ_c. The coefficients A_{mn} and B_{mn} can generally be obtained as in (6.7).

$$f(t) = \frac{A_{00}}{2} + \sum_{n=1}^{\infty} [A_{0n} \cos n(\omega_0 t + \theta_0) + B_{0n} \sin n(\omega_0 t + \theta_0)]$$

$$+ \sum_{m=1}^{\infty} [A_{m0} \cos m(\omega_c t + \theta_c) + B_{m0} \sin m(\omega_c t + \theta_c)]$$

$$+ \sum_{m=1}^{\infty} \sum_{\substack{n=-\infty \\ n \neq 0}}^{\infty} \{A_{mn} \cos[m(\omega_c t + \theta_c) + n(\omega_0 t + \theta_0)] + B_{mn} \sin[m(\omega_c t + \theta_c) + n(\omega_0 t + \theta_0)]\}$$

(6.6)

$$A_{mn} = \frac{1}{2\pi^2} \int_{-\pi}^{\pi} \int_{-\pi}^{\pi} f(\omega_c \tau + \theta_c, \omega_0 t + \theta_0) \cos[m(\omega_c \tau + \theta_c) + n(\omega_0 t + \theta_0)] d(\omega_c \tau + \theta_c) d(\omega_0 t + \theta_0)$$

$$B_{mn} = \frac{1}{2\pi^2} \int_{-\pi}^{\pi} \int_{-\pi}^{\pi} f(\omega_c \tau + \theta_c, \omega_0 t + \theta_0) \sin[m(\omega_c \tau + \theta_c) + n(\omega_0 t + \theta_0)] d(\omega_c \tau + \theta_c) d(\omega_0 t + \theta_0)$$

(6.7)

When Eq. (6.6) is used to represent an AC voltage for a three-phase inverter, the coefficients A_{mn} and B_{mn} in (6.7) are determined by PWM schemes and sampling methods. In [2], it is shown that for naturally sampled sine-triangle PWM (SPWM), the line-to-line voltage v_{ab} can be found as in (6.8); for asymmetrical regular sampled PWM, v_{ab} can be found as in (6.9), where $q = m + n(\omega_0/\omega_c)$. Note in (6.8) and (6.9), M is the modulation index, and J_n is the nth-order Bessel function of the first kind. There can be different definitions for modulation index. In this book, it is defined as in (6.10), where V_{ph-M} is the magnitude or peak value of the fundamental frequency reference phase voltage. For carrier-based PWM without third-order harmonic injection, M is between 0 and 1; for carrier-based PWM with third-order harmonic injection or for space vector modulation (SVM), M can be greater than 1.

$$v_{ab}(t) = \frac{\sqrt{3}}{2} V_{DC} M \cos\left(\omega_0 t + \frac{\pi}{6}\right) + \frac{4V_{DC}}{\pi} \sum_{m=1}^{\infty} \sum_{n=-\infty}^{\infty} \frac{1}{m} J_n\left(m\frac{\pi}{2}M\right) \sin\left[(m+n)\frac{\pi}{2}\right] \sin n\frac{\pi}{3}$$

$$\times \cos\left[m\omega_c t + n\left(\omega_0 t - \frac{\pi}{3}\right) + \frac{\pi}{2}\right]$$

(6.8)

$$v_{ab}(t) = \frac{4V_{DC}}{\pi} \sum_{n=1}^{\infty} \frac{\omega_c}{n\omega_0} J_n\left(\frac{n\omega_0}{\omega_c}\frac{\pi}{2}M\right) \sin n\frac{\pi}{2} \sin n\frac{\pi}{3} \times \cos\left[n\left(\omega_0 t - \frac{\pi}{3}\right) + \frac{\pi}{2}\right]$$

$$+ \frac{4V_{DC}}{\pi} \sum_{m=1}^{\infty} \sum_{n=-\infty}^{\infty} \frac{1}{q} J_n\left(q\frac{\pi}{2}M\right) \sin\left[(m+n)\frac{\pi}{2}\right] \sin n\frac{\pi}{3} \times \cos\left[m\omega_c t + n\left(\omega_0 t - \frac{\pi}{3}\right) + \frac{\pi}{2}\right]$$

(6.9)

$$M = \frac{2V_{ph-M}}{V_{DC}}$$

(6.10)

In addition to avoiding time-domain circuit simulation, the closed-form solutions can provide insights into harmonic characteristics. From Eqs. (6.8) and (6.9), several observations can be made about three-phase inverter AC harmonics. There are no even order baseband harmonics due to lack of baseband harmonics terms in (6.8) and $\sin n\dfrac{\pi}{2}$ term in (6.9). There are also no even order sideband harmonics due to the $\sin\left[(m+n)\dfrac{\pi}{2}\right]$ term in (6.8) and (6.9). There are no triplen (i.e. n is a multiple of 3) or zero-sequence baseband or sideband harmonics due to the $\sin n\dfrac{\pi}{3}$ term. Because of the $\sin\left[(m+n)\dfrac{\pi}{2}\right]$ term, only the sideband harmonics with odd combination of $(m+n)$ exist. In other words, the most important line-to-line voltage sideband harmonics are the even order sideband harmonics of the carrier frequency, such as those at $(\omega_c \pm 2\omega_0)$, $(\omega_c \pm 4\omega_0)$, etc.

One drawback with using Eqs. (6.8) and (6.9) based on DFA is their complexity, involving complicated and computationally time-consuming Bessel functions. Another approach is to use single Fourier series or single Fourier analysis (SFA) proposed in [3]. In this case, the phase voltage v_a is expressed as (6.11). The coefficients A_k and B_k can be obtained through (6.12). Note that the original half-wave symmetry assumption and limitation to include only odd harmonics in [3] are not needed here. As a result, the method described here is more general.

$$v_a = f(\omega_0 t) = \sum_{k=0}^{\infty}[A_k \cos k\omega_0 t + B_k \sin k\omega_0 t] \tag{6.11}$$

$$A_k = \frac{2}{\pi}\int_0^{\pi} f(\omega_0 t)\cos(k\omega_0 t)d(\omega_0 t)$$
$$B_k = \frac{2}{\pi}\int_0^{\pi} f(\omega_0 t)\sin(k\omega_0 t)d(\omega_0 t) \tag{6.12}$$

For asymmetrical regular sampled SPWM, the coefficients in (6.12) can be found through Figure 6.4, which covers one switching or carrier wave cycle T_c, consisting of two equal subintervals: $(2n-1)$th and $(2n)$th subintervals (i.e. each subinterval equal to half of the carrier or switching period). The lines y_{2n-1} and y_{2n} are carrier waves for $(2n-1)$th and $(2n)$th subintervals, respectively; $v_{r(2n-1)}$ and $v_{r(2n)}$ are phase reference voltage values for $(2n-1)$th and $(2n)$th subintervals, respectively, per asymmetrical sampled scheme; and the reference and carrier wave intersection times t_{2n-1} and t_{2n} are switching times for $(2n-1)$th and $(2n)$th subintervals, respectively. During $(2n-1)$th and $(2n)$th subintervals, if $v_{r(2n-1)} > y_{2n-1}$ or $v_{r(2n)} > y_{2n}$, then the switching function will be 1 and $f(\omega_0 t) = V_{DC}$ for a two-level VSI; otherwise, the switching function will

Figure 6.4 Principle of switching angle generation process.

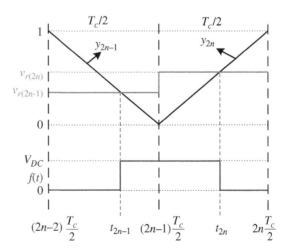

be 0, and $f(\omega_0 t) = 0$. Here, the negative DC bus potential is selected to be zero to simplify the mathematical relations, which will not impact the line-to-line voltage.

Using the relations illustrated in Figure 6.4, Eq. (6.12) can be simplified to Eq. (6.13), where $\alpha_{2n-1} = \omega_0 t_{2n-1}$ and $\alpha_{2n} = \omega_0 t_{2n}$ are the switching angles of the $(2n-1)$th and $(2n)$th switching subintervals, and N is the total number of subintervals in half of the fundamental cycle. Note N is an even integer, implying the carrier frequency ω_c is an integer multiple of the fundamental frequency ω_0, which is of course an approximation in many cases, but a generally acceptable approximation when ω_c is much larger than ω_0. In practice, when ω_c is not much larger than ω_0 in some high-power inverters, ω_c is indeed often selected as an integer multiple of ω_0, for the sake of reduced harmonics.

$$A_k = \frac{1}{\pi}\int_0^{2\pi} f(\omega_0 t)\cos[k\omega_0 t]d(\omega_0 t) = \frac{1}{\pi}\sum_{n=1}^{N}\int_{\omega_0 2(n-1)\frac{T_c}{2}}^{\omega_0 2n\frac{T_c}{2}} f(\omega_0 t)\cos[k\omega_0 t]d(\omega_0 t)$$

$$= \frac{1}{\pi}\sum_{n=1}^{N}\int_{\omega_0 t_{2n-1}=\alpha_{2n-1}}^{\omega_0 t_{2n}=\alpha_{2n}} V_{DC}\cos[k\omega_0 t]d(\omega_0 t) = \frac{V_{DC}}{k\pi}\sum_{n=1}^{N}(\sin k\alpha_{2n} - \sin k\alpha_{2n-1})$$

$$\hspace{8cm}(6.13)$$

$$B_k = \frac{1}{\pi}\int_0^{2\pi} f(\omega_0 t)\sin[k\omega_0 t]d(\omega_0 t) = \frac{1}{\pi}\sum_{n=1}^{N}\int_{\omega_0 2(n-1)\frac{T_c}{2}}^{\omega_0 2n\frac{T_c}{2}} f(\omega_0 t)\sin[k\omega_0 t]d(\omega_0 t)$$

$$= \frac{1}{\pi}\sum_{n=1}^{N}\int_{\omega_0 t_{2n-1}=\alpha_{2n-1}}^{\omega_0 t_{2n}=\alpha_{2n}} V_{DC}\sin[k\omega_0 t]d(\omega_0 t) = \frac{V_{DC}}{k\pi}\sum_{n=1}^{N}(\cos k\alpha_{2n-1} - \cos k\alpha_{2n})$$

For SPWM, the phase a reference voltage is $V_{DC}\dfrac{(1 + M\sin\omega_0 t)}{2}$, where M is the modulation index defined in (6.10). For asymmetrical regular sampled PWM, the reference value for the $(2n-1)$th subinterval in Figure 6.4 is $V_{DC}\dfrac{\left(1 + M\sin\omega_0(2n-1)\frac{T_c}{2}\right)}{2}$, and for the $(2n)$th subinterval is $V_{DC}\dfrac{\left(1 + M\sin\omega_0 2n\frac{T_c}{2}\right)}{2}$. Using these relationships and Figure 6.4, the switching times t_{2n-1} and t_{2n} can be found as in (6.14). Accordingly, the switching angles α_{2n-1} and α_{2n} can be obtained.

$$t_{2n-1} = \left(2n-2+1-v_{r2(n-1)}\right)\frac{T_c}{2} = \left\{2n-2+1-\frac{1}{2}\left[1 + M\sin\left(\omega_0(2n-1)\frac{T_c}{2}\right)\right]\right\}\frac{T_c}{2}$$

$$= \left[2n-1-\frac{1 + M\sin\left(\omega_0(2n-1)\frac{T_c}{2}\right)}{2}\right]\frac{T_c}{2}$$

$$t_{2n} = (2n-1+v_{r2n})\frac{T_c}{2} = \left[2n-1+\frac{1 + M\sin\left(\omega_0 2n\frac{T_c}{2}\right)}{2}\right]\frac{T_c}{2}$$

$$\hspace{8cm}(6.14)$$

From Eq. (6.13), the harmonics can be evaluated for phase a. Similarly, the harmonics for other phases can be obtained. For example, phase b voltage can be in the format of (6.11), and its coefficients for asymmetrical regular sampled SPWM will be as in (6.15), where the switching angles β_{2n-1} and β_{2n} can be found similarly as α_{2n-1} and α_{2n}, with the expression in (6.16).

$$A_k = \frac{V_{DC}}{k\pi} \sum_{n=1}^{N} (\sin k\beta_{2n} - \sin k\beta_{2n-1})$$

$$B_k = \frac{V_{DC}}{k\pi} \sum_{n=1}^{N} (\cos k\beta_{2n-1} - \cos k\beta_{2n})$$

(6.15)

$$\beta_{2n-1} = \left[2n - 1 - \frac{1 + M\sin\left(\omega_0(2n-1)\frac{T_c}{2} - \frac{2\pi}{3}\right)}{2} \right] \frac{\omega_0 T_c}{2}$$

$$\beta_{2n} = \left[2n - 1 + \frac{1 + M\sin\left(\omega_0 2n\frac{T_c}{2} - \frac{2\pi}{3}\right)}{2} \right] \frac{\omega_0 T_c}{2}$$

(6.16)

The line-to-line voltage v_{ab} can be expressed as Fourier series by (6.17), and the coefficients can be determined with (6.18).

$$v_{ab} = v_a - v_b = \sum_{k=0}^{\infty} \left[A_{ab(k)} \cos k\omega_0 t + B_{ab(k)} \sin k\omega_0 t \right]$$

(6.17)

$$A_{ab(k)} = \frac{V_{DC}}{k\pi} \sum_{n=1}^{N} (\sin k\alpha_{2n} - \sin k\beta_{2n} - \sin k\alpha_{2n-1} + \sin k\beta_{2n-1})$$

$$B_{ab(k)} = \frac{V_{DC}}{k\pi} \sum_{n=1}^{N} (\cos k\alpha_{2n-1} - \cos k\beta_{2n-1} - \cos k\alpha_{2n} + \cos k\beta_{2n})$$

(6.18)

The amplitude of the kth harmonic for v_{ab} can be found as in (6.19). The THD up to the K-th order harmonic can be found through (6.20).

$$C_{ab(k)} = \sqrt{A_{ab(k)}^2 + B_{ab(k)}^2}$$

(6.19)

$$THD_{Vab} = \frac{1}{C_{ab1}} \sqrt{\sum_{k=2}^{K} C_{ab(k)}^2}$$

(6.20)

For other PWM schemes, similar approaches can be used to determine voltage harmonics and THD. The sampling method determines the formula for switching times and switching angles. Their values are determined by the phase PWM reference voltages, which are determined by PWM schemes. In general, the PWM reference voltage is a nonsinusoidal periodic function, as a result of sinusoidal fundamental voltage plus an equivalent third-order harmonic injection. Possibly, there will also be a DC offset depending on the zero voltage reference point selection as shown in the SPWM example above with the negative DC bus as the zero voltage reference point. However, there will be no DC offset in line-to-line voltages.

For the popular SVM, Figure 6.5 shows the space vector diagram for the two-level VSC. In this book, the space vector is defined for the phase voltages as in (6.21), where the operator $\vec{a}$ is defined in (6.22). With this definition, the space vector of a symmetric three-phase voltage will have its

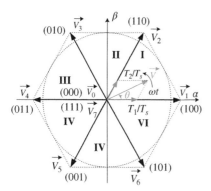

Figure 6.5 Space vector diagram for the two-level VSC.

magnitude $\left\|\vec{V}\right\|$ equal to the phase voltage peak V_{ph-M}, and the phase voltages can be expressed as (6.23).

$$\vec{V} = \left\|\vec{V}\right\| \cdot e^{j\theta} = \frac{2}{3}\left(v_a + \vec{a}\cdot v_b + \vec{a}^2\cdot v_c\right) \tag{6.21}$$

$$\vec{a} = e^{\left(j\frac{2\pi}{3}\right)} = -\frac{1}{2} + j\frac{\sqrt{3}}{2} \tag{6.22}$$

$$v_a = V_{ph-M}\cos\theta = \left\|\vec{V}\right\| \cdot \cos\theta$$
$$v_b = V_{ph-M}\cos\left(\theta - \frac{2\pi}{3}\right) = \left\|\vec{V}\right\| \cdot \cos\left(\theta - \frac{2\pi}{3}\right) \tag{6.23}$$
$$v_c = V_{ph-M}\cos\left(\theta + \frac{2\pi}{3}\right) = \left\|\vec{V}\right\| \cdot \cos\left(\theta + \frac{2\pi}{3}\right)$$

With DC-link voltage V_{DC}, the magnitude of the switching state space vectors (V_1 through V_6 in Figure 6.5) is $\frac{2}{3}V_{DC}$. From Figure 6.6, for a reference voltage in Sector I of the space vector diagram, the duty cycles for vectors V_1, V_2, and V_0 can be found as in (6.24), where M is the modulation index.

$$\begin{cases} d_1 = \frac{\sqrt{3}}{2}M\sin\left(\frac{\pi}{3} - \theta\right) \\ d_2 = \frac{\sqrt{3}}{2}M\sin\theta \\ d_0 = 1 - d_1 - d_2 \end{cases} \tag{6.24}$$

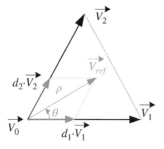

Figure 6.6 A reference voltage vector and its representation in Sector I of the two-level VSC space vector diagram.

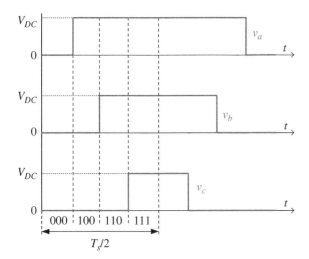

Figure 6.7 Phase voltage switching pattern for continuous SVM for the two-level three-phase VSC.

Equation (6.24) only determines the dwell times or duty cycles for different switching states but not the switching sequence, which is determined by different SVM algorithms. Figure 6.7 shows the switching pattern for phase voltages for the two-level three-phase VSI. The corresponding duty cycles for phase voltages in Sector I are in (6.25). The duty cycle for a phase is defined as the ratio of time spent on positive DC bus within one switching cycle, i.e. the time ratio of the phase voltage equaling to $+V_{DC}$. The duty cycles for other sectors can be obtained similarly. The waveform expressions for phase reference voltages for the whole fundamental cycle are listed in Table 6.1. Figure 6.8 plots phase a reference voltage or modulated phase a reference voltage for the modulation index $M = 1.15$ condition. For other M values, the shape of the reference voltage stays the same but with lower amplitudes. For phase b and phase c, their PWM reference voltage waveforms are the same as that of phase a, except with 120° and 240° phase shift, respectively.

Table 6.1 Phase reference voltage waveforms for continuous SVM (normalized on V_{DC}).

Sector	Phase a	Phase b	Phase c
I: $0 \leq \theta \leq \dfrac{\pi}{3}$	$\dfrac{1}{2} + \dfrac{\sqrt{3}}{4}M\cos\left(\theta - \dfrac{\pi}{6}\right)$	$\dfrac{1}{2} + \dfrac{3}{4}M\sin\left(\theta - \dfrac{\pi}{6}\right)$	$\dfrac{1}{2} - \dfrac{\sqrt{3}}{4}M\cos\left(\theta - \dfrac{\pi}{6}\right)$
II: $\dfrac{\pi}{3} \leq \theta \leq \dfrac{2\pi}{3}$	$\dfrac{1}{2} + \dfrac{3}{4}M\cos\theta$	$\dfrac{1}{2} + \dfrac{\sqrt{3}}{4}M\sin\theta$	$\dfrac{1}{2} - \dfrac{\sqrt{3}}{4}M\sin\theta$
III: $\dfrac{2\pi}{3} \leq \theta \leq \pi$	$\dfrac{1}{2} + \dfrac{\sqrt{3}}{4}M\cos\left(\theta + \dfrac{\pi}{6}\right)$	$\dfrac{1}{2} + \dfrac{\sqrt{3}}{4}M\sin\left(\theta - \dfrac{\pi}{3}\right)$	$\dfrac{1}{2} - \dfrac{3}{4}M\cos\left(\theta - \dfrac{\pi}{3}\right)$
IV: $\pi \leq \theta \leq \dfrac{4\pi}{3}$	$\dfrac{1}{2} + \dfrac{\sqrt{3}}{4}M\cos\left(\theta - \dfrac{\pi}{6}\right)$	$\dfrac{1}{2} + \dfrac{3}{4}M\sin\left(\theta - \dfrac{\pi}{6}\right)$	$\dfrac{1}{2} - \dfrac{\sqrt{3}}{4}M\cos\left(\theta - \dfrac{\pi}{6}\right)$
V: $\dfrac{4\pi}{3} \leq \theta \leq \dfrac{5\pi}{3}$	$\dfrac{1}{2} + \dfrac{3}{4}M\cos\theta$	$\dfrac{1}{2} + \dfrac{\sqrt{3}}{4}M\sin\theta$	$\dfrac{1}{2} - \dfrac{\sqrt{3}}{4}M\sin\theta$
VI: $\dfrac{5\pi}{3} \leq \theta \leq 2\pi$	$\dfrac{1}{2} + \dfrac{\sqrt{3}}{4}M\cos\left(\theta + \dfrac{\pi}{6}\right)$	$\dfrac{1}{2} + \dfrac{\sqrt{3}}{4}M\sin\left(\theta - \dfrac{\pi}{3}\right)$	$\dfrac{1}{2} - \dfrac{3}{4}M\cos\left(\theta - \dfrac{\pi}{3}\right)$

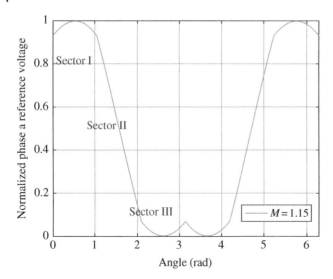

Figure 6.8 Normalized phase *a* reference voltage for continuous SVM for the two-level three-phase VSI.

$$
\begin{cases}
d_a = d_1 + d_2 + \dfrac{1}{2}d_0 = \dfrac{1}{2} + \dfrac{\sqrt{3}}{4}M\cos\left(\theta - \dfrac{\pi}{6}\right) \\[2mm]
d_b = d_2 + \dfrac{1}{2}d_0 = \dfrac{1}{2} + \dfrac{3}{4}M\sin\left(\theta - \dfrac{\pi}{6}\right) \\[2mm]
d_c = \dfrac{1}{2}d_0 = \dfrac{1}{2} - \dfrac{\sqrt{3}}{4}M\cos\left(\theta - \dfrac{\pi}{6}\right)
\end{cases}
\tag{6.25}
$$

With digital implementation, SVM is regular sampled, and often the asymmetrical regular sampling is preferred because of its benefits, especially for high power and relatively low switching frequency converters [2, 4]. In this case, Eqs. (6.13) and (6.15) still apply to determine the phase *a* and phase *b* harmonic coefficients. The switching times and angles can be determined using Eq. (6.26). The references $v_{r2(n-1)}$ and v_{r2n} are from Table 6.1, which change every 60°.

$$
\begin{aligned}
t_{2n-1} &= \left(2n - 2 + \frac{1 - v_{r2(n-1)}}{2}\right)\frac{T_c}{2} \\[2mm]
t_{2n} &= \left(2n - 1 + \frac{1 + v_{r2n}}{2}\right)\frac{T_c}{2}
\end{aligned}
\tag{6.26}
$$

For discontinuous SVM or DPWM (in this case, the popular 60° DPWM), the switching pattern for phase voltages for the two-level three-phase VSC is shown in Figure 6.9. The corresponding duty cycles for phase voltages in Sector I are in (6.27). The duty cycles for other sectors can be obtained similarly. The waveform expressions for phase reference voltages for the whole fundamental cycle are listed in Table 6.2. Figure 6.10 plots the phase *a* reference voltage or modulated phase *a* reference voltage for modulation index $M = 0.5$ and $M = \dfrac{2}{\sqrt{3}} \approx 1.15$ conditions. Unlike the continuous SVM, when *M* decreases from $\dfrac{2}{\sqrt{3}}$, the reference voltage becomes discontinuous and the shape of the waveform changes. For phase *b* and phase *c*, their PWM reference waveforms are again the same as that of phase *a*, except with 120° and 240° phase shift, respectively.

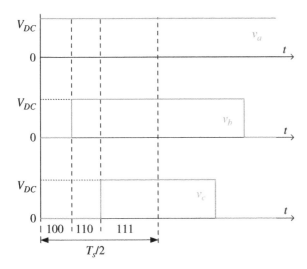

Figure 6.9 DPWM phase voltage switching pattern for the two-level three-phase VSC.

Table 6.2 Phase reference voltage waveforms for 60° discontinuous SVM (normalized on V_{DC}).

Sector (s)	Phase a	Phase b	Phase c
VI&I: $-\dfrac{\pi}{6} \le \theta \le \dfrac{\pi}{6}$	1	$1 - \dfrac{\sqrt{3}}{2}M\cos\left(\theta + \dfrac{\pi}{6}\right)$	$1 - \dfrac{\sqrt{3}}{2}M\cos\left(\theta - \dfrac{\pi}{6}\right)$
I&II: $\dfrac{\pi}{6} \le \theta \le \dfrac{\pi}{2}$	$\dfrac{\sqrt{3}}{2}M\cos\left(\theta - \dfrac{\pi}{6}\right)$	$\dfrac{\sqrt{3}}{2}M\sin\theta$	0
II&III: $\dfrac{\pi}{2} \le \theta \le \dfrac{5\pi}{6}$	$1 + \dfrac{\sqrt{3}}{2}M\cos\left(\theta + \dfrac{\pi}{6}\right)$	1	$1 - \dfrac{\sqrt{3}}{2}M\sin\theta$
III&IV: $\dfrac{5\pi}{6} \le \theta \le \pi - \pi \le \theta \le -\dfrac{5\pi}{6}$	0	$-\dfrac{\sqrt{3}}{2}M\cos\left(\theta + \dfrac{\pi}{6}\right)$	$-\dfrac{\sqrt{3}}{2}M\cos\left(\theta - \dfrac{\pi}{6}\right)$
IV&V: $-\dfrac{5\pi}{6} \le \theta \le -\dfrac{\pi}{2}$	$1 + \dfrac{\sqrt{3}}{2}M\cos\left(\theta - \dfrac{\pi}{6}\right)$	$1 + \dfrac{\sqrt{3}}{2}M\sin\theta$	1
V&VI: $-\dfrac{\pi}{2} \le \theta \le -\dfrac{\pi}{6}$	$\dfrac{\sqrt{3}}{2}M\cos\left(\theta + \dfrac{\pi}{6}\right)$	0	$-\dfrac{\sqrt{3}}{2}M\sin\theta$

$$\begin{cases} d_a = 1 \\ d_b = d_2 + d_0 = 1 - \dfrac{\sqrt{3}}{2}M\cos\left(\theta + \dfrac{\pi}{6}\right) \\ d_c = d_0 = 1 - \dfrac{\sqrt{3}}{2}M\cos\left(\theta - \dfrac{\pi}{6}\right) \end{cases} \tag{6.27}$$

Assuming DPWM is asymmetrical regular sampled, like the continuous SVM, Eqs. (6.13) and (6.15) apply to determine the phase a and phase b harmonic coefficients. The switching times and angles can also be determined using Eq. (6.26). The references $v_{r2(n-1)}$ and v_{r2n} are from Table 6.2, which change every 60°.

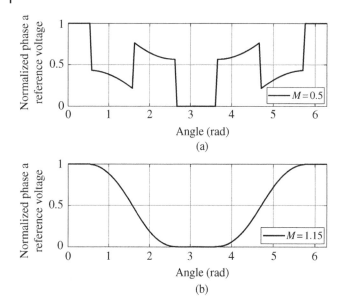

Figure 6.10 Normalized phase *a* reference voltage for discontinuous SVM for the two-level three-phase VSC: (a) modulation index *M* = 0.5; (b) *M* = 1.15.

Table 6.3 Comparison of the voltage THD results.

f_{sw} (kHz)	PWM scheme	Switching model simulation (%)	DFA (%)	SFA (%)	SFA considering noninteger harmonics (%)
4	SPWM	117.443	117.402	117.237	117.414
	SVM	117.597	117.579	117.428	117.580
	DPWM	119.819	120.049	119.846	119.766
6	SPWM	117.431	117.397	117.408	—
	SVM	117.584	117.575	117.576	—
	DPWM	119.885	119.140	119.774	—
8	SPWM	117.444	117.422	117.822	117.406
	SVM	117.588	117.575	116.996	117.575
	DPWM	119.807	119.440	119.003	119.8

Table 6.3 shows the comparison of the THD results using different approaches for three PWM schemes for the example of a 45 kW, 650 V DC to three-phase 460 V RMS, 60 Hz motor drive operating at the modulation index of 0.578. The THD calculation considers the harmonics from second-order harmonic of the fundamental frequency to 10 times of the switching frequency. All THD calculation results match well.

Figures 6.11, 6.12, and 6.13 show the compassion of individual harmonics using different analysis approaches for different PWM schemes. Clearly, when switching frequency f_{sw} is at 6 kHz, an exact integer multiple of the motor fundamental frequency of 60 Hz, the results from both the DFA and SFA methods match almost perfectly as expected. When f_{sw} is not an integer multiple of 60 Hz, the individual harmonics do not perfectly match but the overall accuracy is acceptable. The imperfect match can be compensated by considering the effect of the noninteger harmonics [5].

Table 6.4 compares the computation times of different harmonic analysis approaches. All the times were recorded on a regular desktop PC. The number of harmonic order n calculated in each case depends on the switching frequency, and it is basically equal to half of the switching frequency sideband harmonics between the fundamental frequency and switching frequency. For $f_{sw} = 4$ kHz, $n = 33$ (determined from $0.5 \times 4000/60$). Similarly, for $f_{sw} = 6$ kHz, $n = 49$, and for $f_{sw} = 6$ kHz, $n = 66$. In DPWM case, there is rich sideband harmonics, and therefore

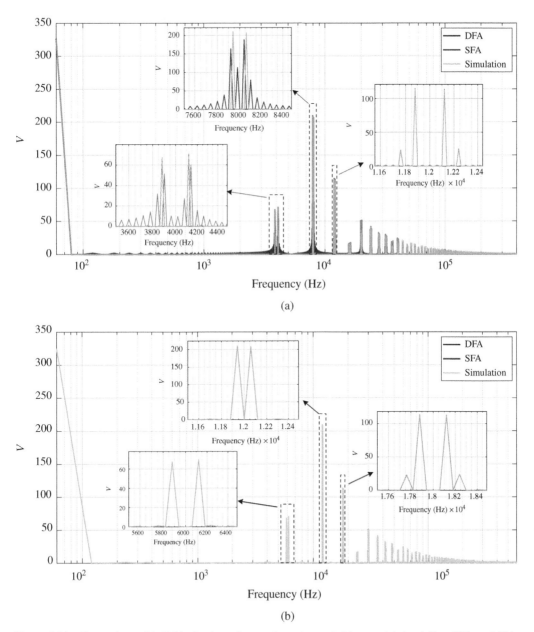

Figure 6.11 Comparison of individual voltage harmonics using switching model simulation, DFA, and SFA: SPWM case. (a) f_{sw} = 4 kHz (b) f_{sw} = 6 kHz (c) f_{sw} = 8 kHz.

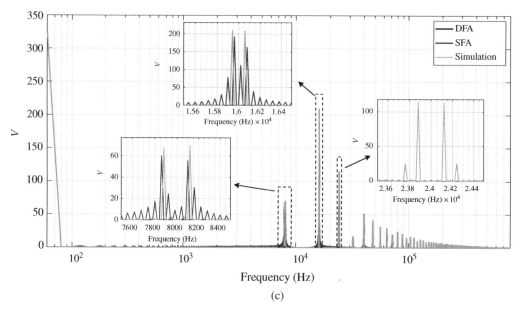

Figure 6.11 (Continued)

larger *n* is needed. Clearly, SFA is much more efficient computationally than its DFA counter-part. Note that in this case, the switching model simulation is based on switching function method, rather than switching circuit method. As a result, it is also quite fast. More details on switching function simulation will be covered in Chapter 7.

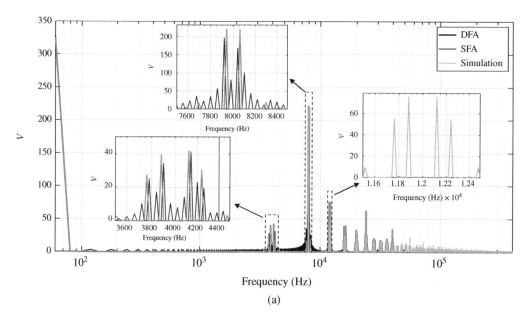

Figure 6.12 Comparison of individual voltage harmonics using switching model simulation, DFA, and SFA: SVM case. (a) f_{sw} = 4 kHz (b) f_{sw} = 6 kHz (c) f_{sw} = 8 kHz.

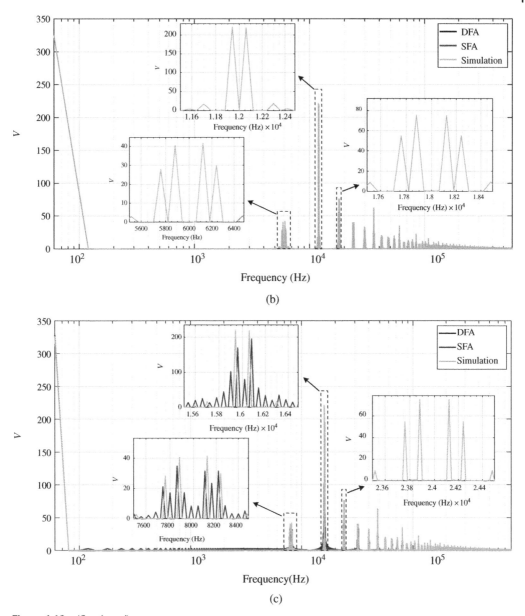

Figure 6.12 (Continued)

It should be pointed out both DFA and SFA ignore DC voltage ripple, dead time, and nonideal-ities of circuit components. Since these effects should be minimized in the design, it is reasonable to neglect them at this stage. The impact of dead time and its compensation will be discussed in Chapters 10 and 12.

Once the AC voltage harmonics are determined, the corresponding harmonic currents can be calculated based on the load impedance. The impedance models for different types of motors will be discussed in Chapter 12.

6.3.2 Inverter Power Loss

The inverter power loss includes all the losses of the inverter components. For the load-side inverter based on two-level VSI, the loss is dominated by the power semiconductor device loss, including the conduction and switching losses. In Chapter 2, the device loss models for a given device at a specific operating point have been presented as functions of the voltage, current, and junction temperature, etc. Equation (6.28) gives the instantaneous device conduction loss power

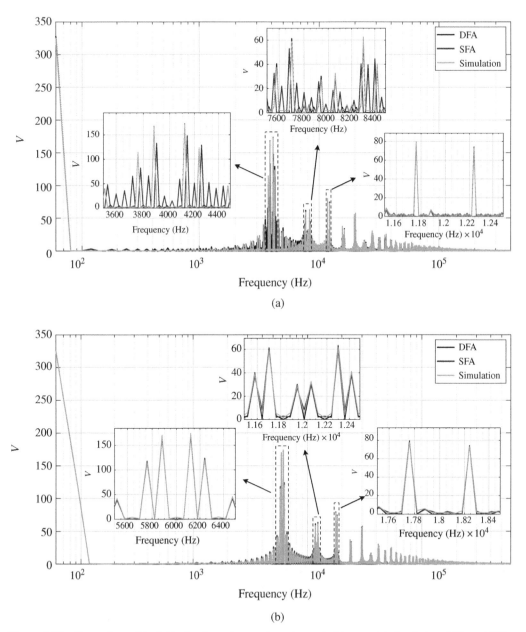

Figure 6.13 Comparison of individual voltage harmonics using switching model simulation, DFA, and SFA: DPWM case. (a) f_{sw} = 4 kHz (b) f_{sw} = 6 kHz (c) f_{sw} = 8 kHz.

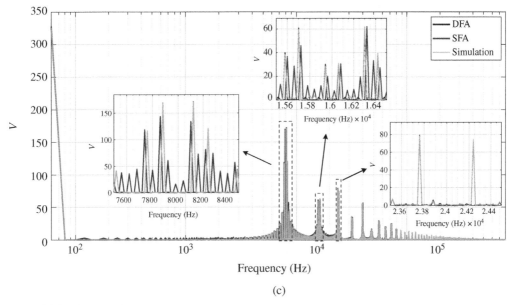

(c)

Figure 6.13 (Continued)

Table 6.4 Comparison of computation times (all units: seconds).

Switching frequency	Modulation	Switching model simulation	DFA	SFA
4 kHz	SPWM	20.73	62.3	6.3
	SVM	19.96	92.4	6.4
	DPWM	23.9	112.5 ($n = 33$)/ 4296 ($n = 100$)	7.6
6 kHz	SPWM	9.52	20.7	3.52
	SVM	9.5	312.9	4.0
	DPWM	9.96	228.1 ($n = 49$)// 1182 ($n = 79$)	3.6
8 kHz	SPWM	20.4	62.7	14.3
	SVM	21.1	491.5	14.3
	DPWM	22.9	471.6 ($n = 66$)// 1337 ($n = 100$)	14.9

p_{con}, in which v_{on} is the on-state device voltage drop (e.g. v_{DSon} for MOSFET and v_{CEon} for IGBT), as the function of current i, junction temperature T_j, and gate voltage V_g (positive gate voltage in conduction loss case). Since the switching is event based, Eq. (6.29) gives the switching energy loss E_{SW} in one switching cycle consisting of turn-on energy loss E_{on} and turnoff energy loss E_{off}. The switching energy is a function of i, T_j, as well as DC-link voltage V_{DC}, and gate resistance R_g and voltage V_g (both positive and negative gate voltages can impact switching loss). It can also be a function of circuit parasitics, denoted by Z here. For a two-level VSI based on IGBT, when the current i is in reverse direction, the conduction loss will occur in the antiparallel diode; for a

VSI based on MOSFET, the reverse conduction loss will occur only in the MOSFET if there is not an antiparallel diode, or in both the MOSFET and antiparallel diode if there is a diode. As a result, v_{on} can be different for forward and reverse conduction cases. In some cases, the turn-on of the switch will cause reverse recovery in the complementary commutating device, resulting in an extra reverse recovery energy loss E_{rr}, which is in general also a function of i, T_j, V_{DC}, R_g, V_g, and Z.

$$p_{con}(t) = v_{on}(i, T_j, V_g) \cdot i(t) \tag{6.28}$$

$$E_{SW} = E_{on}(i, V_{DC}, T_j, R_g, V_g, Z) + E_{off}(i, V_{DC}, T_j, R_g, V_g, Z) \tag{6.29}$$

The functions v_{on}, E_{on}, and E_{off} are part of device characteristics that are already covered in Chapter 2. Given that the thermal time constants of the devices are generally much shorter than the fundamental period for AC load current and voltage, very often, the loss needed for inverter design is the average loss over the fundamental period. Under this assumption (note this assumption will be examined later in Section 6.3.4), the junction temperature T_j can be assumed to be constant for a given operating point, e.g. the rated load condition for the VSI, together with constant V_{DC}, R_g, V_g, and Z.

6.3.2.1 Conduction Loss

To find the average conduction loss of a switching device in the inverter for the whole fundamental period, the total conduction energy loss needs to be determined. Since the switch only conducts part of the time during a switching period, the conduction energy loss for a given switching period is determined by the duty cycle d as in (6.30), where T_{SW} is the switching period. Note that i can be approximated as constant during the conduction period because the switching frequency is generally much higher than the AC operating frequency, or T_{SW} is much shorter than the AC operating period.

$$E_{con} = p_{con}(t) \cdot d(t) \cdot T_{SW} = v_{on}(i, T_j, V_g) \cdot i(t) \cdot d(t) \cdot T_{SW} \tag{6.30}$$

As discussed in Section 6.3.1, in a PWM inverter, the duty cycle $d(t)$ in (6.30) is determined by the reference phase voltage and modulation scheme. The current is determined by voltage, load, and also by the inverter power factor.

The conduction loss model can be analytical, numerical, or through simulation. The simplest analytical model can be closed-form, if the device conduction loss model can be simplified. Assume v_{on} in (6.30) has a linear relationship with device current i as in (6.31), where v_{on_fwd}, V_{0_fwd}, and R_{on_fwd} are forward on-state voltage drop, knee voltage, and equivalent on-state resistance; v_{on_rev}, V_{0_rev}, and R_{on_rev} are reverse direction on-state voltage drop, knee voltage, and equivalent on-state resistance. In the case of an IGBT, reverse conduction will be through its antiparallel diode, and the parameters are diode characteristics; and in the case of MOSFET, reverse conduction will be through the MOSFET channel only if there is no antiparallel diode, or through both the MOSFET channel and antiparallel diode if there is one. For majority carrier devices like MOSFETs, the knee voltages are practically zero.

$$\begin{aligned} v_{on\_fwd} &= V_{0\_fwd} + R_{on\_fwd} \cdot i \\ v_{on\_rev} &= V_{0\_rev} + R_{on\_rev} \cdot i \end{aligned} \tag{6.31}$$

The instantaneous conduction loss, when the device is forward or reverse conducting, can be found as

$$P_{con\_fwd}(t) = V_{0\_fwd} \cdot i(t) + R_{on\_fwd} \cdot i(t)^2$$
$$P_{con\_rev}(t) = V_{0\_rev} \cdot i(t) + R_{on\_rev} \cdot i(t)^2 \tag{6.32}$$

In [6], assuming a positive current direction, the average phase a conduction loss for half a switching period (i.e. half carrier period as in Figure 6.4) can be approximated as in (6.33), where d_a is the duty cycle for phase a, representing the time ratio of the phase conducting through the positive DC bus. Given that phase a current $i_a(t) = I_M \cos(\omega_0 t - \varphi)$, where φ is the power factor angle and I_M is the peak phase current, the average conduction power loss can be obtained via (6.34). Note that the integration is only for half of the fundamental cycle $\left[-\dfrac{\pi}{2} + \varphi, \dfrac{\pi}{2} + \varphi \right]$ when phase a current is positive, but the average is for the whole fundamental cycle of 2π.

$$P_{con\_fwd}(t) = V_{0\_fwd} \cdot d_a(t) \cdot i_a(t) + R_{on\_fwd} \cdot d_a(t) \cdot i_a(t)^2 \tag{6.33}$$

$$
\begin{aligned}
P_{con\_fwd} = V_{0\_fwd} &\left[\frac{I_M}{2\pi} \int_{-\frac{\pi}{2}+\varphi}^{\frac{\pi}{2}+\varphi} d_a(t) \cos(\omega_0 t - \varphi) d(\omega_0 t) \right] \\
&+ R_{on\_fwd} \left\{ \frac{I_M^2}{2\pi} \int_{-\frac{\pi}{2}+\varphi}^{\frac{\pi}{2}+\varphi} d_a(t) [\cos(\omega_0 t - \varphi)]^2 d(\omega_0 t) \right\}
\end{aligned}
\tag{6.34}
$$

Assuming a simple naturally sampled SPWM, $d_a(t) = \dfrac{1 + M \cos \omega_0 t}{2}$ and Eq. (6.34) will yield to (6.35). For the same current in the complementary reverse conducting device (antiparallel diode and/or reverse conducting MOSFET), the average conduction loss is (6.36). Clearly, the loss in forward and reverse directions is different, even if they would have the same loss characteristics (knee voltage and on-state resistance), because of their duty cycle differences. Higher modulation index and higher power factor lead to higher forward conduction loss, indicating in these cases, the current mainly flows in the forward direction, which agrees with the operating principle of the VSI. If the forward and reverse conduction characteristics were the same, then the total conduction loss ($P_{con\_fwd} + P_{con\_rev}$) in a device would be constant, regardless of the M and $\cos \varphi$ values, as expected.

$$P_{con\_fwd} = \frac{V_{0\_fwd} I_M}{2} \left(\frac{1}{\pi} + \frac{M}{4} \cos \varphi \right) + R_{on\_fwd} I_M^2 \left(\frac{1}{8} + \frac{M}{3\pi} \cos \varphi \right) \tag{6.35}$$

$$P_{con\_rev} = \frac{V_{0\_rev} I_M}{2} \left(\frac{1}{\pi} - \frac{M}{4} \cos \varphi \right) + R_{on\_rev} \cdot I_M^2 \left(\frac{1}{8} - \frac{M}{3\pi} \cos \varphi \right) \tag{6.36}$$

For different modulation schemes, different closed-form conduction loss equations can be expected. In [6], it shows that for an SPWM with third-order harmonic injection, the conduction loss for forward and reverse conducting devices is, respectively, (6.37) and (6.38), where M_3 is often selected as a fraction of modulation index M, e.g. $\dfrac{1}{6}, \dfrac{1}{4}$ or $\dfrac{1}{3}$ of M, etc.

$$P_{con\_fwd} = \frac{V_{0\_fwd} I_M}{2} \left(\frac{1}{\pi} + \frac{M}{4} \cos \varphi \right) + R_{on\_fwd} I_M^2 \left(\frac{1}{8} + \frac{M}{3\pi} \cos \varphi - \frac{M_3}{15\pi} \cos 3\varphi \right) \tag{6.37}$$

$$P_{con\_rev} = \frac{V_{0\_fwd} I_M}{2} \left(\frac{1}{\pi} - \frac{M}{4} \cos \varphi \right) + R_{on\_fwd} I_M^2 \left(\frac{1}{8} - \frac{M}{3\pi} \cos \varphi + \frac{M_3}{15\pi} \cos 3\varphi \right) \tag{6.38}$$

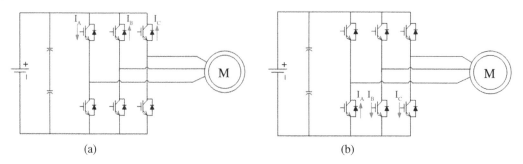

Figure 6.14 A current conduction pattern corresponding to different zero vector placements: (a) use (111) vector and (b) use (000) vector.

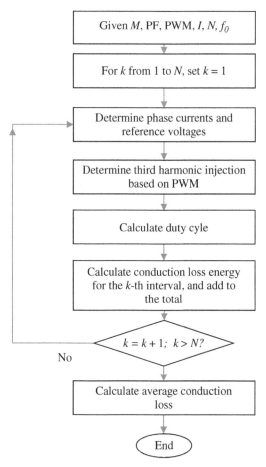

Figure 6.15 Procedure for conduction loss calculation.

Although Eqs. (6.35) and (6.36) for simple SPWM and Eqs. (6.37) and (6.38) for SPWM with third-order harmonic injection are different, the difference in conduction losses due to PWM schemes is marginal with the same fundamental AC current. This is because different PWM schemes only differ in their zero-vector placement. Figure 6.14 illustrates, for a given instant of three-phase current distribution, two different device conduction patterns corresponding to using zero vector (111) and (000), respectively. The fact that $(i_a + i_b + i_c) = 0$ makes the conduction loss difference of the two patterns small. In the case of minority-carrier devices like IGBT, the modulation change, indicated by nonzero M_3 in Eqs (6.35) through (6.38), will not impact the conduction loss part that is due to the knee voltage. In the case of majority-carrier devices like MOSFETs with close forward and reverse conduction characteristics, the impact should also be small.

One drawback with the earlier closed-form equations is that they only consider the dependency of the conduction loss on a single variable, i.e. device current, and also such dependency is only a linearized relationship. As indicated in (6.28), the device conduction loss can be a function of current, junction temperature, and gate voltage; furthermore, the function is generally nonlinear. Here a more general conduction loss calculation procedure will be introduced, which is suitable for use in design optimization. The procedure is illustrated in the Figure 6.15 flowchart with the steps described here.

Step 1: Gather the input information including modulation index M, power factor $\cos \varphi$, PWM scheme, current value (RMS or peak), and AC frequency $f_0 = \dfrac{\omega_0}{2\pi}$.

Step 2: Divide the AC fundamental cycle into N intervals, such that within each interval, the current and voltage can be approximated as constant. For example, based on experience, an N of 1024 is sufficiently large. Each interval will be equal to $\Delta T = 1/(f_0 N)$.

Step 3: Increment k from 1 to N and determine fundamental phase reference voltages u_{a1}, u_{b1}, and u_{c1} as in (6.39). The phase currents can be determined with (6.40), where I_M is the peak current, assuming phase currents are sinusoidal.

$$u_{a1}(k) = M \sin[\omega_0(k-1)\Delta T]$$
$$u_{b1}(k) = M \sin\left[\omega_0(k-1)\Delta T - \frac{2\pi}{3}\right] \tag{6.39}$$
$$u_{c1}(k) = M \sin\left[\omega_0(k-1)\Delta T - \frac{4\pi}{3}\right]$$

$$i_a(k) = I_M \sin[\omega_0(k-1)\Delta T - \varphi]$$
$$i_b(k) = I_M \sin\left[\omega_0(k-1)\Delta T - \varphi - \frac{2\pi}{3}\right] \tag{6.40}$$
$$i_b(k) = I_M \sin\left[\omega_0(k-1)\Delta T - \varphi - \frac{4\pi}{3}\right]$$

Step 4: Determine the third-order harmonic injection reference voltage u_{a3h} based on the PWM scheme. One example for u_{a3h} is (6.41), when a third-order harmonic term is directly injected.

$$u_{a3h}(k) = M_3 \sin[3\omega_0(k-1)\Delta T] \tag{6.41}$$

In many cases, u_{a3h} can be mathematically determined with the general approach explained in [7]. For centered continuous SVM, u_{a3h} can be found as (6.42), where $u_{\max}(k)$ and $u_{\min}(k)$ are defined in (6.43).

$$u_{a3h}(k) = -\frac{u_{\max}(k) + u_{\min}(k)}{2} \tag{6.42}$$

$$u_{\max}(k) = \max[u_{a1}(k), u_{b1}(k), u_{c1}(k)]$$
$$u_{\min}(k) = \min[u_{a1}(k), u_{b1}(k), u_{c1}(k)] \tag{6.43}$$

For 60° and 30° DPWM, u_{a3h} can be found as (6.44) and (6.45), respectively.

$$u_{a3h}(k) = \begin{cases} 1 - u_{\max} & \text{when } |u_{\max}| \geq |u_{\min}| \\ -1 - u_{\min} & \text{when } |u_{\max}| < |u_{\min}| \end{cases} \tag{6.44}$$

$$u_{a3h}(k) = \begin{cases} 1 - u_{\max} & \text{when } |u_{\max}| \leq |u_{\min}| \\ -1 - u_{\min} & \text{when } |u_{\max}| > |u_{\min}| \end{cases} \tag{6.45}$$

For the so-called minimum-loss DPWM, u_{a3h} can be found through (6.46), where $i_{\max}$ and $i_{\min}$ are defined as in (6.47). Different reference voltages for phase a and their corresponding u_{a3h} for different PWM schemes are illustrated in Figure 6.16.

$$u_{a3h}(k) = \begin{cases} 1 - u_{\max} & \text{when } |i_{\max}| \geq |i_{\min}| \\ -1 - u_{\min} & \text{when } |i_{\max}| < |i_{\min}| \end{cases} \tag{6.46}$$

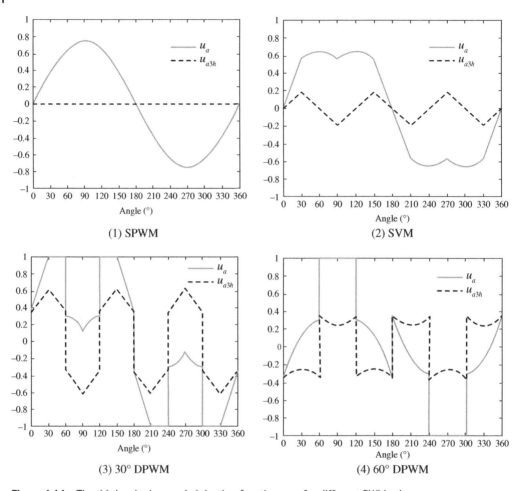

Figure 6.16 The third-order harmonic injection function u_{a3h} for different PWM schemes.

$$i_{\max}(k) = \max[i_a(k), i_b(k), i_c(k)]$$
$$i_{\min}(k) = \min[i_a(k), i_b(k), i_c(k)] \tag{6.47}$$

Step 5: Determine the instantaneous duty cycle

$$d_a(k) = \frac{1 + u_{a1} + u_{a3h}}{2} \tag{6.48}$$

Step 6: Calculate the conduction loss energy for the kth interval with (6.49), where D_{I_a} is the direction function of the current $i_a(k)$. D_{I_a} is 1 when $i_a(k)$ is positive and 0 when $i_a(k)$ is negative. $E_{con\_ap\_fwd}(k)$ and $E_{con\_ap\_rev}(k)$ correspond to the forward and reverse conduction loss energy, respectively, for the upper device of the phase a. To calculate the conduction loss energy of the lower device, Eq. (6.50) should be used.

$$E_{con\_ap\_fwd}(k) = v_{on\_fwd}\big(i_a(k), T_j, V_g\big) \cdot i_a(k) \cdot d_a(k) \cdot \Delta T \cdot D_{I\_a}(k)$$
$$E_{con\_ap\_rev}(k) = -v_{on\_rev}\big(i_a(k), T_j, V_g\big) \cdot i_a(k) \cdot d_a(k) \cdot \Delta T \cdot [1 - D_{I\_a}(k)] \tag{6.49}$$

$$E_{con\_an\_fwd}(k) = -v_{on\_fwd}\big(i_a(k), T_j, V_g\big) \cdot i_a(k) \cdot [1 - d_a(k)] \cdot \Delta T \cdot [1 - D_{I\_a}(k)]$$

$$E_{con\_an\_rev}(k) = v_{on\_rev}\big(i_a(k), T_j, V_g\big) \cdot i_a(k) \cdot [1 - d_a(k)] \cdot \Delta T \cdot D_{I\_a}(k)$$

(6.50)

Step 7: Find the average conduction power loss for phase a upper device using (6.51). The conduction losses for phase a lower device and other phases can be found similarly.

$$P_{con\_ap\_fwd} = f_0 \sum_{k=1}^{N} E_{con\_ap\_fwd}(k)$$

(6.51)

$$P_{con\_ap\_rev} = f_0 \sum_{k=1}^{N} E_{con\_ap\_rev}(k)$$

6.3.2.2 Switching Loss

The average switching loss model of a switching device in the inverter for the whole fundamental period can also be analytical, numerical, or through simulation. Similar to the conduction loss model, the simplest analytical switching loss model is a closed-form solution, if the device switching energy can be simplified to have a linear relationship with device current and DC-link voltage. Assuming that the DC-link voltage is constant as in the case of a VSI, Eq. (6.52) shows the relationship of energy loss on the switch E_{SW} (i.e. the sum of E_{on} and E_{off}) as a linear function of device current i, where V_{DC_rated} and I_{rated} are the rated DC-link voltage and device current (collector current for IGBT or drain current for MOSFET), and E_{on_rated} and E_{off_rated} are turn-on and turnoff energy at the rated condition. Note that the rated condition normally refers to the condition defined in data sheet, with T_j, R_g, and V_g assumed at certain nominal values. The impact of the parasitic impedance Z is neglected. For different T_j and R_g values, they can be assumed constant for a given VSI operating condition, and corresponding E_{on} and E_{off} in (6.52) still only vary with current i. Note that for a hard-switching VSI, a switching commutation generally involves a switch and diode pair (the diode can be the body diode of the complementary switch). The diode can incur switching loss due to reverse recovery. The diode switching energy or reverse recovery loss can also be approximated as a linear function of current as in (6.53).

$$E_{SW} = E_{on}\big(i, V_{DC}, T_j, R_g, V_g, Z\big) + E_{off}\big(i, V_{DC}, T_j, R_g, V_g, Z\big)$$

$$= \frac{V_{DC}}{V_{DC\_rated}} \cdot \frac{i}{I_{rated}} \big(E_{on\_rated} + E_{off\_rated}\big) = \big(k_{on} + k_{off}\big) \cdot i = k_{SW} \cdot i$$

(6.52)

$$E_{diode} = E_{rr}\big(i, V_{DC}, T_j, R_g, V_g, Z\big) = \frac{V_{DC}}{V_{DC\_rated}} \cdot \frac{i}{I_{rated}} \cdot E_{rec\_rated} = k_d \cdot i$$

(6.53)

For a continuous PWM with a high switching frequency f_{SW} relative to the fundamental frequency f_0, the instantaneous switching loss power for a forward conducting switch, e.g. conducting IGBT or forward conducting MOSFET, can be expressed as in (6.54).

$$P_{SW\_fwd} = \big(k_{on} + k_{off}\big) \cdot f_{SW} \cdot i$$

(6.54)

Given current $i = I_M \cos(\omega_0 t - \varphi)$, the average active device switching loss is [6]

$$P_{SW\_fwd} = \frac{I_M}{2\pi} \int_{-\frac{\pi}{2}+\varphi}^{\frac{\pi}{2}+\varphi} P_{SW\_fwd} \cos(\omega_0 t - \varphi) d(\omega_0 t) = \big(k_{on} + k_{off}\big) \cdot f_{SW} \cdot \left(\frac{I_M}{\pi}\right)$$

(6.55)

Similar to the conduction loss case, the integration in (6.55) is only for half of the fundamental cycle $\left[-\dfrac{\pi}{2} + \varphi, \dfrac{\pi}{2} + \varphi\right]$ when phase a current is positive, but the average is for the whole fundamental cycle. This is because P_{SW_fwd} only accounts for the switching loss when the switch forward conducts. In the case of IGBT, this is the total switching loss; in the case of MOSFET, this is only the partial switching loss, as in principle, the body diode can incur loss when the complementary switch turns off due to the reverse recovery. The switching loss of the reverse conducting switch, i.e. the antiparallel diode and/or the body diode, can be obtained as in (6.56).

$$P_{SW\_rev} = k_d \cdot f_{sw} \cdot \left(\frac{I_M}{\pi}\right) \tag{6.56}$$

For DPWM, there is no switching loss during the period when a phase is clamped to positive or negative DC bus. The switching loss depends on the phase angle difference between output voltage and current, i.e. power factor angle φ. As has been derived in [6], for 60° DPWM,

$$P_{SW\_fwd} = \frac{\left(k_{on} + k_{off}\right) \cdot I_M \cdot f_{sw}}{\pi} \left(1 - \frac{1}{2}\cos\varphi\right), \quad \varphi \in \left[0, \frac{\pi}{3}\right]$$

$$P_{SW\_fwd} = \frac{\left(k_{on} + k_{off}\right) \cdot I_M \cdot f_{sw}}{\pi} \frac{\sqrt{3}\sin\varphi}{2}, \quad \varphi \in \left[\frac{\pi}{3}, \frac{\pi}{2}\right] \tag{6.57}$$

For 30° DPWM,

$$P_{SW\_fwd} = \frac{\left(k_{on} + k_{off}\right) \cdot I_M \cdot f_{sw}}{\pi} \left[1 - \frac{(\sqrt{3}-1)}{2}\cos\varphi\right], \quad \varphi \in \left[0, \frac{\pi}{6}\right]$$

$$P_{SW\_fwd} = \frac{\left(k_{on} + k_{off}\right) \cdot I_M \cdot f_{sw}}{\pi} \left(\frac{\sin\varphi + \cos\varphi}{2}\right), \quad \varphi \in \left[\frac{\pi}{6}, \frac{\pi}{3}\right] \tag{6.58}$$

$$P_{SW\_fwd} = \frac{\left(k_{on} + k_{off}\right) \cdot I_M \cdot f_{sw}}{\pi} \left[1 - \frac{(\sqrt{3}-1)}{2}\sin\varphi\right], \quad \varphi \in \left[\frac{\pi}{3}, \frac{\pi}{2}\right]$$

For antiparallel diodes or body diodes, the switching loss for 60° and 30° DPWM will have similar expressions as (6.57) and (6.58), except that $(k_{on} + k_{off})$ should be replaced with k_d.

Like in the case of conduction loss, one drawback with the earlier closed-form equations is that these equations only consider the dependency of switching loss as a linear function of device current. As indicated in Eq. (6.29), the device switching loss characteristics can be a nonlinear function of DC-link voltage, current, junction temperature, and gate resistance and voltages. It can even be influenced by the circuit parasitics, such as parasitic capacitances and inductances [8, 9]. Here a more general switching loss calculation procedure will be introduced, which is suitable for use in design optimization.

The procedure is illustrated in Figure 6.17 flowchart with the steps described here.

Steps 1 to 3 are the same as those of the conduction loss calculation procedure described earlier. One additional piece of information needed in Step 1 is the switching frequency f_{SW}.

Step 4: Determine the instantaneous switching function $S(k)$ value for each phase based on PWM scheme. For continuous PWM schemes, $S \equiv 1$. For 60° DPWM,

$$S_a(k) = \begin{cases} 0 & \text{when } k \in \left(\dfrac{N}{6}, \dfrac{2N}{6}\right) \text{or} \left(\dfrac{4N}{6}, \dfrac{5N}{6}\right), \text{i.e. phase angle} \left(\dfrac{\pi}{3}, \dfrac{2\pi}{3}\right) \text{or} \left(\dfrac{4\pi}{3}, \dfrac{5\pi}{3}\right) \\ 1 & \text{otherwise} \end{cases}$$

$$\tag{6.59}$$

Figure 6.17 Procedure for switching loss calculation.

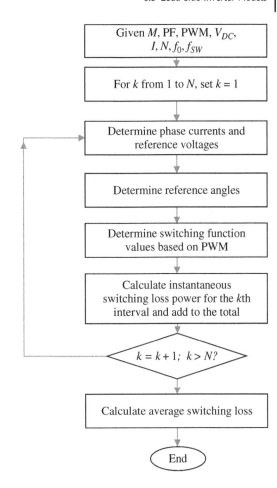

For 30° DPWM,

$$S_a(k) = \begin{cases} 0 & \text{when } k \in \left(\dfrac{N}{12}, \dfrac{2N}{12}\right) \text{or} \left(\dfrac{4N}{12}, \dfrac{5N}{12}\right) \text{or} \left(\dfrac{7N}{12}, \dfrac{8N}{12}\right) \text{or} \left(\dfrac{10N}{12}, \dfrac{11N}{12}\right) \\ 1 & \text{otherwise} \end{cases}$$
(6.60)

where the range $k \in \left(\dfrac{N}{12}, \dfrac{2N}{12}\right)$ or $\left(\dfrac{4N}{12}, \dfrac{5N}{12}\right)$ or $\left(\dfrac{7N}{12}, \dfrac{8N}{12}\right)$ or $\left(\dfrac{10N}{12}, \dfrac{11N}{12}\right)$ corresponds to phase angle range $\left(\dfrac{\pi}{6}, \dfrac{\pi}{3}\right)$, $\left(\dfrac{2\pi}{3}, \dfrac{5\pi}{6}\right)$, $\left(\dfrac{7\pi}{6}, \dfrac{4\pi}{3}\right)$, or $\left(\dfrac{5\pi}{3}, \dfrac{11\pi}{6}\right)$.

Step 5: Calculate the instantaneous switching loss power with (6.61), where D_{I_a} is defined in conduction loss discussion. $P_{SW\_ap\_fwd}(k)$ and $P_{SW\_ap\_rev}(k)$ correspond to the forward and reverse switching power loss, respectively, for the upper device of phase a. To calculate the switching power loss of the lower device, Eq. (6.62) should be used.

$$p_{SW\_ap\_fwd}(k) = \left[E_{on\_ap}(k) + E_{off\_ap}(k)\right] \cdot f_{sw} \cdot D_{I\_a}(k) \cdot S_a(k)$$
$$p_{SW\_ap\_rev}(k) = E_{rr\_ap}(k) \cdot f_{sw} \cdot \left[1 - D_{I\_a}(k)\right] \cdot S_a(k)$$
(6.61)

$$p_{SW\_an\_fwd}(k) = \left[E_{on\_an}(k) + E_{off\_an}(k)\right] \cdot f_{sw} \cdot \left[1 - D_{I\_a}(k)\right] \cdot S_a(k)$$
$$p_{SW\_an\_rev}(k) = E_{rr\_an}(k) \cdot f_{sw} \cdot D_{I\_a}(k) \cdot S_a(k)$$
(6.62)

Step 6: Find the average switching loss power for phase a upper device using (6.63). The switching losses for the phase a lower device and other phases can be found similarly.

$$P_{SW\_ap\_fwd} = \frac{1}{N} \sum_{k=1}^{N} P_{SW\_ap\_fwd}(k)$$

$$P_{SW\_ap\_rev} = \frac{1}{N} \sum_{k=1}^{N} P_{SW\_ap\_rev}(k)$$

(6.63)

The total loss for a device can be found by adding its conduction and switching loss together. The total device loss for the whole inverter is the sum of all device losses. In the two-level VSI, the total device loss includes the loss of six switches and when present, six antiparallel diodes. PWM schemes discussed earlier, including SPWM, continuous SVM, and 60° and 30° DPWM, are all symmetrical switching schemes, and will result in uniform loss distributions. Therefore, the device loss evaluation can be simplified by only considering the loss of one device. Another simplification is to take advantage of time symmetry. For any one device or phase, only half of the cycle or 180° needs to be considered ($\frac{N}{2}$ in the procedure earlier). If we consider all three phases together, only $\frac{1}{6}$ cycle or 60° needs to be considered ($\frac{N}{6}$ in the procedure earlier).

6.3.3 Control Performance

As mentioned earlier in this chapter, the inverter control performance that will impact the power stage design is mainly the dynamic performance characterized through the control bandwidth, which is directly related to switching frequency of the inverter. In the case of the motor drive, the control typically has two control loops, speed regulator loop, and torque regulator loop, as shown in Figure 6.18. Assume the bandwidths of the speed regulator and the torque regulator are ω_{speed} and ω_{torque}, respectively. Then, the relationship in (6.64) should hold, where ω_C is the angular frequency of the PWM carrier, which is related to the inverter switching frequency.

$$\omega_{speed} \ll \omega_{torque} \ll \omega_C$$

(6.64)

For motor drives, torque regulator bandwidth ω_{torque} is roughly equivalent to the current regulator bandwidth ω_I, since the torque is determined by the product of the magnetic flux and current, whereas the magnetic flux in a motor is approximately constant. Practically, selecting ω_C to be at least 5–10 times of current regulator bandwidth ω_I that is in turn 5–10 times of ω_{speed} will be sufficient to guarantee the dynamic performance.

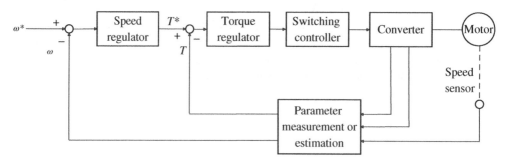

Figure 6.18 Typical motor drive control architecture.

Another requirement is the voltage margin needed for the inverter to support the sudden load change or impact load, which can be characterized by $\frac{dI}{dt}$. Note that I here refers to the magnitude of current, representing the torque. In [10], the relationship between the voltage margin is derived for a synchronous motor load. The motor terminal voltage in per unit can be expressed as (6.65), where ω_b is the base electrical angular frequency, Ψ is the per unit motor stator flux linkage, and δ is the load angle.

$$V = \sqrt{1 + \frac{2\Psi}{\omega_b}\frac{d\delta}{dt} + \frac{1}{\omega_b^2}\left(\frac{d\Psi}{dt}\right)^2 + \frac{\Psi^2}{\omega_b^2}\left(\frac{d\delta}{dt}\right)^2} \tag{6.65}$$

The voltage margin needed can be obtained by evaluating

$$\Delta V^2 = \frac{2\Psi}{\omega_b}\frac{d\delta}{dt} + \frac{1}{\omega_b^2}\left(\frac{d\Psi}{dt}\right)^2 + \frac{\Psi^2}{\omega_b^2}\left(\frac{d\delta}{dt}\right)^2 \tag{6.66}$$

In fact, the per unit voltage margin should be

$$V_{\text{margin}} = \sqrt{1 + \Delta V_{max}^2} - 1 \tag{6.67}$$

Recognizing the maximum voltage margin is likely needed under impact load conditions, the $\frac{1}{\omega_b^2}\left(\frac{d\Psi}{dt}\right)^2$ term can be ignored with $\frac{d\Psi}{dt} \cong 0$. Therefore, the key to determine the voltage margin is to determine $\frac{d\delta}{dt}$. Under the unity power factor condition, which is generally desirable for high-power motor drive, $\frac{d\delta}{dt}$ can be derived as in (6.68), where X_q, T_q'', and T_{q0}'' are q-axis parameters for synchronous motors. X_q is the per unit q-axis synchronous reactance; T_q'' is the q-axis subtransient short-circuit time constant, and T_{q0}'' is the q-axis subtransient open-circuit time constant.

$$\frac{d\delta}{dt} = \frac{\dfrac{IX_q}{\Psi} - \tan\delta + \dfrac{T_q''X_q}{\Psi}\dfrac{dI}{dt}}{T_{q0}'' + T_q''\dfrac{IX_q}{\Psi}\tan\delta} \tag{6.68}$$

Using (6.68) relationship, the maximum $\frac{d\delta}{dt}$ and the maximum voltage for a given $\frac{dI}{dt}$ can be obtained numerically. Under the assumption that current rate of rise is high and δ is small before $\frac{d\delta}{dt}$ reaches its maximum, a further simplification can be achieved as in (6.69), where $k = \frac{dI}{dt}$ and T_{MAX} is the time for I to reach its peak value under the rate of rise of k.

$$\left(\frac{d\delta}{dt}\right)_{MAX} = \frac{kX_q - k\left(X_q - X_q''\right)\cdot e^{-\frac{T_{MAX}}{T_{q0}''}}}{\Psi} \tag{6.69}$$

An example was provided in [10] for a synchronous motor drive under an impact load condition. The conditions are $k = \frac{dI}{dt} = 75$ p.u./s, the speed is 2.0 p.u. (i.e. $\Psi = 0.5$ p.u.), and impact torque is

0.8 p.u. (i.e. impact current $I = 1.6$ p.u.). So, $T_{MAX} = \dfrac{1.6}{75} = 0.021$ seconds. With $X_q = 0.55$ p.u.,

$X_q'' = 0.12$ p.u., and $T_{q0}'' = 0.1343$ seconds, $\left(\dfrac{d\delta}{dt}\right)_{MAX}$ can be calculated as 27.3 rad/s. The calculated

maximum $\dfrac{d\delta}{dt}$ is very close to the value obtained from simulation, which is 25 rad/s. The corresponding peak voltage is 1.27 p.u., close to the simulated result of 1.28 p.u.

6.3.4 Device Maximum Junction Temperature – Maximum Thermal Impedance Requirement

Once we know the power loss, the maximum interface temperature with the thermal management system can be determined based on the thermal characteristics of the components, in order to keep the device junction temperature below T_{jmax}. For steady-state conditions, with periodic voltage and current, all variables associated with the inverter change periodically. As a result, the thermal calculation only needs to consider one single fundamental AC line cycle in time domain, similar to that of the power loss in Section 6.3.2.

If the single-cycle power loss of the device in steady state can be approximated as a constant by its average value P_{avg}, which is the sum of the conduction and switching losses of the device, only the thermal resistance of the device needs to be modeled and also as a constant R_{th-jc}. The temperature difference from the junction to case is

$$T_{jc} = T_j - T_c = P_{avg} \cdot R_{th-jc} \tag{6.70}$$

Given that P_{avg} is the average power loss of one particular device, if the converter has N devices, the total converter power loss will be

$$P_{total} = \sum_{i=1}^{N} P_{avg}(i) \tag{6.71}$$

In the case of two-level VSI with IGBTs and antiparallel diodes and assuming all devices have identical power losses, then

$$P_{total} = 6 \cdot \left(P_{avg-IGBT} + P_{avg-diode}\right) \tag{6.72}$$

Assuming that all the devices share the same heatsink, then the heatsink surface temperature is

$$T_{hs} = P_{total} \cdot R_{th-ha} + T_a \tag{6.73}$$

where T_a is the ambient temperature and R_{th_ha} is the heatsink thermal resistance. The junction temperature for the ith device is

$$T_j(i) = T_{jc}(i) + \Delta T_{TIM}(i) + T_{hs} = P_{avg}(i) \cdot \left[R_{th-jc}(i) + R_{th-TIM}(i)\right] + P_{total} \cdot R_{th-ha} + T_a \tag{6.74}$$

where $\Delta T_{TIM}(i)$ is the temperature drop on the thermal interface material (TIM) between the device baseplate and heatsink, and $R_{th-TIM}(i)$ is the corresponding TIM thermal resistance. Note that in (6.74), we assume each device has its own baseplate. If several devices share a baseplate, e.g. an IGBT/MOSFET with an antiparallel diode, a half-bridge phase leg module, or even a six-pack module, the equation needs to be modified. In the case of an IGBT with an antiparallel diode, the equation becomes

$$T_{j-IGBT} = \Delta T_{jc-IGBT} + \Delta T_{TIM} + T_{hs}$$

$$= P_{avg-IGBT} \cdot R_{th-jc-IGBT} + \left[P_{avg-IGBT} + P_{avg-diode}\right] \cdot R_{th-TIM} + P_{total} \cdot R_{th-ha} + T_a$$

$$T_{j-diode} = \Delta T_{jc-diode} + \Delta T_{TIM} + T_{hs}$$

$$= P_{avg-diode} \cdot R_{th-jc-diode} + \left[P_{avg-IGBT} + P_{avg-diode}\right] \cdot R_{th-TIM} + P_{total} \cdot R_{th-ha} + T_a$$

$$(6.75)$$

In a six-pack case,

$$T_j(i) = \Delta T_{jc}(i) + \Delta T_{TIM}(i) + T_{hs} = P_{avg}(i) \cdot R_{th-jc}(i) + P_{total} \cdot (R_{th-TIM} + R_{th-hs}) + T_a$$

$$= P_{avg}(i) \cdot R_{th-jc}(i) + P_{total} \cdot R_{th-ca} + T_a$$

$$(6.76)$$

where R_{th-ca} is the case-to-ambient thermal resistance.

Given the maximum junction temperature T_{jmax} and assume a six-pack case (or neglect TIM), there should be

$$T_j(i) = P_{avg}(i) \cdot R_{th-jc}(i) + P_{total} \cdot R_{th-ca} + T_a \leq T_{jmax} \qquad (6.77)$$

The corresponding maximum thermal resistance of the thermal management system, or case-to-ambient thermal resistance, should satisfy

$$R_{th\_ca} \leq \frac{T_{jmax} - T_a}{P_{total}} - \frac{\max_i \left[P_{avg}(i) \cdot R_{th-jc}(i)\right]}{P_{total}} = \frac{T_{jmax} - T_{jcmax} - T_a}{P_{total}} \qquad (6.78)$$

where T_{jcmax} is the maximum junction-to-case temperature among all devices on the heatsink.

Since the loss is a function of T_j, if the loss in Section 6.3.2 is determined for T_{jmax}, then (6.78) holds without the need for any iterations. However, in general, the loss can vary within a fundamental cycle even under steady-state conditions. In addition, during the overload case, the thermal time constant must be considered. As a result, the device thermal characteristics need to be modeled using an *RC* network, as discussed in Chapter 2. Generally, devices in an inverter, or even the switch and diode in a device module, can have different losses and thermal characteristics, and therefore, their junction temperatures will be different. When we choose T_{jmax} or some desirable maximum junction temperatures as the calculation temperature for loss for all devices, it is a conservative approach and is often acceptable as a design practice.

If we need to consider temperature variation, to simplify and speed up the calculation, a useful approach is to convert the device power losses for a single cycle in time domain to frequency domain and to obtain the thermal calculation results in the frequency domain. For each thermal calculation case, all the information can be obtained from just a single-cycle loss calculation according to the AC fundamental frequency of the inverter. By means of the fast Fourier transform (FFT), the average power loss and the power loss harmonics for each device are calculated. Injecting the loss harmonics into the device (IGBT, MOSFET, or diodes), thermal impedance networks will result in the temperature harmonics of the junction-to-case temperature rise.

The general procedure of the device loss and junction-to-case temperature rise can be combined as illustrated in Figure 6.19. Steps following Step 6 of the instantaneous device conduction loss and Step 5 of the instantaneous device switching loss described in Section 6.3.2, which are also illustrated in Figures 6.15 and 6.17, are described here.

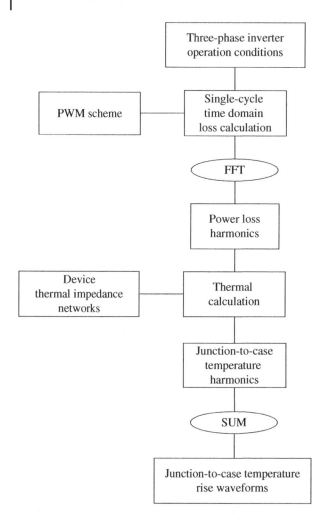

Figure 6.19 Calculation procedure for device junction-to-case temperature.

Step 1: Obtain the total instantaneous device loss by adding the conduction and switching loss from the Step 6 of conduction loss calculation and Step 5 of the switching loss calculation.

Step 2: Perform FFT to obtain loss harmonics. In general, the instantaneous power loss of the device can be expressed as (6.79), where P_{avg} is the average loss of the device, p_n is the nth order power loss harmonic, and N_h is the number of total loss and temperature harmonic orders considered. $N_h = 64$ is sufficient, which is $\frac{1}{16}$ of the total number of points selected for loss calculation in Section 6.3.2.

$$p(t) = P_{avg} + \sum_{n=1}^{Nh} p_n \tag{6.79}$$

The nth order loss harmonic is defined in (6.80), where P_n is its amplitude and θ_n is its phase angle. The fundamental frequency for the device power loss is the electrical fundamental frequency of the AC inverter f_0 ($\omega_0 = 2\pi f_0$).

$$p_n = P_n \cdot \sin(n\omega_0 t + \theta_n) \tag{6.80}$$

Step 3: Determine the junction-to-case temperature harmonics with (6.81). T_{jc-avg} is the average junction-to-case temperature, the same as defined in (6.70); T_{jcn}, γ_n, and t_{jcn} are the amplitude, phase angle, and instantaneous value of the nth harmonic of the junction-to-case temperature, respectively; Z_{th-jcn} is the nth order junction-to-case harmonic thermal impedance.

$$T_{jc-avg} = P_{avg} \cdot R_{th-jc}$$

$$T_{jcn} = P_n \cdot \left| Z_{th-jcn} \right|$$

$$\gamma_n = \theta_n + \tan^{-1}\left(\frac{\text{Im}(Z_{th-cn})}{\text{Re}(Z_{th-cn})}\right) \tag{6.81}$$

$$t_{jcn} = T_{jcn} \cdot \sin(n\omega_0 t + \gamma_n)$$

Step 4: Determine the overall device junction-to-case temperature rise as

$$T_{jc}(t) = T_{jc-avg} + \sum_{n=1}^{N_h} t_{jcn} \tag{6.82}$$

Figure 6.20 shows an example simulation result of the junction temperature variation of an IGBT module in a three-phase VSI during the state-state operation. Both the IGBT and diode junction temperatures vary significantly within one fundamental cycle. Since the case-to-ambient temperature varies little in steady state due to the large time constant of the heatsink, the junction temperature variations in Figure 6.20 can be attributed to the junction-to-case temperature variation.

The peak T_{jc} can be determined for each device of the inverter. The maximum value T_{jcmax} among them should be used to determine the lowest case temperature T_{cmin}, which will in turn determine the maximum allowed TMS thermal resistance, i.e. case-to-ambient thermal resistance R_{th-ca}. In order to keep the device junction temperature below the T_{jmax} limit,

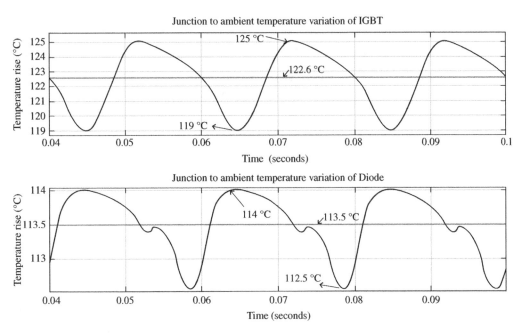

Figure 6.20 Junction-to-ambient temperature variation for an IGBT device module in a VSI.

$R_{th-ca} \leq \dfrac{T_{jmax} - T_{jcmax} - T_a}{P_{total}} = \dfrac{T_{cmin} - T_a}{P_{total}}$ should be satisfied. This agrees with the previous der-

ivation without considering junction temperature variation in steady state, which results in (6.78). Note that P_{total} is the total average power loss of all devices on the heatsink. As mentioned earlier, because the heatsink thermal capacitance is relatively large and the corresponding thermal time constant is relatively long compared with the AC fundamental period, the steady-state heatsink temperature variation in a fundamental period is negligible. Consequently, the average power loss can be used here to determine the required TMS thermal resistance for constant load conditions.

For transient overload conditions, Laplace transform (LT) and inverse Laplace transform (ILT) can be employed in frequency-domain calculation. Since the overload duration in a motor drive is generally much longer than the thermal time constants of devices, the loss and thermal behaviors for the device can be assumed to be in steady state, albeit at a higher "steady-state" load and loss point. The earlier procedure still applies for junction-to case-temperature. For case-to-ambient temperature rise, only the average device losses need to be considered. Assume that the TMS thermal capacitance is C_{th-ca}, the initial maximum total average inverter device loss is the loss at the rated condition $P_{total-rated}$, the overload average inverter device loss is P_{OL}, and the overload duration is t_{OL}. The case-to-ambient temperature rise at end of the overload period is

$$T_{ca}(t_{OL}) = T_c(t_{OL}) - T_a = P_{OL} \cdot R_{th-ca} + (P_{total-rated} - P_{OL}) \cdot R_{th-ca} \cdot e^{-\frac{t_{OL}}{\tau_{ca}}} \tag{6.83}$$

where $\tau_{ca} = R_{th-ca} \cdot C_{th-ca}$ is the TMS thermal time constant. Assuming that the maximum junction-to-case temperature for the overload condition is T_{jc-OL}, which can be calculated using the steady-state procedure illustrated in Figure 6.19, then

$$T_j(t_{OL}) - T_{jc-OL} - T_a = P_{OL} \cdot R_{th-ca} + (P_r - P_{OL}) \cdot R_{th-ca} \cdot e^{-\frac{t_{OL}}{\tau_{ca}}} \tag{6.84}$$

Given that there should be $T_j(t_{OL}) \leq T_{jmax}$, the relationship (6.85) should be satisfied for the thermal management system, where $k_p = \dfrac{P_{OL}}{P_{total-rated}}$. If there is no overload, i.e. $k_p = 1$, then (6.85) reduces to (6.78) for the steady-state rated condition with only R_{th-ca} constraint needed; if the overload period is very long, then (6.85) also can be approximated using (6.78) for the steady-state overload condition with only R_{th-ca} constraint needed. In general, both the TMS thermal resistance R_{th-ca} and capacitance C_{th-ca} have to be considered.

$$R_{th-ca} \cdot \left[k_p \left(1 - e^{-\frac{t_{OL}}{\tau_{ca}}} \right) + e^{-\frac{t_{OL}}{\tau_{ca}}} \right] \leq \dfrac{T_{jmax} - T_{jc\_OL} - T_a}{P_{total-rated}} \tag{6.85}$$

6.3.5 Device Switching Overvoltage

Due to the device and circuit parasitics, there can be overvoltage on the device during switching transients. In principle, higher parasitics (e.g. power loop inductances) and faster switching speed can lead to higher overvoltage. As it is generally desirable to switch fast in an inverter for reduced power loss and smaller passive components, it is important to understand the switching overvoltage relationship with switching speed and circuit parasitics, such that the device peak voltage constraint will not be violated.

6.3.5.1 Turnoff Overvoltage

In Si IGBT-based inverters, the turnoff overvoltage is the main concern. Assuming a DC-link decoupling capacitor, which will be discuss in Section 6.3.6, has already been employed to maintain the DC voltage across the inverter phase leg, reference [11] analyzed the turnoff overvoltage using the simplified model in Figure 6.21. The similar analysis has also been carried out for Si MOSFET-based converter in [12]. In Figure 6.21, V_{bus} represents the voltage on the capacitance part of the decoupling capacitor, which can be considered as a constant during the switching transient; L_{bus} and R_{bus} represent total parasitic inductance and resistance external to the switching device module, including the parasitics of the decoupling capacitor; L_D represents lead inductance of the device drain and source (or collector and emitter); C_{OSS} is the device output capacitance; and R_d represents the equivalent damping resistance related to the device module, whose value is between 0 and gate resistance R_g but not very critical according to [12]. Several other points should be noted about the model in Figure 6.21:

1) The capacitance C_{OSS} for MOSFET and IGBT is generally highly nonlinear, especially during the low-voltage region, as discussed in Chapter 2. However, at high-voltage region, it is relatively small and varies in a limited range. Since the overvoltage condition corresponds to the high-voltage region, C_{OSS} can be considered as a constant.
2) The complementary diode of the device being turned off is not included in the equivalent circuit because the diode is forward biased during the switching transient stage when the voltage overshoot is built up beyond V_{bus} and the diode will not impact the turnoff overvoltage. The difference between the AC load current and the device switching current, i.e. the device drain current, flows through the diode.
3) The device switching current during the turnoff voltage ringing is represented by a current source, whose current is assumed to decrease linearly from the starting load current I_0 to 0 within T_{off} time.
4) The drain-source voltage v_{ds} in Figure 6.21 denotes the voltage across the device module, including the voltage drop on module parasitics L_d and R_d.

As derived in [11] and [12], the voltage overshoot can be first obtained via LT as

$$\Delta v_{ds}(s) = \frac{L_{bus}}{L_{total}} \left[\frac{I_0}{T_{off} C_{OSS}} \frac{(1 + sR_d C_{OSS})}{s\left(s^2 + s\frac{R_{total}}{L_{total}} + \frac{1}{L_{total}C_{OSS}}\right)} \right] \cdot \left[\left(1 - e^{-sT_{off}}\right)\left(1 + \frac{R_{bus}}{sL_{total}}\right) - \frac{R_{bus}T_{off}}{L_{total}} \right]$$

$$(6.86)$$

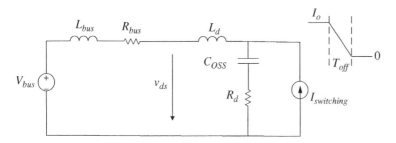

Figure 6.21 Model of turnoff voltage overshoot and resonance. *Source:* Adapted form [11].

where $L_{total} = L_{bus} + L_d$ and $R_{total} = R_{bus} + R_d$. In practice, $sR_dC_{OSS} \ll 1$, $\dfrac{R_{bus}}{sL_{total}} \ll 1$, and $\dfrac{R_{bus}T_{off}}{L_{total}} \ll 1$, such that Eq. (6.86) and its corresponding time-domain equation can be simplified as

$$\Delta v_{ds}(s) = \frac{L_{bus}}{L_{bus} + L_d}\left[\frac{I_o}{T_{off} \cdot C_{out}} \cdot \frac{1}{s\left(s^2 + \dfrac{R_d + R_{bus}}{L_{bus} + L_d} \cdot s + \dfrac{1}{C_{out} \cdot (L_{bus} + L_d)}\right)}\right] \cdot \left[1 - e^{-s \cdot T_{off}}\right]$$

(6.87)

$$\Delta v_{ds}(t) = \frac{I_o L_{bus}}{T_{off}}\left[1 - \frac{1}{\sqrt{1-\zeta^2}} \cdot e^{-\zeta\omega_o t} \cdot \sin\left(\omega_o\sqrt{1-\zeta^2}t + \psi\right)\right]$$

(6.88)

where $\zeta = \left(\dfrac{R_{bus} + R_d}{2} \cdot \sqrt{\dfrac{C_{OSS}}{L_d}}\right)$, $\omega_o = \sqrt{\dfrac{1}{(L_d + L_{bus}) \cdot C_{out}}}$, $\psi = \cos^{-1}\zeta$. From (6.88), the peak over-

voltage can be readily determined. In [12], it is determined that the peak voltage overshoot occurs at approximately

$$T_{peak} = \frac{\pi}{\sqrt{\dfrac{1}{(L_d + L_{bus}) \cdot C_{OSS}} - \left(\dfrac{R_{bus} + R_d}{2(L_d + L_{bus})}\right)^2}}$$

(6.89)

and the corresponding peak voltage overshoot is

$$\Delta v_{ds}\left(T_{peak}\right) = \frac{I_o L_{bus}}{T_{off}}\left[1 - \frac{1}{\sqrt{1-\zeta^2}} \cdot e^{-\zeta\omega_o T_{peak}} \cdot \sin\left(\omega_o\sqrt{1-\zeta^2}T_{peak} + \psi\right)\right]$$

(6.90)

One shortcoming with using (6.88) and (6.90) is that they rely on an assumed constant rate of change for the device current, which in fact can be a complex nonlinear function of the device characteristics and switching process. Another shortcoming is that for fast switching devices, such as WBG devices, the turn-on switching overvoltage can be as large or even larger than the turnoff overvoltage. As a result, an improved and more comprehensive overvoltage model may be necessary in some inverter designs. Simulation-based tools like SPICE can be used for more accurate models for SiC MOSFET as demonstrated in [13] and [14]. However, for design, we still prefer an analytical or numerical model based on parameters from datasheet.

A MOSFET can be modeled as in Figure 6.22 during the switching transient according to [15] and similar previous works. Since the device and circuit parameters are generally highly nonlinear, as discussed in Chapter 2, it is necessary to divide the turn-on and turnoff processes into several stages to facilitate the modeling. As a result, the models can be quite complex and often involve simplifications. Many analytical models have been developed for MOSFET switching process, with different simplifications to balance the complexity and accuracy. Here, a modeling approach based on more recent work for SiC MOSFET in [16] and [17] are introduced, which are suitable for design use.

As shown in Figure 2.18, during the turnoff process, the MOSFET model can be divided into four stages: delay, voltage rise, current fall, and cutoff (or gate voltage fall as in Chapter 2). The peak turnoff overvoltage generally occurs during the current fall stage. However, in order to determine the peak voltage, it is necessary to model the turnoff transient process until the peak overvoltage occurrence.

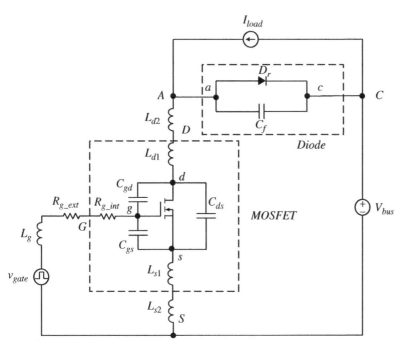

Figure 6.22 Equivalent circuit of a MOSFET for obtaining the model. *Source:* Adapted from [18].

6.3.5.1.1 Turnoff Delay Stage The MOSFET is in ohmic region before turning off. During the turn-off delay stage, the gate drive supply voltage v_{gate} (using v_{gate} instead of v_g to be clearer) is changed from a positive voltage V_{GH} to a negative voltage V_{GL} (or turned off to 0). The gate-to-source voltage v_{gs} drops and the MOSFET remain in ohmic region. This stage ends when v_{gs} reduces to the Miller voltage and the MOSFET enters into saturation region. The equations for this stage are (6.91)–(6.93). In ohmic region, i_d can be approximated as the constant I_{load} and v_{ds} can be neglected. Consequently, the equations can be solved as in (6.94) and (6.95), where p_1 and p_2 are in (6.96), and are both assumed real. If p_1 and p_2 are imaginary, an underdamped solution can be derived for v_{gs} and i_g. The turnoff delay period ends at $t = t_1$ when v_{gs} equals to $(I_{load}/g_{fs} + V_{TH})$, which is called Miller voltage V_{miller}. Note that $C_{ISS} = C_{gs} + C_{gd}$, g_{fs} is the forward transconductance, and V_{TH} is the threshold gate voltage. Note also that $L_s = L_{s1} + L_{s2}$, $L_d = L_{d1} + L_{d2}$, and $R_g = (R_{g\_int} + R_{g\_ext})$ in Figure 6.22.

$$V_{GL} = R_g i_g(t) + v_{gs}(t) + L_g\left(\frac{di_g(t)}{dt}\right) + L_s\left(\frac{di_d(t)}{dt} + \frac{di_g(t)}{dt}\right)$$
$$\approx R_g i_g(t) + v_{gs}(t) + \left(L_g + L_s\right)\left(\frac{di_g(t)}{dt}\right) \tag{6.91}$$

$$i_g(t) = C_{gs}\frac{dv_{gs}(t)}{dt} + C_{gd}\frac{dv_{gd}(t)}{dt} = C_{gs}\frac{dv_{gs}(t)}{dt} + C_{gd}\frac{d\left[v_{gs}(t) - v_{ds}(t)\right]}{dt} \approx C_{ISS}\frac{dv_{gs}(t)}{dt} \tag{6.92}$$

$$v_{gs}(t) = v_{gd}(t) + v_{ds}(t) \tag{6.93}$$

$$v_{gs}(t) = V_{GL} + \frac{(V_{GH} - V_{GL})}{(p_2 - p_1)}\left(p_2 e^{p_1 t} - p_1 e^{p_2 t}\right) \tag{6.94}$$

$$i_g(t) = C_{ISS}p_2 p_1 \frac{(V_{GH} - V_{GL})}{(p_2 - p_1)}\left(e^{p_1 t} - e^{p_2 t}\right) \tag{6.95}$$

$$p_{1,2} = -\frac{R_g}{2(L_g + L_s)} \pm \sqrt{\left(\frac{R_g}{2(L_g + L_s)}\right)^2 - \frac{1}{(L_g + L_s)C_{ISS}}} \tag{6.96}$$

6.3.5.1.2 Voltage Rise Stage In this stage, the MOSFET is about to enter the saturation region from the ohmic region. The channel current decreases from the load current and the device capacitances will be charged; and the drain-source voltage starts to rise and finally reaches the DC bus voltage. In addition to (6.91)–(6.93) (nonapproximated versions), circuit equations also include (6.97)–(6.99). The MOSFET works in saturation region, so its channel current can be modeled as a voltage controlled current source, as in (6.100). There are five independent state variables for these equations, which can be chosen as v_{ds}, v_{gs}, i_d, i_g, and v_{ca}. The equations can be rewritten in a state-space form as in (6.104), where $C_{OSS} = C_{ds} + C_{gd}$, $L_{geq} = L_g + L_s$, and $L_{eq} = L_d + L_s$.

$$v_{ds}(t) = V_{bus} - v_{ca}(t) - R_s i_d(t) - (L_s + L_d)\frac{di_d(t)}{dt} - L_s\frac{di_g(t)}{dt} \tag{6.97}$$

$$i_d(t) = i_{ch}(t) + C_{ds}\frac{dv_{ds}(t)}{dt} - C_{gd}\frac{dv_{gd}(t)}{dt} \tag{6.98}$$

$$C_f\frac{dv_{ca}(t)}{dt} = i_d(t) - I_{load} \tag{6.99}$$

$$i_{ch}(t) = g_{fs} \cdot \left[v_{gs}(t) - V_{TH}\right] \tag{6.100}$$

$$
\begin{bmatrix} \dfrac{dv_{ds}}{dt} \\[2ex] \dfrac{dv_{gs}}{dt} \\[2ex] \dfrac{di_d}{dt} \\[2ex] \dfrac{di_g}{dt} \\[2ex] \dfrac{dv_{ca}}{dt} \end{bmatrix}
=
\begin{bmatrix}
0 & -\dfrac{C_{ISS}g_{fs}}{C_{OSS}C_{ISS} - C_{gd}^2} & \dfrac{C_{ISS}}{C_{OSS}C_{ISS} - C_{gd}^2} & \dfrac{C_{gd}}{C_{OSS}C_{ISS} - C_{gd}^2} & 0 \\[2ex]
0 & -\dfrac{g_{fs}C_{gd}}{C_{OSS}C_{ISS} - C_{gd}^2} & \dfrac{C_{gd}}{C_{OSS}C_{ISS} - C_{gd}^2} & \dfrac{C_{OSS}}{C_{OSS}C_{ISS} - C_{gd}^2} & 0 \\[2ex]
-\dfrac{L_{geq}}{L_{eq}L_{geq} - L_s^2} & \dfrac{L_s}{L_{eq}L_{geq} - L_s^2} & -\dfrac{R_s L_{geq}}{L_{eq}L_{geq} - L_s^2} & \dfrac{R_g L_s}{L_{eq}L_{geq} - L_s^2} & -\dfrac{L_{geq}}{L_{eq}L_{geq} - L_s^2} \\[2ex]
\dfrac{L_s}{L_{eq}L_{geq} - L_s^2} & -\dfrac{L_{eq}}{L_{eq}L_{geq} - L_s^2} & \dfrac{R_s L_s}{L_{eq}L_{geq} - L_s^2} & -\dfrac{R_g L_{eq}}{L_{eq}L_{geq} - L_s^2} & \dfrac{L_s}{L_{eq}L_{geq} - L_s^2} \\[2ex]
0 & 0 & \dfrac{1}{C_f} & 0 & 0
\end{bmatrix}
$$

$$
\cdot
\begin{bmatrix} v_{ds} \\[1ex] v_{gs} \\[1ex] i_d \\[1ex] i_g \\[1ex] v_{ca} \end{bmatrix}
+
\begin{bmatrix}
\dfrac{g_{fs}C_{ISS}V_{TH}}{C_{OSS}C_{ISS} - C_{gd}^2} \\[3ex]
\dfrac{g_{fs}C_{gd}V_{TH}}{C_{OSS}C_{ISS} - C_{gd}^2} \\[3ex]
\dfrac{L_{geq}V_{bus} - L_s V_{GL}}{L_{eq}L_{geq} - L_s^2} \\[3ex]
-\dfrac{L_s V_{bus} - L_{eq}V_{GL}}{L_{eq}L_{geq} - L_s^2} \\[3ex]
\dfrac{-I_{load}}{C_f}
\end{bmatrix}
$$

$$\tag{6.101}$$

The initial conditions for voltage rise stage state variables can be approximated as in (6.102), together with $i_g(t_1)$ that can be determined from (6.95) of the turnoff delay stage. The voltage rise stage ends at $t = t_2$ when v_{ds} reaches V_{bus} or v_{ca} becomes forward biased.

$$
\begin{cases}
v_{ds}(t_1) \approx v_{gs}(t_1) = I_{load}/g_{fs} + V_{TH} \\
i_d(t_1) \approx I_{load} \\
v_{ca}(t_1) \approx V_{bus} - v_{ds}(t_1)
\end{cases}
\tag{6.102}
$$

6.3.5.1.3 *Current Falling Stage*

In this period, the diode will be forward biased after v_{ds} passes V_{bus}. The MOSFET continues to work in saturation region. Equation (6.97) becomes (6.103). Other equations are (6.91), (6.92), (6.93), (6.98), and (6.100). Four independent state variables are chosen as v_{ds}, v_{gs}, i_d and i_g. The equations can be rewritten in a state-space form as in (6.104).

$$
v_{ds}(t) = V_{bus} - R_s i_d(t) - (L_s + L_d)\frac{di_d(t)}{dt} - L_s \frac{di_g(t)}{dt}
\tag{6.103}
$$

$$
\begin{bmatrix} \dfrac{dv_{ds}}{dt} \\[2mm] \dfrac{dv_{gs}}{dt} \\[2mm] \dfrac{di_d}{dt} \\[2mm] \dfrac{di_g}{dt} \end{bmatrix} =
\begin{bmatrix}
0 & -\dfrac{C_{ISS}g_{fs}}{C_{OSS}C_{ISS} - C_{gd}^2} & \dfrac{C_{ISS}}{C_{OSS}C_{ISS} - C_{gd}^2} & \dfrac{C_{gd}}{C_{OSS}C_{ISS} - C_{gd}^2} \\[3mm]
0 & -\dfrac{g_{fs}C_{gd}}{C_{OSS}C_{ISS} - C_{gd}^2} & \dfrac{C_{gd}}{C_{OSS}C_{ISS} - C_{gd}^2} & \dfrac{C_{OSS}}{C_{OSS}C_{ISS} - C_{gd}^2} \\[3mm]
-\dfrac{L_{geq}}{L_{eq}L_{geq} - L_s^2} & \dfrac{L_s}{L_{eq}L_{geq} - L_s^2} & -\dfrac{R_s L_{geq}}{L_{eq}L_{geq} - L_s^2} & \dfrac{R_g L_s}{L_{eq}L_{geq} - L_s^2} \\[3mm]
\dfrac{L_s}{L_{eq}L_{geq} - L_s^2} & -\dfrac{L_{eq}}{L_{eq}L_{geq} - L_s^2} & \dfrac{R_s L_s}{L_{eq}L_{geq} - L_s^2} & -\dfrac{R_g L_{eq}}{L_{eq}L_{geq} - L_s^2}
\end{bmatrix}
\cdot
\begin{bmatrix} v_{ds} \\[2mm] v_{gs} \\[2mm] i_d \\[2mm] i_g \end{bmatrix}
$$

$$
+
\begin{bmatrix}
\dfrac{g_{fs}C_{ISS}V_{TH}}{C_{OSS}C_{ISS} - C_{gd}^2} \\[4mm]
\dfrac{g_{fs}C_{gd}V_{TH}}{C_{OSS}C_{ISS} - C_{gd}^2} \\[4mm]
\dfrac{L_{geq}V_{bus} - L_s V_{GL}}{L_{eq}L_{geq} - L_s^2} \\[4mm]
-\dfrac{L_s V_{bus} - L_{eq}V_{GL}}{L_{eq}L_{geq} - L_s^2}
\end{bmatrix}
\tag{6.104}
$$

The initial conditions for current falling stage state variables can be determined by solutions of (6.101) at $t = t_2$ when $v_{ds} = V_{bus}$. Due to the high order of Eqs. (6.101), (6.104), and nonlinearity of the parameters, especially the capacitances, it is difficult to get a closed-form solution for v_{ds}. Both [16] and [17] use a numerical approach to solve the switching process, which can be illustrated in Figure 6.23. It should be mentioned that most of these previous works focused on switching loss calculation, which requires the whole switching process modeling and calculation. For overvoltage calculation, we only need peak voltage values, which usually occur before the full switching transients complete. Of course, in the design, we are also interested in loss modeling, as explained earlier in Section 6.3.2. In fact, the design tool to be used later in this chapter and also in Chapter 13 has

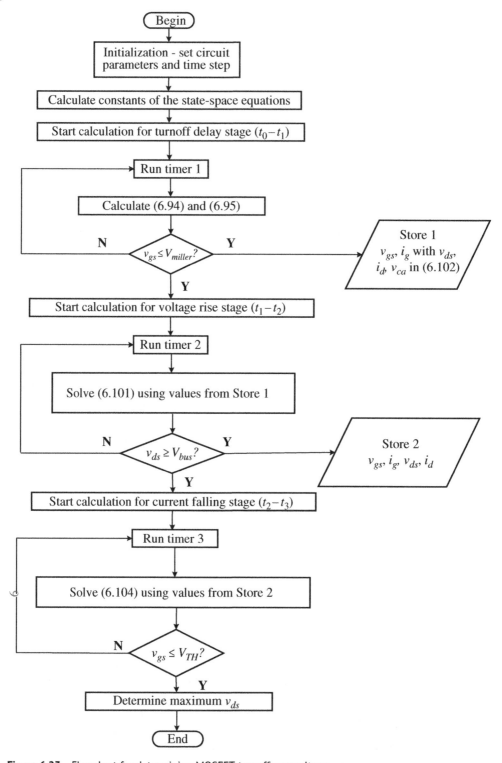

Figure 6.23 Flowchart for determining MOSFET turnoff overvoltage.

a provision to adopt this so-called discrete time modeling based on switching transient stages for switching loss calculation.

Compared with time-domain circuit simulation using physics-based device models, the calculation process in Figure 6.23 only requires datasheet parameters and is therefore preferred. On the other hand, it still involves numerically solving nonlinear differential equations and can be time consuming for the design optimization. To alleviate such issues, simpler analytical solutions have been developed under different simplification assumptions. According to [17], at the start of the current falling stage, there is a substantial difference between i_d and i_{ch}. v_{gs} and i_{ch} remain almost constant throughout this stage, and i_d decreases from its initial value, which will lead to the rise in v_{ds} until it reaches its peak value and $i_d = i_{ch}$. Considering $i_g \ll i_d$ and neglecting the impact of parasitic power loop resistance R_s, Eq. (6.103) is simplified to (6.105). With the approximation of $dv_{dg}/dt \approx dv_{ds}/dt$, (6.106) can be obtained.

$$v_{ds}(t) \approx V_{bus} - (L_s + L_d)\frac{di_d(t)}{dt} \tag{6.105}$$

$$i_d(t) - I_{ch} \approx C_{OSS}\frac{dv_{ds}(t)}{dt} \tag{6.106}$$

The time-domain solution for v_{ds} can be obtained as in (6.107), where ω_a is defined in (6.108). Note that the initial conditions $v_{ds}(t_2) = V_{bus}$ and $I_d(t_2) = I_{load}$ are assumed.

$$v_{ds}(t) = V_{bus} + \omega_a(L_s + L_d)(I_{load} - I_{ch})\sin[\omega_a(t - t_2)] \tag{6.107}$$

$$\omega_a = \sqrt{\frac{1}{(L_s + L_d)C_{OSS}}} \tag{6.108}$$

The channel current I_{ch} can be determined from the previous voltage rise stage and has been derived in [18] as (6.109), where M is defined in (6.110).

$$I_{ch} = \frac{(M-1)I_{load} - g_{fs}V_{TH}}{M} \tag{6.109}$$

$$M = 1 + R_g g_{fs}\frac{C_{gd}}{C_{OSS}} \tag{6.110}$$

Let us consider an example provided in [17] using two SiC MOSFETs from Wolfspeed. The device and circuit parameters are in Table 6.5, where R_{g_int} is the internal gate resistance. The peak voltages under different conditions are shown in Table 6.6, where the external gate resistance R_{g_ext} also includes the gate driver resistance (0.5 Ω in this case). The data for the analytic method 2, simulation, and experiment are obtained from [17], and the analytic method 1 is based on Eq. (6.107). It can be seen that in most cases, the analytic method 1 results are close to the results of the analytic

Table 6.5 Circuit and device parameters.

	L_s (nH)	L_d (nH)	R_{g_int} (Ω)	g_{fs} (Ω^{-1})	C_{OSS} (pF)	C_{ISS} (pF)	C_{gd} (pF)	V_{TH} (V)	V_{bus} (V)
C2M0160120D	6	60	6.5	3.8	55	606	5	2.9	800
C2M0080120D	4	65	3.9	10	92	1130	7.5	2.9	800

Table 6.6 Comparison of maximum v_{ds} in volts using different methods.

| | I_{load} (A) | R_{g_ext} = 9 Ω | | | | R_{g_ext} = 3 Ω | | | |
		Anly1	Anly2	Sim	Exp	Anly1	Anly2	Sim	Exp
C2M0160120D	5	879.4	851.7	847.4	864	917.8	848.6	848.5	880
	10	904.2	891.4	903.9	888	954.6	928.6	924.9	920
	15	928.9	902.0	919.0	912	991.4	951.5	957.9	1000
C2M0080120D	5	876.2	840.3	844.0	872	932.4	937.1	945.2	964
	15	898.6	910.4	931.2	944	971.3	971.9	985.2	990
	25	921.0	927.0	961.7	964	1010.3	1005.7	1030.6	1020

Note: Anly1, analytical method 1; Anly2, analytical method 2; Sim, circuit simulation; Exp, experiment.

method 2, simulation, and/or experiment, and its accuracy should be sufficient for the design. Note that the analytic method results in Table 6.6 used typical values for devices in Table 6.5. To be conservative, the more extreme conditions may be considered in the design. For example, maximum gate threshold voltage and gate source transconductance will result a higher peak voltage.

6.3.5.2 Turn-on Overvoltage

6.3.5.2.1 Turn-on Delay Stage
The MOSFET is in cutoff region before turning on. During the turn-on delay stage, the gate driver supply voltage v_{gate} is changed from a negative or zero voltage V_{GL} to a positive voltage V_{GH}. The MOSFET stays in cutoff region and its input capacitance is charged until v_{gs} reaches the threshold voltage V_{TH}. The equations for this stage are similar to those for the turnoff delay stage. Eq. (6.91) becomes (6.111) with drain current $i_d = 0$; Eqs. (6.92) and (6.93) stay the same. In the cutoff region, v_{ds} is constant and equals to V_{bus} (assuming the voltage drop on the complementary diode negligible) and initial conditions $v_{gs}(0) = V_{GL}$ and $i_g(0) = 0$. Given these conditions, v_{gs} and i_g can be determined as in (6.112) and (6.113), which are similar to (6.94) and (6.95) for the MOSFET turnoff delay stage with the same p_1 and p_2 defined in (6.96). The turn-on delay stage ends at $t = t_1$ when v_{gs} equals to V_{TH}.

$$V_{GH} = R_g i_g(t) + v_{gs}(t) + L_s \frac{di_g(t)}{dt} \tag{6.111}$$

$$v_{gs}(t) = V_{GH} + \frac{(V_{GL} - V_{GH})}{(p_2 - p_1)} \left(p_2 e^{p_1 t} - p_1 e^{p_2 t} \right) \tag{6.112}$$

$$i_g(t) = C_{ISS} p_2 p_1 \frac{(V_{GL} - V_{GH})}{(p_2 - p_1)} \left(e^{p_1 t} - e^{p_2 t} \right) \tag{6.113}$$

6.3.5.2.2 Current Rise Stage
In this stage, the current commutation happens between the MOSFET and its complementary diode in the same phase leg. The MOSFET enters saturation region, and the drain current starts to rise and reaches the load current level at the end of the stage. Here we focus on SiC MOSFET and SiC Schottky diode (or SiC MOSFET body diode), such that the diode

reverse recovery can be neglected. The circuit equations for this stage are similar to those for the current falling stage of the turnoff process. Equations (6.92), (6.93), (6.98), (6.100), and (6.103) still hold. Equation (6.91) becomes (6.114). These equations can be rewritten in state-space form as (6.115), which is almost identical to the state-space equation (6.104) of the turnoff current falling stage, except that V_{GH} replaces V_{GL}.

$$V_{GH} = R_g i_g(t) + v_{gs}(t) + L_s\left(\frac{di_d(t)}{dt} + \frac{di_g(t)}{dt}\right) \tag{6.114}$$

$$
\begin{bmatrix} \dfrac{dv_{ds}}{dt} \\[2mm] \dfrac{dv_{gs}}{dt} \\[2mm] \dfrac{di_d}{dt} \\[2mm] \dfrac{di_g}{dt} \end{bmatrix} =
\begin{bmatrix}
0 & -\dfrac{C_{ISS}g_{fs}}{C_{OSS}C_{ISS}-C_{gd}^2} & \dfrac{C_{ISS}}{C_{OSS}C_{ISS}-C_{gd}^2} & \dfrac{C_{gd}}{C_{OSS}C_{ISS}-C_{gd}^2} \\[3mm]
0 & -\dfrac{g_{fs}C_{gd}}{C_{OSS}C_{ISS}-C_{gd}^2} & \dfrac{C_{gd}}{C_{OSS}C_{ISS}-C_{gd}^2} & \dfrac{C_{OSS}}{C_{OSS}C_{ISS}-C_{gd}^2} \\[3mm]
-\dfrac{L_{geq}}{L_{eq}L_{geq}-L_s^2} & \dfrac{L_s}{L_{eq}L_{geq}-L_s^2} & -\dfrac{R_s L_{geq}}{L_{eq}L_{geq}-L_s^2} & \dfrac{R_g L_s}{L_{eq}L_{geq}-L_s^2} \\[3mm]
\dfrac{L_s}{L_{eq}L_{geq}-L_s^2} & -\dfrac{L_{eq}}{L_{eq}L_{geq}-L_s^2} & \dfrac{R_s L_s}{L_{eq}L_{geq}-L_s^2} & -\dfrac{R_g L_{eq}}{L_{eq}L_{geq}-L_s^2}
\end{bmatrix}
\cdot
\begin{bmatrix} v_{ds} \\[2mm] v_{gs} \\[2mm] i_d \\[2mm] i_g \end{bmatrix}
$$

$$
+
\begin{bmatrix}
\dfrac{g_{fs}C_{ISS}V_{TH}}{C_{OSS}C_{ISS}-C_{gd}^2} \\[4mm]
\dfrac{g_{fs}C_{gd}V_{TH}}{C_{OSS}C_{ISS}-C_{gd}^2} \\[4mm]
\dfrac{L_{geq}V_{bus}-L_s V_{GH}}{L_{eq}L_{geq}-L_s^2} \\[4mm]
-\dfrac{L_s V_{bus}-L_{eq}V_{GH}}{L_{eq}L_{geq}-L_s^2}
\end{bmatrix}
\tag{6.115}
$$

The initial conditions for the current rise stage state variables can be approximated as in (6.116), together with $i_g(t_1)$ that can be determined from (6.113) of the turn-on delay stage. The current rise stage ends at $t = t_2$ when v_{gs} equals to the Miller voltage and the drain current i_d settles at the load current I_{load}.

$$
\begin{cases}
v_{ds}(t_1) = V_{bus} \\[2mm]
i_d(t_1) \approx 0 \\[2mm]
v_{gs}(t_1) = V_{TH}
\end{cases}
\tag{6.116}
$$

6.3.5.2.3 Voltage Falling Stage

In this stage, the diode starts to block. Most circuit equations describing the voltage rise stage of the turnoff transient, (6.92), (6.93), and (6.97)–(6.100), are still valid. Eq. (6.91) is replaced by (6.114) with V_{GL} replaced by V_{GH}. The equations can be rewritten in state-space form as (6.117), which is almost identical to the state-space equation (6.101) of the turn-off voltage rise stage, except that V_{GH} replaces V_{GL}.

$$
\begin{bmatrix} \dfrac{dv_{ds}}{dt} \\[2mm] \dfrac{dv_{gs}}{dt} \\[2mm] \dfrac{di_d}{dt} \\[2mm] \dfrac{di_g}{dt} \\[2mm] \dfrac{dv_{ca}}{dt} \end{bmatrix}
=
\begin{bmatrix}
0 & -\dfrac{C_{ISS}g_{fs}}{C_{OSS}C_{ISS}-C_{gd}^2} & \dfrac{C_{ISS}}{C_{OSS}C_{ISS}-C_{gd}^2} & \dfrac{C_{gd}}{C_{OSS}C_{ISS}-C_{gd}^2} & 0 \\[3mm]
0 & -\dfrac{g_{fs}C_{gd}}{C_{OSS}C_{ISS}-C_{gd}^2} & \dfrac{C_{gd}}{C_{OSS}C_{ISS}-C_{gd}^2} & \dfrac{C_{OSS}}{C_{OSS}C_{ISS}-C_{gd}^2} & 0 \\[3mm]
-\dfrac{L_{geq}}{L_{eq}L_{geq}-L_s^2} & \dfrac{L_s}{L_{eq}L_{geq}-L_s^2} & -\dfrac{R_sL_{geq}}{L_{eq}L_{geq}-L_s^2} & \dfrac{R_gL_s}{L_{eq}L_{geq}-L_s^2} & -\dfrac{L_{geq}}{L_{eq}L_{geq}-L_s^2} \\[3mm]
\dfrac{L_s}{L_{eq}L_{geq}-L_s^2} & -\dfrac{L_{eq}}{L_{eq}L_{geq}-L_s^2} & \dfrac{R_sL_s}{L_{eq}L_{geq}-L_s^2} & -\dfrac{R_gL_{eq}}{L_{eq}L_{geq}-L_s^2} & \dfrac{L_s}{L_{eq}L_{geq}-L_s^2} \\[3mm]
0 & 0 & \dfrac{1}{C_f} & 0 & 0
\end{bmatrix}
\cdot
\begin{bmatrix} v_{ds} \\[2mm] v_{gs} \\[2mm] i_d \\[2mm] i_g \\[2mm] v_{ca} \end{bmatrix}
$$

$$
+
\begin{bmatrix}
\dfrac{g_{fs}C_{ISS}V_{TH}}{C_{OSS}C_{ISS}-C_{gd}^2} \\[4mm]
\dfrac{g_{fs}C_{gd}V_{TH}}{C_{OSS}C_{ISS}-C_{gd}^2} \\[4mm]
\dfrac{L_{geq}V_{bus}-L_sV_{GH}}{L_{eq}L_{geq}-L_s^2} \\[4mm]
-\dfrac{L_sV_{bus}-L_{eq}V_{GH}}{L_{eq}L_{geq}-L_s^2} \\[4mm]
\dfrac{-I_{load}}{C_f}
\end{bmatrix}
$$

(6.117)

The initial conditions for voltage falling stage state variables can be approximated as in (6.118), together with $i_g(t_2)$ and $v_{ds}(t_2)$ that can be determined from (6.115) of the turn-on current rise stage. The voltage falling stage ends at $t = t_3$ when v_{ds} equals to the on-state voltage V_{ds_on}. Note that during the voltage falling stage, the MOSFET goes through saturation to ohmic region.

$$
\begin{cases}
i_d(t_2) = I_{load} \\[2mm]
v_{gs}(t_2) = V_{miller} \\[2mm]
v_{ca}(t_2) = 0
\end{cases}
$$

(6.118)

6.3.5.2.4 Ringing Stage In this stage (termed gate voltage rise stage in Chapter 2), MOSFET is in ohmic region, and the drain current and diode voltage will ring before settling to the steady state. Equations (6.92), (6.93), (6.97), (6.99), and (6.114) can still be used. In addition, Eq. (6.98) becomes (6.119). These equations can be rewritten in state-space form as (6.120), which is almost identical to the state-space equation (6.117) in the previous voltage falling stage, except that the channel current expression changes from $g_{fs}[v_{gs}(t) - V_{TH}]$ to v_{ds}/R_{ds_on}.

$$
i_d(t) = \frac{v_{ds}(t)}{R_{ds\_on}} + C_{OSS}\frac{dv_{ds}(t)}{dt} - C_{gd}\frac{dv_{gs}(t)}{dt}
$$

(6.119)

$$
\begin{bmatrix}
\dfrac{dv_{ds}}{dt} \\[2mm]
\dfrac{dv_{gs}}{dt} \\[2mm]
\dfrac{di_d}{dt} \\[2mm]
\dfrac{di_g}{dt} \\[2mm]
\dfrac{dv_{ca}}{dt}
\end{bmatrix}
=
\begin{bmatrix}
-\dfrac{C_{ISS}}{(C_{OSS}C_{ISS}-C_{gd}{}^2)R_{ds\_on}} & 0 & \dfrac{C_{ISS}}{C_{OSS}C_{ISS}-C_{gd}{}^2} & \dfrac{C_{gd}}{C_{OSS}C_{ISS}-C_{gd}{}^2} & 0 \\[3mm]
-\dfrac{C_{gd}}{(C_{OSS}C_{ISS}-C_{gd}{}^2)R_{ds\_on}} & 0 & \dfrac{C_{gd}}{C_{OSS}C_{ISS}-C_{gd}{}^2} & \dfrac{C_{OSS}}{C_{OSS}C_{ISS}-C_{gd}{}^2} & 0 \\[3mm]
-\dfrac{L_{geq}}{L_{eq}L_{geq}-L_s{}^2} & \dfrac{L_s}{L_{eq}L_{geq}-L_s{}^2} & -\dfrac{R_sL_{geq}}{L_{eq}L_{geq}-L_s{}^2} & \dfrac{R_gL_s}{L_{eq}L_{geq}-L_s{}^2} & -\dfrac{L_{geq}}{L_{eq}L_{geq}-L_s{}^2} \\[3mm]
\dfrac{L_s}{L_{eq}L_{geq}-L_s{}^2} & -\dfrac{L_{eq}}{L_{eq}L_{geq}-L_s{}^2} & \dfrac{R_sL_s}{L_{eq}L_{geq}-L_s{}^2} & -\dfrac{R_gL_{eq}}{L_{eq}L_{geq}-L_s{}^2} & \dfrac{L_s}{L_{eq}L_{geq}-L_s{}^2} \\[3mm]
0 & 0 & \dfrac{1}{C_{ak}} & 0 & 0
\end{bmatrix}
\cdot
\begin{bmatrix}
v_{ds} \\[2mm]
v_{gs} \\[2mm]
i_d \\[2mm]
i_g \\[2mm]
v_{ca}
\end{bmatrix}
$$

$$
+
\begin{bmatrix}
0 \\[2mm]
0 \\[2mm]
\dfrac{L_{geq}V_{bus}-L_sV_{GH}}{L_{eq}L_{geq}-L_s{}^2} \\[3mm]
-\dfrac{L_sV_{bus}-L_{eq}V_{GH}}{L_{eq}L_{geq}-L_s{}^2} \\[3mm]
-I_{load}
\end{bmatrix}
$$

$$(6.120)$$

The initial conditions for ringing stage state variables include $v_{ds}=V_{ds\_on}$. The initial conditions for v_{gs} i_d, i_g, and v_{ca} can be determined with Eq. (6.117) at $t=t_3$.

Similar to the turnoff process, due to the high order of Eqs. (6.115), (6.117), (6.120), and nonlinearity of the parameters, especially the capacitances, it is difficult to get a closed-form solution for diode voltage v_{ca}. Reference [16] uses a numerical approach to solve the switching process, which can be illustrated in Figure 6.24. Note that in this case, it added a calculation procedure for voltage-dependent capacitances.

Like the turnoff transient, although the calculation process in Figure 6.24 only requires datasheet parameters, it involves numerically solving nonlinear differential equations and can be time consuming for the design optimization. To alleviate such issues, simpler analytical solutions are desired but not yet available. There are several observations that can be helpful to determine the peak turn-on overvoltage:

1) As can been seen from Eqs. (6.117) and (6.120), the peak turn-on overvoltage should occur when the diode current is zero or $i_d=I_{load}$, which results in $\dfrac{dv_{ca}}{dt}=0$. Furthermore, before the peak voltage occurs, there should be $i_d>I_{load}$, or $\dfrac{dv_{ca}}{dt}>0$. These can be used in the Figure 6.24 calculation process to determine when the peak turn-on overvoltage occurs and its corresponding value.

2) The peak overvoltage can occur in either the voltage falling stage, when the MOSFET is likely in saturation region, or in the ringing stage, when the MOSFET is in ohmic region. In practice, if the local peaks corresponding to different $i_d=I_{load}$ points start to decrease, the previous peak value can be taken as the overall peak overvoltage. For example, Ref. [14] evaluated experimentally the turn-on overvoltage of a switch pair, consisting of a 900 V/35 A SiC MOSFET (Wolfspeed C3M0065090J) and a 1200 V/33 A SiC Schottky diode (Wolfspeed C4D20120D) under various load currents and gate resistances, and two DC-bus voltages of 400 and 600 V. Figure 6.25 shows the diode voltage and current waveforms when the MOSFET is turned on under various load currents. Figure 6.25a, b correspond to two different DC-bus voltages with

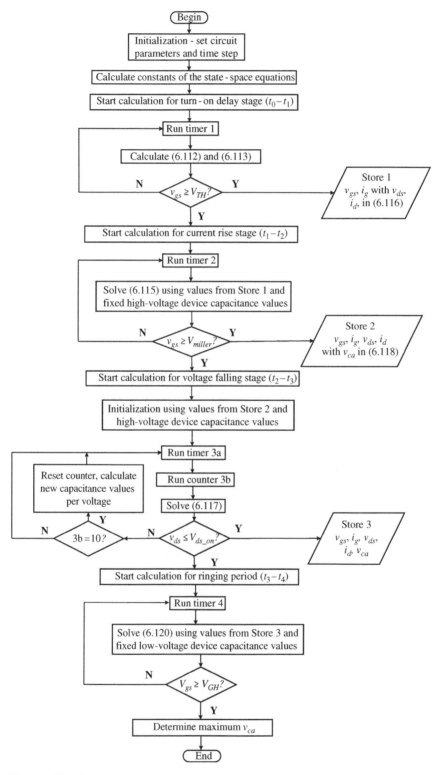

Figure 6.24 Flowchart for determining MOSFET turn-on overvoltage.

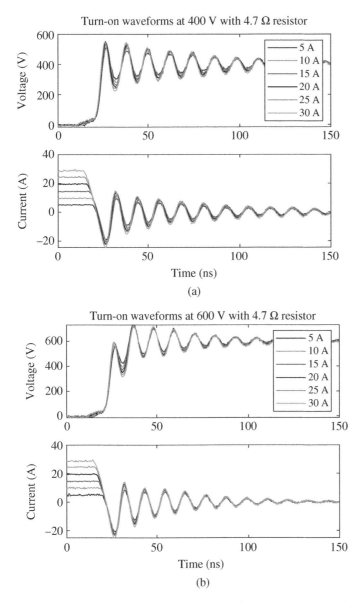

Figure 6.25 Complementary diode voltage and current waveforms when the MOSFET is turned on under different load currents. (a) 400 V DC bus voltage (b) 600 V DC bus voltage.

the same 4.7 Ω gate resistor. It can be seen that for the 400 V DC bus case, the maximum turn-on overvoltage occurs at the first local peak, while for the 600 V DC bus case, it occurs at the second peak.

3) Since the simple and accurate analytical models are not easy to obtain for the turn-on overvoltage, certain data-driven approach may be suitable for design use. Reference [14] used artificial neural network (ANN) to model part of the device characteristics in simulation to evaluate the turn-on overvoltage, achieving very accurate results. It is conceivable that the approach could be extended to solution methods that do not need to involve simulation.

6.3.6 Decoupling Capacitor

As discussed earlier, device switching can cause overvoltage, which is a strong function of power loop inductance. In general, higher power loop inductance will lead to higher switching overvoltage. As shown in Figure 6.1, for a hard-switching VSI, especially a high-power one with relatively large size and large interconnect parasitic inductance, a DC-link decoupling capacitor C_{DEC} is often used to compensate the impact of the interconnect parasitics between the DC source and the devices, and therefore reduce voltage stress on devices. Figure 6.26 shows a more detailed circuit with parasitic inductances included [11]. The DC source represents the DC-link voltage, normally obtained from a rectifier and filtered with a large DC-link capacitor as in the case of the AC-fed motor drive. For high-power inverters, the DC-link capacitor is usually bulky, placed away from the devices, has relatively high parasitic parameters including ESL, and therefore is only suitable for low-frequency harmonic filtering and energy storage. For damping and filtering high-frequency switching transients, the decoupling capacitor should have relatively low ESL and be placed close to switching devices. Compared with other device voltage overshoot suppression schemes such as active and passive snubbers, decoupling capacitors are simpler to implement, less lossy, and less expensive.

To mitigate the impact of parasitic inductance of the interconnect between the DC-link capacitor and the devices and to maintain the DC-link voltage during switching transients, conceptually, a larger C_{DEC} value should be more desirable. However, too large a C_{DEC} may not be necessary and also may have negative effect including cost, size, and excessive energy dump during a short-circuit fault. It is therefore important to select a proper C_{DEC} value to manage the effect of the interconnect parasitics.

As its name indicates, the decoupling capacitor functions through decoupling the switching power loop from the DC source or DC-link capacitor loop that normally has large parasitic inductance, e.g. L_1 in Figure 6.26. The analysis in Section 6.3.5 can be used to understand the impact of the decoupling capacitor. If there are no additional capacitors between the DC source and the inverter, the equations for turnoff overvoltage can be directly applied. As can be seen from (6.107), the turnoff overvoltage is dominated by the resonance between the switching loop inductance, which is represented by $L_{ds} = (L_d + L_s)$ in (6.107), and the MOSFET output capacitance C_{OSS}. A larger total switching loop inductance, for example $L_1 + L_{ds}$ in Figure 6.26, will generally lead to a higher turnoff overvoltage than the case without the additional interconnect inductance L_1. When a decoupling capacitor C_{DEC} is added as in Figure 6.26, there will be two parallel resonances. One is between L_1 and C_{DEC} and the other is between L_{ds} and C_{OSS}. The $L_1 \parallel C_{DEC}$ (i.e. paralleled

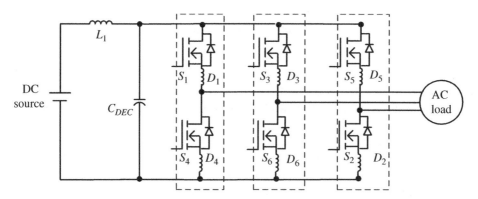

Figure 6.26 Illustration of DC-link decoupling capacitor in an MOSFET-based VSC.

L_1 and C_{DEC}) resonance frequency $\omega_{PR1} = \sqrt{\dfrac{1}{L_1 C_{DEC}}}$ is generally lower than the $L_{ds} \parallel C_{OSS}$ reso-

nance frequency $\omega_{PR} = \sqrt{\dfrac{1}{L_{ds} C_{OSS}}}$. As analyzed in [19], in order for L_1 to be fully decoupled, the decoupling capacitance C_{DEC} needs to be sufficiently large, such that L_1 will not have impact at ω_{PR}. In other words, the equivalent impedance $L_1 \parallel C_{DEC}$ needs to be much smaller than that of L_{ds} at ω_{PR}, such that it does not affect the resonance between L_{ds} and C_{OSS}. This condition can be expressed as

$$\omega_{PR} L_{ds} \gg \left| \frac{1}{j\omega_{PR} C_{DEC} + \dfrac{1}{j\omega_{PR} L_1}} \right| = \frac{\omega_{PR} L_1}{\left| \dfrac{L_1 C_{DEC}}{L_{ds} C_{OSS}} - 1 \right|} \tag{6.121}$$

If we define $L_1 = k_L L_{ds}$ and $C_{DEC} = k_C C_{OSS}$, where k_L and k_C are positive real numbers, and should be both greater than 1.0 in practice, then Eq. (6.121) will lead to (6.122). If we consider k_C to be an integer, then $k_C \gg 2$. According to [19], this means C_{DEC} should be at least 20 times larger than C_{OSS} from the engineering standpoint. Another perspective for this selection means that the equivalent impedance $L_1 \parallel C_{DEC}$ is less than 5% of L_{ds} at ω_{PR}. Since C_{OSS} is usually very small at high v_{ds}, typically less than 1 nF, C_{DEC} can be 100 times higher in practical design, still only in the range of tens to hundreds of nF.

$$k_C \gg 1 + \frac{1}{k_L} \tag{6.122}$$

There are several additional points worth noting for decoupling capacitor selection. First, the decoupling capacitor should be placed as close as possible to the switching devices to minimize L_{ds}. As discussed in Section 6.3.5, in general, L_{ds} will negatively impact the device switching overvoltage. Selection of C_{DEC} to be 50–100 times of C_{OSS} will achieve the decoupling effect and prevent L_1 from impacting at ω_{PR}. However, with the decoupling capacitor, there will also be a low-frequency resonance at ω_{PR1}, which will also cause ringing and contribute to device overvoltage. The impact of the ω_{PR1} resonance on the device turnoff voltage can be analyzed through its impact on the decoupling capacitor voltage [11].

The device switching overvoltage analysis in Section 6.3.5 assumes a constant decoupling capacitor voltage V_{bus}. The selection of the decoupling capacitance value also has this underline goal. However, in practice, V_{bus} can vary during a turnoff switching event because C_{DEC} is not very large. This variation can be explained and analyzed through the low-frequency resonance between C_{DEC} and L_1. An equivalent circuit used in [11] for this low-frequency resonance during the turnoff is shown in Figure 6.27, which also includes the DC-link capacitor parasitic inductance L_{dccap}, the

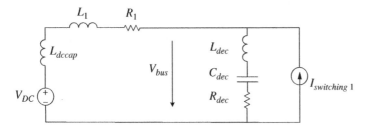

Figure 6.27 Model of voltage overshoot due to low-frequency resonance.

decoupling capacitor parasitic inductance and resistance L_{DEC} and R_{DEC}, and the DC bus parasitic resistance R_1. The voltage on the DC-link capacitance is assumed as a constant voltage source V_{DC} because of its much larger capacitance. The entire inverter phase-leg is modeled as a step-down ramp current source $I_{switching1}$ without explicitly considering the effect of its parasitics (L_{ds}, R_d, and C_{OSS}), which have been considered for the high frequency resonance. $I_{switching1}$ is represented by the current magnitude I_0 and ramp-down time duration T_{off}. The variation of the decoupling capacitance voltage, ΔV_{bus}, can be calculated by (6.123), which is similar to (6.88). Note that

$$\zeta_1 = \left(\frac{R_1 + R_{dec}}{2} \cdot \sqrt{\frac{C_{dec}}{L_{dec}}}\right), \quad \omega_1 = \sqrt{\frac{1}{(L_{dccap} + L_1 + L_{dec}) \cdot C_{dec}}}, \text{ and } \psi_1 = \cos^{-1}\zeta_1.$$

$$\Delta V_{bus}(t) = \frac{I_0 \cdot (L_1 + L_{dccap})}{T_{off}}\left[1 - \frac{1}{\sqrt{1-\zeta_1^2}} \cdot e^{-\zeta_1\omega_1 t} \cdot \sin\left(\omega_1\sqrt{1-\zeta_1^2}t + \psi_1\right)\right] \quad (6.123)$$

The peak value of the second term in (6.123) is approximately $\frac{1}{\sqrt{1-\zeta_1^2}} \cdot e^{-\zeta_1\pi}$. It can be derived that this peak value will decrease with increasing C_{dec} (assuming $\zeta_1 < 1$). In other words, ΔV_{bus} will be reduced when increasing C_{dec}. Another observation is that larger parasitic inductances ($L_1 + L_{dccap}$), larger current I_0, and higher di/dt will all cause higher ΔV_{bus}. All these observations are expected.

The total peak overvoltage is a combined effect of ΔV_{bus} and ΔV_{ds}. Therefore, it is important to minimize L_1 and appropriately increase C_{DEC} from the criterion based on (6.122) to reduce ΔV_{bus}. Based on authors' experience, $C_{DEC} = 250C_{OSS}$ is a good practical choice. If L_1 cannot be reduced, multiple decoupling capacitors can be distributed on the DC bus. In this case, the selection procedure for the decoupling capacitors will be more complicated, but the same analysis approach can be applied.

Since the selection criterion for C_{DEC} as specified in (6.122) is based on the small-signal analysis in [19], some scenarios involving a large-signal event could make it inadequate. One special case is the short-circuit condition for an inverter with several paralleled discrete SiC MOSFETs as described in [20]. In this case, the bus parasitic inductance L_1 is relatively large, the current magnitude I_0 is very large, and also the di/dt is high. The device used in the example is an engineering prototype from GE, which is a 1700 V, 40 mΩ SiC MOSFET rated at 70 A in a TO-268 package. Four such discrete devices are paralleled in a DPT setup for a short-circuit case study for different C_{DEC} and L_1. Table 6.7 shows the corresponding parameters and the peak voltage drops observed in the test. Figure 6.28 shows the corresponding device voltage and current during the short-circuit turnoff process. DC-link voltage for these tests is 600 V, in order to avoid dangerously high overvoltage.

Table 6.7 Case study parameters and results with $V_{DC} = 600$ V.

C_{OSS} (nF)	L_{ds} (nH)	C_{dec} (μF)	L_1 (nH)	f_R (MHz)	f_{R1} (kHz)	C_{dec}/C_{OSS}	I_0 (kA)	T_{off} (μS)	Peak voltage drop ΔV_{ds} (V)
13.3	10	2	40	13.8	563	150	3.1	2	525
13.3	10	7	80	13.8	212	526	3.5	2	250
13.3	10	7	160	13.8	150	526	3.5	2	280
13.3	10	17	80	13.8	136	1278	3.5	2	100

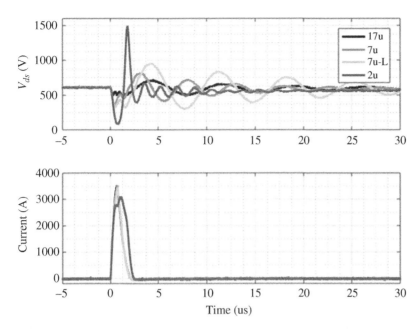

Figure 6.28 Experimental results of short-circuit induced voltage and current with different C_{DEC} and L_1.
Source: From [20] with permission of IEEE.

Note that C_{OSS} is the equivalent output capacitance of the four paralleled devices at 600 V. The
equivalent high- and low-resonance frequencies are simply calculated as $f_R = \dfrac{1}{2\pi}\sqrt{\dfrac{1}{L_{ds}C_{OSS}}}$ and
$f_{R1} = \dfrac{1}{2\pi}\sqrt{\dfrac{1}{L_1 C_{DEC}}}$.

From Table 6.7 and Figure 6.28, it can be seen that the short-circuit currents are over 3 kA, more
than 10 times of the rated current. Moreover, the short-circuit energy can be estimated as high as
0.5 J, about 1000 times of the normal turnoff switching energy. This large current and energy will
cause large variation ΔV_{bus}. In Figure 6.28 example, only 600 V DC-link voltage is used for the 1700
V device, and there is still large room for overvoltage. If a more reasonable 900 V or 1200 V DC-link
voltage is used, the short-circuit current will further increase. It has been shown in [20] that a high
C_{DEC} on the order of 1000 times of C_{OSS} is needed to maintain the voltage within the device's safe
operating area. It can also be seen in Figure 6.28, the low-frequency resonance dominates the device
overvoltage behavior. Clearly, this special case is due to large parasitic inducatnces as a result of
multiple paralleled discrete devices. It is critcal to reduce bus parasitic inductance L_1 as shown
in (6.123) to mitigate this effect.

In the discussion so far, the decoupling capacitor selection has been based on its impact on the
device turnoff overvoltage. The decoupling capacitor can also help suppress the turn-on overvoltage
[21]. Therefore, it is appropriate to consider turn-on overvoltage when selecting the decoupling
capacitor.

A small-signal analysis approach can also be used to analyze the turn-on case. Similar to the turn-
off case, the turn-on process involving the decoupling capacitor will also lead to two resonances.
One is between L_{ds} and diode capacitance C_f with a resonance frequency of $\omega'_{PR} = \sqrt{\dfrac{1}{L_{ds}C_f}}$ and

the other is between L_1 and C_{DEC} with a resonance frequency of $\omega'_{PR1} = \sqrt{\dfrac{1}{L_1 C_{DEC}}}$. Similar to the turnoff case, ω'_{PR} and ω'_{PR1} correspond to the high-frequency resonance and the low-frequency resonance, respectively. As analyzed in [21], which is similar to [19] for the turnoff case, in order for L_1 to be fully decoupled, C_{DEC} needs to be sufficiently large, such that L_1 will not have impact at ω'_{PR}, i.e. the equivalent impedance $L_1 \parallel C_{DEC}$ needs to be much smaller than that of L_{ds} at ω'_{PR}. This condition can be expressed as

$$\omega'_{PR} L_{ds} \gg \left| \frac{1}{j\omega'_{PR} C_{DEC} + \dfrac{1}{j\omega'_{PR} L_1}} \right| = \frac{\omega'_{PR} L_1}{\left| \dfrac{L_1 C_{DEC}}{L_{ds} C_f} - 1 \right|} \tag{6.124}$$

Clearly, (6.124) mirrors (6.121). Since C_f and C_{OSS} represent parasitic capacitances of two complementary switches in an inverter phase leg and therefore are practically equal, the selection of C_{DEC} for the turnoff case should also work for the turn-on case.

Once the C_{DEC} value is determined, a decoupling capacitor can be selected. An important parameter for the capacitor is its parasitic inductance. Typically, multilayer ceramic capacitors are preferred choice for the decoupling capacitors due to their low parasitics. The fact that capacitance of ceramic capacitors usually decreases with voltage is another reason that C_{DEC} is often selected to be hundreds of times of C_{OSS} or C_f rather than only tens of times as required by (6.122).

As shown in [11], increasing the decoupling capacitance will reduce ΔV_{bus}, while also reducing the resonance frequency. This could result in higher EMI noise in lower frequency region, which is undesirable. Measures should be taken, e.g. reselecting C_{DEC} or providing damping, if the decoupling capacitor cause issues in EMI filter design. If the switching overvoltage can be reduced by minimizing parasitic inductance through better mechanical design, e.g. an ultralow inductance laminated busbar, the approach should be adopted.

6.3.7 Decoupling Inductor

In the last subsection, decoupling capacitors are used to compensate for the interconnect parasitic inductances between the inverter DC terminals and the DC-link voltage source. The inverter AC terminals are generally interfaced to an inductive load such as motor. The interconnection, such as cables, as well as the inverter and motor themselves, will introduce parasitic capacitances, which will also impact the inverter performance and may need to be compensated. Figure 6.29 shows measured impedance and a corresponding high-frequency equivalent circuit of a 7.5 kW induction motor with a 2 m cable [22]. In Figure 6.29b, the current on the large inductance L_d that represents the main load inductance can be considered as a current source during the switching transient, while the LRC series resonant networks will affect the switching behavior. During the turn-on transient with high dv/dt, there will be resonant current excited for each LRC series resonant branch (i.e. I_1 to I_4) in Figure 6.29. These additional currents flow into the device and increases the switching current, resulting in slower turn-on speed with larger switching losses. During the turn-off transient, the resonant currents induced by LRC series resonant branches decrease the equivalent inductive current, leading to a longer turn-off time for charging or discharging of the device' output capacitance; hence, the turn-off speed decreases and switching loss increases as well. For the example setup in Figure 6.29, a Wolfspeed 1200 V/20 A SiC MOSFET is used for switching behavior assessment. Due to parasitics, compared to the switching behavior exhibited by using a low parasitic inductor load in a DPT, the tested switching time of the SiC MOSFET increases up to 42%

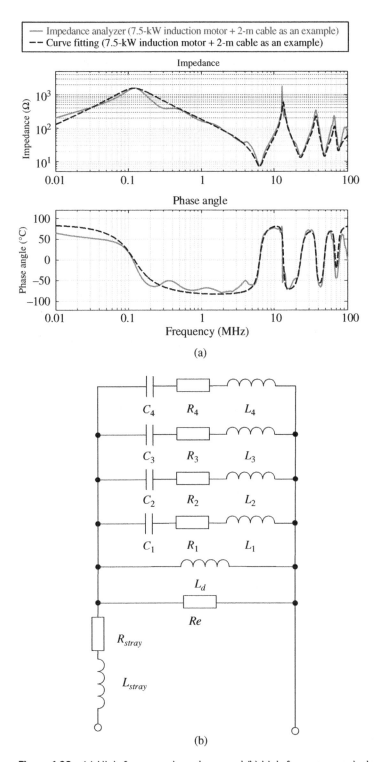

Figure 6.29 (a) High-frequency impedance and (b) high-frequency equivalent circuit model (a 7.5-kW induction motor plus a 2-m cable example).

during turn-on, and doubles during turn-off; an extra 32% of energy loss is dissipated during the switching transient [22].

To mitigate the impact of the inductive load's parasitics on the switching performance of the devices in the VSI, a decoupling inductor L_{DEC} can be placed at the inverter AC terminals. L_{DEC} can be considered as a dual of C_{DEC} on the inverter DC terminals. C_{DEC} decouples the DC source and interconnection parasitic inductances from impacting the device switching and voltage; L_{DEC} decouples the parasitic capacitances of the inductive AC load and interconnection from impacting the device switching and current. As shown in Figure 6.1, in contrast to the shunt connected C_{DEC} on the DC terminals, L_{DEC} is series connected at the AC terminals between the inverter and the AC load. Compared to the DC decoupling capacitor, the AC decoupling inductor for a three-phase VSI needs to cover all three phases.

L_{DEC} functions through the resonant current suppression during the switching transient. There are two basic objectives: (i) minimizing the currents for the LRC branches in Figure 6.29 with high resonant frequencies (f_2 to f_4 in Figure 6.30) and (ii) tuning the resonance period of the LRC branch with the low resonance frequency (f_1 in Figure 6.30) to be much longer than the voltage switching time (i.e. rise or fall time) t_{SW} such that there is almost no response for this series resonant branch to the switching voltage excitation. L_{DEC} can help to modify the high-frequency impedance of the inductive load to meet these two objectives,

As illustrated in Figure 6.30, L_{DEC} can be properly selected to dominate in the high-frequency region, and therefore, the effect of LRC branches corresponding to resonant frequencies f_2 to f_4 will disappear. L_{DEC} will also move the resonant frequency of the $L_1R_1C_1$ branch from f_1 to a lower f_{ring}. If the resonant period $T_{ring} = 1/f_{ring}$ is much longer than t_{SW}, the resonant current I_1 will remain small during the switching transient (Figure 6.31).

Clearly, L_{DEC} should be determined based on f_{ring}, whose selection is directly related to t_{SW}, which is highly dependent on the operating condition. Therefore, when determining f_{ring}, it is important to select a t_{SW} considering the most significant operating conditions such that the switching performance will not be affected by the load under these conditions.

As can be observed in Figure 6.32, t_{SW} (t_{on} or t_{off}) obtained in DPT highly depends on load current I_L, and the longest t_{SW} occurs during the turn-off transient with light load. If f_{ring} were determined based on t_{off} at the light load, the required L_{DEC} would be quite large, but the benefit would be

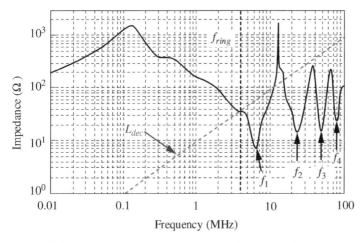

Figure 6.30 High-frequency impedance comparison: the inductive (motor plus cable load) versus the decoupling inductor.

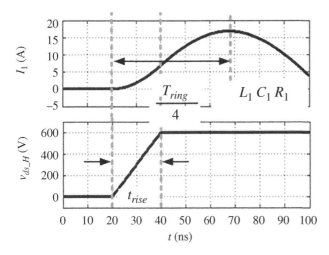

Figure 6.31 Resonant current I_1 during the switching transient.

Figure 6.32 Voltage commutation time dependence on load current measured by DPT.

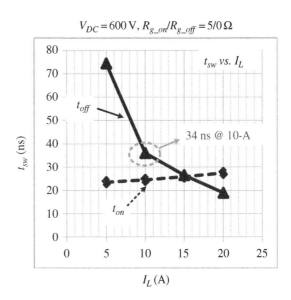

limited since the light load turn-off switching loss is very low. Thus, it is more appropriate to select f_{ring} based on a t_{SW} longer than all of the t_{on}'s and part of the t_{off}'s. Consequently, except for the turn-offs under light load, the switching times and switching losses under other operating conditions will not be affected by the parasitics of the inductive load.

As can be seen in Figure 6.32, the operating condition is selected at 600 V/10 A and the corresponding t_{sw} is 34 ns. The resonance frequency f_{ring} of 3 MHz is selected such that its corresponding resonant period T_{ring} is 333 ns, which is about 10 times of t_{sw}. Combining with the high frequency impedance of the inductive load, a 2 μH decoupling inductor is needed. Figure 6.33 illustrates that after insertion of a 2 μH decoupling inductor, the original resonant frequencies, f_2, f_3, and f_4, are suppressed. Moreover, the resonant frequency f_1 is shifted from 6 to 3 MHz. Considering the three phase, the decoupling inductor per phase is 1.33 μH as the equivalent inductance is

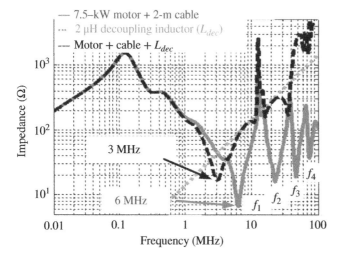

Figure 6.33 High-frequency impedance comparison: inductive load versus decoupling inductor.

determined by the value consisting of one decoupling inductor in series with the two paralleled inductors contributed by the other two phases.

After the L_{DEC} inductance determination, its physical design needs to be carried out. Given its small inductance, and correspondingly small size and low cost, the reasonable design objective is to minimize inductor loss. Because of L_{DEC}'s small inductance, an air-core inductor can be employed. In a three-phase inverter, three single-phase air-core inductors are appropriate.

For an air-core inductor in Figure 6.34a, the design tasks include (i) selection of the wire size based on the current and (ii) determining the diameter d and turns number N to achieve the

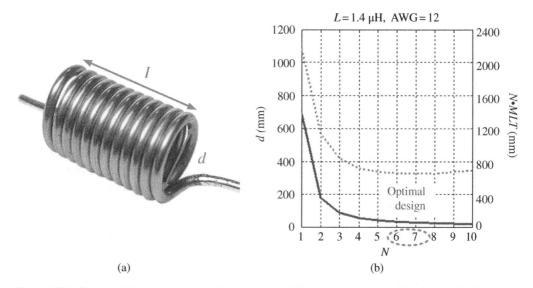

(a)   (b)

Figure 6.34 Design of the air core decoupling inductors. (a) Typical air core inductor. *Source:* Engineering Learn/https://engineeringlearn.com/what-is-inductor-types-of-inductor-uses-function-symbol-complete-details/. (b) Design optimization of air-core inductor: d dependence on N and corresponding $N \times MLT$.

required inductance with the shortest copper wire length (i.e. the smallest $N \times MLT$, where MLT is mean-length-per-turn) for minimum inductor power loss. The inductance can be calculated with (6.125), where L is in μH, and d and the coil length l are in mm.

$$L = \frac{d^2 N^2}{25.4(18d + 40l)} \tag{6.125}$$

In the example [22], based on the 11 A RMS phase current with 4 A/mm^2 maximum current density for copper wires, AWG12 magnet wire with a maximum outside diameter of 2.13 mm is selected. l is approximately $2.13 \times N$ mm, and MLT can be approximated as πd. According to (6.125), the relationship between N and d to achieve a 1.4 μH air-core inductor is illustrated in Figure 6.34b. N and d can be selected to minimize $N \times MLT$. With the AWG12 copper wire, $N = 7$ is appropriate for the $N \times MLT$ close to the minimum. The corresponding d and l are 28.4 mm and 14.9 mm, respectively.

Figure 6.35 shows the switching waveform comparison with three different inductive loads: low-parasitic DPT inductor, induction motor plus 2-m power cable (IM-PC), and IM-PC with 1.4 μH air-core inductors (IM-PC-L_{DEC}). The operating condition is 600 V, 10 A with gate resistance $R_g = 5\,\Omega$ during the turn-on and $R_g = 0\,\Omega$ during the turn-off. In the switching waveforms, a 6.0 MHz ringing can be observed in the switching current for the IM-PC case. After inserting the 1.4 μH air-core inductors, the ringing frequency moves from 6.0 to 3.0 MHz, which agrees with the aforementioned design. Also, as expected, during the turn-on transient, the IM-PC-L_{DEC} case has improved switching performance compared to the IM-PC load only case: dv/dt increases (from 40 to 43 V/ns) and the turn-on switching loss decreases (from 166 to 158 μJ). In fact, the switching behavior with the PC-IM-L_{DEC} load is almost identical to that of the DFT.

Similarly, during the turn-off transient, the decoupling inductor allows the dv/dt with IM-PC load to increase (from 5.6 to 14.0 V/ns), which almost approaches to the dv/dt of the DPT case (16.8 V/ns). Furthermore, the total switching loss of the IM-PC load case decreases (from 224 to 204 μJ) by employing the decoupling inductor, which is nearly the same as the total switching losses achieved with the DPT inductor load (200 μJ).

Clearly, during the switching transient, the decoupling inductor can successfully decouple the interaction between the inductive load and power device, enabling the switching performance with the practical inductive load (e.g. motor plus cable based inductive load in this case study) to achieve the similar fast switching behavior comparable to when the optimally designed DPT inductor is employed.

In addition to mitigating the adverse impact of the inductive load parasitics on the switching behaviors of the VSI, the decoupling inductor is able to reduce dv/dt and overvoltage across the motor terminals. Figure 6.36 illustrates that the decoupling inductor decreases the motor terminal overvoltage slightly (from 1049 to 1018 V) and the dv/dt significantly (up to 44%). The reason for the overvoltage and dv/dt reduction is because the ringing frequency of the voltage across motor terminals becomes lower when L_{DEC} is employed. Figure 6.37 shows the L_{DEC} impact on the motor terminal voltage and dv/dt under different load conditions, which is beneficial to the motor as will be discussed in Chapter 12.

One potential adverse effect of the decoupling inductor is its impact at light load conditions. Based on the design consideration of the decoupling inductor, to minimize the required inductance, there will be a penalty for the turn-off time at the light load (e.g. <10 A in the case study). As can be observed in Figure 6.32, the turn-off time with the motor plus cable load increases as the load current decreases. With the decoupling inductor, the turn-off time becomes even longer at the light load. This is because that the relatively long turn-off commutation time at the light operating

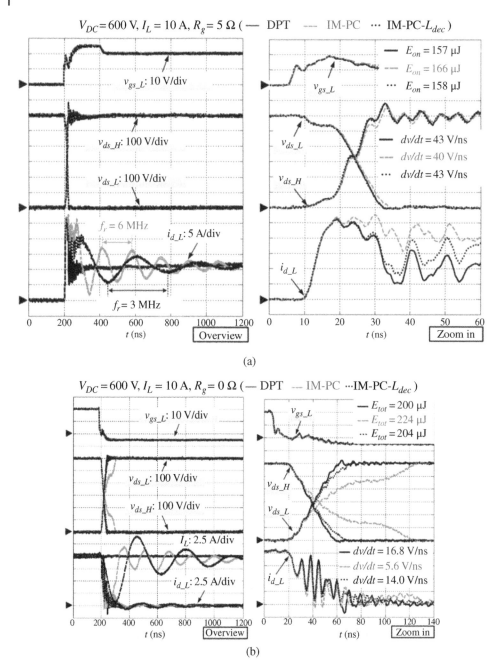

Figure 6.35 Switching waveform comparisons among different inductive loads. (a) Turn-on transient. (b) Turn-off transient.

current cannot avoid the impact of the resonant current induced by the parasitics of the inductive load. In addition, the decoupling inductor reduces the resonant frequency (from 6.0 to 3.0 MHz for the case study); as a result, the duration of the resonant current with this lower resonant frequency becomes longer and the turn-off time increases as well. On the other hand, the tested turn-off switching loss at the light load current stays nearly constant due to the fact that the turnoff loss

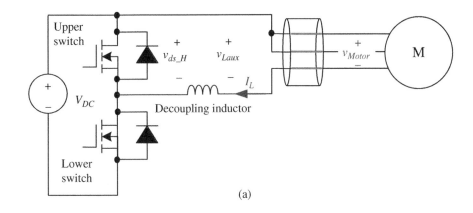

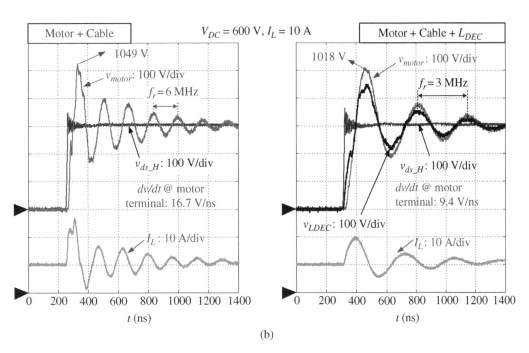

Figure 6.36 Impact of decoupling inductor on motor terminal voltage: v_{ds_H} is the drain-source voltage of the upper switch; v_{motor} is the motor terminal voltage; and v_{Ldec} is the converter terminal voltage (after L_{dec}). (a) Equivalent circuit and measurement terminals. (b) Experimental waveform when lower switch turns on.

is dominated by the energy stored in the junction capacitance of the SiC MOSFET. Therefore, although the turn-off time differences with different inductive loads are significant, their turn-off switching losses are almost identical.

A penalty of the decoupling inductors is the added power loss by the inductors themselves, although the loss may be small with the small air-core inductors. While the decoupling inductors can reduce switching loss in the converter devices, with sufficiently large dv/dt, the cable and motor parasitic capacitances are still charged and discharged by PWM voltages. As a result, the switching power loss due to these capacitances may not be reduced and most likely will dissipate in other parts of the system, e.g. in cable and motor themselves. Therefore, when applying decoupling inductors, the overall system loss could be increased, although reduction of switching loss in VSI is often

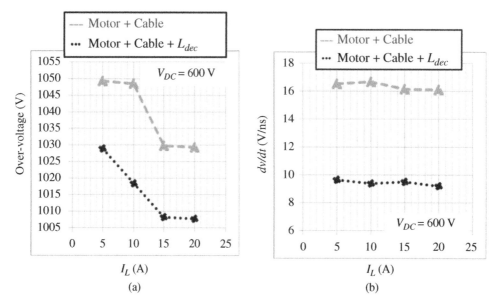

Figure 6.37 Motor terminal over-voltage and its *dv/dt* comparison dependence on load currents. (a) Motor terminal over-voltage dependence on I_L. (b) Motor terminal *dv/dt* dependence on I_L.

desirable for reduced cooling need and better power density. More discussion on cable- and motor-related switching loss can be found in Chapter 12.

6.4 Load-side Inverter Design Optimization

As in the case of the passive rectifier in Chapter 5, after establishing the models among the design constraints, conditions, and variables, as described in Section 6.3, an optimizer can be built for the load-side inverter to carry out the design. The details of design optimization will be described in Chapter 13.

Here an example for aircraft application will be used to illustrate a DC-fed load-side inverter design optimization. Assuming the load is a three-phase AC induction motor. Table 6.8 lists the motor load conditions and parameters.

Other design conditions are listed in Table 6.9, including input parameters from DC source, as well as from the thermal management, mechanical, control, and EMI filter subsystems. The DC

Table 6.8 Induction motor load conditions and parameters.

kVA	Power factor	Rated phase voltage	Rated frequency	Rated speed	Number of pole airs
45 kVA	0.91	230 V RMS	400 Hz	7872 RPM	3
Stator resistance	**Stator leakage inductance**	**Rotor resistance**	**Rotor leakage inductance**	**Magnetizing inductance**	**Slip at full load**
167.2 mΩ	132.2 µH	55.2 mΩ	66.4 µH	3732 µH	0.016

Table 6.9 Other design conditions.

DC source	Thermal management system	Mechanical system	Control	EMI filter
600 V, high impedance to ripple currents	Forced air cool, maximum ambient temperature 65 °C	CM parasitics included	Recommended gate parameters	DC side only with LISN

source with high impedance to ripple current indicates that all ripple currents from the inverter need to be absorbed by DC-link capacitors. The parasitics from the mechanical system include CM parasitic capacitances to be discussed later, although the switching transient impact on device voltage is neglected. Recommended gate parameters refer to the recommended gate resistances and voltages in the datasheet of the selected devices.

Figure 6.38 shows the 45 kVA DC-fed motor drive system used in the design example and the CM parasitic capacitances considered. In this case, for the 45 kVA inverter, it is assumed that the DC-link midpoint to ground capacitance is 6 nF, and the AC terminals to ground parasitic capacitances are 1.5 nF per phase.

Figure 6.39 shows the high-frequency motor and cable models used in the example design. In this case, for the high-frequency motor model, each of the motor RC1 branches has a resistance of 61.5 Ω and a capacitance of 0.97 nF; each of the motor RC2 branches has a resistance of 1.15 Ω and a capacitance of 1.4 nF; each of the motor RL branches has a resistance of 1 Ω and an inductance of 542 μH; and each of motor R branches in parallel with the motor RL branches has a

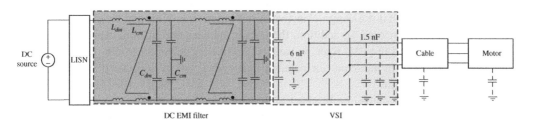

Figure 6.38 The 45 kVA DC-fed motor drive system with CM parasitic parameters.

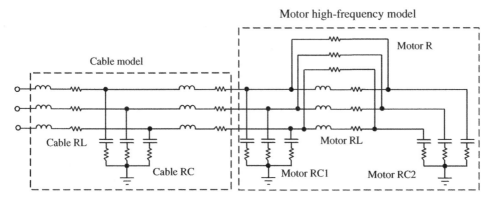

Figure 6.39 High-frequency models of the motor and cable.

Table 6.10 Design constraints.

Efficiency	EMI	DC ripple voltage	AC current THD	Component temperatures	Flux density	Fill factor
>98%	DO-160	<1% peak-to-peak voltage	N/A	<Actual component limits	<80% of actual core saturation limit	<0.8

resistance of 6.97 kΩ. For the T-type AC cable model, each of the cable RC branches has a resistance of 1.0 Ω and a capacitance of 3.4 nF, and each of six cable RL branches has a resistance of 10.5 mΩ and an inductance of 2.1 µH.

The design constraints are listed in Table 6.10. Due to the high switching frequency, the motor current harmonics is generally not a concern, so the AC current THD is not directly used in the design tool as a constraint. A post processor is used to check the AC current THD based on the final design result of the optimizer. For flux density, all magnetic cores have a saturation flux density, a 20% margin is used as design constraint. Note that for the inductor wire selection, current density is not used as a constraint; instead, the thermal model is used to ensure the wire and core temperatures are within their limits. In addition to the listed constraints, the topology is preselected to be a two-level VSC to match the example system used in this chapter.

With all the input design conditions defined and their values given, the design can be carried out to select the design variables that will satisfy the design constraints following the design optimization procedure illustrated in Figure 6.2. Here, the design tool in [23], which will be described in more detail in Chapter 13, is used for the load-side inverter design optimization. Table 6.11 lists the design results. Note that the design objective in this case is the minimum weight, as is typical in the aircraft application case. In addition, the design includes the DC-link capacitor, thermal management system, and the EMI filter, as well as the inverter power stage. On the other hand, no decoupling capacitors or inductors are included, partly because they normally do not contribute much to the overall weight.

In the design example, no inverter AC side EMI filter was included, such that the only grounding point is through LISN on the DC side. The DC DM capacitor for the design result in Table 6.11 is defined as one of the two series connected capacitors as in Figure 6.38. The DC CM capacitor is also part of DC DM capacitor. Thus, in this example, no physically dedicated DC DM capacitors are needed. With the selected two-stage $LCLC$ filter configuration, i.e. two cascaded LC structures as in Figure 6.38, for simplicity, the parameters of the two stages are assumed to be identical. In addition, it can be seen that L_{cm} and L_{dm} are connected together, such that the leakage inductance of the L_{cm} can be used as part of the DM inductance. On the other hand, L_{dm} is also in the CM loop, and therefore can contribute to the overall CM inductance.

One interesting result to note in Table 6.11 is that the DM inductance is 27 µH, larger than the CM inductance of 12.3 µH, which is somewhat counterintuitive. This is because the CM inductance here is only the inductance associated with the physical CM inductor, and it does not include the DM inductance, which is also in the CM loop. Considering the DM inductor, the overall equivalent CM inductance, with contributions from both the CM and DM inductors is still larger than the DM inductance. Another observation is that the DM inductor has a smaller wire gauge than the CM inductor, even though the two carry almost the same current. This is because the inductor design uses thermal models discussed in Chapter 4 rather than the current density as the design constraints.

Table 6.11 Design variable results.

Subsystem	Component	Result
Inverter power stage	Device	Wolfspeed "C3M0065090J," 900 V, 22 A @ 100 °C, 4 in parallel
	PWM scheme	Continuous SVM
	Gate resistance	6 Ω per device (data sheet value with 2.5 Ω external)
	Gate voltage	V_{GH}: 15 V, V_{GL}: − 4 V
	Switching frequency	70 kHz
	DC link capacitor (40 µF)	TDK "B32778P8206" film capacitor, 20 µF, 840 V, 2 in parallel
Thermal management system	Heatsink	Aavid extruded aluminum heatsink, part no: 61325, 153 mm, 3 in parallel
	Fan	NMB "1608VL-04W-B40-B00"
EMI filter	Configuration	Two-stage *LCLC*
	DM capacitor (0.5 µF)	Not needed physically, provided by CM capacitor
	CM capacitor (0.5 µF)	KEMET "R76IN4120(1)30(2)" film capacitor, 1.2 µF, 250 V, 2 in series
	DM inductor (27 µH)	Core: "Magnetics58195" HighFlux; Winding: AWG 10, 13 turns
	CM inductor (12.3 µH)	Core: Vitroperm 500F "T60006-L2019-W838," Winding: AWG 8, 2 turns

Table 6.12 Weight and loss breakdown.

Component	Weight (g)	Loss (W)
DM inductor	1146	131
CM inductor	11	7.6
DM capacitor	0	0
CM capacitor	170	~0
DC link capacitor	223	~0
Devices	412	364 (conduction: 298, switching: 66)
Heatsink and fan	597	—
Total	2559 (specific power: 17.58 kW/kg)	502.6 (efficiency: 98.89%)

Using the design tool, the design objective value and the design constraint values associated with performance can be calculated. In this case, the objective function weight and performance index efficiency are shown in Table 6.12, together with the weight and loss breakdown. Figures 6.40 and 6.41 show the DM and CM EMI noise of the designed inverter, respectively, including both the bare noise and filtered noise. Clearly, with the designed EMI filter, the DM and CM noise are below the required level in DO-160 standard with more than 6 dB margin, satisfying the EMI constraint. It can

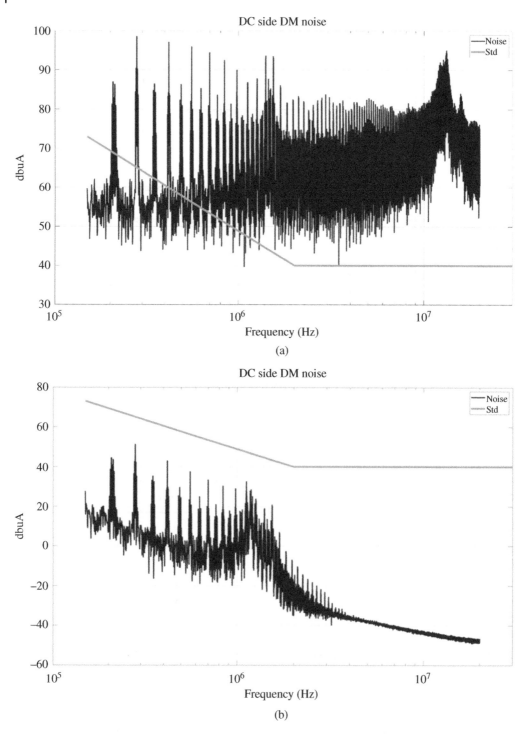

Figure 6.40 DC side DM EMI noise with the designed inverter: (a) bare noise without filter; (b) with the designed filter.

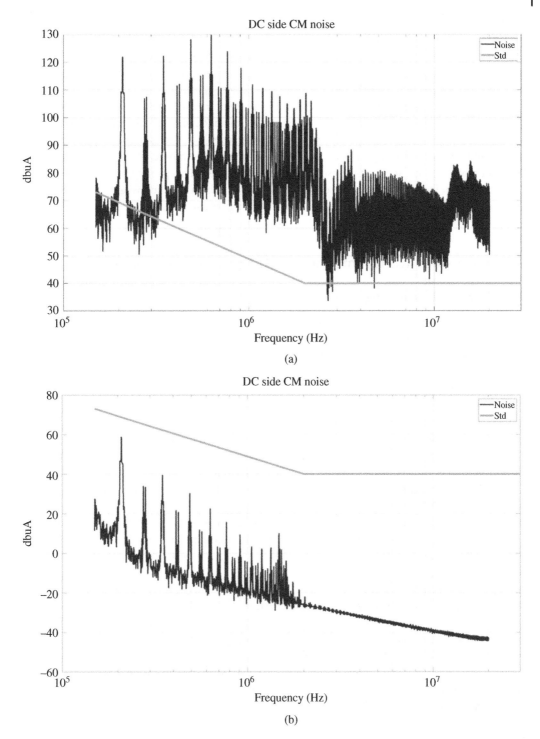

Figure 6.41 DC side CM EMI noise with the designed inverter: (a) bare noise without filter; (b) with the designed filter.

also be observed that the CM bare noise is higher than the DM noise, which justifies the design results with larger equivalent CM filters.

It should be pointed out that all component parts used in the example design are actual off-the-shelf commercial components, including parts for the EMI filter. With the nonideal characteristics, the EMI filter attenuation at high frequency can deteriorate, and there can also be resonance that requires damping. In this case, the CM filter requires damping, so a split capacitor scheme was adopted. The approach is to parallel the CM capacitor with an RC branch, which consists of a small resistor and a capacitor identical (or larger) to the selected CM capacitor. At low frequency, the resistor will not matter; at high frequency, the resistance will dominate and the capacitor in the *RC* branch will not matter, so the resonance can be damped. The penalty is added weight. In Table 6.12, the weight of the CM capacitor is 170 g, which is the weight of 16 KEMET "R76IN4120(1)30(2)" capacitors, rather than eight listed in Table 6.11.

6.5 Load-side Inverter Interfaces to Other Subsystem Designs

6.5.1 General Classifications

In the case of the motor drive, the load-side inverter directly interfaces with these subsystems: rectifier, thermal management system, mechanical system, and control. In some cases, there can be EMI filters on the DC side, AC side, or on both sides. All these interfaced subsystems influence the load-side inverter design through their interface conditions, including both the input conditions and output conditions. All input interface conditions have been described in Section 6.2.3 as part of the load-side inverter design conditions. Here we will examine the output of the load-side inverter needed for the other subsystems.

Rectifier or DC Source – The output from the load-side inverter to its front-end rectifier include: (i) inverter power characteristics, including dynamic characteristics like rate of change; (ii) DC ripple current due to the load-side inverter; and (iii) equivalent DC impedance of the inverter.

EMI Filter – Whether the inverter has EMI filters on DC side, AC side, or both sides, the output of the inverter to the EMI filter include: (i) EMI bare noise, similar to the passive rectifier case. For the DC-side filter, the DC bare noise is needed, and for the AC-side filter, the AC bare noise is needed; (ii) DC and/or AC line current information, also similar to the passive rectifier case. If the time-domain voltage and current waveforms are available, they can be used for both the EMI noise and current information.

Thermal Management System – The output from the inverter to the thermal management system directly associated with the design constraints or variables include: (i) the maximum steady-state power loss of devices; (ii) maximum overload power loss of the devices and its duration; (iii) device or device module thermal characteristics; (iv) device mechanical dimensions; and (v) maximum thermal impedance of the thermal management system.

Mechanical System – The output of the inverter power stage design to the mechanical system include: (i) device dimensions and weight; (ii) passive component dimensions and weight; (iii) current distribution, which refers to the currents on different terminals and wires of the inverter, including the inverter DC currents and AC phase currents; (iv) voltage distribution, which refers to the voltages at all the terminals of electrical parts and components in the inverter; (v) device and component losses, which are needed in mechanical design to consider temperature requirements; and (vi) input and output terminals, including the type, number, and rating. For example, the two-level voltage source inverter in Figure 6.1 has two DC terminals and three AC terminals. If the DC-link midpoint is introduced for grounding or other purposes, there will be an additional DC terminal. The voltage and current ratings of these terminals are also the inverter output to the mechanical design as mentioned earlier.

Control System – Control system has several components that interface with the inverter power stage, including gate drivers, sensors, digital controllers, and protection. For gate drivers, the output from the inverter design will include parameters R_g and V_g (both positive V_{GH} and negative V_{GL}). For sensors, the output (attributes) will include type (e.g. voltage, current, or temperature), range, numbers (e.g. three AC current sensors, two AC voltage sensors, two DC voltage sensors, and a number of temperature sensors for a high-power motor drive), accuracy, bandwidth, and isolation requirement (e.g. $\pm V_{DC}$ or $\pm V_{DC}/2$). For digital controllers, the output (attributes) include the capability requirements (bandwidth, computation power, I/O, etc.) and isolation. For protection, the output includes the types and settings. Note that parameters such as dv/dt and di/dt can influence almost all aspects of control design, and they should also be output to the control system design.

These output interfaces are summarized and classified in Table 6.13. As in the case of Chapter 5, the constraints and design variables are design results. Some of the attributes, such as device or component dimensions and weight, are also design results associated with the design variables, and therefore no more work is needed to obtain them. However, other attributes are not directly

Table 6.13 Output from the load-side inverter to its interfaced subsystems.

Subsystem	Interface type	Item	Value
Rectifier or DC source	Attribute	Inverter power characteristics	Discussion
		DC current including ripple	Discussion
		DC impedance	Discussion
EMI filter	Constraint	AC line current	Design result
	Attribute	EMI bare noise	Discussion
		DC link current	Discussion
Thermal management system	Constraint	Maximum TMS thermal impedance	Design result
	Attribute	Device/component dimensions	Design result
		Maximum steady-state device/component power loss	Design result
		Overload power loss and duration	Design result
Mechanical system	Constraint	Device/component losses	Design result
	Attribute	Device/component dimensions and weight	Design result
		DC input and AC out terminals	Discussion
		Voltage distribution	Discussion
		Current distribution	Discussion
Control system	Design variable	Gate driver R_g, V_g (i.e. V_{GH} and V_{GL})	Design results
	Constraint	Sensor accuracies and bandwidth	Design results
	Attribute	Sensor type, range, number, and isolation	Discussion
		Digital controller capabilities (bandwidth, computation power, and I/O) and isolation	Discussion
		Protection type and settings	Discussion
		Maximum dv/dt and di/dt	Discussion

available from the inverter design results, and therefore need more work. These attributes will be further discussed next.

6.5.2 Discussion

The discussion of the output from the inverter power stage design to other subsystems will be presented by subsystems. Note that the thermal management system has no output that needs further discussion.

6.5.2.1 Rectifier or DC Source

6.5.2.1.1 *Inverter Power Characteristics* The inverter power characteristics include the total peak power and the maximum power change. The peak power can be calculated as the peak load power plus the corresponding inverter power loss, which is part of the design result. If there is short duration overload condition, then the peak overload power for the inverter is the overload power plus the corresponding inverter power loss, which is also part of the design result.

The maximum power change should be related to the dynamic performance requirement of the load. For example, a system motor drive in a steel mill may have a dynamic torque requirement of 15 p.u./s, which can be equated to a $\frac{dI}{dt} = 15$ p.u./s, or $\frac{dP}{dt} = 15$. The dynamic performance will impact the rectifier or DC source design, e.g. the selection of the DC-link capacitors, which will be discussed in Chapter 7.

6.5.2.1.2 *DC Current Including Ripple* Due to switching, the inverter will generate ripple current on the DC link, in addition to an average DC current that corresponds to input power to the inverter. The DC current can be determined by mapping the AC current, which in turn can be determined by harmonic voltages as analyzed in Section 6.3.1. The details for the expression of the inverter DC current will be presented in Chapter 8, where the DC current spectrum is needed for the DC-side EMI filter design.

If the objective is to find the total RMS, DC, and ripple currents on the DC link, analytical models are available. Assuming that the three-phase AC currents are sinusoidal and balanced, for a two-level VSC, the total DC-link RMS current is [24, 25]

$$I_{dc(RMS)} = \left[\frac{3\sqrt{3}}{2\pi} MI_{RMS}^2 \left(1 + \frac{2}{3} \cos 2\varphi \right) \right]^{0.5} \tag{6.126}$$

where M is the modulation index defined as before, I_{RMS} is the RMS value of the inverter AC phase current, and φ is the power factor angle. Neglecting the inverter power loss, the average DC current, i.e. the DC component of the total DC-link current from the inverter, is

$$I_{dc(avg)} = \frac{3}{2\sqrt{2}} MI_{RMS} \cos \varphi \tag{6.127}$$

The ripple RMS current is

$$I_{dc-ripple(RMS)} = \sqrt{I_{dc(RMS)}^2 - I_{dc(avg)}^2} = I_{RMS} \cdot \sqrt{\frac{\sqrt{3}M}{2\pi} + \left(\frac{2\sqrt{3}}{\pi} M - \frac{9}{8} M^2 \right) \cdot \cos^2 \varphi} \tag{6.128}$$

It is also known that the $I_{dc-ripple(RMS)}$ is independent of PWM schemes.

6.5.2.1.3 DC Impedance DC impedance is needed for the DC-link capacitor design as part of rectifier design to ensure small-signal stability, which has been discussed in Chapter 5. The DC impedance can be estimated as follows.

Assuming that the inverter power loss is negligible compared with the motor input power, and the motor drive is a constant power load with a load power P_{load}, we will have

$$v_{DC} \cdot i_{DC} = p = P_{load} \tag{6.129}$$

Differentiating (6.129) with respect to i_{DC} will yield

$$v_{DC} + i_{DC} \cdot \frac{dv_{DC}}{di_{DC}} = \frac{dP_{load}}{di_{DC}} = 0 \tag{6.130}$$

Rearrange (6.130) and we will obtain

$$Z_{INV} = R_{INV} = \frac{dv_{DC}}{di_{DC}} = -\frac{v_{DC}}{i_{DC}} = -\frac{v_{DC}^2}{P_{load}} \approx -\frac{V_{DC}^2}{P_{load}} \tag{6.131}$$

The equivalent inverter DC-link resistance value in (6.131) has been used in Chapter 5 DC-link stability criteria. The assumption that the motor drive is a constant power load is conservative. A real motor drive will have a limited current/power control bandwidth. Accordingly, the equivalent DC impedance should not be a pure negative resistance. On the other hand, Eq. (6.131) has been shown to be a suitable model for motor drive DC-link stability analysis [26].

6.5.2.2 EMI Filter

6.5.2.2.1 DC or AC Bare Noise The EMI bare noise during the design can be obtained through time-domain simulation or analytical analysis, similar to the harmonic current analysis for AC or DC current. The details of the EMI bare noise analysis will be given in Chapter 8.

6.5.2.2.2 DC-Link Current See discussion earlier for the attribute output to rectifier or DC source.

6.5.2.3 Mechanical System

6.5.2.3.1 Voltage Distribution The voltage distribution is a function of the topology and converter grounding. Assuming the inverter in Figure 6.1 is solidly grounded at the DC-link midpoint, then the voltage distribution of the different terminals will be the same as in the case of the passive rectifier in Chapter 5.

6.5.2.3.2 Current Distribution As for the currents, the peak, RMS, and frequency spectrum of the currents through components and inverter terminals will be needed for the busbar and connector design. The AC terminal currents are equal to phase currents, and their calculation has been covered earlier. The DC terminal current is equal to DC average current if we assume all ripple current flowing through DC-link capacitors. In reality, there can be some ripple current into DC source or rectifier too. The DC current including ripple has been discussed earlier for output to the rectifier.

6.5.2.4 Control System

6.5.2.4.1 *Sensor Type, Range, Number, and I/O*
Sensor type here refers to voltage, current, or temperature type for the sensor, not the technology type, which is a design task for the control system. The sensor type really depends on the topology and application. For the three-phase motor drive based on two-level VSI, the AC currents and DC voltage usually need to be measured or estimated for control and protection. In high-power motor drives, AC voltages are also sensed. Very often, there are also some temperature sensors, for example, temperature sensors placed near baseplate or on heatsink, mainly for protection, and possibly also for control. There can also be a DC-link midpoint voltage sensor for protection. The motor drive also can have a speed or position sensor.

The number of sensors depends on topology and on performance requirement. For example, the inverter AC current sensing can use two sensors, since there are only three wires and the three-phase currents should add to zero under normal conditions. However, three sensors are often used in high-power high-performance motor drives, which will enable ground fault detection based on current imbalance. It can also be shown that with three AC current sensors, the current and torque regulation errors in a motor will be smaller than that of two-current sensor case. For multilevel converters, DC voltage sensing for each level will be needed.

The range of the sensors depend on the range of the sensed variables, which need to consider the transient and abnormal conditions. For example, the current and voltage sensors need to consider the maximum overcurrent and overvoltage, respectively.

The isolation requirements for sensors depend on where the sensors will be placed. For example, the voltage sensors need to be placed at the DC or AC terminals and need to bear the voltages of those terminals to ground, where the controllers are electrically placed. The terminal voltages to ground define the isolation requirements for the sensors.

More on sensors can be found in Chapter 10.

6.5.2.4.2 *Digital Controller Capabilities and Isolation Requirements*
Digital controller capabilities here refer to bandwidth, computational/processing power, and I/O. The bandwidth is determined by the switching frequency. The computational power is determined by converter topology, algorithm, and application. For example, a topology with a given number of switches and passive components (i.e. state variable voltages and currents) will require certain switching and voltage/current controls. The number of sensors will require certain data processing. These will determine the computational and processing power. The I/O requirements are also related to number of switches (gates and gate drivers) and sensors. Other auxiliary functions like protection will also need I/O ports.

The isolation requirements for the controllers are related to isolation from sensors, gate drivers, power supplies, and protection circuits.

6.5.2.4.3 *Maximum dv/dt and di/dt*
Control system design needs to consider the noise from the converter. Key indicators for noise level include dv/dt and di/dt. Once the inverter power stage is designed, the maximum dv/dt and di/dt can be estimated from the device switching characteristics. For SiC MOSFETs, maximum dv/dt can typically reach 100 V/ns, for low voltage (e.g. 1200 V) as well as high voltage (e.g. 10 kV) devices. GaN HEMTs can have even faster dv/dt. On the other hand, di/dt depends on the power or current level.

6.6 Summary

This chapter presents the power stage design of AC three-phase PWM inverters for load-side applications.

1) The load-side inverter design is formulated as an optimization problem, with its design variables, constraints, conditions, and objective(s) clearly defined. Using the two-level VSC as an example, the design variables include switching devices, PWM scheme, switching frequency, gate drive parameters, decoupling capacitors, decoupling inductors, and thermal management system performance (i.e. thermal impedance). The design constraints include performance constraints for the inverter (AC power quality, power loss or efficiency, and control performance) and physical constraints associated with the components (maximum temperatures, peak currents, and device peak voltage).

2) The focus of the chapter is on establishing analytical or algebraic models for design constraints as functions of design conditions and variables. The analytical solutions for PWM voltage harmonics are presented using both the DFA method and the more computationally efficient SFA method. Closed-form equations are described for conduction and switching losses of the two-level VSC, when the on-state voltage drop and switching energy are linear functions of the AC line current and the DC-link voltage. For more general cases, numerical algebraic solutions for loss calculation are discussed. Control performance has considered control bandwidth and voltage margin requirements. Based on the device maximum junction temperature constraint and device thermal characteristics, maximum thermal impedance model for the thermal management system is developed considering both the stead-state and overload conditions. Analytical models for peak turnoff and turn-on voltages are developed for MOSFETs based on switching transient stage analysis. Decoupling capacitor selection criteria and decoupling inductor design are both covered.

3) A design optimization example is provided for a DC-fed inverter for minimum weight design. An optimizer was used, which is based on a combination of analytical models and time-domain numerical models.

4) The output from the load-side inverter design as the interface parameters to other subsystem systems are discussed, including interface parameters to the rectifier or DC source, the EMI filter, the thermal management system, the mechanical system, and the control system. Most of the interface parameters are part of the design results. Some are design attributes that have to be determined and modeled based on the design results. The key attribute models that are discussed include DC ripple current, DC impedance, EMI bare noise, component, and converter terminal voltage/current distribution, as well as parameters related to sensor, controller, and protection. The dv/dt and di/dt parameters are also discussed.

References

1 F. Wang, W. Shen, D. Boroyevich, S. Ragon, V. Stefanovic and M. Arpilliere, "Voltage source inverter – development of a design optimization tool," *IEEE Industry Applications Magazine*, vol. 15, no. 2, pp. 24–33, March–April 2009.

2 D. G. Holmes and T. A. Lipo, "*Pulse Width Modulation for Power Converters: Principles and Practice*," IEEE Press, Wiley_Interscience, John Wiley & Sons Inc., 2003.

3 N. Jiao, S. Wang and T. Liu, "Harmonic analysis of output voltage in PWM inverters," in *2018 IEEE International Power Electronics and Application Conference and Exposition (PEAC)*, Shenzhen, China, November 4–7, 2018.

4 D. Zhang, F. Wang, S. El-Barbari, J. Sabate and D. Boroyevich, "Improved asymmetric space vector modulation for voltage source converters with low carrier ratio," *IEEE Transactions on Power Electronics*, vol. 27, no. 3, pp. 1130–1140, 2012.

5 Z. Dong, "High Power Light-Weight Solid-State Circuit Breaker Development and Motor Drive Design for Electrified Aircraft Propulsion," PhD diss., University of Tennessee, Knoxville, 2022.

6 J. W. Kolar, H. Ertl and F. C. Zach, "Influence of the modulation method on the conduction and switching losses of a PWM converter system," *IEEE Transactions on Industry Applications*, vol. 27, no. 6, pp. 1063–1075, 1991.

7 K. Zhou and D. Wang, "Relationship between space-vector modulation and three-phase carrier-based PWM: a comprehensive analysis [three-phase inverters]," *IEEE Transactions on Industrial Electronics*, vol. 49, no. 1, pp. 186–196, 2002.

8 M. Rodríguez, A. Rodríguez, P. F. Miaja, D. G. Lamar and J. S. Zúniga, "An insight into the switching process of power MOSFETs: an improved analytical losses model," *IEEE Transactions on Power Electronics*, vol. 25, no. 6, pp. 1026–1640, 2010.

9 X. Wang, Z. Zhao, K. Li, Y. Zhu and K. Chen, "Analytical methodology for loss calculation of SiC MOSFETs," *IEEE Journal of Emerging and Selected Topics in Power Electronics*, vol. 7, no. 1, pp. 71–83, 2019.

10 F. Wang, "System voltage rating selection for an adjustable speed AC synchronous motor drive considering dynamic performance," in *First IEEE International Electric Machines and Drives Conference*, Milwaukee, Wisconsin, May 1997.

11 Q. Liu, S. Wang, A. C. Baisden, F. Wang and D. Boroyevich, "EMI suppression in voltage source converters by utilizing DC-link decoupling capacitors," *IEEE Transactions on Power Electronics*, vol. 22, no. 4, pp. 1417–1428, July 2007.

12 W. Teulings, J. Schanen and J. Roudet, "MOSFET switching behaviour under influence of PCB stray inductance," in *Conference Record of the 1996 IEEE Industry Applications Conference Thirty-First IAS Annual Meeting*, San Diego, CA, 1996.

13 W. Zhang, Z. Zhang, F. Wang, L. Tolbert, D. Costinett and B. Blalock, "Characterization and modeling of a SiC MOSFET's turn-off overvoltage," in *International Conference on Silicon Carbide and Related Materials (ICSCRM)*, Washington, DC, September 2017.

14 W. Zhang, Z. Zhang, F. Wang, D. Costinett, L. M. Tolbert and B. J. Blalock, "Characterization and modeling of a SiC MOSFET's turn-on overvoltage," in *2018 IEEE Energy Conversion Congress and Exposition (ECCE)*, Portland, OR, September 2018.

15 J. Wang, H. S.-H. Chung and R. T.-H. Li, "Characterization and experimental assessment of the effects of parasitic elements on the MOSFET switching performance," *IEEE Transactions on Power Electronics*, vol. 28, no. 1, pp. 573–590, 2013.

16 M. R. Ahmed, R. Todd and A. J. Forsyth, "Predicting SiC MOSFET behavior under hard-switching, soft-switching, and false turn-on conditions," *IEEE Transactions on Industrial Electronics*, vol. 64, no. 11, pp. 9001–9011, 2017.

17 S. K. Roy and K. Basu, "Analytical model to study hard turn-off switching dynamics of SiC MOSFET and Schottky diode pair," *IEEE Transactions on Power Electronics*, vol. 36, no. 1, pp. 861–875, 2021.

18 M. Rodriguez, A. Rodrıguez, P. F. Miaja, D. G. Lamar and J. S. Zuniga, "An insight into the switching process of power MOSFETs: an improved analytical losses model," *IEEE Transactions on Power Electronics*, vol. 25, no. 6, p. 1626–1640, 2010.

19 Z. Chen, D. Boroyevich, P. Mattavelli and K. Ngo, "A frequency domain study on the effect of DC-link decoupling capacitors," in *2013 IEEE Energy Conversion Congress and Exposition (ECCE)*, Denver, CO, 2013.

20 C. Timms, L. Qiao, F. Wang, Z. Zhang and D. Dong, "New boundary condition for decoupling capacitance selection in SiC phase legs considering short-circuit events," in *2018 IEEE Energy Conversion Congress and Exposition (ECCE)*, Portland, OR, September 2018.

21 L. Wu, J. Zhao, L. Xiao and G. Chen, "Investigation of the effects of snubber capacitors on turn-on overvoltage of SiC MOSFETs," in *2018 1st Workshop on Wide Bandgap Power Devices and Applications in Asia (WiPDA Asia)*, Xi'An, China, May 2018.

22 Z. Zhang, F. Wang, L. Tolbert, B. Blalock and D. Costinett, "Decoupling of interaction between WBG converter and motor load for switching performance improvement," in *2016 IEEE Applied Power Electronics Conference and Exposition (APEC)*, Long Beach, CA, March 20–24, 2016.

23 Z. Dong, R. Ren, F. Wang and R. Chen, "An automated design tool for three-phase motor drives," in *IEEE Design Methodologies for Power Electronics Conference*, Bath, UK, July 14–15, 2021.

24 P. A. Dahono, Y. Sato and T. Kataoka, "Analysis and minimization of ripple components of input current and voltage of PWM inverters," *IEEE Transactions on Industry Applications*, vol. 32, no. 4, pp. 945–950, July/August 1996.

25 J. Kolar, T. Wolbank and M. Schrodl, "Analytical calculation of the RMS current stress on the DC link capacitor of voltage DC link PWM converter systems," in *Ninth International Conference on Electrical Machines and Drives*, 1999.

26 H. Liu, H. Guo, J. Liang and L. Qi, "Impedance-based stability analysis of MVDC systems using generator-thyristor units and DTC motor drives," *IEEE Journal of Emerging and Selected Topics in Power Electronics*, vol. 5, no. 1, pp. 5–13, March 2017.

7

Active Rectifiers and Source-side Inverters

7.1 Introduction

This chapter is on the design of three-phase active PWM AC/DC rectifiers and three-phase source-side PWM DC/AC inverters. Both active rectifiers and source-side inverters have their three-phase AC side connected to the AC source. Although the active rectifier and the source-side inverter have different functions, with the former to convert three-phase AC to DC, and the latter to convert DC to three-phase AC, their hardware portions are very similar. In fact, in some cases, even their operation modes are quite similar. For example, a source-side inverter for a renewable energy source (e.g. PV or wind) may work in the rectifier mode while transferring energy from DC to AC. As a result, it makes sense for the design of the three-phase active rectifier and the source-side inverter to be covered together. To facilitate the discussion, we will mainly use the active rectifier as the example. The source-side inverter will be discussed if certain aspect of its design is distinctly different from that of the active rectifier.

Similar to the passive rectifier and the load-side inverter, this chapter discusses the design of the power stage of the active rectifier and the source-side inverter, as a function block or subsystem of the overall three-phase converter. As in other cases, even when the active rectifier or the source-side inverter works as a stand-alone converter, its power stage is still a subsystem of the whole converter, because there are other subsystems for the whole rectifier or inverter, such as thermal management system, mechanical system, control, and EMI filters.

As for the topology, the active rectifier and the source-side inverter often share the same types of topologies. The two-level PWM voltage source rectifier as the front end of a motor drive will be mainly used in this chapter to facilitate the design discussion. The same principle and approaches can be applied to other topologies and applications. The impact of topology will be discussed in Section 7.5.

7.2 Active Rectifier and Source-side Inverter Design Problem Formulation

We will again use an AC-fed motor drive as the example. Figure 7.1 shows the circuit configuration of the example AC-fed motor drive, with the two-level VSC front-end rectifier and the inverter. It can be seen that the main components for the active rectifier include semiconductor switches, the

Design of Three-phase AC Power Electronics Converters, First Edition. Fei "Fred" Wang, Zheyu Zhang, and Ruirui Chen.
© 2024 The Institute of Electrical and Electronics Engineers, Inc. Published 2024 by John Wiley & Sons, Inc.

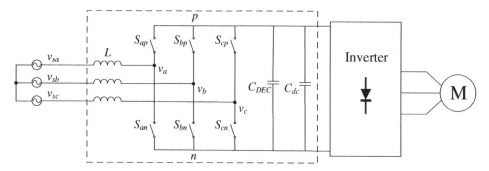

Figure 7.1 Circuit configuration of an AC-fed motor drive with a two-level voltage source rectifier.

DC-link capacitor, and the AC line filter (e.g. an inductor as in Figure 7.1). As explained previously, both the voltage source rectifier and the voltage source inverter require the DC-link capacitor, and its selection needs to consider the requirement of both sides. We have included the DC-link capacitor as a task in the rectifier design, with consideration of the inverter impact through load-side conditions and constraints. If other topologies are used, there may be other passive components; examples include capacitors in different multilevel VSCs in Figure 1.15. For soft-switching topologies, even more complex structures will be involved.

As in the load-side inverter case, Figure 7.1 includes a decoupling capacitor, which is often required for medium- and high-power converters to control switching overvoltage due to parasitic inductances. On the other hand, Figure 7.1 does not include decoupling inductors. With the inductive AC filter, the decoupling inductor is not necessary. The design of the decoupling capacitor is a task in the design of the active rectifier and source-side inverter.

The relationship between the active rectifier or the source-side inverter with its control is similar to that between the load-side inverter and its control. The rectifier or inverter control design related to switching control (i.e. gate drive parameters, PWM schemes, and sometimes switching frequency) will be part of the rectifier or inverter design, while the controller hardware will be covered in Chapter 10. Also similar to the load-side inverter case, the design of the active rectifier or the source-side inverter will not cover the design of the thermal management system, the mechanical system, and EMI filters. However, their interfaces with the rectifier or inverter design will be included. There are also other application-specific auxiliaries such as surge protection devices. The design of these functions or subsystems will be discussed separately in Chapter 12.

7.2.1 Active Rectifier and Source-side Inverter Design Variables

Based on the above discussion, the design for the two-level VSC-based active rectifier and source-side inverter refers to the design and/or selection of switching devices and their control, the AC line filter, and the decoupling capacitor. Therefore, the design variables are mostly the same as those in the load-side inverter case and include the following:

1) Switching devices.
2) Switching scheme: i.e. PWM scheme.
3) Switching frequency: mostly similar to the load-side inverter case. Also, similarly, the switching frequency can be a design condition when the EMI filter is involved. Different from the load-side

inverter, switching frequency can impact control performance of the active rectifier or the source-side inverter in more complex ways since the source-side connection can be more uncertain. For example, the switching frequency can impact control delay, which can impact the stability of the rectifier or source-side inverter.

4) Gate control parameters.
5) DC-link capacitor.
6) AC line filter: for a two-level VSC, the AC side needs inductors for the topology to function as a boost rectifier or buck inverter. Since the inductors also act as filters for current harmonics and ripples, they are also called line filters. Similar to the passive rectifier case in Chapter 5, the AC inductor can be three individual single-phase inductors, or one three-phase inductor. In some cases, a more complex AC line filter topology, like *LCL* filter topology, is used.
7) Decoupling capacitors.
8) Thermal management system: referring to the cooling system thermal impedance, the same as the load-side inverter case.

The same as the load-side inverter case, the impact of the active rectifier or the source-side inverter design on the mechanical system is not treated as a design variable, rather, as a design attribute.

7.2.2 Active Rectifier or Source-side Inverter Design Constraints

The design constraints also include performance constraints related to performance specifications and physical constraints related to component physical capabilities. The common performance specifications include

1) Power quality performance under normal operating conditions, specifically AC input current TDD, the same as the passive rectifier case. Different from the load-side inverter case, where harmonics are a function of the load, the TDD of the active rectifier and the source-side inverter is defined under the assumption that the source is an ideal balanced three-phase voltage source with only positive-sequence fundamental frequency component and zero impedance. In addition to harmonics, negative sequence currents need to be limited as they may cause undesirable effects to equipment in the system, such as torque ripple and loss in generators and motors. DC offset also needs to be limited to avoid saturation in transformers, including transformers often connected between the source and the converter. For the source-side inverter, like PV or battery charger, there will be DC-side power quality requirements as well, such as ripple current through PV or battery.
2) Minimum and maximum DC-link voltages for normal operating conditions, similar to the passive rectifier case.
3) Power loss, the same as the load-side inverter case.
4) Control performance, similar to the load-side inverter case, including both the steady-state and dynamic performance. An important design criterion for the active-rectifier or the source-side inverter is its stability when connecting to the source such as a power grid.
5) Ride-through time or holdup time without input power or with low voltage, similar to the passive rectifier case. In the case of the source-side inverter, such as a PV or a battery inverter, the ride-through requirement is for the inverter not to trip when the grid voltage/frequency is low or high. The voltage or frequency ride-through impact on converter design will be separately considered in Chapter 12.

6) DC-link stability, similar to the passive rectifier case. Sometimes, the active rectifier may have additional large-signal dynamic requirement, which will be discussed in Section 7.3.3.

7) Reliability, often measured by mean-time-between-failures (MTBF) and lifetime. Reliability should be a constraint for any system or subsystem design. It is included in this chapter as an example.

Active rectifiers and source-side inverters often need to comply with EMI/EMC standards on the AC side. Again, since the EMI filter is treated as a separate subsystem, the corresponding EMI requirement will not be included as a performance constraint.

CM voltage or current can also be a concern for the source connected to the rectifier or the inverter, especially, when there is no isolation transformer between the source and the rectifier or the inverter. The CM voltage and current as well as their impact on the rectifier and inverter design will be discussed in Chapter 12.

The physical constraints are those associated with components for the rectifier and the inverter, including switching devices, the DC-link capacitor, the AC line filter, and DC decoupling capacitors. The physical constraints for these components are the same as discussed in previous chapters.

7.2.3   Active Rectifier and Load-side Inverter Design Conditions

The design conditions for the active rectifier and the source-side inverter are mostly similar but also can be substantially different, especially on the DC side electrically. The active rectifier interfaces with the three-phase AC source, the DC load (e.g. a load-side inverter as part of the motor drive) through the DC link, the control, and the thermal and mechanical environment. The source-side inverter interfaces with the DC source (e.g. a DC-interfaced PV, wind or battery) through the DC link, the three-phase AC system source, the control, and the thermal and mechanical environment. Design conditions for both the active rectifier and the source-side inverter include AC source and DC load or source electrical conditions, control conditions, and environmental conditions. The AC source electrical conditions include

1) AC source voltage characteristics under nominal conditions, under which the rectifier or the inverter, and whole converter system (e.g. a motor drive or a PV inverter) need to operate normally and deliver required load power. The characteristics can be defined the same as the passive rectifier case as in Chapter 5.

2) AC source voltage characteristics under transient and abnormal conditions. These include over- and undervoltage conditions and can be defined the same as the passive rectifier case as in Chapter 5.

3) AC source equivalent impedance, which is important during transient conditions like inrush for the passive rectifier case. In the active rectifier or the source-side inverter case, inrush can still occur but can be generally better controlled. On the other hand, source impedance could cause control instability, so the active rectifier or the source-inverter design needs to be tied to certain impedance range.

4) Reactive power, which may be required for the active rectifier and the source-side inverter to provide to the source.

5) Active power, which is often the input requirement for a source-side inverter, e.g. an inverter working in grid-following mode. By the way, if the source-side inverter works as an AC source, it can be treated as a load-side inverter.

Since the DC-side electrical conditions are quite different for the active rectifier and the source-side inverter, they will be discussed separately. The DC side or load electrical conditions for the active rectifier include (most are the same as the passive rectifier case):

1) Nominal load power.
2) Overload.
3) Peak power.
4) Maximum regenerative power, the same as the passive rectifier case if the active rectifier topology can only have unidirectional power flow, such as the Vienna-type rectifier. For the bidirectional topologies like two-level VSC, the maximum regenerative power usually does not impact the rectifier design much, but should still be considered if specified.
5) DC ripple current from the load-side inverter.
6) DC-link undervoltage and overvoltage trip levels.
7) DC load impedance.

The DC side or DC source electrical conditions for the source-side inverter include

1) Minimum DC voltage, including minimum voltages for both normal conditions and abnormal conditions. Normal conditions refer to normal DC voltage range as in the case of PV inverters or battery chargers. Abnormal conditions refer to low-voltage conditions as a result of fault or disturbance, e.g. DC voltage sag as a result of AC or DC source disturbance for the PV inverter or battery charger case.
2) Peak DC voltage, also including voltages under normal and abnormal conditions. The normal conditions refer to normal DC voltage range as in the case of PV inverters or battery chargers. Abnormal conditions refer to overvoltage as a result of DC source fault or disturbance.
3) DC source characteristics, including equivalent impedance. DC sources like PV or battery (together with DC/DC converters connected to these sources) have their own characteristics and need to be considered in the inverter design.

The design conditions of the active rectifier and the source-side inverter related to the thermal management system and the mechanical system are the same as those for the load-side inverter. The EMI filter is often required on the AC source side and may be even required on the DC source side for the source-side inverter. The switching frequency can be either a design variable or design condition depending on if EMI filters are present, as explained in Chapter 6.

Different from the passive rectifier but similar to the load-side inverter, the active rectifier and the source-side inverter are actively controlled and therefore are strongly influenced by the control subsystem or controllers. The design conditions of the active rectifier and the source-side inverter related to the control are also the same as the load-side inverter case.

7.2.4 Active Rectifier and Source-side Inverter Design Objectives and Design Problem Formulation

The design objectives for each subsystem of an AC converter, including the active rectifier and the source-side inverter, should follow those of the whole converter. Based on the selected design objectives, and the design variables, constraints, and conditions, the overall active rectifier design problem can be formulated as the following optimization problem:

Minimize or maximize: $F(X)$ – objective function

Subject to constraints:

$$
G(X,U) = \begin{bmatrix} g_1(X,U) \\ g_2(X,U) \\ g_3(X,U) \\ g_4(X,U) \\ g_5(X,U) \\ g_6(X,U) \\ g_7(X,U) \\ g_8(X,U) \\ g_9(X,U) \\ g_{10}(X,U) \\ g_{11}(X,U) \\ g_{12}(X,U) \\ g_{13}(X,U) \\ g_{14}(X,U) \\ g_{15}(X,U) \\ g_{16}(X,U) \\ g_{17}(X,U) \\ g_{18}(X,U) \\ g_{19}(X,U) \\ g_{20}(X,U) \\ g_{21}(X,U) \\ g_{22}(X,U) \end{bmatrix} = \begin{bmatrix} \text{AC source current TDD} \\ \text{Power loss} \\ -\text{Steady state } V_{DC(\min)} \\ \text{Steady state } V_{DC(\max)} \\ \text{Ride through time} \\ -\text{Control bandwidth} \\ -\text{Voltage margin} \\ -\text{DC link stability margin} \\ -\text{AC source side stability} \\ -\text{Reliability} \\ \text{Device junction temperature} \\ \text{Device peak voltage} \\ \text{Device maximum current} \\ \text{Inductor core window fill factor} \\ \text{Inductor saturation flux density} \\ \text{Inductor temperature} \\ \text{DC and AC capacitor voltage} \\ \text{DC and AC capacitor ripple current} \\ \text{DC and AC capacitor temperature} \\ \text{Decoupling capacitor voltage} \\ \text{Decoupling capacitor temperature} \\ \text{Decoupling capacitor current} \end{bmatrix} \leq \begin{bmatrix} \text{TDD limit} \\ \text{Loss limit} \\ -V_{DC(\min)} \text{ limit} \\ V_{DC(\max)} \text{ limit} \\ t_{rt} \\ -\text{minimum bandwidth} \\ -\text{minimum margin} \\ -\text{phase or gain margin} \\ -\text{phase or gain margin} \\ -\text{MTBF and lifetime} \\ T_{jmax} \\ V_{limit} \\ I_{limit} \\ K_{fill-max} \\ B_{sat} \\ T_{ind(\max)} \\ V_{cap} \\ I_{cap} \\ T_{cap(\max)} \\ V_{cap1} \\ T_{cap1} \\ I_{cap1} \end{bmatrix}
$$

where design variables and design conditions:

$$
X = \begin{bmatrix} x_1 \\ x_2 \\ x_3 \\ x_4 \\ x_5 \\ x_6 \\ x_7 \\ x_8 \\ x_9 \end{bmatrix} = \begin{bmatrix} \text{Devices} \\ \text{PWM scheme} \\ \text{Switching frequency } f_{SW} \text{ (if no EMI filter)} \\ \text{Gate resistance } R_g \\ \text{Gate voltage } V_g \text{ (both high and low voltage } V_{GH} \text{ and } V_{GL}) \\ \text{DC link capacitor} \\ \text{AC side filter} \\ \text{Decoupling capacitors} \\ \text{Maximum TMS thermal impedance} \end{bmatrix}
$$

$$
U = \begin{bmatrix} u_1 \\ u_2 \\ u_3 \\ u_4 \\ u_5 \\ u_6 \\ u_7 \\ u_8 \\ u_9 \\ u_{10} \\ u_{11} \\ u_{12} \\ u_{13} \\ u_{14} \end{bmatrix} = \begin{bmatrix} \text{AC voltage characteristics under nominal conditions} \\ \text{AC voltage characteristics under transient/abnormal conditions} \\ \text{Source impedance} \\ \text{Reactive power} \\ \text{Nominal DC load power} \\ \text{Overload and duration} \\ \text{Peak power} \\ \text{Maximum regenerative energy if unidirectional topology} \\ \text{DC link over and undervoltage trip levels} \\ \text{DC load ripple current} \\ \text{DC load impedance} \\ \text{Ambient temperature } T_a \\ \text{Busbar and connector parasitics} \\ \text{Switching frequency } f_{SW} \text{ (if with EMI filter)} \end{bmatrix}
\tag{7.1}
$$

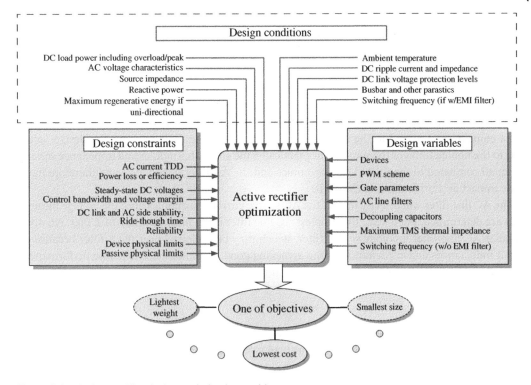

Figure 7.2 Active rectifier design optimization problem.

The design optimization problem for the active rectifier can be illustrated also as in Figure 7.2 [1]. The source-inverter optimization problem can be formulated similarly.

7.3 Active Rectifier and Source-side Inverter Models

In order to solve the optimization problem (7.1), we need the following models for the constraints as functions of design conditions and design variables:

1) AC source harmonic current TDD.
2) Power loss.
3) Control performance.
4) Ride-through time.
5) DC-link stability.
6) Reliability.
7) Device maximum junction temperature.
8) Device peak voltage.
9) AC inductor parameters, which are considered in modeling the AC source harmonics.
10) Decoupling capacitor parameters: capacitance, energy (current), and peak voltage.

This section will focus on how to establish these models, again with focus on models in analytical or algebraic forms.

Ride-through relationship is the same as in Chapter 5; and device loss model, device maximum junction temperature, device peak voltage model, and the decoupling capacitor design are the same as in Chapter 6. As a result, they are not repeated here. The discussion will focus on AC source current harmonics including filter design, control performance, DC-link stability, and reliability.

7.3.1 AC Source Harmonic Current

As explained earlier, the AC source harmonic currents and the TDD depend only on the voltage harmonics of the VSC rectifier or inverter and the AC line filter. Of course, the measured TDD for a converter connected to the source (often the grid) will be different from the design value due to the nonidealities of the source. But practically, the source harmonics and impedance should be small compared to the PWM VSC harmonics and the AC filter impedance, and therefore have little impact on harmonics and the TDD.

The AC line filter will directly impact the AC source current harmonics once the voltage harmonics are determined. For a VSC, the simplest AC line filter is the inductor filter, or L filer. Another popular AC line filter topology is LCL filter, which may be preferred over the L filter because of small inductances and capacitances while achieving good harmonics and dynamic performance. Since the models for these two filters are quite different, they will be discussed separately.

7.3.1.1 *L* Filter

Independent of the AC line filter, the voltage harmonics of the VSC active rectifier or source-side inverter can be evaluated the same way as the load-side inverter, and therefore will not be repeated here. However, the voltage harmonics are functions of the modulation index M, which needs to be determined for the active rectifier and the source-side inverter case. As defined in (6.10), $M = \frac{2V_{ph-M}}{V_{DC}}$. So, the peak phase voltage V_{ph-M} needs to be determined. Taking the active rectifier as the example, with the L filter, the relationship between the source-side phase voltage V_s, the rectifier-side phase voltage V_{ph}, the source-side phase current I_s, and the power factor angle φ can be determined through the phasor diagram in Figure 7.3. Note that $I_X = \omega_0 L I_s$. The relationship is expressed in (7.2), where ω_0 is the fundamental angular frequency and L is the filter inductance. The inductor resistance is assumed negligible.

$$V_{ph} = \sqrt{V_s^2 + \omega_0^2 L^2 I_s^2 - 2\omega_0 L I_s V_s \sin \varphi} \qquad (7.2)$$

For the source-side inverter with L filter, the phasor diagram of the voltage and current is shown in Figure 7.4. The phase voltage expression is the same as the active rectifier case as in (7.2).

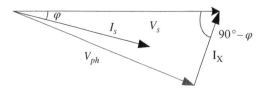

Figure 7.3 Phasor diagram of the voltage and current for the active rectifier with L filter.

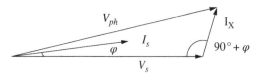

Figure 7.4 Phasor diagram of the voltage and current for the active rectifier with L filter.

In (7.2), V_s, ω_0, and φ are known design conditions, and I_s can be determined from power and reactive power or the kVA rating, which are also known as design conditions. However, inductance L is a design variable (strictly speaking, an intermediate design variable), making the calculation of V_{ph} and the corresponding calculation of the modulation index and voltage harmonics a function of L. In practice, when power factor is near unity, and L value is relatively small, V_{ph} can be approximated to be equal to V_s. For example, with unity power factor and for a 5% or 0.05 p.u. inductance value, at the rated condition of 1 p.u. V_{ph} and I_s, V_{ph} is 1.0012 p.u. Even for low power factor, V_{ph} should be close to V_s.

Once V_{ph} and the corresponding modulation index are determined, the voltage harmonics of the active rectifier or source-side inverter can be calculated using the same approach as in Chapter 6. With the L filter, the current harmonics can be simply determined by (7.3), where n is the harmonic order, $I_{n(RMS)}$ and $V_{n(RMS)}$ are the nth order RMS harmonic current and voltage, respectively, and R is the equivalent resistance of the inductor. In most high-power cases, R is much smaller than $n\omega_0 L$ and can be neglected.

$$I_{n(RMS)} = \frac{V_{n(RMS)}}{\sqrt{R^2 + (n\omega_0 L)^2}} \approx \frac{V_{n(RMS)}}{n\omega_0 L} \tag{7.3}$$

The TDD of AC source-side current can be determined via (7.4), where $I_{1(RMS)}$ is the RMS value of fundamental frequency AC input current at the rated condition. Practically, $I_{1(RMS)}$ can be approximated by the rated RMS current of the active rectifier or source-side inverter, $I_{(RMS)}$. Given that harmonic current standards often specify individual harmonic limits, the expression for individual harmonic current can be found via (7.3).

$$TDD = \frac{\sqrt{\sum_{n=2}^{N} I_{n(RMS)}^2}}{I_{1(RMS)}} \approx \frac{\sqrt{\sum_{n=2}^{N} I_{n(RMS)}^2}}{I_{(RMS)}} \tag{7.4}$$

7.3.1.2 *LCL* Filter

Although L filter is simple, for the source-side inverter, LC filter is often more suitable for better voltage regulation with the capacitor voltage as a state variable. When current harmonics need to be attenuated, which is often the case for both the active rectifier and the source-side inverter, LCL filter is preferred.

Figure 7.5 shows a typical LCL filter topology, where L_1 and L_2 are the converter and source-side inductors, respectively, C_f is the filter capacitor, and R_d is the damping resistor. R_d is necessary to attenuate the resonance between the capacitor and inductors. Active damping schemes are less lossy and can be used in certain applications. However, passive damping is generally more robust, especially in grid-tied applications where damping function must remain effective even when the converter is off. There are other passive damping methods in addition to $R_d C_f$ series approach. The performance improvements of these alternative methods are limited and their design is omitted here.

Similar to the L filter case, to establish the harmonic currents and other relevant models for the LCL filter, the PWM voltages of the VSC must be first determined. The active rectifier phase voltage can be determined in the following sequence. The parasitic resistances of the inductors are neglected. R_d can also be neglected because at the fundamental line frequency, the impedance due to C_f will dominate.

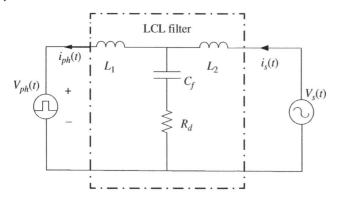

Figure 7.5 Per-phase *LCL* filter equivalent circuit.

The filter capacitor voltage V_{cf} can be found as

$$V_{cf} = \sqrt{V_s^2 + \omega_0^2 L_2^2 I_s^2 - 2\omega_0 L_2 I_s V_s \sin \varphi} \tag{7.5}$$

The filter capacitor current I_{cf} can be found as (neglecting R_d)

$$I_{cf} \approx \omega_0 C_f V_{cf} \tag{7.6}$$

The angle of the V_{cf}, θ_{cf}, with respect to V_s can be found as

$$\theta_{cf} = \cos^{-1}\left(\frac{V_{cf}^2 + V_s^2 - \omega_0^2 L_2^2 I_s^2}{2V_{cf} V_s}\right) \tag{7.7}$$

The rectifier-side current I_{ph} can be found as

$$I_{ph} = \sqrt{I_s^2 + I_{cf}^2 + 2I_s I_{cf} \sin(\theta_{cf} - \varphi)} \tag{7.8}$$

The angle of the I_{ph}, θ_r, with respect to V_s can be found as

$$\theta_r = \cos^{-1}\left(\frac{I_{ph}^2 + I_s^2 - I_{cf}^2}{2I_{ph} I_s}\right) + \varphi \tag{7.9}$$

Finally, the phase voltage V_{ph} can be found as

$$V_{ph} = \sqrt{V_{cf}^2 + \omega_0^2 I_{ph}^2 L_1^2 - 2\omega_0 I_{ph} L_2 V_{cf} \sin(\theta_r - \theta_{cf})} \tag{7.10}$$

Similar to the *L* filter case, since the inductances and capacitance are relatively small, V_{ph} is close to V_s.

Once V_{ph} corresponding modulation index and PWM voltages are determined, the *LCL* filter parameters can be selected based on the TDD and other requirements. Compared to *L* filter with a −20 dB/dec harmonic attenuation, *LCL* filter has a −60 dB/dec attenuation. It is more suitable when higher order harmonics need to be attenuated and can be smaller in size than *L* filter. Since *LCL* filter involves more components, its design needs to consider more constraints and requires more models than the source-side current THD. The constraints additional to current THD include the converter-side current ripple which impact converter switches and inductor L_1, the series fundamental voltage drop, the reactive power or power factor limits, and the resonance frequency. In general, in an *LCL* filter, the converter-side inductor L_1 is constrained by the converter ripple current, the filter capacitor C_f is constrained by power factor, and the source-side inductor L_2 is constrained by the source-side current harmonics.

The maximum capacitance for the filter capacitor can be easily established using the reactive power constraint as in (7.11) and (7.12), where Q_{cf} is the per-phase reactive power on filter capacitor branch, S is the rated kVA of the active rectifier, and k_{cf} is a coefficient. Generally speaking, k_{cf} should be less than 5%, normally in the range of 2–5%.

$$Q_{cf} \approx \omega_0 C_f V_{cf}^2 \approx \omega_0 C_f V_s^2 \leq k_{cf} \frac{S}{3} \tag{7.11}$$

$$C_f \leq k_{cf} \frac{S}{3\omega_0 V_s^2} \tag{7.12}$$

The maximum inductance values of L_1 and L_2 are constrained by the voltage drop limit. Neglecting the filter capacitor because of its small impact at the fundamental frequency, the maximum or minimum converter side voltage can be obtained using (7.2) when the power factor is at the lowest value. The extreme case is when the power factor is 0 (φ is $\pm90°$), and the corresponding maximum or minimum V_{ph} can be found with (7.13). At the rated condition, V_{ph} should be in a close range of V_s, for example, within $\pm5\%$ or $\pm10\%$. As a result, $\omega_0(L_1 + L_2)$ should be within the same percentage of the base impedance (i.e. 1 p.u. impedance).

$$V_{ph} = V_s \pm \omega_0 I_s (L_1 + L_2) \tag{7.13}$$

With *LCL* filter, the voltage on the filter capacitor branch v_{cf} is nearly sinusoidal, and the ripple current in I_r is mainly determined by the converter PWM voltage and the converter-side inductance L_1. Figure 7.6 shows the two-level VSC-based active rectifier represented by switching functions with L filter. The switching functions S_a, S_b, and S_c for phases a, b, and c are defined the same as in Chapter 6.

The source-side three-phase voltages v_{sa}, v_{sb}, and v_{sc} can be assumed as ideal balanced sinusoidal voltage sources. The converter-side three-phase voltages v_a, v_b, and v_c are PWM voltages with both the DM and CM components. Assuming the negative DC bus as the reference point of the system, the phase voltages can be expressed as

$$\begin{bmatrix} v_a \\ v_b \\ v_c \end{bmatrix} = \begin{bmatrix} S_a \\ S_b \\ S_c \end{bmatrix} v_{DC} \tag{7.14}$$

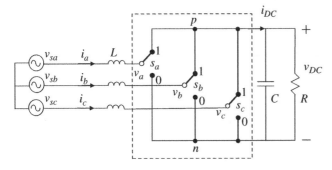

Figure 7.6 Switching function representation of the VSC-based active rectifier.

CM voltage is defined as

$$v_{CM} = \frac{v_a + v_b + v_c}{3} = \frac{S_a + S_b + S_c}{3} v_{DC} \tag{7.15}$$

Figure 7.6 can be redrawn as Figure 7.7, where v_{aN}, v_{bN}, and v_{cN} are the DM components of v_a, v_b, and v_c. Since we are dealing with a three-wire system with $i_a + i_c + i_c = 0$, the CM path N_0 in Figure 7.7 can be considered open ($Z_0 \to \infty$). With symetrical v_{aN}, v_{bN}, and v_{cN}, Eq. (7.16) can be obtained, which can be used to determine the inductor ripple current.

$$\begin{cases} v_{sa} = L_1 \dfrac{di_a}{dt} + v_{aN} = L_1 \dfrac{di_a}{dt} + S_a v_{DC} - \dfrac{v_{DC}}{3}(S_a + S_b + S_c) \\[3mm] v_{sb} = L_1 \dfrac{di_b}{dt} + v_{bN} = L_1 \dfrac{di_b}{dt} + S_b v_{DC} - \dfrac{v_{DC}}{3}(S_a + S_b + S_c) \\[3mm] v_{sc} = L_1 \dfrac{di_c}{dt} + v_{cN} = L_1 \dfrac{di_c}{dt} + S_c v_{DC} - \dfrac{v_{DC}}{3}(S_a + S_b + S_c) \end{cases} \tag{7.16}$$

Two observations can be made from (7.16). First, the ripple current in one phase depends on switching actions of all three phases. For example, $\frac{di_a}{dt}$, which is related to the ripple current in phase a, is a function of S_a as well as S_b and S_c. Second, since swiching functions are determined by PWM scheme, the ripple current is highly dependent on PWM.

With SPWM, example reference voltages for v_a, v_b, and v_c are shown as in Figure 7.8. Only a half line cycle between D0 and D4 is needed due to symmetry. The relationships in (7.16) can be analyzed in detail for each time interval T1 (between D0 and D1), T2 (between D1 and D2), T3 (between D2 and D3), and T4 (between D3 and D4). Due to symmetry, it is sufficeint to analyze T1 nd T2.

In time interval T1, $v_{c\_ref} > v_{a\_ref} > v_{b\_ref}$. These reference voltages can be considered as constant within one switching cycle, as in Figure 7.9, where T_{aon}, T_{bon}, and T_{con} are respective time durations for phases a, b, c to be at the positive DC bus within half of the switching cycle.

In Figure 7.9, when $t \in [0, T_{bon}]$, $S_a = 1$, $S_b = 1$, and $S_c = 1$. From Eq. (7.16) and Figure 7.9, Eqs. (7.17) and (7.18) can be obtained for phase a.

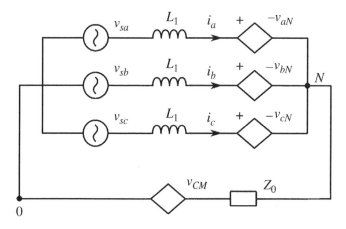

Figure 7.7 An equivalent circuit of the VSC-based rectifier with L filter.

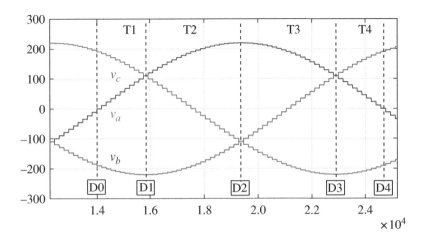

Figure 7.8 SPWM reference phase voltage waveforms for VSC.

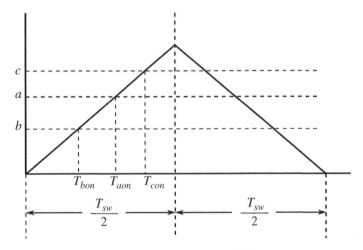

Figure 7.9 Reference phase voltages in one SPWM switching cycle.

$$v_{sa} = L_1 \frac{di_a}{dt} = L_1 \frac{\Delta i_{a\_1}}{T_{bon}} \tag{7.17}$$

$$\Delta i_{a\_1} = \frac{v_{sa} T_{bon}}{L_1} \tag{7.18}$$

When $t \in [T_{bon}, T_{aon}]$, $S_a = 1$, $S_b = 0$, and $S_c = 1$, then

$$v_{sa} = L_1 \frac{di_a}{dt} + \frac{v_{DC}}{3} = L_1 \frac{\Delta i_{a\_2}}{(T_{aon} - T_{bon})} + \frac{v_{DC}}{3} \tag{7.19}$$

$$\Delta i_{a\_2} = \frac{\left(v_{sa} - \frac{v_{DC}}{3}\right)(T_{aon} - T_{bon})}{L_1} \tag{7.20}$$

When $t \in [T_{aon}, T_{con}]$, $S_a = 0$, $S_b = 0$, and $S_c = 1$, then

$$v_{sa} = L_1 \frac{di_a}{dt} - \frac{v_{DC}}{3} = L_1 \frac{\Delta i_{a\_3}}{(T_{con} - T_{aon})} - \frac{v_{DC}}{3} \tag{7.21}$$

$$\Delta i_{a\_3} = \frac{\left(v_{sa} + \frac{v_{DC}}{3}\right)(T_{con} - T_{aon})}{L_1} \tag{7.22}$$

When $t \in \left[T_{con}, \frac{1}{2} T_{SW}\right]$, $S_a = 0$, $S_b = 0$, and $S_c = 0$, then

$$v_{sa} = L_1 \frac{di_a}{dt} = L_1 \frac{\Delta i_{a\_4}}{(T_{SW}/2 - T_{con})} \tag{7.23}$$

$$\Delta i_{a\_4} = \frac{v_{sa}(T_{SW}/2 - T_{con})}{L_1} \tag{7.24}$$

The on-times T_{aon}, T_{bon}, and T_{con} can be expressed as $T_{aon} = d_a T_{SW}/2$, $T_{bon} = d_b T_{SW}/2$, and $T_{con} = d_c T_{SW}/2$, where d_a, d_b, and d_c are the duty cycles for phases a, b, and c, respectively. For SPWM, these duty cycles can be found as in (7.25), where M is the modulation index, and θ_a, θ_a, and θ_a are angles for phases a, b, and c reference voltages.

$$\begin{cases} d_a = \dfrac{1 + M \sin \theta_a}{2} \\ d_b = \dfrac{1 + M \sin \theta_b}{2} \\ d_c = \dfrac{1 + M \sin \theta_c}{2} \end{cases} \tag{7.25}$$

Considering the no-load steady-state condition, $v_{DC} = V_{DC}$, $v_{sa} \approx V_{ph\text{-}M} \sin \theta_a = MV_{DC} \sin \theta_a/2$, $\theta_b = \theta_a - 120°$, and $\theta_c = \theta_a + 120°$. It can be shown that

$$\sum_{k=1}^{4} \Delta i_{a\_k} = \frac{V_{DC}}{3L_1}(-2T_{aon} + T_{bon} + T_{con}) + \frac{v_{sa} T_{SW}}{2L_1} = 0 \tag{7.26}$$

In other words, the average volt-seconds on inductor L_1 is zero, and the current starts with and resets to zero under the no-load condition within the first half of the switching cycle. The above analysis can be extended to the second half of the switching cycle. Referring to Figure 7.9, there are also four subintervals in $\left[\frac{1}{2} T_{SW}, T_{SW}\right]$. When $t \in \left[\frac{1}{2} T_{SW}, T_{SW} - T_{con}\right]$, $S_a = 0$, $S_b = 0$, and $S_c = 0$, the corresponding phase a current change $\Delta i_{a\_5} = \Delta i_{a\_4}$. When $t \in [T_{SW} - T_{con}, T_{SW} - T_{aon}]$, $S_a = 0$, $S_b = 0$, and $S_c = 1$, the corresponding phase a current change $\Delta i_{a\_6} = \Delta i_{a\_3}$. When $t \in [T_{SW} - T_{aon}, T_{SW} - T_{bon}]$, $S_a = 1$, $S_b = 0$, and $S_c = 1$, the corresponding current change $\Delta i_{a\_7} = \Delta i_{a\_2}$. When $t \in [T_{SW} - T_{bon}, T_{SW}]$, $S_a = 1$, $S_b = 1$, and $S_c = 1$, the corresponding phase a current change $\Delta i_{a\_8} = \Delta i_{a\_1}$. Therefore, for the second half of the switching cycle, the average volt-seconds on inductor L_1 is zero, and the current starts with and resets to zero under the no-load condition.

Note that in T1, from their expressions, Δi_{a_1} (Δi_{a_8}), Δi_{a_3}(Δi_{a_6}), and Δi_{a_4} (Δi_{a_5}), are positive, while Δi_{a_2} (Δi_{a_2}) is negative. Their relationship can be qualitatively illustrated in Figure 7.10. In the first half of the switching cycle, the positive peak is equal to Δi_{a_1} and the negative peak is equal to $(-\Delta i_{a_1} - \Delta i_{a\_2})$. It can be shown that in T1 when $\theta_a \in [0, 30°]$, the negative peak is always larger than the positive peak. As a result, the peak-to-peak ripple current can be determined as in (7.27).

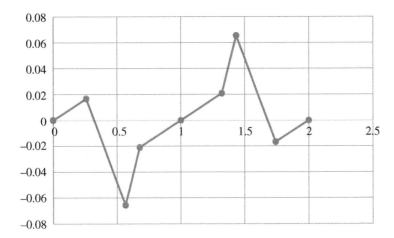

Figure 7.10 Phase *a* current ripple under no-load condition in one SPWM switching cycle.

$$
\begin{aligned}
i_{pp} = 2(-\Delta i_{a\_2} - \Delta i_{a\_1}) &= \frac{2\left(\frac{V_{DC}}{3} - v_{sa}\right)(T_{aon} - T_{bon})}{L_1} - \frac{2v_{sa}T_{bon}}{L_1} \\
&= \frac{2\frac{V_{DC}}{3}(T_{aon} - T_{bon}) - 2v_{sa}T_{aon}}{L_1} \\
&= \frac{MV_{DC}T_{SW}(\sin\theta_a - \sin\theta_b)}{6L_1} - \frac{MV_{DC}T_{SW}\sin\theta_a(1 + M\sin\theta_a)}{4L_1} \\
&= MV_{DC}T_{SW}\frac{\sqrt{3}\cos\theta_a - 3M\sin^2\theta_a}{12L_1}
\end{aligned}
\tag{7.27}
$$

In the time interval T2 of the Figure 7.8, i.e. when $30° \le \theta_a \le 90°$, $v_{a\_ref} > v_{c\_ref} > v_{c\_ref}$. Following the same analysis approach for T1, the inductor current change in each subinterval can be obtained.

For $t \in [0, T_{bon}]$ or $t \in [T_{SW} - T_{bon}, T_{SW}]$,

$$
\Delta i_{a\_1} = \Delta i_{a\_8} = \frac{v_{sa}T_{bon}}{L_1} = \frac{MV_{DC}T_{SW}\sin\theta_a(1 + M\sin\theta_b)}{8L_1}
\tag{7.28}
$$

For $t \in [T_{b\_on}, T_{con}]$ or $t \in [T_{SW} - T_{con}, T_{SW} - T_{bon}]$,

$$
\Delta i_{a\_2} = \Delta i_{a\_7} = \frac{\left(v_{sa} - \frac{v_{DC}}{3}\right)(T_{con} - T_{bon})}{L_1} = MV_{DC}T_{SW}\frac{\left(M\sin\theta_a - \frac{2}{3}\right)(\sin\theta_c - \sin\theta_b)}{8L_1}
\tag{7.29}
$$

For $t \in [T_{con}, T_{aon}]$ or $t \in [T_{SW} - T_{aon}, T_{SW} - T_{con}]$,

$$
\Delta i_{a\_3} = \Delta i_{a\_6} = \frac{\left(v_{sa} - \frac{2v_{DC}}{3}\right)(T_{aon} - T_{con})}{L_1} = MV_{DC}T_{SW}\frac{\left(M\sin\theta_a - \frac{4}{3}\right)(\sin\theta_a - \sin\theta_c)}{8L_1}
\tag{7.30}
$$

For $t \in [T_{aon}, T_{SW}/2]$ or $t \in [T_{SW}/2, T_{SW} - T_{aon}]$,

$$
\Delta i_{a\_4} = \Delta i_{a\_5} = \frac{v_{sa}(T_{SW}/2 - T_{aon})}{L_1} = \frac{MV_{DC}T_{SW}\sin\theta_a(1 - M\sin\theta_a)}{8L_1}
\tag{7.31}
$$

It can be shown that the current resets to zero for both the first and second half of each switching cycle. It can also be seen that Δi_{a_1} (Δi_{a_8}) and Δi_{a_4} (Δi_{a_5}) are positive, and Δi_{a_3} (Δi_{a_6}) is negative. On the other hand, Δi_{a_2} (Δi_{a_7}) can be positive or negative depending on the sign of $\left(M \sin \theta_a - \frac{2}{3}\right)$. If Δi_{a_2} is negative, then the positive peak of the first half cycle is Δi_{a_1} and the negative peak is Δi_{a_4}. It can be shown that when $30° \le \theta_a < 60°$, $\Delta i_{a\_1} < \Delta i_{a\_4}$; when $60° \le \theta_a \le 90°$, $\Delta i_{a\_1} \ge \Delta i_{a\_4}$. If Δi_{a_2} is positive, then the positive peak of the first half cycle is $(\Delta i_{a\_1} + \Delta i_{a\_2})$, which can be larger or smaller than the negative peak Δi_{a_4}, depending on modulation index M and phase angle θ_a.

In summary, there are five cases for the peak inductor ripple current for the SPWM scheme.

Case 1: In T1, the peak-to-peak ripple current i_{pp} can be determined as in (7.27).

Case 2: In T2, when $\Delta i_{a\_2} < 0$ and $30° \le \theta_a < 60°$, the peak-to-peak ripple current i_{pp} is

$$i_{pp} = 2\Delta i_{a\_4} = \frac{MV_{DC}T_{SW} \sin \theta_a (1 - M \sin \theta_a)}{4L_1} \tag{7.32}$$

Case 3: In T2, when $\Delta i_{a\_2} < 0$ and $60° \le \theta_a \le 90°$, the peak-to-peak ripple current i_{pp} is

$$i_{pp} = 2\Delta i_{a\_1} = \frac{MV_{DC}T_{SW} \sin \theta_a (1 + M \sin \theta_b)}{4L_1} \tag{7.33}$$

Case 4: In T2, when $\Delta i_{a\_2} > 0$ and $(\Delta i_{a\_1} + \Delta i_{a\_2}) < \Delta i_{a\_4}$, the peak-to-peak ripple current i_{pp} is the same as in (7.32).

Case 5: In T2, when $\Delta i_{a\_2} > 0$ and $(\Delta i_{a\_1} + \Delta i_{a\_2}) > \Delta i_{a\_4}$, the peak-to-peak ripple current i_{pp} is

$$i_{pp} = 2(\Delta i_{a\_1} + \Delta i_{a\_2}) = \frac{MV_{DC}T_{SW}}{4L_1}\left(\frac{1}{3}\sin \theta_a + M \sin \theta_a \sin \theta_c - \frac{4}{3}\sin \theta_c\right) \tag{7.34}$$

Based on these expressions, phase a peak-to-peak current ripple as the function of θ_a and M can be obtained as shown in Figure 7.11 for ¼ line cycle. The rest of the line cycle will have the same characteristics due to symmetry. It can be seen that the maximum ripple current occurs at 0° (phase voltage 0) and 90° (phase voltage peak). Considering the relatively high modulation index cases for the active rectifier and source-side inverter, i_{pp} at 0° and 90° can be obtained from Case 1 and Case 5

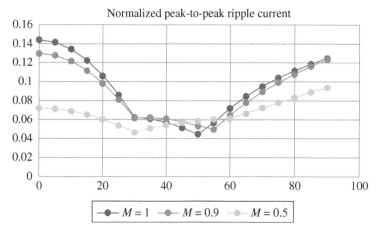

Figure 7.11 Phase a peak-to-peak ripple current for different modulation indices with SPWM switching under no-load condition in ¼ line cycle.

above, respectively, and the results are (7.35) and (7.36). Clearly, i_{pp} at 0° is larger, and the corresponding constraint for the inductance L_1 can be obtained for a given ripple constraint. Considering the worst case at $M = 1$ and a maximum peak-to-peak ripple current Δi_{max}, the inductance constraint can be determined as in (7.37).

$$i_{pp}(\theta_a = 0) = \frac{MV_{DC}T_{SW}}{4\sqrt{3}L_1} \tag{7.35}$$

$$i_{pp}\left(\theta_a = \frac{\pi}{2}\right) = \frac{MV_{DC}T_{SW}}{4L_1}\left(1 - \frac{M}{2}\right) \tag{7.36}$$

$$L_1 \geq \frac{V_{DC}T_{SW}}{4\sqrt{3}\Delta i_{max}} \tag{7.37}$$

For other PWM schemes, similar analysis can be performed. For continuous SVM, the phase reference voltages and duty cycles can be found in Table 6.1, and the corresponding peak to-peak ripple current can be found as in Figure 7.12. Again, for high modulation index cases for the active rectifier and source-side inverter, i_{pp} maximum values can be seen occurring at 0° when $v_a = 0$. The corresponding expression is the same as (7.35). Considering the worst case at $M = \frac{2}{\sqrt{3}}$ for SVM and a maximum peak-to-peak ripple current Δi_{max}, the inductance constraint can be determined as in (7.38).

$$L_1 \geq \frac{V_{DC}T_{SW}}{6\Delta i_{max}} \tag{7.38}$$

For 60° DPWM, the phase reference voltages and duty cycles can be found in Table 6.2, and the corresponding peak-to-peak ripple current can be found as in Figure 7.13. For high modulation index cases, i_{pp} maximum values can be seen again occurring at 0°. The corresponding expression is the same as (7.35) and the inductance constraint is the same as in (7.38) corresponding to $M = \frac{2}{\sqrt{3}}$ for SVM. It can be observed that the maximum ripple can also be reached for some modulation

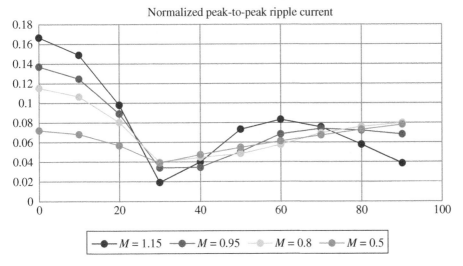

Figure 7.12 Phase peak-to-peak ripple current for different modulation indices with continuous SVM switching under no-load condition in ¼ line cycle.

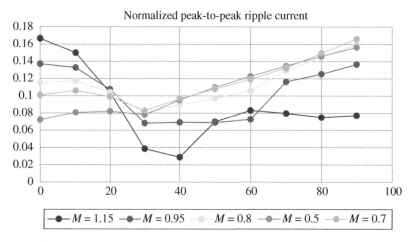

Figure 7.13 Phase peak-to-peak ripple current for different modulation indices with 60° discontinuous SVM switching under no-load condition in ¼ line cycle.

indices at 90°, with the corresponding i_{pp} expression in (7.39). It can be shown that the maximum i_{pp} will occur at $M = \frac{2}{3}$ with the same value as that at 0°.

$$i_{pp} = \frac{MT_{SW}V_{DC}}{2L_1}\left(1 - \frac{3}{4}M\right) \tag{7.39}$$

Δi_{max} in L_1 will impact the converter and inductor loss, as well as the inductor design itself. In principle, Δi_{max} can be treated as an intermediate design variable. For simplicity, sometimes a constraint can be preselected based on experience. As analyzed in [2], a 15% peak-to-peak ripple limit should be sufficient for neglecting the impact of the current ripple on loss.

The third constraint for *LCL* filter is its resonant frequency. For the *LCL* filter as shown in Figure 7.5, the source-side current i_s in s-domain can be generally expressed as in (7.40), where $G_{V1}(s)$ and $G_{V2}(s)$ are the respective transfer functions of the converter-side voltage and source-side voltage to the source-side current. These transfer functions can be obtained as in (7.41) and (7.42), where the angular natural resonance frequency of the *LCL* filter ω_f and its damping coefficient ξ are defined as (7.43) and (7.45).

$$I_s(s) = G_{V1}(s)V_c(s) + G_{V2}(s)V_s(s) \tag{7.40}$$

$$G_{V1}(s) = \left.\frac{I_s(s)}{V_c(s)}\right|_{V_s(s) = 0} = \frac{sC_fR_d + 1}{sL_1L_2C_f\left(s^2 + 2\omega_f\xi s + \omega_f^2\right)} \tag{7.41}$$

$$G_{V2}(s) = \left.\frac{I_s(s)}{V_s(s)}\right|_{V_c(s) = 0} = \frac{s^2L_1C_f + sC_fR_d + 1}{sL_1L_2C_f\left(s^2 + 2\omega_f\xi s + \omega_f^2\right)} \tag{7.42}$$

$$\omega_f = \sqrt{\frac{(L_1 + L_2)}{L_1L_2C_f}} \tag{7.43}$$

$$\alpha = \frac{R_d(L_1 + L_2)}{2L_1L_1} \tag{7.44}$$

$$\xi = \frac{\alpha}{\omega_f} = \frac{R_dC_f\omega_f}{2} \tag{7.45}$$

In general, the resonant frequency f_{res} should be selected to be much higher than the fundamental frequency f_0 but much lower than the switching frequency f_{SW}, typically as in (7.46), where $f_{res} \approx \frac{\omega_f}{2\pi}$ neglecting the impact of the damping resistance R_d.

$$10f_0 \leq f_{res} \leq \frac{1}{2}f_{SW} \tag{7.46}$$

The next constraint is the source-side harmonic currents, which are results of PWM voltages. Based on (7.41), neglecting R_d, the nth order source-side harmonic RMS current $I_{s\_n(RMS)}$ can be determined as (7.47), where $V_{c\_n(RMS)}$ is the nth order harmonic RMS value of the converter-side voltage. The TDD of AC source-side current can be determined via (7.4). $I_{s\_1(RMS)}$ can be found similarly as in the L filter case.

$$I_{s\_n(RMS)} \approx \frac{V_{c_{n(RMS)}}}{n\omega_0(L_1 + L_2)\left|1 - n^2\omega_0^2\frac{L_1L_2}{(L_1 + L_2)}C_f\right|} = \frac{V_{c\_n(RMS)}}{n\omega_0(L_1 + L_2)\left|1 - \left(\frac{n\omega_0}{\omega_f}\right)^2\right|} \tag{7.47}$$

The purpose of R_d is to suppress potential oscillations at the resonance frequency. Therefore, it should be sufficiently large to mitigate the oscillation. On the other hand, it should be small enough to avoid high loss [3]. Hence, a common practice is to set the R_d value to a similar order of magnitude as the C_f impedance at the resonant frequency, e.g. as in (7.48). From (7.45), it can be shown that the condition in (7.48) corresponds to $\xi = \frac{1}{6}$. The power loss on R_d can be estimated as (7.49) using (7.48) and (7.12). $\frac{S}{3}$ is the rated phase kVA rating, k_{cf} is usually less than 5%, and ω_0 is usually less than 1/10 of the ω_f. It can be concluded that the loss due to R_d is negligible. If further loss reduction is desired, more complex passive damping circuit could be adopted, such as RC, RL, and RLC damping [2].

$$R_d = \frac{1}{3\omega_fC_f} \tag{7.48}$$

$$P_R = I_c^2R_d \approx (\omega_0C_fV_s)^2R_d = (\omega_0C_fV_s)^2\frac{1}{3\omega_fC_f} \leq \frac{S}{3}\frac{k_{cf}\omega_0}{3\omega_f} \tag{7.49}$$

In summary, the design of the LCL filter parameters includes several constraints: the maximum inductance $(L_1 + L_2)$ constraint considering voltage drop in (7.13); the maximum filter capacitance C_f constraint considering the power factor in (7.12); the minimum converter-side inductance L_1 constraint considering ripple current limit in (7.37) and (7.38); the filter resonance frequency f_{res} constraint in (7.46); and the damping resistance R_d constraint considering the resonance mitigation and loss in (7.48).

7.3.2 Control Performance

Similar to the load-side inverter case in Chapter 6, the active rectifier and the source-side inverter control performance that will impact the power stage design also involves the dynamic performance characterized through the control bandwidth and voltage margin. In the case of the active rectifier, the control typically has two control loops, the outer DC-link voltage regulator loop, and the inner current regulator loop, as shown in Figure 7.14, where DQ reference frame control is employed. Assume the bandwidths of the voltage regulator and the current regulator are ω_V and ω_I, respectively. Then, the relationship in (7.50) should hold, where ω_C is the angular frequency of the PWM carrier and is related to the inverter switching frequency. Practically, selecting ω_C to be at least 5–10 times of current regulator bandwidth ω_I that is in turn 5–10 times of ω_V will be sufficient to guarantee the dynamic performance.

$$\omega_V \ll \omega_I \ll \omega_C \tag{7.50}$$

In the case of the source-side inverter, there can be two basic working modes, the current mode and the voltage mode. In the current mode, the inverter AC side current is regulated. The current reference can be either from the outer loop DC voltage regulator or the outer loop active power and reactive power regulator. The application scenarios include PV, wind, battery or other energy storage, fuel cell, active power filter inverters, and other grid application inverters working in maximum power point tracking (MPPT), power/current regulation, or grid-following modes. In these applications, the DC-link voltage can be either regulated by the inverter itself, similar to the active rectifier case, or by the energy resources such as PV, wind, fuel cell, and battery. In any case, the current regulator bandwidth ω_I should be significantly higher than the bandwidth of the outer voltage or power loop, and significantly lower than ω_C as shown in (7.50).

In the voltage mode, the inverter AC side voltage is regulated. The typical application scenario is for PV, wind, battery, and other inverter-interfaced energy resources to work in grid-forming mode. A variety of control methods have been developed for the grid-forming inverter and a generalized control structure is shown in Figure 7.15. The inner loop often uses a dual-loop voltage and current control scheme, which consists of an inner filter inductor current control loop and the outer filter capacitor voltage control loop. As a result, the control bandwidth relationship is similar to the inverter in current mode.

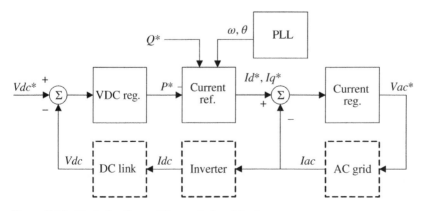

Figure 7.14 Typical active rectifier control architecture.

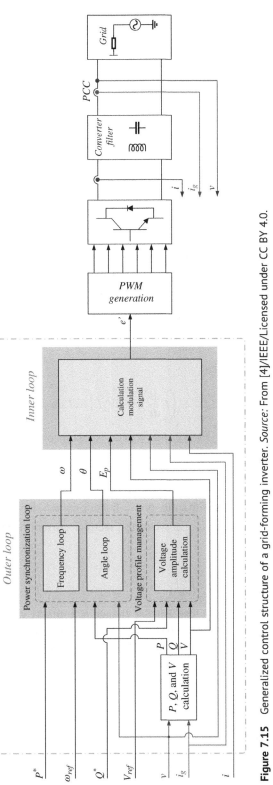

Figure 7.15 Generalized control structure of a grid-forming inverter. *Source:* From [4]/IEEE/Licensed under CC BY 4.0.

Another requirement is the voltage margin needed for the rectifier or the inverter to support the power or current ramp, which can be characterized by $\frac{dI}{dt}$. Similar to the load-inverter case, I here refers to the magnitude of current, representing the power or kVA.

Considering the active rectifier case and using the dq reference frame, the voltage equations are (7.51), where V_{gd} and V_{gq} are the d and q components of the source- or the grid-side voltage, V_d and V_q are d and q components of the converter-side voltage, and I_d and I_q are the d and q components of the current flowing into the rectifier. Note that L is the equivalent total inductance of the line filter, which equals to $(L_1 + L_2)$ in the case of the LCL filter. Note also that in this case, q-axis leads d-axis by 90°.

$$\begin{cases} V_{gd} = V_d - \omega_0 L I_q + L \dfrac{dI_d}{dt} \\ V_{gq} = V_q + \omega_0 L I_d + L \dfrac{dI_q}{dt} \end{cases} \tag{7.51}$$

In general, the d-axis is often selected to align with the source voltage vector, such that $V_{gd} = V_g$ and $V_{gq} = 0$. Under this condition, $I_d = I \cos \varphi$ and $I_q = -I \sin \varphi$ with φ as the power factor angle (defined as the angle the voltage leads the current). In addition, when ramping current, the power factor should be given. Assuming a constant power factor, Eq. (7.51) can be simplified as

$$\begin{cases} V_d = V_g - \omega_0 L I \sin \varphi - L \cos \varphi \dfrac{dI}{dt} \\ V_q = -\omega_0 L I \cos \varphi + L \sin \varphi \dfrac{dI}{dt} \end{cases} \tag{7.52}$$

From (7.52), it can be shown that

$$V^2 = V_d^2 + V_q^2 = V_g^2 + (\omega_0 L I)^2 + \left(L \dfrac{dI}{dt} \right)^2 - 2 V_g L \left(\omega_0 I \sin \varphi + \cos \varphi \dfrac{dI}{dt} \right) \tag{7.53}$$

In per unit, it can be written as (7.54). Note that V_g is assumed to be equal to the base voltage.

$$V_{pu}^2 = 1 + (Z_{pu} I_{pu})^2 + \left(\dfrac{Z_{pu}}{\omega_0} \dfrac{dI_{pu}}{dt} \right)^2 - 2 Z_{pu} \left(I_{pu} \sin \varphi + \dfrac{\cos \varphi}{\omega_0} \dfrac{dI_{pu}}{dt} \right) \tag{7.54}$$

The active rectifier often works with unity power factor or $\varphi = 0$. Eq. (7.54) can be simplified as (7.55). Clearly, when $\frac{dI_{pu}}{dt}$ is negative, i.e. when the rectifier current decreases, higher converter side voltage is needed.

$$V_{pu}^2 = (Z_{pu} I_{pu})^2 + \left(1 - \dfrac{Z_{pu}}{\omega_0} \dfrac{dI_{pu}}{dt} \right)^2 \tag{7.55}$$

For motor drive applications, the load current change is normally within tens of p.u./s. Given that Z_{pu} is around 5% and $\omega_0 = 377$ for a 60 Hz application, the impact on voltage margin is mainly determined by the $Z_{pu} I_{pu}$ term, which is about 5%. In other applications, there can be a requirement for higher current rate of change. In this case, higher voltage margin may be necessary. From (7.53) or (7.54), it can be seen that nonunity power factors could lead to lower voltage margins. In other words, proper reactive power injection could lower the voltage requirement.

7.3.3 DC-Link Stability

The DC-link stability criterion used for the passive rectifier in Chapter 5 still holds for the active rectifier and the source-side inverter. In that case, it is mainly dealing with small-signal stability.

For the small-signal stability, in the case of the active rectifier connecting to an inverter or a DC/DC converter, it is necessary to add the impact of the rectifier impedance and the inverter or DC/DC converter impedance. So, the general equation becomes

$$20 \log Z_{in} - 20 \log Z_{out} \geq Z_m \tag{7.56}$$

The large-signal impact on DC-link capacitance can be determined as in (7.57), where P_{max} is the maximum output/input power, V_{DC} is the DC-link voltage, and ΔV is the allowed voltage change (50 V in [5] for a 650 V DC link). This corresponds to the condition to compensate for power unbalance due to sudden load and source changes. The extreme case would be that in one switching cycle, the rectifier input power drops to zero while the load (e.g. the inverter in a motor drive) keeps the maximum output power, or vice versa. In (7.57), "−" sign corresponds to sudden loss of the rectifier or increase of the inverter load, and "+" sign corresponds to sudden loss of the inverter load.

$$C \geq \frac{P_{max}}{\left(V_{DC}\Delta V \pm \frac{1}{2}\Delta V^2\right)f_{SW}} \tag{7.57}$$

7.3.4 Reliability

Reliability is important to all equipment, including three-phase AC converters. A converter product often has a lifetime and/or a failure rate specification. The latter can be measured in terms of the mean-time-between-failures (MTBF). Reliability should be a constraint for the three-phase AC converter design. While the topic is covered in this chapter, the approach applies to other subsystems as well.

There are two basic approaches in addressing reliability in the converter design [6]. The first one carries out the converter design according to physical constraints of the components and subsystems, expecting to attain an adequate level of reliability as a result of a proper design process. For example, when selecting a capacitor for the rectifier, the designer will ensure the specified operating range of the selected capacitor, including its ratings of voltage, current, and temperature, will cover the operating condition for the capacitor to be used in the rectifier. As a result, the capacitor can achieve its designed reliability as guaranteed by the capacitor vendor.

The second approach, known as reliability-oriented design, takes reliability into consideration in the design process as a design constraint. Here, the second approach is described as it provides the opportunity to include reliability as part of the overall converter design optimization. Following this approach, the detailed reliability models need to be obtained for all system components. These models are then used to assess the reliability of the power converter considering the effect of both operational stresses and the environment on each of the individual components.

It should be pointed out that reliability model for a converter is very complex and challenging, with much ongoing research. The models and information presented here based on Military Handbook MIL-HDBK-217 [7] are somewhat outdated. The intention here is to provide a systematic illustration of the reliability design problem. When better models become available in future, they can replace models here but hopefully the design approach can still be valid.

It should also be noted that in Part 1 of this book on components (Chapters 2 through 4), some reliability (mainly lifetime) models have been discussed for power semiconductor devices, capacitors, and magnetics. Some models there are not the same as models here, reflecting the complexity of the converter reliability modeling and continued progress on the subject.

7.3.4.1 Failure Rate, MTBF, and Lifetime

For system reliability prediction, an important parameter needed is the failure rate of each component, defined as the number of failures per unit time, for example, number of failures per 10^6 hours. Figure 7.16 depicts the well-known bathtub curve of failure rate versus time showing its time dependency. This figure also shows the definition of lifetime, which is the period with constant failure rate corresponding to the life span of a component. The inverse of the failure rate in this constant range is defined as MTBF. Thus, MTBF is not directly related to the lifetime of components. A component could have a very high MTBF (i.e. a very low failure rate), but a relatively short lifetime. Correspondingly, lifetime is defined by the American Society for Testing and Materials (ASTM) as the time at which the time-averaged performance of components degrades below a prescribed or required value, that is, a complete failure or the failure to perform as desired.

In [7], the failure rate model of electronic components is defined as (7.58), where λ is the component failure rate per 10^6 hours, λ_b is the base failure rate, and π_i with $i = 1, ..., m$ are several factors that modify the base failure rate per environmental and operational conditions that affect the component reliability.

$$\lambda = \lambda_b \prod_{i=1}^{m} \pi_i \tag{7.58}$$

In general, factors π_i in (7.58) can include temperature factor (π_T), application-type factor (π_A), power rating factor (π_R), electrical or voltage stress factor (π_S), quality factor (π_Q), and environmental condition factor (π_E). The failure rate λ_s of a system comprised of n components is given by (7.59), where λ_i with $i = 1, ..., n$ corresponds to the failure rate of each individual component. The MTBF of the whole system is then defined as (7.60).

$$\lambda_s = \sum_{i=1}^{n} \lambda_i \tag{7.59}$$

$$\text{MTBF} = \frac{1}{\lambda_s} \tag{7.60}$$

The lifetime LT of a system on the other hand is defined in terms of its n components by (7.61), where LT_i, with $i = 1, ..., n$, corresponds to the lifetime of each individual component. Clearly, the lifetime of a converter is determined by the component with the shortest lifetime.

$$LT = \min\{LT_1, LT_2, ..., LT_n\} \tag{7.61}$$

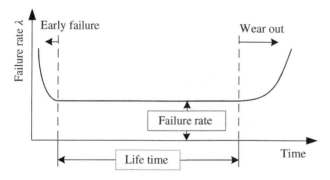

Figure 7.16 Definitions of failure rate and lifetime of components.

7.3.4.2 Reliability Consideration in Design

Although the reliability of the power converter is determined by the reliability of all of its components, including many components in control circuits, here the focus is on the reliability of the main power stage components. The justifications for this approach are (i) many auxiliary components used in the control circuits will be similar for different power stage designs, especially for a given converter topology and power rating; (ii) the reliability consideration for control and auxiliary circuits can be relatively independent of the power stage; and (iii) control circuits are relatively low cost compared with the power stage, especially for the medium- and high-power AC three-phase converters. Understanding or assuming that the reliability differences of the AC three-phase converter designs are determined by the main components of the power stages, including semiconductor devices, capacitors, and inductors, this subsection focuses on the reliability models of these components and their relationships with the converter design and operation parameters. Thermal management system and mechanical system, e.g. cooling fan, can impact reliability but are not considered here.

7.3.4.3 Component Reliability Models

7.3.4.3.1 Device Failure Rate and Lifetime Models
Since the power semiconductor devices have been evolving continually, it is difficult to find a generic model due to the lack of sufficient field data on failures of a particular type or part of a device. One form of the device failure rate model assumes that the power device has the same π factors as low-power transistors, whose data are more abundant. Take Si IGBT as an example. Per [7], the low-power bipolar transistors have factor functions of junction temperature, electrical or voltage stress, type of applications (linear or switching), power rating, quality, and environment. Therefore, a Si IGBT can be assumed to have the same factor functions. Another assumption is that the device base failure rate data can be obtained from generic component failure rates from the field data of the manufacturers. Using the general expression of the failure rate model in [7], the device failure rate λ_{sw} can be modeled as (7.62), where π_{T_j} is the junction temperature factor and other factors have been defined earlier.

$$\lambda_{sw} = \lambda_{b\_sw} \cdot \pi_{T_j} \cdot \pi_A \cdot \pi_R \cdot \pi_S \cdot \pi_Q \cdot \pi_E \text{ Failures/10}^6 \text{ hours} \tag{7.62}$$

For bipolar transistors, the empirical equations for π_{T_j} and π_S are given in [7]. For π_{T_j}, it is (7.63) with junction temperature T_j in °C; and for π_S, it is (7.64), where V_{DC} is the DC-link voltage, and V_{rated} is the device voltage rating. For switching applications, π_A is 0.7. The value of π_E depends on the application environment. For type G_B (ground, benign), i.e. nonmobile, temperature- and humidity-controlled environment accessible to maintenance, π_E is 1; for type G_F (ground, fixed), i.e. moderately controlled environment, π_E is 6; and for type A_{IC} (airborne, inhabited, cargo), it is 13. For normal quality, π_Q can be assumed to be unity. In [7], there are empirical equations for power rating factor π_R but for very low-power devices. For high-power devices, π_R can be assumed as unity for lack of data.

$$\pi_{T_j} = e^{-2114\left(\frac{1}{273+T_j} - \frac{1}{298}\right)} \tag{7.63}$$

$$\pi_S = 0.045 e^{3.1\frac{V_{DC}}{V_{rated}}} \tag{7.64}$$

In [7], the base failure rates are given for many commonly used low-power components. On the other hand, the base failure rates are often unavailable for power electronic devices. In this case, they can be obtained from (7.62) using the field failure rate data of manufacturers and the calculated π factors according to the application specifications under consideration. For example, given a failure rate of 100 FIT (i.e. 100 failures/10^9 hours or 0.1 failures/10^6 hours). FIT stands for failure-in-time and 1 FIT is defined as 1 failure/10^9 hours) for a medium power 1200 V Si IGBT module, and assuming a typical application condition with $T_j = 100\,°C$, $V_{DC} = 600$ V, and type G_F environment, then the base failure rate is found to be $\lambda_{b\_sw} = 0.081$ failures/10^6 hours or 81 FIT.

The lifetime of a power device or device module depends on the lifetime of the materials and package for the device or device module. For example, the lifetime of a Si IGBT module depends primarily on the lifetime of wire bonds and large-area solder joints. Its lifetime can be predicted by accelerated aging tests, where it has been demonstrated that the number of cycles to failure $N_{f\_T_j}$ of the wire bonds is a function of the average junction temperature T_j and the junction-temperature swing ΔT_j. The number of cycles is given by (7.65), where $k_{\Delta T_j} = -5$, $k_{f\_T_j} = 9830$, and $k_{T_j} = 1124$ [6].

$$N_{f\_T_j} = k_{f\_T_j} \cdot \left(\Delta T_j\right)^{k_{\Delta T_j}} \cdot e^{\left(\frac{k_{T_j}}{T_j + 273}\right)} \tag{7.65}$$

Also, the number of cycles to failure of the solder joint is a function of the average case temperature T_c and the case-temperature swing ΔT_c as in (7.66), where $k_{\Delta T_c} = -5$, $k_{f\_T_c} = 9830$, and $k_{T_c} = 337$. T_c is in °C.

$$N_{f\_T_c} = k_{f\_T_c} \cdot \left(\Delta T_c\right)^{k_{\Delta T_c}} \cdot e^{\left(\frac{k_{T_c}}{T_c + 273}\right)} \tag{7.66}$$

In order to predict the lifetime of the selected IGBT module, a mission profile is needed to determine the loss and thermal excursions of the module. These data are used to calculate temperature swings of the junction and the case, and to calculate the number of cycles to failure using (7.65) and (7.66). Generally speaking, the mission profile will generate a varying stress on the component. To determine the actual lifetime of a component, e.g. a Si IGBT, the prediction of the probability of fatigue failure can be approximated by means of linear damage accumulation. Specifically, it has been shown that the damage due to loading with groups of different stress amplitudes accumulates linearly. If the entire mission profile consists of N groups of loadings of different stresses, then the total damage accumulation by one mission profile, D_P, can be determined as in (7.67), where N_{fi} is the number of cycles to failure caused by the ith fixed stress (power cycling or thermal cycling), $D_i = \frac{1}{N_{fi}}$ is the damage ratio of the ith fixed stress in one cycle, and n_i is the number of cycles for the ith fixed stress.

$$D_P = \sum_{i=1}^{N} n_i D_i = \sum_{i=1}^{N} \frac{n_i}{N_{fi}} \tag{7.67}$$

Then, the lifetime t_f can be calculated as (7.68), where t_{D_P} is the time period for one mission profile. Repeated mission profiles will not affect the lifetime calculation as n_i (therefore corresponding D_P) and t_{D_P} increase proportionally.

$$t_f = \frac{t_{D_P}}{D_P} \tag{7.68}$$

The lifetime of the IGBT module can be calculated as the minimum lifetime value between wire bonds $t_{f\_T_j}$ and solder joints $t_{f\_T_c}$ as

$$LT_{IGBT} = \min\left(t_{f\_T_j}, t_{f\_T_c}\right) \tag{7.69}$$

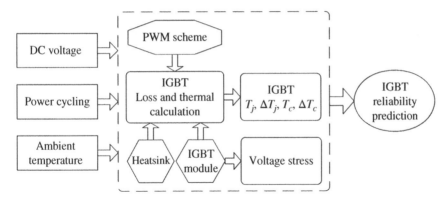

Figure 7.17 Block diagram of the reliability prediction procedures for Si IGBT modules.

Figure 7.17 illustrates the block diagram of the reliability prediction method for IGBT modules. The losses and thermal calculation can follow the approaches in Chapters 6 and 9.

7.3.4.3.2 *Capacitor Failure Rate and Lifetime Models* Capacitors are often seen as the weakest link of the converters from the reliability point of view. With proper use, capacitors can actually have a very low failure rate and long MTBF, but still a relatively short lifetime. The wear-out failure is their most common failure mode. Most such failures are gradual conversions to open circuits as the components become more and more resistive. The reliability of capacitors is very sensitive to the voltage stress and the operating temperature. Referring to [7], the capacitor failure rate λ_C is given in (7.70), where λ_{b_C} is the base failure rate, π_{CV_C} is the capacitance factor, π_{Q_C} is the quality factor, and π_{E_C} is the environmental factor.

$$\lambda_C = \lambda_{b\_C} \cdot \pi_{CV\_C} \cdot \pi_{Q\_C} \cdot \pi_{E\_C} \quad \text{Failures}/10^6 \text{ hours} \tag{7.70}$$

λ_{b_C} is a function of the capacitor voltage stress V and ambient temperature T_A as in (7.71), where V is the sum of the DC voltage and the AC peak voltage, and V_{rated} and T_{rated} are the rated capacitor voltage and temperature, respectively. Note both T_A and T_{rated} are in °C.

$$\lambda_{b\_C} = k_b \cdot \left[\left(\frac{V}{k_{s1}.V_{rated}} \right)^{k_{s2}} + 1 \right] \cdot e^{\left[k_{t1} \cdot \left(\frac{T_A + 273}{T_{rated} + 273} \right)^{k_{t2}} \right]} \quad \text{Failures}/10^6 \text{ hours} \tag{7.71}$$

For film capacitors, $k_b = 9.9 \times 10^{-4}$, $k_{s1} = 0.4$, $k_{s2} = 5$, $k_{t1} = 2.5$, and $k_{t2} = 18$. The capacitance factor π_{CV_C} can be obtained by (7.72), where C is the capacitance value in μF. In normal case, the quality factor π_{Q_C} should be equal to or less than unity. The environmental factor π_{E_C} is 1.0 for type G_B, 2.0 for type G_F, and 6.0 for type A_{IC}.

$$\pi_{CV\_C} = 1.1 \cdot C^{0.085} \tag{7.72}$$

For aluminum electrolytic capacitors, $k_b = 2.54 \times 10^{-3}$, $k_{s1} = 0.5$, $k_{s2} = 3$, $k_{t1} = 5.09$, and $k_{t2} = 5$. The capacitance factor π_{CV_C} can be obtained by (7.73), where C is the capacitance value in μF. In normal case, the quality factor π_{Q_c} should be equal to or less than unity. The environmental factor π_{E_C} is 1.0 for type G_B, 2.0 for type G_F, and 10.0 for type A_{IC}.

$$\pi_{CV\_C} = 0.34 \cdot C^{0.18} \tag{7.73}$$

The operating temperature has a strong effect on the lifetime of capacitors as its lifetime doubles every 10 °C dropped from their maximum operating core temperature (i.e. internal hotspot temperature). The operating voltage also has a significant impact on lifetime. For instance, for a CORNELL DUBILIER aluminum electrolytic capacitor, its operating life can be expressed as (7.74), where M_v is a voltage multiplier, L_b is the expected operating life in hours for full rated voltage and temperature, T is the actual core operating temperature in °C, and T_{rated} is the rated capacitor operating temperature in °C [8]. M_v can be determined by (7.75). Note that (7.74) and (7.75) can be considered as a special case of capacitor lifetime models in Chapter 3.

$$LT_{Cap} = M_v \cdot L_b \cdot 2^{\frac{T_{rated} - T}{10}} \tag{7.74}$$

$$M_v = 4.3 - 3.3 \cdot \frac{V}{V_{rated}} \tag{7.75}$$

Figure 7.18 shows an example of the impact of the core temperature on the lifetime of film capacitors [9].

Figure 7.19 shows a block diagram of the reliability prediction procedure for the DC-link capacitors. All factors affecting reliability are considered, including voltage, ripple current, ESR and its frequency/temperature dependences, losses or heat generation, and thermal characteristics.

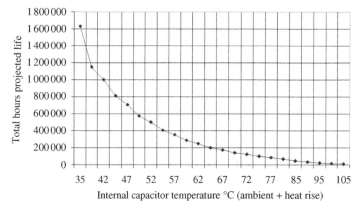

Figure 7.18 Film capacitor life expectancy as a function of temperature. *Source:* From [9]. © [2002] IEEE.

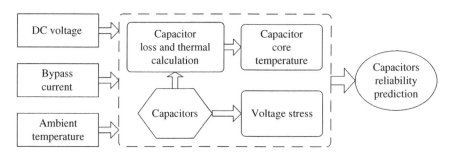

Figure 7.19 Block diagram of the reliability prediction procedure for capacitors.

Table 7.1 Coefficients for different classes of insulation materials.

Insulation class	Maximum operating temperature (°C)	$\lambda_{b\_insulation}$	$k_{T\_insulation}$	$k_{insulation}$
A	105	3.79×10^{-4}	352	14
B	125	3.19×10^{-4}	364	8.7
F	150	3.5×10^{-4}	409	10

7.3.4.3.3 Failure Rate and Lifetime Models for Magnetic Components Because magnetic cores have very high reliability, the most common failure in magnetic components is the winding insulation. In [7], the failure rate of a magnetic component (in this case, inductor) is given by (7.76), where λ_{b_L} is base failure rate of the magnetic component, π_{Q_L} is quality factor, and π_{E_L} is the environmental factor.

$$\lambda_L = \lambda_{b\_L} \cdot \pi_{Q\_L} \cdot \pi_{E\_L} \quad \text{Failures/10}^6 \text{ hours} \tag{7.76}$$

λ_{b_L} can be determined by (7.77), where T_{HS} is the hotspot temperature in °C. The coefficients $\lambda_{b\_insulation}$, $k_{T\_insulation}$, and $k_{insulation}$ for different classes of insulation materials are listed in Table 7.1. Note that the insulation class follows the classification in MIL-PRF-39010E 1997, which is different from the IEC or NEMA classifications.

$$\lambda_{b\_L} = \lambda_{b\_insulation} \cdot e^{\left[\left(\frac{T_{HS} + 273}{k_{T\_insulation}} \right)^{k_{insulation}} \right]} \quad \text{Failures/10}^6 \text{ hours} \tag{7.77}$$

In a normal case, the quality factor π_{Q_L} should be equal to or less than unity. The environmental factor π_{E_L} is 1.0 for type G_B, 4.0 for type G_F, and 5.0 for type A_{IC}.

In [7], the hotspot temperature of the winding is predicted by (7.78), where T_A is ambient temperature and ΔT is the average temperature rise of the inductor or transformer. ΔT can be determined through loss and thermal analysis for magnetics as discussed in Chapters 4 and 5, which can also be directly used to determine the hotspot temperature.

$$T_{HS} = T_A + 1.1 \cdot \Delta T \tag{7.78}$$

According to [10], the lifetime of the inductor, i.e. the lifetime of the magnetic wire insulation for the inductor, can be predicted by (7.79), where V_L is the voltage stress of the wire insulation and T_L is the wire temperature. $f_1(V_L)$ and $f_2(V_L)$ are two linear functions as in (7.80) and $A_1, A_2, B_1,$ and B_2 are constant coefficients, which can be obtained experimentally.

$$LT_L = e^{\left[f_1(V_L) + \frac{f_2(V_L)}{T_L} \right]} \tag{7.79}$$

$$\begin{cases} f_1(V_L) = A_1 + A_2 V_L \\ f_2(V_L) = B_1 + B_2 V_L \end{cases} \tag{7.80}$$

Figure 7.20 shows temperature and voltage stress effects on the lifetime of the magnetic wire. If a wire with high temperature and voltage ratings is used, its lifetime will be very long.

Figure 7.21 shows a block diagram of the reliability prediction procedure for inductors (AC inductors assumed). Similar to the capacitor case, the loss and thermal calculation is a key step.

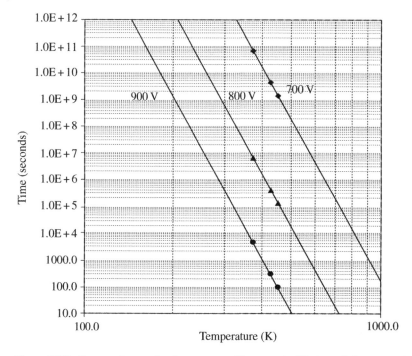

Figure 7.20 Temperature and voltage stress effects on the lifetime of the magnetic wire. *Source:* From [10]. © [2000] IEEE.

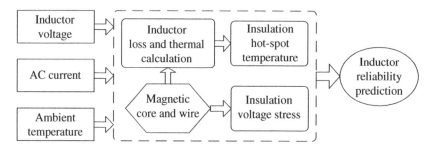

Figure 7.21 Block diagram of the reliability prediction procedure for inductor.

7.3.4.4 Reliability-Oriented Design

The converter reliability improvement can be achieved in two basic ways: from the standpoint of either the system or the components. The former includes minimization of the number of components, which simplifies the converter structure, as well as minimization of stresses on components; the latter involves the employment of the best possible component technologies. In other words, the reliability of the converter can be enhanced by following three simple approaches: (i) simplify the circuit topology to reduce the number of components and simplify its structure; (ii) perform a judicious component selection; and (iii) reduce the stress level of the components.

The reliability-oriented design can be illustrated with a three-phase converter for aircraft applications. The operating conditions are rated power $P = 60$ kW, DC side voltage $V_{DC} = 540$ V (± 270 V), AC side RMS line-to-neutral voltage $V_{AC} = 230$ V, fundamental frequency $f = 400$ Hz, power factor $pf = 0.9$, the maximum ambient temperature $T_A = 65\,°C$, the inlet coolant temperature of the liquid-cooling heatsink is $T_{inlet} = 70\,°C$, and the thermal resistance of the heatsink is 0.0055 °C/W. Note that the power converter is intended for universal use as rectifier and inverter, so the design specifications correspond to the worst-case conditions.

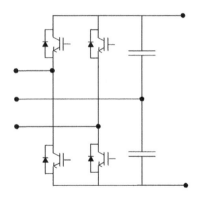

7.3.4.4.1 *Topology* In this book, we focus on two-level VSC topology for convenience and simplicity of the presentation. In applications, many different topologies can often meet the required inverter and rectifier functions. In this specific case, in addition to the two-level VSC, several topology candidates can be identified, including reduced-switch three-phase topologies, multilevel topologies, and Z-source

Figure 7.22 Four-switch VSC.

converter. Most topologies have been introduced in Chapter 1. Figures 7.22 and 7.23 show the reduced 4-switch VSC and Z-source converter. Since aircraft applications can have different voltages, sometimes topologies based on universal modular converters or phase legs may be preferred. Figure 7.24 shows the three-phase convertible voltage-source converter (C-VSC) connected in parallel mode for low-voltage applications and in series or multilevel mode for high-voltage applications.

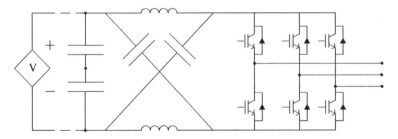

Figure 7.23 Z-source converter.

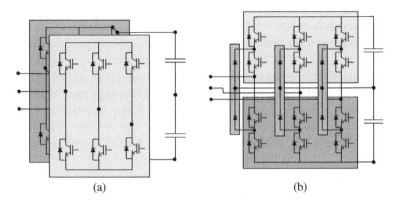

(a)   (b)

Figure 7.24 The three-phase convertible VSC (C-VSC) operating in: (a) parallel mode (low-voltage applications) and (b) series or multilevel mode (high-voltage applications).

Table 7.2 Comparison of converter topologies.

	VSC	4-Sw VSC	Z-source	NPC-VSC	C-VSC
# of switches (IGBT)	6	4	6	12	12
# of diodes	6	4	6	18	18
# of capacitors	2	2	4	2	2
# of inductors	0	0	2	0	3
Total components	14	10	18	32	35
V_{sw}	1	1.73	>1[a]	0.5	0.5
I_{sw}	1	1	1	1	0.5
f_{sw}	1	1	1.15	1	1
Capacitor size	1	~2	1	1	1
NP control need	No	Yes/uncontrolled	No	Yes/limited	Yes/limited
VDF	1	1.67	1	0.5	1 or 0.5
Shoot-through failure	Yes	Yes	No	Yes	Yes

[a] Varies with load.

Table 7.2 shows the comparison of candidate topologies. The first indicator of the reliability is the number of components. The four-switch VSC (4Sw-VSC) with 10 components and the VSC with 14 components present a significant advantage compared to the other topologies, especially to three-level or convertible VSC, which has twice as many switches (IGBTs) and three times the number of diodes. The Z-source converter with 18 components still compares favorably, but requires four capacitors instead of only two, recognized as the least reliable element in power converters. The C-VSC varies its total number of components depending on the usage or not of paralleling inductors for low-voltage applications. In case of employing them, it becomes one of the two topologies requiring inductors together with the Z-source converter. This would certainly increase its parts count and reduce the reliability of this topology.

The next category is on semiconductor device ratings. In Table 7.2, these ratings are presented in per unit, normalized to the two-level VSC topology, in order to facilitate their comparison. The voltage rating determines the DC voltage required to produce 1 p.u. AC side voltage. As shown, the current rating I_{sw} is the same for all topologies except for the C-VSC, which is rated at 50%. Hence, the stress level on the switch for different topologies is mostly differentiated by the voltage magnitude that the switches must withstand. Both NPC-VSC and C-VSC require voltage ratings V_{sw} to be half of the VSC value given their stacked three-level topology, which has the advantage of allowing the usage of 600 V instead of 1200 V IGBT. The 4Sw-VSC requires a significantly higher voltage rating of 1.73 p.u. to generate a balanced three-phase system from only two 60° phase-shifted phase legs. This represents a great limitation for this topology, which would need to operate near its modulation index limit and make its modulator prone to saturation under even minor disturbances. The Z-source converter presents a varying voltage rating, dependent on both the converter load and boost factor being employed. An overrating of 25% would be required for the specific aircraft application.

The third category is on component stress. For comparison purposes, a base switching frequency f_{sw} of 10 kHz is employed. Only the Z-source converter presents different switching frequencies, since it splits the zero vectors into conventional and shoot-through zero states thus increasing its switching frequency by 15%. Also, all topologies present the same frequency location for their harmonic content, centered around f_{sw}. *VDF* is the voltage distortion factor. The 4Sw-VSC presents a significantly higher index. The VSC and Z-source converters present equal values given their identical AC waveforms. The three-level NPC and the C-VSC (when connected in three-level) present better performance indices given their three-level waveforms.

As for capacitor size for different topologies, it is closely related to harmonic currents, as discussed in previous chapters. In Table 7.2, all topologies except the 4Sw-VSC require the same DC-link capacitor. Multilevel structures cannot reduce capacitor size given that for lower modulation indices, they still operate like a two-level VSC. On the other hand, the 4Sw-VSC presents a higher harmonic distortion, and its intrinsic operating mode requires an increased capacitance value in order to limit the DC-link midpoint oscillation.

The two-level VSC and Z-source converter do not require any DC-link midpoint voltage control, while such control is needed for the 4Sw-VSC, the three-level NPC and C-VSC. The NPC and C-VSC are capable of controlling their own DC-link midpoint voltage, but only within a restricted operating region. The 4Sw-VSC cannot control its DC-link midpoint voltage at all, thus representing a significant disadvantage for the topology.

Finally, among all candidate topologies, they all have shoot-through failure mode, except for the Z-source converter, which is an advantage.

Considering all aspects of components count, components stress, and components utilization, the two-level VSC is the most suitable topology. Further analysis can be shown that the C-VSC may be able to achieve a greater device utilization and higher power density than the VSC, as well as a reduced harmonic distortion. However, its complexity and higher number of components are clearly detrimental to its overall reliability. The selection of the popular conventional two-level VSC seems to be a boring outcome. On the other hand, its popularity is justified from the reliability point of view.

7.3.4.4.2 *Switching Frequency and PWM Scheme*

Lower switching frequency leads to lower switching loss, which in turn leads to lower temperature and lower temperature swings, and hence higher reliability for switching devices. However, a lower switching frequency results in higher harmonic distortion and slower dynamic response. A lower switching frequency also translates into bulkier and heavier inductors and capacitors. Consequently, the selection of this parameter is a key design tradeoff in achieving high reliability. For the case aircraft converter design, a minimum control bandwidth of 2 kHz is imposed for the current loop of the converter, as well as a 5% current THD.

Taking advantage of the very high reliability of magnetic components, a lower switching frequency would be preferred using AC inductors to limit AC current harmonics. The constraints to the inductor sizing are the minimum value to limit the AC current THD, and the maximum value to limit the inductor voltage drop, which has been discussed in Section 7.3.1. In the design example, an inductance of 40 μH (i.e. 4.22%) is used with the switching frequency of 20 kHz, which is selected to meet the current control bandwidth requirement of 2 kHz. The 60° DPWM is adopted for lower switching loss.

7.3.4.4.3 *Selection of Components*

The selection of components should be based on reliability data, e.g. published data, or data provided by vendors for the component reliability characteristics. In this example, six-pack IGBT modules are selected since vendors and field data confirm that the failure rate of six-pack modules is lower than that of six individual IGBT modules or three dual-pack modules. The selection of the specific IGBT model is based on the switching and conduction loss calculation and on thermal analysis to predict the maximum junction temperature and its temperature swing. A six-pack Infineon 1200 V, 450 A module FS450R12KE3 is selected among many other available six-pack IGBT modules per performance evaluations.

For the DC-link capacitors, both AEC and film capacitors can be employed. Their pros and cons have been discussed in Chapter 3. To illustrate their different reliability characteristics, both types are selected for further comparison, specifically, Cornell Dubilier AEC 550C332T450DE2B and Electronic Concepts film capacitor UL31Q157K.

7.3.4.4.4 Design Comparison

The design comparison considers the steady-state operation of the power converter. The base failure rate of the Si IGBT module is assumed to be 100 FIT in ground environmental conditions based on the field data report of several IGBT modules. In the reliability prediction, the ambient temperature and load variations are neglected. The lifetime prediction result of the IGBT module is hence mainly determined by the lifetime of the wire bonds. The environmental factor corresponding to A_{IC} in [7] is used for all components considering the aircraft application.

For the capacitor reliability prediction, the capacitor loss calculation is performed with the assumption that the entire high-frequency switching current flows through it. The cooling for the capacitors is natural convection at $T_A = 65$ °C.

For the inductor reliability prediction, high voltage and high-temperature rating (>170 °C) magnet wire is used. The hotspot temperature in the winding is limited to 120 °C, and the voltage stress is limited to 600 V. Because the inductor has obviously higher reliability than the IGBT modules and capacitors, the design focuses on the selection of IGBT modules and capacitors, as well as on modulation scheme in order to reduce stress.

Before comparing the different converter system designs, it is helpful to first compare various components, subsystem configuration, and parameters. Here, the example for capacitor comparison is shown.

Table 7.3 shows the difference in reliability and volume between AEC and film capacitors when constructing the DC-link capacitors bank in Figure 7.25. The AEC solution uses eight Cornell Dubilier units rated at 3300 μF and 450 V, and the film capacitor solution uses eight Electronic Concepts capacitors rated at 150 μF and 500 V. The configuration and the number of the capacitors are determined by the required voltage, current, and capacitance values. Note that in this case, the low capacitance value of the film capacitor is not critical as it is adequate for the application with a proper control design. The larger AEC and its high capacitance value are needed for current capability.

Table 7.3 Comparison of DC-link capacitor solutions.

Capacitors	Film UL31Q157K	Electrolytic 550C332T450DE2B
DC-link voltage ripple (peak-to-peak, V)	9.0	1.1
Loss per capacitor (W)	0.48	7.73
Thermal resistance (°C/W)	12.2	3.26
Internal temperature (°C) (natural convection)	71	90
Failure rate (/10^6 hours)	0.1333	1.858
MTBF (hours)	937 000	67 000
Lifetime (hours)	125 000	120 000
Total volume ($D^2 \times L \times 8$, in^3)	81	378

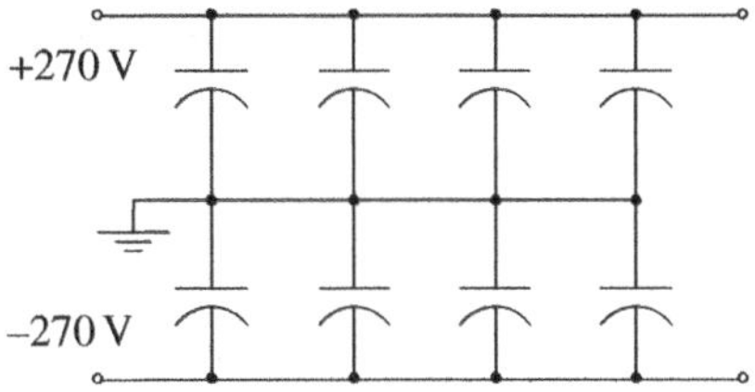

In Table 7.3, the DC-link voltage ripple and loss per capacitor are determined through simulation under the rectifier condition of 60 kW, $V_{DC} = \pm 270$ V, $V_{AC} = 100$ V RMS, $f_{sw} = 12$ kHz, and SPWM scheme. From (7.71) and (7.70), λ_{b_C} and λ_C can be calculated for both the AEC and film capacitors.

For film capacitors,

Figure 7.25 Configuration of the DC-link capacitor.

$$\lambda_{b\_C} = 0.00099 \cdot \left[\left(\frac{270}{0.4 \cdot 500} \right)^5 + 1 \right]$$

$$\cdot e^{\left[2.5 \cdot \left(\frac{65 + 273}{85 + 273} \right)^{18} \right]} = 0.0132 \text{ Failures}/10^6 \text{ hours}$$

$$\lambda_C = 0.0132 \cdot 1.1 \cdot 150^{0.085} \cdot 1 \cdot 6 = 0.1333 \text{ Failures}/10^6 \text{ hours}$$

$$\text{MTBF} = \frac{10^6}{8 \times 0.1333} = 937\,327 \text{ hours}$$

For AEC,

$$\lambda_{b\_C} = 0.00254 \cdot \left[\left(\frac{270}{0.5 \cdot 450} \right)^3 + 1 \right] \cdot e^{\left[5.09 \cdot \left(\frac{65 + 273}{105 + 273} \right)^5 \right]} = 0.127 \text{ Failures}/10^6 \text{ hours}$$

$$\lambda_C = 0.127 \cdot 0.34 \cdot 3300^{0.18} \cdot 1 \cdot 10 = 1.858 \text{ Failures}/10^6 \text{ hours}$$

$$\text{MTBF} = \frac{10^6}{8 \times 1.858} = 67\,262 \text{ hours}$$

From Table 7.3, it can be seen that the film capacitor represents a better solution for its merits on reliability, volume, and losses. Note that the lifetime of the electrolytic capacitor is much longer than its rated value of 15 000 hours due to the low-voltage stress (270 V vs. the rated 450 V) and low core temperature stress (90 °C vs. the rated 108 °C).

Similar comparisons can be performed for modulation scheme, switching frequency, IGBT module configuration, and converter configuration. Finally, a converter configuration with eight film capacitors, one IGBT six-pack module, and one AC inductor is selected. The reliability parameters for the selected design are shown in Table 7.4. More details can be found in [6].

The MTBF and lifetime values in Table 7.4 for the converter power stage correspond to 45 years and 14 years, respectively, with the lifetime determined by the film capacitor lifetime. Although these numbers are not totally unreasonable, as pointed out in [6], the reliability is influenced by many complex factors, and it is extremely difficult to get accurate reliability results by prediction. The numbers shown here are therefore only for reference and their values are best used to compare the different designs.

Table 7.4 Reliability parameters for the selected design.

	IGBT module	Capacitors	Inductor	Power stage
Failure rate (/10^6 hours)	1.45	1.07	0.012	2.532
MTBF (hours)	690 000	937 000	83 333 000	395 000
Lifetime (hours)	458 000	125 000	2 153 000	125 000

7.4 Active Rectifier and Source-side Inverter Design Optimization

Figure 7.2 shows the design optimization problem for the active rectifier and the source-side inverter. With models established among the design constraints, conditions, and variables, the active rectifier or the source-side inverter design optimization can be carried out similarly to the passive rectifier and load-side inverter. The details of design optimization will be described in Chapter 13. An example for an active rectifier design will also be provided there.

7.5 Impact of Topology

Thus far, the three-phase AC converter design in this book has largely used the two-level VSC as the example. While the procedure and general methodology should apply to other topologies, the models will need to be adapted for each specific topology. Analytical or algebraic models can be derived, although in most cases, they will be more complex than the models for the two-level VSC. Simulation can also be conducted. Another important fact is that whether it is through models or through simulation, certain assumptions are made when developing models or building simulations. These assumptions should be examined to ensure their correctness or accuracy for a particular topology. One example is the switching loss energy. In a datasheet, the switching energy is usually obtained through double pulse test on a half-bridge totem-pole set up. Whether the data can be applied to a topology with a different configuration for loss calculation needs to be examined.

In this section, we will examine the impact of topology through two most important aspects: the circuit modeling aspect and the device modeling aspect.

7.5.1 Circuit Modeling for Different Topologies

As presented in Chapter 1, three-phase AC converters can be realized with a variety of topologies. In fact, one of the most important tasks in a converter design is often the topology selection. For example, reference [5] presents a systematic approach to evaluate different topologies for a minimum weight design, and Section 7.3.4 earlier discussed a topology comparison and selection considering reliability, which has also been covered in [6]. Obviously, it is important to include other topologies beyond the two-level VSC in three-phase AC converter designs.

For other different topologies, we can still follow the two-level VSC approach to develop analytical or algebraic models for them. Many analytical models for various three-phase AC converters are available, including harmonics and loss models for three-level and multilevel VSC topologies. On the other hand, it is often difficult, and involves necessary simplifications and approximations to derive analytical models for complex topologies. The time-domain simulation can be more flexible and accurate. To avoid inefficiency of the time-domain simulation, here we will present switching function-based simulation methods for different topologies. Given that the circuit simulation is mainly needed for power loss, harmonics, and EMI noise evaluation, detailed switching transients for devices do not need to be modeled. For example, the equivalent frequency of the sharp switching edges of PWM voltage for WBG devices is beyond 20 MHz, which will have very limited impact on AC or DC voltages and currents of a converter in simulation [11]. Therefore, to simplify simulation and reduce simulation time, all power switch models in the simulation can be replaced with the switching function controlled voltage or current sources. PWM voltages for each phase-leg in a VSC can be synthesized as a function of modulation signals (or gate signals) and the DC-link voltage.

7.5.1.1 Switching Function-Based Simulation for Two-Level VSC

In Chapter 6, we have already presented the switching function models for the two-level VSC and explained how to use them for harmonic analysis in frequency domain. Here, we will further explain how to use the models directly in time-domain simulation.

As can be seen in (7.14), the PWM phase voltages v_a, v_b, and v_c can be considered as controlled voltage sources by switching signals S_a, S_b, and S_c, which are instantaneously obtained from the modulator in time-domain simulation. The phase voltages can be used to determine the line-to-line voltages, which can then be used to determine AC line currents i_a, i_b, and i_c based on load (e.g. motor) model. The DC current can be determined as

$$i_{DC} = S_a i_a + S_b i_b + S_c i_c \tag{7.81}$$

For PWM voltage determination, the switching function method is a steady-state approach and does not involve solving differential equations. When applied to things like loss calculation, it still does not involve solving differential equations, and therefore, in these cases, it is essentially an algebraic numerical method. Computationally, the switching function-based simulation is quite efficient as we already shown in Chapter 6 when comparing different approaches in calculating inverter AC voltage harmonics. On the other hand, when PWM voltages obtained from the switching functions are used for determining harmonic or EMI currents in time domain, i.e. applying voltages to load (e.g. motor) models, it will involve solving differential equations. Even for steady-state calculations, several cycles of simulation or computation may be needed to reach steady state. In this case, the switching function-based approach should still be more computational efficient than the switch-based detailed simulation.

One alternative to further speed up the calculation is to convert the process into frequency domain, if the frequency-domain impedance models are easy to obtain for harmonic and EMI current calculations, the PWM voltages can be converted to frequency domain to perform the calculations. Since both the current THD and EMI noise current are needed in frequency domain, there is even no need to convert everything back to time domain. This approach has been used in loss and thermal analysis in [1] and has been discussed in Chapter 6. It will be further discussed in Chapters 8 and 13.

7.5.1.2 Switching Function-Based Simulation for Multilevel VSC

The same approach can be applied to other topologies. Figure 7.26 shows Phase a of a three-phase three-level ANPC converter, where the switch, antiparallel diode, and parasitic device capacitance are all illustrated. Define switching functions as (7.82), where $j = 1, 2, 3, 4, p, n$.

$$S_j = \begin{cases} 1 \ S_j \text{ switch is on} \\ 0 \ S_j \text{ switch is off} \end{cases} \tag{7.82}$$

Phase a PWM voltage v_{aN} can be expressed as

$$\begin{aligned} v_{aN} = &\left(S_1 \times 1 + S_p \times 0\right)S_2\frac{V_{DC}}{2} \\ &+ [S_4 \times (-1) + S_n \times 0]S_3\frac{V_{DC}}{2} \end{aligned} \tag{7.83}$$

Phase b and c voltages can be derived the same way. With three-phase PWM terminal voltages, the CM and DM voltages of each phase can be determined as

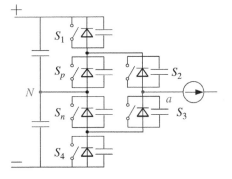

Figure 7.26 Phase a of a three-phase three-level ANPC converter.

$$v_{CM} = \frac{(v_{aN} + v_{bN} + v_{cN})}{3}$$

$$v_{a\_DM} = v_{aN} - v_{CM}$$

$$v_{b\_DM} = v_{bN} - v_{CM} \quad\quad (7.84)$$

$$v_{c\_DM} = v_{cN} - v_{CM}$$

Figure 7.27 shows an equivalent circuit for the three-phase converter, mainly for representing the AC side. This is a general equivalent circuit, valid for all three-phase voltage source inverter topologies. In addition to the voltage sources defined above, the equivalent circuit also contains impedances in the converter circuit, including AC filter inductor L_f; AC parasitic capacitance C_{ACg} (i.e. AC line-to-ground capacitance, which could go through heatsink); motor impedance Z_{motor} and motor neutral-to-ground parasitic capacitance C_g; and DC CM and DM circuit parameters. With this equivalent circuit, the AC-side CM and DM currents, as well as DC-side CM current, can be determined.

The expression of PWM phase voltages v_{aN}, v_{bN}, and v_{cN} for two- and three-level VSCs has been presented as functions of switching functions. In general, for an arbitrary VSC with any number of levels, phase voltages can be obtained in time domain based on PWM modulator output. Different modulation schemes can be automatically considered.

The Figure 7.27 equivalent circuit does not model the DC-side DM circuit. To solve DC DM current, another equivalent circuit, as shown in Figure 7.28, for the ANPC converter can be used, where the DC line model represents DC side connections such as DC cables in an aircraft system. In this case, DC currents have two independent components, which are chosen to be the positive bus current i_{DC1} and neutral wire current i_{DC2}. The negative bus current can be derived using i_{DC1}

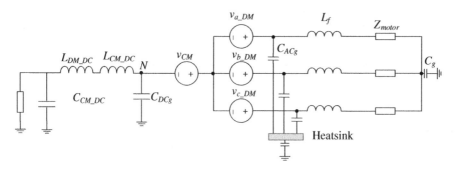

Figure 7.27 General equivalent circuit for a three-phase AC converter system.

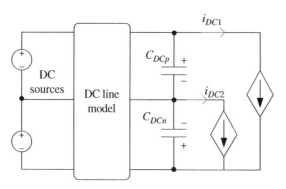

Figure 7.28 DC-side equivalent circuit for a three-phase three-level ANPC converter (note: DC sources can be two or three wires).

and i_{DC2}. Hence, the DC currents can be modeled as two controlled current sources i_{DC1} and i_{DC2} as shown in Figure 7.28. The controlled sources can be obtained through switching functions as (7.85).

$$i_{DC1} = \left(S_{1a}S_{2a}i_a + S_{1b}S_{2b}i_b + S_{1c}S_{2c}i_c\right)$$
$$i_{DC2} = \left(S_{2a}S_{pa} + S_{3a}S_{na}\right)i_a + \left(S_{2b}S_{pb} + S_{3b}S_{nb}\right)i_b + \left(S_{2c}S_{pc} + S_{3c}S_{nc}\right)i_c \tag{7.85}$$

Comparing (7.85) with (7.81), it can be seen that for the two-level VSC, the DC side can be modeled by only one current source, while the three-level ANPC converter needs two current sources due to the neutral wire between the DC-link midpoint and the AC side. The approach can be generalized to other topologies. In this case, two relatively common topologies for high-power and high-voltage applications, flying capacitor clamped and diode-clamped multi-level converters in Figure 7.29 are considered. For AC side, they both can still be modeled by using the equivalent circuit shown in Figure 7.27. For DC side, the DC current of flying capacitor clamped multilevel converter is only determined by the switching actions of the top switch in each phase (e.g. S_{a5}, S_{b5}, S_{c5} in Figure 7.29b), and its equivalent circuit is the same as the two-level VSC as shown in Figure 7.30a. The corresponding i_{DC} can be found as (7.86), where S_{top_j} is the modulation signal of the top switch at phase j.

$$i_{DC} = \sum_{j=a,b,c} S_{top\_j} \cdot i_j \tag{7.86}$$

It is complex to model the DC current of diode-clamped converters with the switching function controlled current sources in Figure 7.30b, since the required number of controlled current sources varies with the voltage levels. It means the simulation circuits need to be manually changed when the voltage level changes. To avoid such a laborious simulation circuit change, a further simplified equivalent circuit similar to the two-level VSC is adopted [12]. By applying Norton's Theorem, the complicated equivalent circuit involving multiple controlled current sources can be simplified as one equivalent current source with an equivalent internal impedance as shown in Figure 7.30b.

By shorting the two DC terminals of Figure 7.30b, the equivalent current i_{DCeq} can be obtained with basic circuit analysis (e.g. the loop current method). The equivalent capacitance C_{eq} can be obtained by opening all current sources i_{DCk} with $k \in (1, n)$, where n is the number of voltage levels minus 1 (or the number of voltage-dividing capacitors), i_{DCk} is the kth current source. For example, for a three-level NPC converter, $n = 3 - 1 = 2$. The results are summarized in (7.87), where C_k is the capacitance value of the kth capacitor on the DC link for the diode-clamped multilevel converters, assuming all capacitors are identical.

$$\begin{cases} i_{DCeq} = i_{DC1} + \dfrac{n-1}{n}i_{DC2} + \dfrac{n-2}{n}i_{DC3} + \cdots + \dfrac{1}{n}i_{DCn} \\[2mm] i_{DCk} = \begin{cases} \displaystyle\sum_{j=a,b,c} \left[S_{j\_k+1} - S_{j\_k}\right]i_j \text{ when } i_j > 0 \\[2mm] \displaystyle\sum_{j=a,b,c} \left[S_{j\_k} - S_{j\_k+1}\right]i_j \text{ when } i_j < 0 \end{cases} \\[2mm] C_{eq} = \dfrac{C_k}{n} \end{cases} \tag{7.87}$$

Figure 7.31 shows that MATLAB/Simulink simulation time comparison between the switching function-based and switch-based simulations for two-level, three-level ANPC, and nine-level flying capacitor VSCs. In this case, since the simulation involves time-domain EMI current calculation, and therefore involves solving differential equations, six line cycles are needed to reach the steady state.

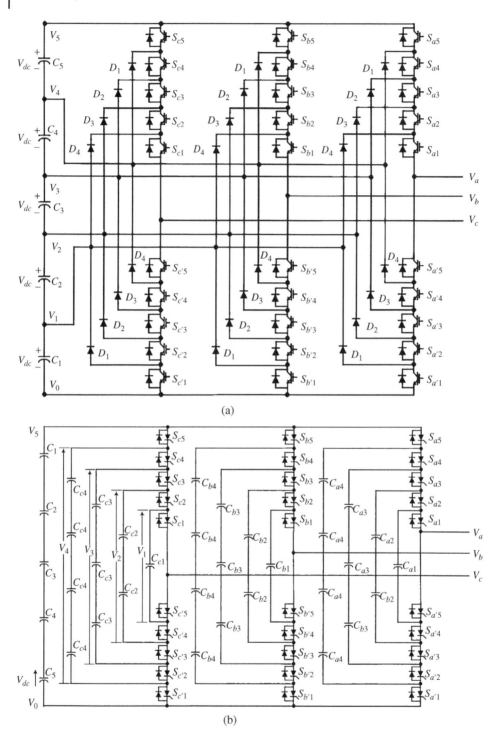

(a)

(b)

Figure 7.29 Two multilevel topologies for three-phase converters. (a) Diode-clamped multilevel converter. (b) Flying capacitor clamped multilevel converter.

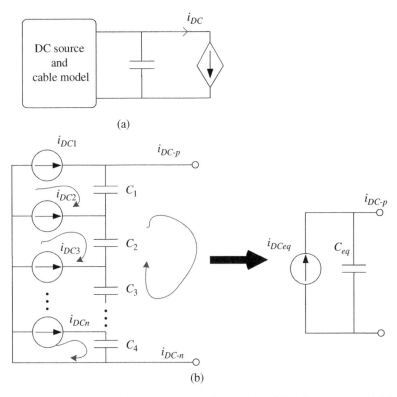

Figure 7.30 General DC side equivalent circuits for multilevel converters. (a) DC side equivalent circuits for flying capacitor clamped converter. (b) DC side equivalent circuit for diode-clamped converter.

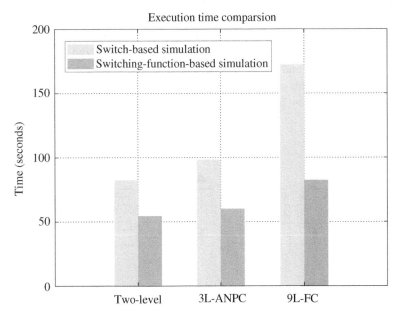

Figure 7.31 Simulation time comparison for two-level, three-level ANPC and nine-level flying capacitor VSCs.

7.5.1.3 Switching Function-Based Loss Calculation for Multilevel VSCs

The switching functions can also be easily used to calculate device losses in time domain. In fact, the loss calculation presented in Section 6.3.2 already has a time-domain numerical approach based on modulation. It can be equivalently implemented in a switching function-based simulation. Basically, the switching function values will determine the on and off status of switches, so the conduction periods and switching actions can be determined. The loss can be calculated, together with the device loss models, i.e. device on-state resistance or on-state voltage drop, switching loss energy E_{on} and E_{off}. The simulation should cover the whole line cycle, although sometimes, a fraction of line cycle is sufficient due to symmetry. For example, 1/6 of a line cycle would be sufficient for a two-level VSC. On the other hand, several line cycles are often used in simulation to have a better average.

7.5.2 Topology Impact on Device Models

One particularly important aspect to examine is related to the parasitics of different converter topologies. Parasitics can make some of the assumptions used in the converter models less accurate or less valid. For example, the devices switching loss model presented in Chapter 6 for the two-level VSC is based on the datasheet switching energy loss E_{on} and E_{off}, which are generally obtained through a DPT on a half-bridge setup. The two-level VSC is based on the half-bridge structure, and therefore, the switching loss calculation based on the datasheet E_{on} and E_{off} is reasonable. On the other hand, for a different topology, the model or parameters may need to be modified.

Here, we will present the impact of topology on switching loss, overvoltage, and PWM voltages, due to parasitic capacitances and inductances for three-level rectifiers or inverters mainly based on the work in [13, 14]. Particularly, we will discuss the impact of extra junction capacitances and switching commutation loops introduced by line-frequency devices (i.e. devices nonactive every other half line-cycle) in three-level rectifiers or inverters.

7.5.2.1 Extra Junction Capacitances and Commutation Loops in Three-Level Converters

7.5.2.1.1 Vienna-Type Rectifiers Figure 7.32 shows three common Vienna-type rectifier variants. Use the variant (c) as the example. For brevity, Figure 7.33 only illustrates its commutation loops for a positive half line-cycle. When the switching commutation takes places between S_1 and D_1, during the turn-on transient, S_1 starts to conduct and its channel current increases. Then, the phase current is commutated from D_1 to S_1. Once this commutation ends, D_1 is blocked. At the same time, junction capacitance C_{D1} of D_1 is charged by DC bus through S_1 channel, and output capacitance

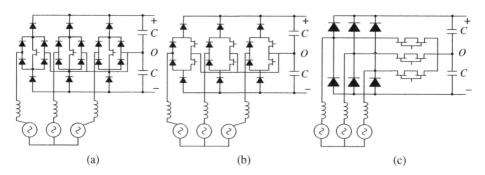

Figure 7.32 Three common Vienna-type rectifier variants.

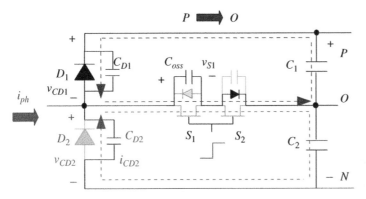

Figure 7.33 Commutation loops with extra capacitance involved in Figure 7.32c Vienna-type rectifier.

C_{oss} of S_1 is discharged. The device loss models for Vienna-type rectifiers generally all consider the commutation between S_1 and D_1, similar to a half-bridge case.

When C_{oss} of S_1 is discharged from $V_{DC}/2$ to zero, the AC terminal voltage (S_1 drain) also drops from V_{DC} to $V_{DC}/2$, as illustrated in Figure 7.34. Therefore, the junction capacitance C_{D2} of D_2 will be discharged through S_1 channel and lost in the channel resistance. This path does not conduct AC phase current for half a line-cycle but the effect also incurs loss and should be considered in device loss model.

The root cause of this effect is that there are three switching branches (positive branch, neutral branch, and negative branch) in the phase leg of three-level converters which forms a T connection, instead of the two positive and negative branches formed by the switch/switch (S–S) or switch/diode (S–D) pair in two-level converters. Consequently, in each half line cycle, although the complementary branch does not conduct, the junction capacitance of this branch is still charged or discharged as long as it has a switching potential (i.e. one terminal potential jumps during switching). This capacitance can be regarded as a paralleled output capacitance across the active switch (i.e. the switching device), as the DC-link voltage is unchanged and the DC-link capacitor can be considered shorted during the process.

The similar analysis can be carried out for the other two Vienna-type rectifiers in Figure 7.32a and b using the commutation loops shown Figure 7.35a and b, respectively [13].

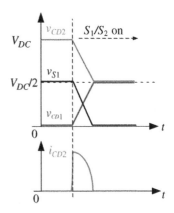

Figure 7.34 Waveform illustration of three junction capacitor voltages and the discharging current of the line-frequency switch during the turn-on transient of the main switch.

7.5.2.1.2 T-Type Three-Level Converters
Figure 7.36 shows a phase leg of a T-type three-level converter. Compared with the Vienna-type rectifier, two diodes are replaced with active switches in the T-type converter, so the commutation can be current independent and in four quadrants. However, similar capacitance effect can still be observed. For transitions from positive level to neutral level in both rectifier and inverter modes, in addition to the conventional switching loop (top), another discharging path from the junction capacitance C_{oss2} of S_2 to the channel of the neutral switch S_{N1} is concurrent (bottom).

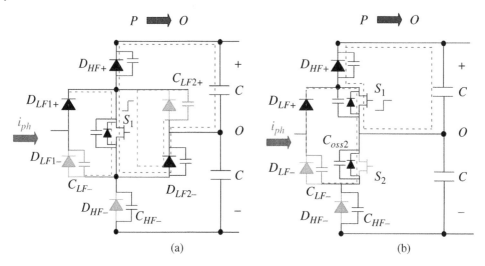

Figure 7.35 Commutation loops with extra capacitance involved in Vienna-type rectifier variants (a) and (b).

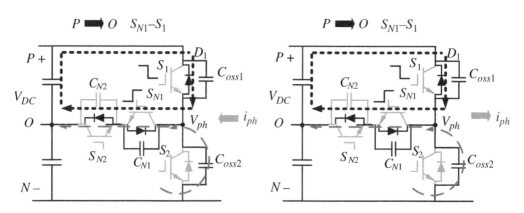

Figure 7.36 Commutation loops with extra capacitances involved in T-type three-level converter. Left: rectifier mode; right: inverter mode.

7.5.2.1.3 I-Shaped Three-Level Converters Commonly used NPC converters adopt I-shaped phase legs. Note that Vienna-type rectifiers (a) and (b) in Figure 7.32 also belong to this category, while the variant (c) belongs to T-type converter.

Figure 7.37 shows the commutation loops of a three-level diode NPC (DNPC) phase leg during half a line cycle. When the converter operates in the first quadrant, phase current is going out (right figure). The outer switch S_1 and clamping diode D_p operate at high frequency, and S_2 stays fully on to form the positive and neutral levels in the positive half line cycle (i.e. outer mode and short loop in Figure 7.37), while S_4 and D_n are fully off. When the converter operates in the second quadrant, phase current is flowing in (left figure). The inner switch S_1/D_1 and S_3 are switched at high frequency to synthesize the positive and neutral levels (i.e. inner mode and long loop in Figure 7.37), while S_2 and S_4/D_4 are fully off and will be active in the other half line cycle.

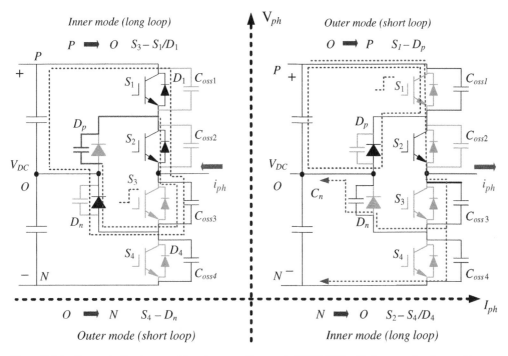

Figure 7.37 Commutation loops with extra capacitances in DNPC VSC with four-quadrant operation (third and fourth quadrants not shown).

There are also extra device junction capacitances and commutation paths for DNPC involving nonactive devices. Assuming the first quadrant operation with outer mode modulation, in addition to the commutation loop (top short dashed loop involving P, O, S_1, and D_p in the right figure), another loop exists (long dashed loops involving P, O, S_1, S_2, S_3 and D_n, or P, N, S_1, S_2, S_3 and S_4 in the right figure). During the S_1 turn-on transient, the switching node voltage at the drain of S_3 increases from $V_{DC}/2$ to V_{DC}, therefore, C_{oss} of the blocked switches S_3 and S_4 as well as C_n of the clamping diode D_n will be also charged, forming extra output capacitances across the main switch S_1.

Similar phenomenon can be seen in the second quadrant operation with the inner mode modulation, as shown in Figure 7.37 (left side). Since the voltage of S_4 remains at $V_{DC}/2$ before and after the turn-on transient of S_3 thanks to D_n clamping, C_{oss4} and C_n do not contribute to extra C_{oss}. However, since the voltage across D_p decreases from $V_{DC}/2$ to zero during the transition, an extra junction capacitance and a switching loop are introduced by D_p in Figure 7.37 (the middle loop involving S_2, S_2, D_p and D_n in the left figure).

In summary, the outer mode modulation of DNPC that only corresponds to a short switching loop in ideal case can introduce extra charged or discharged junction capacitances and another long commutation loop from the complementary branch during the switching commutation. Whereas, the inner mode that corresponds to a long commutation loop owing to the extension of the switching loop into the complementary branch has no extra junction capacitances from the complementary branch; however, there will be extra capacitances and another short commutation loop. Note that a specific operating condition, characterized by its power factor, can be in both outer and inner modes through a line cycle. Only the special

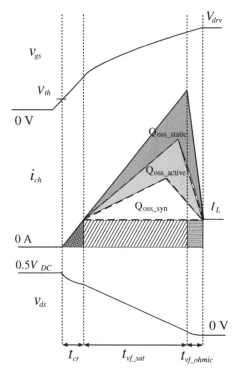

Figure 7.38 Theoretical turn-on transient of the active neutral device in a three-level converter.

case of unity power factor will limit the operation mode of either only outer mode ($PF = 1$) or only inner mode ($PF = -1$) [15].

ANPC has more freedom in modulation because of the flexibility in controlling the clamping switches. The modulation schemes can be outer mode, inner mode, or a combination of inner and outer modes [15]. Since all schemes are rooted in the two basic modes in DNPC, the aforementioned effect and conclusions are also applicable for these different modulations in ANPC.

7.5.2.2 Impact on Switching Loss

The extra commutation loops and device capacitances inevitably introduce switching loss to all types of three-level converters. The first part of the extra loss is the C_{oss_static} energy stored on the nonactive device and dissipated during the turn-on of the active device, as illustrated by charge Q_{oss_static} in Figure 7.38, where Q_{oss_active} and Q_{oss_syn} are the charges of active switch and synchronous switch (i.e. complementary commutating switch of the active device), respectively. The second part is overlap loss caused by the increased duration of saturation region t_{vf_sat} due to this extra charge at turn-on and turnoff. Due to the near-zero voltage drop in the ohmic region, the overlap loss during t_{vf_ohmic} is negligible. In addition to channel current i_{ch}, Figure 7.38 also illustrates the gate voltage v_{gs} and drain-source voltage $v_{ds} \cdot t_{cr}$ is the current rise time to reach the AC current (load current in the inverter case) I_L.

As pointed out earlier, the device loss model for three-level converters in simulation of the Section 7.5.1 still use data obtained from the two-level half-bridge DPT, which cannot consider the extra commutation loops and junction capacitances identified above. In order to have a more accurate loss calculation, a straightforward solution could be to build a DPT phase leg with a three-level structure. However, this is not an economical or even feasible approach in the design stage. A more desirable approach is to still use the switching characteristics in datasheet, e.g. E_{on} and E_{off} data of commercial devices measured based on two-level DPT.

To facilitate the loss calculation taking into account of the extra loss, a method combining device C_{oss} information and conventional DPT data in [13] can be adopted, with step-by-step approach below.

7.5.2.2.1 C_{oss} Loss Calculation The extra C_{oss} energy loss $E_{on\_Coss\_extra}$ contributed by the nonactive device per half line cycle can be calculated with (7.88) as its voltage v_{coss_static} increases from $V_{DC}/2$ to V_{DC}, using the Vienna-type rectifier in Figure 7.33 as an example. Here, T_s and T_{on} denote the switching cycle and turn-on interval, respectively. $C(v_{Coss\_static})$ is the capacitance value at v_{Coss_static}.

$$E_{on\_Coss\_extra} = \int_{T_s}^{T_s + T_{on}} \left(v_{Coss\_static} - \frac{V_{dc}}{2} \right) \cdot i \cdot dt$$

$$= \int_{\frac{V_{DC}}{2}}^{V_{DC}} v_{Coss\_static} \cdot C(v_{Coss\_static}) \cdot dv_{Coss\_static} \qquad (7.88)$$

$$- \frac{V_{dc}}{2} \int_{\frac{V_{DC}}{2}}^{V_{DC}} C(v_{Coss\_static}) \cdot dv_{Coss\_static}$$

Define the energy equivalent capacitance $C_{eq,E}$ for this three-level case as

$$C_{eq,E} = \frac{\int_{\frac{V_{DC}}{2}}^{V_{DC}} v_{ds} \cdot C(v_{ds}) dv_{ds}}{\frac{1}{2} V_{DC}^2} \qquad (7.89)$$

Then, Eq. (7.88) can be reformatted as

$$E_{on\_Coss\_extra} = \frac{1}{2} C_{eq,E,Coss\_static} \left(\frac{V_{DC}}{2} \right)^2 - \frac{V_{DC}}{2} Q_{oss\_static} \qquad (7.90)$$

where $C_{eq,E,Coss\_static}$ is the energy equivalent capacitance of C_{oss_static} and can be calculated from the device voltage-dependent capacitance curve on its datasheet.

7.5.2.2.2 Overlap Loss Calculation

The overlap energy loss $E_{on\_overlap}$ under a given current I_L is simplified as

$$E_{on\_overlap} = E_{on\_cr} + E_{on\_vf} \approx \frac{1}{2} \left(t_{cr} + t_{vf\_sat} \right) V_{ds} I_L \qquad (7.91)$$

where E_{on_cr} is the overlap energy during the current rise time t_{cr}, E_{on_vf} is the overlap energy during the voltage falling interval t_{vf_sat} in the saturation region, and V_{ds} is the steady-state DC blocking voltage on the device before turning on. As will be shown below, t_{vf_sat} can be strongly impacted by the extra junction capacitance of the nonactive device. Therefore, the extra capacitance not only increases the C_{oss} loss but also indirectly increases the overlap loss.

For devices with nonflat gate voltage v_{gs} in the Miller plateau region such as SiC MOSFETs or GaN HEMTs, the total displacement charge Q_{oss_3} in the phase leg of the three-level converter is the sum of three device junction capacitances per phase leg and can be derived based on the triangle area of the channel current i_{ch} curve in Figure 7.38. As analyzed in Chapter 6, the device channel current i_{ch} in the current rise region and saturation region is determined by the gate voltage as

$$i_{ch} = g_{fs} \left(v_{gs} - V_{th} \right) \qquad (7.92)$$

Using (7.92), the slope of i_{ch} can be calculated. The total displacement charge Q_{oss_3} can be obtained as

$$Q_{oss\_3} = Q_{oss\_syn} + Q_{oss\_active} + Q_{oss\_static} = \frac{1}{2} g_{fs} \frac{dv_{gs}}{dt} t_{vf\_sat\_3}^2 \qquad (7.93)$$

where $t_{vf\_sat\_3}$ denotes the voltage falling period of the main active device in the three-level phase leg in the saturation region.

For Si devices, the flat Miller plateau effect should be considered and the reverse recovery charge should be excluded.

From (7.93), the overlap time is proportional to the square root of the total displacement charge as

$$t_{vf\_sat\_3} = \sqrt{\frac{2Q_{oss\_3}}{g_{fs} \frac{dv_{gs}}{dt}}} \propto \sqrt{Q_{oss\_3}} \qquad (7.94)$$

Therefore, the overlap energy $E_{on\_vf\_3}$ of the three-level phase leg in the saturation region can be scaled by the ratio of total displacement charge in three-level phase leg and two-level phase leg, from the overlap energy $E_{on\_vf\_2}$ of the two-level DPT as

$$E_{on\_vf\_3} = \sqrt{\frac{Q_{oss\_T}}{Q_{oss\_H}}} \cdot E_{on\_vf\_2} \tag{7.95}$$

where Q_{oss_2} is the total charge on C_{oss} of the two-level half-bridge.

However, $E_{on\_vf\_2}$ is difficult to be directly measured from DPT and also not available from the device datasheet. Reference [13] established an intrinsic relationship between switching energy and DPT data to realize the loss scaling scheme. The relationship is discussed below.

7.5.2.2.3 Intrinsic Relationship Between Switching Energy and DPT Data
From Figure 7.38, the current rise time t_{cr} can be also derived in (7.96), in addition to the overlap time in (7.94).

$$t_{cr} = \frac{I_L}{g_{fs}\frac{dv_{gs}}{dt}} \tag{7.96}$$

Thus, based on (7.91), (7.94), and (7.96), the total turn-on switching energy in a general setup (i.e. applicable to both two- and three-level phase legs) can be expressed as

$$E_{on} = E_{on\_cr} + E_{on\_vf} + E_{Coss\_total} = \frac{V_{DS}}{2}\frac{1}{g_{fs}\frac{dv_{gs}}{dt}}I_L^2 + \frac{V_{DS}}{2}\sqrt{\frac{2Q_{oss}}{g_{fs}\frac{dv_{gs}}{dt}}}I_L + E_{Coss\_total} \tag{7.97}$$

where E_{Coss_total} consists of both the C_{oss} energy stored in the main active device and the energy stored in the synchronous device during the turnoff of the main active device.

The energy loss E_{on_DPT} and E_{on_DPT} from the conventional DPT can be typically curve-fitted well by a quadratic function

$$\begin{cases} E_{on\_DPT} = k_1 I_L^2 + k_2 I_L + k_3 \\ E_{off\_DPT} = k_4 I_L^2 + k_5 I_L + k_6 \end{cases} \tag{7.98}$$

where k_1 through k_6 are curving-fitting coefficients.

By matching the two expressions (7.97) and (7.98), an interesting and important relationship between the physical device parameters and measured data is revealed

$$\begin{cases} E_{on\_cr} = k_1 I_L^2 = E_{on\_cr\_DPT} \\ E_{on\_vf} = k_2 I_L = E_{on\_vf\_DPT} \\ E_{Coss\_total} = k_3 + k_6 = \sum E_{Coss\_DPT} \end{cases} \tag{7.99}$$

where the constant k_6 in the turnoff transient in DPT represents C_{oss} energy of the active device dissipated during the next turn-on transient, thus is added to E_{Coss_total}.

From (7.99) and (7.95), the total overlap energy $E_{on\_vf\_3}$ of the three-level phase leg can be finally scaled from the linear term $E_{on\_vf\_DPT}$ (i.e. $E_{on\_vf\_2}$) of the two-level half-bridge DPT result. The added overlap energy introduced by the nonactive device is the difference between $E_{on\_vf\_3}$ and $E_{on\_vf\_2}$.

7.5.2.2.4 *DPT Data-Based Loss Calculation, Discussion, and Validation* For the turnoff transient, the actual overlap loss excluding the C_{oss} energy should also be considered. However, for Si MOSFET or WBG devices, this loss is often quite low compared to the turn-on loss, such that the extra turnoff overlap loss caused by the displacement charge of the nonactive device should be even smaller. Therefore, this overlap loss could be omitted to simplify the method.

Finally, the switching loss P_{sw} of the upper half or lower half of a three-level phase leg can be determined based on the two-level DPT data as (7.100), where i_L is the instantaneous phase current, and T_0 is the AC line fundamental period.

$$
P_{sw} = \frac{1}{T_0} \int_0^{\frac{T_0}{2}} \left(k_1 \cdot i_L{}^2 + k_2 \cdot i_L \cdot \sqrt{\frac{Q_{oss\_3}}{Q_{oss\_2}}} + k_4 \cdot i_L{}^2 + k_5 \cdot i_L \right) \cdot f_{sw}
$$
$$
\cdot dt + (k_3 + k_6 + E_{on\_Coss\_extra}) \cdot f_{sw}
$$

(7.100)

Although the presented analysis and proposed methods are conducted on a Vienna-type three-level converter, they can be applied to other three-level converters as well.

The analysis and model have been validated using a T-type Vienna-type rectifier with each phase leg consisting of two 650 V antiseries connected GaN Systems HEMTs and two 1200 V Wolfspeed SiC Schottky diodes. The rectifier operation condition is switching frequency 450 kHz, phase voltage 115 V RMS, DC voltage 650 V, and input power 1.5 kW. The calculated switching losses without and with considering the extra junction capacitance effect are 47.5 W and 55.6 W, respectively, as indicated in Figure 7.39, based on E_{on} and E_{off} data from a conventional half-bridge DPT. The energy equivalent junction capacitances of GaN devices and two diodes are also derived for this estimation. Additionally, associated loss from heatsink-coupled capacitance and other minor losses are included in both cases as the baseline. Clearly, the pure turnoff loss excluding C_{oss} loss is very small, supporting the approach to ignore the extra turnoff loss from the capacitance of the nonactive device. Note that all loss numbers are for the whole three-phase rectifier, including those in Figure 7.39 through Figure 7.41.

The actual power loss of the converter is shown in Figure 7.40. It is evident that the predicted total switching loss after adding the extra C_{oss} of the nonactive diode is 55.6 W, very close to the measured

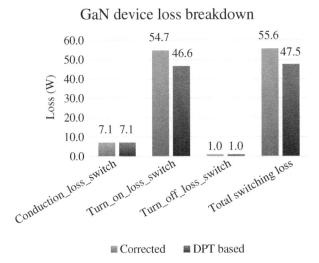

Figure 7.39 Comparison of the calculated device losses in a Vienna-type rectifier with and without correction for extra junction capacitance effect.

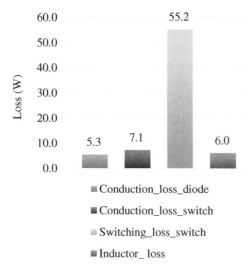

Figure 7.40 Measured converter losses.

switching loss of 55.2 W, and the impact from the nonactive device is 8.1 W. Further breakdown of the turn-on switching loss is given in Figure 7.41, including the overlap loss and the C_{oss} losses of the active GaN devices, the high-frequency synchronous diodes, and the nonactive diodes. It can be seen that the extra overlap loss introduced by the nonactive diodes is only 1.0 W, while the extra C_{oss} loss is 7.1 W.

In general, for fast switching WBG device-based three-level converters, the power loss impact of nonactive devices on the active switch is mainly reflected by the extra C_{oss} loss instead of the extra overlap loss due to the superfast overlap transition. As can be seen in the above example calculation and experiment, this part of loss is as high as 17% of the total switching loss. For slower Si IGBT-based three-level converters, the loss effect will be mainly on extra overlap loss due to the significantly longer switching transition and relatively lower C_{oss} loss.

7.5.2.3 Impact on Device Voltage Stress

In Chapter 6, we have analyzed the turnoff and turn-on overvoltage during switching mainly as a result of parasitic inductances and fast switching. In general, stray inductances from device package, PCB trace, and power switching loop can resonate with the junction capacitances of the devices in a bridge configuration, leading to higher current stress of the active device and higher voltage stress of the synchronous device during the turn-on, and higher voltage stress of the active device during the turnoff.

Here we will examine the impact of added stray inductance associated with the nonactive devices in three-level converters. With participation of the nonactive devices in switching commutation and extra power loops in parallel with the main commutation loop, one more *LC* resonance can occur due to the extra C_{oss} and loop inductance, imposing higher voltage stress on the nonactive devices in the T-type or I-shaped phase leg of three-level converters. This is particularly severe when using fast switching devices.

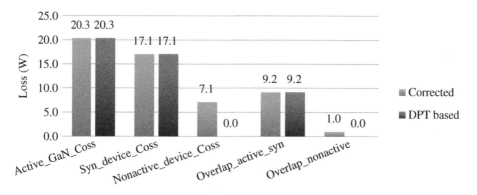

Figure 7.41 Turn-on switching loss breakdown.

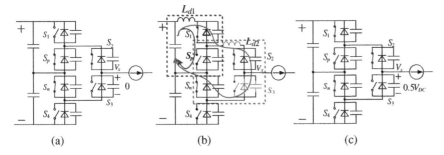

Figure 7.42 Mode analysis for multicommutation loops during positive line cycle: (a) steady-state when $Vx = 0$; (b) switching transition from $Vx = 0$ to $Vx = 0.5V_{DC}$; and (c) steady-state when $Vx = 0.5V_{DC}$.

To illustrate this effect, an example is given for the ANPC inverter. As explained earlier, ANPC has two types of commutation loops, depending on whether it is in the outer or the inner mode. The outer mode is often considered superior under high switching frequency operation in terms of switching loss and device voltage stress since the outer mode corresponds to a shorter loop (see Figure 7.37). However, even for the outer mode, a long commutation loop exists due to the effect from the nonactive switches [16]. Figure 7.42 shows a mode analysis of the multi-loops during a positive half line cycle. First, S_p is turned off, and the load current is transferred to the freewheeling diode of S_p. Then, S_1 is turned on and commutates this load current. Once the current is fully commutated to S_1, the DC (or decoupling) capacitor discharges the junction capacitance of S_1 and charges the junction capacitance of S_p. In the meantime, the junction capacitance of the nonactive switch S_3 is also charged from 0 to $0.5V_{DC}$, since it is in parallel with S_p via S_2 and S_n, and a long commutation loop is formed via switches S_1, S_2, S_3, S_n, and the DC capacitor in Figure 7.42b.

For this extra commutation loop, a resonance can occur between the total loop inductance $L_{d1} + L_{d2}$ and the C_{oss} of S_3, following two intervals during the turn-on transient of the main active switch S_1: (i) the load current I_L is fully commutated from S_p to S_1; (ii) S_2 voltage is charged by the DC bus via S_1 channel until it reaches $0.5\,V_{DC}$. The first resonance cycle can be illustrated by the state trajectory of the stray inductor current i_{Ls} and C_{oss} voltage v_{Coss3} of S_3, as given in Figure 7.43. The first and highest voltage peak appears when i_{Ls} returns to I_L level, indicating the inductive energy stored by the loop inductance is fully transferred to C_{oss} of the nonactive device S_3. Assuming the first resonance amplitude of v_{Coss3} is V_m, and the resonance amplitude of the excitation voltage

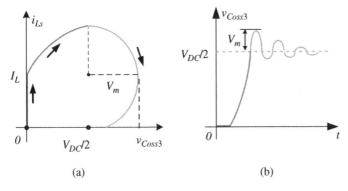

Figure 7.43 Overvoltage analysis of the line-frequency switch S_3 in ANPC phase leg during the turn-on transient of the main switch S_1. (a) State trajectory of current and voltage in the extra commutation loop; (b) voltage ringing and resonance spike.

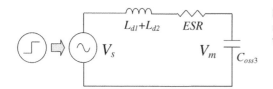

Figure 7.44 A simplified equivalent circuit of the resonance tank during the main switch S_1 turn-on transient.

generated by the turn-on switching action of S_1 is V_s, a simplified equivalent circuit can be obtained as Figure 7.44, where the impact of the original L_{d1} loop is neglected.

From the frequency domain circuit analysis, a simple relationship can be established at the resonance frequency

$$V_m = \frac{V_s}{ESR} Z_0 = \frac{V_s}{ESR} \sqrt{\frac{L_{d1} + L_{d2}}{C_{oss3}}} \tag{7.101}$$

where ESR is the total loop resistance, and Z_0 is the characteristic impedance of the resonance. Therefore, the overvoltage stress is proportional to the square root of the loop inductance.

Figure 7.45 shows the simulation results using PSpice model of Wolfspeed C3M0065090J devices under DC-link voltage of 1000 V. It can be seen that S_2 and S_p will see the same voltage spike during turnoff transition of S_p, assuming $L_{d2} = 0$; when L_{d2} is not zero, the voltage spikes of the two devices become significantly different. As a result of resonance of the long commutation loop inductance and S_3 junction capacitance, the nonactive switch S_3 suffers a much higher voltage spike in the positive half line cycle compared to the active switch S_p.

The impact of switching loops on parasitic inductances and voltage overshoot will be further illustrated through the busbar design for a three-level ANPC converter in Chapter 11.

7.5.2.4 Impact on PWM Voltage and AC Current Distortion

As discussed in Chapter 6 and earlier in this chapter, the PWM voltage harmonic spectrum strongly depends on switching frequency. For ideal switches, which would turn on and off instantaneously, higher switching frequency will push the voltage harmonic frequencies higher, which generally will lead to lower current harmonics for a given load with a given impedance. However, considering the nonideal device characteristics, specifically, the parasitic effect on switching commutation and the corresponding need for the dead time, the PWM voltage quality could deteriorate with higher switching frequency. This is because at a higher switching frequency (therefore, shorter switching period), the fixed commutation transient period will account for a larger portion of the overall switching period. The corresponding voltage errors will cause higher voltage distortion with the switching period. In principle, WBG devices can switch faster with shorter commutation times, and therefore can enable higher switching frequency. However, with certain topology and design, the parasitic effect can be severe to even cause voltage error and current distortion in WBG-based converters.

One such situation is in switch-diode configured converters such as three-level Vienna-type rectifiers. As an example, Figure 7.46 illustrates the turnoff transient of the Figure 7.32c Vienna-type rectifier in postive half AC line cycle. Similar to previous analysis for the turn-on transition, output capacitance C_{oss} of the active switch S_1 is charged from 0 to $V_{DC}/2$, and the junction capacitance C_{D1} of D_1 is discharged from $V_{DC}/2$ to 0. In additon, the junction capacitance C_{D2} of the nonconducting diode D_2 is charged from $V_{DC}/2$ to V_{DC}. Therefore, C_{D2} is also part of the turnoff commutation.

In Figure 7.46, the rectifier phase-leg AC terminal voltage v_x is defined with respect to the DC-link midpoint M, which coincides with the switch S_1 votage v_{ds} during turnoff. Figure 7.47 illustrates the

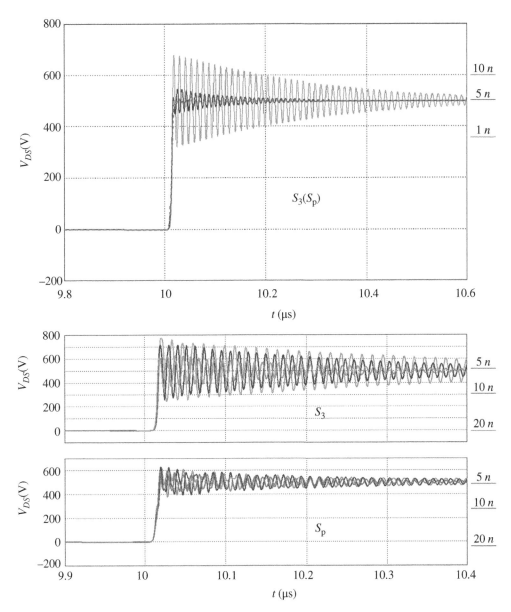

Figure 7.45 Overvoltage in the simulation. Left: L_{d1} = 1 nH, 5 nH, 10 nH and L_{d2} = 0 nH; right: L_{d1} = 5 nH and L_{d2} = 5 nH, 10 nH, 20 nH.

ideal voltage shape of v_x during turnoff vs. the actual voltage shape due to capacitance charing and discharging. Higher equilavent capacitance and/or lower phase current (i.e. source current) will result in more voltage distortion from the ideal case. In Figure 7.46, C_{D2} charging current is part of the phase current i_{ph}, and therefore C_{D2} will slow down the commutation.

To determine the actual voltage distortion, the charging and discharging of all three capacitances C_{oss}, C_{D1}, and C_{D2} need to be considered. On the other hand, the instantaneous values of these capacitance are varying and different for each device during the transient because they are nonlinear and voltage dependent.

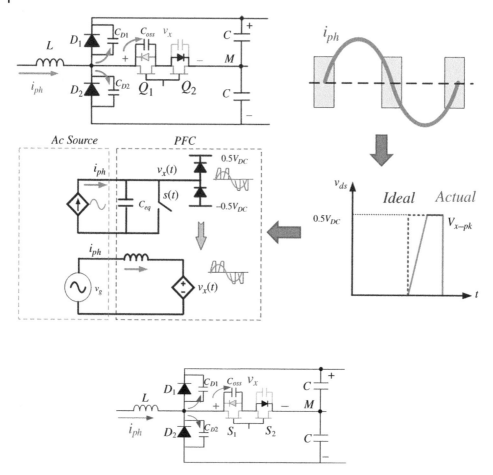

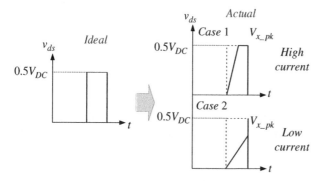

Figure 7.46 Illustration of capacitance charging/discharging during turnoff for one phase leg of T-type Vienna-type PWM rectifier.

Figure 7.47 Illustration of ideal and actual voltage shape during turnoff for T-type Vienna-type rectifier phase leg.

According to [14], it is effective to use charge-based equivalent capacitance to represent the non-linear voltage-dependent total junction capacitances $C_{total}(v_x)$ of the three T-connected devices $(S_1, D_1,$ and $D_2)$. $C_{total}(v_x)$ determines the total voltage ramping and charging interval as

$$t_{charge} = \int_0^{t(V_{x\_pk})} dt = \frac{\int_0^{V_{x\_pk}} C_{total}(v_x)dv_x}{i_{ph}} = \frac{V_{x\_pk}C_{eq,Q}}{i_{ph}} \tag{7.102}$$

where $C_{eq,Q}$ is the charge-based equivalent output capacitance over the voltage range $(0, V_{x\_pk})$ for the three T-connected devices per phase leg.

From the above analysis, C_{D2} not only impacts the switching loss and device stress in the extra power loop but can also introduce extra distortion on PWM voltage and AC current. It is worth noting that for loss evaluation, energy-based equivalent capacitance of the nonactive device should be adopted; for device overvoltage stress analysis, instantaneous and voltage-dependent capacitance value should be used; and for voltage distortion evaluation, the charge equivalent capacitance should be used.

Since the average of the ideal PWM voltage over a switching cycle represents the average fundamental voltage, the distortion voltage due to the slow charging process introduces volt-seconds loss. The phenomenon is similar to the volt-seconds loss due to dead time and therefore can be compensated similarly. The dead-time compensation schemes will be discussed extensively in Chapter 10. Figure 7.48 illustrates the effectiveness of the compensation method through experiment. The test was conducted on the same GaN HEMT-based Vienna-type rectifier used for the loss evaluation above. The operating condition is switching frequency 450 kHz, phase voltage 115 V RMS, DC voltage 600 V, and input power 1.4 kW. After compensation, the current harmonics were significantly reduced, and the current THD drops from 10.3% to 3.0%.

The voltage distortion and current harmonics due to the extra parasitic capacitances can be modeled and quantified analytically. On the other hand, if the proper compensation schemes are adopted, the issues will not significantly impact converter hardware design. As a result, more detailed modeling will be omitted here but can be found in [13, 14].

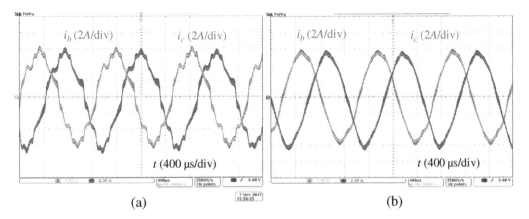

(a)   (b)

Figure 7.48 Experimental comparison of AC phase current quality with and without the compensation scheme. (a) Without compensation; (b) with compensation.

7.6   Active Rectifier and Source-side Inverter Interfaces to Other Subsystem Designs

Since the function of the active rectifier and the source-side inverter is more similar to that of the passive rectifier, their directly interfaced subsystems are also similar to those of the passive rectifier, including the load-side inverter (e.g. the motor drive) or the rectifier (e.g. wind turbine) or the DC/DC converter (e.g. PV, battery, or fuel cell), the thermal management system, the AC-side EMI filter, mechanical system, and control. All input interface conditions have been described in Section 7.2.3. Here we will examine all the output of the active rectifier or the source-side inverter needed for the other subsystems.

Load-side inverter or DC source/load (including DC/DC and rectifier source): In this case, there can be two different scenarios. The first scenario corresponds to the case where the DC-link voltage is regulated from the source side. The output from the active rectifier or the load-side inverter includes (i) maximum and minimum DC voltages during normal operation and (ii) peak DC voltage during abnormal conditions (transients, faults, etc.). These two items are related to the design constraints. The second scenario corresponds to the case where the DC-link voltage is regulated from the DC source side. In this case, the source-side inverter is more like a load-side inverter and its output follows that of the load-side inverter, so will not be further discussed here.

EMI filter: In case there is a source-side EMI filter, the output of the rectifier or the inverter to the EMI filter includes EMI bare noise and AC line current information, similar to the passive rectifier case.

As for the thermal management system, mechanical system, and control, the output from the active rectifier or the source-side inverter are all identical to those of the load-side inverter. The detailed descriptions for each category can be found in Section 6.5.1.

The output interfaces to other subsystems are summarized and classified in Table 7.5. Note that all "discussion" items in Table 7.5 are marked with "∗∗" because all of them are the same as in the load-side inverter case. Therefore, their discussions will not be repeated here.

On item marked with "∗" in Table 7.5 is the peak DC voltage during abnormal conditions. Although it is listed as the design result here, it is not covered in this chapter as part of the active rectifier or the source-side inverter design. Instead, it will be discussed in Chapter 12 when considering the grid requirement impact on the grid-tied converter design.

7.7   Summary

This chapter presents the power stage design of AC three-phase PWM active rectifiers and source-side inverters.

1) The active rectifier or the source-side inverter design is formulated as an optimization problem, with its design variables, constraints, conditions, and objective(s) clearly defined. Using the two-level VSC as an example, the design variables include switching devices, PWM scheme, switching frequency, gate drive parameters, DC-link capacitor, AC line filter, decoupling capacitors, and thermal management system performance (i.e. thermal impedance). The design constraints include performance constraints for the rectifier or the inverter (AC power quality, minimum and maximum DC link voltages for normal operating conditions, power loss or efficiency, control performance, ride through, DC link stability, and reliability) and physical constraints associated with the components (maximum temperatures, peak currents, device peak voltage,

Table 7.5 Output from the active rectifier or the source-side inverter to its interfaced subsystems.

Subsystem	Interface type	Item	Value
Load-side Inverter or DC source/load	Constraints	Max/min DC voltages during normal operation	Design result
		Peak DC voltage during abnormal conditions	Design result*
EMI filter	Constraint	AC line current	Design result
	Attribute	EMI bare noise	Discussion**
Thermal management system	Constraint	Maximum TMS thermal impedance	Design result
	Attribute	Device/component dimensions	Design result
		Maximum steady-state device/component power loss	Design result
		Overload power loss and duration	Design result
Mechanical system	Constraint	Device/component losses	Design result
	Attribute	Device/component dimensions and weight	Design result
		DC input and AC out terminals	Discussion**
		Voltage distribution	Discussion**
		Current distribution	Discussion**
Control system	Design variable	Gate driver R_g, V_g (i.e. V_{GH} and V_{GL})	Design results
	Constraint	Sensor accuracies and bandwidth	Design results
	Attribute	Sensor type, range, number, and isolation	Discussion**
		Digital controller capabilities (bandwidth, computation power, I/O) and isolation	Discussion**
		Protection type and settings	Discussion**
		Maximum dv/dt and di/dt	Discussion**

inductor saturation flux density, and core window fill factor). Note that AC source-side transient conditions are NOT considered and will be covered later in Chapter 12.

2) The focus is on establishing analytical or algebraic models for design constraints as functions of design conditions and variables. Many models similar to those of the load-side inverter can leverage Chapter 6. The AC current harmonics are considered through the L and LCL filter designs. Reliability constraint is considered through the failure rate and lifetime models of major components (devices, capacitors, and inductors). Control performance has considered control bandwidth and voltage margin requirements. Compared to the passive rectifier case, the DC-link stability design is extended to large-signal transient conditions.

3) The impact of the converter topology is discussed. One impact is on converter circuit modeling, with the switching function time-domain calculation or simulation introduced, which is more efficient than the switch-based circuit simulation. Another impact is on switching process and related device models due to added parasitic effects in different topologies. Parasitic impacts on device loss, overvoltage, and PWM voltage quality for three-level converters are discussed. Methods to account for these impacts in the design are presented.

4) The output from the active rectifier or the source-side inverter design as the interface parameters to other subsystem systems are discussed, including interface parameters to the load-side inverter or DC source/load, EMI filter, thermal management system, mechanical system, and control system. These interface parameters are similar either to the passive rectifier case or the load-side inverter case.

References

1 F. Wang, W. Shen, D. Boroyevich, S. Ragon, V. Stefanovic and M. Arpilliere, "Voltage source inverter – development of a design optimization tool," *IEEE Industry Applications Magazine*, vol. 15, no. 2, pp. 24–33, March–April 2009.

2 Y. Jiao and F. C. Lee, "LCL filter design and inductor current ripple analysis for a three-level NPC grid interface converter," *IEEE Transactions on Power Electronics*, vol. 30, no. 9, pp. 4659–4668, September 2015.

3 M. Liserre, F. Blaabjerg and S. Hansen, "Design and control of an LCL-filter-based three-phase active rectifier," *IEEE Transactions on Industry Applications*, vol. 41, no. 5, pp. 1281–1291, September/October 2005.

4 R. Rosso, X. Wang, M. Liserre, X. Lu and O. Engelke, "Grid-forming converters: control approaches, grid-synchronization, and future trends—a review," *IEEE Open Journal of Industry Applications*, vol. 2, pp. 93–109, April 2021.

5 R. Lai, F. Wang, R. Burgos, Y. Pei, D. Boroyevich, T. A. Lipo, B. Wang, V. Immanuel and K. Karimi, "A systematic topology evaluation methodology for high-density PWM three-phase AC–AC converters," *IEEE Transactions on Power Electronics*, vol. 23, no. 6, pp. 2665–2680, November 2008.

6 R. Burgos, G. Chen, F. Wang, D. Boroyevich, W. G. Odendaal and J. D. v. Wyk, "Reliability-oriented design of three-phase power converters for aircraft applications," *IEEE Transactions on Aerospace and Electronic Systems*, vol. 48, no. 2, pp. 1249–1263, April 2012.

7 Military Handbook, "*Reliability Prediction of Electronic Equipment, MIL-HDBK-217F,*" December 1991.

8 J. Sam G. Parler and L. L. Macomber, "Predicting operating temperature and expected lifetime of aluminum-electrolytic bus capacitors with thermal modeling," in *PCIM /Powersystems World*, Chicago, IL, November 1999.

9 J. Lai, H. Kouns and J. Bond, "A low-inductance DC bus capacitor for high power traction motor drive inverters," in *Conference Record of the 2002 IEEE Industry Applications Conference*, Pittsburgh, PA, October 2002.

10 E. Feilat, S. Grzybowski and P. Knight, "Multiple stress aging of magnet wire by high frequency voltage pulses and high temperatures," in *Conference Record of the 2000 IEEE International Symposium on Electrical Insulation*, Anaheim, CA, April 2000.

11 D. Han, S. Li, Y. Wu, W. Choi and B. Sarlioglu, "Comparative analysis on conducted CM EMI emission of motor drives: WBG versus Si devices," *IEEE Transactions on Industrial Electronics*, vol. 64, no. 10, pp. 8353–8363, 2017.

12 R. Ren, "Modeling and Optimization Algorithm for SiC-based Three-phase Motor Drive System," PhD diss., The University of Tennessee, Knoxville, 2020.

13 B. Liu, R. Ren, E. Jones, H. Gui, Z. Zhang, F. Wang and D. Costinett, "Effects of junction capacitances and commutation loops associated with line-frequency devices in three-level AC/DC converters," *IEEE Transactions on Power Electronics*, vol. 34, no. 7, pp. 6155–6170, 2019.

14 B. Liu, R. Ren, E. Jones, F. Wang, D. Costinett and Z. Zhang, "A modulation compensation scheme to reduce input current distortion in GaN based high switching frequency three-phase three-level Vienna type rectifiers," *IEEE Transactions on Power Electronics*, vol. 33, no. 1, pp. 283–298, January 2018.

15 Y. Jiao and F. C. Lee, "New modulation scheme for three-level active neutral-point-clamped converter with loss and stress reduction," *IEEE Transactions on Industrial Electronics*, vol. 62, no. 9, p. 5468–5479, 2015.

16 R. Ren, Z. Zhang, B. Liu, H. Gui, R. Chen, J. Niu, F. Wang, L. M. Tolbert, B. J. Blalock, D. Costinett and B. B. Choi, "Multi-commutation loop induced over-voltage issue on non-active switches in fast switching speed three-level active neutral point clamped phase leg," in *IEEE Energy Conversion Congress and Exposition (ECCE)*, Portland, OR, September 23–27, 2018.

8

EMI Filters

8.1 Introduction

EMI is defined as undesirable electromagnetic noise that corrupts, limits, or interferes with the performance of electronics or an electrical system. It is common to classify EMI into four different groups: conducted susceptibility, radiated susceptibility, conducted emissions, and radiated emissions [1]. The susceptibility or immunity represents the ability of electrical equipment to reject the noise in the presence of external electromagnetic disturbance. The studies of conducted and radiated emissions deal with the undesirable generation of electromagnetic noise from a piece of equipment and the countermeasures that can be taken to reduce it. These emissions are regulated and need to comply with the rule of electromagnetic compatibility (EMC), which is defined as "the ability of a device, unit of equipment or system to function satisfactorily in its electromagnetic environment without introducing intolerable electromagnetic disturbance to anything in that environment" [2]. This chapter focuses on conducted EMI emissions which are generally defined as undesirable electromagnetic energy coupled out of an emitter or into a receptor via any of its respective connecting wires or cables.

The objective of studying conducted EMI is to understand how the noise is generated and to find the solution to reduce its impact in order to meet the EMC requirement for the equipment or system. Two main approaches to dealing with the EMC issue have been used in industry. The first approach is to consider the EMC during the design stage of the equipment by considering all the problems from the start. These might include the EMI noise source, noise propagation, and different topologies that could be used to reduce the EMI generation. The second approach is to wait until the prototype product is built and tested and then add extra components such as an EMI filter or metal shield. For relatively large and expensive equipment like three-phase AC converters, the first approach is generally more cost-effective, resulting in shorter development time and a prototype that will more easily comply with EMC requirements. The second approach may add other interference and might require redesign of the entire system in the worst case.

The basic way to limit EMI emissions is to use filters to attenuate the noise level in the desired frequency range specified in the EMI/EMC standard. The reduction of the EMI filter size, weight, and/or cost is one of the key goals when designing power electronics converters. This chapter is on the design of EMI filters for three-phase AC converters, especially, passive EMI filters. Various EMI/EMC standards, filter topologies, and design methods will be presented. The focus will be on relationship between filter circuit parameters and filter performance as functions of component

Design of Three-phase AC Power Electronics Converters, First Edition. Fei "Fred" Wang, Zheyu Zhang, and Ruirui Chen.
© 2024 The Institute of Electrical and Electronics Engineers, Inc. Published 2024 by John Wiley & Sons, Inc.

characteristics and operating conditions. The nonideal characteristics of filter components will be considered in the design. EMI noise reduction techniques will also be discussed.

8.2 EMI Filter Design Basics

8.2.1 EMI/EMC Standards

Many standards exist to accommodate the wide variety of applications where EMI is an issue. The standards differ in their frequency range of application, the noise amplitude limits, and/or whether the type of noise measured is voltage or current. Figure 8.1 shows the noise limiting level of the commonly used industrial EMC standard CISPR11 (EN55011 and EN61800 standards have the same noise limiting level), aviation standard DO-160 [3], and military standard 461E [4].

Each standard defines its own experimental and noise measurement setup as well as their own line impedance stabilization network (LISN). The military standard 461E is used as an example. The general setup is shown in Figure 8.2. It is composed of a table covered with a ground plane where the LISN and the equipment under test (EUT) are placed. These two pieces of equipment need to be connected through a two-meter power line wire positioned on a nonconductive stand of 5 cm height to avoid any disturbances from the ground. The LISN is defined in accordance with Figure 8.3. For EMI noise measurement, the spectrum analyzer connected to the LISN needs to be set with a certain bandwidth that is dependent on the frequency. The bandwidth dependency on frequency for 461E standard is shown in Table 8.1.

In addition to EMI noise level, other restrictions may exist affecting EMI filter design. For example, some constraints apply to the maximum common mode capacitance allowed in the EMI filter due to the grounding current safety standard. In MIL-STD-461E standard, the line to structure capacitance for each power line is limited to no more than 100 nF for 60 Hz equipment, or 20 nF for 400 Hz equipment. In SAE AS 1831 standard [5], the maximum capacitance value is set to 100 nF per line to ground.

8.2.2 Definition of CM and DM Noise

The conducted EMI noise is usually decoupled and characterized as CM and DM noise. In general, CM noise is between the power line and the ground, while DM noise is between the power lines. It is common to separate the CM and DM noise from the measured noise with a noise separator. The benefit of this approach is because the CM and DM noise are generally decoupled and propagates though different circuit paths and therefore can be characterized and modeled independently. The filter design will be simplified, as the filter for each mode could be designed and trouble shoot independently.

For two-conductor systems (single-phase or DC), as shown in Figure 8.4, the orthogonal CM and DM voltages are defined as:

$$V_{CM} = \frac{V_a + V_b}{2}$$
$$V_{DM} = \frac{V_a - V_b}{2}$$

(8.1)

The CM and DM circuits are usually treated separately. Figure 8.5 shows the decoupled CM and DM circuits with EMI filter inserted to the two-conductor system.

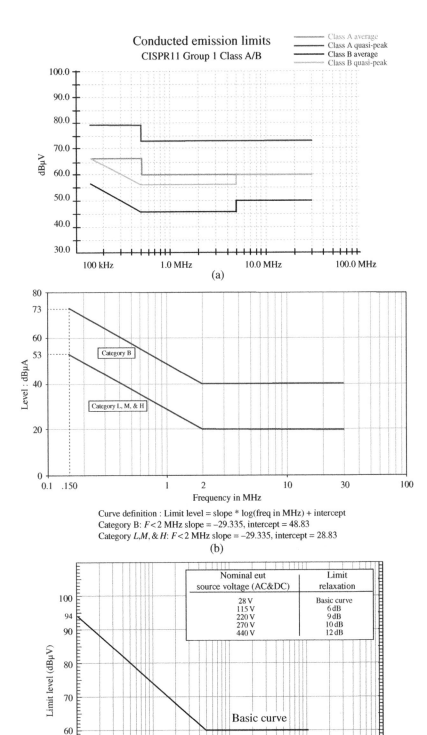

Figure 8.1 Noise level in EMI standards: (a) industrial EMC standard EN55011 (CISPR11); (b) aviation standard DO-160; (c) military standard 461E.

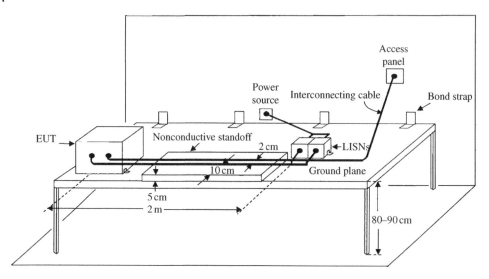

Figure 8.2 Experiment setup of the military standard 461E.

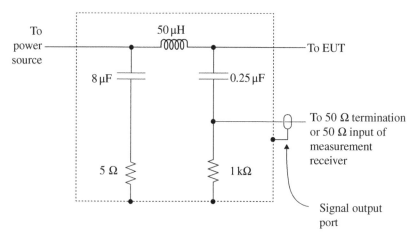

Figure 8.3 LISN circuit.

Table 8.1 Bandwidth dependency on frequency.

Frequency range	6 dB bandwidth
30 Hz to 1 kHz	10 Hz
1 kHz to 10 kHz	100 Hz
10 kHz to 150 kHz	1 kHz
150 kHz to 30 MHz	10 kHz
30 MHz to 1 GHz	100 kHz
Above 1 GHz	1 MHz

Figure 8.4 Single-phase system.

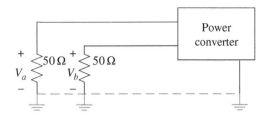

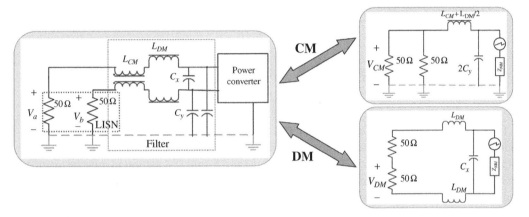

Figure 8.5 Single-phase system and its separated CM and DM circuit.

However, even though V_{CM} and V_{DM} are decoupled through (8.1), the CM and DM circuits are not always decoupled. The equivalent circuit for a single-phase system is illustrated as Figure 8.6. V_1 and V_2 are equivalent noise sources; Z_1 and Z_2 are equivalent propagation path impedances; Z_{L1} and Z_{L2} are equivalent load impedances (i.e. LISN impedance in an EMI network); Z_0 is equivalent ground network impedance.

In general, for LISN, we can assume

$$Z_{L1} = Z_{L2} = Z_L \tag{8.2}$$

Figure 8.6 Single-phase system equivalent circuit.

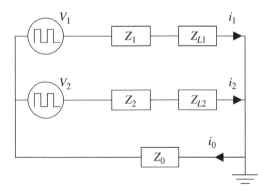

The currents through the load impedances are

$$
\begin{bmatrix} I_1 \\ I_2 \end{bmatrix} = \begin{bmatrix} \dfrac{Z_0 + Z_2 + Z_L}{\Delta} & -\dfrac{Z_0}{\Delta} \\ -\dfrac{Z_0}{\Delta} & \dfrac{Z_0 + Z_1 + Z_L}{\Delta} \end{bmatrix} \cdot \begin{bmatrix} V_1 \\ V_2 \end{bmatrix}
\tag{8.3}
$$

where for simplicity

$$
\Delta = Z_1 Z_2 + Z_1 Z_0 + Z_2 Z_0 + Z_L \cdot (Z_1 + Z_2 + Z_L + 2Z_0)
\tag{8.4}
$$

Clearly, I_1 and I_2 are not decoupled unless $Z_0 = 0$. Using the definition for CM and DM

$$
V_{DM} = \frac{V_1 - V_2}{2}, V_{CM} = \frac{V_1 + V_2}{2}, I_{DM} = \frac{I_1 - I_2}{2}, I_{CM} = \frac{I_1 + I_2}{2}
\tag{8.5}
$$

Then,

$$
\begin{bmatrix} I_{DM} \\ I_{CM} \end{bmatrix} = \begin{bmatrix} \dfrac{4Z_0 + Z_1 + Z_2 + 2Z_L}{2\Delta} & \dfrac{Z_2 - Z_1}{2\Delta} \\ \dfrac{Z_2 - Z_1}{2\Delta} & \dfrac{Z_1 + Z_2 + 2Z_L}{2\Delta} \end{bmatrix} \cdot \begin{bmatrix} V_{DM} \\ V_{CM} \end{bmatrix}
\tag{8.6}
$$

Thus, $Z_1 = Z_2$ must be true for the DM and CM to be decoupled, which implies that the two equivalent propagation paths must be symmetrical. If the symmetrical condition is not met, it is still possible to find two decoupled modes for the system, but the CM and DM definition must be modified. In that case, the physical meaning of CM and DM will no longer be as clear. The derivations also indicate that the CM propagation path characterized by Z_0 will affect DM response. The only exception is when Z_0 is much larger than the other impedances. Note that the sources do not need to be symmetrical for CM and DM decoupling.

For a three-phase system as shown in Figure 8.7, the CM voltage is defined as:

$$
V_{CM} = \frac{V_a + V_b + V_c}{3}
\tag{8.7}
$$

Different from the single-phase case, there are two independent differential modes. Figure 8.8 shows a general equivalent circuit for three-phase converter, where the CM and DM voltages are defined as:

$$
\begin{aligned}
V_{CM} &= \frac{(V_{aN} + V_{bN} + V_{cN})}{3} \\
V_{aN} &= V_{CM} + V_{a\_DM} \\
V_{bN} &= V_{CM} + V_{b\_DM} \\
V_{CN} &= V_{CM} + V_{c\_DM} \\
V_{a\_DM} &+ V_{b\_DM} + V_{c\_DM} = 0
\end{aligned}
\tag{8.8}
$$

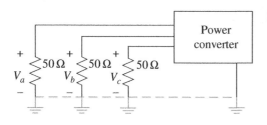

Figure 8.7 Three-phase system.

Figure 8.8 Three-phase system equivalent circuit.

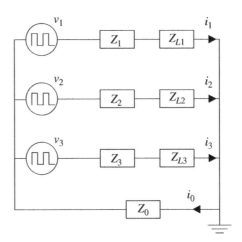

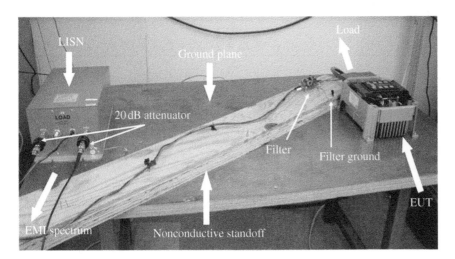

Figure 8.9 Experiment setup [6].

Similar to single-phase, using the equivalent circuit for three-phase converter in Figure 8.8, the parameters are similarly defined as the single-phase case in Figure 8.6, with load (LISN) impedances:

$$Z_{L1} = Z_{L2} = Z_{L3} = Z_L \tag{8.9}$$

The currents through the load impedances are

$$
\begin{bmatrix} I_1 \\ I_2 \\ I_3 \end{bmatrix} =
\begin{bmatrix}
\dfrac{Z_0 Z_2' + Z_0 Z_3' + Z_3' Z_2'}{\Delta_3} & -\dfrac{Z_0 Z_3'}{\Delta_3} & -\dfrac{Z_0 Z_2'}{\Delta_3} \\[3mm]
-\dfrac{Z_0 Z_3'}{\Delta_3} & \dfrac{Z_0 Z_1' + Z_0 Z_3' + Z_3' Z_1'}{\Delta_3} & -\dfrac{Z_0 Z_1'}{\Delta_3} \\[3mm]
-\dfrac{Z_0 Z_2'}{\Delta_3} & -\dfrac{Z_0 Z_1'}{\Delta_3} & \dfrac{Z_0 Z_1' + Z_0 Z_2' + Z_1' Z_2'}{\Delta_3}
\end{bmatrix}
\cdot
\begin{bmatrix} V_1 \\ V_2 \\ V_3 \end{bmatrix}
\tag{8.10}
$$

where for simplicity

$$Z_1' = Z_1 + Z_L, Z_2' = Z_2 + Z_L, \text{and } Z_3' = Z_3 + Z_L$$
$$\Delta_3 = Z_0 Z_1' Z_2' + Z_0 Z_1' Z_3' + Z_0 Z_2' Z_3' + Z_1 Z_2' Z_3' \tag{8.11}$$

Based on the $(\gamma, \delta, 0)$ transformation in (8.12), the CM and DM components can be obtained in (8.13):

$$\begin{bmatrix} V_{DM1} \\ V_{DM2} \\ V_{CM} \end{bmatrix} = \frac{1}{3} \begin{bmatrix} 2 & -1 & -1 \\ -1 & 2 & -1 \\ 1 & 1 & 1 \end{bmatrix} \cdot \begin{bmatrix} V_1 \\ V_2 \\ V_3 \end{bmatrix}, \text{ and } \begin{bmatrix} I_{DM1} \\ I_{DM2} \\ I_{CM} \end{bmatrix} = \frac{1}{3} \begin{bmatrix} 2 & -1 & -1 \\ -1 & 2 & -1 \\ 1 & 1 & 1 \end{bmatrix} \cdot \begin{bmatrix} I_1 \\ I_2 \\ I_3 \end{bmatrix} \tag{8.12}$$

$$\begin{bmatrix} I_{DM1} \\ I_{DM2} \\ I_{CM} \end{bmatrix} = \frac{1}{3}$$

$$\cdot \begin{bmatrix} \dfrac{6Z_0 Z_2' + 3Z_0 Z_3' + 2Z_3' Z_2' + Z_1' Z_2'}{\Delta_3} & \dfrac{(Z_2' - Z_3')(Z_1' + 3Z_0)}{\Delta_3} & \dfrac{2Z_2' Z_3' - Z_1' Z_3' - Z_1' Z_2'}{\Delta_3} \\[2mm] \dfrac{(Z_1' - Z_3')(Z_2' + 3Z_0)}{\Delta_3} & \dfrac{6Z_0 Z_1' + 3Z_0 Z_3' + 2Z_3' Z_1' + Z_1' Z_2'}{\Delta_3} & \dfrac{2Z_1' Z_3' - Z_2' Z_3' - Z_1' Z_2'}{\Delta_3} \\[2mm] \dfrac{Z_2'(Z_3' - Z_1')}{\Delta_3} & \dfrac{Z_1'(Z_3' - Z_2')}{\Delta_3} & \dfrac{Z_1' Z_2' + Z_2' Z_3' + Z_3' Z_1'}{\Delta_3} \end{bmatrix}$$

$$\cdot \begin{bmatrix} V_{DM1} \\ V_{DM2} \\ V_{CM} \end{bmatrix} \tag{8.13}$$

The sufficient condition for decoupling DMs and CM is

$$Z_1 = Z_2 = Z_3 \tag{8.14}$$

Note that, the symmetry condition will not necessarily make the two DMs equal.

8.2.3 EMI Noise Measurement

This section will use a DC-fed motor drive example to illustrate the EMI noise measurement [6]. The experiment setup shown in Figure 8.9 is built to be similar to the experiment setup in the military standard 461E. A table with a copper ground sheet is used, and the LISN and the motor drive (EUT) are screwed to the table. The motor is placed on the ground and connected to the EUT via a 10-m shielded cable. The ground and shield wires are connected on one end to the motor frame and on the other end to the EUT and ground plane. A nonconductive standoff is used to place the 2-meter-long power line connecting the LISN and EUT. The EMI filter is placed close to the EUT. The power source side of the LISN is connected to a DC source, providing 300 V (±150 V), while the signal output of the LISN is connected to a 20-dB attenuator and a spectrum analyzer. The LISN and the measurement connection used complies with the circuit in Figures 8.2 and 8.3.

The standard specifies to use external attenuation of 20 dB. However, in certain cases, the external attenuation has to be increased to protect the spectrum analyzer from the large voltage noise. The EMI spectrum analyzer used for the example measurement is the HP 4195A (10 Hz to 500 MHz) set to the correct bandwidth, while varying the internal attenuation to avoid overload without changing the external attenuator. Changing the internal attenuation does not change the amplitude of the noise when recorded. Only the background noise is affected, and therefore adding too much internal attenuation would make it impossible to see low level changes.

The EUT used for the experiment is the inverter portion of a commercial general-purpose 7.5-hp three-phase motor drive. The switching frequency can be set from 3 to 16 kHz with a rated output frequency of 60 Hz. The drive is used only for DC to AC by injecting the DC voltage across DC bus. Two series capacitors of 47 μF are used to create a midpoint so the total capacitance is 23.5 μF. The load is a 5-hp induction motor wired in a low-voltage configuration (208–230 V).

From Figure 8.3, it can be seen that at low frequency (e.g. line frequency), the LISN can be considered bypassed and will not interfere with the rest of the system. However, at high frequency, the high impedance of the inductor blocks the current noise and the path is provided via the 0.1 μF capacitor to the 50 Ω termination of the spectrum analyzer connected in parallel to the 1 kΩ, as shown in Figure 8.10.

Figure 8.11 is another illustration of the measurement setup given by Mil Std 461E for noise voltage extraction. In addition to the attenuator mentioned earlier to protect the measurement receiver and to prevent overload, a correction factor needs to be applied to the raw data to compensate for the 20-dB attenuator and the voltage drop across the 0.25-μF coupling capacitor. This capacitor is in series with a combination of the 1 kΩ of LISN resistor and the 50 Ω of measurement receiver. The correction factor is shown in Figure 8.12.

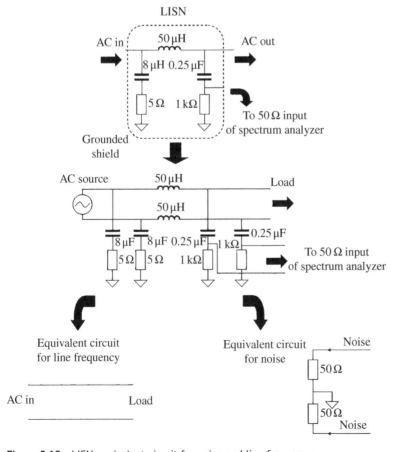

Figure 8.10 LISN equivalent circuit for noise and line frequency.

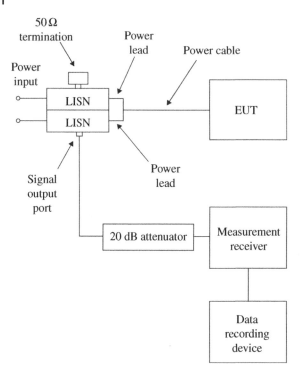

Figure 8.11 Noise voltage measurement illustration.

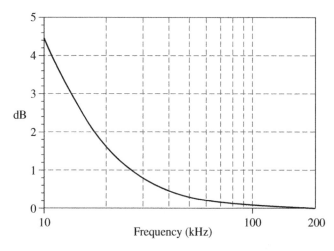

Figure 8.12 Correction factor for LISN capacitor.

To measure the CM and DM noise components directly, a noise separator can be used to convert the voltages from LISN to CM and DM measurements [7]. Alternatively, the CM and DM currents can be measured directly with a current probe, e.g. ETS-LINDGREN model 91550 probe. The decoupling of the two modes is realized simply by changing the wire direction, as shown in Figure 8.13. The current measurement setup is illustrated in Figure 8.14.

In addition to verifying or supplementing the CM and DM voltage measurements, in the case of DO-160 standard, current measurements are directly needed as shown in Figure 8.1b. In order to

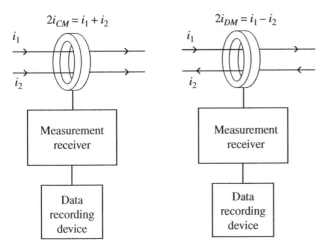

Figure 8.13 CM (left) and DM (right) noise measurement with current probe.

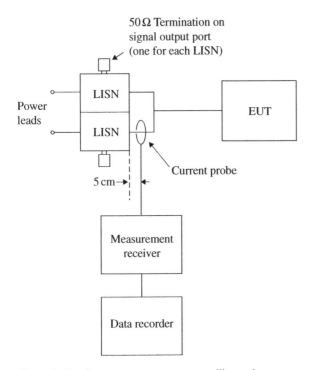

Figure 8.14 Current measurement setup illustration.

accurately determine the current I_P, the reading of the current probe output in microvolts (E_S) needs to be divided by the current transfer impedance Z_T of the probe as:

$$I_P = \frac{E_S}{Z_T} \tag{8.15}$$

The current probe transfer impedance is given in Figure 8.15.

Note that the noise separator is only popular for DC or single-phase case. For three-phase case, it is difficult to build a noise separator, and therefore each phase is usually separately measured. The

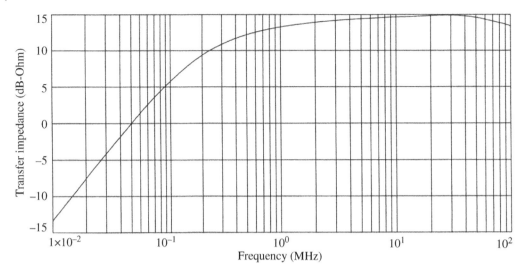

Figure 8.15 Current probe transfer impedance.

CM and DM noise can be obtained using (8.8). If all three phases are symmetrical, then all DM voltages V_{a_DM}, V_{b_DM}, and V_{c_DM} should have the same magnitudes.

8.2.4 Basic EMI Filter Design Method

The DC-fed motor drive in Section 8.2.3 is used as an example to illustrate the basic EMI filter design. The schematic of the IGBT-based motor drive is shown in Figure 8.16, where C_H represents the stray capacitances between IGBTs and heatsink, while C_G is the equivalent capacitance between the motor chassis and the ground. The stray capacitances of cables (not shown in the figure) also need to be considered, which could be quite significant and become predominant for long cables. EMI filter is placed at the DC input side.

8.2.4.1 Filter Attenuation
Figure 8.16 includes a basic yet popular EMI filter to attenuate both CM and DM noise, which is re-illustrated in Figure 8.17. The filter equivalent circuit is further separated for CM and DM,

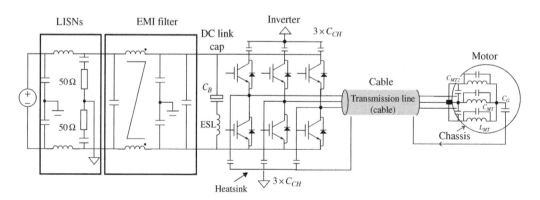

Figure 8.16 A DC-fed motor drive schematic for EMI filter design.

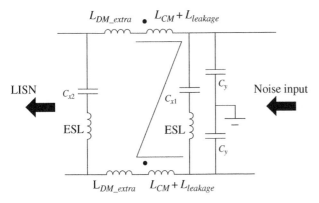

Figure 8.17 A basic EMI filter.

Figure 8.18 CM filter equivalent circuit.

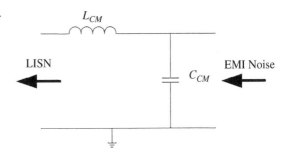

Figure 8.19 DM filter equivalent circuit.

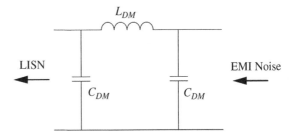

respectively, in Figures 8.18 and 8.19. The CM filter is an LC filter, while the DM filter is a Π-type CLC filter.

Note that the LISN is characterized for each line by a 50-Ω resistor and is approximated by a 25-Ω or 100-Ω resistor for CM and DM correspondingly (two resistors in parallel or in series). The equivalent CM noise source can be represented by a noise voltage source in series with an equivalent impedance Z_g. The equivalent circuit for the CM filter attenuation derivation is shown in Figure 8.20 and the corresponding equations are

$$CM_{atteuation} = \frac{v_{LISN(without\ filter)}}{v_{LISN(with\ filter)}} \approx \frac{I_{S\_CM(25\ \Omega)}}{I_{O\_CM(25\ \Omega)}}$$

$$\frac{I_{S\_CM(25\ \Omega)}}{I_{O\_CM(25\ \Omega)}} \approx \frac{Z_{L_{CM}} + Z_{C_{CM}}}{Z_{C_{CM}}} = \frac{j\omega L_{CM} + \dfrac{1}{j\omega C_{CM}}}{\dfrac{1}{j\omega C_{CM}}} = 1 - \omega^2 L_{CM} C_{CM} \qquad (8.16)$$

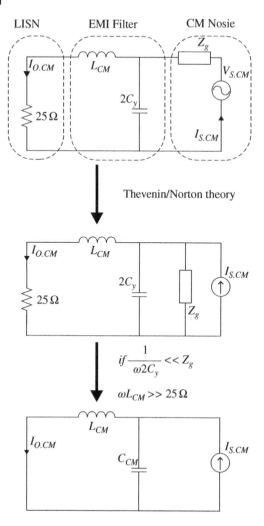

LISN EMI Filter CM Nosie

Thevenin/Norton theory

$$if \frac{1}{\omega 2C_y} \ll Z_g$$

$$\omega L_{CM} \gg 25\,\Omega$$

Figure 8.20 Equivalent circuit for the derivation of CM filter attenuation.

Therefore, the theoretical CM filter attenuation can be approximated by a second-order LC filter with a -40 dB/decade slope passing through the resonance frequency f_{CM} of L_{CM} and C_{CM} ($f_{CM} \approx \frac{1}{2\pi\sqrt{L_{CM}C_{CM}}}$ with $C_{CM} = 2C_y$).

The same analysis can be done for the DM filter. The DM noise source is more complex to define due to the complex characteristic of the DM noise impedance. For simplicity and without loss of generality for a voltage source inverter (VSI), the equivalent noise source is assumed to be a voltage source in series with low impedance Z_T. The equivalent circuit for the DM filter attenuation derivation is shown in Figure 8.21 and the DM attenuation is given by:

$$DM_{attenuation} = \frac{v_{LISN(without\ filter)}}{v_{LISN(with\ filter)}} \approx \frac{V_{S\_DM(100\,\Omega)}}{V_{O\_DM(100\,\Omega)}}$$

$$\frac{V_{S\_DM(100\,\Omega)}}{V_{O\_DM(100\,\Omega)}} = \frac{Z_{C_{DM}} + Z_{L_{DM}}}{Z_{C_{DM}}} = 1 - \omega^2 L_{DM}C_{DM} \qquad (8.17)$$

$$f_{DM} \approx \frac{1}{2\pi\sqrt{L_{DM}C_{DM}}} \text{ with } C_{DM} = C_{X1} = C_{X2} \text{ and } L_{DM} = 2L_{leakage}$$

Figure 8.21 Equivalent circuit for derivation of DM filter attenuation (no extra DM inductors, only CM choke leakage).

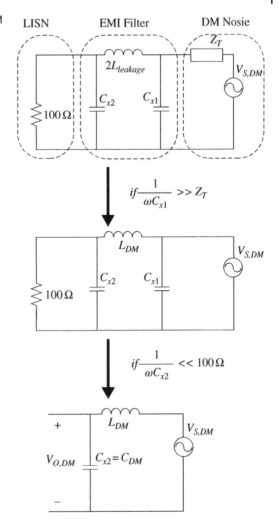

Therefore, the theoretical DM filter attenuation can also be approximated by a second-order LC filter with a -40 dB/decade slope passing through the resonance frequency f_{DM} ($f_{DM} \approx \dfrac{1}{2\pi\sqrt{L_{DM}C_{DM}}}$ with $C_{DM} = C_{X1} = C_{X2}$ and $L_{DM} = 2L_{leakage}$). Note that in this case C_{X1} is not needed due to low Z_T, although the assumption may not be always valid as will be discussed below.

8.2.4.2 Filter Corner Frequency

The next step is to find the corner frequencies for both CM and DM filters. When these frequencies are known, the components' values can be determined, and the filter can be designed.

From the bare CM and DM noise measured with the noise separator (or from the model) as in Figure 8.22, the highest peaks are plotted separately with the EMI standard limits as in Figure 8.23. The required attenuation for the CM and DM filter can be calculated as in (8.18). The 6-dB margin is needed because the standard limits are for total noise and not for CM or DM noise.

$$\left(V_{attenuation\_required\_CM}\right)_{dB} = \left(V_{original\_CM}\right)_{dB} - \left(V_{standard}\right)_{dB} + 6\,\text{dB}$$

$$\left(V_{attenuation\_required\_DM}\right)_{dB} = \left(V_{original\_DM}\right)_{dB} - \left(V_{standard}\right)_{dB} + 6\,\text{dB}$$

$$(8.18)$$

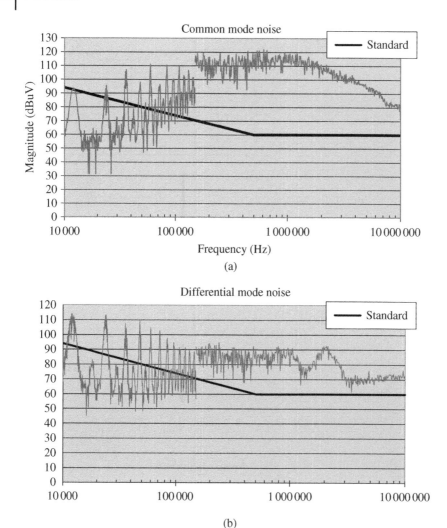

Figure 8.22 Bare noise: (a) CM and (b) DM [6].

The vertical lines in Figure 8.24 represent the difference between the noise peaks and the standard limit, while the upper pink dots are the final required attenuation when margin is applied. The corner frequency f_c is simply found by drawing a 40 dB/decade slope line that is tangent to the required CM or DM attenuation and crossing the 0-dB axis, or by using the formula (8.19), where f_{pk} is the frequency point on the 40 dB/dec slope line that determines the required attenuation Att_req in dBμV. For example in Figure 8.24, $f_{c(CM)}$ and $f_{c(DM)}$ are 6552 Hz and 2444 Hz, respectively:

$$f_{c(CM\ or\ DM)} = \frac{f_{pk}}{10^{\frac{Att\_req}{40}}} \tag{8.19}$$

The above discussion assumes a one-stage LC filter with a -40 dB/decade attenuation. The approach applies to other filter topologies with a different attenuation (slope). Also, this method is to meet the low-frequency specification, since at high frequencies, the attenuation will not be ideal with real inductors and capacitors.

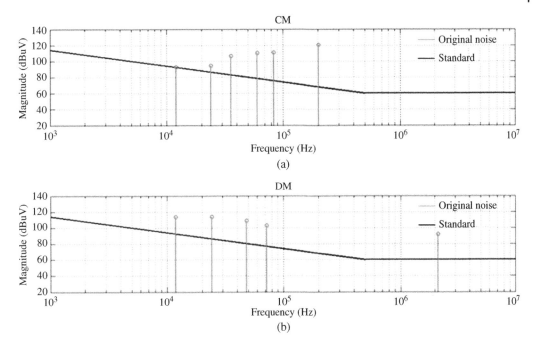

Figure 8.23 Bare noise peaks: (a) CM and (b) DM [6].

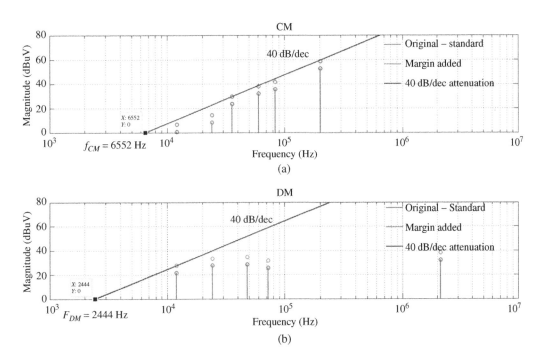

Figure 8.24 Required attenuation and corner frequency for (a) CM and (b) DM [6].

8.2.4.3 Parameter Selection and Physical Design/Selection

When both CM and DM filter corner frequencies are known, the inductance and capacitance values can be determined. For an *LC* filter, another equation is needed, e.g. a maximum capacitance constraint given by the leakage current limit. For higher-order filters, there are more freedoms in parameter selections. With these values determined, physical design and selection of inductors and capacitors can be carried out. The procedure can follow Chapters 3 and 4 with more discussions later in this chapter.

8.2.5 EMI Filter Topology

It is common to use one- or two-stage low-pass EMI filter topologies, which lead to many cell possibilities by arranging the capacitors and inductors differently. A general approach is presented when the noise source impedance is known.

8.2.5.1 Basic Approach to Choose Filter Structure

The basic rule for the topology selection is to maximize impedance mismatch between the source (i.e. converter) and the filter (input side) as well as between the load (i.e. LISN) and the filter (output side). The basic cell topologies are given in Figure 8.25 and differ depending on the source and load impedances.

For the basic filter topology in Figure 8.16 or Figure 8.17, the CM choke inductance is high while its leakage inductance is small; and the DM capacitance is high whereas the CM capacitance is limited and small. Therefore, the 25-Ω equivalent CM impedance of the LISN can be considered low compared to the CM choke, while the 100-Ω equivalent DM impedance of the LISN considered high compared against the leakage inductance. For the source impedance (i.e. input impedance of the motor drive), it is common for the CM to have high input impedance. Given the high-source and low-load CM impedances, topology 2 in Figure 8.25 is preferred. The DM source impedance is more complex since it has the DC-link capacitor and other impedances that will affect the total impedance. In the example for Figure 8.16 system, tests were performed and the Π-topology 1 was found suitable, which is for high source and load impedances. Even though the attenuation from

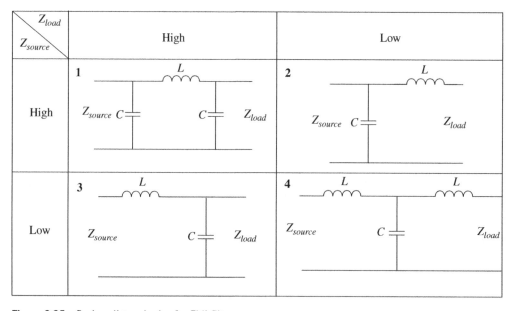

Figure 8.25 Basic cell topologies for EMI filters.

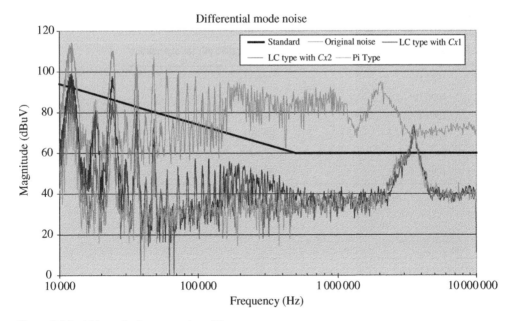

Figure 8.26 DM topologies comparison [6].

C_{X2} (see Figure 8.21) is more important, the extra capacitor C_{X1} helps to reduce the DM noise. The measurements of the EMI noise were compared between the Π-topology 1 and LC topologies 2 and 3. Both topologies "2" and "3" were tested when either C_{X1} or C_{X2} only was used. From Figure 8.26, topology 3 with C_{X2} only performed better than topology 2 with C_{X1} only, while Π-topology 1 achieves the best attenuation.

The impact of using either C_{X1} or C_{X2} only is different due to the impedance mismatch for the input or output of the filter. Here, the attenuation of C_{X1} is smaller than C_{X2} since it is in parallel with the DC-link capacitor of the drive. The EMI noise when C_{X2} only is used almost meet the standard at low frequency.

Note that in all cases in Figure 8.26, noise in high-frequency region violates the EMI standard, illustrating the nonidealities of the components and/or the impact of noise impedances, which must be considered in the EMI filter design.

8.2.5.2 Two-stage and Multistage Filter

The use of a two-stage or multistage filter to further increase the attenuation and reduce the size of the filter can be useful in some applications. The same analysis as Section 8.2.4 to calculate the corner frequency can be done by assuming a new attenuation slope depending on the two-stage filter topology. Then, all components values can be determined and compared to the single stage. An example is shown in Table 8.2.

Generally, when a two-stage filter is used, the corner frequency is pushed to higher frequencies due to the higher attenuation. In this case in Table 8.2, the corner frequencies of the one-stage filter for the CM and DM are 3780 Hz and 2090 Hz, respectively, while they are 7000 Hz and 5860 Hz respectively for the two-stage filters, assuming 80 dB/dec attenuation for both modes. The CM corner frequency is doubled while the DM is almost tripled. However, to compare the gain, the components values have to be calculated and the maximum CM capacitance is still limited to 100 nF per line [5], so 200 nF for C_{CM} for one-stage and only 50 nF per line or 100 nF for C_{CM} for two-stage. In this example, the leakage is assumed to be 0.5% of the CM choke made of nanocrystalline cores.

Table 8.2 Components comparison between one- and two-stage filters.

	Single-stage 40 dB/dec attenuation	Two-stage 80 dB/dec attenuation
$f_{corner(CM)}$	3780 Hz	7000 Hz
$f_{corner(DM)}$	2090 Hz	5860 Hz
If C_{CM}	200 nF	100 nF
L_{CM}	8.9 mH	5.2 mH
Total $L_{leakage}$	44.5 µH	26 µH
C_{DM}	130 µF	29 µF

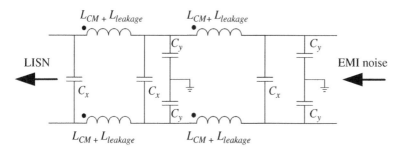

Figure 8.27 Two-stage EMI filter for the same frequency range.

The reduction of the DM capacitance is from 130 to 29 µF, while the CM inductance is reduced from 8.9 to 5.2 mH. But to compare them it is important to remember that two CM chokes and three DM capacitors are needed for a two-stage filter as shown in Figure 8.27, so the final size/weight/cost might not be reduced.

The two-stages in the Figure 8.27 filter cover the same (low to high) frequency range. A two-stage filter can also be designed such that one stage is used to attenuate the low-frequency noise, while the other one attenuates the high-frequency noise. Multistage filters can provide even more flexibility with more complexity.

8.3 EMI Filter Design Problem Formulation

8.3.1 EMI Filter Design Variables

The design for the EMI filter refers to the design and/or selection of EMI filtering strategy (active vs. passive filters, passive filter topology), filter inductors, and capacitors. For a passive filter with a given topology, the design variables include:

1) Switching frequency: as explained in Chapters 6 and 7, switching frequency directly impacts harmonics, switching loss, CM voltage and current, EMI, and control performance. It is selected to be a design variable for EMI filter because its strong non-monotonic relationship with EMI filter corner frequency, which will be explained further in Section 8.4. Using the EMI filter design to select preferred switching frequencies can speed up overall converter design optimization.

2) EMI filter inductors: magnetic core, wire, and other associated characteristics.
3) EMI capacitors: type, capacitance, voltage and current rating, and other associated characteristics.
4) Thermal management system: in this case, it specifically refers to the thermal impedance of the cooling system.

8.3.2 EMI Filter Design Constraints

The EMI filter design constraints include performance constraints related to performance specifications and physical constraints related to component physical capabilities. They are as follows:

1) EMI noise attenuation requirements: This is to meet the EMI/EMC standards.
2) Leakage current requirement: For system and personnel safety consideration, the maximum allowed leakage current to ground may be defined for applications, which sets a limit on the filter component values, e.g. Y capacitors.
3) Loss: The converter usually has efficiency requirement, which puts a limit to the total loss of all subsystems, including the EMI filters.
4) Filter components physical constraints: These include inductor temperature, core saturation flux density, core fill factor, voltage capability, and reliability, and capacitor temperature, voltage and current capability, and reliability.

8.3.3 EMI Filter Design Conditions

Depending on applications, EMI filters for three-phase converters can be on the side of AC source, AC load, DC source, and DC load, or their combinations. So, the EMI filter can interface with converter AC or DC side as its noise source, and with LISN as its load. In the case without LISN, e.g. the load-side EMI in a motor drive, the motor and cables become the filter load. The EMI filter also interfaces with the thermal management system. In summary, the EMI filter design condition can be grouped as:

1) Converter electrical operating conditions: As the EMI filter interfaces with the load/source and the converter, the converter operating conditions (i.e. voltage and current) will be also EMI filter design condition. For AC or DC source-side filter, the conditions include source voltage characteristics under nominal conditions (rated, range, unbalance, and harmonics), transient and abnormal conditions (over- and undervoltages); current characteristics (RMS, peak, harmonics, and unbalance). For AC or DC load-side filter, the conditions include DC-link voltage, load power (nominal, overload, and peak), and load current (RMS, peak, harmonics, and unbalance).
2) EMI bare noise and system high-frequency impedance: As discussed in Chapters 5–7, the EMI bare noise is an input for the EMI filter design from the rectifier/inverter. In addition, the system high-frequency impedances are needed, including: load (i.e. LISN, motor, and cables) CM and DM impedances in the EMI frequency range, converter parasitic capacitances to the heatsink or ground, passive components (e.g. DC-link capacitor) parasitics, and grounding impedances.
3) Thermal management system condition: mainly the ambient temperature. It can also impact parasitic capacitances as mentioned earlier and in Chapter 9.

The mechanical and control systems can also impact parasitic capacitances (See Chapter 11). The mechanical system can also impact temperature range of the EMI filter.

8.3.4 EMI Filter Design Objectives and Design Problem Formulation

The design objective for each subsystem of an AC converter, including the EMI filter, should follow that of the whole converter. Based on the selected design objective(s), and the design variables, constraints, and conditions, the overall EMI filter design problem can be formulated as the following optimization problem:

$$G(X, U) \text{Minimize or maximize}: \quad F(X) - \text{objective function}$$

Subject to inequality constraints:

$$G(X, U) = \begin{bmatrix} g_1(X, U) \\ g_2(X, U) \\ g_3(X, U) \\ g_4(X, U) \\ g_5(X, U) \\ g_6(X, U) \\ g_7(X, U) \\ g_8(X, U) \\ g_9(X, U) \end{bmatrix} = \begin{bmatrix} \text{EMI noise} \\ \text{Leakage current} \\ \text{Filter power loss} \\ \text{Inductor core window fill factor} \\ \text{Inductor saturation flux density} \\ \text{Inductor temperature} \\ \text{Capacitor voltage} \\ \text{Capacitor ripple current} \\ \text{Capacitor temperature} \end{bmatrix} \leq \begin{bmatrix} \text{Standards defined limit} \\ \text{Standards defined limit} \\ \text{Loss limit} \\ K_{u(max)} \\ B_{sat} \\ T_{ind(max)} \\ V_{cap} \\ I_{cap} \\ T_{cap(max)} \end{bmatrix}$$

where design variables and design conditions:

$$X = \begin{bmatrix} x_1 \\ x_2 \\ x_3 \\ x_4 \end{bmatrix} = \begin{bmatrix} \text{Switching frequency} \\ \text{EMI inductor} \\ \text{EMI capacitor} \\ \text{Thermal impedance of cooling} \end{bmatrix}$$

$$U = \begin{bmatrix} u_1 \\ u_2 \\ u_3 \end{bmatrix} = \begin{bmatrix} \text{Electrical operating conditions} \\ \text{Noise and impednaces} \\ \text{Ambient temperature} \end{bmatrix}$$

(8.20)

The optimization problem can also be illustrated as in Figure 8.28.

8.4 EMI Filter Models

To solve the optimization problem in (8.20) and Figure 8.28, models are needed for the constraints as functions of design conditions and design variables. These models include:

1) EMI noise with EMI Filters: This model will require the EMI noise source and propagation path impedance models, as well as the EMI filter models. Since the EMI standards are based on EMC test receiver, the receiver model is also needed.
2) Leakage current: This can generally be obtained with the operating conditions and the EMI filter low-frequency models.
3) Loss model.
4) Inductor and capacitor physical constraints models.

Given the complexity of the EMI noise model with filters, this section will separate the discussion into three models: (i) EMI noise source model; (ii) EMI propagation path component models; and

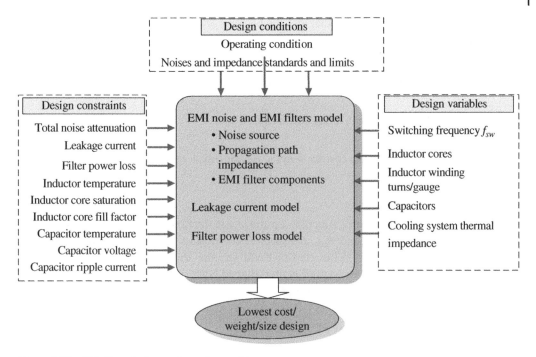

Figure 8.28 EMI filter design optimization problem.

(iii) EMI filter attenuation model. With the EMI component models, the leakage current model, the loss model, and the physical constraints models, including the thermal and reliability models, can be obtained.

8.4.1 EMI Noise Source Model

As discussed in Chapters 5–7, EMI noise source should be the design results of the inverter or the rectifier. While practical EMI filter design often relies on measurements and detailed circuit simulation to obtain the EMI noise source, analytical or algebraic models can be established similar to low-frequency voltage/current harmonics, using double Fourier analysis (DFA) or switching function-based simulation. In these models, ideal switch is assumed as the switching transient will only impact the harmonics/noises in the very high frequency range.

The DFA method provides insights into harmonic noise characteristics; however, the integral limits need to be derived for each pulse-width modulation (PWM) scheme, and also, the sideband harmonics needs to be calculated to high orders for good accuracy, which involves complicated computations. The switching function-based model is more flexible and efficient. Both of these methods will be discussed here. The single Fourier analysis method generally does not accurately model the sideband harmonics and is not suitable for EMI noise modeling.

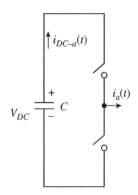

Figure 8.29 Phase leg a of a two-level VSC.

8.4.1.1 DFA-Based Model

For a voltage source converter (VSC), the EMI noise source generally includes AC-side voltage and DC-side current sources. For the Phase a phase leg in Figure 8.29, assuming a constant DC-link voltage V_{DC}, the AC voltage between the AC terminal and the negative DC bus can be expressed as a double Fourier series:

$$
\begin{aligned}
v_{aN}(t) &= V_{DC} \cdot s_a(t) \\
&= \frac{A_{00}}{2} + \sum_{n=1}^{\infty} [A_{0n} \cos(ny) + B_{0n} \sin(ny)] \\
&\quad + \sum_{m=1}^{\infty} [A_{m0} \cos(mx) + B_{m0} \sin(mx)] \\
&\quad + \sum_{m=1}^{\infty} \sum_{\substack{n=-\infty \\ (n \neq 0)}}^{\infty} [A_{mn} \cos(mx+ny) + B_{mn} \sin(mx+ny)] \\
x &= \omega_c t, \ y = \omega_0 t
\end{aligned}
\tag{8.21}
$$

where $S_a(t)$ is the Phase a switching function, ω_0 and ω_c are the fundamental and carrier angular frequencies, respectively. A_{mn} and B_{mn} are spectral coefficients defined by a double Fourier integral in (8.22), which is only a function of PWM scheme and modulation index M. For a two-level VSC, with the commonly used center-aligned SVM, the double integral limits for C_{mn} calculation is summarized in Table 8.3 [8].

$$
C_{mn} = A_{mn} + jB_{mn} = \frac{V_{DC}}{2\pi^2} \int_{y_r}^{y_f} \int_{x_r}^{x_f} e^{j(mx+ny)} dx dy
\tag{8.22}
$$

Then, the AC EMI noise voltage source of the three-phase VSC for CM and DM voltages can be determined as:

$$
V_{CM} = \frac{V_{aN} + V_{aN} + V_{aN}}{3} = \frac{1}{3} \sum_{m=1}^{\infty} \sum_{n=-\infty}^{\infty} C_{mn} \cos(mx+ny) \left[1 + 2\cos\left(n\frac{2\pi}{3}\right) \right]
\tag{8.23}
$$

$$
V_{DM\_a} = V_{aN} - V_{CM}
\tag{8.24}
$$

From Figure 8.29, the DC-side current contributed by Phase a current can be expressed as:

$$
i_{DC-a}(t) = s_a(t) \cdot i_a(t)
\tag{8.25}
$$

Transform (8.25) to frequency domain using convolution:

$$
I_{DC-a}(\omega) = S_a(\omega) \otimes I_a(\omega) = \int_{-\infty}^{+\infty} S_a(\mu) I_a(\omega - \mu) d\mu
\tag{8.26}
$$

Table 8.3 Double Fourier integral limits for two-level SVM (Phase a).

y_r	y_f	x_r	x_f
0	$\dfrac{\pi}{3}$	$-\dfrac{\pi}{2}\left[1 + \dfrac{\sqrt{3}}{2}M\cos\left(y - \dfrac{\pi}{6}\right)\right]$	$\dfrac{\pi}{2}\left[1 + \dfrac{\sqrt{3}}{2}M\cos\left(y - \dfrac{\pi}{6}\right)\right]$
$-\pi$	$-\dfrac{2\pi}{3}$		
$\dfrac{\pi}{3}$	$\dfrac{2\pi}{3}$	$-\dfrac{\pi}{2}\left[1 + \dfrac{3}{2}M\cos y\right]$	$\dfrac{\pi}{2}\left[1 + \dfrac{3}{2}M\cos y\right]$
$-\dfrac{2\pi}{3}$	$-\dfrac{\pi}{3}$		
$\dfrac{2\pi}{3}$	π	$-\dfrac{\pi}{2}\left[1 + \dfrac{\sqrt{3}}{2}M\cos\left(y + \dfrac{\pi}{6}\right)\right]$	$\dfrac{\pi}{2}\left[1 + \dfrac{\sqrt{3}}{2}M\cos\left(y + \dfrac{\pi}{6}\right)\right]$
$-\dfrac{\pi}{3}$	0		

Phase a current can be considered as sinusoidal with a magnitude I_M and a power factor $\cos \varphi$:

$$i_a(t) = I_M \cos(\omega_0 t + \varphi) \tag{8.27}$$

Then, (8.26) becomes [9]

$$I_{DC-a}(\omega) = \frac{I_M}{2} \left[e^{j\varphi} S_a(\omega - \omega_0) + e^{-j\varphi} S_a(\omega + \omega_0) \right] \tag{8.28}$$

Therefore, the spectral content of Phase a contribution to the DC current is simply the superposition of its switching function spectrum with frequency shifts of $\pm \omega_0$. The same can be found for Phases b and c. The total DC current from the VSC is

$$i_{DC}(t) = S_a \cdot i_a(t) + S_b \cdot i_b(t) + S_c \cdot i_c(t) \tag{8.29}$$

The corresponding frequency domain expression is

$$
\begin{aligned}
I_{DC}(\omega) = I_{DC-a}(\omega) + I_{DC-b}(\omega) + I_{DC-c}(\omega) &= \frac{I_M}{2} \left[e^{j\varphi} S_a(\omega - \omega_0) + e^{-j\varphi} S_a(\omega + \omega_0) \right] \\
&+ \frac{I_M}{2} \left[e^{j\left(\varphi - \frac{2\pi}{3}\right)} S_b(\omega - \omega_0) + e^{-j\left(\varphi - \frac{2\pi}{3}\right)} S_b(\omega + \omega_0) \right] \\
&+ \frac{I_M}{2} \left[e^{j\left(\varphi + \frac{2\pi}{3}\right)} S_b(\omega - \omega_0) + e^{-j\left(\varphi + \frac{2\pi}{3}\right)} S_b(\omega + \omega_0) \right]
\end{aligned} \tag{8.30}
$$

Since the EMI noise limit is in frequency domain, (8.30) can be directly used as the noise source. If needed, substituting the Fourier transform of (8.21) into (8.28) and transforming back to time domain will yield [9]:

$$
\begin{aligned}
i_{DC-a}(t) = \frac{\hat{A}_{00}}{2} &+ \sum_{n=1}^{\infty} \left[\hat{A}_{on} \cos(nx) + \hat{B}_{on} \sin(nx) \right] \\
&+ \sum_{m=1}^{\infty} \sum_{n=-\infty}^{\infty} \left[\hat{A}_{mn} \cos(my + nx) + \hat{B}_{mn} \sin(my + nx) \right]
\end{aligned} \tag{8.31}
$$

where the harmonic coefficients are defined as:

$$
\begin{aligned}
\hat{A}_{00} &= \frac{I_M}{2V_{DC}} [A_{01} \cos \varphi + B_{01} \sin \varphi] \\
\hat{A}_{mn} &= \frac{I_M}{2V_{DC}} [(A_{m,n-1} + A_{m,n+1}) \cos \varphi + (B_{m,n-1} - B_{m,n+1}) \sin \varphi] \\
\hat{B}_{mn} &= \frac{I_M}{2V_{DC}} [(B_{m,n-1} + B_{m,n+1}) \cos \varphi - (A_{m,n-1} - A_{m,n+1}) \sin \varphi]
\end{aligned} \tag{8.32}
$$

Note that with the assumption for i_a in (8.27) and balanced three-phase currents, the DC-side current source models are only for DM.

8.4.1.2 Switching Function Simulation-Based Model

The switching function-based approach to evaluate PWM AC voltages has been presented in Chapters 6 and 7 for two- and multilevel VSCs. In general, for an arbitrary VSC with any number of levels, v_{aN}, v_{bN}, and v_{cN} can be obtained in time domain based on PWM modulator output. Different modulation schemes can be automatically considered. After obtaining phase voltages, v_{aN}, v_{bN}, and v_{cN}, the CM and DM voltages of each phase can be determined using Eq. (8.8). Then, FFT can be used to obtain the CM and DM voltage harmonics models in frequency domain to facilitate EMI noise evaluation.

Similarly, the DC-side currents can be considered as controlled current sources by switching function signals and AC currents. For a two-level VSC, the DC current is (8.29). For multilevel converters, the DC current source currents can be obtained following the discussion in Chapter 7. After the time domain DC-side currents are determined, the DM current harmonics models in frequency domain can be obtained through FFT. Note again that the DC current models for VSC generally only apply to DM.

8.4.1.3 Noise Sources from Two Converters

The three-phase AC converter system often has two or more relatively independently controlled or operated converters. For example, an AC-fed motor drive consists of a front-end passive or active rectifier and an inverter. In this case, both the rectifier and inverter can generate EMI noise. Unlike the low-frequency harmonics, which can be analyzed and modeled separately, the EMI noise generated in the inverter may significantly impact the noise in the rectifier, and vice versa. Moreover, the two noise sources may need to be considered together, e.g. in the CM case when the rectifier and inverter share part of the CM noise propagation path. Therefore, the EMI noise source models need to consider two converter case.

For the motor drive with the diode rectifier case as shown in Figure 8.30, the AC input side EMI noise needs to be evaluated. Since the diodes do not switch and contribute little to EMI noise, the main EMI DM noise source is the DC-side current from the inverter represented by $i_{DC\text{-}INV}$, and the CM noise source is still the inverter CM voltage.

The diode rectifier mainly provides a propagation path for EMI noise from the inverter. The conduction of the diode will determine not only the normal operation current loop but also the flow path of high-frequency conducted noise currents. Figure 8.31 shows the diode rectifier CM and DM paths when two diodes are conducting, which exhibits higher noise as verified by measured phase current waveforms [10]. Therefore, the diode rectifier can be conservatively modeled by two-phase conducting in EMI analysis and modeling.

For a motor drive with active rectifier, with a relatively large DC-link capacitor, the AC side DM noise for the source and load will be mainly determined by the rectifier and the inverter, respectively, and therefore can be separately modeled. On the other hand, the CM noise will share the propagation path and need to be analyzed together. Figure 8.32 shows the equivalent CM circuit, where V_{CM-REC} and V_{CM-INV} are CM noise voltage sources generated by the rectifier and the inverter switching, respectively; Z_{CM-REC} and Z_{CM-INV} are the respective equivalent CM

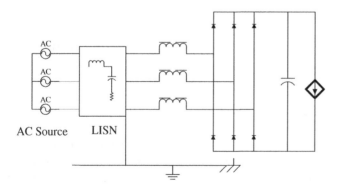

Figure 8.30 A motor drive with the diode rectifier.

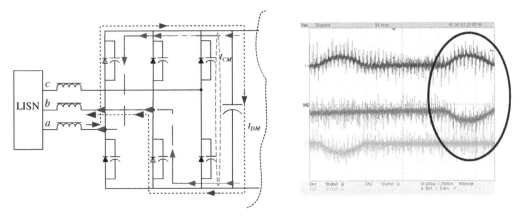

Figure 8.31 Diode rectifier DM and CM noise conduction paths with two diodes conducting and corresponding measured phase current waveforms.

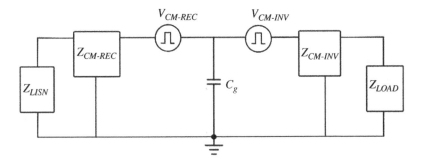

Figure 8.32 Equivalent CM circuit for a motor drive with active rectifier.

impedances; C_g represents DC-side capacitance to ground. In many cases, C_g is dominated by relatively small parasitic capacitances, and therefore the CM noise on LISN and load sides cannot be decoupled.

8.4.1.4 Noise Source Impedance

With the EMI filter topology selection based on impedance mismatch, the noise source impedance is obviously very important. Unfortunately, the converter equivalent impedance is difficult to model accurately, and therefore we will have to rely on estimation or measurement. In general, for a VSC, a high impedance is assumed for the CM noise source (low parasitic capacitance to ground and therefore high impedance at low frequencies), while a low impedance is assumed for the DM noise source (impact of high DC-link capacitance). However, as shown in Figure 8.26, these assumptions may not be valid.

Noise source impedance can be measured using the insertion loss method [11]. Figure 8.33a shows the measured CM noise impedance of a 7.5-kW diode front-end motor drive with the AC-side EMI filter. Figure 8.33b shows the corresponding CM noise measurements. It can be seen that the CM noise source impedance is very low in the MHz range, rather than high as assumed for the EMI filter design, which directly contributed to the high noise peak around 3.5 MHz. This observation reinforces the need to consider the noise source impedance in the EMI filter design for the whole frequency range, even though it is difficult.

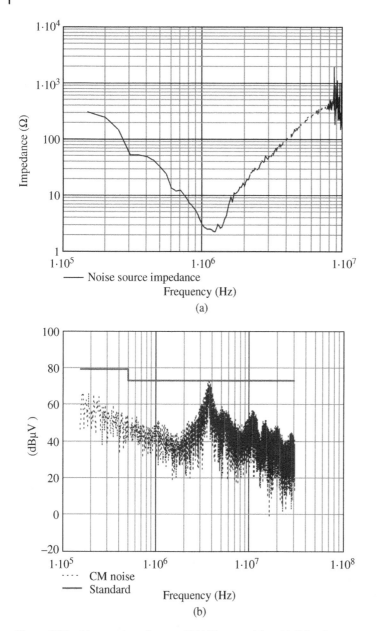

Figure 8.33 Measured results on a 7.5-kW motor drive. (a) CM noise source impedance and (b) CM noise vs. standard limit.

8.4.2 EMI Propagation Path Impedance Model

As mentioned earlier, the propagation path impedance includes the load (e.g. motor and connecting cable), EMI filters, converter equivalent impedance (i.e. noise source impedance), LISN, and DC link. The noise source impedance has been discussed earlier. LISN impedance characteristics are defined by EMI standards and can also be directly measured. Thus, only the EMI filter components and motor and cable models are discussed here. The EMI test receiver model is also introduced.

8.4.2.1 CM Inductor

8.4.2.1.1 CM Inductor Equivalent Circuit A three-winding CM inductor (choke) can usually be modeled with the equivalent (CM) circuit shown in Figure 8.34 in a wide frequency range. In this model, L_{cm} is the self-inductance of the CM inductor and plays the most important role in expected CM attenuation; R is the combination of winding AC resistance and the core loss equivalent resistance and mainly acts as the damping to the CM noise. Both L_{cm} and R are nonlinear and frequency-dependent; The inter-winding

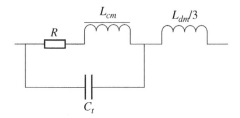

Figure 8.34 Equivalent CM circuit of a three-winding CM inductor.

capacitance C_t will cause resonance together with L_{cm} and affect the performance of the CM inductor; leakage inductance of the CM choke, L_{dm}, is used for DM noise attenuations. Since it is related to the leakage field of the inductor, L_{dm} can be treated as linear for the frequency range of interest.

The self-inductance is expressed as:

$$L_{cm}(f) = \mu_C(f)\mu_0 \frac{A_e}{l_e}(3N_t)^2 \tag{8.33}$$

where A_e is the effective cross-sectional area of the core, l_e is the effective loop length of the core (circumference of the toroidal choke), N_t is the turns number for one winding, and $\mu_C(f)$ is the frequency-dependent relative permeability of the core material.

Resistance consists of the equivalent core loss resistance and winding AC resistance. The detailed core loss and winding loss models are discussed in detail in Chapter 4. Note that the flux in the CM choke includes both the flux generated by CM currents and the leakage flux generated by DM currents. The leakage flux often dominates, so the core loss should be mainly determined by the leakage flux.

The winding loss equivalent resistance can also be determined as [12]:

$$R_W(f) = R_{WDC} \cdot \frac{A(f)}{2\pi f}\left\{\frac{e^{2A(f)} - e^{-2A(f)} + 2\sin[2A(f)]}{e^{2A(f)} + e^{-2A(f)} - 2\cos[2A(f)]} + \frac{2\left(n_l^2 - 1\right)}{3} \cdot \frac{e^{A(f)} - e^{-A(f)} + 2\sin[A(f)]}{e^{2A(f)} + e^{-A(f)} - 2\cos[A(f)]}\right\} \tag{8.34}$$

where R_{WDC} is the DC winding resistance at a given temperature, n_l is number of winding layers, and A for a round wire is

$$A(f) = \left(\frac{\pi}{4}\right)^{3/4} \cdot \frac{d^{3/2}}{\delta t^{1/2}} \tag{8.35}$$

where d is the conductor diameter, t is the distance between the centers of the two adjacent conductors, and δ is the skin depth of the wire at frequency f.

The inter-winding capacitance and the leakage inductance models have been discussed in Chapter 4.

8.4.2.1.2 CM Inductor Core Saturation and Temperature Rise It has been explained in Chapter 4 that the CM inductor core saturation can be due to leakage flux generated by DM currents. The flux density constraint is given by (4.63). In addition, the CM voltage can also cause CM inductor saturation through flux or volt-seconds. The constraint can be specified as [13]:

$$A_e \cdot l_e \geq \left(\frac{|VS_{Lcm}|}{B_{max}}\right)^2 \cdot \frac{\mu_0\mu_C}{L_{cm}} \tag{8.36}$$

where most terms have been defined in (8.33) and VS_{Lcm} is the peak CM volt-seconds on the CM inductor. The CM path can be modeled to include the system CM voltage V_{cm} in series with L_{cm}, an equivalent CM capacitance C_{cm} (primarily the filter capacitor C_y), and an equivalent resistance R_{eq}. For continuous SVM,

$$|VS_{Lcm}| = \frac{\pi V_{cm-M} \cdot L_{cm}}{\sqrt{R_{eq}^2 + \omega^2 L_{cm}^2 \left[\left(1 - \frac{\omega_{res}^2}{\omega^2}\right)\right]^2}} \tag{8.37}$$

where V_{cm-M} is the peak system CM voltage (e.g. $V_{DC}/2$ for a two-level VSC), ω is the AC fundamental angular frequency, and $\omega_{res} = \frac{1}{\sqrt{L_{cm} \cdot C_{cm}}}$. Reference [14] further analyzed the PWM impact CM inductor volt-seconds. In general, low modulation index and DPWM correspond to worse-case. The filter should also be designed such that the resonance frequency ω_{res} is much higher than the operating frequency ω.

The temperature rise can be determined by using the models in Chapter 4.

8.4.2.1.3 CM Inductor Impedance in Filter Design

In the basic EMI filter design method discussed in Section 8.2.4, the EMI filter corner frequencies are determined based on the first several EMI noise peaks in low-frequency region. Then, the EMI filter and its corresponding component parameters, including CM inductances, are selected, assuming inductors and capacitors are ideal inductances and capacitances. With the ideal component assumption, there would be no need to check whether a resulted filter design would meet the EMI noise attenuation requirement in high-frequency regions.

A real CM inductor is not an ideal inductance. From the equivalent circuit in Figure 8.34, its equivalent CM impedance is

$$Z_{CM}(f) = \frac{[R(f) + j2\pi f \cdot L_{CM}(f)] \cdot \frac{1}{j2\pi f \cdot C_t}}{[R(f) + j2\pi f \cdot L_{CM}(f)] + \frac{1}{j2\pi f \cdot C_t}} + j2\pi f \cdot \frac{L_{DM}}{3} \tag{8.38}$$

Even with constant R and L_{CM}, the equivalent Z_{CM} would not be a linear function of frequency due to the parasitic capacitance C_t. As discussed in Chapter 4, for commonly used ferrite and nanocrystalline materials for CM inductors, both the magnetic permeability and loss are frequency-dependent, which can be characterized through the real part μ' and imaginary part μ'' of the frequency-dependent complex permeability $\bar{\mu}$. Thus, R and L_{CM} will be frequency-dependent. Therefore, the CM inductance determined based on EMI noise peaks at low frequencies, and the corresponding impedance and attenuation may not be adequate at high frequencies as μ' decreases with frequency. Clearly, the EMI filter attenuation performance in the entire EMI frequency range should be examined. A DC-fed motor drive with a one-stage AC-side CM filter based on nanocrystalline core is used as an example to illustrate this design consideration.

Figure 8.35 shows the equivalent CM circuit without and with the CM filter, where V_{cm} is the CM voltage; $Z_{cm(DC)}$ is the DC-side CM impedance; C_{sc} is the AC-side parasitic capacitance; $L_{dm(AC)}$ is the AC-side DM inductance or the leakage inductance of the CM inductor; and Z_m represents motor and cable impedance. Without CM filter, the AC-side CM bare noise current magnitude is

$$I_{cm-bare}(f) = \left| \frac{V_{cm}\left\{[1/(j2\pi f C_{sc})] \middle\| \left(\frac{j2\pi f L_{dm(AC)}}{3} + \frac{Z_m}{3}\right)\right\}\left[1/(j2\pi f C_{sc}) \middle\| \left(\frac{j2\pi f L_{dm(AC)}}{3} + \frac{Z_m}{3}\right)\right]}{Z_{cm(DC)} + \left\{[1/(j2\pi f C_{sc})] \middle\| \left(\frac{j2\pi f L_{dm(AC)}}{3} + \frac{Z_m}{3}\right)\right\}\left[1/(j2\pi f C_{sc}) \middle\| \left(\frac{j2\pi f L_{dm(AC)}}{3} + \frac{Z_m}{3}\right)\right]} \right| \tag{8.39}$$

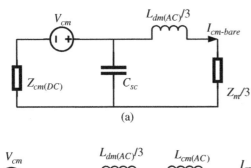

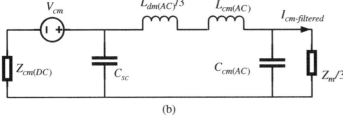

Figure 8.35 Equivalent CM circuit (a) without CM filter and (b) with CM filter.

With the CM filter inserted, the AC-side CM noise current magnitude becomes $I_{cm-filtered}(f)$, which can still be solved analytically as a function of V_{cm} and all the impedances including the filter impedances.

The insertion loss of the CM filter, IL_{cm}, should be no smaller than the required attenuation Att_{req} in the entire EMI frequency range:

$$IL_{cm}(f) = 20\log\left|\frac{I_{cm-bare}(f)}{I_{cm-filtered}(f)}\right| \geq Att_{req}(f), \quad f \in (\text{all EMI frequency range}) \tag{8.40}$$

In order to obtain $IL_{cm}(f)$ at all key frequency points, the corresponding impedance of the CM inductor at these points need to be calculated to check the requirement (8.40). For toroidal nano-crystalline cores, the datasheet usually provides A_L (inductance per square of turns numbers) at 100 kHz. Then,

$$L_{cm(AC)}(f) = A_{L\_100\,kHz}\frac{\mu'(f)}{\mu'(100\,kHz)}(3N_t)^2 \tag{8.41}$$

If the designed CM inductance cannot provide enough impedance in the high frequency range due to decreased permeability, the number of turns for the inductor needs to be increased to increase the impedance. If the turns number will lead to flux density exceeding B_{max}, then a larger core must be selected. In the design, the process needs to cover the whole EMI frequency range. Practically, the validation usually needs to cover up to 5 MHz. Beyond 5 MHz, the EMI noise is greatly impacted by parasitics and coupling effects and will be not accurate.

In addition to frequency dependency, which can be observed and characterized with small signal near zero magnetic field, the permeability associated with the CM inductor is also a function of large signal current level, similar to the DC current biased permeability discussed in Chapter 4 and the characteristics for some DM inductors to be discussed next. Figure 8.36 shows the CM inductor impedance measurement setup and results for a motor drive front-end diode rectifier under different currents. Clearly, the impedance value in low-frequency region, which is dominated by L_{cm}, decreases considerably with increasing current, indicating a permeability drop due to current.

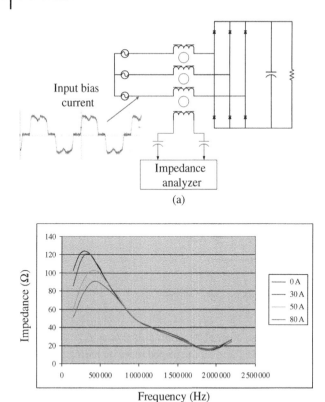

Figure 8.36 CM impedance changes with current level and frequency. (a) Measurement setup and (b) measured impedance vs. frequency at different currents.

The physical design for the EMI CM inductor can follow the same procedure in Chapter 4. Compared to the low-frequency gapped inductors, the differences include core type, geometry, fill factor definition, and flux density calculation (considering both DM leakage and CM volt-seconds).

8.4.2.2 DM Inductor

The DM inductor in an EMI filter can be the leakage inductance of a CM choke, an air core inductor if the required inductance is small, a gapped core inductor with high permeability core material, or a distributed air gap core-based inductor. The needed models for calculation of inductance, self-capacitance, core saturation, core loss and winding loss, and temperature rise have all been covered in Chapter 4. The physical design of DM inductor is also covered there.

A special case is that powder or amorphous core with distributed air gap is used for the DM inductor. Due to the current-dependent permeability of these materials as discussed in Chapter 4, DM inductor inductance drop occurs at relatively high operating current. It needs to be checked that the DM inductor could provide enough impedance to achieve required DM noise attenuation in the worst case. A common practice is to guarantee that the DM inductance cannot drop more than 40% at peak current compared to the value at zero current. Also, the unequal DM impedances of three phases due to the unequal instantaneous three-phase currents can convert DM noise to CM noise, which may impact the CM filter design as will be discussed later in Section 8.5.4. A current bias-dependent DM inductor model and an EMI filter design flowchart considering this current dependent permeability are presented in [15].

8.4.2.3 CM and DM Capacitors

EMI filter capacitors are generally modeled as an ideal capacitor in series ESL and ESR. Capacitor equivalent circuit, loss, thermal, and lifetime models can be found in Chapter 3.

8.4.2.4 Motor and Cable

8.4.2.4.1 Motor Conventional motor equivalent circuit models with only resistances, inductances, and back-EMF are low-frequency models and not valid in EMI frequency range, where parasitic capacitances within windings, among windings, and between windings and frame need to be considered. In Chapter 6 design example, a motor high-frequency model is shown in Figure 6.39 and the corresponding parameters are given for a 45-kVA motor. In general, the high-frequency motor model is not available from datasheet. Estimation or scaling from a known motor can be performed based on the rated power, voltage, frequency, and speed. Some discussion on motor parasitic parameters are provided in Chapter 12.

If desired, motor models can be obtained through relatively simple measurements. Reference [16] describes an induction motor model based on measurements. The basic idea is to model the three-phase motor by three magnetically coupled per-phase circuit model as shown in Figure 8.37. Two measurements are performed: (i) CM impedance measurement as illustrated on the left side of Figure 8.38, where all three motor terminals are shorted so three per-phase circuits are connected in parallel; (ii) DM impedance measurement as illustrated on the right side of Figure 8.38, where two per-phase circuits are connected in parallel and both are connected in

Figure 8.37 Motor model based on three coupled per-phase models [16].

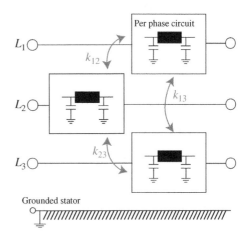

Figure 8.38 Measurement setup illustrations [16].

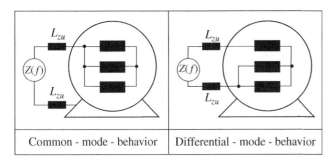

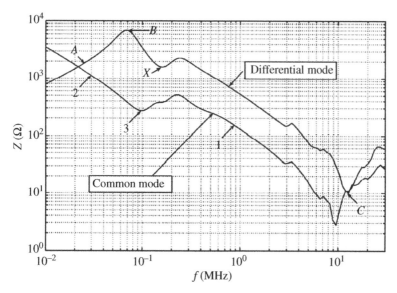

Figure 8.39 Impedance measurement results of 15-kW induction machine [16].

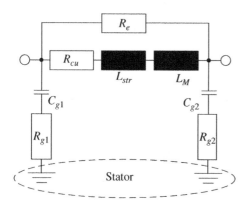

Figure 8.40 Per-phase high-frequency equivalent circuit for induction motor.

series with the third. Figure 8.39 shows the CM and DM measurement results of a 15-kW induction motor.

Based on the characteristics of the measured CM and DM impedance results, the per-phase circuit model can be synthesized as in Figure 8.40, where the inductance includes a stray inductance L_{str} and a mutual inductance M which fully couples with the other two phases (i.e. coupling coefficients $k_{12} = k_{12} = k_{12} = 1$ in Figure 8.37); R_e and R_{CU} account for iron and copper loss, respectively; and C_{g1} and R_{g1}, and C_{g2} and R_{g2} are parasitic parameters. All these parameters can be determined with the measured CM and DM impedance curves in Figure 8.39 [16].

Note that the Figure 8.40 model is similar to the model in Figure 6.39, except the latter does not have mutual coupling between phase inductances.

8.4.2.4.2 Cable
For EMI filter design, cable models need to consider line-to-line and line-to-ground capacitances as shown in Figure 6.39. For long cables, segmented or distributed models may be needed. The cable parameters are usually available from the datasheet or can be calculated with simple formulas. Chapter 12 provides some discussion on cable impact and modeling.

8.4.2.5 EMI Test Receiver
FFT is usually applied to the time-domain voltage waveforms on the LISN to obtain the EMI spectrum for EMI prediction and analysis. However, because EMI standards have specific requirements on the resolution bandwidth (RBW), the envelope detector, the peak detector, the quasi-peak

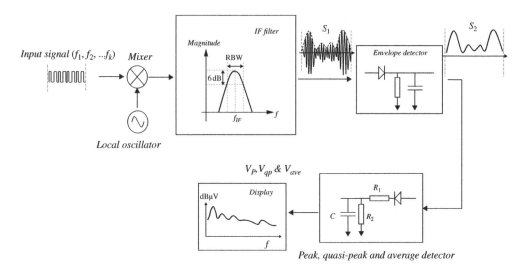

Figure 8.41 Operating principle of the peak, quasi-peak, and average noise measurement in a spectrum analyzer (150 kHz to 30 MHz) [17].

detector, and the average detector for spectrum analyzer, the FFT results, which do not consider the influence of these requirements, may not agree with actual measurement results.

A typical quasi-peak EMI measurement in a spectrum analyzer is illustrated in Figure 8.41 based on CISPR 16 [17]. The input EMI noise spectrum is mixed with signal generated by a local oscillator. The output of the mixer is fed to the intermediate-frequency (IF) filter. The IF filter has 9-kHz 6-dB RBW. The IF filter has zero attenuation to the signal at f_{IF}. For the signal not at f_{IF}, the attenuation can be approximately characterized with a Gaussian function. The IF signal is fed to an envelope detector. The envelope detector can follow the fastest possible changes of the envelope of the IF signal but not the instantaneous value of the IF waveform itself. The envelope signal is further fed to a quasi-peak detector. The quasi-peak detector has a charging time constant τ_C and discharging time constant τ_D defined in EMI standards. For average or peak noise measurement, the quasi-peak detector is replaced with an average or a peak detector, while other components stay the same. The difference between quasi-peak detector and the other two detectors is the charging and discharging time constants. For example, the average detector is basically a low-pass filter, which retains the DC component in the output of the envelope detector. On the other hand, the peak detector has a very small charging time constant.

The IF filter model, envelope detector model, and the peak/quasi-peak/average detector model are developed in [18, 19] based on the operating principle of EMC spectrum analyzers, which is discussed next. The conventional FFT results can be different from the measured peak/quasi-peak/average values using spectrum analyzer and the difference could be significant in some cases. Generally, the FFT results are no more than the average EMI values.

8.4.2.5.1 IF Filter Modeling The IF filter's transfer function can be expressed by a Gaussian function:

$$G_{IF}(f,c) = \exp\left(-(f-f_{IF})^2/c^2\right) \tag{8.42}$$

where f is the noise signal frequency and f_{IF} is the intermediate frequency of the IF filter and c is given by:

$$c = \frac{RBW}{2\sqrt{\ln 2}} \tag{8.43}$$

8.4.2.5.2 Envelope Detector Modeling The output of the envelope detector at a given frequency can be modeled with the frequency component with the maximum magnitude within a set of frequency components at the output of the IF filter. However, these frequency components have phases. So, in time domain, the magnitude, which determines the envelope, of the vector sum of all of these frequency components is not equal to the maximum magnitude of these frequency components. To accurately model the envelope detector, the magnitude of the vector sum of these frequency components must be considered.

The signal including different frequency components fed to the envelope detector from the IF filter is expressed as:

$$s(t) = \sum_{i=1}^{p} A_i \cos(\omega_i t + \theta_i) \tag{8.44}$$

where, A_i, ω_i, and θ_i are the magnitude, frequency, and phase of the i-th frequency component, respectively, and p is the number of frequency components.

Taking frequency ω_1 and phase θ_1 as the frequency and phase references, (8.44) becomes

$$s(t) = \sum_{i=1}^{p} A_i \cos(\omega_1 t + \alpha_i) \tag{8.45}$$

where $\alpha_i = (\omega_i - \omega_1)t + \theta_i$. (8.45) can be further simplified to

$$s(t) = c(t) \cos(\omega_1 t + \alpha_1) \tag{8.46}$$

$c(t)$ in (8.46) represents the envelope and it is given by:

$$c(t) = \sqrt{\left(\sum_{i=1}^{p} A_i \cos(\alpha_i)\right)^2 + \left(\sum_{i=1}^{p} A_i \sin(\alpha_i)\right)^2} \tag{8.47}$$

In (8.47), the AC and DC components of the envelope can be represented as:

$$c(t) = \sqrt{\sum_{i=1}^{p} A_i^2 + \left[\sum_{k=1}^{p}\sum_{j=1}^{p} A_k A_j \cos\left(\omega_k t - \omega_j t + \theta_k - \theta_j\right)\right]} \tag{8.48}$$

In the equations aforementioned, for switching mode power electronics applications, if f_{sw} is the switching frequency of a constant frequency and constant-duty cycle waveform, the period T_{sw} of the envelope is $1/f_{sw}$. For any harmonic components ω_k and ω_j, condition $\frac{\omega_k - \omega_j}{2\pi} = nf_{sw}$ is always met, where $n = |k - j|$, and $j \neq k$. For a SPWM waveform, $\frac{\omega_k - \omega_j}{2\pi} = nf_{mod}$, where $n = |k - j|, j \neq k$, and f_{mod} is the modulation frequency. Equation (8.48) indicates that the period T_{sw} of the envelope is $1/f_{sw}$ or $1/f_{mod}$. $c(t)$ is equal to the square root of the sum of a DC component, which is equal to the sum of A_i^2, and AC components that include all of the cross-product terms in (8.47). The magnitude of the AC components in (8.48) is a function of both the magnitudes and phases of the frequency components in (8.44).

8.4.2.5.3 Peak, Quasi-Peak, and Average Detection Based on the envelope waveform, the peak, quasi-peak, and average values can be determined by the detector circuit with different charging and discharging time constants. In the process of obtaining the peak and average values of the envelope in one period, the peak value of the envelope is the maximum value of these N sampled data V_i:

$$V_{peak} = \max_{i=1,\dots,N}(V_i) \tag{8.49}$$

The average value is

$$V_{avg} = \frac{\sum_{i=1}^{N} V_i}{N} \tag{8.50}$$

Table 8.4 Charging and discharging time constants of quasi-peak detectors.

	Band A: 9–150 kHz	Band B: 150 kHz to 30 MHz	Band C/D: 30 MHz to 1 GHz
RBW	200 Hz	9 kHz	120 kHz
T_D	500 ms	160 ms	550 ms
T_C	45 ms	1 ms	1 ms

For quasi-peak detection, CISPR standards define the charging and discharging time constants in Table 8.4 for the quasi-peak detector circuit in Figure 8.42a. R_c, R_d, C, $V_{envelope}$, and $V_{quasi-peak}$ are charging resistance, discharging resistance, charging capacitor, input envelope voltage, and output quasi-peak voltage of the quasi-peak detector, respectively. Ignoring small ripples, the steady-state quasi-peak value $V_{quasi-peak}$ can be derived based on charge balance on the capacitor C. The equation used to calculate the quasi-peak value based on the sampled data in Figure 8.42b is derived as (8.51). Δt_i is the i-th time interval when the envelope waveform $V_{envelope}$ is larger than $V_{quasi-peak}$. Q is the number of the sampled data that are larger than $V_{quasi-peak}$.

$$\frac{\sum_{i=1}^{Q}\left(V_i - V_{quasi-peak}\right)}{R_c} = \frac{V_{quasi-peak} \times N}{R_d} \tag{8.51}$$

Based on (8.49), (8.50), and (8.51), $V_{peak} \geq V_{quasi-peak} \geq V_{avg}$. The equality holds when the output voltage V_i of the envelope detector has DC component only.

Figure 8.42 (a) Quasi-peak detector circuit and (b) Quasi-peak value based on capacitor charge balance.

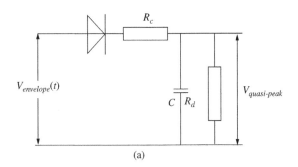

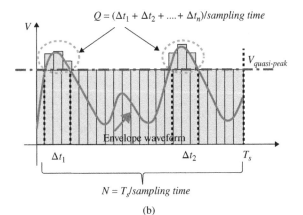

8.4.3 EMI Filter Corner Frequency vs. Switching Frequency

As explained earlier, corner frequency is an important intermediate value in the EMI filter design. This subsection establishes the relationship between the EMI filter corner frequency and switching frequency.

After the EMI noise source and EMI propagation path impedances are modeled, the EMI noise can be calculated and the required attenuation can be obtained. The CM filter of a DC-fed motor drive for aircraft application is used as an example (i.e. use DO-160 Standard).

The required CM noise attenuation can be determined as:

$$I_{RA(f_{index})} = 20 \log \left[\frac{V_{CM(f_{index})} \times 10^6}{Z_{CM(f_{index})}} \right] - I_{std(f_{index})} + \text{Margin (dB}\mu\text{A)} \tag{8.52}$$

where $I_{RA(f_{index})}$, $I_{std(f_{index})}$, $V_{CM(f_{index})}$, and $Z_{CM(f_{index})}$ are the noise current required attenuation, EMI standards defined noise current limit level, CM noise voltage, and CM noise propagation path impedance at frequency f_{index}, respectively. The calculation of the CM noise voltage and CM noise propagation path impedance have been discussed in Sections 8.4.1 and 8.4.2.

After the required noise attenuation is calculated and the EMI filter type is selected, the noise peak frequency f_{pk}, which determines the EMI filter corner frequency, can be identified with the flowchart in Figure 8.43. f_{pk} should satisfy

$$I_{RA(f_{index})} \geq k \cdot 20 \log \frac{f_{index}}{f_{pk}} + I_{RA}\left(f_{pk} \right) \text{ (dB}\mu\text{A)} \tag{8.53}$$

where k indicates the filter type ($k = 2$ and $k = 4$ represent single- and two-stage LC filters, respectively).

The CM filter corner frequency can be calculated with

$$\log f_c = \log f_{pk} + \frac{I_{std(f_{pk})} - 20 \log \left[\frac{V_{CM(f_{pk})} 10^6}{Z_{CM(f_{pk})}} \right]}{20k} \tag{8.54}$$

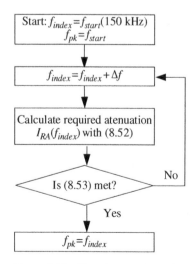

Start: $f_{index} = f_{start}$(150 kHz)
$f_{pk} = f_{start}$

$f_{index} = f_{index} + \Delta f$

Calculate required atenuation $I_{RA}(f_{index})$ with (8.52)

Is (8.53) met? — No

Yes

$f_{pk} = f_{index}$

Figure 8.43 Flowchart to determine the noise peak frequency f_{pk}.

Since f_{pk} is related to noise voltage V_{CM} and thus related to switching frequency f_{sw}, the CM filter corner frequency f_c as a function of f_{sw} can be obtained. Figure 8.44 shows a case study calculation result for a three-level ANPC inverter with commonly used nearest three space vector (NTSV) modulation when one-stage LC filter and two-stage LC filter are assumed.

From Figure 8.44, several findings are observed. When $20 \text{ kHz} < f_{sw} < 150 \text{ kHz}$, f_c and f_{sw} have a nonlinear (and even non-monotonic) relationship, some preferred f_{sw} exists such as 140 kHz. When $150 \text{ kHz} < f_{sw} < 2 \text{ MHz}$, if a single-stage LC filter is applied, f_c has slight change as f_{sw} increases. If two- or multiple-stage LC filters are applied, f_c increases fast as f_{sw} increases. The findings provide a guideline for switching frequency and EMI filter-type selection.

The nonlinear relationship between f_c and f_{sw} in the 20–150 kHz range can be understood because the noise peak frequency and CM voltage amplitude are not linearly changed when switching frequency increases. The different

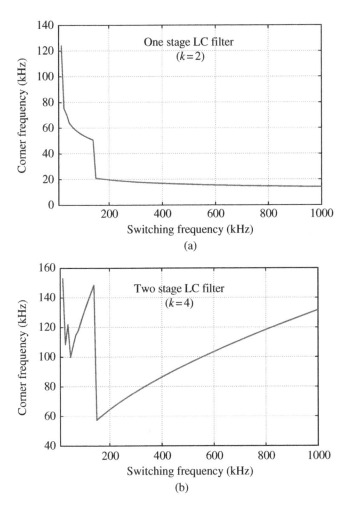

Figure 8.44 Calculated CM filter corner frequency vs. switching frequency: (a) one-stage *LC* filter and (b) two-stage *LC* filter.

trends of filter corner frequency between a one-stage and a multiple-stage filter when switching frequency is in the 150 kHz to 2 MHz range can also be explained. As switching frequency increases, the *LC* filter impedance increases, and the DO-160 EMI standard limit decreases at a slow rate of around 30 dB/decade. For single-stage *LC* filter, the benefit gained from the increase of *LC* filter impedance is almost cancelled out by the decrease of the EMI standards limit. As a result, f_c has slight change as f_{sw} increases. For multiple-stage *LC* filters, the filter impedance increases much faster as switching frequency increases and outweighs the decrease of EMI standard limit. Thus, f_c increases fast as f_{sw} increases.

8.5 EMI Filter Design Optimization and Some Practical Considerations

Figure 8.28 shows the design optimization problem for the EMI filter. After establishing the models among the design constraints, conditions, and variables, an optimizer can be built for the EMI filter to carry out the design process. The details of design optimization will be described in Chapter 13.

Here, we will discuss some practical issues in EMI filter design and implementation that have not been fully considered in previous sections.

8.5.1 Grounding Effect

Grounding is an important aspect of the EMI filter design and implementation. The CM noise propagates through power circuit and ground. Changing filter connections to ground will alter the filter CM impedance.

8.5.1.1 Grounding Impedance

For the simple filter shown in Figure 8.17, the purpose of the C_y capacitors is to bypass the high-frequency CM noise. Their high-frequency performance is a function of the grounding impedance. For a typical EMI filter PCB layout, the C_y capacitors are directly soldered on the copper plane of the filter. The filter ground is then connected to the converter ground. In Figures 8.45 and 8.46, measurements were made with the same filter only changing its connection to the ground plane [6]. Specifically, the long wire trace corresponds to the case where a 50-cm-long wire is used to connect to ground; the copper foil trace is to ground the filter using a 15-cm copper; and long and short wire plus bolted-to-heatsink (HS) cases ground the filter via a wire either long (50 cm) or short (15 cm), and simultaneously also bolt it down directly to the converter (i.e. DC-fed motor drive) heatsink. For the cases with two connections, two relatively far parallel paths are provided for the CM noise to flow, which effectively reduces the equivalent inductance. We can notice however, that the copper foil (one path) measurement is almost identical to the best case two points grounding case, indicating that copper foil connection provides low grounding impedance and the best solution.

8.5.1.2 One-point vs. Multipoint Grounding for Multistage Filter

For multistage filter, there are two possible grounding patterns: one-point grounding and multipoint grounding. Analysis shows that the CM capacitors should be grounded separately [20]. Because of the large difference in current between grounding paths, the mutual inductance

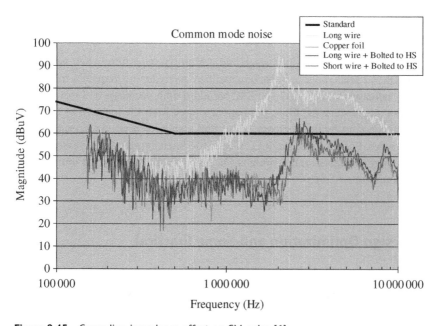

Figure 8.45 Grounding impedance effect on CM noise [6].

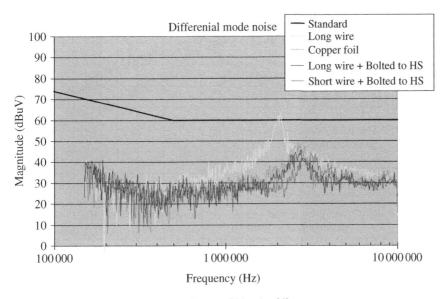

Figure 8.46 Grounding impedance effect on DM noise [6].

between two grounding paths (CM noise loops) are detrimental to the CM filter's performance. Thus, the mutual inductance should be as small as possible.

Experiments are carried out on a DC-fed motor drive using a two-stage filter in Figure 8.47 to verify the analysis [20].

The measured CM noise results for the original unfiltered case and four different filter grounding cases are compared in Figure 8.48. Case 1: All CM capacitors share the same ground plane on the PCB. The ground plane on the PCB is then grounded to the true ground in the EMI measurement setup. It shows a noise reduction up to 28 dB at low frequencies but not very good reduction at high frequencies. Case 2: The ground planes on both the top and bottom sides of the PCB are divided into two separate parts. Two PCB ground plane parts are grounded to one point on the true ground with two separate grounding paths close to each other, so the mutual inductance is large. There is only several dB improvement at low frequencies and no improvement at high frequencies. Case 3: When the two grounding paths in (Case 2) are moved far away from each other, the mutual inductance between them is greatly reduced and the noise reduction is significant. There is a 20–30 dB improvement in the entire measured frequency range as compared with the results in Case 2. Case 4: The

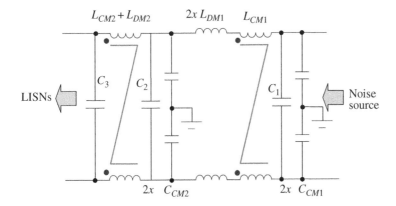

Figure 8.47 A two-stage LC filter.

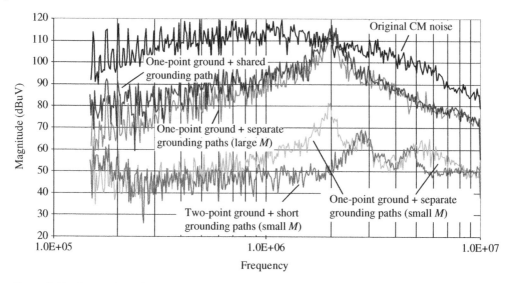

Figure 8.48 Comparison of measured CM noise [20].

capacitors are grounded at two different points of the true ground, and the grounding paths are shortened (40% of the previous cases) so that both the mutual inductance and the impedance of the two grounding paths are smaller. There is another improvement of up to 30 dB from the range of 300 kHz to 2.4 MHz, and up to a 15-dB improvement from 5 to 8 MHz as compared with the result in Case 3.

8.5.2 EMI Filter Coupling

As indicated in the grounding discussion aforementioned, the parasitic couplings between multi-stage filters can significantly impact EMI filter attenuation performance. Two basic parasitic coupling effects are discussed here.

8.5.2.1 Inductive Coupling in a Π-Type Filter
Figure 8.49 shows a *CLC* Π-type CM filter where L is the CM choke, and C1 and C2 represent the filter capacitors. Lp1 and Lp2 are used to describe the parasitic inductances of input and output traces, respectively. The parameters of mutual inductances M1–M10 are used to describe the

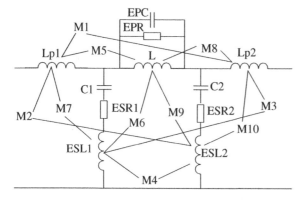

Figure 8.49 Parasitic couplings in a Π-type CM filter [21].

Figure 8.50 Crucial couplings in a Π-type CM filter [21].

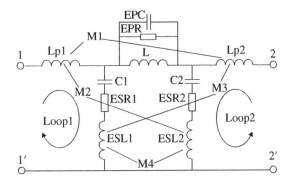

near-field inductive couplings in the filter. The inductive coupling can be divided into five categories: (i) coupling between input and output traces: M1; (ii) coupling between traces and capacitors: M2, M3, M7, and M10; (iii) coupling between the two capacitors: M4; (iv) coupling between CM inductor and input and output traces: M5 and M8; and (v) coupling between CM inductor and the capacitors: M6 and M9.

Based on the analysis conducted in [21], four types of couplings have no influence on the performance of the Π-type CM filter: the coupling between CM choke and the capacitors; the coupling between CM choke and the input trace or output trace; the coupling of the input trace with its nearby capacitor; and the coupling of the output trace with its nearby capacitor. The crucial couplings are from inductive couplings of circuit branches, which belong to two loops, input loop and output loop of the filter. In other words, M1–M4 are important and M5–M10 can be ignored in Figure 8.49, which can be simplified as Figure 8.50.

Notice that the sum of M1, M2, M3, and M4 represents the mutual inductance between input loop (Loop1) and output loop (Loop2), and it is possible to consider the four mutual inductances as a whole. A simplified high-frequency model of Π-type CM filter is shown in Figure 8.51, where inductance M plays the role of the mutual inductance between the input loop and the output loop of the filter with

$$M = M1 + M2 + M3 + M4 \tag{8.55}$$

Figure 8.51 A simplified high-frequency model of Π-type CM filter [21].

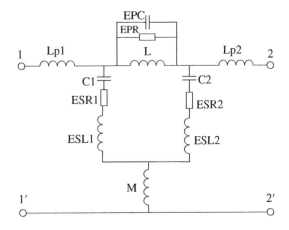

8.5.2.2 Capacitive Coupling in a T-Type Filter

In the conventional equivalent circuit analysis of *LCL* T-type filter, only the self-parasitics of two inductors L_1 and L_2 and bypass capacitor C are considered. Other parasitic couplings from signal traces and loops are often ignored in this structure due to the high impedance at both input and output ends. However, the filter performance could be much degraded compared to its ideal performance with only self-parasitics included. The cause is mutual capacitance coupling from the input of L_1 to the output of L_2 possibly formed by the signal traces and space distance. A small mutual capacitance can have a significant impact on the T-type filter.

Figure 8.52a shows the equivalent circuit considering this mutual capacitance C_p. Two inductor impedances are defined as Z_1 and Z_2. The transfer gain of the T-type filter can be derived from the ratio of the output voltage over the input voltage. As clearly seen, this circuit has a Δ-connection from Z_1, Z_2 to C_p. Using the Δ/Y transformation, an equivalent circuit can be derived as in Figure 8.52b [22].

The three Y-connected impedances are as follows:

$$\begin{cases} Z_{Y1} = \dfrac{Z_1 Z_{cp}}{Z_1 + Z_2 + Z_{cp}} \\[2mm] Z_{Y2} = \dfrac{Z_2 Z_{cp}}{Z_1 + Z_2 + Z_{cp}} \\[2mm] Z_{Y3} = \dfrac{Z_1 Z_2}{Z_1 + Z_2 + Z_{cp}} \end{cases} \tag{8.56}$$

Since C_p is assumed no more than several pF, $Z_{cp} \gg Z_1$, $Z_{cp} \gg Z_2$ at low and medium frequencies, e.g. below 10 MHz. Equation (8.56) can be approximated as:

$$\begin{cases} Z_{Y1} = \dfrac{Z_1 Z_{cp}}{Z_1 + Z_2 + Z_{cp}} \approx Z_1 \\[2mm] Z_{Y2} = \dfrac{Z_2 Z_{cp}}{Z_1 + Z_2 + Z_{cp}} \approx Z_2 \\[2mm] Z_{Y3} = \dfrac{Z_1 Z_2}{Z_1 + Z_2 + Z_{cp}} \approx \dfrac{Z_1 Z_2}{Z_{cp}} = sC_p Z_1 Z_2 \end{cases} \tag{8.57}$$

Therefore, through this transformation, the major change is an inserted impedance Z_{Y3} in series with capacitor branch. It indicates that Z_{Y3} act as an inductance and thus contributes to the ESL of the capacitor branch, potentially degrading this bypass branch performance. This series impedance as a product of Z_1, Z_2, and C_p is strongly frequency-dependent and has remarkable differences between CM and DM cases, depending on the impedance characteristics and magnitude of the two inductors L_1 and L_2. And especially in the CM filter case, this impact becomes more significant due to the high impedance of CM choke at low frequency.

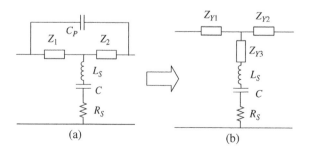

Figure 8.52 Mechanism of capacitance coupling in a T-type filter. (a) Y and (b) Δ equivalent circuits.

8.5.2.3 Coupling in Multistage Filters

From the analysis of *CLC* Π-type filter and *LCL* T-type filter, one important conclusion can be drawn is that in the *CLC* structure the dominant coupling effect is from inductive coupling between the input and output loops, whereas in the *LCL* structure the dominant coupling effect is from capacitive coupling between the input and output terminals. This presents a duality feature between the two basic structures: Π-type and T-type.

In a general cascaded filter structure, both the inductive and capacitive couplings exist and the dominant couplings that influence the filter attenuation are marked in Figure 8.53 using lumped components.

In general, all the loop currents can induce mutual inductive coupling through the magnetic field. In practice, however, the input and output loops have the strongest inductive coupling effect. This can be well understood because the input loop current noise has the highest strength, and output loop current noise is most attenuated. Similarly, the input and output terminals have the strongest capacitive coupling effects. The coupling effects of commonly used filter structure are summarized in Table 8.5.

It can be seen that *LCLC* and *CLCL* have the combined inductive and capacitive coupling effect. They suffer from the capacitive coupling effect due to the *LCL* T-joint, and from the inductive coupling effect due to coupling between the input and the output loops. For these cases with both coupling effects, some general behaviors can be observed. With mutual capacitance C_p at pF level, the capacitive coupling dominates and occurs from low frequency, such as hundreds of kHz. Whereas the inductive coupling effect in CM cases usually occurs at frequencies above MHz and could occur at hundreds of kHz in DM cases, mainly related to mutual inductance value and filter capacitance value. With mutual C_p at 100 fF level or less, both capacitive coupling and inductive coupling are visible and mingles together. With mutual C_p at several fF levels, the inductive coupling dominates and can occur from MHz or 100s of kHz if mutual inductance is less than several nH, whereas the capacitive coupling becomes trivial and occurs above MHz.

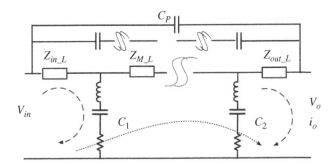

Figure 8.53 General coupling including both capacitive and inductive effects in a multicomponent filter.

Table 8.5 General dominant coupling effects in typical filter structures.

Filter structure	Inductive coupling	Capacitive coupling
CLC	Highest	Lowest
LCL	Lowest	Highest
LCLC	Medium	One T-joint, high
CLCL	Medium/lower	One T-joint, high

8.5.3 Mixed-Mode Noise

EMI CM and DM filter designs are usually carried out separately based on CM and DM noises and propagation paths, given that the two modes are usually well decoupled. However, mixed-mode (MM) noise exists, causing difficulties on the noise recognition and influencing the filter design. The MM noise was first identified as the non-intrinsic DM noise in the switching mode power supply. Here, the AC-side MM noise is discussed for a DC-fed motor drive system based on the three-phase two-level VSI in Figure 8.54, where the solid lines are power current conducting paths, and dashed lines represent grounding loops through parasitic capacitances between the inverter switches and the heatsink, as well as between the motor windings and motor chassis [23].

Considering switching state <100>, which means that the top switch of phase a and bottom switches of phases b and c are turned on. As shown in Figures 8.55 and 8.56, during this time period,

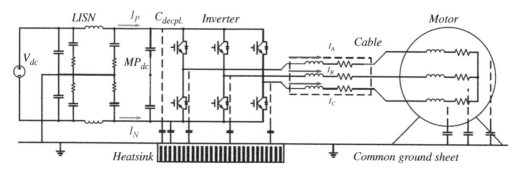

Figure 8.54 A DC-fed three-phase two-level VSI motor drive system.

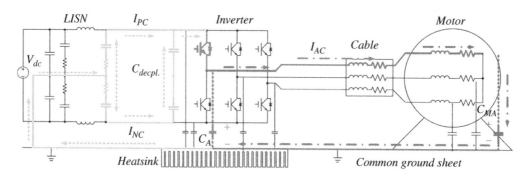

Figure 8.55 Charging of parasitic grounding capacitance during <100>.

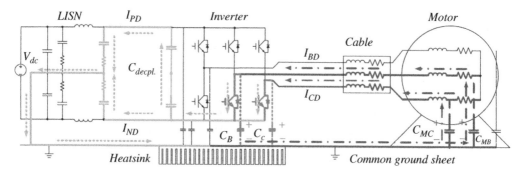

Figure 8.56 Discharging of parasitic grounding capacitance during <100>.

the motor parasitic grounding capacitance of phase a C_{MA} is being charged while phase b and c capacitances C_{MB} and C_{MC} are being discharged. These charging and discharging currents are part of the grounding current, not the intrinsic DM noise due to the line-to-line voltage applied on the line impedance of cable and motor yet captured by the AC DM noise measurement.

Assuming a balanced three-phase motor and cable, the impedance on the charging or discharging loops through each phase should be equal to each other. Hence, the phase a charging current I_{AC} has the same magnitude but opposite direction with phase b discharging current I_{BD} or phase c I_{CD}. Assuming the magnitude of the charging or discharging current as I_0 and the charging direction of I_{AC} as positive, the CM noise can then be deduced as:

$$I_{CM} = \frac{I_{AC} + I_{BD} + I_{CD}}{3} = -\frac{I_0}{3} \tag{8.58}$$

With the definition of DM current, the charging and discharging currents are calculated as:

$$I_{DM\_a} = I_0 - I_{CM} = \frac{4I_0}{3} = I_{MM\_a}$$

$$I_{DM\_b} = I_{DM\_c} = I_{BD} - I_{CM} = I_{CD} - I_{CM} = -\frac{2I_0}{3} = I_{MM\_b} = I_{MM\_c} \tag{8.59}$$

In (8.59), the acquired noise is the MM noise, which is observed in DM measurement but not the intrinsic noise for each phase. The equation shows that in the time period of <100>, in the measurement of AC DM noise of phase a, there is part of the noise going through the ground with the magnitude of four times of the CM noise. For phases b and c, the grounding noise is two times the CM noise.

Similar analysis can be carried out for all other switching states. Table 8.6 summarizes the output MM noise of each phase during every possible voltage vector condition.

Due to the existence of the MM noise, DM filter can be ineffective to reduce DM noise. Figure 8.57 shows the experimental EMI noise measurement with different AC-side DM LC filters. With the 140 µH DM inductors and 50 nF capacitors, attenuation of the AC DM noise is expected beyond the corner frequency of 60 kHz. When the DM inductors are increased to 160 µH and DM capacitors to 68 nF and 100 nF, resulting in corner frequencies of 48 kHz and 40 kHz, the noise level is similar to previous results of 60 kHz corner frequency.

However, when a simple CM LC filter is added, change in the DM noise spectrum occurs. A CM choke of 200 µH is series connected with the 140 µH DM inductor, and the DM capacitors are grounded making them also CM capacitors. Thus, a CM LC filter is realized. Considering the leakage inductance of 3 µH from the CM choke, it has no significant influence on the corner frequency of the DM LC filter. Based on conventional EMI filter design, this new filter structure will only filter the AC CM noise. However, as shown in Figure 8.58, the DM noise is further attenuated. It indicates the fact that the CM filter is helping attenuate the DM noise, which actually validates the existence of the MM noise in the motor drive system.

Table 8.6 CM and MM noise summary for all possible switching states of the two-level VSI.

	<000>	<100>	<010>	<001>	<110>	<101>	<011>	<111>
I_{CM}	I_0	$-I_0/3$	$-I_0/3$	$-I_0/3$	$I_0/3$	$I_0/3$	$I_0/3$	I_0
I_{MM_a}	0	$-4I_{CM}$	$2I_{CM}$	$2I_{CM}$	$2I_{CM}$	$2I_{CM}$	$-4I_{CM}$	0
I_{MM_b}	0	$2I_{CM}$	$-4I_{CM}$	$2I_{CM}$	$2I_{CM}$	$-4I_{CM}$	$2I_{CM}$	0
I_{MM_c}	0	$2I_{CM}$	$2I_{CM}$	$-4I_{CM}$	$-4I_{CM}$	$2I_{CM}$	$2I_{CM}$	0

dBμA

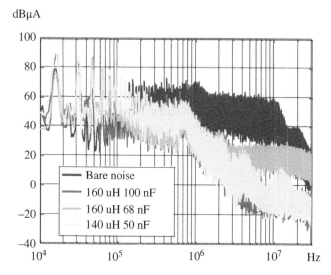

Figure 8.57 AC DM noise with different AC DM *LC* filters [23].

dBμA

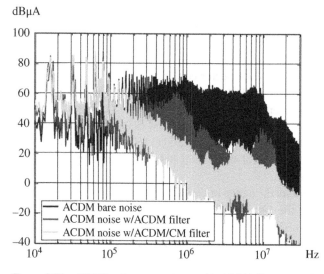

Figure 8.58 AC DM noise comparison: with AC DM filter vs. with AC CM and DM filters [23].

As discussed, the measured DM noise by definition is not only intrinsic AC DM noise. Part of it is the MM noise going through the ground. Considering its characteristics, no so-called MM filter is needed to attenuate it. Typical CM filters either increasing or bypassing the output grounding impedance will help suppress the MM noise. If CM noise is well attenuated, MM noise will not cause total noise exceeding standards. Meanwhile, DM capacitors will create low impedance path among phases, which also suppress the MM noise.

In order to avoid overdesign of the DM filter in the case of high level MM noise emission, it is recommended to implement both the CM and DM filter when checking if the DM noise is attenuated by the designed DM filter. When the CM noise is suppressed by the CM filter, the influence of the MM noise on DM bare noise measurement will be minimized.

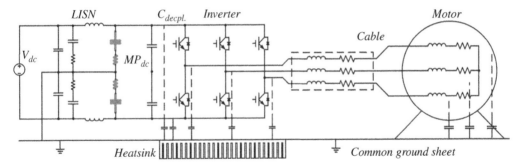

Figure 8.59 A DC-fed three-phase motor drive system.

8.5.4 EMI Noise Mode Transformation Due to Propagation Path Unbalance

Another type of MM noise is due to the propagation path unbalance as already analyzed in Section 8.2.2. It can lead to transformation of noise mode between CM and DM, which could also cause filter overdesign or unnecessary design iteration. The phenomenon will be discussed also for a three-phase DC-fed motor drive system.

8.5.4.1 DC CM Noise Propagation Unbalance and Mode Transformation

Figure 8.59 shows again the structure of a DC-fed motor drive system. The dashed lines with capacitors indicate the path of CM noise propagation, including the parasitic grounding capacitances. For the DC-side propagation path, the LISNs are also included. Although the input impedance of the LISN is usually defined in EMI standards, the grounding branches of the highlighted parts in Figure 8.59 are not strictly regulated. For commercial products, the tolerance of the resistor and capacitor could vary from 10% to 20%, which means there is an expected difference between the positive and negative grounding impedances. If the grounding branches of the LISNs are different, the charging or discharging currents of the positive and negative buses will not be identical. As a result, the total grounding currents flowing through the positive bus and negative bus will be different. The DC DM noise measurement will include this grounding current difference. Therefore, part of the CM noise is transformed to the DM spectrum, which is also called MM noise.

8.5.4.2 DM Noise Propagation Unbalance and Mode Transformation

For the DM noise propagation path, the discrepancies among the three-phase components of the DM filter are the main contributors to the noise propagation path unbalance. The inevitable discrepancies in inductors and capacitors would result in the unbalance of the DM noise path impedance, which converts DM noise to CM noise. Moreover, the powder or amorphous core with distributed air gap is commonly used for DM inductors, which has current-bias-dependent permeability. Due to the unequal instantaneous currents for three phases, current-bias-dependent permeability will cause the unbalanced DM inductance. This unbalance of the DM noise path will also convert DM noise to CM noise [14].

8.6 EMI Noise and Filter Reduction Techniques

This section overviews several techniques for EMI noise and/or EMI filter size or weight reduction.

8.6.1 Switching Frequency

Switching frequency is a key design parameter for power converters. Generally, increasing switching frequency will reduce passive components and increase switching device power loss. Therefore, for a specific application, there would be an optimal switching frequency. When considering the EMI filter, the relationship between the switching frequency and the filter weight/size is not linear due to the noise spectrum and the standard requirement as discussed in Section 8.4.3. Higher switching frequency does not necessarily lead to higher filter corner frequency. There are some preferred switching frequencies (e.g. 40 kHz, 70 kHz, 140 kHz, or above 500 kHz) in most cases. Note that the weight of an inductor is also not only determined by the inductance value/corner frequency but also related to the volt-second on the inductor. For example, although the corner frequency is small at some switching frequencies, the volt-seconds on the inductor may be amplified if the filter resonant frequency is close to the switching frequency, resulting a larger inductor.

In addition to selecting an optimized fixed switching frequency, random PWM (RPWM) or varible frequency PWM (VFPWM) can be considered to further reduce EMI noise. The principle is to spread the energy concentrated around the harmonics of the switching frequency to a wider frequency range and thus reducing the EMI noise peak [24].

8.6.2 Modulation Scheme

Modulation schemes affect the EMI noise source and thus can be utilized for EMI noise reduction.

The main idea for CM volatge reduction is to utilize the redudant vectors with reduced CM volatge magnitude. For two-level voltage source converter, the active zero state PWM (AZSPWM) and near state PWM (NSPWM) are popular modulation schemes for CM volatge reduction. AZSPWM is a variation for SVM but inverses the aligment of the medium-length duty and reduces the CM voltage peak to be $\pm V_{DC}/6$. Similarly, NSPWM is a variation for DPWM and reduces both the switching loss and CM voltage, but it could not work with a low modulation index.

Figure 8.60 displays the four typical PWM methods gate logics and the corresponding CM voltage. Although AZSPWM and NSPWM could effectively reduce the CM voltage peak, they may increase the CM voltage harmonics at several frequencies. Figure 8.61 shows the voltage spectrum comparison between SVPWM and AZSPWM, DPWM and NSPWM [13]. In Figure 8.61a, the CM voltage magnitude of AZSPWM is reduced to 1/3 of SVM, but some harmonics of switching frequency is higher with AZSPWM than with SVM. CM current could be amplified with certain CM resonant frequency. In Figure 8.61b, NSPWM generates more third-order harmonics of switching frequency than DPWM. Therefore, the improved modulation schemes could reduce the CM voltage amplitude but also increase high-frequency components. The benefit of CM voltage reduction could not result in CM current improvement if CM resonant frequency is close to these frequency components. The application of modulation scheme for CM noise reduction should also consider CM loop resonant frequency. The CM EMI noise current is measured in four different PWM cases. Figure 8.62 is the comparison between SVPWM and AZSPWM, and Figure 8.63 is the comparison between NSPWM and DPWM [13]. Obvious CM current noise attenuation has been achieved around switching frequency, but the improvement at higher frequencies is limited.

8.6.3 EMI Filter Topology

In a three-phase inverter or rectifier, the CM EMI noise on DC and AC sides share the propoagtion path. The EMI noise coulpling between DC and AC sides makes the filter design more complicated as adding EMI filter on one side will change the propagation impedance and EMI noise level on

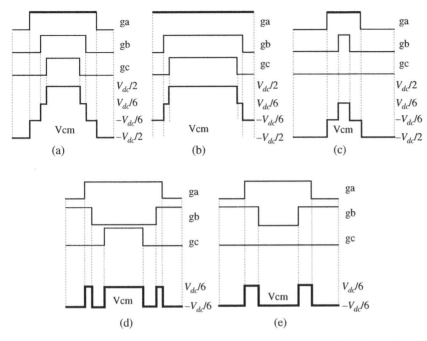

Figure 8.60 Gate logic and CM voltage for (a) SVM, (b) DPWM (clamped to positive bus), (c) DPWM (clamped to negative bus), (d) AZSPWM, and (e) NSPWM.

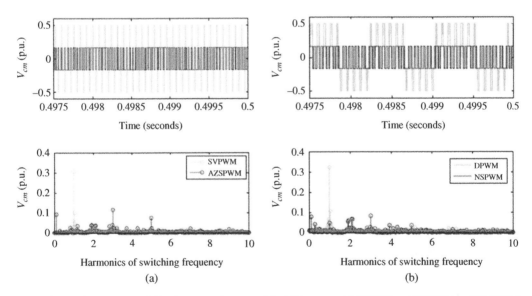

Figure 8.61 Comparison of CM voltages and spectra ($M = 0.9$) between (a) SVPWM and AZSPWM and (b) DPWM and NSPWM.

both sides. On the other hand, it provides an opportunity to use some filter topologies that connect DC and AC sides with better overall attenuation.

One approach is connecting the DC and AC midpoint, and thus confining the CM noise within the converter and supressing both AC and DC CM noise. Figure 8.64 shows a filter topology, where

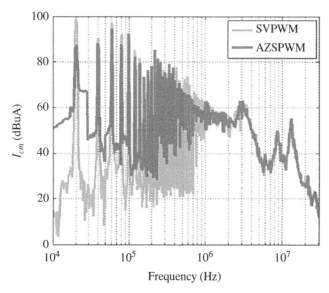

Figure 8.62 CM EMI noise of SVPWM and AZSPWM [13].

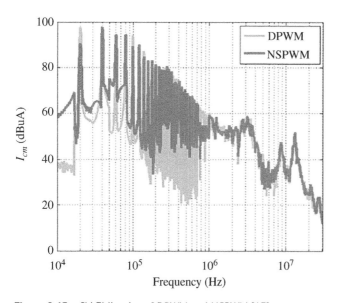

Figure 8.63 CM EMI noise of DPWM and NSPWM [13].

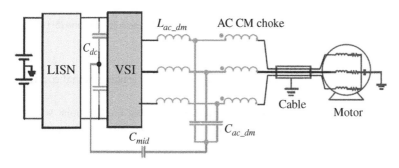

Figure 8.64 Filter topology connecting DC midpoint and AC neutral.

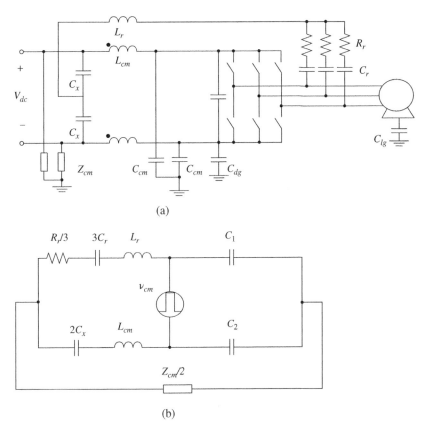

Figure 8.65 Impedance balancing circuit for DC-side CM noise reduction: (a) topology and (b) equivalent circuit.

the AC-side DM capacitor neutral is connected with the DC-link capacitor midpoint which creats a bypass branch for CM noise.

Another approach is impedance balancing concept. Figure 8.65 shows the circuit scheme for DC-side CM noise reduction [25]. An impedance bridge is formed by adding a L_r-R_r-C_r branch from the inverter AC side to DC side. In the EMI frequency range, the impedance of R_r and C_r is much smaller than L_r, and impedance C_x is much smaller than L_{cm}. The motor CM impedance is modeled as capacitance C_1 and device parasitic impedance is modeled as C_2, so the CM current in LISN branch is zero if (8.60) is satisfied:

$$\frac{L_r}{L_{cm}} = \frac{C_1}{C_2} \tag{8.60}$$

Figure 8.66 shows a circuit scheme that can reduce CM noise for both DC and AC sides [26]. Two auxiliary branches are added to the conventional CM filter: one branch is an inductor L_r connecting the AC side and the ground (C_r is used to block the fundamental and low-frequency components, and R_r is used to damp the LC resonance); and the other branch is a capacitor C_{return}, which connects the AC and DC sides through the AC DM capacitors neutral and DC-link capacitor midpoint. Similarily, the CM current flowing through the motor can be theoretically zero if (8.61) holds:

$$\frac{L_r}{L_{cmac}} = \frac{C_{return}}{2C_{cmdc}} \tag{8.61}$$

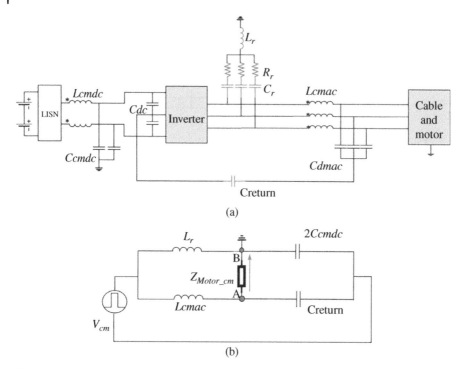

Figure 8.66 Impedance balancing circuit for both DC- and AC-side CM noise reduction: (a) topology and (b) simplified equivalent circuit.

Define the impedance balancing circuit ratio as:

$$n = \frac{L_r}{L_{cmac}} = \frac{C_{return}}{2C_{cmdc}} \tag{8.62}$$

The CM current flowing to the DC-side LISNs will be smaller than the conventional CM filter without impedance balancing circuit if $n > 1$.

With larger n, DC-side CM current will be smaller. With $n > 1$, the AC-side motor CM current is not ideally cancelled, but the cancelling effect still exsits and AC CM noise can be significantly reduced. Figure 8.67 shows the experimental results when $n = 4$. Around 10–20 dB noise attenuation is achieved on the AC side and 5–15 dB noise attenuation on the DC side in the 150 kHz to 1-MHz frequency range. In the higher EMI frequency range, the CM noise attenuation is not obvious. This is because in the MHz frequency range, the power stage parasitic capacitances begin to influence balancing perforamnce of the impedance bridge. Also, the parasitics of the EMI filter in the impedance bridge (such as EPC of the CM inductor) cannot be neglected in the high frequency range.

In general, filter topologies that properly connect DC and AC sides can create extra CM noise propagathion path to achive bypassing or cancelling effect thus reducing CM EMI noise. Since the added branches will not carry the load current, the filter size and weight could be much smaller than regular filter in the main current conduction path.

8.6.4 Active/Hybrid Filter

The active or hybrid passive/active filter could be an alternative to the bulky passive filter. Due to bandwidth limitations, the active filter is more effective in reducing low-frequency EMI noise and a passive filter will be needed to reduce high-frequency noise. Still, the total size and weight of the hybrid filter could be much smaller than a conventional passive EMI filter.

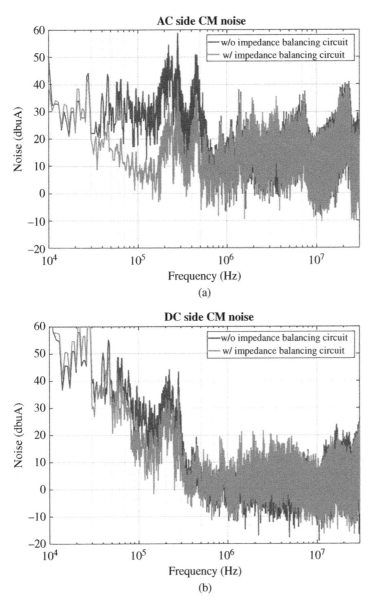

Figure 8.67 (a) AC-side and (b) DC-side CM EMI noise with and without the impedance balancing circuit when n = 4 [26].

Active filters are based on cancelling concept. Both CM voltage and CM current cancellation with feedforward and feedback control can be used [27]. Figures 8.68 and 8.69 show the CM noise voltage cancellation scheme. Figures 8.70 and 8.71 show the CM noise current cancellation scheme.

In Figure 8.68, a voltage source is generated by a cancellation circuit and is in series with CM noise source. Figure 8.69 shows three approaches for CM noise cancellation. For the feedforward cancellation in Figure 8.69a, the CM noise voltage V_{CM} is sensed and amplified $A(s)$ times, and then injected into the DC bus. The CM current of the LISNs is given by:

$$I_{CM}(s) = [1 - A(s)]I_{CMO}(s) \tag{8.63}$$

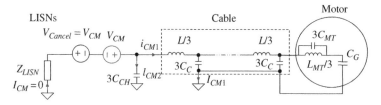

Figure 8.68 CM noise voltage cancellation circuit.

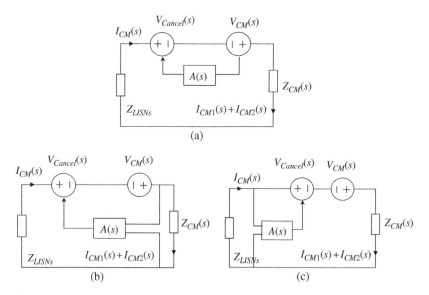

Figure 8.69 CM noise voltage cancellation approach: (a) feedforward; (b) feedback 1; and (c) feedback 2.

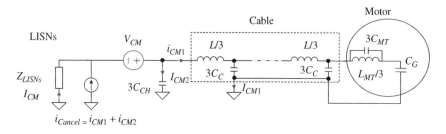

Figure 8.70 CM noise current cancellation circuit.

$I_{CMO}(s)$ represents the original CM noise flowing through LISNs without any EMI filters inserted

$$I_{CMO}(s) = V_{CM}(s)/[Z_{CM}(s) + Z_{LISN}(s)] \qquad (8.64)$$

$A(s)$ should be unity gain to obtain the best cancellation. For the feedback cancellation in Figure 8.69b, the CM noise voltage drop on impedance $Z_{CM}(s)$ is sensed and amplified $A(s)$ times and then injected to the DC bus. The CM current of LISNs is given by:

$$I_{CM}(s) = \frac{1}{1 + A(s)/[1 + Z_{LISN}(s)/Z_{CM}(s)]} I_{CMO}(s) \qquad (8.65)$$

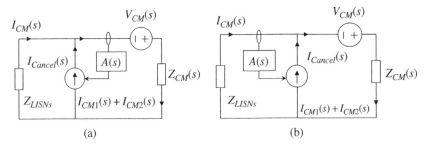

Figure 8.71 CM current cancellation approach: (a) feedforward and (b) feedback.

Equation (8.65) suggests that voltage gain $A(s)$ should be very high, and the feedback cancellation is more effective if the magnitude of impedance $Z_{CM}(s)$ is not smaller than the LISNs' impedance $Z_{LISN}(s)$:

$$|Z_{CM}(s)| \geq |Z_{LISN}(s)| \tag{8.66}$$

For the CM noise source, (8.66) can usually be met in low EMI frequency range since the impedance of the CM parasitic capacitance is usually very high there. For the feedback cancellation in Figure 8.69c, the CM noise voltage drop on the LISNs is sensed and amplified $A(s)$ times and then injected into the DC bus. The CM current of the LISNs is given by:

$$I_{CM}(s) = \frac{1}{1 + A(s)/[1 + Z_{CM}(s)/Z_{LISN}(s)]} I_{CMO}(s) \tag{8.67}$$

Also, $A(s)$ should be very high, and the feedback cancellation is more effective if the magnitude of impedance $Z_{CM}(s)$ is not larger than $Z_{LISN}(s)$:

$$|Z_{CM}(s)| \leq |Z_{LISN}(s)| \tag{8.68}$$

which usually cannot be easily satisfied in low EMI frequency range.

In Figure 8.70, a current is generated between the ground and the DC bus. In Figure 8.71a, the CM noise is sensed on the noise source side. The sensed current is amplified $A(s)$ times and injected to the DC bus with reference to the ground. The CM noise current of the LISNs is given by:

$$I_{CM}(s) = \frac{1 - A(s)}{1 - A(s)/[1 + Z_{CM}(s)/Z_{LISN}(s)]} I_{CMO}(s) \tag{8.69}$$

Equation (8.69) shows that the current gain $A(s)$ should approach unity and the feedforward cancellation is more efficient if $|Z_{CM}(s)| \geq |Z_{LISN}(s)|$. In Figure 8.71b, the CM noise is sensed on the LISNs' side. The sensed current is amplified $A(s)$ times and injected to the DC bus from ground. The CM noise current of the LISNs is given by:

$$I_{CM}(s) = \frac{1}{1 + A(s)/[1 + Z_{LISN}(s)/Z_{CM}(s)]} I_{CMO}(s) \tag{8.70}$$

Equation (8.70) shows that the current gain $A(s)$ should be very high, and this feedback is more efficient if $|Z_{CM}(s)| \geq |Z_{LISN}(s)|$.

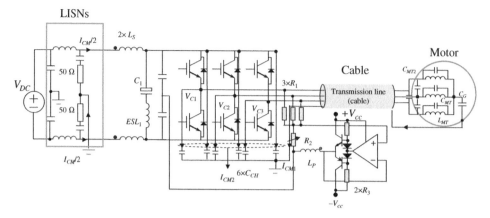

Figure 8.72 Implementation of feedforward CM noise voltage cancellation.

Figures 8.72 and 8.73 show the implementation of active CM filters. In Figure 8.72, the voltage between the AC neutral point and the DC bus is sensed using resistors and injected to the DC side through a CM transformer. There is a disadvantage for the CM noise voltage cancellation in high-voltage applications. The transformer's primary side inductance cannot be small to avoid large magnetizing current. Then the inductance of the secondary side will be large due to the high turn ratio of the voltage transformer. Hence, the voltage transformer's weight/size will be large. In Figure 8.73, a current transformer is used to sense CM noise current on the DC bus. The cancellation current is then injected to the common ground via injection capacitor C. The transformer's primary winding has only one turn. Therefore, the current transformer's weight/size is much smaller than that of the voltage transformer.

As mentioned earlier, passive filter is usually used together with the active filter to reduce high-frequency noise and to reduce the di/dt of the CM current so that active filter can work properly. Figure 8.74 shows the measured CM noise using hybrid filter with current feedforward cancellation. Around 50-dB attenuation is achieved by the active filter at the first noise peak.

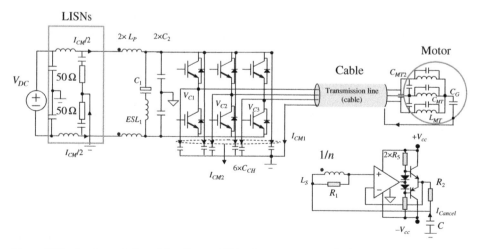

Figure 8.73 Implementation of feedforward CM noise current cancellation.

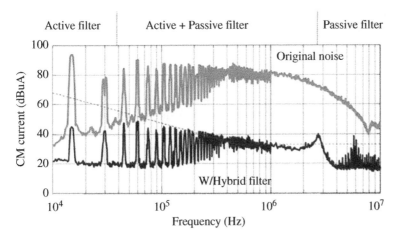

Figure 8.74 Comparison of the measured CM noise [27].

8.6.5 Paralleled Converters Interleaving Angle Optimization

By shifting the phase angle of the carriers used for PWM in paralleled inverters, the switching frequency related harmonics at the AC output terminal can be reduced. Interleaving actually does not eliminate any harmonics, it converts certain harmonics from the output into circulating components, and as a result, the output EMI noise and ripple currents are reduced. Interleaving angle can be optimized to maximize the EMI filter corner frequency [28].

When the objective is EMI noise and EMI filter reduction, symmetric interleaving may not be the most effective way to reduce certain harmonic peaks that determine the EMI filter corner frequency.

Considering the two paralleled inverter case, when symmetric interleaving ($\gamma = \pi$) is applied, the average harmonic coefficient of the two inverters using (8.22) can be evaluated as:

$$C_{mn,avg} = C_{mn}\left|\cos\left(\frac{m\pi}{2}\right)\right| = \begin{cases} C_{mn}, & m \text{ is an even number} \\ 0, & m \text{ is an odd number} \end{cases} \tag{8.71}$$

When asymmetric interleaving ($\gamma \neq \pi$) is applied to two interleaved inverters,

$$C_{mn,avg} = \frac{1}{2}\left|\left(C_{mn,1} + C_{mn,2}e^{jm\gamma}\right)\right| = C_{mn}|\cos(m\gamma/2)| \leq C_{mn} \tag{8.72}$$

Hence, symmetric interleaving will eliminate the odd number carrier harmonics while keeping the even number harmonics amplitude unchanged. Asymmetric interleaving could reduce the amplitude of all the harmonic components.

Following the CM and DM voltage harmonics calculation, noise propagation path impedance calculation, and the noise peak frequency identification illustrated in previous sections, the relationship between interleaving angle and CM filter corner frequency for two paralleled interleaved inverters can be expressed as (8.73), where most variables are defined in (8.54). The relationship between interleaving angle and AC-side DM filter corner frequency can be derived similarly. From (8.73), it can be seen that EMI filter corner frequency is a function of switching frequency, modulation index, interleaving angle, EMI filter type, and EMI noise propagation path impedance.

$$\log f_c = \log f_{pk} + \frac{I_{std(f_{pk})} - 20\log\left[\dfrac{V_{CM(f_{pk})}\cos\left(\dfrac{m\gamma}{2}\right)10^6}{Z_{CM(f_{pk})}}\right]}{20 \cdot k} \tag{8.73}$$

Whether the switching frequency is in the EMI frequency range or not will influence the trend how the EMI filter corner frequency varies as interleaving angle changes. Two case studies are provided here.

The first case is that the switching frequency is $f_{sw} = 70$ kHz (<150 kHz). The CM and DM filter corner frequencies are calculated as a function of modulation index and interleaving angle. Assume a three-level ANPC converter with NTSV modulation, and a one-stage filter ($k = 2$). The results are shown in Figure 8.75 [28]. From Figure 8.75a, the CM filter corner frequency peak is achieved when interleaving angle is around 180°, and the second optimal interleaving angle range is 30–90°. From Figure 8.75b, the DM filter corner frequency peak is obtained when interleaving angle is around 30–60°, and the second optimal interleaving angle range is around 120–180°.

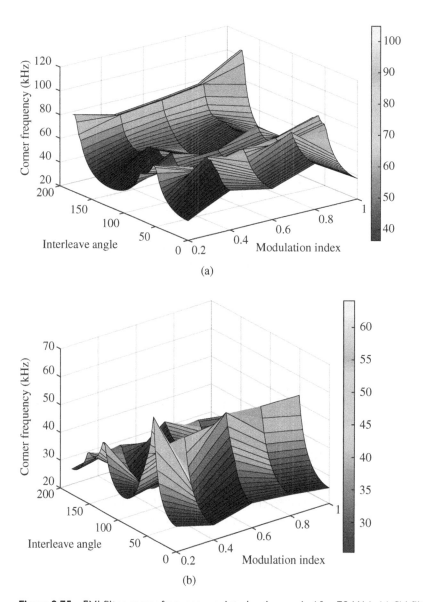

Figure 8.75 EMI filter corner frequency vs. interleaving angle (f_s = 70 kHz): (a) CM filter; (b) DM filter.

The second case is when switching frequency $f_s = 200$ kHz (>150 kHz). In this case, the first and second carrier harmonics are in the EMI frequency range. The analysis is similar to the first case. The CM and DM filter corner frequencies are also calculated, and the results show that the CM filter corner frequency peak is achieved when interleaving angle is around 180° while the DM filter corner frequency peak is achieved when interleaving angle is in the 90–150° range.

8.6.6 EMI Filter Integration

Integration techniques can reduce the size of passive components by implementing multiple functions into one component.

Different types of inductors can be integrated to form one inductor that achieve different functions. The leakage inductance of the CM choke can act as a DM inductor to suppress DM noise. In interleaved topologies, the interphase inductors and DM inductors can be integrated together, which will limit system circulating current and filtering DM noise at the same time. In [29], CM choke and DM inductor integration is achieved by placing a low-permeability core in the window area of the high-permeability CM choke core. The integrated choke can achieve higher DM inductance.

In medium- and low-power applications, planar EMI filters are used and capacitors can also be implemented within inductors. The main passive integration technologies include low-temperature co-fired ceramic (LTCC) technology and planar ferrite magnetic and ceramic dielectric passive integration [30, 31]. Inductors and capacitors have also been integrated as transmission line EMI filters into busbars [32].

8.7 Interface to Other Subsystem Designs

In general, the EMI filter in a three-phase converter system interfaces directly with the converter power stage, whether the filter is on AC side(s), DC side, or both. The EMI filter also directly interfaces with thermal management system and mechanical system. Note that the EMI filter is also directly interfaced with the source and/or the load. The active or hybrid EMI filter will also need control, while the passive EMI filter will not need any control. In some relatively low power applications, EMI nose generated by the switching mode power supplies (e.g. for gate drivers) in the converter can be a relatively important contributor to the overall EMI noise.

The input from the interfaced subsystems has been discussed in Section 8.5. Here, we will examine the output from the EMI filter. Table 8.7 summarizes all the output from EMI filter to its interfaced other subsystems. As in the previous chapters, the "discussion" items are further explained here.

8.7.1 Voltage Distribution

The voltage distribution is a function of the converter and filter topology and grounding. The EMI filter is interfaced with converter and source/load. When the converter and filter topology and grounding method are determined, the EMI filter terminal voltage potentials can be readily determined.

8.7.2 Current Distribution

As for the currents, the peak, RMS, and frequency spectrum of the currents through components and terminals will be needed for the busbar and connector design. The AC terminal currents are

Table 8.7 Output from EMI filter to its interfaced subsystems.

Subsystem	Interface type	Item	Value
Converter	Constraint	Switching frequency	Design result
Thermal management system	Attribute	Components dimensions	Design result
		Components loss	Design result
		Components thermal impedance	Design result
Mechanical system	Attribute	Components dimensions and weight	Design result
		Voltage distribution	Discussion
		Current distribution	Discussion
		Input/output terminals	Discussion
Load	Attribute	*dv/dt*	Discussion

equal to phase currents. The DC terminal current is equal to DC average current if we assume all ripple current flowing through the DC-link capacitor. The EMI noise source and propagation path impedance models are discussed and current harmonics of each filter components can be obtained.

8.7.3 Input/Output Terminals

Input/output terminals will be needed for mechanical system design, including the type (AC or DC), number and rating. Since the EMI filter usually has one side connecting to the converter input or the output, and the other side connecting to the source or the load, some terminals are external and others are internal. Even within the EMI filters, there can also be internal terminals.

8.7.4 Load-side *dv/dt*

Loads like motor often has a *dv/dt* limit. With an AC-side EMI filter between the inverter and the motor, the *dv/dt* at the motor terminal is expected to be reduced even though the design objective of the EMI filter is not on *dv/dt* reduction. With the given *dv/dt* at the inverter terminal and the EMI (DM) filter impedance, the motor-side *dv/dt* can be determined. More discussion on motor *dv/dt* can be found in Chapter 12.

8.8 Summary

This chapter presents the EMI filter design.

1) The EMI filter design for the three-phase converter is formulated as an optimization problem with its design variables, constraints, conditions, and objective defined. The focus is on establishing models as functions of design conditions and variables. The EMI noise source models (i.e. AC voltage harmonics and DC current harmonics) are introduced with both DFA method and switching function simulation-based method. FMI filter components models are presented, considering their nonideal characteristics. Other important EMI noise propagation path impedance models including EMC test receiver model and high-frequency motor model are also introduced. Relationship between EMI filter corner frequency and switching frequency is established.

2) Practical EMI filter design and implementation issues are covered. Grounding impact, parasitic coupling effects including inductive coupling and capacitive coupling and their impact on filter performance, mixed CM/DM mode and its impact on EMI filter design, have been discussed.

3) State-of-the-art EMI noise and filter reduction techniques covering switching frequency selection, modulation schemes, filter topologies, active/hybrid filters, interleaving paralleled converters, and filter integration are introduced.

References

1 C. R. Paul, *"Introduction to Electromagentic Compatibility,"* New York: Wiley, 1992.

2 IEC, "IEC61000: Electromagnetic Compatibility," 2000.

3 SAE, *"Environmental Condition and Test for Airborne Equipment (in DO-160D),"* RTCA, Inc., 2004.

4 Department of Defense, USA, "Requirement for the Control of Electromagentic Interference Characteristics of Subsystems and Equipment (in Military Standard 461E)," 1999.

5 SAE, *"Aerospace Standard (in SAE AS 1831),"* The Engineering Society for Advancing Mobility Land Sea Air and Space, 1997.

6 Y. Y. Maillet, "High-Density Discrete Passive EMI Filter design for DC-fed Motor Drives," Master Thesis, Virginia Polytechnic Institution and State University, Blacksburg, VA, 2008.

7 S. Wang, F. Lee and W. Odendaal, "Characterization, evaluation, and design of noise separator for conducted EMI noise diagnosis," *IEEE Transaction on Power Electronics*, vol. 20, no. 4, pp. 974–982, 2005.

8 D. G. Holmes and T. A. Lipo, *"Pulse Width Modulation for Power Converters: Principles and Practice,"* Piscataway, NJ: IEEE Press, 2003.

9 B.P. McGrath and D.G. Holmes (2009). "A general analytical method for calculating inverter DC-link current harmonics," *IEEE Transactions on Industry Applications*, vol. 45, no. 5, pp. 1851–1859.

10 W. Shen, F. Wang and D. Boroyevich, "Conducted EMI characteristic and its implications to filter design in 3-phase diode front-end converters," in *IEEE IAS Annual Meeting*, Seattle, WA, 2004.

11 D. Zhang, D. Y. Chen, M. J. Nave and D. Sable, "Measurement of noise source impedance of off-line converters," *IEEE Transactions on Power Electronics*, vol. 15, no. 5, pp. 820–825, 2000.

12 M. Bartoli, A. Reatti and M. Kazimierczuk, "Minimum copper and core losses power inductor design," in *IAS Annual Meeting*, San Diego, CA, 1996.

13 F. Luo, S. Wang, F. Wang, D. Boroyevich, N. Gazel, Y. Kang and A. C. Baisden, "Analysis of CM volt-second influence on CM inductor saturation and design for EMI filters in three phase DC-fed motor drive systems," *IEEE Transactions on Power Electronics*, vol. 25, no. 7, pp. 1905–1914, 2010.

14 D. Jiang, F. Wang and J. Xue, "PWM impact on CM noise and AC CM choke for variable-speed motor drives," *IEEE Transaction on Industry Applications*, vol. 49, no. 2, pp. 963–972, 2013.

15 R. Ren, B. Liu, Z. Dong and F. Wang, "Current-bias Dependent Permeability of Powder and Amorphous Core Induced Unbalanced DM Impedance and Mixed-mode Noise," in *IEEE ECCE*, Baltimore, MD, 2019.

16 M. Schinkel, S. Weber, S. Guttowski, W. John and H. Reichl, "Efficient HF Modeling and Model Parameterization of Induction Machines for Time and Frequency Domain Simulations," in *IEEE APEC*, Dallas, TX, 2006.

17 Keysight, *"Spectrum Analyzer Basics (Application Note 150),"* Keysight, 2014.

18 Z. Wang, S. Wang, P. Kong and F. C. Lee, "DM EMI noise prediction for constant ON-time, critical mode power factor correction converters," *IEEE Transaction on Power Electronics*, vol. 27, no. 7, pp. 3150–3157, 2012.

19 L. Yang, S. Wang, H. Zhao and Y. Zhi, "Prediction and analysis of EMI spectrum based on the operating principle of EMC spectrum analyzer," *IEEE Transaction on Power Electronics*, vol. 35, no. 1, pp. 263–275, 2020.

20 S. Wang, Y. Maillet, F. Wang, R. Lai, R. Burgos and F. Luo, "Investigating EMI filter's grounding of EMI filters in power electronics systems," in *IEEE PESC*, Rhodes, Greece, 2008.

21 H. Chen, Z. Qian, Z. Zeng and C. Wolf, "Modleing of parasitic inductive couplings in a PI-shaped common mode EMI filter," *IEEE Transaction on Electromagnetic Compatibility*, vol. 50, no. 1, pp. 71–79, 2008.

22 B. Liu, R. Ren, F. F. Wang, D. Costinett and Z. Zhang, "Capacitve coupling in EMI filters containing T-shaped joint: mechanism, effects, and mitigation," *IEEE Transaction on Power Electronics*, vol. 35, no. 3, pp. 2534–2547, 2020.

23 J. Xue and F. Wang, "Mixed-mode EMI noise in three-phase DC-fed PWM motor drive system," in *IEEE ECCE*, Denver, CO, 2013.

24 D. Jiang and F. Wang, "Variable switching frequency PWM for three-phase converters based on current ripple prediction," *IEEE Transactions on Power Electronics*, vol. 28, no. 11, pp. 4951–4961, 2013.

25 L. Xing and J. Sun, "Conducted common-mode EMI reduction by impedance balancing," *IEEE Transaction on Power Electronics*, vol. 27, no. 3, pp. 1084–1089, 2012.

26 R. Chen, Z. Zhang, R. Ren, J. Niu, H. Gui, F. Wang, L. M. Tolbert, D. J. Costinett and B. J. Blalock, "Common-mode noise reduction with impedance balancing in DC-fed motor drives," in *IEEE APEC*, San Antonio, TX, 2018.

27 S. Wang, Y. Y. Maillet, F. Wang, D. Boroyevich and R. Burgos, "Investigation of hybrid EMI filters for commom-mode EMI supression in a motor drive system," *IEEE Transaction on Power Electronics*, vol. 25, no. 4, pp. 1034–1045, 2010.

28 R. Chen, J. Niu, H. Gui, Z. Zhang, F. Wang, L. M. Tolbert, B. J. B. D. J. Costinett and B. B. Choi, "Modleing, analysis, and reduction of harmonics in parallel and interleaved three-level neutral point clamped inverters with space vector modulation," *IEEE Transaction on Power Electronics*, vol. 35, no. 4, pp. 4411–4425, 2020.

29 R. Lai, Y. Maillet, F. Wang, S. Wang, R. Burgos and D. Boroyevich, "An integrated EMI choke for differential-mode and common-mode noise supression," *IEEE Transaction on Power Electronics*, vol. 25, no. 3, pp. 539–544, 2010.

30 M. H. F. Lim, "Low temperature Co-fired Ceramics Technology for Power magnetics Integration," PhD Diss., Virginia Polytechnic Institution and State University, Blacksburg, VA, 2008.

31 R. Chen, "Integrated EMI Filters for Switch Mode Power Supplier," PhD Diss., Virginia Polytechnic Institution and State University, Blacksburg, VA, 2004.

32 F. Luo, A. C. Baisden, D. Boroyevich, K. Ngo, F. Wang, P. Mattavelli, L. Coppola, N. Gazel and Y. Kang, "An improved design for transmission line busbar EMI filter," in *IEEE Energy Conversion Congress and Exposition (ECCE)*, Atlanta, GA, 2010.

9

Thermal Management System

9.1 Introduction

Thermal management is essential to power electronics systems, including three-phase AC converters. Appropriate cooling is necessary in order for electronic devices and other components to maintain a safe operating temperature. It is estimated that 55% of all electronic system failures are caused directly by thermal-related issues [1]. Additionally, the thermal management system takes up a large percentage of the total system size, weight, and/or cost. Figure 9.1 shows the weight breakdown of a typical back-to-back three-phase VSI system. It can be seen that about 30% of the converter weight comes from the heatsink and the fan. In the Chapter 6 design example, if we exclude the EMI filter, which often dominates the weight in aircraft applications, the heatsink and fan are the heaviest parts. There are a number of cooling technologies available for power electronics systems, such as air cooling, liquid cooling, and phase-change cooling, with different cooling performance, maintenance needs, size, weight, and cost. Therefore, for power electronics system design, including the three-phase AC converter design, it is important to select a suitable cooling strategy and design a thermal management system to help achieve the overall system design objective, e.g. the smallest size or the lowest cost, while meeting the system cooling requirement.

This chapter covers the design of thermal management systems commonly used for three-phase AC converters. The focus is on forced-air and forced-liquid convection cooling systems, although natural convection cooling and some advanced cooling techniques including phase change cooling (e.g. heat pipes) will also be discussed. The focus is on the design optimization of the thermal management system as a subsystem of the overall converter.

9.2 Cooling Technology Overview

The basic equation for describing heat transfer by convection is [1]:

$$q = h \cdot A \cdot \Delta T \tag{9.1}$$

where q is the rate of heat transfer (i.e. thermal power) with the unit of W, h is the heat transfer coefficient with the unit of W/(cm^2 · °C) or W/(cm^2 · K), A is the surface area with the unit of cm^2, and ΔT is the temperature difference between the heat source (e.g. a device) and the air or fluid with the unit of °C or K. In this book, we use °C as the preferred temperature unit but will also use K where appropriate. A technique that can provide a larger h can transfer more heat for a given

Design of Three-phase AC Power Electronics Converters, First Edition. Fei "Fred" Wang, Zheyu Zhang, and Ruirui Chen.
© 2024 The Institute of Electrical and Electronics Engineers, Inc. Published 2024 by John Wiley & Sons, Inc.

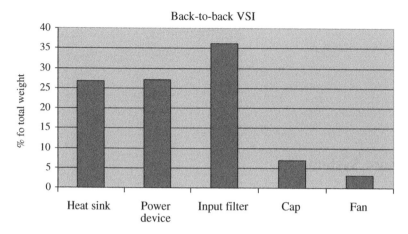

Figure 9.1 Typical weight distribution for a back-to-back VSI converter.

temperature difference or can maintain a lower temperature difference for a given transferred heat power.

Usually, power electronics applications employ basic or conventional cooling techniques, including natural convection, forced-air convection, and forced-liquid convection. These techniques can achieve thermal densities on the order of 100 W/cm^2. In most cases, they are sufficient for high-power converter applications. For certain high-density applications, higher thermal densities as high as 1 kW/cm^2 can be expected [2]. In these cases, thermal densities may be beyond the capabilities of the basic cooling methods. As a result, some advanced cooling techniques that the microelectronics industry has first developed, such as heat pipes, liquid immersion, and jet impingement, have started to be applied in power electronics. The general trend is to move the cooling medium (i.e. fluid) closer to the heat source (i.e. semiconductor) itself [1].

This section overviews the basic conventional cooling methods, as well as some of the more advanced cooling techniques. The different types of heatsinks and thermal materials are also discussed.

9.2.1 Basic Conventional Cooling Methods for Power Electronics

The most basic cooling method is natural air convection. With this method, the heat-generating power electronic components, especially, the power semiconductor devices, are placed on a heat spreading component, commonly a finned heatsink. While this approach is very simple, the finned heatsink can be very large without the help of airflow. To reduce the size of the heatsink, one can add a fan, which forces cooler air across the heatsink surface, helping heat transfer to achieve forced-air convection. With the same principle, cold liquid can be flown through a heat spreader, e.g. a cold plate, to remove the heat, achieving forced-liquid convection. These basic cooling techniques are discussed in this section.

9.2.1.1 Natural Convection

Natural convection employs a thermally conductive metal heatsink, as shown in Figure 9.2. Aluminum is a popular heatsink material. The fins of the heatsink serve to increase the surface area where the heat may be exchanged from the heatsink to air. Limiting factors in this technique include: (i) limited surface area where the flat side of the heatsink contacts the heat-generating

Figure 9.2 Example of a finned heatsink for natural convection. *Source:* Priatherm.

device; (ii) limited thermal conductivity of the heatsink material; and (iii) limited ability of the heatsink to exchange heat with the surrounding air. The heat transfer coefficient h of natural convection is limited to about 0.003 W/(cm^2 · °C) [3].

9.2.1.2 Forced-Air Convection

Forced-air convection is the next step up from natural convection. With this method, a finned heatsink is also used, but with the addition of a fan or a blower that circulates air across the surface of the heatsink fins. The heat extraction from the heat-generating device to the heatsink is the same as for natural convection, but the capability for the heatsink to exchange heat with the surrounding air is enhanced. The heat transfer coefficient h for this technique is directly proportional to the velocity of the airflow. With forced-air convection, h is limited to about 0.03 W/(cm^2 · °C) [3], which is about an order of magnitude of an improvement over natural convection. Compared with the natural convection, which only needs a heatsink, forced-air convection needs additional fan(s) or blower(s), and sometimes air ducts to control the airflow, which all add to the size, weight, and cost of the thermal management system. The fan or blower also needs to be powered and can incur power loss, albeit the loss is relatively small in high-power converter cases.

9.2.1.3 Forced-Liquid Convection

With forced-liquid convection, a liquid is used as an intermediary material to carry heat away from the heat-generating components, e.g. devices. The devices are mounted on a cold plate with embedded pipes, as shown in Figure 9.3. The cold plate transfers heat from the heat source to the liquid passing through. The liquid must then be sent to an external heat exchanger, where the heat is released into the surrounding ambient. With this method, both the velocity and the physical properties of the liquid determine the heat transfer capability.

Figure 9.3 Cold plate used in forced-liquid convection. *Source:* Boyd.

Table 9.1 shows a comparison of the properties of water and several fluorochemical coolants. As can be seen from the table, water not only provides the best heat transfer properties but also has the lowest weight per unit volume of the selected coolants. For this reason, water is an excellent choice in many applications. In some applications, including electric vehicles and electrified

Table 9.1 Thermal properties of selected liquid coolants at 1 atm. of pressure.

Fluid	Saturation temperature (boiling point) T_{sat} (°C)	Liquid density ρ_f (kg · m³)	Liquid specific heat c_{pf} (J/kg/K)	Vapor density ρ_g (kg · m³)	Latent heat of vaporization h_{fg} (kJ/kg)	Surface tension $\sigma \times 10^3$ (N/m)
FC-72	56.6	1600.1	1102.0	13.43	94.8	8.35
FC-87	32.0	1595.0	1060.0	13.65	87.93	14.53
PF-5052	50.0	1643.2	936.3	11.98	104.7	13.00
Water	100.0	957.9	4217.0	0.60	2256.7	58.91

Source: From [3]/with permission of IEEE.

aircraft, however, water cannot be used because it may freeze at low temperatures. Fluorochemical coolants have lower freezing points than water, but they generally have reduced heat transfer capability. The oil used for cooling in vehicles and aircraft systems has properties similar to these fluorochemical coolants. The heat transfer coefficient for forced-liquid convection is limited to about 1 W/(cm² · °C) for fluorochemical coolants, and about 10 W/(cm² · °C) for water [3].

In general, the forced-liquid convection thermal management system includes cold plate, liquid coolant, pump, liquid piping system, and liquid-to-ambient condenser or heat exchanger. It should be noted that in some design examples in literature, only the cold plate is considered, because the rest of the liquid cooling system is assumed to be already available. This is often the case in hybrid electric vehicles and more electric aircraft, where liquid cooling system is in place for mechanical system needs, and no additional condensers or heat exchangers are needed. However, in pure electric vehicles and electrified aircraft, the whole liquid cooling system may need to be counted as part of the electrical system.

9.2.2 Advanced Cooling Techniques

To overcome the heat transfer performance limitations of forced convection, the microelectronics community has developed cooling techniques including the use of phase change and direct liquid cooling. Phase change techniques involve the cooling liquid boiling, or changing its phase to gas, as it extracts heat from the heat-generating components. The coolant used in phase change methods must have a boiling point below the desired component operating temperature, typically 10–40 °C below the component operating temperature [3]. In direct liquid cooling, the heat-generating components are either directly submerged in the coolant, or the coolant is forced onto or across the surface of the component. Direct liquid cooling can also be combined with phase change; in this case, the coolant directly touches the component and boils as a result of the heat absorption. Because of the high amount of energy required to boil a liquid, phase change is a very effective cooling technique. Methods such as jet impingement and spray cooling are designed to meet the cooling needs of 1 kW/cm². This subsection summarizes several advanced cooling techniques that incorporate phase change, direct liquid cooling, or their combinations.

9.2.2.1 Liquid Immersion and Pool Boiling

Liquid immersion is the simplest form of direct liquid cooling. A well-known case of the liquid immersion is oil-filled transformers or circuit breakers in electric power systems, where oil is used both as a coolant and as electrical insulation. For semiconductor device cooling, liquid immersion method is usually used when the semiconductor die can be submerged; as a result, coolant must

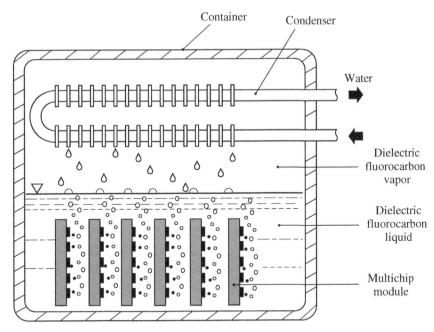

Figure 9.4 Diagram of a pool boiling system. *Source:* From [3]/with permission of IEEE.

provide electrical isolation, as in the case of the transformer oil. Although water has superior heat removal properties, it will not work as a dielectric. If this technique also employs phase change, it is then referred to as pool boiling. For this to occur, the fluid must be chosen appropriately. In either case, the coolant may be either passively circulated over the device by the effect of heated coolant rising away from the device and drawing cooler fluid up or by an active system that circulates coolant with force. Pool boiling cooling needs a condenser to convert the coolant back into liquid, thereby releasing thermal energy to the ambient air, or to another medium, as shown in Figure 9.4, which is a diagram of the pool boiling with no forced flow [3]. Figure 9.5 is a picture of an actual cooling system using pool boiling with a forced flow [4].

9.2.2.2 Heat Pipes

Heat pipes are a form of phase change cooling, where the coolant is enclosed in a pipe that is either embedded in the device to be cooled or in a cold plate. The entire pipe is a closed system, as such, the coolant is not pumped through the pipe; instead, the coolant boils near the heat source and travels toward the condenser end of the heat pipe, where the liquid condenses. Since these systems cannot always rely on being oriented such that gravity pulls the coolant back down to the heat source, they often incorporate a wicking material enclosed in the pipe. The wick draws the condensed coolant back toward the heat source [1]. Figure 9.6 shows a cutaway picture of a system employing heat pipes for cooling. The devices to be cooled are packaged on the left. After undergoing the phase change process, the vapor coolant travels to the condenser section on the right. The condenser

Figure 9.5 Picture of pool boiling with forced flow. *Source:* From [4].

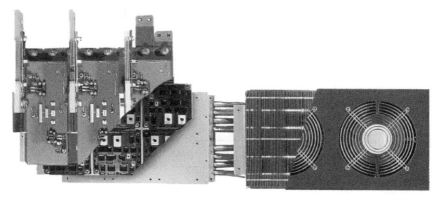

Figure 9.6 Cutaway picture of a heat pipe cooling system.

exchanges heat from the coolant in the pipes into the air, condensing the coolant so that it may be wicked back to the devices on the left. Microelectronics applications have also used wickless micro heat pipes, where the diameter of the pipe is defined as being less than 1 mm.

9.2.2.3 Jet Impingement

Jet impingement is a cooling method where the liquid coolant is forced out of a small nozzle under high pressure onto the heat source. In most cases, the jet is directed onto the semiconductor die itself. The jet can pass through a gaseous, enclosed environment (as shown in Figure 9.7) or onto a submerged die. A jet impingement system can also incorporate phase change if the coolant is designed such that it vaporizes as it impacts the die. A disadvantage of jet impingement is the possibility of surface erosion due to the high impact force [3].

9.2.2.4 Spray Cooling

Spray cooling is similar to jet impingement, except that the liquid is broken up into a wide spray of droplets before impacting the surface. Sprays can be formed either by high-pressure liquid forced through a wide opening (Figure 9.8) or by atomizing the liquid using high-pressure air. While atomized sprays achieve better performance, they are extremely difficult to realize in a practical system. Unlike jet impingement, spray cooling does not erode the device surface. The disadvantage,

Figure 9.7 Jet impingement directly onto the semiconductor die. *Source:* From [4].

Figure 9.8 Spray cooling using a high-pressure nozzle. *Source:* From [4].

however, is that the spray nozzles are more likely to clog [3]. Spray cooling can be very effective, as [5] suggests that a heat flux as high as 1 kW/cm^2 can be achieved even without phase change.

9.2.2.5 Microchannel/Mini-channel Liquid Convection

Using channels on the order of 10 μm in diameter, microchannel cooling pumps liquid coolant directly onto the heat-generating device, or through a heatsink attached to the device. There is also mini-channel cooling, where the channel diameter is on the order of 1 mm. As with most advanced cooling techniques, microchannels can be combined with phase change, where the liquid coolant is boiled inside the channel, close to the heat source [3]. One potential issue with micro- or mini-channel cooling is its requirement on purity of the coolant to avoid clogging.

Most of the advanced phase change and direct cooling methods are yet to be widely used for power electronics, especially, not yet for high-power three-phase AC converters. One exception is heat pipes, which have been used successfully in high-power converter applications, such in the GE Innovation Series motor drives in late 1990s. Practically, all advanced cooling techniques involve liquid cooling. Many of these cooling techniques require relatively bulky (often outdoor) heat exchangers or condensers. Use of these techniques makes more sense when the existing cooling system is already available and can be leveraged, or when power electronics equipment is indoor.

9.2.3 Comparison of Cooling Technologies

Figure 9.9 and Table 9.2 summarizes the heat transfer coefficient ranges for different categories of cooling methods. Figure 9.10 depicts temperature rise ΔT between the heat-generating device and the ambient temperature versus the dissipated power q for various thermal resistance R_{th} values. The curves in Figure 9.10 is based on the simple equation of (9.2). Using this figure, one can quickly determine what range of thermal resistance needs to be achieved between the device and ambient for an effective thermal system design.

$$\Delta T = \frac{q}{R_{th}} \tag{9.2}$$

For an example Si IGBT module (1200 V/200 A Infineon FF200R12KT3 dual-pack module) that may be used in a three-phase AC power converter, its surface area is 7.65 cm^2. If this module dissipates a maximum of 500 W, a cooling method is needed to provide a surface heat flux of at least

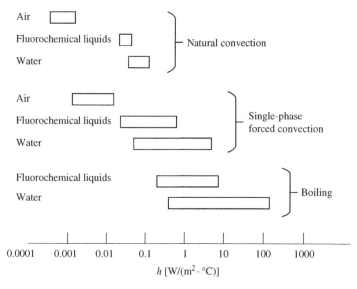

Figure 9.9 Heat transfer coefficient ranges for various cooling methods. *Source:* From [3]/with permission of IEEE.

Table 9.2 Typical *h* value ranges for different cooling schemes.

Cooling type	Heat transfer coefficient [W/(cm² · °C)]	Comments
Radiation	<0.1	Black body, 120 °C to room temperature environment
Air, free convection	0.0003–0.0012	Typical about 0.0005
Air, forced convection	0.001–0.01	Typical about 0.005
Liquid, forced convection	0.02–0.2	Fluorocarbons
Liquid, forced convection	0.2–0.7	Water and water/glycol
Boiling	0.2–0.7	Fluorocarbons

Source: Adapted from [3].

65.4 W/cm². If the heatsink temperature is limited to 85 °C, and assuming an ambient temperature of 60 °C, then the temperature rise for the heatsink will be 25 °C. According to Figure 9.11, forced-air convection will provide adequate cooling capability. If higher temperature rise is allowed, less cooling will be needed. However, the true limit for temperature rise is the device junction temperature. Assuming the 500-W loss is evenly generated by the four devices, two IGBTs and two diodes, then the maximum junction temperature will be seen by the diodes. Given that the thermal resistance from the junction to heatsink for the diodes is 0.26 °C/W, the temperature rise from the heatsink to junction for the diodes is 32.5 °C. As a result, the maximum junction temperature for the diodes is 117.5 °C, which is just slightly below the maximum allowable junction temperature of 125 °C. Since the IGBTs have a lower junction to heatsink thermal resistance, their maximum junction temperature is lower. In order to improve the power density and dissipate more heat, a more powerful cooling method will be needed, e.g. forced-liquid convection cooling or phase change cooling.

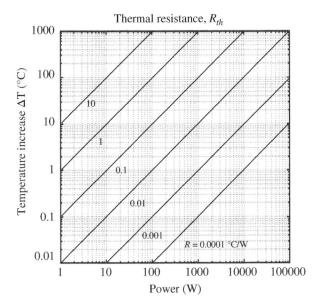

Figure 9.10 Temperature difference between device and ambient versus power dissipated for various thermal resistance values.

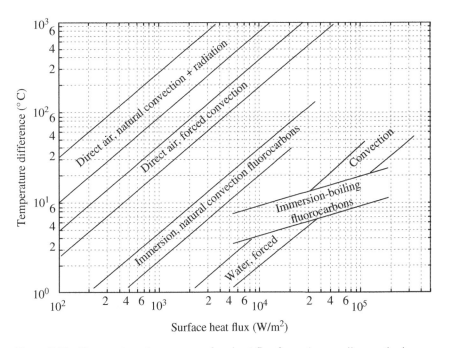

Figure 9.11 Temperature rise versus surface heat flux for various cooling methods.

9.2.4 Heatsinks and Other Components

9.2.4.1 Heatsink Types and Configurations

For a given cooling method, heatsinks to large extent determine the cooling performance. Both heatsink configurations and materials are important for their performance. For air cooling heatsinks, metal structures with fins are typical configurations for increased heat transfer. Heatsinks

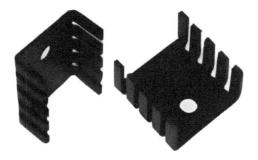

Figure 9.12 Stamping heatsinks.

can be classified by the way they are fabricated. Figure 9.12 shows the stamped heatsinks, which are made by stamping copper or aluminum sheets into desired shapes. Stamped heatsinks are a low-cost solution to low-density thermal needs. Figure 9.13 shows casting heatsinks, which use sand, lost or fusible core, or die casting to produce high-density aluminum or copper/bronze pin fins. Casting heatsinks can yield with excellent performance for impingement cooling. They are more expensive but can be made into more sophisticated shapes. Figure 9.14 shows the example of heatsinks made through extrusion of metals or metal alloys. Extrusion allows the formation of elaborate 2D shapes capable of dissipating large heat loads. As a result, the extrusion heatsinks are the most popular heatsink type for power converters. The design limits for extrusion heatsinks are usually set by: fin height-to-gap aspect ratio, minimum fin thickness-to-height, and maximum base-to-fin thickness. A fin height-to-gap aspect ratio up to 6 and a minimum fin thickness of 1.3 mm are attainable with standard extrusion, while a 10 to 1 aspect ratio and a fin thickness of 0.8 mm are attainable with special design. Aluminum alloys are used in both casting and extrusion heatsink; however, the aluminum alloys for casting heatsinks typically have lower thermal conductivity than aluminum alloys used for extrusion heatsinks.

Figure 9.15 shows heatsinks with bonded/fabricated fins. Thermally conductive aluminum filled epoxy is used to bond planar fins on a grooved extrusion plate. With this type of heatsinks,

Figure 9.13 Casting heatsinks.

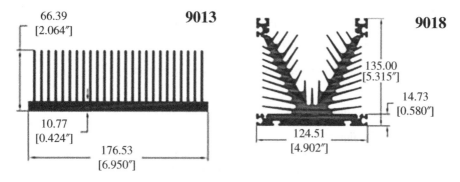

Figure 9.14 Extrusion heatsinks [products of R-theta].

manufacturing technology limits for extruded profiles like fin spacing can be overcome, and greater
fin height-to-gap aspect ratio can be achieved
(~20–40). The surface of the heatsink is increased
by a higher number of fins through keeping the same
size of the baseplate, and the cooling capacity is
increased without increasing the volume. However,
bonding materials like epoxy may have a much lower
thermal conductivity, and their impact on overall
thermal resistance needs to be considered.

Figure 9.16 shows the heatsinks with folded fins.
In this case, aluminum or copper sheet metal is
folded into fins and then attached to a baseplate or
directly to the heat surface via brazing or epoxying.
Due to the availability and fin efficiency limitations,
folded fins are not suitable for high-profile heatsinks.
Folded fins allow the fabrication of high-
performance heatsinks when extrusion or bonded

Figure 9.15 Bonded/fabricated fins. *Source:*
Bhoomi Heatsinks.

fins are not able to meet certain performance requirements, e.g. weight requirement.

With the advancement of additive manufacturing, 3D printing can be a promising approach to
make heatsinks with sophisticated geometry to achieve better cooling performance. Figure 9.17
shows a 3D-printed aluminum air-cooled heatsink for a 50-kW three-phase inverter with a
1000-V DC link. With a power density constraint 100 W/in^3, the design goal of the heatsink was
to minimize the maximum junction temperature of the semiconductor devices [6].

In addition to classification by manufacturing methods, heatsinks can also have different fin
geometry. Fins can be rectangular, square, or round (i.e. pin fins). In forced convection cooling,
generally, rectangular fins have better performance than square fins, whose back edges have poorer
airflow passing through. Rectangular fins also have better performance than round fins; however,
pressure drop is also higher for rectangular fins and therefore requires more powerful fans. Round
fins are good if the airflow direction is uncertain or if airflow may not be straight through the heatsink. The performance of pin fin is being debated. Usually in liquid cooling, pin fin has more benefit, as well as good manufacturability of combined pin fin and baseplate. In recent years, pin fin
heatsinks get more used due to their better performance than flat fins in certain applications,
although the true heatsink performance must be determined considering airflow.

For liquid cooling, metal cold plates with embedded cooling loops are generally used, as shown in
Figure 9.3. The geometry of the liquid loops can be simple circles or squares but can also be designed

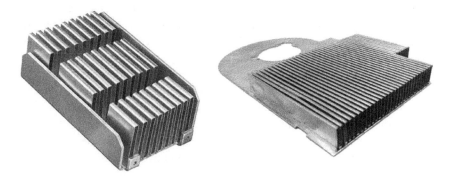

Figure 9.16 Folded fins.

Figure 9.17 A 3D-printed aluminum air-cooled heatsink. *Source:* From [6]/with permission of IEEE.

to be more sophisticated for improved heat transfer. For example, fin structures, which can help to create turbulent flow for more efficient heat transfer, have been used in the liquid cooling loops for some electric vehicle inverters. There are also micro- and mini-channel structures, as already mentioned earlier.

9.2.4.2 Thermal Materials

Materials with good thermal properties may be used throughout the construction of a power converter, from substrate of a semiconductor device module, to heat spreaders, thermal interfaces, and heatsinks. For effective removal of heat, it is desirable to use materials with high thermal conductivity. However, an equally critical consideration is matching the coefficient of thermal expansion (CTE) at material interfaces. Poorly matched CTEs can result in excess mechanical stresses due to thermal or power cycling, leading to premature failures.

In [7], thermal materials are classified as first-, second-, or third-generation materials. A summary of materials and their typical properties is presented in Table 9.3. First-generation materials consist mainly of the traditionally used aluminum, copper, and their alloys. While copper offers high thermal conductivity, it also has a high weight density, resulting in heavy heatsinks. The rather high CTE of copper is an additional concern. A lighter weight alternative is aluminum. Its weight density is 30% of that of copper. Its drawback is lower thermal conductivity and even greater CTE. Still, aluminum has been a popular low-cost choice for lightweight applications.

To reduce the size and weight of thermal management solutions, more advanced materials with higher conductivity and lower density have been developed. The bulk of second- and third-generation materials are composites, in which a matrix material is used for structure and the reinforcement material is used to manipulate the thermal properties. Also included in third-generation materials are exotic materials such as chemical vapor deposition (CVD) diamond, highly oriented pyrolytic graphite (HOPG), and natural graphite. One thermal material that has received considerable attention is graphite. Other promising thermal materials include carbon nanotube. Many of these advanced materials, including graphite, do not have the same thermal conductivity in all directions, unlike copper and aluminum. Instead, their thermal conductivity is very high in the in-plane direction through the material but considerably lower in the direction normal to that plane. In Table 9.3, the "in-plane" thermal conductivity is the higher conductivity, while the "through" thermal conductivity is the lower conductivity in a single direction. This characteristic adds some limitations on the material use; on the other hand, it can actually be quite useful with proper design. The orientation of the material can be chosen when fabricating the heat spreader, heatsink, etc., such that the thermal conductivity is extremely high in two dimensions and low in

Table 9.3 Typical properties of thermal materials.

Reinforcement	Matrix	Thermal conductivity [W/(m · °C)] In-plane	Through	CTE[a] (ppm/°C)	Density (g/cm³)
First-generation materials					
	Copper	400		17	8.9
	Aluminum	218		23	2.7
Second-generation materials					
Natural graphite	Epoxy	370	6.5	−2.4	1.94
Continuous carbon fibers	Polymer	330	10	−1	1.8
Continuous carbon fibers	SiC	370	38	2.5	2.2
Discontinuous carbon fibers	Copper	300	200	6.5–9.5	6.8
SiC particles	Copper	320		7–10.9	6.6
Carbon foam	Copper	350		7.4	5.7
Third-generation materials					
	CVD diamond	1100–1800		1–2	3.52
	HOPG	1300–1700	10–25	−1.0	2.3
	Natural graphite	150–500	6–10	−0.4	
Continuous carbon fibers	Copper	400–420	200	0.5–16	5.3–8.2
Continuous carbon fibers	Carbon	400	40	−1.0	1.9
Graphite flake	Aluminum	400–600	80–110	4.5–5.0	2.3
Diamond particles	Aluminum	550–600		7.0–7.5	3.1
Diamond and SiC particles	Aluminum	575		5.5	
Diamond particles	Copper	600–1200		5.8	5.9
Diamond particles	Cobalt	>600		3.0	4.12
Diamond particles	Silver	400–600		5.8	5.8
Diamond particles	Magnesium	550		8	
Diamond particles	Silicon	525		4.5	
Diamond particles	SiC	600		1.8	3.3

[a] CTE is the in-plane CTE for materials with different in-plane and through characteristics.
Source: Adapted from [7].

the third dimension. This can be useful in evenly spreading the heat from multiple sources, or even directing the heat flow to a cooler area.

Currently, the advanced second- and third-generation materials are mostly used in the baseplate for power semiconductor devices or module packages, or in heat spreaders. For lightweight applications, the promising materials include natural graphite, natural graphite/epoxy, HOPG, continuous carbon fiber/carbon, and graphite flake/aluminum. All of these materials have a weight density lower than aluminum but a thermal conductivity greater than copper.

Figure 9.18 shows the examples of heatsinks with graphite fins from Dreyer Systems. Graphite has a thermal conductivity of 370 W/(m · °C), equal to copper, but 75% lighter than copper, and

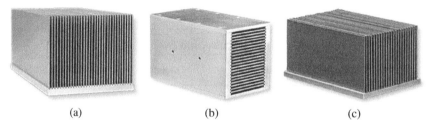

(a)   (b)   (c)

Figure 9.18 Heatsinks with natural graphite fins and different bases: (a) nickel-plated copper; (b) aluminum; and (c) copper. *Source:* From [8].

even 25% lighter than aluminum. One issue with the graphite is its negative CTE, which may be compensated with metal meshed with graphite.

9.2.4.3 Thermal Interface Materials

Since both the devices and heatsinks have solid surfaces, their contacts cannot be perfect, which can cause poor thermal conduction and poor cooling performance. It is therefore necessary to apply a soft compliant thermal interface material (TIM) between the device baseplate and heatsink, which will fill gaps and imperfections to minimize the interface thermal resistance. Since any gaps and poor surface contacts can cause great degradation of cooling performance, TIMs are critical to an efficient thermal management system design.

There are several different types of TIMs, including phase change materials, gap fillers, thermal grease, and thermal pads. Phase change materials are solid at room temperature and begin to soften and become a viscous liquid at an elevated temperature. They can fill the microscopic imperfections between the heat source and the heatsink, reducing the thermal interface resistance. Gap-filling materials are soft and compliant, allowing them to fill any large gaps between the heat source and heatsink. Thermally conductive grease can be applied in thin layers to help with the thermal transfer between surfaces that are typically smooth and flat, such as the surfaces between a device baseplate and a heatsink. Thermal pads are used between the device baseplate and heatsink with added capability of providing electrical insulation.

All these types of TIMs have been used in power converters. Gap-filling materials are seen more in passive components and certain device packages to help thermal transfer. Thermal grease is most popular for device cooling. Its advantage includes superior thermal conductivity and far less sensitivity to mounting pressure than other thermal interface materials. The thermal conductivity of thermal grease ranges from less than 1 W/(m·°C) to close to 10 W/(m·°C).

The thickness of the thermal grease will impact its thermal resistance. Typically, a thickness of around 0.1 mm is applied between the device baseplate and the heatsink to ensure it is thick enough for good contact, while not too thick to sacrifice thermal performance. It is difficult to apply an even layer of thermal grease manually. A stencil can be used to improve the process. One example of thermal grease use is for a 10-kV discrete SiC device, which has a baseplate dimension of 26.7 mm by 47.2 mm. The device was used in a medium-voltage converter to be described further in Chapter 11. The thermal grease S606P-50 from T-Global Technology was used, which has a thermal conductivity of 8 W/(m·°C). With a thickness of 0.1 mm, the estimated thermal resistance is

$$R_{th-TIM} = \frac{0.1 \text{ mm}}{8 \text{ W/(m} \cdot {}^\circ\text{C}) \times (26.7 \text{ mm} \times 47.2 \text{ mm})} \approx 0.01 \, {}^\circ\text{C/W}$$

Sometimes, it is necessary to electrically insulate semiconductor devices from the heatsink. One example is devices in a TO-220 package, which is non-isolated. In this case, thermal grease may not be suitable, since it is difficult to eliminate voids in the grease layer, even though the grease is electrically nonconductive. Alternatively, thermally conductive but electrically insulating thermal pads can be used as the TIM.

Thermal pads are less compliant. However, with proper mounting pressure, their thermal performance can be quite good, close to that of thermal grease. For example, thermal pad TG-A486G-150-150-0.5-1A also from T-Global Technology is 0.5 mm thick and its thermal conductivity is 6.3 W/(m · °C). Its thermal conductivity can be as good as the thermal grease, but it is thicker. For a device in a TO-247 package, the thermal resistance is 0.212 °C/W, which is still quite good. In this case, the thermal pad can provide good electrical insulation (13 kV/mm). Thermal pads are also easy to use. There are also phase change compound-based thermal pad, which can be much thinner. For example, CD-02-05-247-N from Wakefield-Vette can provide insulation up to 9.2 kV and has a thickness of 0.076 mm. For a TO-247 package, the thermal resistance is 0.1844 °C/W.

9.2.4.4 Fan/Blower, Pump, Duct, Pipe, and Exchanger

For forced convention cooling, heatsink is only one part of the thermal management system. For forced-air cooling, fan/blower and air duct will impact the thermal performance, and at the same time contribute to the cost, size, and weight of the thermal management system. For forced-liquid cooling, it is the pump, pipe, and heat changer. Very often, there is a trade-off relationship among different components within a thermal management system.

Figure 9.19 illustrates a typical fan performance curve and a heatsink convection curve. The intersection point of the two curves determines the operating point of the thermal management system. The heatsink convection curve shows that higher airflow will require higher pressure, with nearly a linear relationship. The fan performance curve is a plot of pressure rise and power required versus volumetric flow rate. Pressure of the fan decreases as airflow increases, which means that in order to achieve a higher power flow, a fan with higher pressure is needed. The shape of the power curve depends on the blade type, which usually can be expressed as a fifth-order polynomial [9].

Clearly, in forced-air cooling system, the fan can be a crucial component affecting both convection and thermal management system objective function, e.g. size or weight. Figure 9.20 shows airflow curves for a group of standard tube-axial fans weighing from 7.5 to 20 g. The fan laws [10, 11] indicate the relationship between fan volume, airflow, and static pressure. On the other hand, if the weight is the design target, there is no explicit equation describing the relationship between fan weight and airflow curve.

The dimension of the air duct is usually larger than those of the heatsink and fan, and there will be a certain amount of bypass flow. This is strongly related to the cross-sectional geometry of the duct and the pressure drop across the heatsink. With some simple empirical equations in [12] and some of the approximations, equations have been derived to predict the airflow produced by the heatsink, fan, and duct. In general, an analytical thermal model can be built to calculate thermal resistance, as will be presented in Section 9.4.1.

Figure 9.19 Fan operating point in relation to the heatsink convection and fan performance curves.

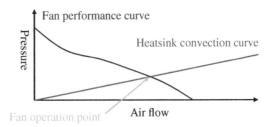

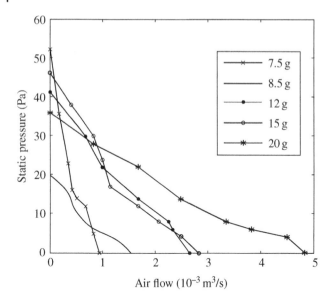

Figure 9.20 Typical relationship of fan weight, airflow, and static pressure. *Source:* From [9]/IEEE.

9.3 Thermal Management System Design Problem Formulation

As discussed earlier, the forced-air and forced-liquid convections are still the most popular cooling methods for medium- and high-power converters, including three-phase AC converters. The forced-air convection cooling is even more popular due to its simplicity, low cost, and low maintenance needs. Therefore, forced-air cooling thermal management system will be used here as the example for design discussion. The simplest and most popular forced-air cooling system is a plate-fin heatsink with horizontal inlet cooling stream, as shown in Figure 9.21, with a suitable fan mounted to face the fins, a vent in the opposite direction of the fan, and a simple box duct.

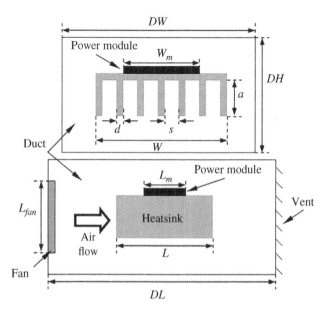

Figure 9.21 Geometrical configuration of heatsink, fan, and duct (top: front view; bottom: side view).

The thermal management system design for a forced-air cooling system needs to select or design the heatsink, fan, duct, and TIM. If commercial off-the-shelf parts are used, we simply need to select them based on types (including materials and geometry for the heatsink) and size. Here, the size refers to physical size and/or rating. If a custom heatsink is used, we will also need to select material and design the detailed geometry as shown in Figure 9.21.

9.3.1 Thermal Management System Design Variables

Based on aforementioned discussion, the design variables include the following:

1) Heatsink: for off-the-shelf commercial products, this means the type and dimension for a given geometry of the heatsink; for custom heatsinks, this means material, dimensions as shown in Figure 9.21, including fin height a, base thickness b, number of fins n, fin width d, channel width s, and heatsink width W, and heatsink length L. Note that b and n are not marked in Figure 9.21. For high-power converters, there can be many power device modules. As a result, sometimes more than one heatsink will be used for better cooling design. In this case, the number of heatsinks is also a design variable.
2) Fan or blower: this means the fan type, rating, and number. The type refers to specific manufacturer and type (e.g. DC vs. AC, metal vs. plastic). The rating refers to power and pressure. The number refers to how many fans will be needed, as sometimes multiple small fans are preferred to one large fan.
3) Air duct: air duct design includes its width, length, height, and material.
4) TIM: this includes type and thickness.

9.3.2 Thermal Management System Design Constraints

The thermal management system design constraints also include performance constraints and physical constraints. For a forced-air cooling system, the common performance specifications include:

1) Maximum thermal impedance $Z_{th-TMS(max)}$ for thermal management system, which will help to keep the junction temperatures of the devices below their maximum permissible value T_{jmax}. The thermal impedance should include the thermal impedance of the heatsink-to-ambient Z_{th-ha} and the thermal impedance of the device baseplate-to-heatsink Z_{th-bh}, which is usually dominated by TIM. For a converter without transient performance requirements, e.g. without overload, thermal resistances are sufficient; if transient thermal performance needs to be considered, then thermal capacity or thermal capacitance will also be needed. Since TIM is usually a very thin layer, only its thermal resistance R_{th-bh} needs to be considered. In some device datasheet, R_{th-bh} is specified or included as part of the device junction-to-heatsink thermal resistance R_{th-jh}.
2) Maximum heatsink temperature T_{hs-max}. This is an optional constraint. In some converters, especially, low-voltage low-/medium-power converters (e.g. 100s of volts and 10s of kW level), the heatsink is usually grounded to the converter chassis, which could be in touch with other equipment or even would be touched by humans. In this case, it makes sense to limit the maximum heatsink temperature, for example, below 80 °C.
3) Minimum heatsink surface dimension. As can be seen in Figure 9.21, the heatsink needs to accommodate the device. The width and length of the heatsink need to be larger than the width and length of the device baseplate, respectively.

9.3.3 Thermal Management System Design Conditions

The thermal management system interfaces directly with the inverter and/or rectifier power stage, the mechanical system, and the ambient. The design conditions are from these subsystems and the ambient. The design input from the inverter or the rectifier power stage include:

1) Maximum steady-state power loss of the devices, including the conduction loss and switching loss, which have been determined in Chapter 6. Note that power loss is generally not constant for AC three-phase converters over a line cycle, even under the steady-state conditions. It is usually necessary to have the instantaneous power loss over a line cycle for the thermal management system design. Also, the power loss per individual device or device module will be needed.
2) Maximum overload power loss of the devices and its duration. Many inverters such as motor drives have overload requirement for certain period of time. The corresponding power loss and duration will influence the cooling design and therefore are needed as design input.
3) Device or device module thermal characteristics
4) Device mechanical dimensions. These will be used to set the heatsink dimension constraint.

The ambient temperature T_a is an input condition to the thermal management system design. Usually, the maximum ambient temperature is a key parameter.

There are no input parameters from the mechanical system to the thermal management system.

It should be pointed out that passive components, especially magnetic components, also need cooling. On the other hand, the cooling design for the magnetic components are less standard in power converters. As a result, the design conditions from the EMI filter and the passive components of the inverter/rectifier are not included in this section. The cooling design of the passives will be discussed in Section 9.7.2.

9.3.4 Thermal Management System Design Objectives and Design Problem Formulation

As in other cases, the design objectives for the thermal management system should follow those of the whole converter. If the design objective for the target motor drive is minimum cost, size or weight, the thermal management system design objective should also be associated with cost, size, or weight correspondingly.

Based on the selected design objectives, and the design variables, constraints, and conditions, the overall thermal management system design problem can be formulated as the following optimization problem:

$$\text{Minimize or maximize}: \quad F(X) - \text{objective function}$$

Subject to constraints:

$$G(X,U) = \begin{bmatrix} g_1(X,U) \\ g_2(X,U) \\ g_3(X,U) \end{bmatrix} = \begin{bmatrix} \text{TMS thermal resistance} \\ \text{TMS thermal impedance} \\ -\text{Heatsink dimesnions} \end{bmatrix} \leq \begin{bmatrix} R_{th-TMS(max)} \\ Z_{th-TMS(max)} \\ -\text{Device dimensions plus margins} \end{bmatrix}$$

where design variables and design conditions:

$$
X = \begin{bmatrix} x_1 \\ x_2 \\ x_3 \\ x_4 \end{bmatrix} = \begin{bmatrix} \text{Heatsink} \\ \text{Fan} \\ \text{Air duct} \\ \text{TIM} \end{bmatrix}
$$

$$
U = \begin{bmatrix} u_1 \\ u_2 \\ u_3 \\ u_4 \\ u_5 \end{bmatrix} = \begin{bmatrix} \text{Maximum steady state power loss for devices} \\ \text{Maximum overload power loss and duration for device} \\ \text{Device thermal charateristics} \\ \text{Device dimensions} \\ \text{Ambient temperature} \end{bmatrix} \tag{9.3}
$$

The thermal management system design optimization problem can be illustrated also as in Figure 9.22.

It should be mentioned that the optional constraint of the heatsink temperature is not included in the optimization problem (9.3) and Figure 9.22. Also, there can be some important intermediate design variables, including airflow velocity and fan pressure.

In addition to design variables, there are also attributes as a result of the thermal management system design. In this case, the heatsink can result in parasitic capacitance, which can impact CM current and EMI noise, and even loss in some higher voltage converters. The design attributes will be further discussed in Section 9.6.

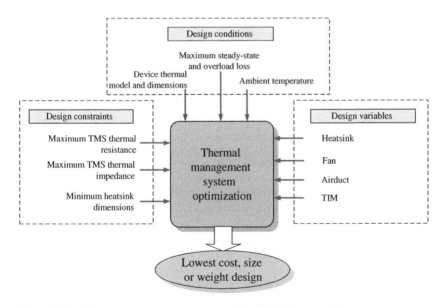

Figure 9.22 Thermal management system design optimization problem.

9.4   Thermal Management System Models

In order to solve the thermal management system design optimization problem, the relationships must be established among the design variables, constraints, conditions, and objectives. Specifically, design constraints (output) G as functions of design conditions (input) U and variables X need to be established. The set of design variable X values making up a design will be associated with one value of the objective function $F(X)$.

Based on Eq. (9.3) and Figure 9.22, we need the following models for the constraints as functions of design conditions and design variables:

1) Thermal impedance
2) Heatsink dimensions

The detailed parameters to be used in a heatsink and fan for forced-air cooling system are listed in Table 9.4. Note that most of dimension parameters for the heatsink, duct, and fan are illustrated in Figure 9.21. Also, in this case, weight is assumed as the design objective.

9.4.1   Thermal Impedance

In an actual design of the thermal management system, there are two common methods. One method is based on commercial off-the-shelf heatsinks, and the other is based on custom-designed heatsinks. We will discuss these two methods separately.

9.4.1.1   Commercial Off-the-Shelf Heatsink Case

For commercial off-the-shelf heatsinks, their thermal characteristics are usually given in the datasheet. Figure 9.23 shows an example of heatsink characteristics in a datasheet. These characteristics include thermal resistance of different heatsinks as function of air velocity, as well as the pressure drop as function of air velocity. Note that all of these heatsinks have a base of 130 mm by 70 mm and 0.55 mm elliptical fins. Their differences lie in the height of the heatsinks. From top to bottom, the curves correspond to heatsink heights of 8, 10, 12, 15, 20, 25, and 30 mm, respectively.

The heatsink selection or design is to find a heatsink large enough to hold all the devices and at the same time can have the thermal resistance smaller than $R_{th-TMS(max)}$ subtracting the TIM thermal resistance. Since the heatsink (i.e. heatsink-to-ambient) thermal resistance R_{th-ha} is a function of air velocity, for a given commercial heatsink, a minimum air velocity has to be maintained. For example, for the lowest heatsink (i.e. the heatsink with the shortest 8 mm fins) represented in Figure 9.23, in order to achieve a heatsink thermal resistance $R_{th-ha} = 0.6\,°C/W$, a 3 m/s air velocity is needed. Given that the pressure drop for these heatsinks is 4 mm H_2O or 0.01 psi, a fan will be

Table 9.4   Detailed parameters for forced-air cooling system thermal model.

Category	Parameters
Heatsink geometry	Fin height, baseplate thickness, number of fins, fin width, channel width, and heatsink length
Duct geometry	Duct width, duct length, and duct height
Fan	Airflow curve, fan weight, and fan dimension
Material properties	Thermal conductivity of heatsink material, thermal conductivity of airflow, and density of heatsink material

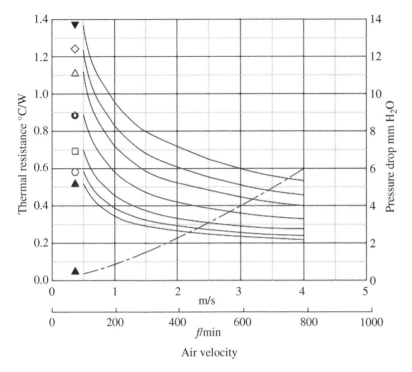

Figure 9.23 Example of heatsink characteristics in a datasheet [Alpha UB Series].

needed with a performance curve intersecting with the heatsink convection curve at a point beyond 3 m/s. The procedure to generate the thermal resistance of a commercial-based thermal management system design is shown in Figure 9.24. In this case, the minimum weight is assumed to be the design objective, and the thermal resistance of the heatsink corresponding to the minimum weight (including heatsink and fan) will be the selection result.

Considering TIM, the total thermal resistance of the thermal management system is

$$R_{th-TMS} = R_{th-ha} + R_{th-TIM} \tag{9.4}$$

The thermal capacitance of the thermal management system is dominated by the heatsink, which can be determined via the weight of the heatsink as in (9.5), where c_{hs} is the specific heat capacity and W_{hs} is the weight of the heatsink. c_{hs} for aluminum and copper, together with some other useful properties, can be found in Table 9.5.

$$C_{th-ha} = c_{hs} W_{hs} \tag{9.5}$$

9.4.1.2 Custom-Designed Heatsink Case

Much of the discussion in this part is based on [9], which utilizes other previous works, including references [12, 13]. For a custom-designed heatsink with the geometry in Figure 9.21, its total thermal resistance considering forced-air convection is the combination of the thermal resistances of the heatsink base and fins. Considering the fin effect, the total thermal resistance can be written as (9.6), where k is the thermal conductivity of the heatsink material; b, W, L, and s are heatsink dimension parameters defined in Figure 9.21; n is the number of heatsink fins; h is the convective heat transfer coefficient; and R_{th-fin} is the thermal resistance of a single fin:

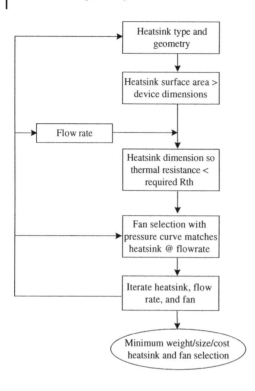

Figure 9.24 Design or selection procedure for commercial heatsink and fan.

Table 9.5 Selected thermal properties of aluminum and copper.

	Aluminum	Copper
Density (kg/m³)	2700	8960
Thermal conductivity (W/(m · °C))	238	400
Specific heat capacity (J/(kg · °C))	900	385
Volumetric heat capacity (J/(m³ · °C))	2.43×10^6	3.45×10^6

$$R_{th-ha} = \frac{b}{k \cdot L \cdot W} + \frac{1}{\frac{n}{R_{th-fin}} + h \cdot (n-1) \cdot s \cdot L} \tag{9.6}$$

From [14], R_{th-fin} can be expressed as in (9.7), where m is the fin parameter and can be approximated as (9.8). In Eqs. (9.7) and (9.8), P_c and A_c are the perimeter and the area of the heatsink side as can be seen from Figure 9.21, and can be found as $P_c = 2(a + L)$ and $A_c = (a \cdot L)$, respectively.

$$R_{th-fin} = \frac{1}{\sqrt{h \cdot k \cdot P_c \cdot A_c}} \frac{1}{\tan h(m \cdot a)}, \tag{9.7}$$

$$m \approx \sqrt{\frac{h \cdot P_c}{k \cdot A_c}} \tag{9.8}$$

As discussed earlier, the convection heat transfer coefficient h [unit: W/(m² · °C)] represents the effectiveness of a cooling system and therefore is one of the most important parameters for the heatsink thermal model. The h coefficient can be determined using the model developed in [15] as in (9.9), where k_f represents the fluid thermal conductivity [unit: W/(m·°C)] and Nu is Nusselt number:

$$h = \frac{Nu \cdot k_f}{s} \tag{9.9}$$

Nu is dimensionless and represents the ratio of convective to conductive heat transfer across the convection boundary. It can be calculated from the fully developed and developing channel flow presented in [15] as in (9.10), where Re_s^* is the adjusted channel Reynolds number and P_r is the Prandtl number. Both Re_s^* and P_r are dimensionless.

$$Nu = \left[\left(\frac{Re_s^* \cdot P_r}{2} \right)^{-3} + \left(0.664 \cdot \sqrt{Re_s^*} \cdot P_r^{\frac{1}{3}} \cdot \sqrt{1 + \frac{3.65}{\sqrt{Re_s^*}}} \right)^{-3} \right]^{-\frac{1}{3}} \tag{9.10}$$

P_r represents the ratio of viscosity and thermal diffusivity. For air, $P_r = 0.7$. Re_s^* considers the channel parameters and is defined as in (9.11), where Re_s is Reynolds number and represents the ratio of inertial forces to viscous forces. Re_s can be determined as in (9.12), where V_{ch} is the average channel velocity of the air in m/s and ν is the kinematic viscosity of the air in m²/s. ν can be obtained by (9.13), where μ and ρ are viscosity [unit: kg·s/m] and density [unit: kg/m³] of the air, respectively.

$$Re_s^* = \frac{Re_s \cdot s}{L} \tag{9.11}$$

$$Re_s = \frac{s \cdot V_{ch}}{\nu} \tag{9.12}$$

$$\nu = \frac{\mu}{\rho} \tag{9.13}$$

The average channel velocity V_{ch} can be determined through the free stream or approaching velocity V_f by (9.14) based on conservation of mass. V_f can be considered equal to the duct airflow velocity of the fan.

$$V_{ch} = V_f \cdot \left(\frac{s + d}{s} \right) \tag{9.14}$$

The required pressure drop of the heatsink can be obtained [16] as in (9.15), where f_{app} is the apparent friction factor for a hydrodynamically developing flow, K_c is the coefficient of a sudden contraction of heatsink channel, and K_e is the coefficient of a sudden expansion of heatsink channel. K_c and K_e are defined in (9.16) and (9.17) [14, 16]. f_{app}, K_c, and K_e are all dimensionless.

$$\Delta P = \left[f_{app} \cdot \frac{n \cdot L \cdot (2a + s)}{a \cdot W} + K_c + K_e \right] \cdot \frac{\rho \cdot V_{ch}^2}{2} \tag{9.15}$$

$$K_c = 0.42 \left[1 - \left(1 - \frac{n \cdot d}{W} \right)^2 \right] \tag{9.16}$$

$$K_e = \left[1 - \left(1 - \frac{n \cdot d}{W} \right)^2 \right]^2 \tag{9.17}$$

f_{app} for a rectangular channel can be obtained by using (9.18) developed in [16], where Re_{ch} is the channel Reynolds number, and L^* is the normalized or dimensionless length of the channel, and f

is the friction factor. Re_{ch}, L^*, and f are all dimensionless. Re_{ch} and L^* can be found as in (9.19) and (9.20), where D_h is the hydraulic diameter of the channel and is defined as in (9.21).

$$f_{app} = \frac{1}{Re_{ch}} \left[\left(\frac{3.44}{\sqrt{L^*}} \right)^2 + (f \cdot Re_{ch})^2 \right]^{\frac{1}{2}} \tag{9.18}$$

$$L^* = \frac{L}{D_h \cdot Re_{ch}} \tag{9.19}$$

$$Re_{ch} = \frac{D_h \cdot V_{ch}}{v} \tag{9.20}$$

$$D_h = 2 \frac{s \cdot a}{s + a} \tag{9.21}$$

The friction factor f can be approximated as in

$$f = \frac{1}{Re_{ch}} \left[24 - 32.527 \left(\frac{s}{a} \right) + 46.721 \left(\frac{s}{a} \right)^2 - 40.829 \left(\frac{s}{a} \right)^3 + 22.954 \left(\frac{s}{a} \right)^4 - 6.089 \left(\frac{s}{a} \right)^5 \right] \tag{9.22}$$

Generally, the required air pressure of the fan is related to the pressure drop of the heatsink and duct geometry. As in [9], an empirical expansion coefficient can be adopted to depict the relationship as in (9.23), where K_{e-fan} is the coefficient of a sudden expansion of the fan, and L_{fan} is the length of the fan frame as illustrated in Figure 9.21.

$$K_{e-fan} = \left[1 - \left(\frac{L_{fan}^2}{DH \cdot DW} \right)^2 \right]^2 \tag{9.23}$$

Since the situation when airflow expands from the fan to the duct is very similar to the air expansion case after the channel, Eqs. (9.23) and (9.17) exhibit the same format. Verified by FEA simulation for different forced-air cooling system designs, Eq. (9.23) is found to describe the airflow expansion well with a relatively high accuracy. The error of the analytical calculation results compared with the FEA simulation results is <10%.

The airflow of the fan V_{fan} [unit: m^3/s] can be approximately obtained as in (9.24). The pressure of the fan P_{fan} can be found as in (9.25).

$$V_{fan} = V_d \cdot DH \cdot DW \tag{9.24}$$

$$P_{fan} = \Delta P + \frac{K_{e-fan} \cdot \rho \cdot V_{fan}^2}{2} \tag{9.25}$$

P_{fan} and V_{fan} determine the operating point of the fan and must be on the fan airflow curve. With the aforementioned equations and the given heatsink geometry, channel geometry, and airflow curve of the fan, equations can be solved and the total thermal resistance can be obtained. An example will be provided in Section 9.5.

For the custom-designed heatsink, the thermal capacitance can be determined the same way as in the commercial off-the-shelf heatsink case using Eq. (9.5). In this case, the total weight of the heatsink can be obtained by (9.26), where ρ_{hs} is the heatsink material weight density.

$$W_{hs} = \rho_{hs} \cdot (b \cdot L \cdot W + n \cdot a \cdot d \cdot L) \tag{9.26}$$

In addition to the thermal resistance model in (9.6) and (9.7), other types of models have been developed. One such example is based on the set of Eqs. (9.27)–(9.30) presented in [11]. R_{th-d},

R_{th-fin}, and R_{th-A} are thermal resistances of the base, fin, and air convection, respectively, and are illustrated in Figure 9.25. Compared with the models in (9.6) and (9.7), this set of models are simpler. The difference for the thermal resistance calculation is less than 10% between the two models for the design example to be covered in Section 9.5. The same example has been used in [9, 17] with the former using the more complex method and the latter using the simpler method. It is expected that the difference of the two methods could grow beyond 10% when the heatsink size becomes large.

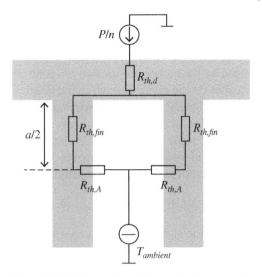

Figure 9.25 Illustration of thermal resistances of different parts in a heatsink.

$$R_{th-d} = \frac{b}{k \cdot W \cdot L} \tag{9.27}$$

$$R_{th-fin} = \frac{a}{k \cdot d \cdot L} \tag{9.28}$$

$$R_{th-A} = \frac{1}{h \cdot L \cdot a} \tag{9.29}$$

$$R_{th-ha} = R_{th-d} + \frac{1}{2n}\left(R_{th-fin} + R_{th-A}\right) \tag{9.30}$$

9.4.2 Heatsink Dimensions

Both for the cooling and mechanical support needs, the heatsink surface area must be larger than the baseplate of the heat-generating components, e.g. power devices or device modules, placed on top of the heatsink. As can be seen in Figure 9.21, there simply should be $W > W_m$, and $L > L_m$, where W_m and L_m are width and length of the power module, respectively.

In the case of multiple devices or device modules, a predefined layout can be used to establish dimension constraints for the heatsink. For example, if there are n devices, we can assume all of them are placed in one row, or multiple rows, with a certain orientation. The length of the device module can be in the direction of the heatsink length or width. In general, any layout-related design optimization is quite difficult to automate. Care must be taken in programming to ensure a feasible design can be found. In the one row case, there can be two options: (i) $W > n \cdot W_m$, and $L > n \cdot L_m$; or (ii) $L > n \cdot W_m$, and $W > n \cdot L_m$. If multiple rows are used, there can be more combinations of options. Sometimes, margins are needed for installation and tolerance.

9.5 Thermal Management System Design Optimization

After establishing the models among the design constraints, conditions, and variables, as described in Section 9.4, the thermal management system design can be carried out to select a set of variables under the given conditions to achieve the lowest cost, size, and/or weight, while meeting the design constraints. As in the case of the other subsystems like inverters and rectifiers, an optimizer can be built for the thermal management system to carry out the design process, as will be described in Chapter 13.

9.5.1 Design Optimization Example

Here, a heatsink-fan cooling design example is given by using the analytical models in Section 9.4. In this case, the maximum heatsink thermal resistance is the design constraint, which is determined by the maximum junction temperature of the device (note that only one device module is assumed), the device thermal characteristics, and the ambient temperature. The task is to determine the optimal heatsink geometry and optimal fan selection to achieve a minimum total weight. The equation for the total weight is given in (9.31), which is equal to the sum of the heatsink and the fan weight while the duct weight is neglected. The design objective, design variables, design constraints, and design input are listed in Table 9.6. The determination of maximum case-to-ambient thermal resistance $R_{th-ca(\max)}$ constraint will be explained later.

$$W_{total} = W_{hs} + W_{fan} = \rho_{hs} \cdot (b \cdot L \cdot W + n \cdot a \cdot d \cdot L) + W_{fan} \tag{9.31}$$

In addition to design variables related to heatsink dimensions and fan, the average channel airflow velocity V_{ch} is an important intermediate variable. The dimensions of the heatsink are constrained by the dimensions of the duct (i.e. heatsink dimensions should be smaller than the duct), as well as by the dimensions of the power module (i.e. heatsink dimensions should be larger than the power module). Fin thickness d and channel width s are limited by manufacturability (both assumed to be >1 mm).

Table 9.7 lists design conditions. Since the duct dimensions are usually confined by other electrical and mechanical components, a fixed duct of 60 mm × 60 mm × 60 mm is used here. Restricted

Table 9.6 Heatsink-fan cooling system design parameters.

Design objective	Minimum total weight W_{total}
Design variables	$n, d. W, L, a$, fan
Design constraints	n is integer, $n > 1$
	$DW > d > 1$ mm
	$DW > W > W_m$
	$DL > L > L_m$
	$DW > s > 1$ mm
	$DH > a > 0$
	$V_{ch} > 0$
	$R_{th-total} < R_{th-ca(\max)} = 2.5\,°C/W$
	P_{fan} and V_{fan} should be on fan curve

Table 9.7 Design conditions.

Item	Description
Power module (high temperature)	20 mm × 20 mm × 2 mm SiC devices Power loss: 80 W Junction temperature: 250 °C Case bottom temperature: 230 °C
Heatsink material	Aluminum ($k = 200$ W/m · °C)
Duct dimension	60 mm × 60 mm × 60 mm
Ambient temperature	30 °C

by manufacturability, the base thickness b of the heatsink is set as a constant value of 4 mm. From the design conditions, the maximum case-to-ambient thermal resistance $R_{th-ca(max)}$ can be determined by case temperature T_c, ambient temperature T_a, and total device power loss P as:

$$R_{th-ca(max)} = \frac{(T_c - T_a)}{P} = \frac{(230 - 30)}{80} = 2.5°C/W$$

Table 9.8, which is based on the datasheets of a popular fan vendor (NMB Technologies Corporation), lists the main properties of each type of fans. It can be seen that fans of the same size (i.e. frame area) usually have the same weight but different characteristic flow curves. The maximum airflow and maximum static pressure will simultaneously increase when the input power and noise increase. Thus, given an input power and acoustic noise limit (3.3 W and 40 dB in this case), the most powerful fan can be directly selected as a candidate for a certain fan size, which will lead to a smaller heatsink.

On the other hand, fans of different sizes may have the same weight but different fan characteristic curves. For example, in Table 9.8, the fans labeled 6, 7, 8, and 9 have the same weight (25 g) but vary widely in maximum airflow and maximum static pressure. They should all be part of database as candidates because any of them can be preferred choice with certain specific heatsink geometry and duct geometry. More details about fans can be found in vendors' product information.

Using the thermal models in Section 9.4, the design procedure can be established and optimization can be carried out. To simplify the procedure, as discussed in Section 9.2.4, the airflow characteristic curve of the fan can be approximately expressed as a fifth-order polynomial. In this design example, (9.32) can approximate the fan curve, where the unit for V_{fan} is in liter/s:

$$P'_{fan} = 3.6653 - 29.0837V_{fan} + 80.9036V_{fan}^2 - 90.002V_{fan}^3 + 15.4747V_{fan}^4 + 40.9982V_{fan}^5 \tag{9.32}$$

For a feasible solution, P'_{fan} must be equal to P_{fan} in (9.25). A procedure to find an optimal solution can be illustrated in a flowchart as in Figure 9.26. In this example, the optimization is conducted

Table 9.8 Candidate fan properties sorted by weight.

Fan part number	Weight (g)	Max airflow ×10^{-3} (m^3/min)	Max static pressure (pa)	Size (mm)	Fan label
1004KL04WB50	7.5	0.057	50.6	25 × 10	1
1204KL04WB50	8.5	0.110	54.0	30 × 10	2
1404KL04WB50	12	0.160	41.0	35 × 10	3
1604KL04WB50	15	0.17	46.0	40 × 10	4
2004KL04WB50	20	0.29	35.0	30 × 10	5
1608KL04WB50	25	0.27	69.4	40 × 20	6
1204KL04WB50	25	0.39	109.7	40 × 28	7
2106KL04WB50	25	0.4	35.0	50 × 15	8
2404KL04WB50	25	0.54	30.1	60 × 10	9
2406KL04WB50	45	0.52	51.2	30 × 10	10
2410RL04WB80	65	0.81	69.1	30 × 10	11

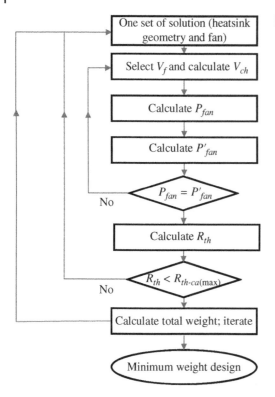

Figure 9.26 Flowchart for optimization.

with the optimization toolbox of MATLAB. One of the iterative optimization methods, the Newton method, is used to solve this nonlinear problem. In the optimization procedure, it evaluates the Hessian matrix to improve the rate of convergence.

Figure 9.27 shows the optimization results for different fans. It can be seen that a light fan with a small flow capacity requires a large heatsink, which also has a relatively heavy weight. As a result, the total weight will be high when the heatsink weight dominates. On the other extreme, a small

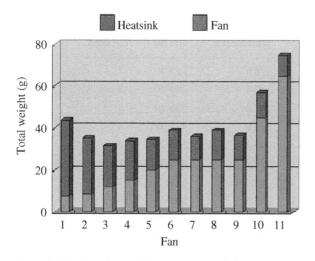

Figure 9.27 Optimized weight with the module operated at 250 °C junction temperature.

Table 9.9 Heatsink geometry corresponding to optimal design with Fan label 3.

n	d (mm)	s (mm)	W (mm)	L (mm)	a (mm)
13	1	1.42	30.0	20	23.4

heatsink will require a strong airflow, and therefore a large fan with a large flow capacity, which leads to a heavy fan. As a result, the total weight will also be high with the fan weight dominating. The lowest weight design falls somewhere in the middle, when both the fan and the heatsink account for appreciable portion of the total weight. A minimum value for a total weight of 31.3 g is achieved by using Fan label 3. Table 9.9 shows the detailed heatsink geometry design results corresponding to the optimal design. Note that it would not change the design method and trend if fans from other vendors were chosen.

Some of the corresponding intermediate design results are $V_f = 0.702$ m/s, $V_{ch} = 1.1964$ m/s, $V_{fan} = 2.52$ liter/s or $0.15\,\text{m}^3$/min, $P_{fan} = P'_{fan} = 2.6135$ Pa. Using (9.6) and (9.7), the thermal resistance of the heatsink R_{th-ha} is $2.24\,°C$/W, and using the simplified formulae (9.27)–(9.30), the thermal resistance R_{th-ha} is $2.49\,°C$/W, both of which meet the constraint of $2.5\,°C$/W.

It should be mentioned that the impact of the TIM is neglected in this example as the heatsink thermal resistance is relatively large.

9.5.2 Design Verification

The aforementioned model-based design has been verified through simulation and experiment [9]. Figure 9.28 shows the simulation results using FEA software I-DEAS. In the simulation, all input are the same as design conditions in Table 9.7, and the 80 W power loss is assumed to be dissipated from the top surface of the SiC dies. In Figure 9.28, the maximum junction temperature is very close to the design input of $250\,°C$. With simple calculation $R_{th-ha} = \frac{P}{(T_C - T_A)}$, the heatsink thermal resistance is $2.52\,°C$/W, which is close to the design result.

To further evaluate the performance of the designed cooling system, a verification test was conducted using the SiC diodes as heat source. Both a thermocouple and an infrared camera were used to measure the temperature. A DC current was applied to the SiC diodes to generate an 80 W loss. The ambient temperature was $30\,°C$. The module temperature was measured close to $250\,°C$, indicating a successful design. From Figure 9.29, the thermal resistance can be calculated as $2.47\,°C$/W, which is also close to the design optimization result.

9.6 Thermal Management System Interface to Other Subsystems

9.6.1 General Classification

Essentially all other subsystems in the converter will interface with the thermal management system. In the case of the motor drive, these include subsystems that are heat generating: the rectifier, the inverter, and the EMI filter. The mechanical system and the control system will not generate much heat that will

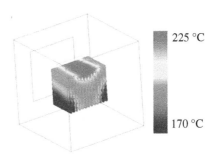

225 °C

170 °C

Figure 9.28 Results of simulation by I-DEAS.

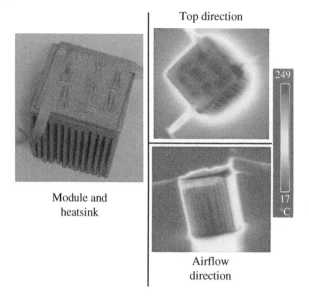

Top direction

Module and
heatsink

249

17
°C

Airflow
direction

Figure 9.29 Power module, heatsink, and experiment results. *Source:* From [9]/with permission of IEEE.

require cooling; however, they will either have to accommodate the need of the thermal management system or will be impacted by it.

The input conditions from other subsystems in a three-phase AC converter have been discussed in Section 9.3.3. As we have discussed with other subsystem in previous chapters, the thermal management system design also generates output that will be needed for other subsystems. Here, we will examine the output needed one by one for other subsystems.

9.6.1.1 Rectifier and Inverter

The key output from the thermal management system to the inverter and rectifier are the thermal impedance of the thermal management system and the device junction temperatures. The other less obvious output is the parasitic capacitance of the heatsink, which could induce extra switching loss. The parasitic loss impact can be important and will be discussed in Chapter 12 for grid applications.

9.6.1.2 EMI Filter

The EMI filter can incur loss and therefore needs cooling. If the EMI filter is based on passive components, then the cooling system design result will be the output to the EMI filter. The passive component cooling is quite different from the semiconductor device cooling. As a result, it will be separately discussed in the next section. Another impact of the thermal management system to EMI filter is the parasitic capacitance of the heatsink, which can impact the CM noise.

9.6.1.3 Mechanical System

The output of the thermal management system to the mechanical system are mainly the dimensions, weight, and materials of the heatsink, duct, and fan. The operating temperatures of these components should also be passed to the mechanical system design. The electrical potentials of the cooling system components are also important.

Table 9.10 Output from thermal management system to its interfaced subsystems (assuming forced-air cooling).

Subsystem	Interface type	Item	Value
Rectifier/inverter	Constraint	TMS thermal impedance	Design result
	Attribute	Device maximum junction temperatures	Discussion
	Attribute	Heatsink parasitic capacitances	Discussion
EMI filter	Constraint	Cooling system performance	Design result
	Attribute	Heatsink parasitic capacitances	Discussion
Mechanical system	Attribute	Dimensions/size/weight of heatsink, duct, and fan	Design results
		Heatsink temperature	Discussion
		Heatsink voltage	Discussion
Control system	Attribute	Fan power	Design result

9.6.1.4 Control System

The thermal management system, specifically, the forced convection cooling, will require air or liquid flow. In the case of the air cooling system, the fan power needs to be provided by the auxiliary power from the control system. Another related item is the temperature sensor to monitor the thermal management system, which is really part of the control system design.

9.6.2 Discussion

Table 9.10 summarizes all the output from the thermal management system to its interfaced other subsystems within the overall converter system (e.g. in this case, an AC-fed motor drive). As can be seen, there are several items that require further discussion.

9.6.2.1 Voltage Distribution

For mechanical system, it is the voltage distribution. For low-voltage (100s of volts or lower) three-phase converters, the heatsink (as well as duct and fan) is generally directly mounted on the enclosure, which should be grounded. For higher voltage converters, the heatsink can be floating, especially, in multilevel converters. In this case, the heatsink voltage distribution needs to be determined for electrical insulation design in the mechanical system.

9.6.2.2 Device Maximum Junction Temperatures

Maximum permissible junction temperature T_{jmax} for converter components, specifically, the power semiconductor devices, are component constraints. Although T_{jmax} can be used as a constraint in the thermal management system design, it is not necessary since the maximum thermal impedance constraints are derived based on T_{jmax}, as discussed in Chapter 6. Therefore, it is more appropriate to treat the actual maximum junction temperature T_{jmax} as an attribute of the thermal management system design result.

Once the thermal impedance of the thermal management system is determined, including the thermal impedance of the heatsink Z_{th-hs} and the thermal resistance of the TIM R_{th-TIM}, the device junction temperatures can be determined based on device power loss and device thermal characteristics (i.e. device thermal impedance).

In the steady-state case, we only need to consider the thermal resistances. Given the maximum device power loss P_{max} and device junction-to-case thermal resistance R_{th-jc}, the maximum junction temperature can be found as:

$$T_{jmax} = P_{max} \cdot \left(R_{th-jc} + R_{th-TIM} + R_{th-hs}\right) + T_a \tag{9.33}$$

Equation (9.33) applies to the case of a single device per heatsink, which is rarely the case in high-power converters. More commonly, there will be multiple devices on one heatsink, including multiple device dies in a device module. The dies in the module can also be different, for example, an IGBT module with IGBT and diode dies in it. In the case of multiple devices on a heatsink, the heatsink surface temperature T_{hs} can be considered uniform. Assuming that the maximum total power loss is $P_{total-max}$, and the maximum power loss for the i-th device is P_{max-i}, the maximum junction temperature for the i-th device is

$$T_{jmax-i} = P_{max-i} \cdot \left(R_{th-jc-i} + R_{th-TIM-i}\right) + P_{total-max} \cdot R_{th-hs} + T_a = P_{max-i}$$
$$\cdot \left(R_{th-jc-i} + R_{th-TIM-i}\right) + T_{hs} \tag{9.34}$$

where $R_{th-jc-i}$ and $R_{th-TIM-i}$ are the junction-to-case thermal resistance and TIM thermal resistance of the i-th device, respectively.

If the transient condition, such as the overload condition, has to be considered, then thermal capacitance must be included to determine T_{jmax}. In this case, the frequency domain approach used in Section 6.3.4 can be used. The equivalent thermal impedance network for the device will be needed, e.g. the Cauer model discussed in Chapter 2, to help the calculation. The procedure will not be repeated here.

9.6.2.3 Heatsink Temperature

The heatsink temperature T_{hs} should be considered in the mechanical system design. The value can be obtained in the process of determining the device junction temperature as in (9.34).

9.6.2.4 Parasitic Capacitances

When a device module is placed on a heatsink, there can be parasitic capacitances between the device terminals and the heatsink. If the heatsink area is relatively large, the capacitance can be appreciable. For example, for a heatsink for a 10 kV, 20 A SiC MOSFET half-bridge, the parasitic capacitance due to the heatsink is approximately 30 pF, which led to more than 5% switching loss increase at the rated converter operating condition [18]. When the heatsink is grounded, the parasitic capacitance is part of the device to ground capacitance, which can contribute to CM current, especially in converters based on fast-switching wide bandgap (WBG) devices.

The accurate determination of the parasitic capacitances is difficult without detailed layout information. In the design optimization stage, it is usually a good idea to estimate the capacitances based on experience and rating of the converter, and then, use FEA or other simulation-based approach in detailed design.

9.7 Other Cooling Considerations

9.7.1 Force-Liquid Convection Cooling

So far, the discussion of the thermal management system design has focused on the forced-air convection cooling. In high-power three-phase AC converters, the forced-liquid convection is also often used for its more effective cooling. As mentioned earlier, the liquid cooling can result in higher power density, especially, in applications where cooling systems are already in place.

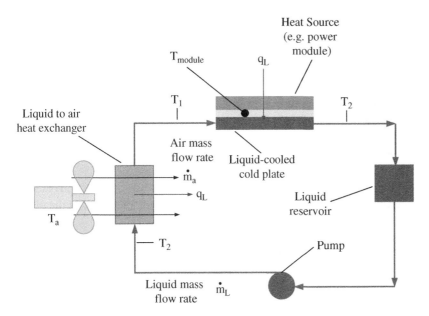

Figure 9.30 A closed-loop forced liquid cooling system.

There are two general types of forced-liquid cooling systems, the open system and the closed-loop system. Figure 9.30 illustrates a closed-loop liquid cooling system. In order for this system to function, there are several components needed: the cold plate, the liquid coolant, liquid pipes, the pump, and liquid-to-air heat exchanger. In this case, the thermal management system includes the selection and/or design of all these components as well as the operating conditions of coolant temperature flow. One example is the closed-loop cooling for a liquid-cooled medium voltage converter that requires deionized water.

For an open cooling system, the coolant flow and temperature are controlled externally, such that the coolant inlet temperature is not impacted by the converter system itself. Examples are liquid-cooled converters using facility cooling water directly and liquid-cooled converters using an available chiller. In a laboratory setting, an open cooling system is often used. In this case, the cooling system design involves the design of the cold plate and flow rate for a given inlet coolant temperature.

In an open cooling system, the inlet coolant temperature T_1 is known, which can be considered as an environmental condition. Other design conditions include the devices to be cooled by the system, their individual losses, and the total loss. If we assume TIM and its corresponding thermal resistance for individual devices are known, the design constraint is the maximum cold plate surface temperature T_{cp}, which can be determined by the maximum permissible junction temperature T_{jmax}, device thermal impedances, the TIM thermal resistance for devices, and power loss. Considering the steady-state condition, take the conservative definition for the cold plate thermal impedance as the impedance between the cold plate surface temperature T_{cp} and the coolant outlet temperature T_2:

$$T_{cp} = P_{total} \cdot R_{th-cp} + T_2 \tag{9.35}$$

The coolant inlet and outlet temperature satisfy the relation:

$$T_2 = T_1 + \frac{P_{total}}{m \cdot c_{lq}} \tag{9.36}$$

where m is the mass flow rate and c_{lq} is the specific heat of the liquid coolant. Combining (9.35) and (9.36) yields

$$T_{cp} = P_{total} \cdot \left(R_{th-cp} + \frac{1}{m \cdot c_{lq}} \right) + T_1 \tag{9.37}$$

To keep T_{cp} below its maximum temperature constraint $T_{cp(max)}$, we need to select a cold plate and a mass flow rate such that its thermal resistance will satisfy

$$R_{th-cp} \le \frac{T_{cp(max)} - T_1}{P_{total}} - \frac{1}{m \cdot C_{lq}} \tag{9.38}$$

Figure 9.31 shows an example cold plate thermal resistance and pressure drop as function of mass flow rate. Clearly, the thermal resistance will decrease as the mass flow rate increases. At the same time, the pressure drop will also increase.

A design example is given in Chapter 12 for the liquid cooling system of a 1-MW medium voltage converter. The ATS-CP-1004 cold plate is used in that case, with its characteristics shown in Figure 9.31.

In a closed-loop system, T_1 is unknown, and a heat capacity C_{HX} can be introduced to characterize the liquid-to-ambient (e.g. air) heat exchanger. The corresponding relationship is

$$T_2 = T_a + \frac{P_{total}}{C_{HX}} \tag{9.39}$$

C_{HX} can be regarded as the inverse of the equivalent thermal resistance of the heat exchanger $R_{th-HX} = 1/C_{HX}$. Note that here C_{HX} has a unit of W/°C, different from common definition of thermal/heat capacity or capacitance with the unit of J/°C. Combining (9.35) and (9.39) yields

$$T_{cp} = P_{total} \cdot R_{th-cp} + \frac{P_{total}}{C_{HX}} + T_a = P_{total} \cdot \left(R_{th-cp} + \frac{1}{C_{HX}} \right) + T_a \tag{9.40}$$

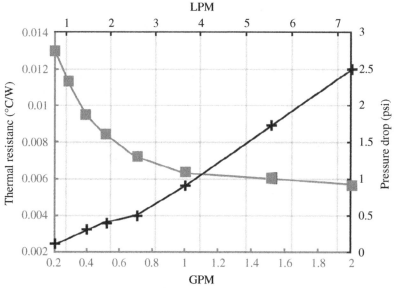

Figure 9.31 Example cold plate thermal resistance and pressure drop as function of mass flow rate.

In the closed-loop cooling case, to keep T_{cp} below its maximum temperature constraint $T_{cp(\max)}$, we need to select a cold plate and heat exchanger such that their combined thermal resistance will satisfy

$$R_{th-cp} + \frac{1}{C_{HX}} \leq \frac{T_{cp(\max)} - T_a}{P_{total}} \tag{9.41}$$

Although there is no explicit role for flow rate in (9.41), it is not excluded, as the thermal resistances already include its effect. In other words, we do not need to calculate liquid flow rate or liquid temperature in determining the needed thermal resistances of the cold plate and heat exchanger. However, in selecting these components, the needed flow rate must be designed, and corresponding pump needs to be selected to provide the flow. Note that the aforementioned derivations assume that there is no heat gain or loss from pump or connecting pipes.

9.7.2 Cooling for Passives

The discussion so far in this chapter has focused on the cooling system design for semiconductor devices. Passive components, especially, the magnetic components, can incur significant losses. Inductors and transformers can have both core loss and winding loss and therefore need to be cooled. With faster switching WBG-based converters and higher power density requirements, the loss density and heat density in passives can be quite high and pose considerable design challenges for their thermal management.

By design, the semiconductor devices pass their heat through one side of the package, mostly through their bases or baseplates. In certain cases, semiconductor devices may also pass heat through their top sides. As a result, it is relatively easy to remove heat and cool semiconductor devices by using heatsinks on one side, or in some cases, on both sides for double-sided cooling. On the other hand, the inductor or transformer geometry configurations do not lend themselves for easy heat removal and cooling. The heat can be generated throughout the cores and windings and must be dissipated through their surfaces, which are generally three-dimensional or at least two-dimensional.

Since it is quite difficult to cool a two- or three-dimensional heat-generating component, there is no standard approach for passive cooling. Here, we will discuss some commonly used approaches.

9.7.2.1 Improving Internal Thermal Conduction

The real design limit is the hot spot temperature within a passive component, similar to the junction temperature of devices. It is important to reduce the thermal resistance of the passive component itself. Any air gaps are undesirable, since air is a poor thermal conductor. It is often a good idea to fill the gaps between inductor/transformer windings, between windings and cores, and even between wires in a winding. The gap filler materials have been discussed earlier as TIM, and they generally need to be good thermal conductors as well as good electrical insulators. A traditional material is the transformer oil used in oil-filled transformers, circuit breakers, reactors, and even capacitors. The drawback with oil is its flammability and high maintenance requirement. The dry-type transformers and inductors use polymer-based potting materials to fill the gaps that improve both electrical insulation and thermal conduction. These materials include silicone gel, epoxy, etc.

Some more exotic approaches, mainly used in low-power high-density prototypes, include embedding metal conductors in magnetic cores to improve the heat removal.

9.7.2.2 Air Convection Cooling

As we already know, most AC three-phase converters are air-cooled. Naturally, the passive components in these converters are air-cooled as well. In forced-air cooling system, there will be airflow, which is mainly designed for semiconductor device cooling. If possible, in the mechanical system design, the layout should be designed in such a way that the passives, especially, the inductors and transformers, are in the path of the airflow. Consequently, the thermal performance for these passives can be improved. If it is not possible to place the passives along the path of airflow, the passive thermal performance will have to be evaluated under the natural cooled condition. Figure 9.32 shows the ANSYS simulation of a 6.25 kV/1 kV, 50 kVA, 10 kHz single-phase transformer intended for a DC/DC converter within the Figure 11.1 converter. The transformer windings are insulated with silicone elastomer as potting material. Without forced-air cooling, the maximum temperature

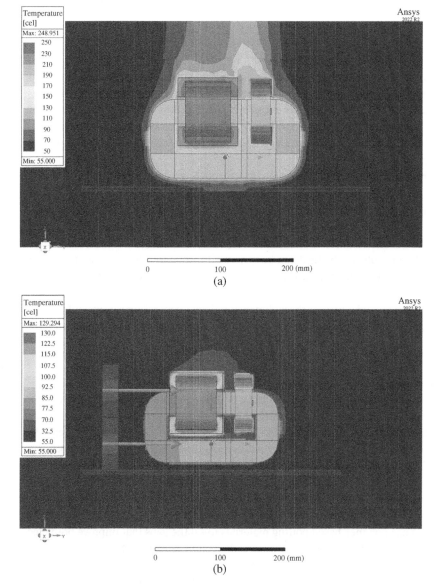

Figure 9.32 Comparison of thermal performance of transformers without or with airflow. (a) Without airflow and (b) with a 4 m/s airflow.

in the winding could approach 250 °C; with an average of 4 m/s air velocity, which are achieved with four dedicated fans, the maximum temperature will drop to more acceptable range of <130 °C. The specified ambient temperature is 55 °C.

In Chapters 4 and 5, simple analytical thermal models have been presented for inductors considering natural and forced-air cooling conditions. For detailed design, FEA simulation is often used to design and evaluate the inductor or transformer thermal performance.

Capacitors generally have very low loss. Still, airflow can be helpful in maintaining the ambient temperature surrounding capacitors below their maximum working temperatures.

9.7.2.3 Liquid Convection Cooling

In forced-liquid convection cooling system, there is often no forced air. In this case, the passive components either will simply be without forced cooling (i.e. natural cooling) or will have to be liquid cooled as well. As pointed out earlier, since the passives have to be cooled in more than one dimension (i.e. multiple sides), it is quite difficult to have a cold plate that can cover multiple dimensions. In this case, several approaches can be taken:

1) Liquid Immersion: As mentioned earlier, oil has been used in transformers, which does not have liquid flow. Some of the advanced liquid immersion techniques discussed in Section 9.2.2 could also be used for passive components. Figure 9.33 shows a prototype for a 30 kW, 600 V to 10 kV pulsed power supply [19]. The right side box contains a 600 V to 10 kV transformer and two high-voltage mica DC capacitors filled with transformer oil. A metal cooling pipe with its connection terminals on the right side of the converter is placed in the transformer oil. When cooling oil from the chiller passes through the pipe, it will help cool the transformer oil and remove heat from the transformer and capacitors.

2) Cooling Surface Creation: A cooling surface can be created to facilitate using cold plate to cool passives. Figure 9.34a shows a CM choke for an AC-side EMI filter in a 100-kW DC-fed three-phase motor drive. In order to use liquid cooling, the choke is placed in an aluminum case and potted to improve the thermal conductivity. Then the whole case is screwed onto the cold plate, as shown in Figure 9.34b. The EMI filter has two AC CM chokes, one DC CM choke, and an AC DM inductor. All four inductors are liquid cooled as shown in Figure 9.34c. The final EMI filter assembly is shown in Figure 9.34d, where two cold plates are used for both sides to improve the overall cooling performance [20].

3) Embedded Cooling: Liquid can be flown into the magnetic components to cool them. Figure 9.35a shows a prototype 3 kV/3 kV, 350 kVA, 10 kHz transformer by ABB [21]. It adopts coaxial cable windings with hollow inner tubes, which can be used to pass de-ionized water for cooling. Figure 9.35b shows a prototype AC CM choke for a 1-MW cryogenically cooled inverter [22].

Figure 9.33 A 30-kW, 600-V to 10-kV pulsed power supply prototype. *Source:* From [19]/with permission of IEEE.

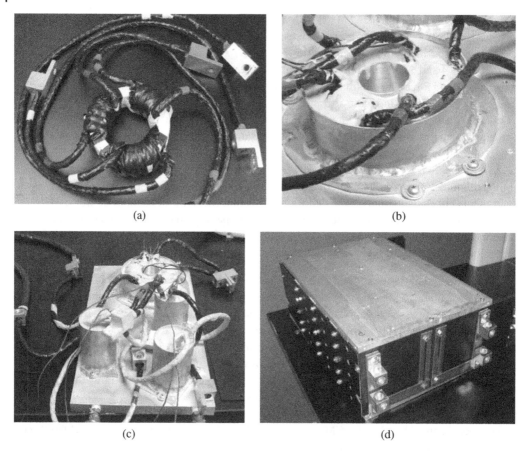

(a)　　　　　　　　　　　　　　(b)

(c)　　　　　　　　　　　　　　(d)

Figure 9.34 A protype EMI filter for a 100-kW motor drive: (a) AC CM choke; (b) AC CM choke in aluminum case with potting screwed onto cold plate; (c) all filter inductors on cold plate; (d) integrared liquid cooled EMI filter assembly.

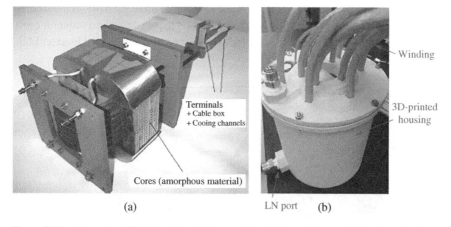

(a)　　　　　　　　　　　　　　(b)

Figure 9.35 Examples of the liquid cooled magnetics: (a) a transformer with hollow windings as cooling loops [21]/with permission of IEEE; (b) a CM choke with 3D-printed housing enabling liquid nitrogen (LN) cooling.

A 3D-printed housing is built to accommodate liquid nitrogen (LN) cooling as well as cores and windings. The housing design allows the copper windings to be cooled close to LN temperature, while keeping nanocrystalline cores to be close to room temperature, since copper will have lower loss at cryogenic temperature while magnetic core will have lower loss near the room temperature.

9.8 Summary

This chapter presents the thermal management system design for AC three-phase converters, although most of the methods apply to other types of converters, especially, high-power converters. The main points covered and key takeaways are the following:

1) Forced-air and forced-liquid convection cooling still dominate high-power converter applications. Advanced phase change and/or direct liquid cooling techniques are developed but not yet widely applied. One exception is the heat pipe technology, which has been successfully used in high-power converters.
2) Heatsink materials are largely based on aluminum or copper. Graphite-based materials are promising for high-density converters due to their lightweight, while care must be taken on their non-isotropic thermal conductivity and negative CTE. Heatsinks with extrusion remain to be popular due to their low manufacturing cost and good thermal performance. 3D-printing can be an effective way in building high-performance heatsinks.
3) In addition to heatsinks (cold plate for liquid cooling), other components including fan/blower and duct (pump, pipe, and heat exchanger for liquid cooling) are integral part of the thermal management system. TIM is important for overall cooling performance. Thermal grease and thermal pads are the two popular TIMs, with thermal pads used when electrical insulation is needed between the device and heatsink.
4) Thermal management system design can be formulated as a design optimization problem. For forced-air cooling, the design variables include design or selection of the heatsink, the air duct, and the fan. The numbers of fan and heatsinks can also be design variables. An important intermediate design variable is air velocity in the heatsink channel. The design constraints include thermal impedance of the thermal management system and the heatsink surface dimension to accommodate heat-generating devices.
5) The main model needed for thermal management system design is the thermal resistance model. For off-the-shelf heatsinks, it is simply the heatsink characteristics as function of airflow velocity from datasheet. For custom heatsink, analytical models have been presented as functions of airflow, and heatsink geometry and materials, which have been validated through FEA simulation and experiment for a single module case.
6) Forced-liquid cooling system can be an open or a closed-loop system. The design of the open system involves the selection of the cold plate with sufficient mass flow rate; and the design of the closed system involves the selection of the cold plate and the heat exchanger with small enough combined thermal resistance. In all cases, pumps are needed to provide needed flow.
7) Passive components dissipate heat through their surfaces, which are three- or two-dimensional, and their cooling design is less standard. Common methods to improve thermal performance and cooling include gap filling and potting with materials that are good thermal conductors and electrical insulators, and introducing airflow. In liquid cooled systems, their thermal performance can be improved with immersion, cooling surface creation, and embedded coolant loops.

References

1 L. Yeh and R. C. Chu, *"Thermal Management of Microelectronic Equipment: Heat Transfer Theory, Analysis Methods, and Design Practices,"* New York: ASME Press, 2002.

2 M. C. Shaw, J. R. Waldrop, S. Chandrasekaran, B. Kagalwala and X. Jing, "Enhanced thermal management by direct water spray of high-voltage, high power devices in three-phase, 18-hp AC motor drive demonstration," in *IEEE Intersociety Conference on Thermal and Thermomechanical Phenomena in Electronic Systems*, San Diego, CA, 2002.

3 I. Mudawar, "Assessment of high-heat-flux thermal management schemes," *IEEE Transactions on Components and Packaging Technologies*, vol. 24, no. 2, pp. 122–141, 2001.

4 I. Mudawar and T. Tope, "Purdue University International Electronic Cooling Alliance," [Online]. Available: http://www.ecn.purdue.edu/PUIECA/main.html. [Accessed 26 September 2022].

5 C. Xia, "Spray/Jet cooling for heat flux high to 1 kW/cm^2," in *Eighteenth Annual IEEE Semiconductor Thermal Measurement and Management Symposium*, San Jose, CA, March 12–14, 2022.

6 T. Wu, B. Ozpineci, M. Chinthavali, Z. Wang, S. Debnath and S. Campbell, "Design and optimization of 3D printed air-cooled heat sinks based on genetic algorithms," in *2017 IEEE Transportation Electrification Conference and Expo (ITEC)*, Chicago, IL, June 22–24, 2017.

7 C. Zweben, "Thermal Materials Solve Power Electronics Challenges," *Power Electronics Technology*, February 2006.

8 Dreyer-System GmbH, "Dreyer system," [Online]. Available: https://www.dreyer-system.eu/highpower-heat-sinks.html. [Accessed 26 September 2022].

9 P. Ning, F. Wang and K. Ngo, "Forced air cooling system design under weight constraint for high temperature SiC converter," *IEEE Transactions on Power Electronics*, vol. 29, no. 4, pp. 1998–2007, 2014.

10 U. Drofenik, G. Laimer and J. W. Kolar, "Theoretical converter power density limits for forced convection cooling," in *PCIM Europe*, Nuremberg, Germany, June 2007.

11 U. Drofenik and J. W. Kolar, "Analyzing the theoretical limits of forced air-cooling by employing advanced composite materials with thermal conductivities >400 W/mK," in *Proc. of 4th Int. Conf. on Integrated Power Systems (CIPS'08)*, Naples, Italy, June 2008.

12 R. Hossain, J. R. Culham and M. M. Yovanovich, "Influence of bypass on flow through plate fin heatsinks," in *23rd IEEE SEMI-THERM Symposium*, San Jose, CA, March 18–22, 2007.

13 W. A. Khan and M. M. Y. J. R. Culham, "Effect of bypass on overall performance of pin-fin heat sinks," in *9th AIAA/ASME Joint Thermophysics and Heat Transfer Conference*, San Francisco, California, June 2006.

14 J. R. Culham and Y. S. Muzychka, "Optimization of plate fin heatsinks using entropy generation minimization," *IEEE Transactions on Components and Packaging Technologies*, vol. 24, no. 2, pp. 159–165, 2001.

15 P. Teertstra, M. M. Yovanovich, J. R. Culham and T. Lemczyk, "Analytical forced convection modeling of plate fin heatsinks," in *Proceedings of 15th IEEE SEMI-THERM Symposium*, San Diego, CA, March 9–11, 1999.

16 S. Lee, "Optimum design and selection of heatsinks," in *Proceedings of 11th IEEE SEMI-THERM Symposium*, San Jose, CA, February 7–9, 1995.

17 P. Ning, G. Lei, F. Wang and K. D. T. Ngo, "Selection of heatsink and fan for high-temperature power modules under weight constraint," in *Twenty-Third annual IEEE Applied Power Electronics Conference and Exposition (APEC 2008)*, Austin, TX, February 2008.

18 X. Huang, S. Ji, J. Palmer, L. Zhang, L. M. Tolbert and F. Wang, "Parasitic capacitors' impact on switching performance in a 10 kV SiC MOSFET based converter," in *IEEE 6th Workshop on Wide Bandgap Power Devices and Applications (WiPDA)*, Atlanta, GA, October 31–November 2, 2018.

19 H. Sheng, W. Shen, H. Wang, D. Fu, Y. Pei, X. Yang, F. Wang, D. Boroyevich, F. C. Lee and C. W. Tipton, "Design and implementation of a high power density three-level parallel resonant converter for capacitor charging pulsed-power supply," *IEEE Transactions on Plasma Science*, vol. 39, no. 4, pp. 1131–1140, April 2011.

20 J. Xue and F. Wang, "A practical liquid-cooling design method for magnetic components of EMI filter in high power motor drives," in *Proceedings of the IEEE Energy Conversion Congress and Exposition (ECCE)*, Milwaukee, WI, September 2016.

21 L. Heinemann, "An actively cooled high power, high frequency transformer with high insulation capability," in *Seventeenth Annual IEEE Applied Power Electronics Conference and Exposition (APEC 2002)*, March 2022.

22 R. Chen, J. Niu, R. Ren, H. Gui, F. Wang, L. Tolbert, B. Choi and G. Brown, "A cryogenically cooled MW inverter for electrified aircraft propulsion," in *AIAA/IEEE Electrified Aircraft Symposium (EATS)*, August 26–28, 2022.

10

Control and Auxiliaries

10.1 Introduction

This chapter includes the design of control and auxiliaries. Although they generally are not the main contributors to the design objectives (e.g. cost, size, and weight) in a medium- or high-power three-phase converter, compared to other components and subsystems in previous chapters, the control and auxiliaries, as illustrated in Figure 10.1, are essential parts of the converter. They cover many details and, if not designed properly, will cause incomplete functionalities of power converter systems.

In detail, this chapter will discuss the controller architecture, design and selection of digital controller hardware, isolation strategy and isolation components (signal isolator and isolated power supply), and gate driver design. It will also include sensor selection, including high-voltage considerations for the medium-voltage application and conditioning circuit design under the high-frequency high-noise environment. The protection will be focused, first on device-level protection, such as short circuit and crosstalk, and then converter-level protection, including overcurrent, over-/undervoltage, and overtemperature. Printed circuit boards (PCBs), as a critical electrical and mechanical interconnection means, will also be discussed. Finally, the deadtime setting and compensation, considering the unique characteristics of wide bandgap (WBG) power semiconductors and WBG-based converters, will be covered.

10.2 Control Architecture

Figure 10.2 displays a common control architecture defined in IEEE Std 1676-2010 [1]. It consists of a system control layer that determines the overall mission of the system, an application control layer responsible for maintaining overall functions of the power electronics, a converter control layer that implements common functions of converters, a switching control layer that handles the switching logic/sequence, and the hardware control layer that manages everything specific to the power hardware. Each layer has characteristic processing and communication speed requirements, irrespective of the final applications. Considering recent WBG-based applications with faster switching speed and higher switching frequency as compared to the Si-based solutions, the response times associated with hardware control and switching control could be even faster.

10.2.1 System Control Layer

All functions associated with the determination of the system duties and mission of the power electronics system are considered in this layer. Any human–machine interfaces are also included. From the system controller's perspective, the lower control layers execute all power

Design of Three-phase AC Power Electronics Converters, First Edition. Fei "Fred" Wang, Zheyu Zhang, and Ruirui Chen.
© 2024 The Institute of Electrical and Electronics Engineers, Inc. Published 2024 by John Wiley & Sons, Inc.

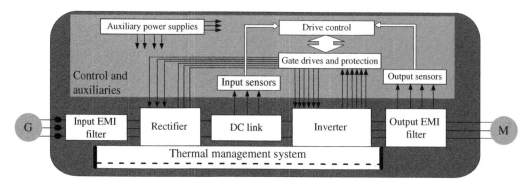

Figure 10.1 Overview of control and auxiliaries in three-phase power converter system.

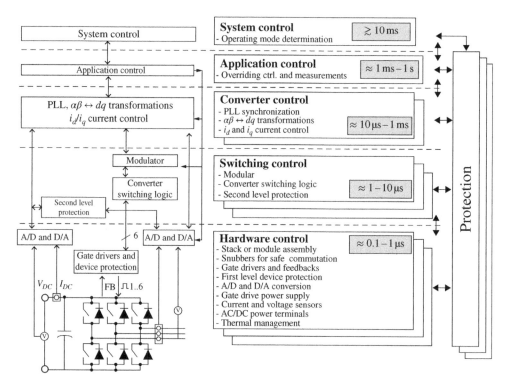

Figure 10.2 Recommended architecture for power electronics applications.

electronic system functions that are necessary to fulfill the system mission. Using the motor drive in a rolling steel mill as example, the system control pertains to control meeting steel rolling requirements.

10.2.2 Application Control Layer

The application control layer determines the operation of the power electronics system to meet the mission command from the system control. From the viewpoint of the application controller, the lower control layers enable the power electronics system to be either controlled current source or controlled voltage source, as illustrated in Figures 10.3 and 10.4. The boundary between the

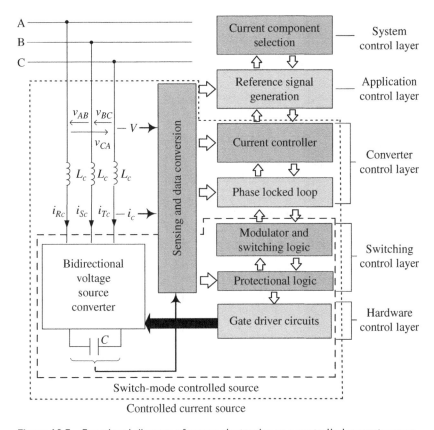

Figure 10.3 Functional diagram of power electronics as a controlled current source.

application control layer and lower control layers (dashed line in Figures 10.3 and 10.4) is determined by those control subsystems that enable the power electronics to behave as a controlled current or voltage source. Therefore, the output signal for the application control layer at the interface with the converter control layer should contain either current or voltage reference signals. In the motor drive example, the application control pertains to speed and torque control.

10.2.3 Converter Control Layer

The primary characteristic of the converter control layer is the feedback control system, as shown in Figures 10.3 and 10.4, while other components support the input and output requirements of the feedback control system. The converter control layer implements many of the common functions of converters such as synchronous timing, phase-locked loop (PLL), current and voltage measurement filtering, and feedback control calculations.

10.2.4 Switching Control Layer

The switching control layer, together with the lower hardware control layer, enables the power electronics to behave as a switch-mode-controlled source and includes modulation control and pulse width modulation (PWM) generation. These functions are common and independent of the final application. Here, the switch-mode-controlled source is defined as the power electronics hardware operation with the PWM modulation control.

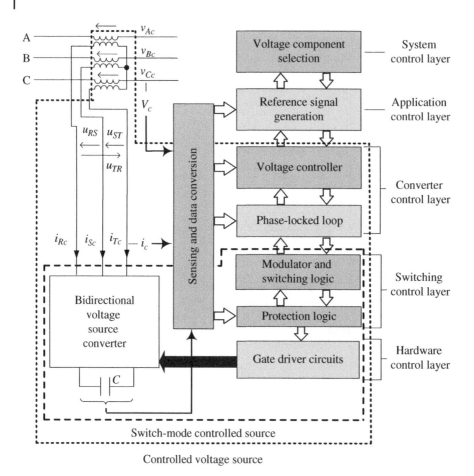

Figure 10.4 Functional diagram of power electronics as a controlled voltage source.

10.2.5 Hardware Control Layer

The hardware control layer directly interfaces with converter power stage and manages everything specific to power devices, such as gate driver, galvanic isolation, limits of di/dt and dv/dt, and device-level protections. The functions of this layer are common for any applications.

10.3 Control Hardware Selection and Design

Different control layers have different requirements for the control hardware selection and design. Typically, for system and application control layers, programmable logic controller (PLC) and digital signal processor (DSP) are adopted considering relatively lower response time requirements (>1 ms). At these layers, power electronics applications do not pose unique requirements for the control hardware selection; thus, the general selection criteria, such as processor, memory, communication interface, are sufficient and will not be discussed in detail.

At the converter and switching control layers, DSP, field-programmable gate array (FPGA), and even application-specific integrated circuit (ASIC) are used for sub-microsecond response time. Usually, one microcontroller will handle the control tasks for both converter and switching layers.

If so, beyond the general considerations, like processors, memory, and communications, several unique functions to support power electronics converter are necessary to be integrated, such as PWM generation and analog-to-digital conversion (ADC). For certain power electronics topologies, like modular multilevel converter (MMC), cascaded H-bridge, or converter system consisting of multiple paralleled/series converters, excessive PWM channels are needed, which is challenging for the general-purpose microcontroller. Therefore, a combination of microprocessor and FPGA is also common where microprocessor is primarily responsible for fulfilling the converter control functions and modulation, and PWM generations are taken care of by FPGA. Additionally, for certain applications with high sensing accuracy requirements, such as the healthcare and medical power supplies, the embedded ADC in the microcontroller might not be adequate. In this case, a peripheral ADC unit with higher resolution will be introduced, cooperating with microcontroller to accomplish the control. For the even faster response hardware control layer, such as gate driver, device-level protection, and switching transient (e.g. di/dt, dv/dt) management, dedicated analog or mixed-mode circuits need to be designed. For the hardware control layer, the design of gate driver, fast response protection, etc., will be discussed in Sections 10.5 and 10.7.

It is noted that temperature is an attribute that can affect the selection of the suitable controller, for example, in high-temperature or cryogenically cooled applications.

10.4 Isolation

Isolation is essential between the controller and power stage. As illustrated in Figure 10.5, it includes signal isolators, sensors, protection circuitries, as well as power supplies for all circuitries tied with the power stage (i.e. located on the secondary side of the isolation layer). Notably, sensors may integrate the isolation layer, such as Hall effect-based current transducer, etc. They are separately discussed in Section 10.6.

10.4.1 Signal Isolator

As illustrated in Figure 10.6, signal isolator is to provide: (i) galvanic isolation between the ground of controller and the ground of circuitries tied with the power stage (e.g. gate driver) and (ii) safe transfer of the control, fault, and sensing signals between the primary side and secondary side of

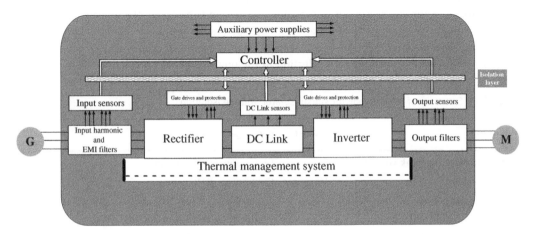

Figure 10.5 Isolation needs in power converter design.

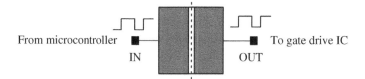

Figure 10.6 Signal isolator (gate driver as an example).

signal isolator. Galvanic isolation capability and common-mode transient immunity (CMTI) are the key criteria for signal isolator selection.

Galvanic isolation capability indicates the maximum potential difference to be handled between the primary and secondary sides of the signal isolator. In a phase-leg-based voltage source converter (VSC), the worst case is when the lower switch in the phase leg turns off, the potential difference across the signal isolator associated with circuitries tied with the upper switch. In steady state, this potential difference is approximately equal to the DC-link voltage; considering the switching transition, and its resultant parasitic ringing and overvoltage across the lower switch, the dynamic potential difference across the upper side signal isolator can be even higher.

In the datasheet of the signal isolator, multiple terminologies can be found to define the galvanic isolation capability in different manners. Here, a clarification is given, and the correct value for signal isolator in power converter design is recommended according to [2].

First is the basic isolation vs. reinforced isolation. Basic isolation refers to isolation that provides sufficient protection against electrical shock as long as the insulation barrier is intact. However, practically, some safety regulations require the basic isolation to be supplemented with a secondary isolation barrier for redundancy so that the additional barrier provides shock protection, even if the first barrier fails. This is called double isolation. To make systems compact and save cost, it is desirable to have only one level of isolation that has the required electrical strength, reliability, and shock protection of two levels of basic isolation. This is called reinforced isolation. Therefore, with a given signal isolator, the basic isolation voltage rating is always higher than the corresponding reinforced isolation voltage.

Second, multiple terminologies are used to describe isolation voltage capability, including maximum repetitive peak voltage, working voltage, maximum transient isolation voltage, isolation withstand voltage, and maximum surge isolation voltage. Different terminologies indicate the isolation voltage capability of the signal isolator according to different voltage profiles, as summarized in Table 10.1. For the signal isolator in the converter design considering the worst-case scenario discussed earlier, the whole DC-link voltage is applied across the upper side signal isolator continuously. Thus, the maximum repetitive peak voltage should be selected as the required galvanic isolation capability for the signal isolator selection in the power converter.

Table 10.1 Summary of terminology for signal isolator voltage isolation.

Terminology	Symbol	Definition
Maximum repetitive peak voltage	V_{IORM}	Continuous, peak voltage
Working voltage	V_{IOWM}	Continuous, RMS voltage
Maximum transient isolation voltage	V_{IOTM}	Transient (in a range of seconds), peak voltage
Isolation withstand voltage	V_{ISO}	Transient (in a range of seconds), RMS voltage
Maximum surge isolation voltage	V_{IOSM}	Transient (in a range of micro-seconds), peak voltage

Figure 10.7 CMTI of the signal isolator (high side gate driver as an example).

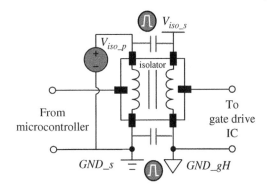

Table 10.2 Minimum CMTI of state-of-the-art commercially available signal isolators.

Isolation techniques	Optocoupler	Capacitive	Transformer	RF-based
Minimum CMTI	30 kV/μs	100 kV/μs	75 kV/μs	200 kV/μs

Source: Adapted from [3–6].

In addition to the galvanic isolation, CMTI is another important selection criterion for the signal isolator. As illustrated in Figure 10.7, for the signal isolator associated with the upper side circuitries in the phase leg, the ground of secondary side is connected with the source terminal of the upper switch, which is the middle point of the phase leg. During the switching transient of the lower switch, the potential of phase-leg middle point varies between the positive and negative DC bus with a high dv/dt. Considering the inevitable coupling capacitance (input-to-output capacitance) across the signal isolator, which provides the parasitic paths, dv/dt induced current through the parasitic paths may lead the isolator to lose control by false triggering a function or causing erroneous feedback. This also occurs for the current transducer located at the converter AC terminal before the filter. CMTI illustrates the maximum dv/dt that the single isolator can endure without potential malfunction.

Table 10.2 lists the minimum CMTI of the state-of-the-art commercial signal isolators with different isolation technologies, including optical (i.e. optocoupler), capacitive, magnetic (i.e. transformer), and radio-frequency (RF)-based approaches. It is the minimum CMTI rather than the typical CMTI in the datasheet that should be used to identify the capability of the signal isolator.

Beyond the basic isolation functions, the design of signal isolator may also consider other requirements posed by the converter design, such as signal transmission frequency range to cover the desired switching frequency, propagation delay and distortion to ensure control accuracy, and so on. This will be discussed in Section 10.5 as part of the gate driver design.

10.4.2 Isolated Power Supply

Like the signal isolator, isolated power supply also should be capable of providing sufficient galvanic isolation and CMTI of which the design considerations are similar to the signal isolator discussed earlier and will not be repeated. In most of the isolated power supply datasheet, input-to-output capacitance instead of CMTI is provided. Equivalently, small input-to-output capacitance indicates high CMTI capability.

To be noted, in addition to the isolation requirements, power rating, input voltage range, number of output channels, and the corresponding voltage value(s) are also critical design parameters. Their design criteria are very application-specific and generally straightforward; thus, it will not be discussed here. Some case studies, e.g. gate driver-related isolated power supply design, will be covered in the specific sections.

10.4.3   Discussion on Isolation Strategies for Low-Power Converter Design

To save cost, in some low-power low-voltage three-phase converter cases, the isolation design is significantly simplified. For instance, as illustrated in Figure 10.8, the negative DC bus is defined as the common grounding point for both control/protection/sensing circuitries and power stage. In principle, no isolation is needed between the control/power boards as they all share the same grounding point. This is also true for the lower side gate drivers. In the meantime, the boot-strap-based gate driver (details will be presented in Section 10.5) could be adopted for the upper side devices without dedicated isolation.

Though the cost reduces, potential noise problems, such as high common mode dv/dt, inductive and capacitive couplings between any floating nodes and common ground point, etc., could worsen the converter performance. Some design challenges on the sensing and signal processing circuits will be discussed in Section 10.6.5.

10.5   Gate Driver

Gate driver is the main link between the world of control and the world of power for power conversion system. The world of control is the world of 1.8 V, 3.3 V, or 5 V with limited current generated by microcontroller, while the world of power is the world of up to thousands of volts and thousands of amperes delivered by power devices. A good gate driver is not just something nice to have but something must have for a power electronics converter. As clearly illustrated in Figure 10.9, which uses motor drives as an example, the gate driver serves as the interface between logic-level control signals and power devices.

In this section, we focus on the gate driver design in a phase-leg-based VSC, as the example circuit shown in Figure 10.10. First, gate driver fundamentals and design considerations are presented. Second, key power device characteristic parameters used for gate driver design are highlighted. Third, detailed gate driver design is introduced based on the basic functional blocks. The unique

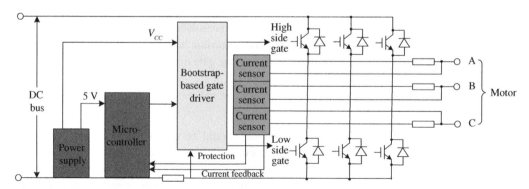

Figure 10.8   Example of isolation strategy for low-power converter design.

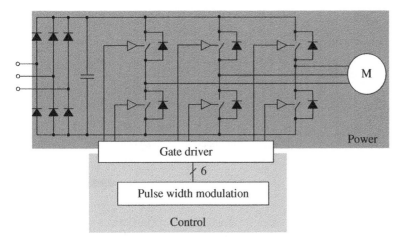

Figure 10.9 Gate drive as interface between the world of control and the world of power (motor drives as an example).

challenges posed by fast-switching WBG devices are also emphasized and the corresponding design considerations are discussed.

10.5.1 Gate Driver Fundamentals

A good gate driver is able to efficiently and reliably control power devices under static, dynamic, and fault conditions. Specifically, in the static state, a good gate driver can keep the device in on-state with minimized conduction loss and keep the device safely in off-state to prevent spurious change of the switch state due to external or internal disturbances. During the dynamic transient, gate driver is capable of driving the device from on to off and off to on states with fast speed and low switching losses, acceptable electromagnetic inter-

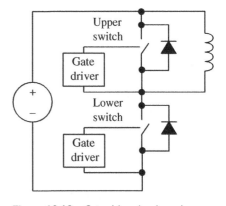

Figure 10.10 Gate driver in phase-leg configuration.

ference (EMI), and low-voltage spike and parasitic ringing. In case of any hazardous situations, gate driver can quickly and reliably protect the device, including typically shoot-through, overcurrent, overvoltage, and overtemperature faults. This section focuses on the gate driver design for the normal control of power devices, and protection function will be discussed in Section 10.7.

To achieve the basic function of a gate driver, several functional blocks are necessary. As illustrated in Figure 10.11, they include gate drive IC/buffer circuit, signal isolator, and isolated power supply. Gate drive IC is to amplify the driving capability of logic-level control signal from microcontroller to drive power device. In case the driving capability provided by the gate drive IC is not sufficient, an extra buffer circuit is added. This can be commonly observed in gate driver design of power modules for high-power applications. Signal isolator is to provide galvanic isolation between the world of control and the world of power due to their potential differences and noise interference. Isolated power supply is to power the secondary side of the signal isolator, gate drive IC, and buffer circuit if applicable. Detailed introduction and design criteria will be presented in

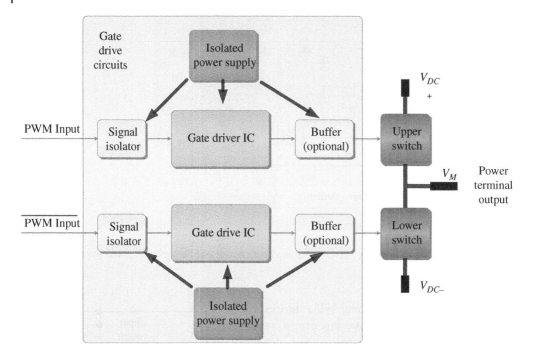

Figure 10.11 Basic functional blocks of a gate driver.

Section 10.5.3. Note that the gate driver isolation strategy and architecture presented here represent the most popular scheme used in power converters. The scheme is also adequate and convenient for gate drive needs of WBG power devices. Other isolation strategies like using transformers for both the signal and power isolation are also used for converters.

As already discussed in previous chapters, high-voltage application, high dv/dt, di/dt environment, severe parasitic ringing, and high-temperature operation become unique for WBG devices. These features pose new requirements for WBG device gate driver, as it is the neighboring component of power devices. The special requirements include high galvanic isolation capability, high common-mode transient and ringing immunity capability, and high operating temperature capability. Dedicated design criteria have to be considered for the WBG gate driver design.

10.5.2 Gate Driver-Related Key Device Characteristics

An ideal gate driver should be designed catering for the specific device it is to drive. That is, the gate driver should be carefully designed, depending on specific device characteristics. In practice, people may prefer a more generic gate driver design for various devices to save time and cost. However, it is desirable for gate driver to be matched with devices, especially for the new WBG devices that are fast and potentially more sensitive to driving control. Thus, it is critical to understand the impact of the device on the gate driver design and identify the gate driver-related key device characteristics before the design of gate driver. As discussed earlier, gate driver should meet static and dynamic requirements imposed by the device. Thus, it is necessary to review the static and dynamic characteristics of power devices and identify the key device parameters needed for gate driver design. Note again that the dynamic characteristics here refer to parameters obtained from static characterization that will impact device switching characteristics.

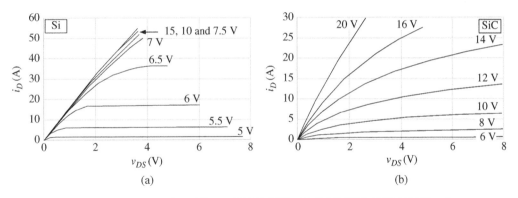

Figure 10.12 Output characteristics of Si and SiC MOSFETs. (a) Si MOSFET APT34M120J and (b) SiC MOSFET C2M0080120D. *Source:* Adapted from [7, 8].

10.5.2.1 Gate Driver Design Considering Device Static Characteristics

Starting from the static state, to minimize the on-state resistance or on-state voltage and conduction loss when the device is on, and in the meantime, not overstress device gate, device static characteristics significantly affect the on-state gate voltage selection. Specifically, output characteristics of the device are critical to identify the proper on-state gate voltage. For example, Figure 10.12 illustrates output characteristics of two example devices, one is Si MOSFET (APT34M120J from Microsemi) and the other is SiC MOSFET (C2M0080120D from Wolfspeed/Cree). It shows clearly that for Si MOSFET, the change of V–I curves as the gate voltage increases from 7.5 V to 15 V becomes smaller, indicating it is not necessary to further increase the gate voltage of the Si MOSFET beyond 15 V to reduce the on-state resistance/voltage even if the maximum allowable positive gate voltage of this Si MOSFET is up to 30 V (see Table 10.3). In contrast, for SiC MOSFET, according to the V–I curve, it is obvious that as the gate voltage increases up to 20 V, the on-state resistance/voltage can still reduce significantly. However, further increase of the gate voltage is limited by its maximum positive gate voltage rating, which is 25 V for this specific case (see Table 10.3).

Additionally, when device is in off-state, to prevent spurious change of the switch state due to external or internal disturbances, it is preferred to introduce negative off-state gate voltage, especially for fast-switching WBG devices. However, concerning the reverse conduction characteristics along with the negative gate voltage rating of WBG devices, off-state gate voltage selection becomes a trade-off between power loss, reliability, and noise immunity capability. For example, as shown in Figure 10.13 for a GaN HEMT, due to the unique "diode like behavior," the negative off-state gate voltage amplitude increase will lead to more energy loss during the dead time after the switching transient, which cannot be negligible as compared to the device switching energy loss. On the other hand, the higher negative gate voltage is better with regard to the noise immunity capability.

For SiC MOSFET, its intrinsic body diode dominates the reverse conduction characteristics, which is regardless of the off-state gate voltage. However, due to the relatively high-voltage forward

Table 10.3 Gate voltage ratings of Si and SiC MOSFETs.

Type	Manufacturer	Model	Gate voltage maximum ratings
Si MOSFET	Microsemi	APT34M120J	+30 V/−30 V
SiC MOSFET	Wolfspeed/Cree	C2M0080120D	+25 V/−10 V

Source: Adapted from [7, 8].

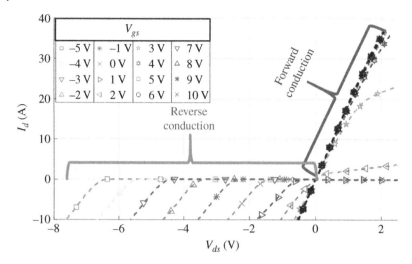

Figure 10.13 Output characteristics of GaN HEMT (GS66508P from GaN Systems). *Source:* Adapted from [9].

drop of SiC MOSFET's body diode, it is interesting to observe that as the off-state gate voltage decreases from 0 V, the reverse conduction voltage drop slightly increases, leading to additional reverse conduction loss. Thereby, similar to the GaN transistor, this phenomenon affects the selection of off-state gate voltage for the SiC MOSFET.

Furthermore, for both GaN HEMTs and SiC MOSFETs today, the maximum allowable negative gate voltage is limited as compared to their Si counterparts. For example, in Table 10.3 as compared to -30 V negative gate voltage rating of Si MOSFET, it is only -10 V for SiC MOSFET. Similar situation can be found for GaN HEMTs. Thus, the negative gate voltage rating is also an important device characteristic for the off-state gate voltage selection in the gate drive design.

In summary, with respect to device static characteristics, output characteristics, including forward and reverse conduction performance, as well as the gate voltage rating of the power device play a significant role for the gate driver design.

10.5.2.2 Gate Driver Design Considering Device Dynamic Characteristics

Device dynamic characteristics are primarily determined by gate driver, device itself, and operating condition. Therefore, a good gate driver design should be based on consideration of those device parameters that affect dynamic characteristics. The key device parameters that affect the switching performance and are critical for gate driver design include Miller capacitance C_{rss} (i.e. C_{GD} in Figure 10.14), input capacitance C_{iss} (i.e. sum of C_{GD} and C_{GS} in Figure 10.14), output capacitance C_{oss} (i.e. sum of C_{GD} and C_{DS} in Figure 10.14), internal gate resistance of the device $R_{G(in)}$, threshold voltage V_{th}, and transconductance g_{fs}. Specific impact of these device parameters on the gate driver design will be discussed in Section 10.5.3.

10.5.3 Gate Driver Design

Based on gate driver basic functional blocks in Figure 10.11, the typical gate driver is illustrated in Figure 10.15. In addition to signal isolator, isolated power supply, and gate drive IC, gate resistor and decoupling capacitor are necessary components for a gate driver. Detailed introduction of individual components is presented as follows.

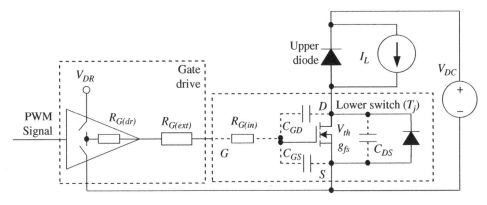

Figure 10.14 Device dynamic performance dependence on device, gate drive, and operating condition.

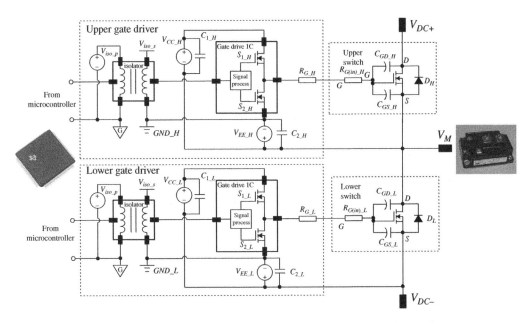

Figure 10.15 Typical gate driver circuit with detailed components highlighted. *Source:* Octopart.

10.5.3.1 Signal Isolator

Using the traction drive of a gasoline powered car as an analogy for a gate driver, signal isolator is the "steering wheel" of a gate driver. As discussed in Section 10.4.1, galvanic isolation capability and CMTI are critical design parameters, in particular, for WBG devices.

Regarding the galvanic isolation capability, considering the turn-off transition, and its resultant parasitic ringing and overvoltage across the lower switch, the dynamic potential difference across the upper signal isolator can potentially be higher than the DC bus voltage. With the conservative consideration, the galvanic isolation capability of the signal isolator for the gate drive design is preferred to be greater than the voltage rating of the power device.

For the CMTI, it is determined by the dv/dt during the switching transient. Apparently, fast-switching WBG devices pose increased challenges on the selection of signal isolator. Table 10.4 lists the tested peak dv/dt of several representative devices. Typically, more than 100 kV/μs peak dv/dt can be easily observed for fast-switching devices. Note that instead of average dv/dt, peak dv/dt

Table 10.4 Tested peak *dv/dt* for WBG devices.

Device under test	Model number	Typical *dv/dt*
SiC discrete switch	CMF20120D from Wolfspeed/Cree	~30 kV/μs
SiC power module	CPM212000025B from Wolfspeed/Cree	~80 kV/μs
GaN e-mode HEMT	GS66508P from GaN Systems	~200 kV/μs

should be used here as the worst-case scenario. Accordingly, signal isolator should be selected with CMTI capability greater than the peak *dv/dt* of the device under test.

According to Tables 10.2 and 10.4, it can be seen that the CMTI of state-of-the-art signal isolators barely meets the requirement. Considering the advancement of emerging WBG devices, CMTI capability of signal isolator should be further improved.

Additionally, signal transmission frequency range, propagation delay, and distortion of the signal isolator should also be taken into account based on the converter design and control requirement. Specifically, maximum and minimum signal transmission frequencies should cover required switching frequency range of the power converter under design. Propagation delay and distortion introduced by the signal isolator should be sufficiently short to not introduce unacceptable control error. In general, these two selection criteria are relatively easy to meet according to the requirement of the state-of-the-art power electronics converters.

10.5.3.2 Isolated Power Supply

Power supply is the "gas tank" of a gate driver. It provides necessary voltage for the operation of gate driver circuit. Also, it offers necessary voltage and power for driving the switch into on-state and off-state.

Similar to the signal isolator, isolated power supply also should be capable of providing sufficient galvanic isolation and CMTI. Specifically, with conservative consideration, galvanic isolation capability should be greater than the voltage rating of the power device. In the meantime, CMTI should be greater than peak *dv/dt* during the switching transient. In most of the isolated power supply datasheet, input-to-output capacitance instead of CMTI is provided. Equivalently, small input-to-output capacitance indicates high CMTI capability. Practically, for fast-switching WBG devices, this capacitance should be limited in a range of several pF.

Additionally, output voltage and power rating are another two significant design criteria for the isolated power supply. Output voltage of the isolated power supply is determined by the turn-on and turn-off gate voltages for the power device, while the turn-on and turn-off gate voltages are designed based on the device static and dynamic requirements. As discussed in Sections 10.5.1 and 10.5.2, the static requirement includes the minimization of on-state resistance or on-state voltage and conduction loss, and the suppression of spurious change of switch state due to the external or internal disturbance. During the switching transient, gate voltage should be selected to turn on and turn off power device fast, and, in the meantime, maintain sufficient margin to not exceed gate voltage maximum ratings of the device. Gate voltages for both turn-on and turn-off should be selected based on the trade-off between power loss, noise immunity, and gate terminal reliability (i.e. voltage margin). Device output characteristics and gate voltage ratings significantly affect the selection of the gate voltage. Practically, the device datasheet usually provides recommended turn-on and turn-off gate voltages, which should be a good reference for the output voltage determination of the isolated power supply.

Also, in practice, it is not easy to find an isolated power supply with the output voltage directly meets the design requirement. In this case, another non-isolated voltage regulator (e.g. low-dropout (LDO) regulator) can be introduced to tune the output voltage of the isolated power supply. Additionally, a dedicated 5 V or 3.3 V non-isolated voltage regulator can be employed to power the secondary of the signal isolator and logic gates, if necessary.

Power rating requirement of the isolated power supply is determined by the needed power for the secondary side of the signal isolator, gate drive IC, and charging/discharge input capacitance of power devices, which can be expressed as:

$$P_{out} > P_{iso} + P_{gd} + P_{sw} \tag{10.1}$$

where P_{iso} refers to the power dissipated by secondary side of isolator which can be obtained from the datasheet of signal isolator, P_{gd} indicates the power dissipated by gate drive IC which can be found based on the datasheet of the gate drive IC, and P_{sw} is the output power to drive the power device for an intended converter application. Detailed derivation for P_{sw} calculation is illustrated as follows.

Assume the gate charge of the device under test is Q_g. During the turn-on as shown in Figure 10.16a, input capacitance of the device C_{iss} is charged from V_{EE} to V_{CC}, where V_{EE} is the off-state gate voltage, which is zero or a negative value, and V_{CC} is the on-state gate voltage. Then turn-on energy supplied by the power supply E_{PS_on} is

$$E_{PS\_on} = V_{CC} \times Q_g \tag{10.2}$$

Similarly, during the turn-off as shown in Figure 10.16b, C_{iss} is discharged from V_{CC} to V_{EE}. Then turn-off energy supplied by the power supply E_{PS_off} is

$$E_{PS\_off} = V_{EE} \times (-Q_g) \tag{10.3}$$

In total, during one switching cycle, the energy provided by power supply, E_{PS_tot}, is

$$E_{PS\_tot} = (V_{CC} - V_{EE}) \times Q_g \tag{10.4}$$

Therefore, the needed power to drive the device P_{sw} with switching frequency f_s is

$$P_{sw} = (V_{CC} - V_{EE}) \times Q_g \times f_s \tag{10.5}$$

Note that P_{SW} is a function of V_{CC}, V_{EE}, Q_g, and f_s, and is regardless of the gate resistance R_G. Besides, during one switching cycle, there is no net energy stored in C_{iss}. Therefore, according to the energy conservation, power provided by the isolated power supply to drive the device is dissipated primarily by R_G. Thus, the power dissipation by gate resistor P_{rg} is the same as P_{sw} and also independent of the gate resistance.

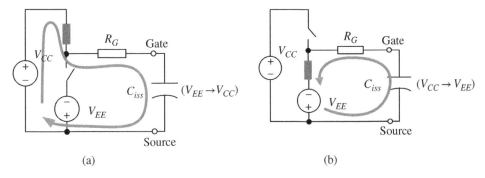

Figure 10.16 Output power to drive the device under test. (a) Turn-on and (b) turn-off.

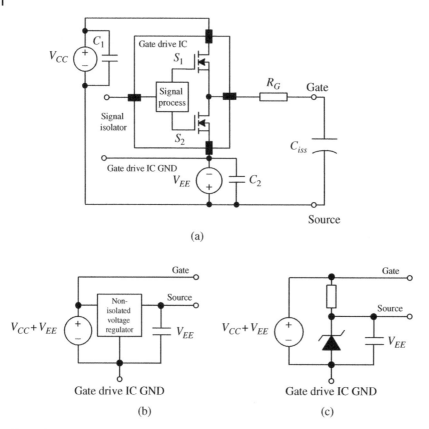

(a)

(b)   (c)

Figure 10.17 Isolated power supply configuration with negative turn-off gate voltage. (a) Two isolated power supplies, (b) one isolated power supply with non-isolated voltage regulator, and (c) one isolated power supply with Zener diode.

Typically, negative gate voltage is preferred for the fast turn-off switching speed and, in the meantime, provide sufficient voltage margin to prevent the external or internal disturbance. In practice, it is not necessary to design two isolated power supplies for both positive gate voltage V_{CC} and negative gate voltage V_{EE}. One isolated power supply along with a non-isolated power supply is adequate. Starting from a two-isolated power supply solution in Figure 10.17a, two options are illustrated in Figure 10.17b and in Figure 10.17c for V_{CC} and V_{EE} generation with one isolated power supply: one is to use the non-isolated voltage regulator, e.g. LDO regulator, and the other is to employ Zener diode.

10.5.3.3 Gate Drive IC

Gate drive IC is the "engine" of a gate driver. It provides necessary voltage to drive the switch into on-state and off-state, sufficient current to charge and discharge the switch input capacitance, and low loop impedance with quick response for fast switching. Accordingly, the primary selection criteria of the gate drive IC include adequate operating voltage range, sufficient peak source/sink driving current, and small pull-up and pull-down resistances of the gate drive IC buffer output.

Specifically, the operating voltage range of gate drive IC V_{GDIC} should be higher than the potential difference between the turn-on gate voltage V_{CC} and turn-off gate voltage V_{EE}, yielding

$$V_{GDIC} > V_{CC} - V_{EE} \tag{10.6}$$

Second, the peak source and sink driving current of the gate drive IC $I_{source/sink}$ should be larger than the peak gate current I_{g_peak} during the turn-on and turn-off transient. In the gate driver to charge the capacitive load, as shown in Figure 10.18, the initial gate current has the maximum amplitude, and its value is determined by:

$$I_{G\_peak} = \frac{V_{CC} - V_{EE}}{R_G} \quad (10.7)$$

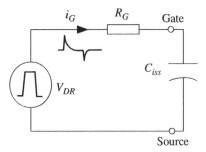

Figure 10.18 Transient gate current in gate drive.

where R_G is the total gate loop resistance, including the pull-up or pull-down resistance of the gate drive IC $R_{G(dr)}$, external gate resistance $R_{G(ext)}$, and internal gate resistance of the device $R_{G(in)}$, as shown in Figure 10.19.

In addition, different turn-on and turn-off gate resistances are usually used to optimize the switching performance during both the turn-on and turn-off transients. Thus, the required peak source and sink currents of the gate drive IC can be different and therefore should be calculated separately. Note that (10.7) is only valid without considering the gate loop parasitic inductance. Practically, due to this inductance, the peak gate current would be smaller. Therefore, the peak source and sink driving current selection based on (10.7) should be for the worst-case scenario with sufficient margin to satisfy the practical need.

Third, the pull-up and pull-down resistances of the gate drive buffer stage should be adequately smaller than the desired gate resistance; otherwise, the switching speed would be limited. Practically, it is recommended to select the gate drive IC with the pull-up and pull-down resistances much smaller than the internal gate resistance of the device. Then the contribution of the pull-up and pull-down resistances to the total gate resistance is small, and its impact on the device switching dynamics is negligible.

Similar to the signal isolator, two additional selection criteria of the gate drive IC should be considered: signal transmission frequency range, and propagation delay and distortion. Specifically, maximum and minimum signal transmission frequencies should cover required switching frequency range of the power converter under design. Propagation delay and distortion introduced by the gate drive IC should be sufficiently small to avoid introducing unacceptable control error.

10.5.3.4 Gate Resistor

Gate resistor is the "gas pedal" of a gate driver. It controls the speed of switching transients. Although gate resistor with small resistance is preferred, gate resistor must be designed properly to accommodate trade-offs among various switching behaviors, such as switching speed, switching loss, crosstalk suppression, and switch stresses.

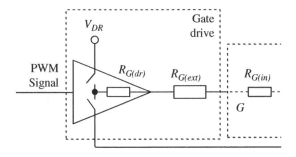

Figure 10.19 Total gate loop resistance.

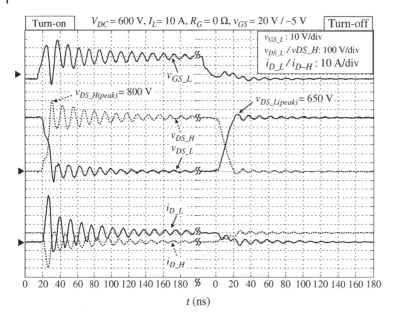

Figure 10.20 Measured switching waveforms of SiC devices.

Furthermore, usually it is difficult to achieve the optimal turn-on and turn-off behaviors based on a single-gate resistance. Figure 10.20 shows measured switching waveforms of Wolfspeed/Cree 1200-V/20-A SiC MOSFETs operating at 600 V/10 A with gate resistance of $0\,\Omega$. Obviously, the overvoltage and ringing during the turn-on transient is more severe as compared to that during the turn-off transient. Thus, it is highly recommended to design the gate driver circuit capable of tuning the turn-on and turn-off gate resistances, separately. Figure 10.21 illustrates a common approach to set different turn-on and turn-off gate resistances by using diodes. Note that low forward voltage drop and fast reverse recovery are the two key selection criteria for the diode. Additionally, some commercial gate drive ICs have the independent turn-on and turn-off output pins; as a result, different turn-on and turn-off gate resistances can be easily set.

In addition to the gate resistance, power rating of the gate resistor is another selection criterion to consider. As discussed earlier, power dissipation by gate resistor P_{rg} is the same as P_{sw} and is proportional to gate voltage ($V_{CC} - V_{EE}$), gate charge of the device Q_g, and switching frequency f_s as in (10.5). It is noted that the power dissipation of the gate resistor is independent of the gate resistance.

10.5.3.5 Decoupling Capacitor

Different from its namesake in Chapter 6, decoupling capacitor here acts as the "fuel injector" of the gate driver. It provides the energy for the gate current pulse during switching transient. Gate driving current originates from the power supply V_{CC} and V_{EE}. However, physically the power supply cannot be placed in close proximity to the gate driver and power device due to the relatively large volume. As a result, large parasitics are inevitably introduced, as shown in Figure 10.22. Therefore, it is necessary to place the decoupling capacitor locally with small footprint and low equivalent series inductance (ESL) to bypass the power supply resultant parasitics and improve the gate driver dynamic performance.

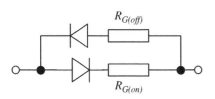

Figure 10.21 Using diode to set different turn-on and turn-off gate resistances.

Three design criteria should be considered for the decoupling capacitor selection, including sufficient voltage rating, adequate capacitance, and minimized ESL. Specifically, the voltage rating should meet

$$V_{C1} > V_{CC}, V_{C2} > |V_{EE}| \qquad (10.8)$$

where V_{C1} is the voltage rating of turn-on-related decoupling capacitor C_1 and V_{C2} is the voltage rating of turn-off-related decoupling capacitor C_2.

Second, the capacitance should be sufficient to ensure the stable gate voltage during the switching transient. In other words, with the conservative assumption, i.e. all the gate charge during the switching transient is provided by the decoupling capacitor, the voltage across the capacitor should remain almost constant. Therefore, the required capacitance should satisfy

$$C_1 > \frac{Q_g}{\Delta V_{CC}} = \frac{Q_g}{k_{VCC} \times V_{CC}}, C_2 > \frac{Q_g}{|\Delta V_{EE}|}$$
$$= \frac{Q_g}{k_{VEE} \times |V_{EE}|}$$
$$(10.9)$$

where ΔV_{CC} and ΔV_{EE} are the turn-on and turn-off gate drive output voltage variations, respectively; k_{VCC} and k_{VEE} are coefficients to indicate the two voltage variation percentages, respectively. Practically, 1–5% is selected for k_{VCC} and k_{VEE}.

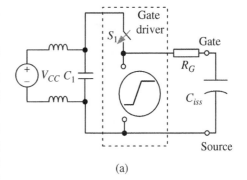

(a)

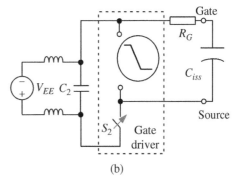

(b)

Figure 10.22 Decoupling capacitance for dynamic gate current. (a) Turn-on transient and (b) turn-off transient.

Third, to minimize the ESL introduced by the decoupling capacitor, surface mount ceramic capacitor is usually utilized. Also, multiple capacitors with small capacitance in parallel are preferred since ESL can be further reduced due to this paralleling configuration.

10.5.4 Bootstrap Gate Driver

Beyond the typical isolation-based gate driver design, a cost-efficient bootstrap gate driver may also be considered, especially for the low-power low-voltage converters, e.g. the example illustrated in Figure 10.8.

Figure 10.23 displays the circuit diagram of the bootstrap gate driver. In a phase-leg configuration, taking MOSFET's terminology as an example, drain terminal of the lower switch remains the same as the negative bus; therefore, practically, it is feasible to design the lower side gate driver sharing the same negative bus terminal as the ground; thus, the isolation is not needed. In the meantime, the power supply of the upper side gate driver is provided by the stored charge of the bootstrap capacitor C_B along with the isolation diode D_B and damping resistor R_B.

The operating principle is illustrated in Figure 10.24. When the lower switch in the phase-leg configuration is on, as shown in Figure 10.24a, the lower side gate drive power supply V_{CC}, through the bootstrap diode D_B and damping resistor R_B together with the short path provided by the lower switch, would (i) feed the upper gate driver and (ii) charge the upper bootstrap capacitor C_B for the charge compensation. Therefore, during the off-state of the lower switch in Figure 10.24b, the charge stored in C_B alone is used to power the upper gate driver, while D_B is reverse-biased to isolate the high bus voltage.

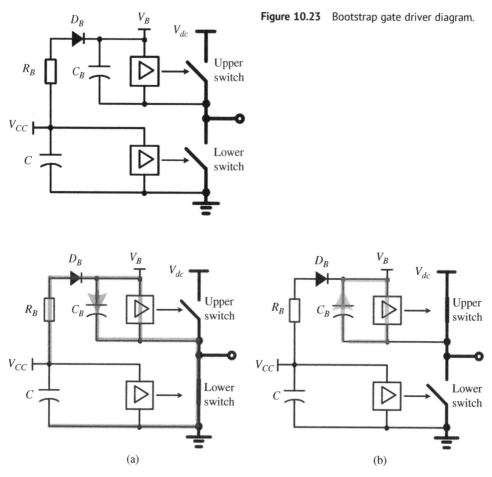

Figure 10.23 Bootstrap gate driver diagram.

Figure 10.24 Bootstrap gate driver operating principle. (a) Lower switch is on and (b) lower switch is off.

Bootstrap gate driver is a simple and cost-efficient solution. However, there are several limitations for the bootstrap gate driver: (i) for the high-voltage high-power application, it is challenging from the perspective of the practical implementation to find the proper bootstrap diode D_B, as D_B should have the breakdown voltage higher than the bus voltage, and it is preferred to have low forward voltage drop (as this forward voltage drop could reduce the V_{CC} to charge C_B) and minimized reverse recovery when the lower switch transitions from on to off; (ii) it is almost impossible to maintain the charge of C_B as the duty cycle of the upper switch close to 1. In other words, the usage of bootstrap gate driver poses limitation on the duty cycle control; and (iii) it is not straightforward to provide the negative gate voltage, which is preferred for the high-power converter design.

10.6 Sensors and Measurements

In the power converter design, sensors are essential for the controller and protection. In general, four types of sensors are adopted: current sensors, voltage sensors, temperature sensors, and application-oriented sensors, such as speed and position sensors for the motor drive application. From

Table 10.5 Summary of voltage sensor technologies.

Technology	Descriptions
Hall effect sensors	With isolation, DC and AC with low measurement bandwidth (~10 kHz)
Potential transformers (PTs)	With isolation, AC only (from several Hz with good measurement bandwidth)
Resistor dividers	Without isolation, DC and AC with good measurement bandwidth. Additional isolation or alternative means (e.g. differential op. amp. with adequate common-mode voltage handling capability) are needed

the converter designer's viewpoints, several questions need to be answered for the sensor design and selection. This includes (i) what types of sensors are needed? (ii) how many sensors are needed? (iii) what technologies are preferred with the consideration of steady-state accuracy, measurement bandwidth, cost, and size/weight? The answers to these questions are highly dependent on applications and the resultant specification and requirements. As this is case by case scenario and the corresponding selection usually is straightforward. So, in this section, a brief overview of sensor types and technologies is provided. Meanwhile, two special considerations, including high-voltage sensing technology for medium-voltage design and conditioning circuit design considering high-frequency high-noise environment for WBG-based converters, will be highlighted.

10.6.1 Voltage Sensors

In the three-phase AC converter, typically, DC-link voltage is the primary sensing target. It is noted that for multilevel and modular topologies, more than one DC voltage sensor is required for the DC-link voltage control, submodule voltage balancing control, and over-/undervoltage protections. Sometimes, three-phase AC voltages also need to be measured, such as the grid voltage for the grid-tied inverter system, which usually needs three line-to-ground or two line-to-line sensors.

Table 10.5 summarizes the voltage sensor technologies, including Hall effect sensors, potential transformers (PTs), and resistor dividers. In practice, resistor-based sensing circuits are most common considering low cost and adequate measurement bandwidth. The drawback is the extra consideration on the isolation. Typically, the conditioning circuit with operational amplifier is introduced to cope with it, which will be discussed in detail in Section 10.6.5.

10.6.2 Current Sensors

Typically, three-phase AC currents are the primary sensing target. For a three-phase balanced system, only currents associated with two phases are needed, although three sensors are often used in high-power high preperformance converters for added accuracy and protection. The sensed current information is essential for the AC current control (amplitude, phase angle, and fundamental frequency) and overcurrent protection. Sometimes, DC current is also measured for the overcurrent protection or control on the DC side.

Table 10.6 summarizes the current sensor technologies, including Hall effect sensors, current transformers (CTs), and resistor shunts. In practice, Hall effect sensors are most common in medium- and high-power converters considering adequate measurement bandwidth for both DC and AC together with the isolations.

Table 10.6 Summary of current sensor technologies.

Technology	Descriptions
Hall effect sensors	With isolation, DC and AC with good measurement bandwidth (10s to 100s of kHz)
Current transformers (CTs)	With isolation, AC only (from several Hz with good measurement bandwidth)
Resistor shunts	Without isolation, DC and AC with good measurement bandwidth. Additional isolation or alternative means (e.g. differential op. amp. with adequate common-mode voltage handling capability) are needed

10.6.3 Temperature Sensors

Temperature sensors are usually included for the converter control, power derating, and overtemperature protection. In practice, an embedded temperature sensor within the power module or a sensor on the heatsink can be used as an indicator of the device temperature. Note that this sensed temperature is lower than the device junction temperature; therefore, a thermal model presented in Chapter 9 together with the measured temperature can be adopted to estimate the device junction temperature.

Typically, an NTC thermistor is embedded within the power module where NTC stands for "Negative Temperature Coefficient," meaning that the resistance decreases with increasing temperature. NTC thermistors are generally made of ceramics or polymers, suitable for use within a temperature range between −55 and 200 °C. There are special families of NTC thermistors that can be used at temperatures approaching absolute zero (−273.15 °C). The temperature sensitivity of an NTC sensor is expressed as "percentage change per °C (or K)." Depending on the materials used and the specifics of the production process, the typical values of temperature sensitivities range from −3% to −6%/°C.

In addition to the NTC thermistor, resistance temperature detectors (RTDs) and thermocouples are also popular temperature sensors. RTDs are made from metals with the most accurate measurements (±0.5% measurement tolerance). Also, RTDs have a much wider temperature range between −200 and 800 °C as compared to NTC thermistor. Regarding the thermocouple, it has the best working temperatures as high as 2500 °C. Also, thermocouple is more cost effective with faster response time as compared to RTDs.

Table 10.7 compares different temperature sensor technologies. In summary, for the power converter typically operating in the temperature ranges from −40 to 150 °C, NTC thermistor is preferred due to smaller size, faster response, greater resistance to shock/vibration, and lower cost.

10.6.4 High-Voltage Sensors

For the medium-/high-voltage converter design, special considerations is needed for voltage sensors. There are several options for voltage measurement in these converters. Beyond Hall effect transducers, PTs, and resistor dividers discussed in Section 10.6.1, capacitor dividers and resistor–capacitor ladders are also included. Though resistor divider, in theory, has unlimited bandwidth, practically, it requires an output filter to limit the bandwidth, which will cause a delay due to the RC time constant. Capacitor dividers block the DC voltage component. RC ladders are also a good solution and have been used for device dynamic voltage measurement.

The main challenges of the high-voltage sensor design include signal fidelity, isolation design, and compactness. The generalized voltage sensor design example is given in Figure 10.25. It consists of voltage divider, RC filter, follower, voltage-to-frequency converter (VFC), and fiber optic

Table 10.7 Summary of temperature sensor technologies.

Technology	Descriptions	Examples
NTC thermistor	Moderate temperature ranges from −55 to 200 °C; smaller size, faster response, greater resistance to shock and vibration; slightly less precise than RTDs; similar precision to thermocouples, poorer than RTDs; suitable for power converter applications	MEAS series from TE connectivity
RTDs	Wide temperature ranges from −200 to 800 °C; most accurate measurements; relatively slower response and higher cost	PT100 RTD
Thermocouple	Best working temperature ranges up to 2500 °C; similar precision to NTC, poorer than RTDs; adequate response and low cost	Different types (e.g. J and K) dependence on temperature ranges and accuracy

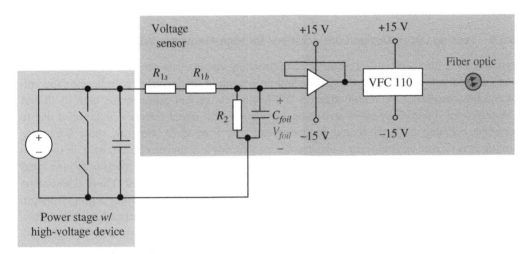

Figure 10.25 Schematic of test setup and generalized voltage sensor.

transmitter. For the voltage divider, several resistors are used to satisfy UL 347A standard and achieve the proper resistance while maintaining a low-profile design with surface mount (SMD) resistors. The follower provides a high impedance input and voltage source for the VFC, while the VFC provides a digital signal that can be isolated from the main power loop with the fiber optic.

The main error sources in the measurement arise due to the ADC, transducer, and a ground bounce or EMI due to high dv/dt. A sigma-delta-based VFC is chosen to digitize the analog signal from the resistor divider, so it can be transmitted to control via optic fiber. The rising and/or falling edges are counted over a period, T_{samp}. The VFC gives good resolution with 2 V/kHz with a maximum frequency of 4 MHz. The resolution should not be an issue here, since this is well below the 1% design target. However, due to the floating nature of the voltage sensors in some cases, e.g. submodules in the MMC, which may cause a ground bounce or EMI that impacts the operation of the VFC. Therefore, the special attention needs to be given to the grounding of each voltage sensor in the actual converter circuit with the details presented in [10].

The other remaining error sources include the transducer or resistor divider. To reduce the noise impact on the divider from surrounding circuitries, the following methods are recommended: increasing the signal-to-noise ratio (SNR), reducing bandwidth, layout improvements, and shielding.

1) *Increasing Signal-to-Noise Ratio:* the SNR is increased by reducing the input resistance. This will increase the power loss, but often this is necessary to maintain the fidelity of the measurement.
2) *Reducing Bandwidth:* by increasing C_{foll} in Figure 10.25, the bandwidth of the voltage sensor is reduced. However, this will also increase the measurement delay due to the increased RC time constant.
3) *Layout Improvements:* signal integrity and layout go hand in hand. However, the high-voltage nature of the voltage sensor can make this nontrivial. Additionally, it is ideal to have the sensor be low-profile and as compact as possible.
4) *Shielding:* shielding of high-speed traces is common, and often it is desirable to form a transmission line to reduce noise coupling. It is common in current sensing and has been implemented with Rogowski coils. However, this can cause insulation concerns and may even add capacitance between traces, which are both highly undesirable.

An actual design example for the high-voltage sensing circuit design based on MMC, one of the most popular multilevel topologies for the medium-/high-voltage applications, is documented in detail in [10].

10.6.5 Sensing Circuit Design Considerations for High-Frequency WBG Converters

Beyond sensors discussed earlier, sensing and processing circuits between sensors and microcontroller are also critical. Sometimes, sensing and processing in the feedback path become the vulnerable link, especially for high-frequency WBG-based converters experiencing high dv/dt and di/dt transients due to high-speed switching. In other words, signal integrity issues arise due to capacitive coupling from dv/dt and inductive coupling from di/dt. Notably, these issues are commonly perceived by engineers, and some are well known in signal electronics field. In this subsection, a summary to explain some of the phenomena and effects that are specifically associated with high-frequency noise and transients in power electronics is described. These phenomena include nonlinear DC or low-frequency shift on the sensing signals. If without a proper design, the measurement tolerance induced by the sensing and conditioning circuitries can lead to inaccurate control as well as voltage/current distortion. For the sake of better illustration, in the following discussion, some test waveforms from a GaN-based three-phase rectifier are presented.

10.6.5.1 DC Bias from Analog Signal Isolation Amplifier and Mitigation
10.6.5.1.1 Observed DC Bias on Isolated Voltage Sensing When isolation is mandatory, voltage or current sensing is often realized by isolated schemes shown in Figure 10.26. An analog isolation amplifier (iso-amp) is the key device to achieve the isolation between control and power domains. At high-power side of an iso-amp, the voltage will be sampled by a resistor divider in a single-ended form or by a divider pair in a differential form. And the current will be sampled through a shunt resistor. Then, the isolated sensing signal is processed by downstream conditioning amplifiers and filters before transmitting to ADC. However, operated in high dv/dt conditions, this scheme is observed to suffer severe distortion, i.e. a nonlinear DC shift.

High dv/dt during the switching transient emits high level of noises either conductively or radiationally to its surrounding. Although it is understandable that certain high-frequency noise might be coupled onto the sensing signal of the output voltage, it is surprising to observe a significant DC bias in the sensed DC voltage. As can be seen in Figure 10.27 and Figure 10.28, the DC error could be considerable.

Although this section focuses on sensing circuits, because of the sensing voltage shift ΔV_{DC}, the actual voltage regulated by feedback closed-loop control is different from the reference value as in

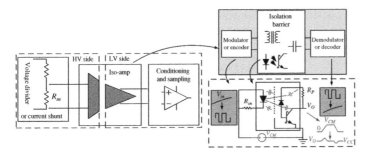

Figure 10.26 Iso-amp function block and formation of DC bias through CM noise penetration and demodulation.

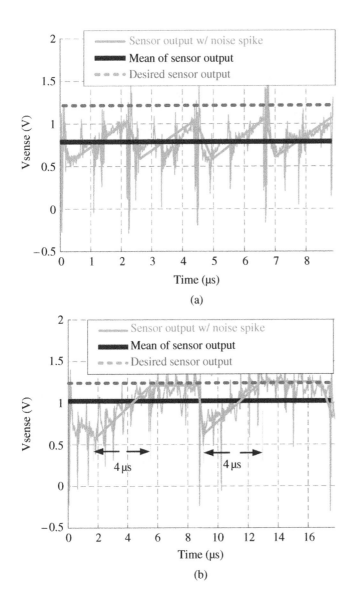

Figure 10.27 Voltage sensing distortion (black/solid line profiles the actual sensor output signal; dashed line denotes the reference voltage). (a) 450 kHz switching and (b) 112.5 kHz switching.

(10.10). This DC error is even larger when the DC-link voltage increases as if a nonlinear voltage-dependent sampling gain exists.

$$V_{actual} = V_{ref} - \Delta V_{DC} \tag{10.10}$$

10.6.5.1.2 Origin of DC Bias

The mechanisms causing the observed DC bias from analog signal isolation amplifier is due to the basic structure of an analog iso-amp in Figure 10.26, where the analog input is modulated into digital format, e.g. via a sigma-delta ADC at the primary side, and the signal is reconstructed to analog output via a demodulator at the secondary side. An intermediate digital transmitting medium is implemented across the isolation barrier of the iso-amp. As a result, a high CM voltage exists due to the different grounding potentials. Taking optical iso-amp as an example, when a rising $v_{CM,r}$ pulse is imposed onto an iso-amp, the internal coupling capacitance of the iso-amp induces CM current pulses penetrating the barrier, and a positive differential voltage spike is generated at the output through the pull-up resistor. In addition, the turn-off pulse also induces a negative spike $v_{CM,f}$ at the output. These spikes may falsely toggle the logic level, jeopardize the signal quality of effective signal, and finally demodulate analog spikes. Thus, a DC bias can consequently be generated after the downstream filter through the average of these positive and negative analog spikes. This relationship is expressed in (10.11). It should be noticed that the polarity of the spikes could be reversed depending on internal logic of an iso-amp IC.

$$\Delta v_{DC\_out} = \int_0^{T_s} \left(v_{CM\_r} - v_{CM\_f} \right) dt / T_s = f_s \int_0^{T_s} \left(v_{CM\_r} - v_{CM\_f} \right) dt \tag{10.11}$$

10.6.5.1.3 Impacting Factors of dv/dt Induced DC Bias

1) CM dv/dt in power converter

    The magnitude of CM isolation voltage across the iso-amp is determined by the CM impedance network distribution considering parasitics of converter (e.g., transformer inter-winding capacitance, if applicable) and sensing circuits (e.g., isolation capacitance of the iso-amp). The edge slopes of this voltage are related to dv/dt of device turn-on and turn-off, determined by the specific gate driver, gate resistance, switching speed, and DC-link voltage. The turn-off dv/dt is also strongly impacted by load current level.

2) Asymmetric of dv/dt transient

    Assuming the turn-on dv/dt and turn-off dv/dt are the same, it seems reasonable that a minimum DC bias might be induced per (10.11). However, in practice, the two dv/dt values might be different due to the junction capacitor charging effect, load current, and different gate drive on/off speed control. Taking a GaN-based application as an example, this dv/dt disparity becomes larger. The turn-on transient of 650 V e-mode GaN device is usually less than 10 ns and dv/dt can be above 100 kV/µs, while its turn-off transient could last for hundreds of ns at AC current zero crossings. Thus, turn-on dv/dt becomes the primary contributor to the CMTI issue instead of the turn-off dv/dt at low current level or zero crossing. As shown in Figure 10.27, the turn-on dv/dt induced negative spikes, which are below the desired voltage sensing level, are almost the same in case (a) and (b), whereas no positive peaks show up except some noises.

3) Transient speed of iso-amp

    The aforementioned results also highlight another important distortion factor that was not considered previously, i.e. the output signal rise/fall time of an iso-amp IC. Ideally, the output spike due to dv/dt should only occur during the switching transient. However, it is noticed that at different switching frequencies, the profiles of the output ripples in Figure 10.27a, b have rising ramps which increase linearly from the same negative valley. This is caused by the relatively slow rise time of the iso-amp ACPL-C87B output, roughly 4 μs. For example, in 450-kHz switching frequency case, the dv/dt induced spike is reset by the next switching event before it can reach the steady-state level within its 2.2 μs cycle. Whereas in 112.5 kHz case, the dv/dt-related ripple can jump back to the desired DC voltage in one switching cycle.

4) Switching frequency

Switching frequency will not impact the peak spike of an iso-amp output, but it does impact the DC bias level. This is because: (i) the same noise spike pattern will be averaged over different switching periods; thus, the bias is proportional to f_s, as shown in (10.11); (ii) the noise spike of an iso-amp may not reach the correct sensing level if the switching cycle is too short as analyzed earlier. As a result, high frequency induces more DC sensing drift.

From experimental results in Figure 10.28 recorded by DSP controller, the sampling output voltage is compared with the actual voltage at two different frequencies. In the steady state, the output voltage sampling begins to suffer high DC shift, and it is clear that this DC shift becomes much higher when three-phase rectifier is operated at higher switching frequency.

In summary, the speed of iso-amp, i.e. the rise/fall time, impacts the sensing output distortion induced by CM spikes. More instantaneous distortions and DC bias will be introduced when higher switching frequency, lower speed sensing iso-amp, and higher speed devices are adopted.

10.6.5.1.4 Mitigation Solutions The aforementioned analysis has shown that the DC bias issue is not just by the switching noise propagation, but it is at the basis of the demodulation of the interference. With high CMTI and fast speed iso-amp, this effect could be suppressed. Unfortunately, the CMTI of the commercial analog iso-amp is relatively low in comparison with device dv/dt, especially for fast WBG devices. Table 10.8 lists the state-of-the-art commercial iso-amps. As it can be seen, the CMTI of some products is not precisely defined. Detailed information of iso-amp error levels vs. CM voltage or dv/dt is also unknown, which further increases the difficulty of sensing circuit design. The state-of-the-art commercial voltage sensing iso-amp was from Avago before year 2018, typical with 15 kV/μs CMTI [11] using optocoupler isolation, and current sensing iso-amp has up to 75 kV/μs CMTI [12] using complementary metal–oxide–semiconductor (CMOS) isolation.

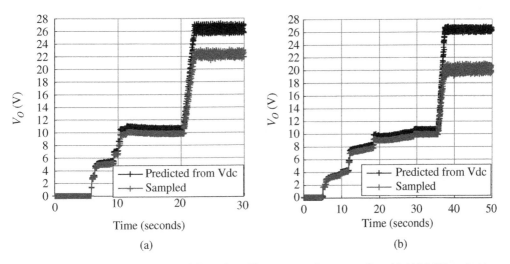

Figure 10.28 Recorded experimental data of rectifier output voltage sampling. (a) 112.5 kHz switching and (b) 450 kHz switching.

Table 10.8 Key parameters of state-of-the-art commercially available analog isolation amplifiers.

Amplifier name	Isolation technique	Bandwidth (kHz)	CMTI (kV/μs)	t_r and t_f (μs) (10–90%)
ACPL-C87B	Optical	100	[a]15 (typ), 10 (min)	2.7 (typ), 4.0 (max)
ACPL-790B/A/0	Optical	200	[a]15 (typ), 10 (min)	1.7 (typ)
Si8920A/B	CMOS	950	[b]75 (typ), 50 (min)	0.42 (typ)
AMC1311B	SiO_2	275	[c]140 (typ), 75 (min)	1.3 (typ)
AMC1311	SiO_2	220	[c]30 (typ), 15 (min)	1.3 (typ)

[a] CMTI failure is defined as the output error of 200 mV deviation from the average output voltage persisting for at least 1 μs, tested at $V_{CM} = 1$ kV.
[b] CMTI failure is defined as the output error of more than 100 mV persisting for at least 1 μs, tested at $V_{CM} = 1.5$ kV.
[c] CMTI failure definition is not specified, tested at $V_{CM} = 1$ kV.

Thus, CMOS-based iso-amp is a better option currently, since it shows higher CMTI due to RF transmission. To be noted, an even higher CMTI voltage sensing iso-amp AMC1311B from Texas Instruments was released in 2018 [13].

Optionally, CM filters may be possible to use at the high side of the iso-amp to limit the CM voltage across the iso-amp. This will be similar to the approach for CMTI enhancement of the gate driver discussed in Section 10.5. However, the power supply lines should also be in series with the CM filter to avoid high CM voltage stress across the high side pins of the iso-amp.

10.6.5.2 DC Bias from RF Interference Rectification in Op-amp and Mitigation

For Si device switching at 10s of kHz, a moderate filter implemented in the conditioning circuitry is sufficient to guarantee the integrity of sensed signals. However, when using the similar design principles in high-frequency WBG-based converters, the experimental results reveal that the voltage and current sensing signals along the conditioning circuitry could see a pronounced DC offset or low-frequency distortion, which further grows as DC-link voltage or switching frequency increases. An example of AC current sensing at 450 and 112.5 kHz switching frequencies in an GaN-based Vienna-type rectifier is provided in Figure 10.29, where the sampling signal outputted by digital-to-analog converter (DAC) has much difference from the current probe result. At higher switching frequency, this low frequency or DC shift becomes larger. It should be mentioned the poor quality of the AC current is due to several distortions, including the sensing distortion, sampling distortion, and the PWM voltage loss during the current zero crossings.

10.6.5.2.1 Identification of Observed Sensing Noise Pattern
To identify the cause of this distortion, the output signal of the Hall sensor is checked. High-switching transient noises are observed in the sensor output as shown in Figure 10.30. Since the onboard sensor is placed close to the main power loop, the noise is found induced by high di/dt of switching transients. And noise at turn-on seems more pronounced than that at turn-off, which can be explained from the fact that at turn-off, commutation from off-device to on-device is largely determined by load current charging.

The sensor output, ADC sampling, and probe measurement are further compared. From results in Figure 10.31, both the sensor output and final ADC sampling have DC shift distortion in switching period scale and low-frequency distortion in line period scale. Although the coupled noise and DC shift patterns may be different at different operation points, making the analysis intricate, some key observations are captured: (i) due to the asymmetric nature of the coupled high frequency (HF) noise as mentioned earlier, the mean value of sensor output over a switching period can be shifted

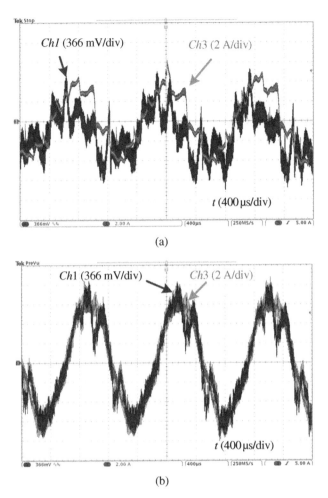

Figure 10.29 Experimental waveforms of AC current sampling (*Ch1*: sampled signal, *Ch3*: current probe result). (a) 450 kHz switching and (b) 112.5 kHz switching.

from the original level and falsely conveyed to the downstream conditioning amplifier; (ii) this DC shift is thus influenced by switching frequency; (iii) the fact that an offset also exists between the envelops of the sensor output and probe result indicates the sensor itself is also affected by the HF noise.

10.6.5.2.2 *Origin of DC Bias and Low-Frequency Distortion* It is found that this severe DC distortion is determined by the amount of high di/dt-induced noise picked up by signal-ground traces between sensor output and conditioning amplifier input. And the cause of this DC bias is the radio-frequency interference (RFI) rectification effect on op-amp, a phenomenon seldom considered in power electronics before. With this effect, when an op-amp sees high-frequency out-of-band noises at its input, its output can induce an instantaneous DC shift due to the nonlinear I–V characteristics of differential transistor pairs at its input stage. And the strength and envelop of the time-domain bias are dependent on the amplitude, frequency and continuity of the input coupled out-of-

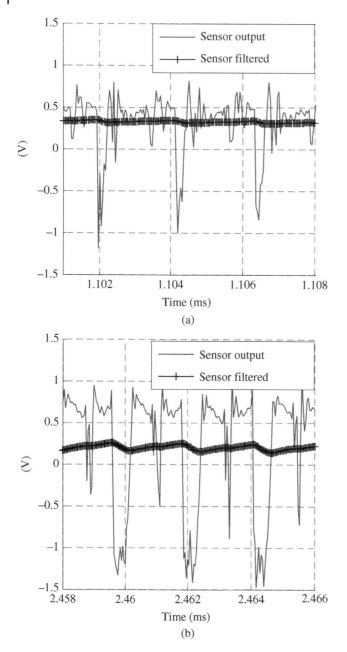

(a)

(b)

Figure 10.30 AC current sensor output waveforms with 450 kHz PFC (due to 25 MHz low bandwidth voltage probe used in the example, only the noise envelop is captured where the higher frequency oscillations are not accurate). (a) Higher line current instant and (b) lower line current instant.

band noise. Furthermore, noise coupled at the output of an op-amp can also contribute to the input noise via its feedback path. This explains why DC shift at current sensor output was observed, as an internal op-amp exists at the output stage of the current sensor to provide desired quiescent bias.

To help understand this rectification effect for readers in power engineering domain, a brief introduction is provided. RFI rectification was firstly observed in naval research on the susceptibility of transistors to microwave energy [14, 15]. Since then, the effects and mechanisms of RFI distortion

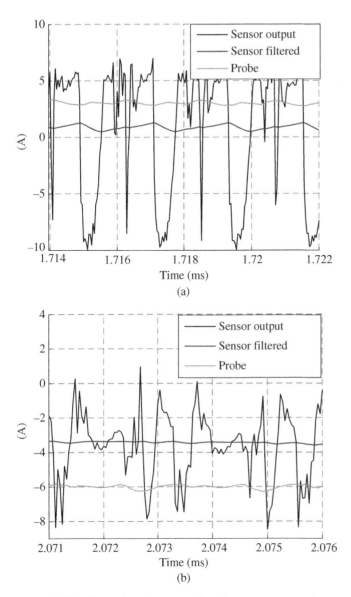

Figure 10.31 Comparison between onboard sensor output and current probe result at different current levels with 450 kHz PFC in switching cycle scale. (a) current instant at the positive line cycle and (b) current instant at the negative line cycle.

on analog and digital ICs have been investigated experimentally and theoretically, in both time domain and frequency domain. Analog electronics, in particular, the amplifiers are most susceptive to the EMI noise. Technical notes from industry have attributed the DC bias effect on op-amp to the nonlinear transfer characteristics of its input transistors, e.g. emitter-coupled bipolar junction transistor (BJT) has exponential I–V curve and source-coupled junction-gate field-effect transistor (JFET) or CMOS have quadratic I–V curve. Although this theory can confirm the observation that BJT-based amplifiers typically suffer higher RFI distortion than other types, it fails to explain why this effect only occurs at out-of-band frequencies and why the DC offset varies with frequencies. The frequency-dependent nonlinearity and DC bias level were later analytically modeled in [16] for a CMOS op-amp, indicating that the parasitic capacitances C_{GS} of input differential pairs and their total source-to-ground capacitance are the determinants. The mechanism of negative feedback on DC bias was also revealed in [17] for CMOS op-amp. It is noticed that some models are only valid at low-level RF input, i.e. two transistors of the input differential pairs are both in saturation region, while other models may be feasible for both saturation region and subthreshold region where large signal noise switches the transistor on and off.

The susceptivity of an amplifier to RFI/EMI can be evaluated by a rejection ratio (EMIRR), from which the corresponding DC bias can be estimated as (10.12) (All voltage units in mV). However, this is only valid at low-amplitude RFI/EMI [18].

$$\Delta V_{DC} = \frac{1}{10^{\left(\frac{EMIRR_{IN}\ (dB)}{20}\right)}} \left(\frac{V_{RF\_peak}^2}{100\ mV}\right) mV \tag{10.12}$$

10.6.5.2.3 Mitigation Approaches A straightforward approach to mitigate the RFI effect is attenuating EMI noise before the input pins of an op-amp. RC-based CM and DM filters have been built in in some commercial amplifiers [19]. However, the filter corner frequency is usually set above 10s of MHz to serve general purposes, beyond interested frequency range of the state-of-the-art GaN-based power converters. Therefore, a set of external EMI filters should be provided for converter sensing design, as shown in Figure 10.32.

For single-ended signals, RC or LC filter can be applied whose bandwidth should be selected between the current control bandwidth frequency and the high switching frequency, e.g. in the middle range, without impacting obvious delay to the feedback control. For differential sensing signals, both CM and DM filters should be adopted following similar design consideration, where

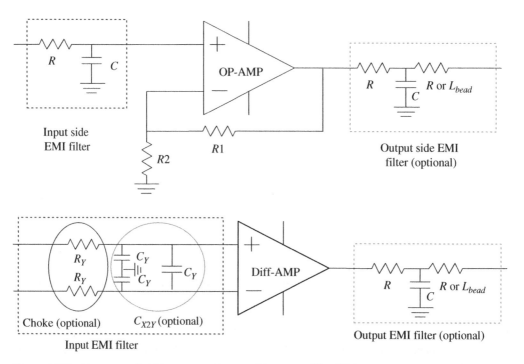

Figure 10.32 Recommended filter approaches to mitigate amplifier RFI issue: upper for general op-amp and lower for differential amplifier.

the DM filter corner frequency is determined by (10.13) and the CM filter corner frequency determined by (10.14) can be slower as it does not affect the feedback control:

$$f_{DM} = \frac{1}{2\pi R_Y (2C_X + C_Y)} \tag{10.13}$$

$$f_{CM} = \frac{1}{2\pi R_Y C_Y} \tag{10.14}$$

However, unlike signal processing, power conversion deals with high voltage and current. Particularly, in WBG-based converters, high di/dt and dv/dt and high frequency could likely induce strong noise coupling. Relying purely on filter itself is not effective enough to eliminate RFI noise without sacrificing sensing bandwidth. In fact, as reported in [20], the prefiltered voltage sensing circuit can be still distorted once PWM-based hard switching operation is activated, leading to the design compromise by moving the local sensing circuit of the power board far away and placing it onto another board. This can prevent integration of sensing functions in power modules/circuits and even lower power density.

Additionally, the direct control of the noise source and associated coupling path is more effective and straightforward. Taking current sensing distortion as an example, if tracing back to the source of the Hall sensor noise, it is induced by high di/dt current circulated in the switching loop and then coupled onto the signal side of the sensor through space magnetic coupling. Given the fixed di/dt, smaller power loop means less diffusing magnetic field generation, thus reducing the magnetic coupling. However, less loop inductance can adversely induce higher di/dt and dv/dt. Therefore, signal loop layout for sensing design becomes another critical step. In high power density converter design, surface-mount Hall sensor chip is preferred due to its low profile and weight. When it is placed on the power board close to converter terminals, longer traces of control power supply nets and output nets are inevitable to interface with the control board; thus, it is prone to taking in the high di/dt-induced voltage noise. Therefore, an effective solution is to layout the sensor circuit and the conditioning circuit with the lowest loop area, in addition to the front-end filters. The basic concept is shown in Figure 10.33, where three traces including control ground, control power line, and sensing signal should all be tightly placed on both the power board and the control board.

To implement this scheme, a vertical trace arrangement is preferred to minimize the area of two types of signal loops. In addition, the Hall sensor interface loop and power supply loops should be paralleled to the flux orientation and orthogonal to the switching loop so that the coupling could also be reduced.

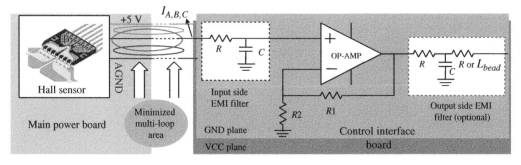

Figure 10.33 RFI noise suppression approach through PCB layout for op-amp-based sensing circuit. EMI filter can be optional.

10.6.5.2.4 Discussion of Other Noises and Mitigation Since the magnetic field strength is inversely proportional to the distance, another conservative but effective approach is to move Hall sensors far away from the power loop, which is also experimentally verified but at the cost of larger board space.

Apart from di/dt inductive coupling noise, other noises also exist for Hall sensors. For example, dv/dt capacitive coupling from high voltage side to signal side of the Hall sensor IC could also superpose significant noise onto the low-level sensor output via the isolation capacitance. Recommendations are: (i) place Hall sensors at non-dv/dt nodes, such as after filter inductors which absorb voltage dv/dt in the first place; (ii) choosing Hall sensor chips with internal capacitive shielding can also avoid this type of distortion; and (iii) adding proper filter for this HF noise is also suggested. However, solutions (i)–(iii) alone may be ineffective, since a DC offset can still be produced if dv/dt patterns at turn-on and off are not symmetric, following the mechanism discussed earlier.

10.6.5.3 DC Bias Due to RFI Rectification in In-amp and Mitigation

Aside from op-amp, instrumentation amplifier (in-amp) is typically used for high-voltage sensing where strong CM signals can be eliminated by the differential structure. However, the typical common-mode rejection ratio (CMRR) of in-amps rolls off at 200 Hz, with only 20–30 dB remaining at 100 kHz, and the in-amp has only up to 20 kHz bandwidth [19]. If a strong RFI noise is coupled at the input and is not effectively suppressed, the DC rectification effect happens as well, similar to the RFI issue in the op-amp case. Choosing an in-amp with higher EMI robustness is always the first design choice; however, decent immunity at low-voltage signal level can barely cover the high noise strength in power conversion systems.

More practical mitigation in power electronics applications should be again to reduce the noise from its source. The first typical solution is inserting external EMI filter pairs at inputs of the in-amp, with both CM and DM filters implemented by RC filters, chokes, or X2Y capacitors, as shown in Figure 10.34.

Note that the divider impedance network should also be considered for the input EMI filter design; otherwise, improper filter characteristic may affect the control bandwidth of the voltage regulator due to the enlarged phase delay of the filter. Based on the Thevenin Theorem, the equivalent source impedance from the sensing voltage side is $R_H//R_L$; thus, corner frequencies for CM and DM filters are, respectively:

$$f_{CM} = \frac{1}{2\pi(R_H//R_L + R_Y)C_Y} \tag{10.15}$$

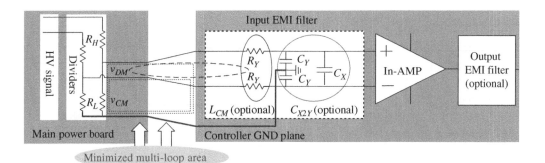

Figure 10.34 Combined RFI noise suppression approach through PCB layout and EMI filter for in-amp-based sensing circuit.

$$f_{DM} = \frac{1}{2\pi(R_H//R_L + R_Y)(2C_X + C_Y)} \qquad (10.16)$$

Clearly, neglecting the divider impedance $R_H//R_L$ in the filter will lead to selection of higher R_Y and cause underestimated phase delay and much slower voltage controller dynamics than one would expect. Similarly, failure to model it in the control transfer function should also be prohibited. In the case of RC filter, since $R_H//R_L$ is often above 10s of kΩ, R_Y and C_X could be removed to simplify the conditioning circuit. If LC filters are used, a second-order RLC circuit analysis would be needed to properly determine the cut-off frequency and damping factors.

Another solution is to minimize the signal coupling loop. Different from the single-ended signal, the differential sensing circuit shown earlier contains two types of signal trace loops between the sensing divider and the in-amp: the DM signal loop and the CM signal-GND loop. When it comes to PCB trace layout for the differential signal, the CM ground loop is often overlooked. In fact, it is as important as the DM loop. Since the high di/dt can induce significantly high CM and DM noise voltage on the control interface circuit, the CM noise voltage will also be imposed on the in-amp pins. High stress of such CM noise not only causes clamped 0-V output of the in-amp (since its allowable input CM voltage is limited within its power supply voltage and decreases as its DM voltage increases) but also introduces heavier RFI effect as the RFI rectification involves CM and DM noises simultaneously. Reducing both noises is ultimate, and minimizing both types of trace loops is a must. This requires delicate layout on the power board to minimize not only the gap between the differential signal pairs as short as possible but also the gap between signal pairs and the signal ground, where the ground wire can be vertically placed through the multilayer PCB design. The signal connection from the power board to the control interface board should also be short and compact until the three signal traces safely land onto the controller ground plane. On the interface board, the differentia pair should continue to follow the aforementioned best practice to reduce the possible coupling noise.

10.7 Protection

As described in control architecture of Figure 10.2, two layers of protections are included: device-level protection with the faster response time (100s of ns to several µs) and converter-level protection with relatively longer response time (sub-µs for electrical protection to sub-ms for thermal protection).

10.7.1 Device-Level Protection

Regarding the device-level protection, there are two types of protection techniques: one is to prevent the fault based on the understanding of the device behavior in power converters and the other is to reliably clear the fault with fast response protection circuit. Normally, the first type of techniques relies on the feedforward control with the predefined command, while the latter one is achieved by the feedback control in collaboration with the online monitoring technique. Crosstalk suppression is one representative of the first type of protection techniques, while short-circuit protection belongs to the second type of protection techniques.

10.7.1.1 Short-Circuit Protection
In the widely applied voltage source converter, shoot-through failure and short-circuit protection are crucial. Notably, the requirement of fast short-circuit and overcurrent protection for WBG

devices has been shown to be much stricter than conventional Si devices. This can be explained by the much smaller die size of WBG devices, resulting in reduced thermal capacity, and the potential for a more rapid spike in junction temperature. In comparison with the Si devices, which generally guarantee at least 10 μs short-circuit withstand time, 1 μs is the typical time for SiC devices. The situation is even worse for GaN transistors. Recent literature has shown that a healthy 650-V GaN HEMT can be destroyed by a short-circuit pulse as brief as 400–600 ns at 400 V, and in some cases even only 200 ns [20]. This short-circuit withstand capability increases significantly with lower bus voltages because of lower leakage current and loss after the device turns off. In any case it is desirable to turn off the devices as quickly as possible under short-circuit conditions, as sustained short-circuit conditions will likely lead to overcurrent and overtemperature, causing device degradation.

Perhaps the most common overcurrent protection scheme is desaturation protection. Desaturation protection requires sensing of the drain-source voltage. It triggers the gate driver to turn off, when this voltage indicates that the device has left its normal working region (ohmic region for a MOSFET or saturation region for a bipolar device like IGBT) and entered the abnormal high current region (saturation region for a MOSFET or active region for a bipolar device). In fact, the desaturation protection gets its name from its use in bipolar devices. The basic operating principle is illustrated in Figure 10.35 using IGBT's terminology as an example.

Details of design and implement desaturation protection circuit, as illustrated in Figure 10.36, have been documented in [21], particularly the special considerations for fast-switching high-noise WBG devices and will not be repeated here.

Desaturation protection scheme has several limitations and shortcomings, especially about its potential use in WBG device protection. First, a diode must be used to protect the sensing circuit from the much higher drain-source voltage, when the device is in off state, which reduces the response time and adds capacitive loading to the output capacitance of the device. Second, the overcurrent threshold is set by the saturation current level corresponding to the selected gate driver voltage and cannot be selected independently. Third, with the given gate voltage, the saturation current is a strong function of junction temperature, and in general, for GaN HEMT, it drops as temperature increases while the opposite trend can be observed for SiC MOSFET; as a result, a fixed setting of the desaturation protection cannot predictably respond to overcurrent faults of WBG devices in a wide range of operating temperatures. More importantly, desaturation protection

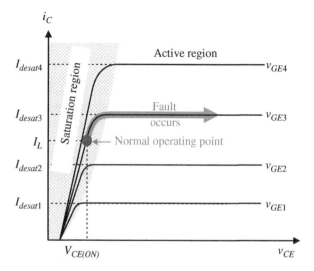

Figure 10.35 Operating principle of the desaturation protection.

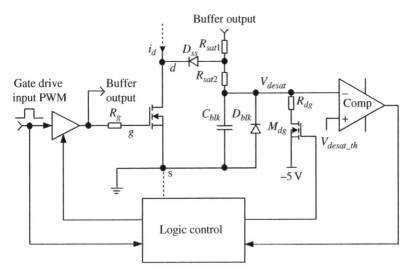

Figure 10.36 Schematic of a desaturation overcurrent protection scheme.

replies on having sufficient knowledge on the device to be protected, which is a challenge for emerging devices, like WBG power semiconductors. Once properly designed, desaturation overcurrent protection has been shown to function well for both SiC and GaN with a total response time of ~200 ns. But the desaturation detection circuit adds capacitive loading to the drain of the WBG device, increasing both the response time and the switching loss. In addition, the protection design requires prior knowledge of the device dynamic characteristics discussed in Chapter 2.

Figure 10.37 shows another overcurrent protection scheme, which was developed for Si IGBTs and applied to SiC MOSFETs. This scheme is based on sensing the voltage between the power source and Kelvin source terminals of the device. The voltage across the parasitic inductance L_S between them represents $L_s \dfrac{di_D}{dt}$ where i_D is the device drain current. By passing the voltage signal

Figure 10.37 Schematic of a *di/dt* integration overcurrent protection scheme.

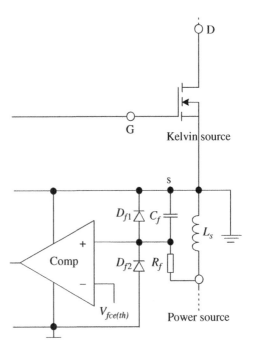

through an analog integrator, i_D can then be sensed by a comparator for overcurrent fault detection. This scheme relies on a sufficient parasitic inductance existing between the source and Kelvin source terminals to enhance the sensitivity. With near-chip scale GaN FETs or low parasitic packaging-based SiC devices and modules, the inductance may not be adequate to implement this scheme. This issue can be resolved by using an external Rogowski coil to pick up the switching loop di/dt with the controllable coil inductance to guarantee the measurement sensitivity. Usually, a custom pick-up coil integrated with PCB is preferred to fit with the package of WBG devices, in the meantime, to improve the density of the whole measurement system. This concept has been utilized for SiC power module and GaN transistors for high-speed dynamic current measurement and short-circuit and overcurrent protection. The limitation of this approach is that the custom pick-up coil design depends on the footprint of the specific device; as a result, the generality of the approach is lost.

10.7.1.2 Crosstalk Suppression

In the phase-leg configuration, the switching process of the lower switch will affect the operating condition of the upper one, which in turn might worsen the switching performance of the lower switch and the reliability of upper switch. This interaction between the two switches is called crosstalk and illustrated in Figure 10.38.

During turn-on transient of the lower switch, the drain-source voltage rise on the upper switch v_{DS_H} induces a current that is coupled through its Miller capacitance C_{GD_H}. This induced current generates a voltage drop across the gate resistance R_{G_H}. If this positive spurious gate voltage v_{GS_H} exceeds the threshold voltage of the upper switch, the upper switch will partially turn on. As a result, a shoot-through current will flow from the upper switch to the lower one, leading to excessive switching losses in both switches. Meanwhile, based on the transfer characteristics of power devices, this additional shoot-through current will increase the channel current of the lower switch, affecting the gate voltage. The increased gate voltage of the lower switch will reduce the gate current, resulting in the decrease of dv/dt. Thus, the turn-on switching speed of the lower switch would be limited by this self-regulating mechanism.

During the turn-off transient of the lower switch, similarly, a negative spurious upper switch gate voltage would be induced. This negative spurious gate voltage will not generate a shoot-through problem, but it may degrade the upper switch's reliability if its magnitude exceeds the maximum allowable negative-biased gate voltage acceptable to the semiconductor device itself.

Consequently, in order to mitigate crosstalk, the fast-switching speed capability of power devices in the phase leg has to be sacrificed. Based on the aforementioned analysis, the core element of crosstalk is the spurious gate voltage of the upper switch v_{GS_H}. The peak value of this spurious gate voltage is shown as:

$$v_{GS\_H(max)} = \frac{dv}{dt} \cdot R_{G\_H} C_{GD\_H} \cdot \left(1 - e^{-\frac{V_{DC}}{\frac{dv}{dt} \cdot C_{iss\_H} R_{G\_H}}}\right) \tag{10.17}$$

where dv/dt is the slew rate of drain-source voltage of the lower switch v_{DS_L}, partially representing the switching speed. R_{G_H}, C_{GD_H}, and C_{iss_H} refer to the gate resistance, Miller capacitance, and input capacitance of the upper switch, respectively. Thus, dv/dt, R_{G_H}, C_{GD_H}, C_{iss_H}, and V_{DC} determine the peak value of spurious gate voltage of upper switch.

Assuming dv/dt approaches ∞, this yields

$$v_{GS\_H(max)}\Big|_{\frac{dv}{dt} = \infty} = \frac{C_{GD\_H} V_{DC}}{C_{GS\_H} + C_{GD\_H}} = \frac{V_{DC}}{1 + C_{GS\_H}/C_{GD\_H}} \tag{10.18}$$

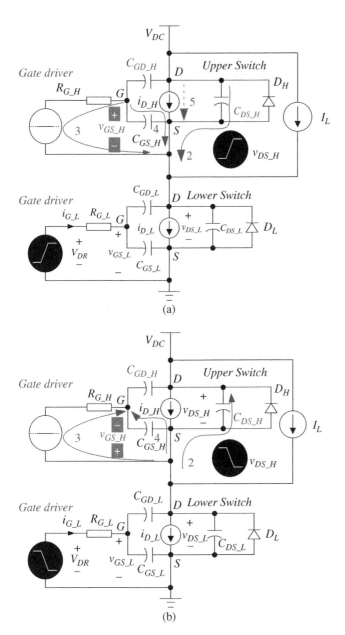

Figure 10.38 Mechanism causing crosstalk interference. (a) Turn-on transient of lower switch and (b) turn-off transient of lower switch.

Hence, the limit of this peak value is merely dominated by the junction capacitance and operating voltage of the upper switch. If this value exceeds the threshold voltage or maximum allowable negative-biased gate voltage of the upper switch, then theoretically there is a switching speed limitation placed on the lower switch. Therefore, different devices have different crosstalk impact due to their unique characteristics.

To suppress the crosstalk, the key is to minimize the spurious gate voltage of the nonoperating switch in a phase leg, and its amplitude is determined by the portion of displacement current from the Miller capacitance that is used to charge/discharge the gate-source capacitance of the affected

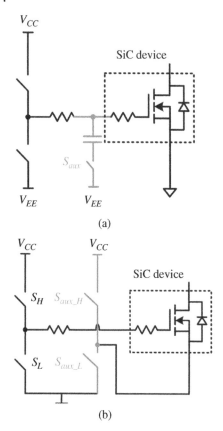

(a)

(b)

Figure 10.39 Examples of crosstalk suppression taking SiC MOSFET as an example. (a) Gate impedance regulation and (b) gate voltage control.

device [22]. Once its gate impedance becomes small during the switching transient, most displacement current will be bypassed by the gate loop, resulting in less current that would induce spurious gate voltage, and thus crosstalk is mitigated. This approach is entitled gate impedance regulation, and one implementation solution is illustrated in Figure 10.39a.

In addition to regulating gate loop impedance for crosstalk mitigation, a more direct solution to avoid crosstalk is to actively control the gate voltage of the affected device during the switching transient of the opposite switch in a phase leg. Specifically, we can negatively precharge the gate-source capacitance of the upper switch before the turn-on transient of the lower switch so that $C\dfrac{dv}{dt}$-induced positive spurious gate charge will be canceled by this prestored negative gate charge. Once the prestored gate-source charge of the upper switch is sufficient, crosstalk during the turn-on transient of the lower switch can be eliminated. In this case, the gate voltage control circuit should be designed, with an example implementation given in Figure 10.39b.

10.7.2 Converter-Level Protection

Converter-level protection includes overcurrent protection, overvoltage and undervoltage protections, and overtemperature protection. The ground fault detection and protection are also considered. Typically, hardware-based protections are designed for the electrical parameters (e.g. current and voltage) with sub-μs response time, while the software protection is adopted to manage the thermal-related protection (e.g. temperature and I^2t) with sub-ms response time. Software-based protection also exists for the overcurrent and under-/overvoltage protection as a backup.

10.7.2.1 Overcurrent Protection

There are two types of hardware-based detection schemes: peak current detection and di/dt detection. In general, peak current detection based on current transducer and the measured absolute current value is the straightforward approach, while the measurement bandwidth of the current transducer may limit the response time. If so, di/dt detection could be a good complementary approach, offering a more sensitive indicator with faster response time. Sometimes, integration of both detection schemes is adopted to achieve the better trade-off between measurement accuracy and fast response time.

If the overcurrent protection is implemented in the software, the response time is slower, typically, at least with one switching cycle delay. Also, to avoid false triggering, a digital filter is adopted for the better noise immunity. One unique scenario, which is worth mentioning, is that if the converter is overloaded or experiences a high impedance short-circuit fault, a thermal-related I^2t protection will be triggered to ensure converter's operation under the safe temperature range. As I^2t

protection is less time-sensitive but with relatively complex algorithm, a software-based approach is typically implemented.

Once the overcurrent protection is detected, usually, all active switches will be turned off, in the meantime, the circuit breakers, if applicable, will be opened.

10.7.2.2 Overvoltage and Undervoltage Protections

Generally, overvoltage and undervoltage protections are associated with the DC-link. Overvoltage protection level is set by the DC-link capacitor bank and other voltage limits. Once the detected peak voltage exceeds the upper limit, the overvoltage protection will be triggered. Similar to the overcurrent protection, a software-based protection may also be considered with a digital filtering for enhanced noise immunity with the penalty of slower response time.

In general, undervoltage protection is also essential with the consideration that the low DC-link voltage may cause excessive inrush current and overmodulation. Therefore, the threshold level is set by maximum allowable inrush current and minimum tolerable operating voltage. Typically, hardware-based protection with faster response time and software protection as a backup are implemented.

For both overvoltage and undervoltage protections, once the fault is detected, all active switches will be turned off, in the meantime, the circuit breakers, if applicable, will be opened.

10.7.2.3 Overtemperature Protection

As discussed in Section 10.6.3, usually the case temperature or NTC temperature of the power devices is measured, while the actual device junction temperature is estimated based on the loss calculation and thermal model. Once the estimated junction temperature approaches the maximum operating temperature of the power device, unlike the overcurrent and over/undervoltage protections, the typical practice is not to shut down the converter, instead derating the converter with lower operating power, guaranteeing the safe operation, in the meantime, maximizing the availability of the power converter. This scheme is very useful for certain applications, e.g. locked rotor condition or low-speed motor operating condition.

10.7.2.4 Ground Fault Detection

Ground fault detection is another important converter-level protection. Depending on the fault type, current-based approach may not be effective, such as the high impedance fault. Therefore, multiple thresholds can be used for different level of actions. Additionally, for systems with high impedance ground fault, CM voltage detection can be used as another effective mean. Practically, a dedicated box outside the power converter can also be used for the ground fault detection. Some discussion on ground fault detection is provided in Chapter 12 for motor drive applications.

10.7.2.5 Common Protection Equipment Applied in Converter Systems

Beyond the power converter itself, several protection equipment is commonly used. It includes: (i) circuit breakers to physically isolate the fault, (ii) fuses as an effective passive protection for the short-circuit fault, and (iii) metal oxide varistor (MOV) for the lightning strike and other voltage dynamics. Relays and contactors are also often used.

Some related auxiliary circuits are precharge circuits and voltage suppression circuits like dynamic braking and crowbar. All AC/DC converters require precharge to establish DC-link voltage first before connecting to AC source voltages to avoid excessive inrush currents. More discussions on precharge and dynamic braking can be found in Chapter 12.

10.8 Printed Circuit Boards

PCBs play significant roles on converter design, particularly for the control and auxiliaries. Generally, almost all the functions of control and auxiliaries are implemented by PCBs. In practice, multiple PCBs will be designed to achieve different functionalities and later integrate both electrically and mechanically. Figure 10.40 displays a typical example of the PCBs' partitioning for a general converter design, primarily including microcontroller board, interface board, gate driver board, and power stage board. It is noted that power stage board is essential for the low-voltage low-power converters while for the medium-voltage high-power applications, other means, such as busbar with mechanical integration is introduced, causing PCB an optional solution. Digital controllers are generally grounded, but many other PCBs can be at high-voltage potentials, such as gate drivers, isolated power supplies, and sensors.

Several examples of the PCB boards associated with different functional blocks are given in Figure 10.41.

PCBs, as the general tool applied in power converter design, turn out to be one of the reliability bottlenecks. Key failure modes, failure mechanisms, and critical stressors are summarized in Table 10.9. It is noted that temperature capabilities and vibration tolerance of the PCBs are crucial, which may also impact the mechanical layout and enclosure design.

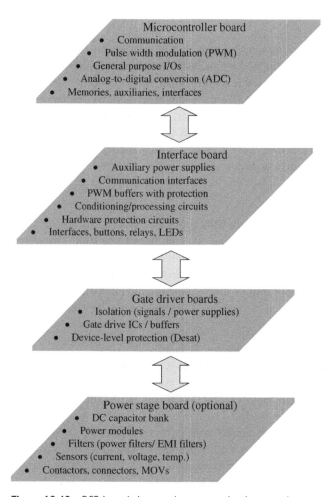

Figure 10.40 PCB boards in actual converter implementation.

(a)

(b)

(c)

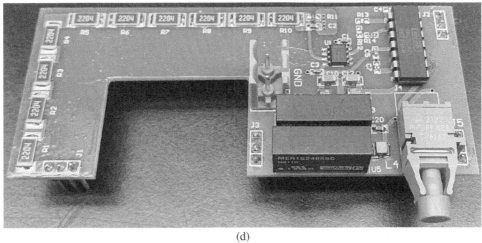

(d)

Figure 10.41 Examples of PCB prototypes in actual medium-voltage converter. (a) Microcontroller and interface board, (b) gate driver board, (c) desaturation protection board, (d) DC-link voltage sensing board, (e) AC current sensing board and (f) power stage board with DC-link capacitor banks.

(e)

(f)

Figure 10.41 (Continued)

Table 10.9 Failure mode and effect anaylsis of PCB boards.

Failure modes	Failure mechanisms	Critical stressors
Open circuit	Solder joint fatigue	Temperature and vibration
	Trace fatigue	
	Through hole plating cracks (not z-axis fatigue)	
	Via cracks (where via interfaces with a layer, getting isolation where plaiting did not adhere, or fatigue)	
Short circuit	Conductive filament formation	Relative humidity and current
	Electrochemical migration or corrosion	
	Tin Whisker formation: not that many people use emersion Sn, but Sn-plated leads are used.	Pressure, vibration, temperature, and use of pure Tin

In summary, to support the converter design, particularly the mechanical design, types, numbers, dimensions, and weight of PCBs are critical inputs.

10.9 Deadtime Setting and Compensation

Phase-leg configuration is the basic cell for many power converter types. The phase-leg configuration consists of two series-connected power semiconductor switches with a DC voltage source across the lower/upper terminals and a DC/AC current source through its middle point. To avoid short-circuit of the DC voltage source, then dead time, a small interval during which both the upper and lower switches in a phase leg are off, is introduced. Dead time is also called blanking time.

This blanking time, as a design variable associated with the control, affects the overall converter performance primarily from two aspects. One is output voltage disturbance and the resultant low-order harmonics distortion. This is a well-known issue, especially for high-voltage IGBT-based voltage source inverters. The other is the extra energy dissipation due to the freewheeling diode conduction during the superfluous dead time. This has been extensively discussed for one of the typical DC/DC voltage source converters, synchronous buck converter employing lower voltage devices that are capable of third-quadrant operation, such as MOSFETs or HEMTs.

Beyond the well-known challenges for the deadtime setting, for WBG-based converter, several unique challenges should be considered for deadtime setting due to devices' inherent properties. First, turn-off time t_{off}, which determines the dead time t_{dt}, is significantly sensitive to the operating conditions and characteristics of inductive loads. As can be observed in Figure 10.42 using SiC MOSFET as an example, compared to the tested t_{off} at the operating current of 20 A, t_{off} at 5 A increases by a factor of 5. Thus, the traditional fixed dead time depending on t_{off} at the worst operating point (i.e. 600 V/20 A in Figure 10.42) is not suitable. Moreover, under the same operating condition, t_{off} with the motor load is doubled as compared to that when an optimally designed inductor load is employed, as illustrated in Figure 10.43.

Second, reverse conduction-induced extra energy loss E_{dt} during the superfluous dead time is significantly higher than the loss dissipated in the switch channel, as can be observed in Figure 10.44, where the conduction loss of a SiC MOSFET is displayed and compared when the current flows through the device's channel vs. the freewheeling diode [23]. Moreover, this reverse conduction loss is dissipated during each switching cycle and is significant compared to the

Figure 10.42 Turn-off time t_{off} dependence on inductive load current I_L.

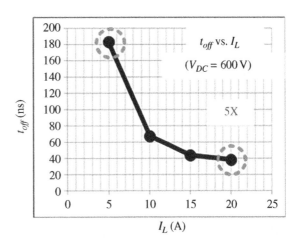

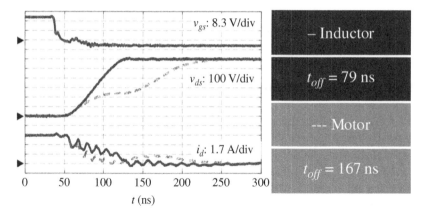

Figure 10.43 Turn-off time t_{off} dependence on inductive loads.

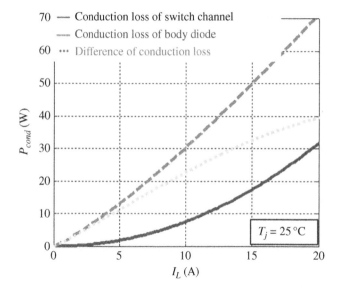

Figure 10.44 Conduction loss comparison.

switching loss E_{sw}. Figure 10.45 shows the ratio between E_{dt} and E_{sw}. It can be found that an additional 15% E_{sw} would be dissipated at 600 V/15 A during the 500 ns deadtime interval.

Accordingly, the method to set the proper dead time is critical. Additionally, the deadtime compensation might be even more challenging for high-frequency converter enabled by WBG devices as the accumulated volt-second error over a fundamental cycle is proportional to the switching frequency. Therefore, deadtime setting and compensation are discussed as follows.

10.9.1 Deadtime Setting

To better set the proper dead time, turn-off behaviors under various operating conditions must be understood; then deadtime setting model can be derived to identify the optimal dead time under different turn-off transition modes.

The following analysis, based on the phase-leg configuration in Figure 10.46a, focuses on the turn-off transition of both switches in the phase leg (i.e. t_1 to t_2 and t_3 to t_4 in Figure 10.46b), since

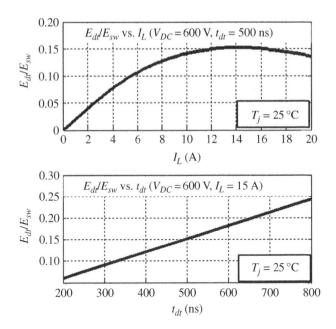

Figure 10.45 E_{dt}/E_{sw} vs. I_L and t_{dt}.

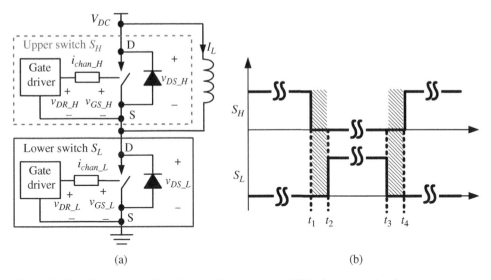

Figure 10.46 Phase-leg configuration. (a) Schematics and (b) logic control signals.

it directly affects the deadtime setting. It is noted that as the basic cell, the analysis based on the phase-leg configuration can be extended to different types of converters because of their similar switching transitions.

Assuming the inductive load current I_L flows into the middle point of the phase leg, as shown in Figure 10.46a, the lower switch is the operating device while the upper one is the synchronous device. Before t_1 in Figure 10.46b, the upper switch is on, I_L flows through the channel of the upper switch. From t_1 to t_2, the upper switch turns off and the lower switch remains in the off state; hence, the only flowing path of I_L is the freewheeling diode of the upper switch. Thus, I_L only transfers from the device's channel to its freewheeling diode. Therefore, from t_1 to t_2 in Figure 10.46b, the turn-off command of the upper switch will not result in the switching commutation between

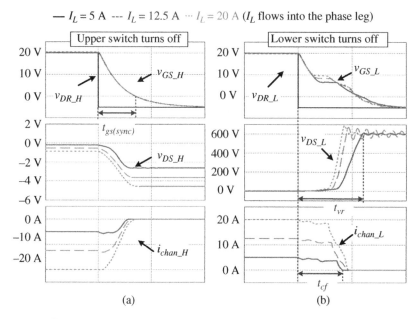

Figure 10.47 Typical switching waveforms during the turn-off transient: (a) upper switch turns off and (b) lower switch turns off.

lower and upper switches in the phase leg. Also, during this interval, the drain-source voltage of the upper switch v_{DS_H} only changes from on-state voltage of the device's channel to the voltage drop of its freewheeling diode. Due to this limited variation of v_{DS_H}, the power loop parasitics and operating conditions have almost no effect on the gate voltage of the upper switch v_{GS_H}, which is verified by simulation waveforms in Figure 10.47a. Hence, once v_{GS_H} drops below the threshold of the upper switch (or with conservative consideration v_{GS_H} drops to zero since threshold voltage is temperature-dependent and hard to practically fix), the lower switch can be turned on; otherwise, more energy will be dissipated by the freewheeling diode, as shown in Figure 10.48.

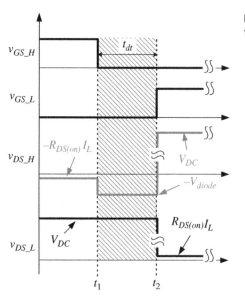

Figure 10.48 Schematic diagram of switch waveforms from $[t_1, t_2]$.

Note that v_{GS_H} is merely determined by gate loop parameters. The second-order gate loop equivalent circuit in Figure 10.49 can be simply solved to obtain the duration $t_{GS(sync)}$ when v_{GS_H} drops from the on-state gate voltage to zero, where R_G, L_G, L_{CM}, and C_{iss} in Figure 10.49 represent resistance associated with the gate loop, gate loop inductance, common source inductance shared in both gate and power loops, and power devices' input capacitance, respectively. In the end, the optimal dead time $t_{dt(opt)}$ from t_1 to t_2 should be set to the gate voltage fall time $t_{GS(sync)}$.

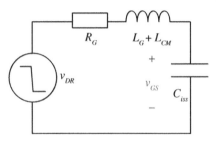

Figure 10.49 Gate loop equivalent circuit from $[t_1, t_2]$.

From t_3 to t_4 in Figure 10.46b, the hard turn-off transition occurs between the lower switch and the upper freewheeling diode. There are two current sources for turn-off voltage commutation: one is the current in the gate loop via charging the Miller capacitance and the other is the inductive load current from the power loop to charge/discharge the power devices' output capacitances. Therefore, the hard turn-off switching transition has two modes: (i) gate-loop-dominated turn-off and (ii) power-loop-dominated turn-off. As compared to the gate current, which stays almost the same in the given gate drive circuit, the inductive load current will be changed under different operating currents. Thus, the turn-off transition mode will be shifted as the operating condition varies. According to the simulation waveforms during the turn-off transient from t_3 to t_4 in Figure 10.47b, the switching trajectories with different I_L are plotted in Figure 10.50. At light load (e.g. 5 A), dv/dt is limited by the power loop. All I_L is used to charge/discharge output capacitances of lower and upper switches. Channel current of the lower switch i_{chan_L} drops to 0 A before the drain-source voltage of the lower switch v_{DS_L} rises to V_{DC}, namely channel current fall time t_{cf} becomes shorter than drain-source voltage rise time t_{vr}. This is the power loop dominated turn-off mode. In contrast, at heavy load (e.g. 20 A), dv/dt is limited by the gate loop. A portion of I_L for charging/discharging output capacitances is sufficient to follow the gate loop dominated dv/dt. Therefore, i_{chan_L} keeps the rest of I_L until v_{DS_L} rises to V_{DC} so that t_{cf} is

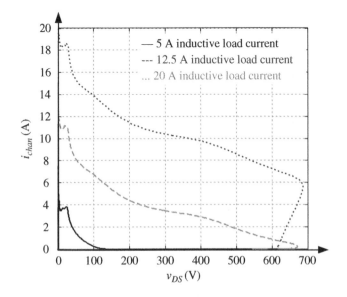

Figure 10.50 Switching trajectory during turn-off from $[t_3, t_4]$.

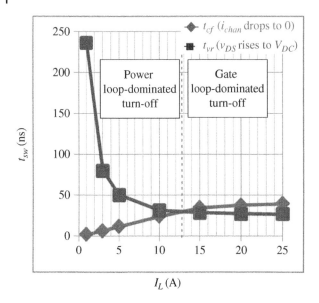

Figure 10.51 t_{cf} vs. t_{vr} dependence on operating current.

longer than t_{vr} during the gate loop dominated turn-off mode. Figure 10.51 illustrates the relation between t_{cf} and t_{vr} under different turn-off transition modes with various I_L.

According to the aforementioned analysis, the optimal dead time $t_{dt(opt)}$ from t_3 to t_4 greatly depends on the turn-off transition modes. For gate loop-dominated turn-off, $t_{dt(opt)}$ prefers to be t_{cf}. If $t_{dt} < t_{cf}$, obviously, a shoot-through failure is incurred; on the contrary, $t_{dt} > t_{cf}$ causes longer diode conduction and more energy dissipation (see Figure 10.52a). For power loop dominated turn-off, $t_{dt(opt)}$ prefers to be t_{vr}. If $t_{cf} < t_{dt} < t_{vr}$, original synchronous turn-on transient of the upper switch at t_4 turns to be hard switching (see Figure 10.52b), and this unexpected hard switching event

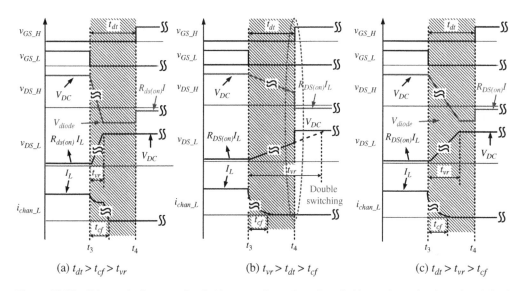

Figure 10.52 Schematic diagram of switching waveforms from [t_3, t_4]: (a) gate loop-dominated and (b–c) power loop-dominated.

Table 10.10 $t_{dt(opt)}$ depedence on different turn-off transitions when current flows into phase leg.

Optimal dead-time $t_{dt(opt)}$	Turn-off of synchronous switch (upper switch in Figure 10.46a)	Turn-off of operating switch (lower switch in Figure 10.46a)	
		Power loop dominated	Gate loop dominated
	Gate voltage fall time $t_{gs(sync)}$	Drain-source voltage rise time t_{vr}	Channel current fall time t_{cf}

increases the total switching loss; to avoid this double switching, a longer t_{dt} should be adopted which leads to more loss due to the diode conduction (see Figure 10.52c).

Table 10.10 summarizes the optimal dead time $t_{dt(opt)}$ dependence on different turn-off transitions when inductive current flows into the middle point of the phase leg. Similarly, $t_{dt(opt)}$ for the opposite current direction can be derived.

10.9.2 Deadtime Compensation

Dead time t_{dt} adds voltage distortion at the midpoint of the phase leg. The ideal midpoint voltage of a phase leg is meant to mirror the input PWM signal; however, the device parasitic capacitance and nonideal characteristics of the inductive load along with the previously mentioned dead time cause the midpoint voltage to become distorted from the input PWM signal.

For the phase leg in Figure 10.53, the effect of the dead time is either to cause voltage gain or voltage loss compared with the voltage reference, as illustrated in Figure 10.54.

Figure 10.54a shows gate signals for phase A positive switch S_{A+} and negative switch S_{A-}. With a dead time t_d inserted between the signals to prevent shoot-through, there will be a delayed turn-on for both switches. If phase A current $i_A > 0$ with its polarity defined in Figure 10.53, during switching transient, the current will flow through the positive switch S_{A+} or the negative diode D_{A-}, which can be the body diode or the antiparallel diode of the negative switch. When S_{A+} is turned off, the current will be immediately commutated to D_{A-} and the dead time will not have any impact. In contrast, when S_{A-} is tuned off, the current will continue to remain in D_{A-} and will not be commutated back to S_{A+} until S_{A+} is turned on after the dead time t_d. Hence, when $i_A > 0$, the dead time will cause a net loss for phase A voltage v_{AN} (defined in Figure 10.53) compared with the ideal no deadtime case, as illustrated in Figure 10.54b. When $i_A < 0$, the situation is reversed and the dead time will cause a net gain for phase A voltage v_{AN}, as illustrated in Figure 10.54c.

In other words, during the dead time, both positive and negative switches in a phase leg are off, and the phase voltage v_{AN} depends on the direction of current i_A. If we define an error voltage

$$v_e = (v_{AN})_{\text{ideal}} - (v_{AN})_{\text{actual}} \tag{10.19}$$

Figure 10.53 Phase A phase leg in a two-level voltage source converter.

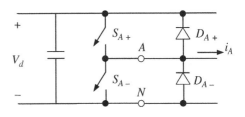

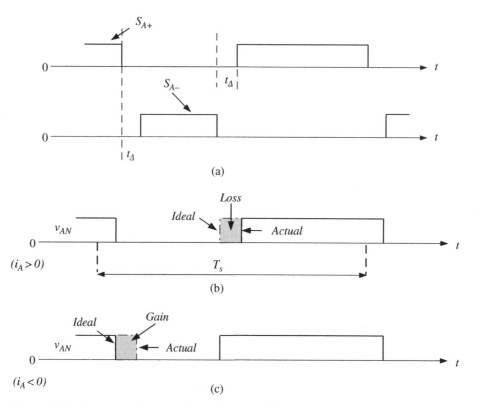

Figure 10.54 Illustration of the deadtime impact on VSC phase-leg voltage. (a) Gate signals of switches in phase leg A with the dead time t_d, (b) phase A current $i_A > 0$ case, and (c) phase A current $i_A < 0$ case.

The average error voltage over one switching period T_s due to t_d is

$$\Delta V_{AN} = \begin{cases} +\dfrac{t_d}{T_s} & i_A > 0 \\[2mm] -\dfrac{t_d}{T_s} & i_A < 0 \end{cases} \tag{10.20}$$

Note that the error does not depend on current magnitude but on current direction. The same analysis can be applied to phase B to yield

$$\Delta V_{BN} = \begin{cases} -\dfrac{t_d}{T_s} & i_A > 0 \\[2mm] +\dfrac{t_d}{T_s} & i_A < 0 \end{cases} \tag{10.21}$$

Note that in (10.21), we have assumed $i_B = -i_A$ as shown in Figure 10.55. With this the average line-to-line voltage error can be found as in (10.22). The actual vs. the ideal average line-to-line voltage as a function of the reference voltage is illustrated in Figure 10.56. Figure 10.57 illustrates the impact of the dead time on the sinusoidal line-to-line output voltage. It can be seen that the voltage distortion occurs when the current changes direction at the zero crossing.

$$\Delta V_o = \Delta V_{AN} - \Delta V_{BN} = \begin{cases} +\dfrac{2t_d}{T_s} & i_A > 0 \\[2mm] -\dfrac{2t_d}{T_s} & i_A < 0 \end{cases} \tag{10.22}$$

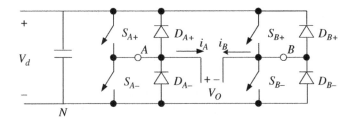

Figure 10.55 Phase A and B phase legs in a two-level VSC.

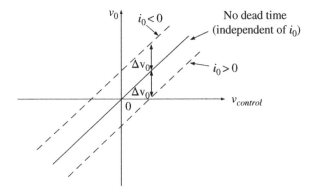

Figure 10.56 Actual vs. ideal average line-to-line voltage as function of reference voltage.

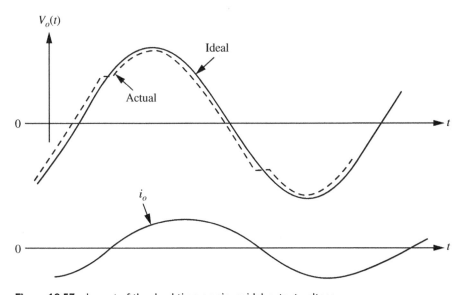

Figure 10.57 Impact of the dead time on sinusoidal output voltage.

From the aforementioned analysis, several observations can be made regarding the voltage distortion due to the dead time.

1) The voltage distortion is synchronous with the load current and therefore will generate low-order harmonics.
2) The voltage distortion is the ratio of the dead time t_d to the PWM switching period T_s. Therefore, the distortion will be more severe with higher switching frequencies and a given dead time. Increasing switching frequency without taking care of the dead time will make the voltage distortion and corresponding harmonics worse, not better.
3) The distortion is independent of modulation index. Hence, the impact is more severe at low modulation index (low motor speed in motor drive case).

Clearly, the deadtime effect should be corrected to mitigate the low-order harmonics, and in motor drive case, torque error and potential instability issues. A straightforward deadtime correction method is compensation for voltage error. Assuming that phase A reference voltage $v^* = V_{ph-M}\sin(\omega_e t)$, the average phase A voltage corresponding to Figure 10.53 in a switching period T_s is

$$\bar{v}_{AN} = \frac{t_+}{T_s}V_{DC} = V_{DC}\left(\frac{1}{2} + \frac{v^*}{V_{DC}}\right) = V_{DC}\left[\frac{1}{2} + \frac{M}{2}\sin(\omega_e t)\right] \tag{10.23}$$

where t_+ is the on-time for the positive switch and $M = \dfrac{2 \cdot V_{ph-M}}{V_{DC}}$ is the modulation index as defined in Chapter 6. Equation (10.23) can be rewritten as:

$$t_+ = \frac{T_s}{2} + \frac{T_s M}{2}\sin(\omega_e t) \tag{10.24}$$

According to the earlier analysis, with dead time t_d, the actual phase A on time is

$$t_+ = \frac{T_s}{2} + \frac{T_s M}{2}\sin(\omega_e t) - t_{d \cdot}\,\mathrm{sgn}(i_A) \tag{10.25}$$

where $\mathrm{sgn}(i_A)$ is the sign function for i_A, returning 1 for positive i_A and -1 for negative i_A. The average voltage considering the dead time becomes

$$\bar{v}_{AN} = \frac{t_+}{T_s}V_{DC} = V_{DC}\left[\frac{1}{2} + \frac{M}{2}\sin(\omega_e t) - \frac{t_{d \cdot}}{T_s}\mathrm{sgn}(i_A)\right] \tag{10.26}$$

which clearly is not the intended value in (10.23). The straightforward approach to compensate for the error is to add a correction item in phase A positive switch on time t_+ such that it becomes

$$t_+ = \left[\frac{T_s}{2} + \frac{T_s M}{2}\sin(\omega_e t) + t_{d \cdot}\,\mathrm{sgn}(i_A)\right] - t_{d \cdot}\,\mathrm{sgn}(i_A) \tag{10.27}$$

The term $\left[\dfrac{T_s}{2} + \dfrac{T_s M}{2}\sin(\omega_e t) + t_{d \cdot}\,\mathrm{sgn}(i_A)\right]$ in (10.27) is the modulation with the correction, where the $t_{d \cdot}\,\mathrm{sgn}(i_A)$ term will compensate the $-t_{d \cdot}\,\mathrm{sgn}(i_A)$ term as a result of the dead time and eliminate the error caused by the dead time. For positive i_A, t_+ will increase, and for negative i_A, t_+ will decrease. The compensation method in (10.27) is simple and straightforward and has been successfully adopted in some industry motor drive applications. However, there can be a few issues that may affect its accuracy and effectiveness.

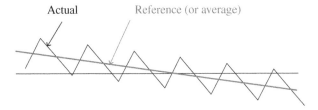

Figure 10.58 Current zero crossing with switching ripple.

1) Current polarity detection. It can be seen that the correction term depends on the sign of the current. Near current zero crossing or at low current level, the detection can be challenging, especially, when there is current ripple due to switching. 10.58 illustrates (with exaggeration) an AC current with switching ripple around its zero crossing. If the actual current measurement is used, the correction term $t_d \cdot \text{sgn}(i_A)$ will toggle back and forth in this region, which is definitely undesirable. To prevent this from happening, there are two simple approaches. The first approach is not to wait for zero crossing to toggle the deadtime correction. Instead, it toggles near the zero current (below certain threshold). As a result, the flat spot can be eliminated at zero crossing at the expense of small vertical distortions. The second approach is to use current reference instead of measurement for compensation as shown in Figure 10.58. In [24], the AC current magnitude and phase angle are determined from the measurement, and then a sinusoidal waveform is reconstructed, which is free of switching ripple and can be used for current polarity detection.

2) Switching transitions. The aforementioned analysis has assumed that the switching transition and current commutation are instantaneous, which can be a good approximation when current is high, and the deadtime time is relatively long (much longer than the device parasitic capacitance charging and discharging periods). This assumption is reasonable for slow-switching Si devices but less so for fast-switching WBG devices, where the capacitance charging/discharging plays a significant part in switching transition. Figure 10.59 illustrates switching transition under different current conditions due to the parasitic capacitance effect. Assume that $i_A < 0$ and initially, the bottom (negative) switch conducts, so that when the bottom switch turns off, the current immediately commutates to the top (positive) diode. When the current is high or the parasitic capacitance is relatively small, the top switch/diode capacitance is discharged quickly (and correspondingly the bottom switch/diode capacitance is charged quickly). In this scenario, v_{AN} can be considered to rise almost instantaneously from 0 to V_{DC}, as shown in Figure 10.59b, which also corresponds to the simplified case in Figure 10.54. If the current is low or the parasitic capacitance is not negligible, the top device capacitance discharge and bottom device charge will take some time so that v_{AN} will take some time to rise from 0 to V_{DC}, as shown in Figure 10.59c. In the case of very low current, the device capacitance discharge/charge may not finish by the end of the dead time, as shown in Figure 10.59d.

To more accurately compensate for the dead time, other schemes can be used beyond the simple open-loop approach in (10.27). One approach is the closed-loop voltage regulation, including closed-loop modulation, which has been used in industry application. The closed-loop approach will also be able to compensate for voltage errors caused by nonidealities other than switching transitions, including voltage drops on switches. The limitation of this approach is mainly in high-frequency applications because of the feedback response and control bandwidth requirements.

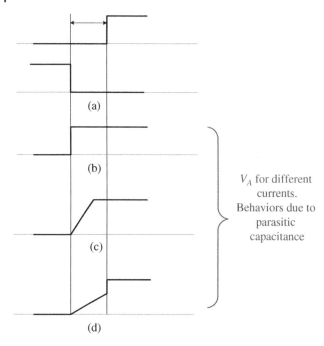

Figure 10.59 Illustration of switching transition during dead time: (a) turn-off and turn-on signals for two switches in phase A; (b) v_{AN} under high current and negligible capacitance; (c) v_{AN} under low current or nonnegligible capacitance; and (d) v_{AN} under very low current.

Another scheme is using the volt-seconds balance theory to directly consider the impact of the parasitic capacitances and nonideal transitions [25, 26]. This approach generally needs a high bandwidth current transducer with inaccurate dynamic current sensing and often still relies on open-loop model, which can limit its accuracy.

One promising approach for an accurate deadtime compensation is to use online monitoring of switching transients and compensate for the deadtime volt seconds with details summarized in [27] and will not repeat here.

10.10 Interface to Other Subsystems

As already covered in other chapters, control and auxiliaries in this chapter interface with other subsystems, including rectifier/inverter, thermal management system, mechanical system, and EMI filter (especially when active/hybrid filter is used). The rectifier/inverter provides input on gate driver, power supplies, sensors, and protection; thermal management systems provides input on ambient or cooling conditions, and required cooling (e.g. fan) power; mechanical system provides input on temperature range; and EMI filter provides requirements on control (in case of active/hybrid filters).

Table 10.11 summarizes the output from the control and auxiliaries subsystem to others. The only item needing discussion is the voltage distribution of the various controller and auxiliary components, which can be determined based on their location and isolation design.

Table 10.11 Output from the control and auxiliaries to their interfaced subsystems.

Subsystem	Interface type	Item	Value
Rectifier/inverters	Design variables/ attribute	Gate driver	Design result
		Sensors	Design result
		Controllers	Design result
		Protections	Design result
		Isolation/isolators	Design result
		PCBs	Design result
Thermal management system	Attribute	Component loss	Design result
	Design variable	Temperature sensor	Design result
Mechanical system	Attribute	PCBs	Design result
	Attribute	Voltage distribution (gate drivers, power supplies, sensors etc., and their PCBs)	Discussion
	Design variable	Controller architecture and isolation design	Design result

10.11 Summary

In this chapter,

- A general control architecture is introduced with five layers: system control layer, application control layer, converter control layer, switching control layer, and hardware control layer.
- Depending on the requirements of each layer, control hardware includes PLC, DSP, FPGA, and ASIC.
- Isolation, including signal isolator and isolated power supply, is usually needed, particularly for medium-voltage high-power appliances.
- Gate driver design is highly dependent on the characteristics of power devices and significantly affects the device performance in the actual converter.
- Basic sensing and measurement techniques are overviewed. Special attention should be given to sensing circuit design, especially for high-frequency WBG-based converter.
- Device-level protection and converter-level protection are also summarized where at the device level, usually, shoot-through protection and crosstalk suppression should be included. Fast response is the key. At the converter level, overcurrent, overvoltage, undervoltage, overtemperature, and ground fault should be considered with a mix of hardware-based and software-based schemes.

- PCBs play significant roles on converter design, particularly for the control and auxiliaries. Generally, almost all the functions of control and auxiliaries are implemented by PCBs. Attention should also give to the reliability of PCBs.
- Finally, deadtime setting for converter operational performance optimization and deadtime compensation for improved power quality are discussed. The corresponding models are summarized in Table 10.10 and Eq. (10.27).

References

1 WG18 – Power Electronics Building Blocks (PEBB), "IEEE Guide for Control Architecture for High Power Electronics (1 MW and Greater) Used in Electric Power Transmission and Distribution Systems," in *IEEE Std 1676-2010*, pp. 1–47, 11 February 2011.

2 Texas Instruments, "High-voltage reinforced isolation: definitions and test methodologies," August 2019. [Online]. Available: http://www.ti.com/general/docs/lit/getliterature.tsp?baseLiteratureNumber=slyy063 (accessed February 2023).

3 BROADCOM, "Avago ACPL-4800 high CMR intelligent power module and gate drive interface optocoupler datasheet," 2007. [Online]. Available: https://docs.broadcom.com/docs/AV01-0193EN (accessed February 2023).

4 Texas Instruments, "Texas Instruments ISO7841x high-performance, 8000-VPK reinforced quad-channel digital isolator datasheet," 2017. [Online]. Available: http://www.ti.com/lit/ds/symlink/iso7841.pdf (accessed February 2023).

5 Analog Devices, "ADuM120N/ADuM121N datasheet, Rev. C," 2017. [Online]. Available: http://www.analog.com/media/en/technical-documentation/data-sheets/ADuM120N_121N.pdf (accessed February 2023).

6 Silicon Labs, "Isolated gate drives," 2016. [Online]. Available: https://www.silabs.com/products/isolation/isolated-gate-drivers (accessed February 2023).

7 Microsemi, "APT34M120J Si MOSFET datasheet, Rev. C," 2011. [Online]. Available: https://www.microsemi.com/existing-parts/parts/60368#resources (accessed February 2023).

8 Wolfspeed, "Cree C2M0080120D SiC MOSFET datasheet, Rev. C," 2015. [Online]. Available: https://www.wolfspeed.com/media/downloads/167/C2M0080120D.pdf (accessed February 2023).

9 F. Wang, Z. Zhang and E. Jones, "Wide bandgap device characterization," in *IEEE Applied Power Electronics Conference and Exposition*, Long Beach, CA, March 2016.

10 Lai, R., Wang, F., Burgos, R., Y. Pei, D. Boroyevich, B. Wang, T. A. Lipo, V. D. Immanuel, K. J. Karimi, "A systematic topology evaluation methodology for high-density PWM three-phase AC–AC converters," *IEEE Transactions on Power Electronics*, vol. 23, no. 6, pp. 2665–2680, 2008.

11 Broadcom, "Precision optically isolated voltage sensor," 2017. [Online]. Available: https://www.broadcom.com/products/optocouplers/industrial-plastic/isolation-amplifiers-modulators/isolation-amplifiers/acpl-c87b (accessed February 2023).

12 Silicon Labs, "Isolated amplifier for current shunt measurement," 2021. [Online]. Available: https://www.silabs.com/documents/public/data-sheets/si8920-datasheet.pdf (accessed February 2023).

13 Texas Instruments, "AMC1311x high-impedance, 2-V input, reinforced isolated amplifiers," 2022. [Online]. Available: http://www.ti.com/lit/ds/symlink/amc1311.pdf (accessed February 2023).

14 R. A. Amadori, V. G. Puglielli and R. E. Richardson, "Prediction methods for the susceptibility of solid state devices to interference and degradation from microwave energy," in *Proc. 1973 Int. Symp. Electromagn. Compat.*, 1973, pp. 1–17.

15 R. E. Richardson, V. G. Puglielli and R. A. Amadori, "Microwave interference effect in bipolar transistor," *IEEE Transactions on Electromagnetic Compatibility*, vol. EMC-17, pp. 216–219, November 1975.

16 F. Fiori and P. S. Crovetti, "Prediction of high-power EMI effects in CMOS operational amplifiers," *IEEE Transactions on Electromagnetic Compatibility*, vol. 48, no. 1, pp. 153–160, February 2006.

17 F. Fiori, "A new nonlinear model of EMI-induced distortion phenomena in feedback CMOS operational amplifier," *IEEE Transactions on Electromagnetic Compatibility*, vol. 44, no. 4, pp. 495–502, November 2002.

18 M. B. Aimonetto and F. Fiori, "On the effectiveness of EMIRR to qualify OpAmps," in *Proc. 2015 Int. Symp. Electromagn. Compat.*, Dresden, 2015, pp. 40–44.

19 C. Kitchin and L. Counts, "*A Designer's Guide to Instrumentation Amplifiers,*" Third Edition, Analog Devices Inc., 2006.

20 L. Xue, D. Boroyevich and P. Mattavelli, "Driving and sensing design of an enhancement-mode-GaN phase leg as a building block," in *Proc. IEEE 3rd Workshop Wide Bandgap Power Devices Appl. (WiPDA)*, November 2015, pp. 34–40.

21 F. Wang, Z. Zhang and E. Jones, "*Characterization of Wide Bandgap Power Semiconductor Devices,*" Institution of Engineering and Technology, September 2018.

22 Z. Zhang, F. Wang, L. M. Tolbert and B. J. Blalock, "Active gate driver for crosstalk suppression of SiC devices in a phase-leg configuration," *IEEE Transactions on Power Electronics*, vol. 29, no. 4, pp. 1986–1997, April 2014.

23 Wolfspeed, "C2M0080120D SiC MOSFET datasheet, Rev. C," 2019. [Online]. Available: http://www.wolfspeed.com/downloads/dl/file/id/167/product/10/c2m0080120d.pdf (accessed February 2023).

24 A. Muñoz and T. Lipo, "On dead-time compensation technique for open-loop PWM-VSI drives," *IEEE Transactions on Power Electronics*, vol. 14, no. 4, pp. 683–689, July 1999.

25 Z. Zhang and L. Xu, "Dead-time compensation of inverters considering snubber and parasitic capacitance," *IEEE Transactions on Power Electronics*, vol. 29, no. 6, pp. 3179–3187, 2014.

26 Y. Wang, Q. Gao and X. Cai, "Mixed PWM for dead-time elimination and compensation in a grid-tied inverter," *IEEE Transactions on Industrial Electronics*, vol. 58, no. 10, pp. 4797–4803, 2011.

27 J. Dyer, Z. Zhang, F. Wang, D. Costinett, L. M. Tolbert and B. J. Blalock, "Online condition monitoring based dead-time compensation for high frequency SiC voltage source inverter," in *IEEE Applied Power Electronics Conference and Exposition*, San Antonio, TX, 2018.

11

Mechanical System

11.1 Introduction

Although power electronics converters, including three-phase AC converters, are functionally electrical equipment, physically, they are integrated electro-thermal-mechanical equipment. In Chapter 9, we have presented the design of the critically important thermal management system. This chapter focuses on the mechanical system design, which is also an essential part of power electronics equipment. Here, the mechanical system refers to the mechanical structure and associated parts in a converter, mainly including busbars, enclosure (cabinets), connectors, and other mechanical supports.

While providing the essential mechanical structural support function to a power electronics converter, the mechanical system is also important to the performance, cost, size, and/or weight of the overall converter. For example, the busbar is very important to the switching performance of the AC three-phase converter. As discussed in Chapter 6, the busbar parasitic impedance plays a critical role in impacting the device transient overvoltages. In some converters, due to its manufacturing complexity, the busbar can also be a significant contributor to the converter cost, even comparable to the cost of the semiconductor devices. Converter enclosures or cabinets are usually grounded and serve as safety shields, and they are also significant cost items. In addition, mechanical parts are often heavy and can occupy large space, and therefore, are important to the converter power density.

It should be clarified that since this book is intended for power electronics professionals with background in electrical engineering and authors are all electrical engineers, the mechanical system design described here is mainly done as the design of a supporting subsystem for the overall converter. We will not perform actual mechanical structure design and analysis. Instead, we will describe the converter mechanical design procedure to select appropriate mechanical parts to meet the electrical system needs, while considering the basic mechanical system integrity by following standard design practice. Examples of electrical system needs include: the proper spacing between parts and components to satisfy electrical insulation requirement; the shape of mechanical parts with no sharp edges to avoid high electric field and corona; and the laminated multilayer busbar with adequate conductor and insulation layer thicknesses to meet current density, insulation, and parasitic inductance needs. Examples of standard practice to meet basic mechanical system integrity include cabinet with proper mechanical strength to meet weight requirements and proper placement of parts to ensure adequate center of gravity.

Figure 11.1a shows the circuit of a 100-kVA, 13.8-kV medium-voltage (MV) three-phase AC to 850-V low-voltage (LV) DC power electronics converter, which consists of two AC/DC H-bridges

Design of Three-phase AC Power Electronics Converters, First Edition. Fei "Fred" Wang, Zheyu Zhang, and Ruirui Chen.
© 2024 The Institute of Electrical and Electronics Engineers, Inc. Published 2024 by John Wiley & Sons, Inc.

plus an isolated dual active bridge (DAB) DC/DC converter module in each phase. The modules are cascaded on the MV AC side and paralleled on the LV DC side. Figure 11.1b and c show the pictures of the side view and top view, respectively, of one of the modules. Figure 11.1d shows the pictures of the whole converter cabinet with and without the front panel. The whole converter contains six modules as shown in Figure 11.1b and c for all three phases, together with a bottom shelf housing the filter inductors, sensors, and auxiliary power supplies. There is also a controller box attached to the cabinet.

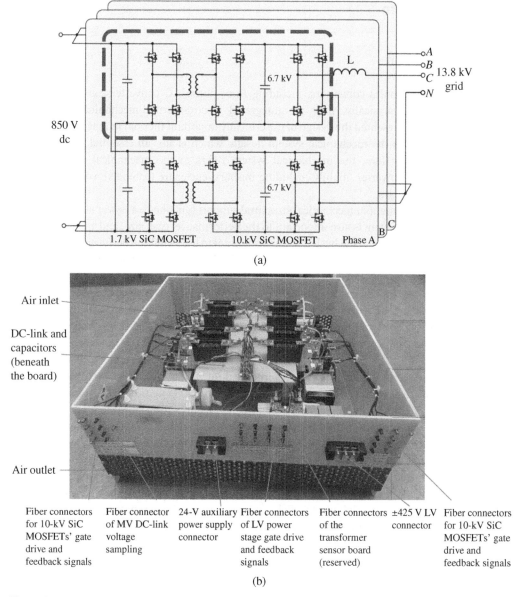

(a)

(b)

Figure 11.1 A medium-voltage three-phase AC/DC/DC power converter: (a) circuit topology; (b) side view of one H-bridge AC/DC plus a DAB module; (c) top view of one H-bridge AC/DC plus a DAB module; and (d) whole converter.

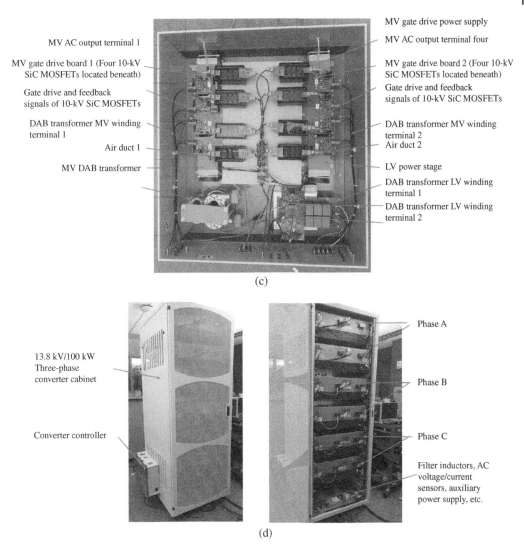

MV gate drive power supply
MV AC output terminal four
MV AC output terminal 1
MV gate drive board 1 (Four 10-kV SiC MOSFETs located beneath)
MV gate drive board 2 (Four 10-kV SiC MOSFETs located beneath)
Gate drive and feedback signals of 10-kV SiC MOSFETs
Gate drive and feedback signals of 10-kV SiC MOSFETs
DAB transformer MV winding terminal 1
DAB transformer MV winding terminal 2
Air duct 1
Air duct 2
MV DAB transformer
LV power stage
DAB transformer LV winding terminal 1
DAB transformer LV winding terminal 2

(c)

13.8 kV/100 kW Three-phase converter cabinet
Converter controller

Phase A
Phase B
Phase C
Filter inductors, AC voltage/current sensors, auxiliary power supply, etc.

(d)

Figure 11.1 (Continued)

From Figure 11.1, the following mechanical parts can be identified:

1) Cabinet: The metal cabinet in Figure 11.1d provides an enclosure to the converter and the mechanical structural support for the modules in Figure 11.1b and c, and all other parts. For a high-power converter cabinet, it usually has shelves for better utilization of space. In the example case, instead of shelves, the cabinet has six sliding rails for the six modules, as well as the bottom shelf for the inductors and auxiliary parts. With the sliding rails, the modules can be more easily accessed, removed, serviced, and replaced. The cabinet panels have vents for this air-cooled converter. Note that the cabinet can have optional casters for easy moving.

2) Drawer: The optional drawer that contains the modules in Figure 11.1 can have multiple functions. In the example case, the drawer is made of garolite boards, which provide electrical insulation among modules and between modules and other parts of the converters (e.g. the metal cabinet frame), in addition to providing mechanical support. The drawer wall also supports many different types of connectors, and it has vents for airflow.

Terminal of the DC-
link capacitor 3

Terminal of the DC-
link capacitor 4

DC-link voltage
sampling daughter
board

Three connectors
(DC+, DC–, and
DC+) between the
gate drive board 1
and the DC-link

Terminal of the DC-
link capacitor 2

Voltage balancing
resistor

Terminal of the DC-
link capacitor 1

Three connectors
(DC+, DC–, and
DC+) between the
gate drive board 2
and the DC-link

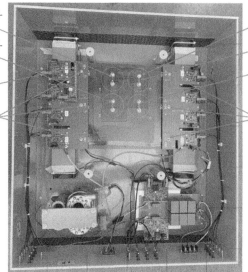

MV AC output terminal 2

DC+ terminal

Wire connecting the DC+
terminal and the 10-kV SiC
MOSFET

Wire connecting the two 10
10-kV SiC MOSFETs

10-kV SiC MOSFET's source
terminal

10-kV SiC MOSFET's drain
terminal

Figure 11.2 Illustration of the busbar and its connection to SiC devices and DC-link capacitors.

3) Busbar: The busbar connects DC-link capacitors. In Figure 11.1b and c, the busbar is hidden beneath the garolite board supporting the isolated power supplies. It can be more clearly seen in Figure 11.2. Since the current is relatively low for this MV converter (AC RMS current is 4.2 A), a PCB-based busbar with sufficient insulation strength is a good choice. For the same reason, the 10-kV SiC MOSFETs are not on the PCB busbar; instead, as shown in Figure 11.2, they are on another PCB with gate drivers close to the busbar. For higher current converters, copper- or aluminum-based laminated busbars are often employed.

4) Supporting boards and spacers/standoffs: In Figures 11.1 and 11.2, there are a number of electrical components and parts that require additional mechanical support and/or electrical isolation. Examples include the garolite board above the busbar for supporting and insulating the eight isolated power supplies, and the structure to hold the MV DAB transformer and to attach it to the bottom of the garolite drawer. Note that the air duct is part of the thermal management system, and in this case, it is also part of the mechanical system, providing mechanical support as well as electrical isolation for heatsinks (therefore, for SiC devices). In this example converter, the 10-kV SiC MOSFETs are in a non-isolated package. Each device has its individual heatsink,

and therefore no isolation is provided between the device and heatsink. For the purpose of electrical insulation, the heatsinks are placed 10 mm apart. Note that, four heatsinks share a common fan and air duct as shown in Figure 11.1. Therefore, an air duct made of fiberglass is used to provide the electrical isolation as well as mechanical support for the heatsinks.

5) Connectors: As can be seen in Figure 11.1, there are many connectors for the example converter. The design and selection of these connectors are part of the mechanical system design. The connectors include power connectors and signal connectors. The former includes AC and DC power terminals, and LV power supply connectors, and the latter includes connectors for gate drive input and feedback signals and sensor signals.

6) Fasteners (bolts, nuts, screws, and washers): They are necessary for fixing converter components and parts to the converter structure. Although in some rare cases, welding and soldering can be used, more commonly, fasteners like bolts, nuts, and screws are used with washers to provide necessary protection/cushion. Both metal and plastic fasteners are used in power electronics, with plastic ones used when electrical insulation is needed and high mechanical strength is not required.

7) Electrical/optical cables, wires, and buses: Figure 11.1b and c show a number of wires and cables, both for power and signal transmission. Although their functions are electrical, their selection and layout are considered as part of mechanical system design. For high-power and high-voltage converters, fiber optic cables are often used for isolation and noise immunity as in the example converter. Figure 11.1 does not contain metal buses as the power terminals, which may be used in high-current cases.

11.2 Mechanical System Design Problem Formulation

As discussed earlier through the example, the mechanical system for an AC three-phase converter consists of a number of parts and components. The mechanical system design should include the selection and design of these parts and components. Moreover, the design needs to include the placement and layout of these parts and components. Clearly, the design should be based on the needs of the converter and the other subsystems in the converter. In principle, the mechanical system design can follow the same approach as the other subsystems, that is, to formulate the design problem by clearly identifying the variables, constraints, conditions, and objectives for the design. Then the design problem becomes an optimization problem.

11.2.1 Mechanical System Design Variables

Based on the earlier discussion for the example system of the MV converter in Figure 11.1, the design variables include the following parts and components:

1) Busbar(s), for connecting power devices, DC-link capacitors, and DC and AC terminals. Sometimes, filter components are also connected to (like EMI filter capacitors) or integrated with (like CM chokes) the busbars.

2) Cabinet(s) or enclosure(s). Multiple cabinets and enclosures may be needed for high-power and high-voltage converters. Not all cabinets need to be identical. For example, some high-power converters can have dedicated cabinets for controller and/or liquid cooling system. In fact, Figure 11.1 converter has a dedicated small enclosure for its controllers, which is attached to the main cabinet as shown in Figure 11.1d.

3) Drawers and/or shelves, which are usually needed for high-voltage and high-power converters but will not be needed for LV converters, such as in the case of most two-level VSC-based motor drives.
4) Supporting boards and spacers or standoffs, which are always needed in certain form.
5) Connectors, including power and signal connectors.
6) Fasteners.
7) Electrical/optical cables, wires, and buses.

It should be pointed out the mechanical system design is not limited to selection or design of the earlier mentioned components and parts. The mechanical system layout is critically important for both the performance and design objectives. For example, the layout will impact the voltage distribution and size. On the other hand, it is quite challenging to systematically or analytically optimize the mechanical system layout. Here, the layout design will be limited to parts and components, for example, the busbar layout in relation to the device modules and the DC-link capacitors. Often, the layout design is treated in a practical and qualitative way. In Chapter 13, virtual prototyping related to the system layout will be briefly introduced.

11.2.2 Mechanical System Design Constraints

The mechanical system design constraints also include performance constraints and physical constraints. The common performance specifications related to the converter include:

1) Insulation level, which includes the voltage withstand level or dielectric strength as often characterized by the DC or 50/60 Hz breakdown voltage as measured by a hi-pot tester, and partial discharge inception voltage (PDIV) level and partial discharge extinction voltage (PDEV) level characterized by a PD tester. The PDIV and PDEV level settings depend on the converter operating condition and the application. For instance, for the example system in Figure 11.1, the rated module DC-link voltage is 6.75 kV, and there are two modules cascaded per phase. As a result, the peak voltage can be 13.5 kV. Therefore, the insulation withstand voltage and PDIV must be beyond this value with sufficient margin following IEEE and other standards [1]. In addition, this converter is for grid-tied application and should meet the grid operation requirement. Grid overvoltage conditions, such as lightning, switching, and temporary overvoltage, may occur as will be discussed in Chapter 12. Therefore, mechanical system design needs to consider more insulation constraints. Since the insulation capability is a function of environment, including temperature and pressure, its design needs to consider the operating environment.
2) Thermal performance requirements. The mechanical system needs to consider both the ambient temperature as well as the operating temperature of the converter. It must be capable of operating over the temperature range needed by the converter. It also needs to consider the airflow condition requirement. For example, in Figure 11.1, the cabinet and drawer must accommodate airflow needed for the fan and air duct of the forced-air cooling system. The layout design should also consider to provide some airflow through the MV transformer.
3) Environment constraint beyond temperature capability, including outdoor conditions, where a weather-proof enclosure may be needed. One example is the NEMA-3-rated enclosure intended for outdoor use primarily to provide protection against windblown dust, rain, sleet, and external ice formation. Another condition is low-pressure condition, such as in a high-altitude environment as in aerospace applications.
4) Application size and weight constraints, which are related to the overall converter. For example, converters are often height and weight limited due to the building constraints. The size and weight limitations can also be due to transportation constraints.

5) Inherent size and strength constraints, which are due to the dimensions and weight of the electrical, thermal, and even mechanical parts and components that the mechanical system supports. For example, the enclosure needs to be sufficiently large and strong to house and support all the converter parts and connectors. The physical constraints for the mechanical system design are related to the intrinsic limits of the mechanical parts and components themselves. For example, the busbar conductor should not exceed its temperature limit while carrying the rated current. The enclosure, cabinet, and other mechanical supports should have sufficient strength to bear the weight of the components.

11.2.3 Mechanical System Design Conditions

The mechanical system interfaces directly with all other subsystems of the converter, as well as the ambient. The design conditions are from these subsystems and ambient.

The design input from the inverter or the rectifier power stage include:

1) device dimensions and weight;
2) passive component dimensions and weight;
3) current distribution, which refers to the currents on different terminals and wires. Examples include DC currents and AC phase currents for an inverter or a rectifier;
4) voltage distribution, which refers to the voltage potentials at all the terminals of electrical parts and components, as discussed in Chapters 5 through 10;
5) component losses, which are needed to determine temperature range of all electrical parts and components, such that the temperature range can be considered in mechanical system design.
6) input and output terminals, including the type, number, and rating. For example, the converter module in Figure 11.1b and c has two MV AC terminals and two LV DC terminals. Their voltage and current ratings are also the input to the mechanical design. There can also be internal terminals, e.g. terminals to DC link and EMI filters.

The design input from the thermal management system include:

1) dimensions and weight of fans, heatsinks, and air duct;
2) voltage distribution of fans, heatsinks, and air duct. Heatsinks and other parts can be grounded or floating, and their voltage potentials will impact the mechanical system design;
3) temperature and flow requirements, which include temperature range of the thermal management system parts/components, and the need for airflow inlets and outlets.

The EMI filter is essentially the extension of the rectifier or inverter power stage, and therefore its input to the mechanical system are similar to those of the rectifier or inverter power stage. Assuming the EMI filter contains only passive components, its input to the mechanical system design includes:

1) passive component dimensions and weight;
2) current distribution;
3) voltage distribution;
4) component losses;
5) input and output terminals, including the type, number, and rating. Since the EMI filter usually has one side connecting to the converter input or the output, and the other side connecting to the source or the load, some terminals are external and others are internal. Even within the EMI filters, there can also be internal terminals.

The design input from the controllers and auxiliaries include:

1) types, numbers, dimensions, and weight of PCBs for controllers and auxiliaries. As discussed in Chapter 10 and shown in Figure 11.1, high-power converters usually have multiple PCBs, including gate drive boards, power supply boards, sensor boards, and controller boards;
2) voltage distribution, which refers to the voltage levels of different PCBs and controller and auxiliary parts/components. As discussed in Chapter 10, digital controllers are generally grounded, but many other PCBs can be at high-voltage potentials, such as gate drives, isolated power supplies, and sensors;
3) temperature capabilities of the PCBs and components, which may impact the mechanical layout and enclosure design;
4) noise immunity capability of PCBs and components, which may also impact the mechanical layout. For example, a commercial IC has certain common mode transient immunity (CMTI), which has to be considered.

11.2.4 Mechanical System Design Objectives and Design Problem Formulation

The same as other converter subsystems, the design objectives for the mechanical system should also follow those of the whole converter. In principle, the overall mechanical system can be designed following the same optimization problem formulation as other subsystems, based on the selected design objectives, and the design variables, constraints, and conditions. However, as mentioned earlier, that will inevitably involve the layout optimization of the whole converter, which does not yet have a mature method to realize. As a result, we will adopt an approach to separately design and determine different variables associated with the mechanical system. For example, the design of the busbar will be considered as an independent design problem. The whole converter mechanical system design will be carried out following a procedure with proper iterations to achieve a good if not "optimal" design, while satisfying various constraints.

The design procedure for the mechanical system based on the example converter in Figure 11.1 is described here. The sequence for some of these parts can be altered as this is an iterative process.

11.2.4.1 PCBs

The first step for mechanical design is to have full inventory of components and subsystems, especially, PCBs. In Chapter 10, we have discussed partition strategy used for high-power three-phase converters and different types of PCBs needed. Some PCBs will be at signal level voltage potentials and grounded, some will be at high potentials, and others will be connected between two terminals with large voltage difference. The types of boards often depend on power rating, voltage rating, and application, while number of boards depends on application and topology. For the MV converter in Figure 11.1, there are 14 different types of boards as listed in Table 11.1 with their functions and voltage potentials indicated.

11.2.4.2 DC Busbar

As will be discussed in detail later, the busbar design needs to consider: (i) connections to the devices, DC-link capacitors, capacitor voltage balancing resistors and/or bleeder resistors, and other components on the DC link; (ii) insulation requirements; and (iii) stray inductances. In the example converter, due to the relatively low current, the PCB-based busbar is adopted as shown in Figure 11.3. DC-link capacitors are directly mounted to the PCB busbar. Insulation distances are determined following the IPC-2221 Standard, and slots are adopted to increase the creepage

Table 11.1 Types of PCBs in a medium-voltage converter.

	Type	Description		Type	Description
1	Converter controller	In controller box, ground	8	MV AC voltage sensor	For AC phase voltage sensing
2	Optical fiber	Three different types, connecting to LV DC, MV DC, and MV AC at different potentials	9	MV AC current sensor	For AC line current sensing
3	LV power	Power stage board for LV DC (850 V)	10	Auxiliary power supply connection	Two boards for 24 V to modules and to devices
4	LV gate drive	For 1.7-kV SiC MOSFET	11	MV GDPS	Two boards for primary and secondary of MV gate drive power supply
5	LV DC sensor	LV DC voltage sensing	12	MV GDPS to gate drive DC/DC	Power supply from 24 V to gate drive voltages
6	MV DC-link busbar	Two types, connecting devices and capacitors	13	MV gate drive	For 10-kV SiC MOSFET
7	MV DC voltage sensor	For H-bridge DC voltage	14	MV desaturation protection	Drain-source voltage detection and short-circuit protection

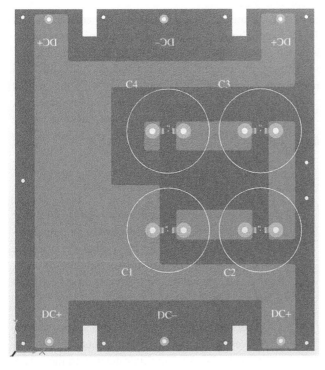

Figure 11.3 PCB-based busbar.

Source

Gate

Drain

Figure 11.4 A 10 kV/20 A SiC MOSFET from Wolfspeed.

distance. The stray inductance is estimated with FEA simulation, which is considered as part of the stray inductance in the switch commutation loop.

11.2.4.3 Device and Gate Drive

Once the busbar is designed, the gate drive board and its connection to devices need to be designed. The design needs to consider: (i) the connection between the gate drive board and the device; (ii) layout for minimum gate loop; and (iii) insulation requirements.

For the example converter in Figure 11.1, the following measures are adopted.

1) Minimize device commutation loop. The switching commutation loop of the half-bridge is made as small as possible to reduce the switching overshoot, which requires putting the devices (see Figure 11.4) and the DC bus close, while maintaining the required clearance and creepage distances.

2) Minimize gate loop. To reduce the gate voltage or current ringing, the gate loop should be small. Therefore, the gate drive board (see Figure 11.6) is directly mounted to the device. The gate drive circuit can be seen in Figure 11.5.

3) Considering the flexibility of test and debug, as well as the PCB cost, the gate drive board is made separately, rather than placed on the DC busbar PCB.

4) Considering the insulation requirements, slots are added between the gate drive circuits of two devices in each half-bridge to reduce the PCB size.

5) The connection between the device and the DC busbar PCB is through screw connectors and wires considering the device's special package. Those connections also introduce stray inductance, and the total commutation loop inductance consists of contributions from the busbar, the DC-link capacitor, the capacitor connection, and the device connection. With the total inductance estimated, the switching voltage overshoot can be estimated. With the desired switching speed, if the overshoot is too high, either the design needs to be improved to reduce the inductance or decoupling capacitors can be added close to the devices.

11.2.4.4 Device Cooling

Device cooling mechanical design needs to consider: (i) heatsink connection and layout; (ii) air duct structure; (iii) insulation; and (iv) cooling performance verification (e.g. with FEA simulation). For Figure 11.1 converter, the following have been considered in the design.

1) As discussed in Chapter 9, the semiconductor devices can be cooled with one heatsink per device or one heatsink for multiple devices. In the case of non-isolated SiC MOSFET in Figure 11.4, since the device baseplate is its drain terminal, multiple

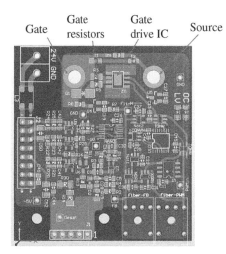

Gate

Gate resistors

Gate drive IC

Source

Figure 11.5 The 10 kV SiC MOSFET gate drive circuit.

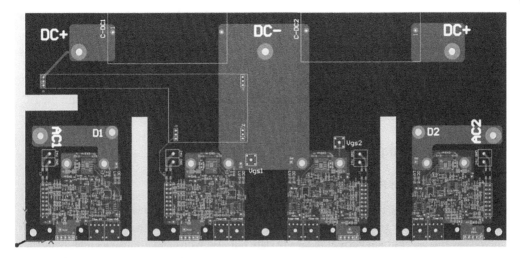

Figure 11.6 The 10 kV SiC MOSFET gate drive board.

devices sharing the same heatsink require insulation between the device and the heatsink. Considering the high voltage, a thick insulation layer would be needed, which would increase the thermal resistance between the device and the heatsink, and require more cooling. In addition, the coupling capacitance could be large between adjacent devices, and between the device and ground, through the heatsink, which would increase the CM current as well as the loss (see Chapter 12 for further discussion). Therefore, one heatsink per device is adopted.

2) Since the heatsinks are not isolated from the devices on them, the distance between two heatsinks needs to be sufficient for insulation. To save space, multiple devices are located within one air duct for forced-air cooling. Heatsinks are floated from the air duct to increase the creepage distance and allow a smaller distance between the heatsinks (see Figure 11.7).

3) Slots on the air duct side panels are used for electrical wire connection, and the air duct has no lid, which results in air leakage. In this case, FEA simulation with the fan model is conducted to verify the cooling performance and the temperature range.

11.2.4.5 Mechanical Layout and Support

Mechanical layout and support need to consider: (i) component layout; (ii) mechanical support and fixing; and (iii) insulation. Specific measures are as follows.

1) The height of the busbar, gate drive, device, and heatsink, as well as the air duct, need to be coordinated. The connector between the busbar and the gate drive board is determined such that the heatsink has a certain distance (10 mm) from the air duct, and standoffs for the busbar and the gate drive boards are available. The air duct is selected to be at least 20 mm

Figure 11.7 The heatsink floating and not touching the air duct on bottom or sides for increased creepage distance between two adjacent heatsinks.

wider than the heatsink so that the distance between the heatsinks and the two sides of the air duct is at least 10 mm.

2) Since the power supply for the fan is low voltage (LV) to the ground, the distance between the fan and the nearest high-voltage components, such as the devices and heatsink, needs to be sufficient to meet the creepage and clearance requirement (>100 mm in Figure 11.1 converter).

3) Standoffs are used to support the busbar and gate drive, and screws are used to mount the DC-link capacitor, air duct, and standoffs to the garolite board.

11.3 Busbar Design

11.3.1 Busbar Design Problem Formulation

In high-power converters, the busbar is the commonly used connector for power device modules, DC-link capacitors and their voltage balancing/bleeder resistors, and filters. The busbar not only bears high voltage and carries high current in steady state but also withstands/transmits high-frequency voltage and current during switching transients. The resonance between the parasitic inductance of the busbar and the parasitic capacitances of devices can cause voltage and current overshoot. A poor design of the busbar with high parasitic inductance can result in high-voltage overshoot as analyzed in Chapter 6, which can damage the devices. One example for such voltage overshoot can be seen later in Section 11.3.3. To avoid its occurrence, the devices must be switched slower or additional components such as snubbers must be added, and neither solution is desirable.

Clearly, the parasitic loop inductance of the busbar is critically important, especially for fast-switching converters based on WBG devices. Moreover, the busbar design has to meet a number of other important criteria, including dimensions, current density, and insulation. Converter topologies, ratings, and applications determine the types, ratings, and numbers of devices and passives, which will impose requirements for the geometry and dimensions of a busbar. The number of DC and AC input/output terminals can dictate the mechanical complexity. The current level, including average and RMS currents as well as the low- and high-frequency components, are responsible for the busbar thickness and number of connections, which determine the current density. Equation (11.1) shows the formulation for the design optimization problem. The design constraints, variables, and conditions are self-explanatory. One constraint on current distribution for symmetry is really a design rule or practice for magnetic field reduction, which will lead to low inductance and will be explained later.

Minimize: $L(X)$ – busbar parasitic loop inductance

Subject to constraints:

$$G(X, U) = \begin{bmatrix} g_1(X) \\ g_2(X) \\ g_3(X) \\ g_4(X) \end{bmatrix} = \begin{bmatrix} - \text{dimensions} \\ - \text{insulation capability} \\ \text{current density} \\ - \text{current distribution} \end{bmatrix}$$

$$\leq \begin{bmatrix} - \text{device and capacitor dimensions} \\ - (\text{converter voltage rating} + \text{required marging per application}) \\ \text{maximum current density per temperature capability} \\ - \text{asymmetry} \end{bmatrix}$$

where design variables X and design conditions U:

$$X = \begin{bmatrix} x_1 \\ x_2 \\ x_3 \\ x_4 \\ x_5 \\ x_6 \end{bmatrix} = \begin{bmatrix} \text{Busbar geometry} \\ \text{Conductor material} \\ \text{Conductor thickness} \\ \text{Insulator material} \\ \text{Insulator thickness} \\ \text{DC and AC connections} \end{bmatrix} \tag{11.1}$$

$$U = \begin{bmatrix} u_1 \\ u_2 \\ u_3 \\ u_4 \\ u_5 \end{bmatrix} = \begin{bmatrix} \text{device numbers and dimensions} \\ \text{passive numbers and dimensions} \\ \text{current} \\ \text{voltage} \\ \text{DC and AC terminals} \end{bmatrix}$$

11.3.2 Busbar Design Procedures and Considerations

Although the busbar design problem can be theoretically formulated as in Eq. (11.1), some of the relationships among the design objective, variables, constraints, and conditions are difficult to be expressed or modeled analytically. Here, we will describe these relationships based on combination of physical principles, analytical models, and simulation. A design procedure based on [2] in a systematic but qualitative way will be presented. The corresponding flowchart is shown in Figure 11.8.

11.3.2.1 Component Selection and Busbar Geometry

According to Figure 11.8, the first step for the busbar design is to select components and design/select busbar geometry or layout. For a VSC, the main components that influence busbar layout include devices, DC-link capacitors, and to some extent, heatsink/cold plate, which are based on the inverter/rectifier design. For the given design problem, the design criteria for busbar geometry should be for low inductance, while also considering power density and complexity.

Once the components are determined, the terminals need to be identified based on the topology. For a two-level three-phase VSC, there are two DC terminals (positive and negative terminals), and three AC terminals for a, b, and c phases, respectively. These terminals are marked in Figure 11.9. All these terminals are at different voltage potentials, and therefore each terminal needs to have its own bus or bar. In practice, since AC conductors in a VSC are not in the switching loops of the power devices and their parasitic inductances are not critical, sometimes only DC terminals are connected using busbars. Since it is generally desirable to achieve magnetic field cancelation and minimize busbar parasitic inductance, a two-layer busbar is commonly used with a positive DC bus layer and a negative DC bus layer. In other cases, AC terminals are also included in the busbar, where a multilayer structure may be needed. For example, a three-layer busbar can be used with the middle layer connected to AC terminals, corresponding to the circuit structure in Figure 11.9. Note that the three AC buses and terminals in the middle layer are separated electrically from one another, unlike the DC terminal cases, where positive and negative DC bus layers are both whole piece conductors.

In addition to external terminals, busbar connects with device modules and capacitors via internal terminals, often the screw holes. In high-power converters, the device modules can have

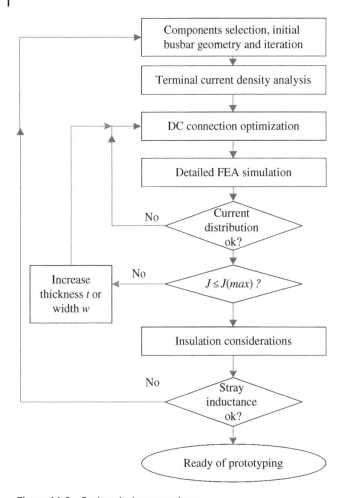

Figure 11.8 Busbar design procedure.

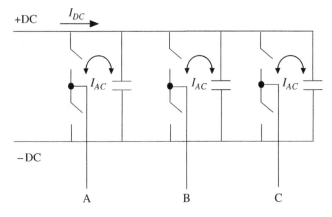

Figure 11.9 Illustration of the AC and DC terminals and current flows in a three-phase two-level VSI.

multiple terminals, and there are often multiple DC-link capacitors. Consequently, there can be many internal terminals (holes) on a busbar. Although they take space, the distributed terminals can be beneficial for current distribution and sharing, as will be explained next.

The busbar geometry design needs to consider the type (DC or AC) and number of terminals, structure (e.g. layered structure), number of layers, and positions of the terminals relative to other terminals, power device modules, and DC-link capacitors. One important geometry design principle is related to the location of DC input/output terminals. In principle, it is better to have the terminals closer to the devices, rather than to the DC-link capacitors. The DC link carries both the average (i.e. DC) current and high-frequency AC switching current. The DC current flows through devices and input/output terminals, while the AC current flows between devices and DC-link capacitors. The relationship is illustrated in Figure 11.9. By placing DC input/output terminals closer to devices, the overlap of the DC and AC currents in the busbar will be reduced, which will reduce the current density and loss in the busbar. Following the similar principle, for the three-phase structure, to have three sets of capacitors with each set close to one phase and the power device modules on the phase can often be beneficial, as the capacitor current only flows to devices as indicated in Figure 11.9. Note that, sometimes, one may not have the freedom to locate the external DC or AC terminals due to the overall converter constraint.

There are a number of additional design principles for the busbar geometry:

- DC input/output terminal connections should be symmetrically positioned relative to the power device modules. The symmetry among DC terminals and power device modules will help DC current distribution.
- For the AC current distribution, the symmetry between DC-link capacitors and power device modules is desirable, given that the AC current flows through capacitors and device modules. High-power applications, such as three-phase AC applications, often contain more than one DC-link capacitor. To ensure that each capacitor operates under similar conditions, the AC current distribution and the current sharing among the capacitors should be balanced. Phase currents, and consequently the current through the power devices, are defined by the load. For a three-phase balanced load, currents through the power devices should be balanced.
- Sharp corners and bends can cause eddy currents and consequently losses and heat. In converters with relatively high voltages (even above several hundreds of volts), they can also cause corona, leading to potential loss, noise, and insulation problems. To avoid these effects, corners and bends should have a sufficient large radius, as illustrated in Figure 11.10.

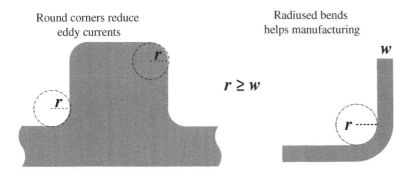

Figure 11.10 Busbars with round corners and bends. *Source:* From [2]/with permission of IEEE.

11.3.2.2 Busbar Dimensions, Current Density, and Current Distribution

11.3.2.2.1 Minimum Dimension and Current Density Requirements The second step in Figure 11.8 busbar design flowchart is the dimension design based on current density and distribution analysis. For a given geometry (shape), the busbar dimensions, specifically, the conductor thickness should be designed to satisfy the current density requirement to prevent overheating in any part of the busbar. A typical maximum current density value for passive cooled busbars is 5 A/mm² for copper busbar [2]. Current density analysis needs to be conducted especially for the terminals, which usually are the narrowest part of the busbar.

Assuming that mechanical strength is not the concern, the minimum busbar thickness is determined by the input and output terminal connectors. As illustrated in Figure 11.11, given the maximum current density J and the width w, which can be determined by the size of the components on the busbar and terminals, the thickness t can be determined to keep the current density within its specification.

If copper conductor is used, the minimum total cross-sectional area A needed for the busbar to carry a DC current I_{DC} can be calculated by (11.2), where I_{DC} is in amperes and N is the number of conductors in the busbar assembly [3]. It indicates that if the busbar is built with conductors laminated together, then a 5% safety margin is needed for each additional conductor to compensate for the compounding heat gain within the conductors. Equation (11.2) shows that the maximum current density is approximately 5 A/mm² for $N = 1$ and lower for higher N cases. Note that (11.2) is only valid for $I_{DC} < 300$ A. For I_{DC} above 300 A, the ampacity table can be used to determine the area [4]. The ampacity is a function of ambient temperature and temperature rise. Correlating with the ampacity table, Eq. (11.2) corresponds to a temperature of 70 °C or a temperature rise of 30 °C with an ambient temperature of 40 °C. Equation (11.2) applies to the rectangular conductor. From A and w, the thickness t of the conductor can be determined.

$$A = 0.2025 \cdot I_{DC} \cdot [1 + 0.05 \cdot (N - 1)] \ \text{mm}^2 \tag{11.2}$$

To account for the high-frequency AC current, the skin effect must be considered to calculate the required cross-sectional area. The skin effect can be considered using the skin depth. The equation to approximate skin depth for a rectangular conductor while accounting for the thickness is given in (11.3), where f is AC frequency, σ is electrical conductivity, and μ_0 is the permeability of free space. For the three-phase AC converter, the dominant AC frequency is the switching frequency. For copper, $\sigma = 5.96 \times 10^7$ S/m at 20 °C with a thermal coefficient of -4.04×10^{-3} (1/°C), and for aluminum, $\sigma = 3.77 \times 10^7$ S/m at 20 °C with a thermal coefficient of -3.90×10^{-3} (1/°C).

$$\delta = \frac{1}{\sqrt{\pi f \sigma \mu_0}} \left(1 - e^{-\sqrt{\pi f \sigma \mu_0}}\right) \tag{11.3}$$

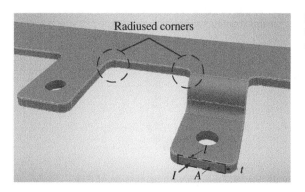
Radiused corners

Figure 11.11 Busbar thickness design considerations based on maximum current density. *Source:* Adapted from [2].

Figure 11.12 illustrates the current distribution in a rectangular connection terminal. The DC current distributes evenly on the whole conductor area, while the AC current only distributes on the surface within the skin depth δ. Typically, the thickness t should be greater than 2δ. The total current density ($J_{DC} + J_{AC}$) should be below the maximum density, e.g. $5\,A/mm^2$ for passive cooling case [2]. A higher current density would mean that a smaller cross-sectional area is allowed, leading to a size reduction for the busbar thickness. Typically, the busbar conductors are sized for a 30 °C temperature rise, though the proper temperature rise should be designed based on the specific application ambient conditions.

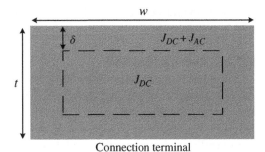

Figure 11.12 Current distribution illustration in a connection terminal.

11.3.2.2.2 *Current Distribution Analysis*

To determine the current density, the current distribution needs to be analyzed. In a three-phase converter, the DC-link current contains DC and AC components, which have been derived as (6.130). In high-power and -current cases, the DC input/output terminals can be split into several connectors in order to improve the current distribution with acceptable conductor thickness. This reduces ohmic losses and evenly spread the heat across the busbar, which reduces the hot spots. With more than one set of DC and AC input/output terminals and multiple DC-link capacitors, the current distribution on busbars should be evaluated for balanced current sharing among terminals and among DC-link capacitors [5].

The current distribution analysis can be done with equivalent circuits or through FEA simulation. With equivalent circuits, DC and AC current distribution analyses need to be conducted separately, since DC and AC currents will have different paths and different circuit parameters. DC analysis is to evaluate current flows through the busbar DC terminals and the devices, while AC analysis is to evaluate the current distribution among the DC-link capacitors and the power devices. If the busbar is symmetric, DC and AC analysis can each use a single-loop RL equivalent circuit [2]. Multiple branch equivalent circuit can be used in the case of asymmetry. The equivalent circuit parameters can be obtained through analytical or simulation approach.

The equivalent circuit parameter extraction will require a 3D model, and very often the practical busbar design will involve asymmetries. FEA simulation is often preferred, since it avoids the need for the busbar equivalent circuit parameter extraction and can automatically consider asymmetry. It can also model the skin effect and proximity effect.

There are two common issues in busbar current distribution. One is the high current density in some vulnerable paths, e.g. the terminals, and the other is the current imbalance among paralleled components. A good method to avoid these issues is to design a symmetrical busbar with all the paths symmetric and of the same length. This, however, is not always possible or feasible. Hence, in some cases, it becomes necessary to guide or reroute currents (both DC and AC) by adding features in busbar, such as adding intentional obstacles (holes/slots) to the current flow path to make it an unattractive route. In [2], it shows an example, in which by adding incisions on the busbar, the originally unbalanced current distribution among three paralleled IGBT modules is mitigated. Two different types of symmetrical cuts are tried, triangular and round cuts. The current density, while not drastically different, has been reduced. The incision dimensions can be designed through simulation to find a current distribution with acceptable current density in the busbar and current level in components.

11.3.2.3 Electrical Insulation

Electrical insulation is an essential part of the busbar, since busbar conductors can be at different voltages and must be insulated electrically. The insulation design includes the electrical insulation material selection and insulation configuration design. The design of the electrical insulation is driven by the operating voltage, the operating temperature, and the environmental conditions. The operating voltage and the insulation materials, which have their corresponding dielectric strength, determine the thickness of the insulation layer. The common insulation materials used in power converters are mostly polymer-based and can be in the form of film (e.g. Nomex, Kapton, Tedlar, Mylar, and Ultem), curable materials (e.g. resins like Valox), coating materials (e.g. epoxy powder coating), or some their combinations [2]. Only in rare low-voltage cases, the busbar insulation relies on air. In many cases, especially the relatively low current cases, PCB based busbars are used, which often use materials like FR4 as the insulation material, which is made from a flame-retardant epoxy resin and glass fabric composite.

In general, thinner insulation layer in a busbar is desirable for smaller size and weight. In addition, a thinner insulation layer can result in a higher capacitance, which can act as a filter to reduce the EMI noise. It is therefore important to select insulation materials with high dielectric strength. The choice of the insulation material and its thickness in a busbar is not only driven by the electrical requirement but can also be driven by the mechanical requirement such as match with conductors. For example, for a laminated busbar, it needs to survive the temperature cycling as a result of load current and temperature changes. A thinner and more flexible insulation material with its CTE better matched with conductors is more desirable.

Busbar insulation is to prevent short circuit between busbar conductors. The short circuit can happen as a result of the insulation breakdown internally, and it can also happen through the insulation surface externally. Similar to the case of a PCB, it is important to consider the creepage distance in the busbar design. The required creepage distance between two conductors depends on the busbar geometry, the voltage difference of the conductors, the exterior material of the insulation, and the environment. If the surface of the busbar is susceptible to moisture or contamination, a larger creepage distance will be needed to avoid flashover due to surface tracking.

Other factors in selection and design of insulation include ability to resist pollutants and moisture, and temperature capability. For example, the popular insulation material Nomex is susceptible to moisture and therefore should not be used in a humid environment to avoid dielectric strength degradation and possible delamination. Every insulation material is rated for a certain temperature range. It is well known that the lifetime of an insulation material typically decreases by half for every 10 °C temperature rise, so it is important to apply the material within its upper temperature limit. The temperature capability of the busbar is usually limited by its insulation materials, similar to the cases of film capacitors, cables, and to large extent, inductors, transformers, and electric machines. The low temperature is usually not a problem for insulation materials, although in some extreme cases, the mechanical integrity can be a concern. One example is that certain insulation materials may crack at very low temperature and fail electrically afterward. For aerospace applications and cryogenically cooled environment, the low-temperature capability of the insulation materials should also be considered.

Some applications such as aerospace applications involve high altitude and low pressure. It is well known that the breakdown electric field of the air will decrease with lower pressure. As a result, the creepage distance needs to be increased as compared to the normal pressure condition. Other options include potting, conformal coating, and powder coating of polymer materials on the exterior of the busbar. In addition to having better dielectric strength than air, the coating materials can be resistant to moisture, dust, dirt, and salt, etc.

11.3.2.4 Busbar Parasitics

The busbar design optimization problem in (11.1) has the minimum busbar loop parasitic inductance as the design objective. Clearly, the parasitic loop inductance is one of the most important busbar parameters. In addition, the parasitic resistance and capacitance can also impact the performance of the busbar, and therefore should also be evaluated.

11.3.2.4.1 Parasitic Inductance In addition to significantly impacting the device switching overvoltage as discussed in Chapter 6, in the case of multiple DC-link capacitors, if the connections for capacitors and devices are asymmetric, the parasitic inductance difference in each current path may affect current sharing.

The busbar parasitic inductance can be estimated analytically or through simulation. Figure 11.13 shows a simple two-layer busbar consisting of two paralleled conductor plates separated by an insulation layer, where w, t, and l represent width, thickness, and length of each conductor, and d is the thickness of the insulator or the distance between two conductors.

The parasitic inductance of a plate includes two parts: self-inductance L_s and mutual inductance M. The total parasitic inductance of the busbar considering the same current flowing in two plates in opposite directions can be determined as:

$$L_{total} = 2(L_s - M) \tag{11.4}$$

An effective inductance reduction technique is to utilize magnetic field canceling effect. For the two plates in Figure 11.13 with identical dimensions, the magnetic fields generated by the currents in opposite directions on two plates have a canceling effect. In other words, the mutual inductance can cancel part of the self-inductance. Clearly, M should be maximized in order to minimize the busbar total inductance. Many analytical formulas have been developed with various approximations and accuracies for L_s and M (e.g. [6]). In general, an effective way to reduce L_{total} is to decrease the distance d between the two plates. Increasing the overlap area of the two plates (w in Figure 11.13) should also help. Therefore, the laminated structure with the conductors closely placed on top of the each other and separated only by a thin layer of insulation is preferred in the busbar design.

The parasitic inductance of the busbar will change with frequency as well, mainly due to skin effect. On the other hand, the impact of the frequency on parasitic busbar inductance is relatively

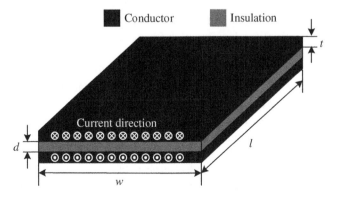

Figure 11.13 A simple two-layer busbar structure.

small for the frequency range of interest to power converters. For example, in [2], the differences of the simulated inductances at 10 and 100 kHz for five different busbars are all around 10%. Besides, the low-frequency inductances are higher than those at high frequencies due to skin effect, and therefore the calculated value with DC analysis is a conservative result.

11.3.2.4.2 Parasitic Capacitance and Resistance

Between busbar conductors, there are voltage differences and corresponding electric fields. Consequently, there will be corresponding parasitic capacitances. Unlike parasitic inductance, busbar parasitic capacitance generally can benefit the converter by reducing the total impedance of the busbar and providing a filter for high-frequency noise, including EMI noise. In addition, it will help reduce resonance and voltage overshoot during switching.

If we neglect the busbar resistance, the total impedance of the busbar can be approximated by (11.5), where C_{total} is the equivalent capacitance of the busbar. Obviously, to achieve low impedance, the capacitance should be designed as large as possible while minimizing the inductance.

$$Z = \frac{L_{total}}{C_{total}} \tag{11.5}$$

The parasitic capacitance of a laminated busbar is determined by the geometry of conductors and insulators as well as the insulation material property. For a simple two-layer busbar in Figure 11.13, the capacitance C_{total} can be obtained by (11.6), where ε is the permittivity of the insulation material.

$$C_{total} = \varepsilon \frac{w \cdot l}{d} \tag{11.6}$$

From (11.6), larger capacitance can be achieved by increasing the overlapping area of the two busbar conductors and/or by reducing d. However, the busbar area will also impact inductance, size, weight, and cost. Thus, an effective method to increase the capacitance is to minimize the thickness of the insulation layer d while satisfying insulation requirement. As discussed earlier, smaller d is also beneficial in reducing parasitic inductance.

The busbar parasitic resistance is determined by materials and shape of the conductors. Generally, with the proper design of the busbar to meet the current density or temperature requirement, the busbar parasitic resistance is quite small and has minimum impact on converter loss. One important characteristic of the busbar resistance is its frequency dependency. An example in [2] shows that the busbar resistance more than doubles when frequency changes from DC to 100 kHz, and at the same time, the power loss also almost doubles, though still quite small. This frequency-dependent resistance can be beneficial for damping high-frequency ripples and noise, while without causing appreciable extra loss.

Although analytical models can be derived to estimate the parasitic parameters for simple busbar structures, for practical busbars with complex geometry, 3D model-based FEA simulation is often necessary to calculate the parasitic parameters of the busbar. Simulation can also more easily consider frequency dependency, which is important for both the parasitic inductance and resistance.

As shown in Figure 11.8, the busbar design optimization is an iterative process, where the geometry, terminal connections, and dimensions can be varied to meet the requirements on current density (or thermal temperature limits), current distribution, and electric field, while minimizing parasitic inductance. 3D FEA simulation is an integral part of the busbar design optimization. An example design is presented as follows.

11.3.3 Busbar Layout Design Example for a Three-Level ANPC Converter

Chapter 6 has shown that for a two-level VSI, the peak device voltage during switching transient can be significantly higher than the steady-state voltage as a result of the switching loop inductances. Chapter 7 explained that this voltage overshoot becomes even more severe and complicated in a three-level topology, where multiple switching loops exist during the switching transient [7]. The overvoltage can be mitigated by using snubbers. However, passive snubbers introduce extra energy loss and decrease the switching speed of the device, while active snubbers increase the cost and complexity and reduce the reliability of the converter.

A busbar with low loop inductance will help mitigate the overvoltage, and in the case of WBG-based converters, to preserve their high-switching speed performance. Note that low loop inductance could lead to fast device current rise during short circuit. However, with proper design, the short-circuit fault can be detected even under high di/dt condition with the fault current being limited to acceptable value through protections such as desaturation protection. Since the device failure due to over current mainly occurs in saturation region after the current is limited, the low loop inductance will have little impact on short-circuit withstand capability of the converter. In other words, low busbar parasitic inductance is beneficial and desirable for VSCs.

General design guideline and procedure for busbars have been presented earlier in this section. However, the impact of switching loops on busbar design has not been considered, which are critical for topologies like three-level converters. This subsection will present a design example for a three-level ANPC converter with much of the materials here based on [8].

11.3.3.1 Switching Loops in a Three-Level ANPC Converter

To optimize the busbar layout of a three-level converter, it is necessary to have a good understanding of the switching loops and parasitic circuit parameter distribution. According to the structure of the switching loops, three-level converters can be categorized into two main groups: converters with symmetric switching loops (e.g. T-type converters) and converters with asymmetric switching loops (e.g. NPC converters). Since the asymmetric structure makes the busbar design more complex, we will focus on ANPC converters. The T-type converter switching loop evaluation can be found in [8].

The circuit configuration of one phase-leg with DC-link capacitors in the three-level ANPC converter is illustrated in Figure 11.14. Different busbar parts, parasitic inductances, and switching loops are highlighted. There are six distinct conducting bars: the positive DC-link bar, the negative DC-link bar, the AC terminal bar, and the DC neutral bar, as well as two middle bars, which can be connected to the AC bar through the clamping switches S_{2H} and S_{2L} or can be floating.

A set of switching function signals are also plotted in Figure 11.14, which correspond to a low-loss modulation scheme [9]. The switching patterns are divided into four states for one line cycle as listed in Table 11.2. S_{1H}, S_{3H}, S_{1L}, and S_{3L} switch at switching frequency, while S_{2H} and S_{2L} switch at line frequency. During the half-line cycle with negative AC output voltage, the operation state changes between N and O^-, which means that S_{1L} and S_{3L} operate at switching frequency while S_{2L} and S_{3H} stay ON, and S_{2H} and S_{1H} are OFF.

As shown in Figure 11.14 and also analyzed in Chapter 7, there are two switching loops corresponding to the negative half-line cycle operation. One goes through S_{1L} and S_{3L} in the bottom half of the phase-leg, and it is the short commutation loop. The parasitic inductance of the switching loop resonates with the output capacitance of S_{1L} or S_{3L}, depending on which one acts as the synchronous switch during the switching. The other switching loop, marked by the green dashed line, is the long commutation loop. The parasitic inductance of the switching loop resonates with the output capacitance of S_{2H}.

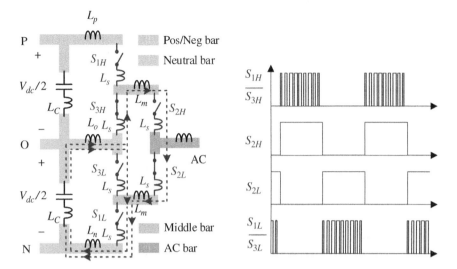

Figure 11.14 A phase leg with DC-link capacitors of a three-level ANPC converter and a set of switching function signals corresponding to a low-loss modulation scheme.

Table 11.2 Operation states of one phase of a three-level ANPC converter.

State	S_{1H}	S_{3H}	S_{2H}	S_{2L}	S_{3L}	S_{1L}
P	On	Off	On	Off	On	Off
O^+	Off	On	On	Off	On	Off
O^-	Off	On	Off	On	On	Off
N	Off	On	Off	On	Off	On

Assuming each busbar part is independent and is not coupled with other busbar parts, the total parasitic inductance of the two loops can be evaluated. For the short commutation loop, it includes the lower-half DC (decoupling) capacitor, the neutral bar, the negative bar, and two switches S_{1L} and S_{3L}. Thus, the total short loop parasitic inductance L_{sl} is (11.7), where L_C is the ESL of the capacitor, L_o and L_n are the inductances of the neutral and negative busbar parts, and L_s is the inductance of each switch:

$$L_{sl} = L_C + L_o + L_n + 2L_s \tag{11.7}$$

The long switching loop consists of the lower-half DC (decoupling) capacitor, the neutral bar, two parts of the middle bar, the negative bar, and four switches S_{1L}, S_{3H}, S_{2H}, and S_{2L}. Thus, the total long loop parasitic inductance L_{ll} is (11.8), where L_m is the inductance of the middle bar:

$$L_{ll} = L_C + L_o + L_n + 2L_m + 4L_s \tag{11.8}$$

Clearly, the long switching loop has larger parasitic inductance than the short one and should be paid more attention in busbar design.

11.3.3.2 Busbar Design Procedure Considering Multiple Switching Loops

In Section 11.3.2, the magnetic field canceling effect and its benefit for busbars have been explained. Based on this principle, a procedure can be devised as a guideline for designing two-layer laminated busbars for three-level converters. The procedure is illustrated in Figure 11.15 and is described as follows.

1) Place device modules. The placement of the power modules significantly influences the busbar layout. In most cases, all the power modules are located in the same plane. Hence, this step can be accomplished in a 2-D planar layout. The primary design criteria include two aspects: (i) for symmetric switching or commutation loops, an identical layout should be kept to achieve the same parasitics for all loops; and (ii) the connection between the busbar and modules should be designed for ease of manufacturing. In a layered busbar, holes have to be drilled on the bottom layers such that the connectors on the upper layer can go through and touch the power module. As a result, the effective area of the bottom layers decreases, and the current distribution and current density can be affected.

2) Determine the needed busbar parts and find the parts shared by multiple switching loops. The composition of each switching loop should be examined to understand the involved busbar parts. For example, the ANPC converter busbar parts for the two lower-half switching loops can be easily identified from Figure 11.14, and the shared busbar parts for the two loops include the negative DC bar and the negative middle bar.

3) Place the busbar parts. This step needs to be conducted in 3D. The returning path should be selected from the shared parts by multiple switching loops, and it should be placed in one layer.

4) Adjust the area of the returning path to fully overlap with the outgoing paths. Based on the magnetic field cancelation principle, the outgoing and returning paths should be overlapped to maximize the cancelation effect. Usually, there is only one returning path and multiple outgoing paths based on step 3 of this design procedure. Therefore, the area of the returning path usually should be increased to cover all the outgoing paths.

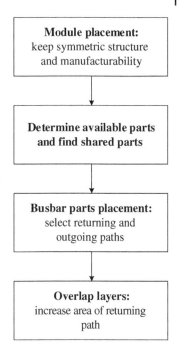

Figure 11.15 The design procedure of a two-layer laminated busbar for three-level converters.

11.3.3.3 Busbar Design Implementation Example

The Figure 11.15 flowchart and the aforementioned description provide a general procedure for busbar layout. The detailed layout is highly dependent on the geometry of the components such as device modules, DC-link capacitors, and heatsink or cold plate. Here, a busbar layout design example following the design methodology is demonstrated for a three-phase, three-level ANPC inverter. The inverter is rated for 500 kVA with cryogenic cooling and derated to 200 kVA with room temperature water cooling. The input DC voltage is 1 kV, the output fundamental frequency is 3 kHz, and the switching frequency is 60 kHz.

11.3.3.3.1 Power Module Selection Power module selection is an input to the mechanical system design, including the busbar design. As described in Section 11.3.1 when formulating the busbar design problem, the power module information needed for the mechanical system and the busbar

 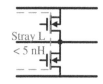

Figure 11.16 Wolfspeed HT-3000 series SiC MOFET module and equivalent circuit. *Source:* From [8]/with permission of IEEE.

design includes dimensions, and rated voltage and current. In addition, power modules also contribute to the parasitic circuit parameters, including the inductance of the switching loop(s). Commercial power modules can be in different forms, such as single-device module, half-bridge module, phase-leg module, and three-phase module. Based on commercial availability, half-bridge modules are used for this three-level ANPC converter, since they will have reduced external connections between switches compared with the single-switch modules. Three half-bridge modules are required for each three-level ANPC phase-leg, one for S_{1H} and S_{3H}, one for S_{1L} and S_{3L}, and one for S_{2H} and S_{2L}. For the example converter, the Wolfspeed HT-3000 series half-bridge module is selected, with its voltage rating of 900 V and on resistance of 1.2 mΩ. Figure 11.16 shows a picture of the module package. According to the manufacturer, the stray inductance of the module is <5 nH.

11.3.3.3.2 Capacitor Selection Film capacitors are selected for the DC-link capacitors due to their higher current capability and lower ESL compared with electrolytic capacitors. There are two options for the capacitor selection. One is to choose a single large capacitor, and the other is to use multiple small capacitors in parallel.

Based on the inverter design results, the total required capacitance of the three-phase converter is 300 μF. Table 11.3 shows the comparison of the two options, 100 μF large capacitor and 10 μF small capacitor. Since the total weight of 10 small capacitors is less than that of a single large capacitor, paralleling small capacitors is adopted in this minimum weight inverter design example. Also, the ESL of one large capacitor is >20 nH, which would require decoupling capacitors, while no decoupling capacitors are needed for small capacitors in parallel since the overall ESL is sufficiently low. Nevertheless, paralleling capacitors requires the layout to guarantee good current distribution.

11.3.3.3.3 Busbar Layout Design The detailed design steps is presented here following the layout design procedure in Figure 11.15.

1) Place the power modules. To keep the symmetric structure of the upper and lower side of the converter, there are mainly three ways as shown in Figure 11.17 to place the power modules in one plane. The colored lines represent the busbar connections among the terminals of the power modules and the capacitors.

    Options (a) and (b) are the two cases of side-by-side placement. For option (a), the main issue is the conflict between busbar parts and the gate-source terminals of the modules (marked as "GS" in Figure 11.17). To implement the gate drive boards, it is inevitable to drill holes on the busbar parts, especially, for the module in the middle. Drilling holes decreases the effective

Table 11.3 Capacitor comparison.

Part	Value (μF)	ESL (nH)	Weight (g)	Price ($)	Qty
KYOCERA AVX FFVE6K0107K	100	25	350	60.00	1
Vishay KP1848C61060JK2	10	15	15	4.20	10

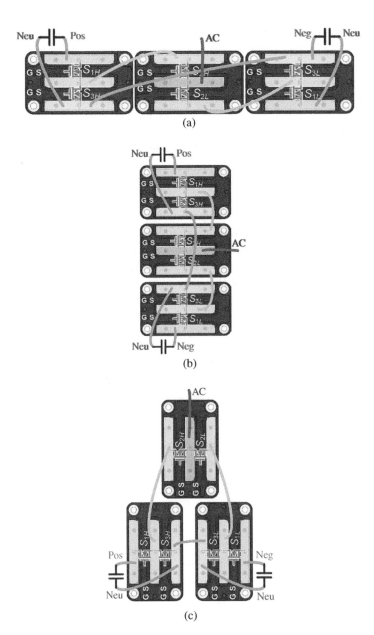

Figure 11.17 Three options of symmetric power module placement.

conductor cross-sectional area and makes the loops less symmetric. In addition, the terminal for the AC busbar part is in the center and surrounded by other terminals, which makes it difficult for the connection of the AC output. Therefore, option (a) is not a good placement.

For option (b), the AC output terminal can be extended easily. However, there is a conflict between the neutral bar (lines that connect to Neu) and the middle bars (lines only connected internally not to any terminals). Assuming the neutral bar is in the upper layer while the middle bars are in the lower layer and they overlap with each other, holes have to be drilled on the middle bars so that the neutral bar can be connected with the module terminals. Compared with

(a) and (b), option (c) is more like the schematic of a three-level ANPC converter. Because the line frequency power module is not in the same row with the other two modules, the conflict between busbar parts and module terminals can be avoided. Therefore, option (c) is selected as the placement for further busbar layout. Although the length of the middle bars seems longer than (a) and (b), it is not a critical issue because of the magnetic cancelation effect.

2) Determine the available busbar parts and find the shared parts by multiple switching loops. Here, the loops in the lower half of the phase leg as highlighted in Figure 11.14 are taken as the design example. In terms of the short switching loop, it only includes two busbar parts: the neutral bar and the negative bar. The long switching loop is more complicated and more critical, which consists of four parts: neutral bar, negative bar, and two middle bars. The neutral and negative bars are shared by two switching loops.

3) Place the busbar parts. Since the short switching loop only includes the neutral and negative bars, only one of them can be selected as the returning path, and they should be laminated and located in two layers. Considering that the neutral bar is also shared by the two loops in the upper half of the phase leg, the neutral bar is selected as the returning path, and it is placed in one layer. All the other busbar parts are located in the other layer.

4) Increase the area of returning path to fully overlap with the outgoing paths. Figure 11.18 shows the conceptual 3D view of the busbar layout for the long switching loop. The middle bars (two larger bars on the top layer) and the negative bar (the smaller bar on the top layer) are placed in the same layer. The neutral bar (the bar on the bottom layer) is a whole layer and serves as the returning path of the switching loop. With such a design, the busbar parts are coupled, and their magnetic fields can cancel each other with the opposite current flowing directions, resulting in low loop inductance. The layout of the switching loops in the upper half of the phase leg can follow the same procedure.

Figure 11.19 shows the equivalent circuits of the two switching loops considering the busbar structure. For the short switching loop, the negative and neutral bars are coupled, and their mutual inductance is M_{on}. For the long switching loop, the negative and middle bars are coupled with

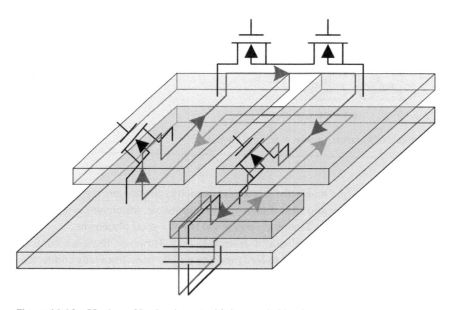

Figure 11.18 3D view of busbar layout with long switching loop.

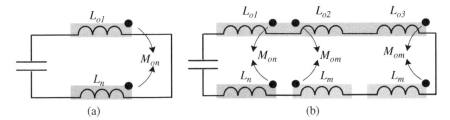

Figure 11.19 Equivalent circuits of the switching loops: (a) the short switching loop and (b) the long switching loop.

the neutral bar. The mutual inductance between the middle and neutral bars is M_{om}. Note that the overlapped area between the negative and neutral bars only accounts for a small portion of the total busbar area. Thus, the effective self-inductance of the neutral bar in the short loop (L_{o1}) is smaller than that in the long loop ($L_o = L_{o1} + L_{o2} + L_{o3}$).

With Figure 11.19, the total loop inductances in (11.7) and (11.8) are modified as:

$$\begin{cases} L_1 = L_C + L_{o1} + L_n - 2M_{on} + 2L_s \\ L_2 = L_C + L_o + L_n + 2L_m - 2M_{on} - 4M_{om} + 4L_s \end{cases} \tag{11.9}$$

In terms of implementation, Kapton sheets and epoxy are selected as the insulation materials. Kapton sheets are mainly used between two layers, while the epoxy is used to seal the edge of the holes and bars. Figure 11.20 shows the cross section of the laminated busbar structure. The distance between two layers is 0.5 mm. The dielectric strength of the Kapton sheets and epoxy is 300 kV/mm and 15.7 kV/mm, respectively, which is sufficient for 1 kV voltage. However, if a higher voltage rating is required, the distance may need to increase, resulting in a larger loop inductance.

Next the capacitor placement is considered. As recommended in [10], the adjacent two paralleled capacitors are placed oppositely as shown in Figure 11.21a. There are 10 capacitors arranged in two rows of five. The top plate is the negative/positive bars, while the bottom one is the neutral bar. Figure 11.21b sketches the top view of the placement of the capacitors on the busbar plate. The solid lines represent the current on the top layer (positive/negative bars), and the dashed lines represent the current on the bottom layer (neutral bar), while the dashed dotted lines are the current inside the capacitors. For the left-side capacitor, the current flows into the capacitor on the bottom layer, while the current flows out of the right-side capacitor through the top layer. Therefore, currents on two laminated plates are in opposite directions, which enhances the magnetic cancelation, reduces the total inductance, and helps even the current distribution.

The finalized exploded view of the busbar design for one phase leg of the three-level ANPC inverter is shown in Figure 11.22. The neutral busbar is on the top layer and covers the area of the bottom layer, where the positive, negative, middle, and AC busbars are placed laterally.

Figure 11.20 Cross section of the laminated busbar layers.

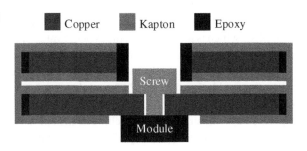

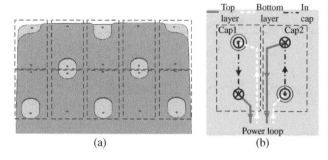

Figure 11.21 Paralleled DC-link capacitors on busbar: (a) designed busbar parts for capacitors; and (b) top view of the busbar with capacitor placement and current flow directions.

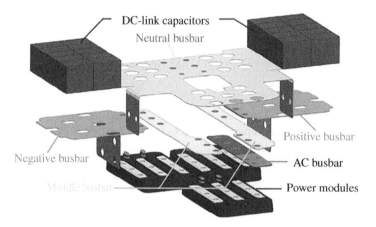

Figure 11.22 Designed laminated busbar for one phase-leg.

11.3.3.3.4 *Busbar Design Validation*   As mentioned earlier, the parasitic parameters of the busbar can be estimated analytically based on equations [6]. However, because of the complicated structure in Figure 11.22 and the irregular shape of the busbar parts, it is difficult to get an accurate result with analytical models. Using FEA tools like Ansys Q3D, the parasitic inductances can be extracted from the 3D drawing. Based on the simulation result, the busbar inductance of the long switching loop is found to be 12.6 nH when the applied frequency is 20 MHz.

Figure 11.23 illustrates the surface current density of the neutral busbar with two designs. Notably, as shown in Figure 11.23a, the current does not flow along the shortest path A in the designed busbar layout; instead, it follows path B, which overlaps with the middle bars. If the area of the neutral busbar is reduced and does not fully overlap with the other bars as shown in Figure 11.23b, the high-frequency current on the returning path (neutral busbar) cannot follow the outgoing path (middle bars). The simulated inductance of the long switching loop is 79.4 nH at 20 MHz, which is more than six times of the value of the designed busbar layout. This comparison verifies the effectiveness of the magnetic field cancelation for the designed busbar layout.

The equivalent circuit of the busbar with parasitics is extracted from Ansys Q3D and simulated in Saber along with the SiC MOSFET module model. The stray inductance of each MOSFET in the half-bridge module is assumed to be 2.5 nH, and we also assume that there is no mutual inductance. Figure 11.24a illustrates the switching transient waveforms of the SiC MOSFET drain-source voltage. The resonant frequency of the short switching loop is 34.5 MHz, while that of the long

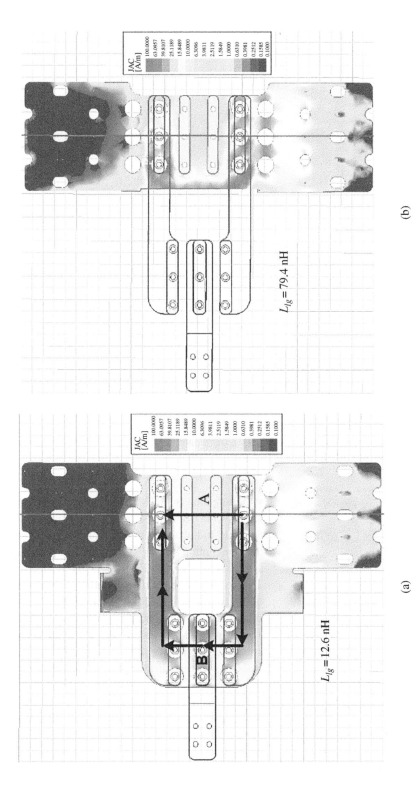

$L_{lg} = 12.6$ nH

(a)

$L_{lg} = 79.4$ nH

(b)

Figure 11.23 Simulated surface current density on neutral busbar: (a) designed busbar layout and (b) layout without full overlap.

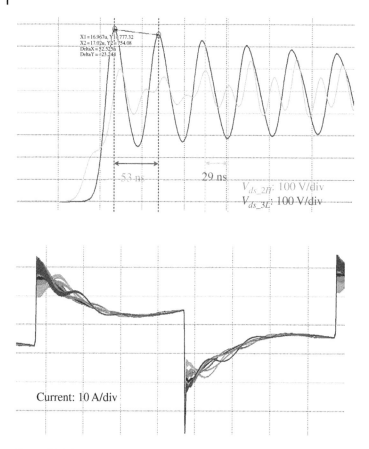

Figure 11.24 Simulated waveforms with module, capacitor, and busbar models: (a) device voltages during switching transient and (b) currents of paralleled DC-link capacitors.

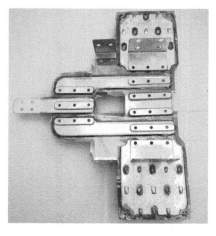

Figure 11.25 A prototype busbar.

switching loop is 18.9 MHz. Based on the output capacitance value from the device module datasheet, the total parasitic inductances of the short and long switching loops including busbar, capacitors, and power modules are calculated as 6.5 nH and 17.5 nH, respectively.

Figure 11.24b plots the simulated current waveforms of 10 paralleled DC-link capacitors in one switching cycle at full load conditions with the busbar model. The current RMS value of each capacitor is about 8 A, and the largest RMS difference among the capacitors is 0.6 A, which indicates good current sharing among the capacitors.

The busbar design has also been validated through prototyping and testing. Figure 11.25 shows a busbar prototype based on the design example. First, the prototype passed the insulation test. The impedance of the busbar loops including the DC-link capacitors but without the power modules is measured with an impedance analyzer. The measured short and long switching loop inductances of the busbar are 2.5 nH and 10 nH, respectively. Considering the inductance of one power module is around 4 nH (2.5 nH per MOSFET assumed earlier), the result matches with the simulation in Figure 11.24.

Figure 11.26 shows the power stage prototype of one phase of the three-phase, three-level ANPC inverter equipped with the fabricated busbar. When the inverter was tested under the rated conditions of 1 kV and 200 kVA at the room temperature with water cooling, the highest temperature of the DC-link capacitors was 38 °C, and the temperature difference among the capacitors was within 5 °C. This indicates good current sharing among capacitors.

Sample measured switching transient waveforms are shown in Figure 11.27. The resonant frequencies of the short and long switching loops are 35.7 MHz and 17.2 MHz, respectively, which

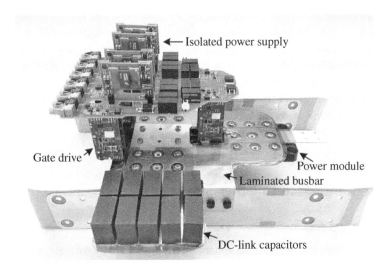

Figure 11.26 Power stage prototype of one phase-leg of the 500-kVA three-level ANPC inverter.

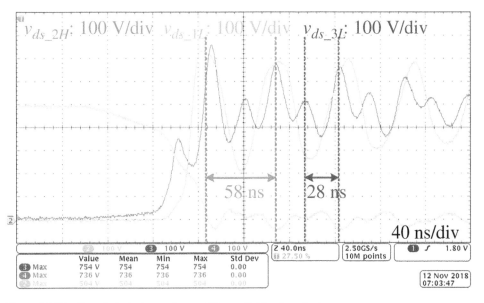

Figure 11.27 Sample tested waveforms of the switching transient.

Table 11.4 Loop parasitic inductance comparison.

References	[8]	[9]	[12]	[13]	[14]	[15]	[16]
Power (kVA)	500	200	1000	750	N/A	N/A	475
DC bus (kV)	1	1.2	2.4	2	N/A	N/A	1.1
Short loop (nH)	6.5	55	N/A	78	96	47.9	95
Long loop (nH)	17.5	135	115	208	150	76.2	118

N/A, not available.

are very close to the simulation results in Figure 11.24. With the low loop parasitic inductances, high dv/dt of the SiC MOSFETS can be tolerated without causing excessive voltage overshoot. The peak observed drain-source voltage of S_{2H} is 736 V and S_{3L} is 754 V, which is lower than the voltage rating of the SiC MOSFET module (900 V). In fact, even operating this inverter at 500 kVA with cryogenic cooling, the voltage overshoot was below 900 V [11].

Table 11.4 compares the loop inductances of the designed and built busbar, which is detailed in [8], with those reported in other NPC-type converters. The designed and built busbar achieves significantly lower parasitic inductances for both short and long switching loops.

11.4 Mechanical System Interface to Other Subsystems

11.4.1 General Classifications

Similar to thermal management system, the mechanical system interface essentially with all other subsystems in the converter, since they all need to be accommodated by the mechanical structure of the converter.

The input conditions to the mechanical system from other subsystems in a three-phase AC converter have been discussed in Section 11.2.3. Here, we will examine the output of the mechanical system needed for the other subsystems.

11.4.1.1 Rectifier and Inverter

The key output from the mechanical system to the inverter and rectifier are busbar parasitic parameters, including parasitic inductances, capacitances, and resistances. Among them, the inductances are design results, while others are design attributes. Another attribute is the temperature range, which refers to surrounding temperature of the inverter/rectifier parts. So far, the inverter/rectifier design directly uses the ambient temperature, which is reasonable when considering device junction temperature and cooling system design. However, many components inside a converter are not in direct contact with the ambient or directly cooled on a heatsink or a cold plate. Understanding surrounding temperature range of these components is necessary for correct selection and design of these components and converter. This is especially important in some applications, e.g. outdoor applications, high-temperature or cryogenically cooled applications.

11.4.1.2 EMI Filter

The EMI filter design will be impacted by parasitics of the converter. For example, the parasitic capacitances of device modules to ground in a three-level ANPC can impact the CM noise, and correspondingly the EMI filter design. The parasitic inductances can also impact EMI noise with their impact on switching transients. Temperature range is also an attribute.

11.4.1.3 Thermal Management System

Certain mechanical system parts can incur loss, specifically, the busbars which carry currents, and can generate loss and heat as discussed earlier. These losses and their distribution should be considered in the overall thermal mechanical system design.

As mentioned in Chapter 10, there is no output from the mechanical system to the control system. On the other hand, the temperature is an attribute that can help select right controller and auxiliary components, e.g. for high-temperature or cryogenically cooled applications.

11.4.2 Discussion

Table 11.5 summarizes all the output from the mechanical system to its interfaced subsystems within the converter system. In essence, there are only four items, parasitic inductances, parasitic capacitances, busbar current distribution and power loss, and temperature for different subsystems and components. They are further discussed here.

11.4.2.1 Busbar (and Loop) Parasitic Inductances

In aforementioned discussion, the parasitic inductance is the design objective for the busbar, therefore, a design result. For simple busbar structure such as the two-layer structure in Figure 11.13, the parasitic inductance can be analytically calculated [6]; for a more practical complex structure, usually, 3D FEA simulation will be needed. However, at the early design stage, FEA simulation approach may be impractical. In this case, an estimated value based on general dimensions, or power and voltage ratings, can be a reasonable alternative. For example, based on the device and capacitor dimensions, the busbar area can be estimated. Based on the voltage rating, the insulation layer thickness can be estimated. Together, the parasitic inductance can be estimated. In lieu of dimensions, power or current rating can be used to estimate the parasitic inductances based on some known designs. The eventual detailed design accurately models the actual busbar, whose design will greatly impact the parasitic inductance.

Table 11.5 Output from mechanical system to its interfaced subsystems.

Subsystem	Interface type	Item	Value
Rectifier/inverter	Constraint/objective	Busbar (and loop) parasitic inductances	Design result and discussion
	Attribute	Parasitic capacitances	Discussion
	Attribute	Temperature range	Discussion
EMI filter	Constraint/objective	Busbar (and loop) parasitic inductances	Design result and discussion
	Attribute	Parasitic capacitances	Discussion
	Attribute	Temperature range	Discussion
Thermal management system	Constraint	Current distribution	Design result
	Attribute	Busbar power loss	Discussion
Control system	Attribute	Temperature range	Discussion

It should be emphasized that the parasitic inductance of the busbar is only part of the switching loop inductance, which also includes the stray inductances of devices, capacitors, and other components in the loop. For simple calculation, we can add up all these inductances to find the total loop inductance. However, the layout can impact the actual value due to mutual inductances among components. One example is the capacitor layout in Figure 11.21 and the corresponding canceled/reduced capacitor parasitic inductances. The simple sum of all parasitic inductances in the loop without considering the cancelation or enhancing effect of mutual inductances can lead to erroneous results.

11.4.2.2 Parasitic Capacitances

Parasitic capacitances are even more difficult to determine analytically, especially the parasitic capacitances between a conductor in converter and the ground. In this case, FEA simulation can be used to determine these parasitics, or they can be estimated based on known converters for different power and voltage ratings.

11.4.2.3 Busbar Current Distribution and Power Loss

In the discussion earlier, we have covered the need to consider the current distribution, not only to achieve a more symmetric distribution which usually leads to lower inductance but also to satisfy the maximum current density constraint which is related to power loss and temperature rise. The busbar power loss, albeit a small portion of the overall converter power loss, should be considered in the overall thermal management system. In some applications, the busbar may be actively cooled such that a smaller (thinner or narrower) busbar can be adopted, such as in a cryogenically cooled converter [8, 11].

In order to determine the power loss from the busbar, simple estimation based on RMS current and busbar DC resistance can be employed. More accurate results can be obtained using FEA simulation, which can consider the actual current distribution and frequency dependency due to skin effect and proximity effect.

11.4.2.4 Temperature Range

Since the temperature distribution depends on mechanical structure and layout, temperature ranges for various components and subsystems can also be most easily determined with FEA simulation.

11.5 Summary

This chapter discusses the selection and design of mechanical parts to complement and support the electrical and thermal components and subsystems, from the viewpoint of an electrical engineer. These mechanical parts for converters include busbars, enclosures (cabinets), shelves and/or drawers, supporting boards and spacers or standoffs, connectors, fasteners, and cables/busses/ wires. Although PCBs are generally not considered as mechanical parts, they play important roles in system layout and therefore, mechanical system design. The main points covered in this chapter are summarized as follows.

1) In addition to their structural support functions, mechanical parts can also impact the converter electrical and/or thermal performance.
2) In principle, the mechanical system design can be formulated as a design optimization problem, similar to other converter subsystems. However, since the mechanical system design almost always involves layout, it is very difficult to be modeled analytically or even algebraically. 3D

FEA simulation is often needed. As a result, the converter mechanical system design often follows a procedure, considering step by step the PCBs, busbars, devices and gate drives, cooling, mechanical layout and support. Given the difficulty to analytically model the overall mechanical system, different mechanical parts are often selected or designed separately.

3) One of the key mechanical parts is the busbar. Although the design goal for busbar is often to minimize its parasitic inductances, the busbar design needs to consider geometry, current density, and electrical insulation. Since busbar layout is critical, FEA simulation, together with analytical models, are needed for the busbar design. A key design method for busbar parasitic inductance reduction is magnetic field cancelation.

4) One busbar layout design example is presented for a three-level ANPC converter also considering the switching loop impact. Utilizing the magnetic field cancelation effect in component placement and geometry design, a low inductance two-layer busbar can be achieved. FEA simulation, analytical inductance model, and circuit simulation are all useful tools for the busbar design.

5) Busbar parasitic capacitances and resistances can be beneficial in damping the switching and EMI noise.

It should be pointed out again that we have not considered mechanical structure design and analysis, which is essential part of the mechanical system design, and could add further constraints on components and parts. For example, when considering structural strength of a busbar, a critical parameter is short circuit current, which can create mechanical force on the busbar.

References

1 IEEE Std C57.12.91-2020. IEEE Standard Test Code for Dry-Type Distribution and Power Transformers.

2 A. D. Callegaro, J. Guo, M. Eull, B. Danen, J. Gibson, M. Preindl, B. Bilgin and A. Emadi, "Bus bar design for high-power inverters," *IEEE Transactions on Power Electronics*, vol. 33, no. 3, pp. 2354–2367, 2018.

3 Mersen, "Design guide formulas," [Online]. Available: https://ep-fr.mersen.com/en/products/engineering/design-guide-formulas. Accessed in 3 October 2022.

4 Copper Development Association, "Ampacities and mechanical properties of rectangular copper busbars: Table 1. Ampacities of copper No. 110," [Online]. Available: https://www.copper.org/applications/electrical/busbar/bus_table1.html. Accessed in 3 October 2022.

5 R. J. Pasterczyk, C. Martin, J.-M. Guichon and J. L. Schanen, "Planar busbar optimization regarding current sharing and stray inductance minimization," in *2005 European Conference on Power Electronics and Applications*, Dresden, Germany, September 11–14, 2005.

6 Z. Piatek, B. Baron, T. Szczegielniak, D. Kusiak and A. Pasierbek, "Inductance of a long two-rectangular busbar single-phase line," *Przegląd Elektrotechniczny (Electrotechnical Review)*, vol. 89, no. 6, pp. 290–292, 2013.

7 H. Gui, R. Chen, Z. Zhang, J. Niu, R. Ren, B. Liu, L. M. Tolbert, F. Wang, D. Costinett, B. J. Blalock and B. Choi, "Modeling and mitigation of multiloops related device overvoltage in three-level active neutral point clamped converter," *IEEE Transactions on Power Electronics*, vol. 35, no. 8, pp. 7947–7959, 2020.

8 H. Gui, R. Chen, Z. Zhang, J. Niu, L. M. Tolbert, F. Wang, B. J. Blalock, D. J. Costinett and B. B. Choi, "Methodology of low inductance busbar design for three-level converters," *IEEE Journal of Emerging and Selected Topics in Power Electronics*, vol. 9, no. 3, pp. 3468–3478, June 2021.

9 Y. Jiao, S. Lu and F. C. Lee, "Switching performance optimization of a high power high frequency three-level active neutral point clamped phase leg," *IEEE Transactions on Power Electronics*, vol. 29, no. 7, pp. 3255–3266, 2014.

10 Cree, "Design considerations for designing with Cree SiC modules part 2. Techniques for minimizing parasitic inductance," 2014. [Online]. Available: https://www.mouser.com/pdfDocs/Cree-Design-Considerations-for-Designing-with-Cree-SiC-Modules-part-2.pdf. Accessed in 6 October 2022.

11 R. Chen and F. Wang, "SiC and GaN devices with cryogenic cooling," *IEEE Open Journal of Power Electronics*, vol. 2, pp. 315–326, April 2021.

12 D. Zhang, J. He and S. Madhusoodhanan, "Three-level two-stage decoupled active NPC converter with Si IGBT and SiC MOSFET," *IEEE Transactions on Industry Applications*, vol. 54, no. 6, pp. 6169–6178, 2018.

13 J. Wang, B. Yang, J. Zhao, Y. Deng, X. He and Z. Xu, "Development of a compact 750kVA three-phase NPC three-level universal inverter module with specifically designed busbar," in *IEEE Applied Power Electronics Conference (APEC)*, Palm Springs, CA, February 21–25, 2010.

14 L. Popova, T. Musikka, R. Juntunen, M. Lohtander, P. Silventoinen, O. Pyrhönen and J. Pyrhönen, "Modelling of low inductive busbars for medium voltage three-level NPC inverter," in *2012 IEEE Power Electronics and Machines in Wind Applications (PEMWA)*, Denver, CO, July 16–18, 2012.

15 L. Popova, R. Juntunen, T. Musikka, M. Lohtander, P. Silventoinen, O. Pyrhönen and J. Pyrhönen, "Stray inductance estimation with detailed model of the IGBT module," in *IEEE Eur. Conf. Power Electron. Appl*, 2013 15th European Conference on Power Electronics and Applications (EPE), Lille, France, September 2–6, 2013.

16 F. He, S. Xu and C. Geng, "Improvement on the laminated busbar of NPC three-level inverters based on a supersymmetric mirror circulation 3D cubical thermal model," *Journal of Power Electronics*, vol. 16, no. 6, pp. 2085–2098, 2016.

12

Application Considerations

12.1 Introduction

In previous chapters thus far, the general aspects related to the design of three-phase AC converters have all been covered, without focusing on applications. Very often, the specific applications also need to be considered as the application conditions may impact the design of a converter. This chapter considers these impacts on three-phase AC converter design. Given that the two most popular applications of the three-phase AC converters are motor drives and grid interface converters, the focus will be on these two applications. For the motor drive, the design of the motor-side filters and other techniques to limit the impact of high dv/dt, CM voltage, and long cables will be discussed in relation with the three-phase PWM inverter. For grid interface applications, in addition to the source-side requirements and conditions similar to the AC-fed motor drive case, others such as grounding, abnormal condition ride-through, and transients-induced overvoltage/overcurrent will also be considered. EMI filters have been covered in Chapter 8 and therefore will be excluded here.

12.2 Motor Drive Applications

Bulk of the materials presented in this book have used motor drives as the application example to facilitate the discussion. A typical three-phase AC-fed motor drive system configuration is shown here again in Figure 12.1. Clearly, in addition to the three-phase rectifier and inverter, the motor drive system also includes the motor load and connecting wires (usually cables) and the three-phase power source. For a DC-fed motor drive, the power source is DC.

The AC three-phase motor load can be an induction motor or a synchronous motor, including the wound field synchronous motor (WFSM) and the permanent magnet (PM) synchronous motor. Whether it is an induction motor or a synchronous motor, the motor drive controls the motor in a similar way, that is, controlling the speed through the drive inverter frequency and controlling the torque through voltage (i.e. motor flux linkage) and current. In addition, both types of the motors are similar in structure, consisting of a stator with stator windings and a rotor with rotor field (windings or PM), even though they realize their rotor field differently. As a result, the motor requirements to and interactions with the drive inverter are similar.

All three-phase motors are designed to expect a symmetrical and balanced three-phase sinusoidal AC voltage or current source, except in the case of some small so-called brushless DC motors (i.e. DC-fed inverter driven motors), where trapezoidal voltages/fluxes are employed. On the other hand, the load-side inverters as designed in Chapter 6 are voltage source inverters and generate

Design of Three-phase AC Power Electronics Converters, First Edition. Fei "Fred" Wang, Zheyu Zhang, and Ruirui Chen.
© 2024 The Institute of Electrical and Electronics Engineers, Inc. Published 2024 by John Wiley & Sons, Inc.

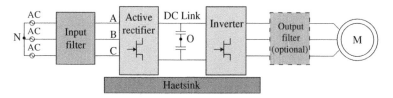

Figure 12.1 A typical AC-fed AC/DC/AC motor drive system.

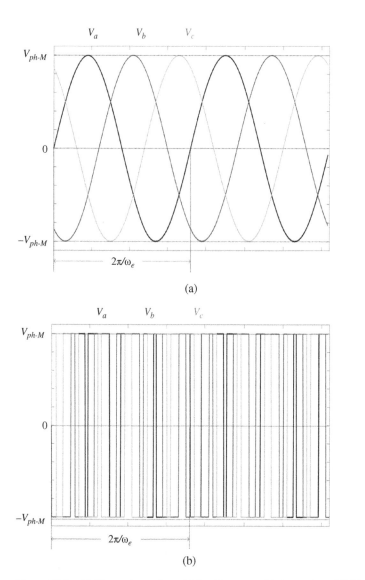

Figure 12.2 (a) Three-phase sinusoidal voltages vs. (b) three-phase PWM voltages.

three-phase PWM voltages as shown in Figure 12.2, where the three-phase voltages are defined as the voltages from the phase terminals to the DC-link midpoint O.

In Figure 12.2a, the phase voltages can be expressed as in (12.1), where V_{ph-M} is the magnitude of the phase voltage and ω_e is the motor electrical frequency. Clearly, the three-phase voltages in

Figure 12.2a are (i) sinusoidal with only one fundamental frequency component, (ii) symmetrical since phase *a* voltage leads phase *b* voltage by $\frac{2\pi}{3}$, which leads phase *c* voltage by $\frac{2\pi}{3}$, and (iii) balanced since $(v_a + v_b + v_c) = 0$:

$$
\begin{cases}
v_a = V_{ph-M} \cdot \sin(\omega_e t) \\
v_b = V_{ph-M} \cdot \sin\left(\omega_e t - \frac{2\pi}{3}\right) \\
v_c = V_{ph-M} \cdot \sin\left(\omega_e t + \frac{2\pi}{3}\right)
\end{cases}
\tag{12.1}
$$

On the other hand, the PWM phase voltages in Figure 12.2 are certainly not sinusoidal and furthermore are also not balanced since in general $(v_a + v_b + v_c) \neq 0$. Under normal steady-state operations, the PWM phase voltages can be considered symmetrical since their fundamental voltages are symmetrical.

With the nonsinusoidal PWM voltages, they can generate harmonic currents in motors; with the unbalanced PWM voltages, they can generate CM voltages. In addition, the pulse-shaped voltage waveforms can have high dv/dt, compared with sinusoidal voltages. All these can adversely impact motors and therefore need to be considered in the motor drive design.

The PWM voltages may also adversely impact cables with harmonics and high dv/dt. In addition, cables tend to have high capacitances, which will be charged and discharged during PWM switching and can lead to extra loss. These also need to be considered in the motor drive design.

This section will discuss the strategies and corresponding designs to mitigate the adverse effect of the PWM voltages on motors and cables.

12.2.1 Harmonics in Motors

12.2.1.1 Harmonic Current
When harmonic voltages are directly applied to the motor terminals, harmonic currents will occur. The adverse effects of the motor harmonic currents include extra winding loss and harmonic torque ripple. Consequently, the harmonic currents must be managed. Typically, motors are designed to tolerate 4–5% of current THD at the rated load.

Figure 12.3 shows a steady-state harmonic equivalent circuit for the induction motor [1], where V_n and I_n are RMS voltage and current of the n-th order harmonic respectively; R_s and L_{ls} are the stator winding resistance and leakage inductance, respectively; R_r and L_{lr} are the equivalent rotor winding resistance and leakage inductance referred to the stator, respectively; L_m is the magnetizing inductance; n is the harmonic order; and s_n is the slip corresponding to the n-th harmonic. s_n can be found as in (12.2), where ω_m is the angular speed of the motor and p is the motor pole pair number.

$$
s_n = \frac{n\omega_e - p\omega_m}{n\omega_e}
\tag{12.2}
$$

It can be seen from Figure 12.3 that the harmonic currents are determined mainly by the leakage inductances. Typically, the total leakage inductance at the rated motor frequency account for about 0.1–0.3 p.u. impedance. For example, a 1-MW, 4160-V, 60-Hz induction motor used in

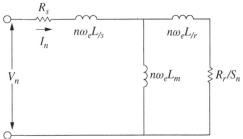

Figure 12.3 Steady-state harmonic equivalent circuit of the induction motor.

a cement plant drive has $X_{ls} = \omega_e L_{ls} = 0.127$ p.u. and $X_{lr} = \omega_e L_{lr} = 0.100$ p.u., where $\omega_e = 2\pi \cdot 60$ rad/s, so the total leakage reactance is 0.227 p.u. at 60 Hz. For the motor drive inverter switching at a modest 1080 Hz, the harmonic number in Figure 12.3 corresponds to $n = 18$. The equivalent impedance due to the total leakage will be approximately 4.0 p.u. at the switching frequency. From Chapter 6, it can be determined that the voltage harmonics near the switching frequency for commonly used PWM scheme is below 10%. Therefore, the corresponding harmonic current is below 2.5%. The voltage harmonics magnitude can increase at high multiples of the switching frequency, but the equivalent impedance will also increase, so the current harmonics will remain small. Moreover, at high frequencies, the winding resistance will also significantly increase due to the skin effect. Therefore, the current harmonics in an induction motor due to PWM switching are generally not a concern, especially, with the trend of employing WBG devices and increased switching frequencies.

For a WFSM, its equivalent circuit model in *dq* reference frame can be given as in Figure 12.4. Note that the fundamental frequency current will be DC for this rotating reference frame model. The harmonic impedance is determined by the subtransient reactance X_d'' and X_q'', which are defined as in (12.3). Although X_d'' and X_q'' are relatively small compared with the synchronous reactance X_d and X_q, their per unit values are generally large enough to limit harmonic currents. For example, a 12 MW, 6600-V WFSM used in a steel mill has a $X_d'' = 0.2530$ p.u. and $X_q'' = 0.3204$ p.u. The motor has a base frequency of 24 Hz and top frequency of 44 Hz. Even though the corresponding Si IGCT motor drive inverter switches at only 540 Hz, the three-level NPC topology makes the equivalent switching frequency doubled to 1080 Hz, which is more than 24 times of the top frequency. Similar to the induction motor example aforementioned, the harmonic currents generally should not be a concern for WFSM.

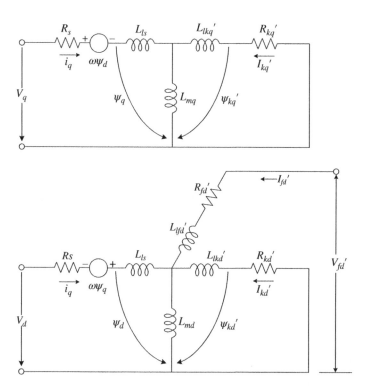

Figure 12.4 Equivalent circuit of the WFSM (*k* and *f* subscripts indicate damper and field windings, respectively, and ' superscript indicates referred to the stator side).

One potential issue worth mentioning is that the harmonic currents in a WFSM are mostly going through the damper windings. Unlike induction motor rotor windings (or rotor bars in the case of the most popular squirrel-cage induction motor), which are designed for the continuous duty, the WFSM damper windings are often designed for transient duty based on generator applications and therefore may have very limited current capability. It is important to control harmonic currents in this case. In most cases, WFSMs used in motor drive applications can get rid of the damper windings if possible.

$$X_d'' = X_{ls} + \frac{X_{md} \cdot X_{lfd}' \cdot X_{lkd}'}{X_{md} \cdot X_{lfd}' + X_{md} \cdot X_{lkd}' + X_{lfd}' \cdot X_{lkd}'} \approx X_{ls} + \frac{X_{lfd}' \cdot X_{lkd}'}{X_{lfd}' + X_{lkd}'}$$

$$X_q'' = X_{ls} + \frac{X_{mq} \cdot X_{lkq}'}{X_{md} + X_{lkq}'} \approx X_{ls} + X_{lkq}' \qquad (12.3)$$

In the case of the PM synchronous motor, there are no damper windings or field windings, and the equivalent circuit in dq reference frame becomes as in Figure 12.5. In this case, the harmonic current will see the synchronous inductance, the same as the fundamental current. On the other hand, the synchronous reactance values in per unit for PM motors are generally smaller than those of WFSMs due to the larger PM motor air gaps. For example, a commercial PM motor has a $X_d = 0.1454$ p.u. and $X_q = 0.1402$ p.u., which are even smaller than X_d'' and X_q'' values of the WFSM aforementioned. However, the PM motor drives are relatively small and switch at relatively high frequency, such that the harmonic currents are also not a big concern.

Figure 12.5 Equivalent circuit of the PM motor.

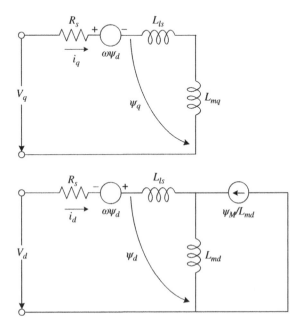

12.2.1.2 Dead-Time Harmonics

While the harmonic currents in motors due to PWM inverter switching is generally not a concern, one exception is the harmonics caused by phase-leg switching dead times. The necessity of a dead time or blanking time in a VSI has been explained in Chapter 10. Also, the mechanism causing harmonics because of dead time has been discussed there and will not be repeated here.

In addition to voltage error and corresponding low-order current harmonics in motor, the dead-time caused voltage distortion can also result in instability and torque ripple in a motor drive [2]. Here, we will introduce the torque ripple issue.

Assume that the motor voltage has an error ΔV with respect to the refence value V^*. Since the torque T is proportional to the voltage square for an induction motor, then we can have

$$T \propto (V^* + \Delta V)^2 \tag{12.4}$$

which, with $\Delta V \ll V^*$, can be approximated to

$$T \approx T_0 \left(1 + \frac{2\Delta V}{V^*} \right) \tag{12.5}$$

where T_0 is the torque corresponding to the case without any voltage error. Therefore, the torque error is

$$\Delta T = T - T_0 \approx T_0 \frac{2\Delta V}{V^*} \tag{12.6}$$

Since V^* is proportional to motor frequency, the torque error is worse at low frequency or low speed.

The dead-time effect should be corrected to mitigate the low-order harmonics, torque error, and potential instability issues. The dead-time compensation methods have been covered in Chapter 10.

12.2.1.3 Harmonic Core Loss

In addition to current harmonics, PWM voltage harmonics can also cause extra core loss in motors. Motor cores and windings for stators and rotors are similar to those for primary and secondary of magnetic transformers. With nonsinusoidal voltages, the flux density in magnetic cores can also be nonsinusoidal, which can lead to extra loss as discussed in Chapter 4 for magnetics.

Reference [3] provides an experimental evaluation of the core loss under PWM voltages in comparison with sinusoidal and square voltages. Since the core loss is largely independent of load, the evaluation was conveniently done at the no-load condition. By measuring the total no-load loss $P_{no-load}$, and calculating stator and rotor winding losses based on current measurement, the core loss can be determined as:

$$P_{core} = P_{no-load} - P_{winding-stator} - P_{winding-rotor} \tag{12.7}$$

The experiment was conducted on a standard induction motor rated at 7.5 kW, 380/220 V, 50 Hz. The core material is silicon steel. The impact of the PWM voltage modulation index, switching frequency, and modulation scheme is evaluated. The switching frequency of the inverter is varied from 1 to 20 kHz. The core loss increase with PWM voltages is found to be from 10 to 80% under various operating conditions, compared to the sinusoidal voltages. The core losses with six-step square voltages are close to that of the sinusoidal voltages. It is further concluded that:

- The modulation index for the PWM voltage has a strong influence on the motor core loss increase. Generally, higher modulation index can lead to lower loss increase.

- The switching frequency does not strongly influence core losses. In fact, with the switching frequency increase, a small reduction of the core losses is observed. Obviously, high switching frequency could lead to high losses in other categories.
- The modulation scheme does not strongly influence the core loss increase. The three commonly adopted modulation schemes (SPWM, SPWM plus third harmonic injection, and SVM) do not change the motor core losses.

Therefore, in motor drive design and applications, the motor core loss increase due to PWM voltages should be considered.

12.2.1.4 Harmonics in Cables

In many motor drive applications, the drive and motor may be far apart, and relatively long cables need to be used. With harmonic currents, extra copper loss due to skin effect needs to be considered. Since the cable carries the same currents as the motor windings, the motor harmonic currents can be used to evaluate the cable copper loss, which also needs to consider the cable harmonic resistances.

Cables usually have relatively high parasitic capacitances both between phase conductors and between conductors and ground. With PWM voltages, the AC cable parasitic capacitances will be charged and discharged every switching cycle, which will lead to extra switching loss. The total power loss for a three-phase cable will be proportional to the switching frequency f_{sw} as:

$$P_{sw-cable} = \frac{K}{2} \cdot f_{sw} \cdot C_{c-ll} \cdot V_{ll}^2 \tag{12.8}$$

where C_{c-ll} is the line-to-line capacitance of the cable and V_{ll} is the line-to-line peak voltage. $K = 6$ for continuous three-phase PWM with six commutations per switching cycle and $K = 4$ for discontinuous PWM with four commutations per switching cycles. In general, $V_{ll} \geq V_{DC}$, as will be discussed later due to the cable transmission line effect. In Si IGBT motor drives, the cable switching loss is relatively small due the low switching frequencies. In WBG device drives, the switching frequency may be significantly increased for other benefits, such as reduced EMI filters. In this case, the cable loss can be important. Another scenario is in medium-voltage (MV) drives with SiC devices, where both the voltage and switching frequency can be high.

Equation (12.8) only applies to cables with three symmetrical phase-conductors. If the cable also has a symmetrical ground conductor, the equation becomes

$$P_{sw-cable} = \frac{K}{2} \cdot f_{sw} \cdot C_{c-ll} \cdot V_{ll}^2 + \frac{K}{4} \cdot f_{sw} \cdot C_{c-lg} \cdot V_{ll}^2 \tag{12.9}$$

where C_{c-lg} is the line-to-ground capacitance of the cable. Note that some of the power in (12.9) may be recovered due to the soft-switching of some of the devices. The remaining will be lost in the inverter switches, load, cables, and other parts of the circuit. The decoupling inductors presented in Chapter 6 can reduce the loss in the inverter, but as already pointed out there, the total loss due to the cable capacitances will not be reduced.

12.2.2 CM Voltage in Motors

CM voltages have been discussed in previous chapters. Specific to motors, CM voltages can have detrimental effects in the form of shaft voltage and bearing current. These effects must be considered in the motor drive application design, either in the inverter, the motor, or through an external filter.

As discussed earlier, the three-phase PWM voltages are generally symmetrical but unbalanced. By inspecting the motor drive system configuration in Figure 12.1, the DC-link midpoint O is either the actual balancing point of the drive system or can be considered as the balancing or reference point. The PWM drive inverter, in generating a balanced three-phase positive-sequence voltage, which is DM voltage, also inherently produces a zero-sequence voltage, which is CM voltage. For a given load voltage level at a given frequency, the load ABC phase voltages V_{AO}, V_{BO}, and V_{CO} (and their waveforms) are determined with respect to the DC-link midpoint O and can be decomposed into

$$
\begin{aligned}
V_{AO} &= V_{AN} + V_{NO} \\
V_{BO} &= V_{BN} + V_{NO} \\
V_{CO} &= V_{CN} + V_{NO}
\end{aligned}
\tag{12.10}
$$

where V_{AN}, V_{BN}, and V_{CN} are balanced positive-sequence voltages that always sum to zero, and V_{NO} is the equivalent CM zero-sequence voltage source of the system and can be determined through

$$
V_{NO} = \frac{V_{AO} + V_{BO} + V_{CO}}{3}
\tag{12.11}
$$

The magnitude and waveform of V_{NO} depend on the topology and PWM strategy. For example, for a two-level VSI, its peak CM voltages are positive DC-link potential and negative DC-link potential, which correspond to the two zero states of (ppp) or (+++) and (nnn) or (−−−). For a three-level NPC VSI, its space vector diagram is shown in Figure 12.6. As can be seen, in low modulation index region, or for low-frequency operation for a motor drive, corresponding to inner hexagon of the space vector diagram, each PWM switching cycle goes through zero states origin (+++) or (−−−), resulting in a peak CM voltage equal to positive DC-link potential or negative DC-link potential according to (12.11). For high-frequency operation, or in high modulation index region, corresponding to outer hexagons in Figure 12.6, the maximum peaks for CM voltages are two-thirds of the positive or negative link potential, corresponding to the six small vector redundant switching states: (0−−), (++0), (−0−), (0++), (−−0), and (+0+). In other words, multilevel converters can

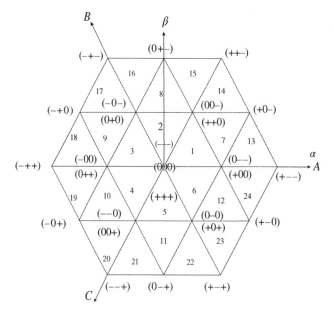

Figure 12.6 Space vector diagram of the three-level PWM modulator.

reduce CM voltages under some conditions compared to the two-level converter; however, in motor drive applications, the peak CM voltage will not decrease because of the low-frequency region, as long as the DC-link voltage stay constant throughout the frequency or speed range.

Based on (12.10) and (12.11), the drive inverter-motor side can be electrically represented by Figure 12.7, where the motor is represented by a balanced three-phase (stator) winding. With the inverter DC-link midpoint O grounded through an impedance Z_{NO} while the motor neutral nominally ungrounded, the midpoint potential to ground practically equals to the CM voltage V_{NO}, generated by the PWM inverter. Consequently, the motor neutral point potential largely reflects how much CM voltage actually falls on the motor.

The stator CM voltage can be coupled to the stator or motor frame and the rotor through the parasitic or stray capacitances as illustrated in Figure 12.8. The motor frame is usually grounded for safety reasons, so no CM voltages can appear on the stator frame, although the frame can provide a CM current path, which can contribute to EMI.

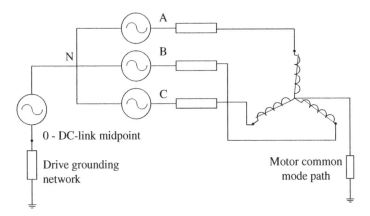

Figure 12.7 CM circuit of the motor drive system (inverter side).

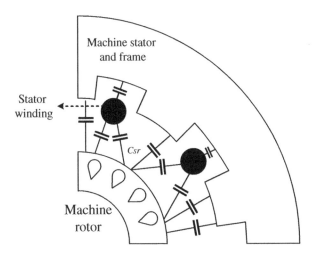

Figure 12.8 Motor stator winding capacitance coupling to stator frame and rotor.

On the other hand, CM voltage can appear on the rotor through capacitive coupling. Although the coupling is through air gap and generally quite weak, the voltage level can be significant, since the peak value of the system CM voltage V_{NO} can be equal to the positive and negative DC-link potential as analyzed earlier.

Since the rotor is directly mounted on the motor shaft, which is usually made of steel, the voltage on the rotor is also the shaft voltage. On the shaft, there are also mechanical loads, gearboxes, etc. If the shaft is exposed and the shaft voltage is sufficiently high, it may cause arcing and sparks, which can be hazardous for certain environments, such as in mining applications.

Shaft voltage can also cause bearing currents. Motor bearings are attached to the motor frame and the motor shaft. If there is sufficient voltage buildup across motor bearing between motor shaft and frame, it is possible to have bearing currents that can damage the bearing. Figure 12.9 illustrates the motor stator, rotor, shaft, and bearings in a setup to measure the shaft voltage and bearing current. Note that the bearing insulators are not always the part of the motor design and are added here to facilitate measurement.

Motor shaft voltage and bearing current problems have long been recognized even under 50- or 60-Hz sinusoidal voltages with earlier concerns mainly on circulating bearing currents due to magnetic asymmetry in motors, which led to the practice of insulating one bearing in large motors [4] (e.g. >200 kW). The circulating bearing currents refers to the currents circulating between the two bearings through the stator frame and the rotor shaft. As seen in Figure 12.9, one bearing insulator is sufficient to block circulating bearing current. The PWM VSI-generated CM voltages result in new sources and mechanisms of producing shaft voltages and bearing currents. As summarized in [5, 6], there are three general types of motor bearing currents associated with PWM VSI drives:

1) Bearing current due to discharge of equivalent capacitance between the motor shaft (rotor) and the motor frame (stator) when bearing lubricant oil film breaks down, also termed as electric discharge machining (EDM) current;
2) Bearing current due to dv/dt in CM voltages when the bearing provides a low capacitive impedance path, i.e. displacement current;
3) Circulating bearing current due to magnetic flux resulting from CM currents.

The bearing related failures are reported to account for a large percentage of all motor failures (51% according to [7]), and damages and failures due to excessive bearing currents have been reported in

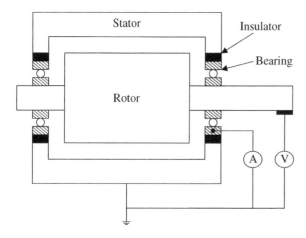

Figure 12.9 Illustration of shaft voltage and bearing current measurement.

PWM VSI drive applications. With motor drives being used in more applications requiring higher power, higher frequency, and/or higher voltage, while also adopting faster switching WBG devices, the potential shaft voltage and bearing current issues can become more pronounced. It is important to understand these potential issues and apply corresponding mitigations in the design and operation of the motor drive system.

12.2.2.1 Shaft Voltage Evaluation

The first step to understand the shaft voltage and bearing current issue is to evaluate the shaft voltage. Given the system CM voltage V_{NO} determined in (12.11), the shaft voltage can be evaluated based on the equivalent circuit model in Figure 12.10 [8, 9]. In this model, C_b and Z_b represent bearing; V_{ms} is the CM voltage on the motor stator winding (i.e. the neutral); C_{sf}, C_{sr}, and C_{rf} are mutual capacitances between stator and frame, stator and rotor, and rotor and frame, respectively; Z_{dg} is the drive DC-link midpoint grounding impedance; and Z_{pr} is the optional paralleled grounding impedance at the motor terminal. This model, together with the CM circuit model in Figure 12.7, depicts the shaft voltage and bearing current phenomenon in a three-phase PWM VSI motor drive system.

In order to determine the shaft voltage, it is necessary to know the motor parasitic capacitances C_{sf}, C_{sr}, and C_{rf}. In principle, these capacitances are mainly related to the motor geometry, which is related to the power and speed rating. Figure 12.11 shows the calculated parasitic capacitances of motors of different power ratings from 5 to 900 Hp, based on design data for four-pole, 460-V induction motors and associated bearing dimensions [10].

In an example, the shaft voltage and bearing current of a motor driven by a three-phase three-level NPC inverter were measured using the setup in Figure 12.9 [11]. The motor used is a 60-Hz, six-pole, 2000-Hp, 2300/4160-V induction motor whose windings can be either Δ- or Y-connected. As noted earlier, high-power, MV motors usually have one of their bearings insulated from the stator frame to avoid circulating bearing currents, which is the case for this test motor. So, the focus can be on the noncirculating-type bearing currents, i.e. types (1) and (2) as identified earlier. To facilitate bearing current measurement, the second bearing was also insulated in this test, as shown in Figure 12.9. The shaft voltages were measured from the shaft to the grounded motor frame, while the bearing currents were measured through grounding leads of the bearing. Table 12.1 summarizes some of the important parameters for the test. In order to understand the CM voltage distribution, the motor stator winding neutral voltage needs to be measured. When the motor neutral is inaccessible, an artificial neutral point can be used to access the motor neutral potential. In the test, the neutral point of a three-phase *RC* filter installed at the motor terminal was used for this purpose.

Figure 12.10 CM circuit model for noncirculating bearing current.

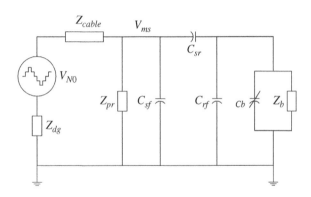

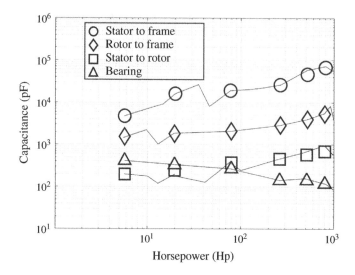

Figure 12.11 Motor parasitic capacitances for different ratings. *Source:* From [10]/ with permission of IEEE.

Table 12.1 Nominal test conditions and parameters.

Motor	60-Hz, six-pole 2000-Hp, 2300/4160 induction motor
Drive inverter	2300-V, 2-MW, three-level NPC based on 1200 A Si IGBTs
DC link	Rectifier supplied, 3500 V nominal
Switching	Continuous SVM, 1500 Hz
Grounding	DC-link midpoint impedance grounded

The cabling between the drive and motor and the system grounding also affect CM voltage distribution and CM current path. In the test, a 5-kV nonshielded three-phase cables were used. The grounding wire of the cable and its sheath were grounded at both the drive and motor ends. The drive inverter itself has a single grounding point through DC-link midpoint grounding impedance, which will be discussed later in this section. The motor was isolated electrically from ground with only its frame grounded.

The first important observation is the relation between the shaft voltage and the motor neutral voltage. The motor shaft voltage results from direct capacitive coupling from the stator winding CM voltage, i.e. motor neutral voltage in this case. This is evident from the identical shapes of shaft voltage and the corresponding motor neutral voltage waveforms in Figure 12.12. These waveforms corresponded to the motor operating at 3 Hz, i.e. modulation index equal to 0.05. Clearly, with this low modulation index, the NPC operates as a two-level inverter, i.e. within the inner hexagon of the space vector diagram as in Figure 12.6. Given that the PWM inverter has a DC-link voltage of 3500 V and switches at 1500 Hz, the total system CM voltage V_{NO} should have a peak-to-peak voltage around 3500 V and a frequency of 750 Hz. Referring to Figures 12.7 and 12.10, since the drive DC-link midpoint is impedance grounded and the motor neutral is open, it can be expected that most of the V_{NO} should fall on the motor neutral. It can be seen from Figure 12.12 that the motor neutral voltage has a peak-to-peak value around 2500 V at a frequency of 750 Hz. As explained earlier, this low modulation index operating condition corresponds to the worst-case scenarios in terms of CM voltage level. The peak-to-peak shaft voltage is around 120 V.

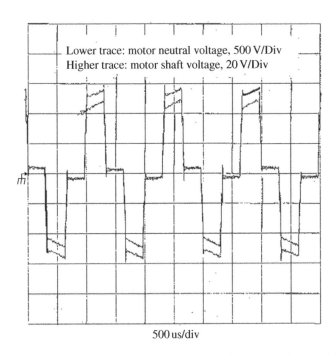

Lower trace: motor neutral voltage, 500 V/Div
Higher trace: motor shaft voltage, 20 V/Div

500 us/div

Figure 12.12 Shaft voltage vs. motor neutral voltage at low modulation index case.

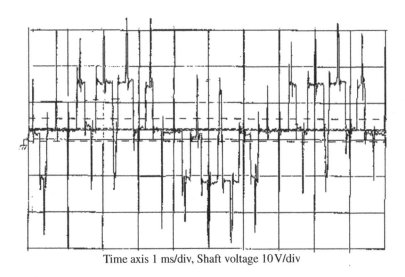

Time axis 1 ms/div, Shaft voltage 10 V/div

Figure 12.13 Shaft voltage at a high modulation index case.

At higher modulation index, the CM voltages were observed lower. Figure 12.13 shows the motor shaft voltage waveform at 57 Hz (modulation index 0.95), with its peak-to-peak around 60 V. Given that the peak CM voltage at high modulation index should be two-thirds of that at low modulation index, the actual peak-to-peak shaft voltage should be higher than 60 V in this case. However, with very high modulation index, the dwell times at the small vectors are very short and their peak voltages are easily missed during the measurement. Notice the distinct three-level steps in this waveform.

12.2.2.2 Bearing Current Evaluation

As discussed earlier, there are three types of bearing currents associated with the CM voltages generated by PWM voltages of the VSI, of which, the noncirculating bearing currents, i.e. the EDM and dv/dt bearing currents, are more of a concern for high-power motors.

The basic mechanism for harmful bearing current is the existence of equivalent capacitance C_b as shown in Figure 12.10. When the motor rotates at a speed, the bearing resistance will be high in the megaohm range [9]; and, as speed increases, lubricant oil can form a film, creating an insulating boundary between the bearing race and the ball. Therefore, a capacitance C_b is created, and the shaft voltage or bearing voltage can build up. When the shaft voltage is above a threshold to breakdown the insulating lubricant oil, the EDM bearing current can occur. Even without breakdown, high dv/dt on C_b can produce bearing current, which, even though at lower magnitude than the EDM current, can cause lubricant chemical change, leading to increased bearing wear and eventual failure. In general, there are three conditions for bearing currents, especially, EDM currents to occur: (i) a source, which is provided by the CM voltage on motor neutral; (ii) a capacitive coupling mechanism, accomplished by C_{sr}; and (iii) sufficient rotor voltage buildup on rotor depending on the existence of C_b.

The threshold voltage V_{rf-th} depends on the lubricant thickness and bearing surface roughness [5]. V_{rf-th} is 0.2–1 V under 60-Hz sinewave operation [6]. For PWM voltages, V_{rf-th} was observed to be from around 10 to 35 V peak, depending on the lubricant film thickness, which is related to the operating condition, such as temperature. Higher temperature leads to thinner film, and therefore lower V_{rf-th}. Hence, to avoid the harmful EDM bearing current, the shaft voltage should be controlled to be below 10 V peak.

The shaft voltage depends on motor parasitic capacitances, which are defined in Figure 12.10. Among them, the bearing capacitance C_b is a function of the lubricant film thickness and therefore is a variable. In general, C_b is a function of bearing load, velocity, and temperature, as well as lubricant permittivity and viscosity. The exact value of C_b is difficult to determine but can be estimated within a range. Since low range C_b will result in high shaft voltage, it can be used to estimate the shaft voltage as a conservative approach.

The bearing current effect on bearing life depends on the current magnitude as well as contact area. In fact, it is found that the bearing life degradation is related to current density. Since EDM currents generally have higher peaks, they tend to have higher current density and are more detrimental to bearing life. In [12], it states that dv/dt currents have much lower peaks and do not degrade bearing life. It also recommends that the bearing current peak density to be below 0.8 A/mm^2 to avoid the EDM current. On the other hand, the conclusions were based on test results from slow IGBTs in 1990s, and faster dv/dt from using WBG devices can potentially increase the dv/dt current peaks to be on the same order as those of the EDM currents if no mitigation measures are taken.

Here, some examples of the bearing current test results are shown for the same drive-motor setup for the aforementioned shaft voltage tests. Figure 12.14 shows the waveform of bearing current due to the bearing oil film breakdown. This measurement was taken when the motor ran at 3 Hz, corresponding to the same worse-case scenario as in Figure 12.12: low modulation index, and high CM and shaft voltages. As observed by others in low-voltage (LV) PWM VSI drives, the discharge bearing currents are random in nature and can have very high peak values and decay very rapidly in an oscillatory manner, indicating a second-order LC resonance between the equivalent capacitance and equivalent inductance. As described in [12], the equivalent capacitance is mainly the motor CM equivalent capacitance, which is equal to $C_{sf} \| (C_{sr} + C_{rf} \| C_b)$, i.e. the stator-to-frame capacitance in parallel with the series combination of the stator-to-rotor capacitance and the parallel combination of the rotor-to-frame and bearing capacitances; the equivalent inductance

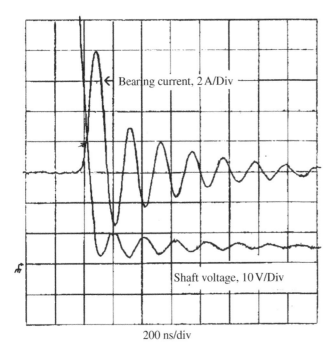

Figure 12.14 shows the bearing current, 2 A/Div and Shaft voltage, 10 V/Div at 200 ns/div.

Figure 12.14 EDM bearing current vs. shaft voltage.

consists of the stator winding CM inductance and other CM inductances in the loop, e.g. CM choke of the EMI filters (Chapter 8) and the decoupling inductors (Chapter 6). The current peak in Figure 12.14 is around 8 A, which is above the level known to cause bearing degradation.

Figure 12.15 shows a more complete picture of the shaft voltage behavior when discharge bearing currents occur. It first quickly goes to zero and then reverses its polarity, apparently no longer following the motor neutral voltage. In fact, this is a voltage level shift phenomenon due to the bearing impedance change as a result of discharge bearing currents, which was also observed with two-level LV drives [8, 13]. It is also worth noting that this case corresponds to a low impedance grounding at the inverter DC-link midpoint, leading to a higher motor neutral voltage than the case of Figure 12.12, with the peak-to-peak voltage close to the full DC-link voltage of 3500 V, compared with 2500 V in Figure 12.12.

Figure 12.16 shows the dv/dt bearing currents together with the motor neutral and shaft voltages. Clearly, the currents are a result of the IGBT switching. The peak values of these capacitive currents are on the order of tens to hundreds of milliampere, much smaller than discharge currents, though they occur every time an IGBT switches. If higher dv/dt WBG devices are used, the dv/dt bearing currents could also cause bearing life degradation.

12.2.2.3 Shaft Voltage and Bearing Current Mitigation

The aforementioned analysis and test results have shown that motor shaft voltages and bearing currents need to be considered, and the potential issues need to be mitigated in the design and operation of motor drive systems. Understanding the mechanisms for shaft voltage and bearing current generation, there are five general approaches in solving the motor shaft voltage and bearing current problems: (i) eliminate or reduce system CM voltage V_{NO} by using modulators or topologies with lower CM voltages, which in general will add complexity and/or sacrifice other performance;

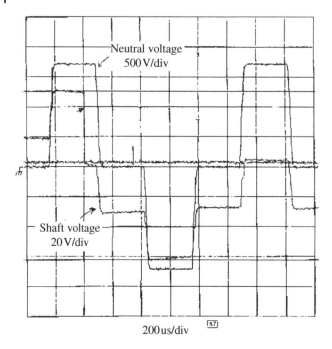

Figure 12.15 Shaft voltage collapse as EDM current occurs.

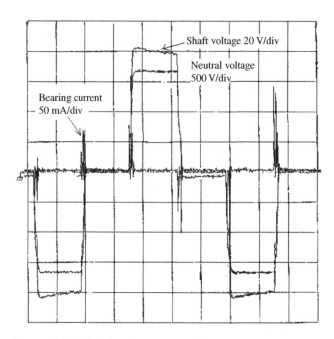

Figure 12.16 *dv/dt* bearing currents with motor neutral and shaft voltages.

(ii) eliminate or reduce coupling capacitance C_{sr} such as using electrostatic shields between the stator and the rotor, which is a motor-based solution and requires special motors; (iii) eliminate motor shaft voltage V_{mr} by grounding the motor shaft, which is very effective in eliminating shaft voltage and bearing current but would require often undesirable grounding brushes; (iv) eliminate bearing

Table 12.2 Results with various bearing current reduction schemes (all values are peak values).

Cases	Motor neutral ground	Motor filter present	Motor filter neutral ground	Drive midpoint ground	Shaft ground brush	Motor neutral voltage	Motor shaft voltage	Motor bearing current
1	No	No	N/A	Impedance	No	1650 V	66.4 V	>10 A
2	No	Yes	No	Impedance	No	1400 V	54 V	10 A
3	No	No	N/A	Impedance	Yes	1650 V	30 mV	75 mA
4	Solid	Yes	No	No	No	0	25 V	100 mA
5	3.3 kΩ resistor	Yes	No	No	No	1200 V	40 V	2.4 A
6	No	Yes	Solid	No	No	600 V	28 V	300 mA
7	No	Yes	No	No	No	1050 V	50 V	1.1 A

current by insulating both motor bearings, which requires a special motor but may be a relatively inexpensive approach, though the shaft voltage issue remains; and (v) eliminate or reduce motor neutral voltage by redesigning CM circuitry. The last approach includes: (a) solidly grounding the motor neutral or grounding the motor neutral through an impedance (e.g. properly chosen resistor or capacitor), which will require motor neutral accessibility; (b) grounding other neutrals in the system, e.g. grounding the motor terminal Y-connected RC filter neutral; and (c) increasing the drive grounding impedance.

To better illustrate their effectiveness, test results for some of these approaches on the same three-level NPC-driven motor described earlier are summarized in Table 12.2, where the first two cases are baseline cases without any mitigations. Note that even solidly grounding the motor winding neutral cannot completely eliminate the motor shaft voltage and bearing currents. On the other hand, it may not be necessary to totally eliminate the shaft voltage and bearing currents as long as they are below the threshold to pose operation hazard (e.g. arcing) or degrade bearing life.

12.2.3 Switching *dv/dt* Impact on Motors

In addition to the impact on CM currents and bearing currents, high *dv/dt* voltages can also adversely impact the motor and cable insulation. There are two main mechanisms for *dv/dt* to impact insulation, one is to cause overvoltage at the motor terminal and along the cable, and the other is to create nonuniform voltage distribution in motor windings.

12.2.3.1 Motor Terminal Overvoltage

When the inverter is connected to the motor through a cable, the voltage and current travels between the inverter and the motor through the cable as electromagnetic waves. From the transmission line theory, there will be reflection and refraction phenomena at the motor terminal due to the characteristic impedance difference between the cable and the motor.

Figure 12.17 illustrates the voltage and current traveling wave reflection and refraction at the boundary between two transmission media, with characteristic impedance Z_A and Z_B, respectively.

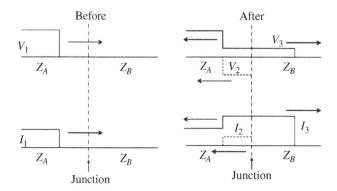

Figure 12.17 Illustration of voltage and current traveling wave reflection and refraction.

With the incoming voltage V_1 and current I_1, the reflection voltage V_2, reflection voltage I_2, refraction voltage V_3, and refraction current I_3 can be determined as in (12.12):

$$V_2 = \left(\frac{Z_B - Z_A}{Z_A + Z_B}\right) V_1$$

$$I_2 = \left(\frac{Z_A - Z_B}{Z_A + Z_B}\right) I_1$$

$$V_3 = \left(\frac{2Z_B}{Z_A + Z_B}\right) V_1 \tag{12.12}$$

$$I_3 = \left(\frac{2Z_A}{Z_A + Z_B}\right) I_1$$

If the cable terminal is shorted, then $Z_B = 0$, the reflection voltage $V_2 = -V_1$, and the terminal voltage $V_3 = 0$. If the cable terminal is open, then $Z_B \to \infty$, the reflection voltage $V_2 = V_1$, and the terminal voltage $V_3 = 2V_1$, corresponding to a full reflection. In the case of the cable connected with the motor, the motor winding impedance is inductive and much higher at high frequency than the cable characteristic impedance, making the condition close to a cable terminal open circuit. Therefore, with a motor driven by an inverter with a connecting cable, the motor terminal voltage can double the inverter terminal voltage, so-called the voltage doubling effect. If we define the reflection coefficient Γ as in (12.13), $\Gamma = 1$ indicates full reflection and terminal voltage doubling. In reality, Γ will be less than 1 for a cable and a motor case, and the terminal voltage can be determined as in (12.14).

$$\Gamma = \left(\frac{Z_B - Z_A}{Z_A + Z_B}\right) \tag{12.13}$$

$$V_3 = V_1 + V_2 = (1 + \Gamma) \tag{12.14}$$

The aforementioned discussion is for instantaneous voltage and current. For a pulse voltage with a certain peak at the inverter terminal, the corresponding peak voltage at the motor terminal depends on the cable length and the dv/dt. Assuming that the cable length is l_c, and voltage traveling wave velocity is v, then the traveling time from the inverter terminal to the motor terminal will be $t_p = \frac{l_c}{v}$. Figure 12.18 shows the motor terminal voltage traveling waves with different rise times. The voltage is normalized based on the inverter terminal voltage, and the rise time t_r is measured in terms of t_p. Note that $\Gamma = 1$ is assumed.

Figure 12.18 shows four cases. In both $t_r = t_p$ and $t_r = 2t_p$ cases, the motor terminal voltage reaches 2.0 before the first reflected wave starts to come back negative due to the low impedance at inverter terminal (the inverter is voltage-controlled and therefore its terminal can be considered as short circuited). In $t_r = 3t_p$ case, the first reflected wave makes the peak motor terminal start to drop at $t = 3t_p$, and therefore the peak motor terminal voltage is $2 \times \frac{2}{3} = \frac{4}{3}$. In $t_r = 4t_p$ case, the first reflected wave also makes the peak motor terminal start to drop at $t = 3t_p$; but in this case, the peak motor terminal voltage is $2 \times \frac{1}{2} = 1$, i.e. no overvoltage. In summary, for a motor connected to a two-level inverter, the peak motor terminal line-to-line voltage can be found by (12.15), which is

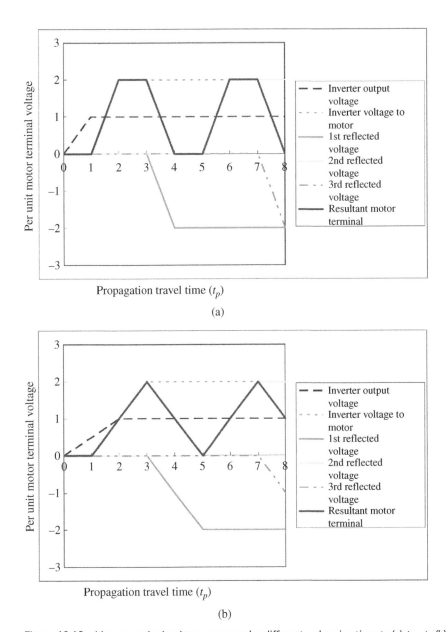

Figure 12.18 Motor terminal voltage wave under different pulse rise time t_r. (a) $t_r = t_p$ (b) $t_r = 2t_p$ (c) $t_r = 3t_p$ (d) $t_r = 4t_p$.

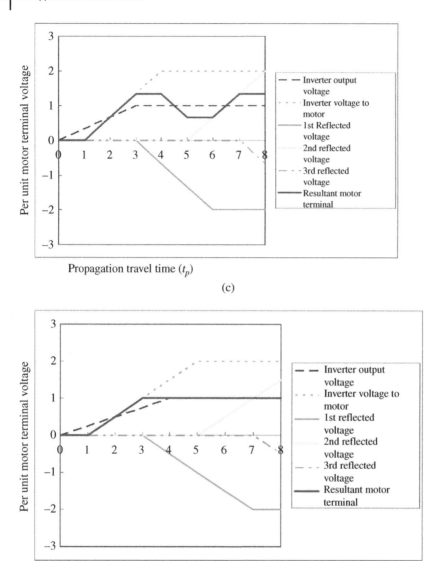

Figure 12.18 (Continued)

slightly different from those given in some references such as in [14], as the traveling waves are considered here beyond $t = 3t_p$. Note that (12.15) has a discontinuity at $t_r = 4t_p$ if $\Gamma < 1$ due to approximations involved in wave reflection and refraction analysis. Essentially, we should not worry about overvoltage when $t_r > 4t_p$.

$$V_{ll-pk} = \begin{cases} (1 + \Gamma), & \text{for } t_r \leq 2t_p \\ \dfrac{2l_c}{v \cdot t_r}(1 + \Gamma) = \dfrac{2t_p}{t_r}(1 + \Gamma), & \text{for } 2t_p < t_r \leq 4t_p \\ 1, & \text{for } t_r > 4t_p \end{cases} \tag{12.15}$$

From Figure 12.18 and Eq. (12.15), it is clear that for a given t_p, which is determined by the cable length for a given v, V_{ll-pk} is determined by t_r. In other words, for a given t_r, there is a critical cable length for the voltage doubling effect. As an example, assuming $\Gamma = 0.9$, $v = \frac{c}{3}$ (c is speed of light), and $t_r = 0.5$ μs, the critical cable length for a full-voltage doubling effect is

$$l_c = \frac{v \cdot t_r}{2} = \frac{0.5 \times 10^{-6} \times 10^8}{2} = 25 \text{ m}.$$

If we wish to have the voltage overshoot to be less than 20% for this 25 m cable, then the rise time should be

$$t_r \geq \frac{2l_c}{v \cdot V_{ll-pk}}(1 + \Gamma) = \frac{2 \times 25}{10^8 \cdot 1.2}(1 + 0.9) = 0.79 \text{ μs}.$$

For the two-level converter, the line-to-line voltage step for each switching is the full DC-link voltage V_{DC}. For a N-level converter, the voltage step is $V_{DC}/(N-1)$. The voltage doubling effect will be just for doubling the voltage step. For a three-level converter, the peak motor terminal line-to-line voltage becomes

$$V_{ll-pk} = \begin{cases} \left(1 + \dfrac{1}{2}\Gamma\right), & \text{for } t_r \leq 2t_p \\ \dfrac{2l_c}{v \cdot t_r}\left(\dfrac{1}{2} + \dfrac{1}{2}\Gamma\right) + \dfrac{1}{2} = \dfrac{t_p}{t_r}(1 + \Gamma) + \dfrac{1}{2}, & \text{for } 2t_p < t_r \leq 4t_p \\ 1, & \text{for } t_r > 4t_p \end{cases} \tag{12.16}$$

A smaller t_r or higher dv/dt will result in a shorter critical cable length, and therefore will more easily cause overvoltage problems.

Once the pulse voltage appears at the motor terminal, it will stress the motor insulation. Motor insulation systems consist of ground wall insulation and winding turn insulation. The stress on ground wall insulation is determined primarily by winding voltage level (magnitude), while the stress on turn insulation is also a function of voltage rise time. The turn insulation is more vulnerable to voltages with steep fronts due to initial nonuniform voltage distribution in winding coils. In recognizing these facts, ANSI/IEEE Standard 522-2004 specifies a motor winding coil voltage capability by an envelope as shown in Figure 12.19 [15], where the lower voltage capability with fast

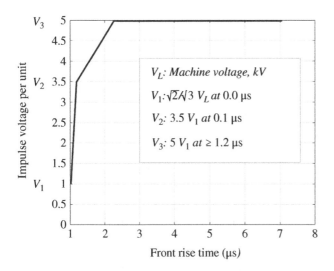

Figure 12.19 Coil impulse voltage withstand envelope.

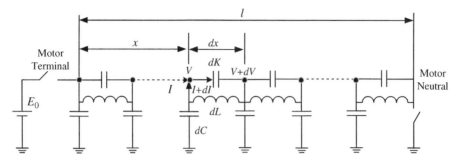

Figure 12.20 Distributed winding circuit model.

fronts actually concerns with turn insulation, while the higher voltage capability (>5 μs rise time) defines the ground wall insulation [16].

The initial voltage distribution in a motor winding can be analyzed according to the well-developed model for transformer windings [17]. Figure 12.20 shows a distributed circuit model for a motor winding with distributed winding inductance L, winding to ground capacitance C, and winding turn-to-turn capacitance K. dL, dC, and dK represent elemental values corresponding to an elemental winding length $dx \cdot dV$ and dI represent elemental winding-to-ground voltage V and winding current I changes over elemental dx.

The basic equations are

$$dL = L_0 \, dx, dC = C_0 \, dx, dK = \frac{K_0}{dx} \tag{12.17}$$

where L_0, C_0, and K_0 are per unit length values. Assume that a voltage E_0 is suddenly applied to a winding of length l. Initially, inductance branch can be assumed open. Figure 12.21 shows the equivalent circuit for determining the initial voltage distribution, where the effect of the winding inductance and resistance are both ignored, and the switch at the neutral point is to indicate different connection conditions.

At a point x on the wining, the voltage and current on the capacitance $\frac{K_0}{dx}$ has the relation:

$$I = -\frac{K_0}{dx} \frac{\partial (dV)}{\partial t} = -K_0 \frac{\partial}{\partial t} \left(\frac{dV}{dx} \right) \tag{12.18}$$

The voltage and current relation for the capacitance ($C_0 \cdot dx$) is

$$dI = -C_0 \cdot dx \cdot \frac{\partial V}{\partial t} \quad \text{or} \quad \frac{dI}{dx} = -C_0 \frac{\partial V}{\partial t} \tag{12.19}$$

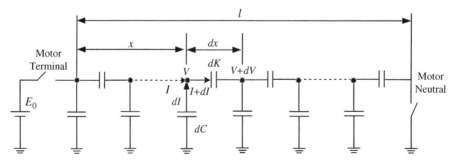

Figure 12.21 Equivalent winding circuit for determining the initial voltage distribution.

Differentiating (12.18) with respect to x yields

$$\frac{dI}{dx} = \frac{d}{dx}\left[-K_0\frac{\partial}{\partial t}\left(\frac{dV}{dx}\right)\right] = -K_0\frac{\partial}{\partial t}\left(\frac{d^2V}{dx^2}\right) \tag{12.20}$$

Combining (12.19) and (12.20) to eliminate current i, we can obtain

$$\frac{d^2V}{dx^2} = \frac{C_0}{K_0}V \tag{12.21}$$

The general solution for (12.21) is

$$V = Ae^{\alpha x} + Be^{-\alpha x} \tag{12.22}$$

with

$$\alpha = \sqrt{\frac{C_0}{K_0}} \tag{12.23}$$

The coefficients A and B can be found with boundary conditions. For example, if the end of the winding, i.e. the motor neutral is grounded (the switch at the neutral terminal in Figure 12.21 is closed), the boundary condition is

$$\begin{cases} x = 0, & V = E_0 \\ x = l, & V = 0 \end{cases} \tag{12.24}$$

The initial voltage distribution in the motor winding can be found as:

$$V = E_0\frac{e^{-\alpha l}}{e^{\alpha l} - e^{-\alpha l}}\left[e^{\alpha(l-x)} - e^{-\alpha(l-x)}\right] = E_0\frac{\sinh[\alpha(l-x)]}{\sinh(\alpha l)} \tag{12.25}$$

With the open neutral, the boundary condition becomes

$$\begin{cases} x = 0, & V = E_0 \\ x = l, & I = 0 \text{ or } \dfrac{dV}{dx} = 0 \end{cases} \tag{12.26}$$

The corresponding initial voltage distribution in the motor winding can be found as:

$$V = \frac{E_0}{e^{\alpha l} + e^{-\alpha l}}\left[e^{\alpha(l-x)} + e^{-\alpha(l-x)}\right] = E_0\frac{\cosh[\alpha(l-x)]}{\cosh(\alpha l)} \tag{12.27}$$

Recognizing $\alpha l = \sqrt{\frac{C_0}{K_0}}l = \sqrt{\frac{C_0 \cdot l}{K_0/l}}$, we can conclude that the initial voltage distribution in a winding depends on the ratio between the total winding-to-ground capacitance and the total winding turn-to-turn capacitance. Figure 12.22 shows the initial voltage distributions as function of αl. Given that the typical αl value for a transformer winding is between 5 and 15, and the value for

Figure 12.22 Initial voltage distribution in a winding with a steep front pulse voltage as function of αl for the grounded neutral case (left) and the open neutral case.

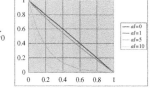
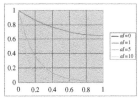

a motor winding is probably higher given the slot structure, it can be seen that the winding coils near the motor terminal will bear the highest voltage, regardless of the neutral connections.

The voltage distribution aforementioned is for winding-to-ground voltage, which stresses the main winding-to-frame insulation that is usually not the major concern. The winding turn-to-turn insulation is more of a weak point due to the slot constraint. The turn-to-turn voltage can be characterized by $\frac{dV}{dx}$. For the grounded neutral case,

$$\frac{dV}{dx} = -\alpha E_0 \frac{\cosh[\alpha(l-x)]}{\sinh(\alpha l)} \tag{12.28}$$

For the open neutral case,

$$V = -\alpha E_0 \frac{\sinh[\alpha(l-x)]}{\cosh(\alpha l)} \tag{12.29}$$

Figure 12.23 shows the initial $\frac{dV}{dx}$ distribution in a motor winding as function of αl. The value is normalized by $-\alpha E_0$. Clearly, if αl is sufficiently high (e.g. >5), the motor terminal winding coil will see the highest initial turn-to-turn voltage, as expected.

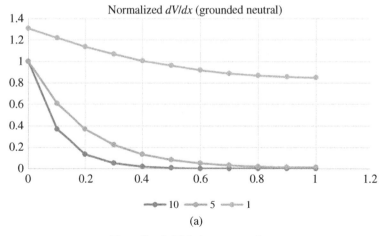

(a)

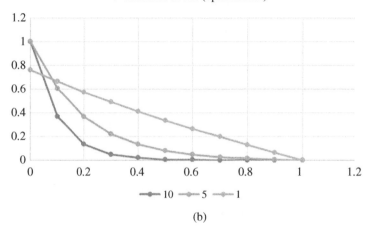

(b)

Figure 12.23 Normalized initial $\frac{dV}{dx}$ distribution in a motor winding as function of αl: (a) grounded neutral and (b) open neutral.

Another overvoltage phenomenon observed is the greater than 2 p.u. voltages at motor terminals in PWM VSI drive applications [18]. The cause for such phenomenon is due to the narrow switching pulses for line-to-line voltages. By properly eliminating such narrow pulses in the modulator, this issue can be avoided.

To mitigate the high dv/dt-related problems on motors, external filters have been developed. Although there are many different filter approaches, they can be divided into three categories: motor terminal filter, dv/dt filter, and sinewave filter.

12.2.3.2 Motor Terminal Filter

Motor terminal filters function based on the principle of impedance matching. As can be seen in (12.13), if $Z_B = Z_A$, then $\Gamma = 0$, and therefore there will be no voltage overshoot. For a lossless cable, which can be used to model a high-power cable as the resistive loss is very small, its characteristic impedance is a pure resistance. In order to fully eliminate the overvoltage, the motor terminal filter should be a pure resistor. However, pure resistor filter will generate unacceptably high loss. Consequently, an RC filter is usually adopted as the motor terminal filter, with a typical configuration as shown in Figure 12.24.

To select the RC terminal filter parameters, their relationship with the motor terminal voltage and the cable charateristic impedance needs to be derived. Take phase A of the three phases in Figure 12.24 as an exmaple. The incident voltage and current waves at the motor filter terminal are related by (12.30), and the refected voltage and current waves by (12.31):

$$v_{A1} = Z_0 \cdot i_{A1} \tag{12.30}$$

$$v_{A2} = -Z_0 \cdot i_{A2} \tag{12.31}$$

Assuming that the motor impedance is high and can be approximated as an open-circuit branch, the voltage and current in the RC filter are equal to the refraction waves as:

$$v_{A3} = v_{A1} + v_{A2} \tag{12.32}$$

$$i_{A3} = i_{A1} + i_{A2} = \frac{1}{Z_0}(v_{A1} - v_{A2}) \tag{12.33}$$

The capacitor current can be expressed as in (12.34), which can be rewritten as (12.35) by utilizing (12.32) and (12.33):

$$C\frac{dv_C}{dt} = C\frac{d(v_{A3} - R \cdot i_{A3})}{dt} = i_{A3} \tag{12.34}$$

$$C\frac{d(v_{A1} + v_{A2})}{dt} - \frac{C \cdot R}{Z_0}\frac{d(v_{A1} - v_{A2})}{dt} = \frac{1}{Z_0}(v_{A1} - v_{A2}) \tag{12.35}$$

Figure 12.24 Typical motor terminal filter configuration.

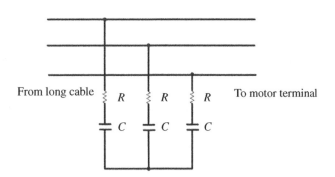

From long cable R R R To motor terminal

C C C

Rearranging Eq. (12.35), v_{A2} can be expressed in terms of v_{A1} as:

$$\frac{dv_{A2}}{dt} + \frac{1}{C \cdot (R + Z_0)} v_{A2} = \frac{(R - Z_0)}{(R + Z_0)} \frac{dv_{A1}}{dt} + \frac{1}{C \cdot (R + Z_0)} v_{A1} \tag{12.36}$$

Assuming v_{A1} is an ideal square wave pulse with an amplitude V and applying Laplace transform to Eq. (12.36) will yield

$$sV_{A2}(s) + \frac{1}{C \cdot (R + Z_0)} V_{A2}(s) = \frac{(R - Z_0)}{(R + Z_0)} V + \frac{V}{sC \cdot (R + Z_0)} \tag{12.37}$$

Defining $C \cdot (R + Z_0) = \tau$, $V_{a2}(s)$ can be expressed as:

$$V_{A2}(s) = \frac{(R - Z_0)}{(R + Z_0)} \frac{V}{\left(s + \dfrac{1}{\tau}\right)} + \frac{V}{s\tau\left(s + \dfrac{1}{\tau}\right)} = \frac{(R - Z_0)}{(R + Z_0)} \frac{V}{\left(s + \dfrac{1}{\tau}\right)} + \frac{V}{s} - \frac{V}{\left(s + \dfrac{1}{\tau}\right)} \tag{12.38}$$

Applying inverse Laplace transform will yield

$$v_{A2} = V - \left[1 - \frac{(R - Z_0)}{(R + Z_0)}\right] V \cdot e^{-\frac{t}{\tau}} \tag{12.39}$$

The terminal voltage of the RC filter and the motor will be

$$v_{A3} = v_{A1} + v_{A2} = 2V - \left[1 - \frac{(R - Z_0)}{(R + Z_0)}\right] V \cdot e^{-\frac{t}{\tau}} \tag{12.40}$$

From (12.40), we can examine some special cases. In the case of a pure resistor filter (i.e. $C \to \infty$), $\tau \to \infty$, the equation will reduce to (12.41), which agrees with (12.12). In the case of a very large R, or equivalently no terminal filter, $v_{A3} \to 2V$, corresponding to the open-circuit case. In the case of a very small C (i.e. $C \to 0$), $\tau \to 0$, then also $v_{A3} \to 2V$. In other words, too small a capacitance will make the filter ineffective.

$$v_{A3} = V + \frac{(R - Z_0)}{(R + Z_0)} V = \frac{2Z_0}{(R + Z_0)} V \tag{12.41}$$

On the other hand, too large a capacitance or pure resistance will make the filter too lossy, especially, in higher voltage applications. For example, the total power loss for the three-phase filter in Figure 12.24 for a two-level VSI can be found as (12.42), where C_Δ is the capacitance for the equivalent Δ-connected RC filter with $C_\Delta = \frac{C}{3}$, V_{DC} is the DC-link voltage, ϵ is the ratio of the terminal overvoltage between 0 and 1, and f_{sw} is switching frequency. Clearly, the loss is proportional capacitance value C.

$$P_{filter} = 3 \times \frac{1}{2} C_\Delta (1 + \epsilon)^2 V_{DC}^2 \cdot f_{sw} = \frac{f_{sw}}{2} C (1 + \epsilon)^2 V_{DC}^2 \tag{12.42}$$

For the Figure 12.24 filter, R is usually selected to be equal to the cable characteristics impedance Z_0 (typically between 20 and 100 Ω), and selection of C is a trade-off between filter power loss and voltage overshoot coefficient ϵ. If we wish the peak overvoltage ratio to be below ϵ, the peak of the reflected voltage v_{A2} should be less than ϵV. According to (12.39), the relation (12.43) should be satisfied, under the condition $R = Z_0$. From Figure 12.18, the terminal voltage starts to decrease at $t = 3t_p$ if t starts when the pulse voltage is generated at the inverter terminal; therefore, the time instant should be $t = 2t_p$ if t starts when pulse reaches the motor and filter terminal as defined for Eqs. (12.39) and (12.40). Consequently, $t = 2t_p$ is used in (12.43) for determining v_{A2} peak.

$$V - V \cdot e^{-\frac{2t_p}{\tau}} \le \epsilon V \tag{12.43}$$

Relation (12.43) can be manipulated to yield (12.44), where $t_p = \frac{l_c}{v}$, $v = \frac{1}{\sqrt{L_0 C_0}}$, and $Z_0 = \sqrt{\frac{L_0}{C_0}}$ have been used. L_0 and C_0 are per unit length inductance and capacitance of the cable.

$$C \geq -\frac{t_p}{Z_0 \ln(1-\epsilon)} = -\frac{l_c}{v Z_0 \ln(1-\epsilon)} = -\frac{l_c C_0}{\ln(1-\epsilon)} \tag{12.44}$$

Note that C in (12.44) depends on cable length l_c, which is inconvenient for the general-purpose design. In reality, if C is selected large enough, it should work for relatively long cables. For example, a 45-kW inverter with 650-V DC bus drives a motor through a 150-m cable with a 50 Ω characteristic impedance. Assume $v = 10^8$ m/s. If we wish to keep the overvoltage below 20%, then according to (12.44), C should be

$$C \geq -\frac{l_c}{v Z_0 \ln(1-\epsilon)} = -\frac{150}{10^8 \times 50 \times \ln 0.8} = 0.134\,\mu\text{F}$$

With $C = 0.134\,\mu\text{F}$, each switching cycle results in 0.068 J loss according to (12.42). If the switching frequency is 20 kHz, the power loss will be 1.36 kW. For a 45-kW drive, this loss is clearly too high. Therefore, the RC terminal motor filter is more suitable for higher power and lower switching frequency applications.

One question related to the motor terminal filter is how to determine the characteristics of a three-phase cable, including the characteristic impedance Z_0 and wave traveling speed v. In the single-phase (two conductors, such as coaxial cable) case, the definition is clear since there is only one impedance. The termination design, etc., can all be based on this impedance value. In the case of three-phase cables with multiple conductors (at least four including ground), there are multiple impedances. It is important to understand which of them are the more critical parameters, and what is the right approach to measure or calculate them.

12.2.3.2.1 *Three-Phase Three-Conductor Cable Characteristics*   Let's start with the simplest case of a three-phase cable, a cable with only three phase-conductors, and no ground wires, sheaths, or armors. Assume all three conductors are symmetrical with respect to ground and all voltage potentials are referred to ground. In addition, a lossless cable is assumed, which is acceptable for high-current cables.

The general charge and flux linkage equations for the three conductors are (12.45) and (12.46), where q_j and ψ_j are phase charges and flux linkages, C_{jk} and L_{jk} are self and mutual capacitances and inductances of phase conductors, and u_j and i_j are phase voltages and currents, respectively, with $j, k \in (A, B, C)$. Based on the symmetry assumption, all phase-conductor self-capacitances are the same as $C1$, $C2$ is the mutual capacitance between any two phase-conductors, $L1$ is the self-inductance for a phase, and $L2$ is the mutual inductance between two phases. Note that $C1 > C2$ and $L1 > L2$ should hold, and all quantities are for per unit length.

$$\begin{bmatrix} q_A \\ q_B \\ q_C \end{bmatrix} = \begin{bmatrix} C_{AA} & C_{AB} & C_{AC} \\ C_{BA} & C_{BB} & C_{BC} \\ C_{CA} & C_{CB} & C_{CC} \end{bmatrix} \begin{bmatrix} u_A \\ u_B \\ u_C \end{bmatrix} = \begin{bmatrix} C1 & C2 & C2 \\ C2 & C1 & C2 \\ C2 & C2 & C1 \end{bmatrix} \begin{bmatrix} u_A \\ u_B \\ u_C \end{bmatrix} \tag{12.45}$$

$$\begin{bmatrix} \psi_A \\ \psi_B \\ \psi_C \end{bmatrix} = \begin{bmatrix} L_{AA} & L_{AB} & L_{AC} \\ L_{BA} & L_{BB} & L_{BC} \\ L_{CA} & L_{CB} & L_{CC} \end{bmatrix} \begin{bmatrix} i_A \\ i_B \\ i_C \end{bmatrix} = \begin{bmatrix} L1 & L2 & L2 \\ L2 & L1 & L2 \\ L2 & L2 & L1 \end{bmatrix} \begin{bmatrix} i_A \\ i_B \\ i_C \end{bmatrix} \tag{12.46}$$

The charge and current, and flux linkage and voltage satisfy (12.47) and (12.48), respectively:

$$\begin{bmatrix} i_A \\ i_B \\ i_C \end{bmatrix} = \frac{\partial}{\partial t} \begin{bmatrix} q_A \\ q_B \\ q_C \end{bmatrix} \tag{12.47}$$

$$\begin{bmatrix} u_A \\ u_B \\ u_C \end{bmatrix} = \frac{\partial}{\partial t} \begin{bmatrix} \psi_A \\ \psi_B \\ \psi_C \end{bmatrix} \tag{12.48}$$

Differentiating the charge and flux equations (12.45) and (12.46), and rearranging them will yield the current and voltage wave equations (12.49) and (12.50):

$$\frac{\partial^2}{\partial x^2} \begin{bmatrix} i_A \\ i_B \\ i_C \end{bmatrix} = \begin{bmatrix} C1 & C2 & C2 \\ C2 & C1 & C2 \\ C2 & C2 & C1 \end{bmatrix} \begin{bmatrix} L1 & L2 & L2 \\ L2 & L1 & L2 \\ L2 & L2 & L1 \end{bmatrix} \frac{\partial^2}{\partial t^2} \begin{bmatrix} i_A \\ i_B \\ i_C \end{bmatrix} \tag{12.49}$$

$$\frac{\partial^2}{\partial x^2} \begin{bmatrix} u_A \\ u_B \\ u_C \end{bmatrix} = \begin{bmatrix} C1 & C2 & C2 \\ C2 & C1 & C2 \\ C2 & C2 & C1 \end{bmatrix} \begin{bmatrix} L1 & L2 & L2 \\ L2 & L1 & L2 \\ L2 & L2 & L1 \end{bmatrix} \frac{\partial^2}{\partial t^2} \begin{bmatrix} u_A \\ u_B \\ u_C \end{bmatrix} \tag{12.50}$$

Clearly, even with symmetrical cables, the three-phase waves are coupled together, making it difficult to characterize using one parameter (e.g. characteristic impedance). The common practice to overcome the aforementioned coupling difficulty is to use modal transformation such as Karenbauer transformation in (12.51) to decouple the waves:

$$T = \frac{1}{3} \begin{bmatrix} 1 & 1 & 1 \\ -2 & 1 & 1 \\ 1 & -2 & 1 \end{bmatrix} \tag{12.51}$$

$$T^{-1} = \begin{bmatrix} 1 & -1 & 0 \\ 1 & 0 & -1 \\ 1 & 1 & 1 \end{bmatrix} \tag{12.52}$$

With Karenbauer transformation, the modal voltage and current can be defined as in (12.53) and (12.54):

$$\begin{bmatrix} u_{m1} \\ u_{m2} \\ u_{m0} \end{bmatrix} = T^{-1} \begin{bmatrix} u_A \\ u_B \\ u_C \end{bmatrix} = \begin{bmatrix} 1 & -1 & 0 \\ 1 & 0 & -1 \\ 1 & 1 & 1 \end{bmatrix} \begin{bmatrix} u_A \\ u_B \\ u_C \end{bmatrix} = \begin{bmatrix} u_A - u_B \\ u_B - u_C \\ u_A + u_B + u_C \end{bmatrix} \tag{12.53}$$

$$\begin{bmatrix} i_{m1} \\ i_{m2} \\ i_{m0} \end{bmatrix} = T^{-1} \begin{bmatrix} i_A \\ i_B \\ i_C \end{bmatrix} = \begin{bmatrix} 1 & -1 & 0 \\ 1 & 0 & -1 \\ 1 & 1 & 1 \end{bmatrix} \begin{bmatrix} i_A \\ i_B \\ i_C \end{bmatrix} = \begin{bmatrix} i_A - i_B \\ i_B - i_C \\ i_A + i_B + i_C \end{bmatrix} \tag{12.54}$$

Clearly, modes 1 and 2 are differential modes (positive and negative sequence), and mode 0 is the CM. In modal 1-2-0 reference frame, the wave equations become (12.55) and (12.56):

$$\frac{\partial^2}{\partial x^2} \begin{bmatrix} i_{m1} \\ i_{m2} \\ i_{m0} \end{bmatrix} = \begin{bmatrix} (L1-L2)(C1-C2) & 0 & 0 \\ 0 & (L1-L2)(C1-C2) & 0 \\ 0 & 0 & (L1+2L2)(C1+2C2) \end{bmatrix} \frac{\partial^2}{\partial t^2} \begin{bmatrix} i_{m1} \\ i_{m2} \\ i_{m0} \end{bmatrix}$$

$$\tag{12.55}$$

$$\frac{\partial^2}{\partial x^2} \begin{bmatrix} u_{m1} \\ u_{m2} \\ u_{m0} \end{bmatrix} = \begin{bmatrix} (L1-L2)(C1-C2) & 0 & 0 \\ 0 & (L1-L2)(C1-C2) & 0 \\ 0 & 0 & (L1+2L2)(C1+2C2) \end{bmatrix} \frac{\partial^2}{\partial t^2} \begin{bmatrix} u_{m1} \\ u_{m2} \\ u_{m0} \end{bmatrix}$$

(12.56)

The three modal waves are decoupled, and their behaviors can be analyzed the same way as the single-phase case. If we know the characteristic impedance of the cable for each mode, the related design can be done similarly to the single-phase case. Define modal characteristic impedance using vector notation through

$$\vec{U}_m = \begin{bmatrix} u_{m1} \\ u_{m2} \\ u_{m0} \end{bmatrix} = [Z_m] \cdot \vec{I}_m = [Z_m] \begin{bmatrix} i_{m1} \\ i_{m2} \\ i_{m0} \end{bmatrix}$$

(12.57)

From wave equations (12.55) and (12.56), it can be found

$$[Z_m] = \begin{bmatrix} \sqrt{\dfrac{(L1-L2)}{(C1-C2)}} & 0 & 0 \\ 0 & \sqrt{\dfrac{(L1-L2)}{(C1-C2)}} & 0 \\ 0 & 0 & \sqrt{\dfrac{(L1+2L2)}{(C1+2C2)}} \end{bmatrix}$$

(12.58)

Therefore,

$$Z_{m1} = Z_{m2} = \sqrt{\frac{(L1-L2)}{(C1-C2)}}$$

$$Z_{m0} = \sqrt{\frac{(L1+2L2)}{(C1+2C2)}}$$

(12.59)

In the normal operation of a three-phase PWM drive, only DM overvoltages are of concern. The motor terminator design should be concerned with only characteristic impedance Z_{m1}. Since modal voltages and currents are phase quantities, the fully matched terminator should be a line-to-neutral resistance of Z_{m1}.

One common way to measure Z_{m1} is to measure the open- and short-circuit impedance between two phase-conductors with the third one floating. Based on (12.45) and (12.46), for the short-circuit condition $i_C = 0$, and $i_B = -i_A$, the measured AC impedance at angular frequency ω is

$$Z_{SC} = \frac{u_A - u_B}{i_A} = 2\omega \cdot (L1-L2) \cdot l_c$$

(12.60)

The cable length l_c is included here since $L1$ and $L2$ are per unit length values. For the open-circuit measurement, $q_C = 0$ and $q_B = q_A$, the measured AC impedance at angular frequency ω is

$$Z_{OC} = \frac{u_A - u_B}{i_A} = \frac{2}{\omega \cdot (C1-C2) \cdot l_c}$$

(12.61)

The equivalent characteristic impedance from these measurements is Z_{eq} in (12.62), which is twice of the DM characteristic impedance of the cable as in (12.59):

$$Z_{eq} = \sqrt{Z_{SC} \cdot Z_{OC}} = 2\sqrt{\frac{(L1 - L2)}{(C1 - C2)}} = 2Z_{m1} \tag{12.62}$$

According to modal wave equations (12.55) and (12.56), the traveling velocity for DM waves is

$$v_{m1} = v_{m2} = \frac{1}{\sqrt{(L1 - L2)(C1 - C2)}} = \omega \cdot l_c \cdot \sqrt{\frac{Z_{OC}}{Z_{SC}}} \tag{12.63}$$

12.2.3.2.2 Characteristic Impedance for Unshielded Cables of More Than Three Conductors

For many three-phase cables, there are more than three conductors. Three of these conductors are phase conductors, and the rest are ground wires, sheaths, and/or armors. In the case of unshielded cables, the aforementioned analysis for the three-conductor case is basically applicable. The difference is that there will be more than three modes. Assuming a cable of three conductors ABC with one ground wire G and all conductors are symmetrical, the modal transformation matrix has the form of

$$T^{-1} = \begin{bmatrix} 1 & -1 & 0 & 0 \\ 1 & 0 & -1 & 0 \\ 1 & 0 & 0 & -1 \\ 1 & 1 & 1 & 1 \end{bmatrix} \tag{12.64}$$

The four modes are: (a) from A to B; (b) from A to C; (c) from A to G; and (d) from A, B, C, and G to ground. Again, the first two modes are differential modes. The fourth mode is the CM. The third mode is negligible if the ground wire is kept floating and becomes part of CM if grounded.

If there are three symmetrical ground wires G1/G2/G3, there will be two additional modes from B to G2 and C to G3. In this case, CM currents will be minimized if ABC phases are balanced. The effect of armor can be treated the same as the ground wires.

The characteristic impedance for two differential modes can be obtained similarly as in the three-conductor case. Strictly speaking, however, the grounding schemes for the armor and ground wires will affect the impedance. The right approach is to connect the grounding wires the same way as they will be used and then measure the characteristic impedance. The difference, though, may be small in this case of the unshielded cables.

12.2.3.2.3 Characteristic Impedance for Three-Phase Shielded Cables

First consider a three-phase symmetrical shielded cable without any armor or ground wires. In this case, there should be six modes. Numbering the conductors in the order of A, As, B, Bs, C, and Cs and recognizing the shielding effects, the modal transformation matrix is

$$T^{-1} = \begin{bmatrix} 1 & -1 & 0 & 0 & 0 & 0 \\ 0 & 1 & 0 & 0 & 0 & -1 \\ 0 & 0 & 1 & -1 & 0 & 0 \\ 0 & 0 & 0 & 1 & 0 & -1 \\ 0 & 0 & 0 & 0 & 1 & -1 \\ 0 & 1 & 0 & 1 & 0 & 1 \end{bmatrix} \tag{12.65}$$

Clearly, there are three types of modal currents or voltages: coaxial modes 1 (A to As), 3 (B to Bs), and 5 (C to Cs); differential modes 2 (As to Bs) and 4 (As to Cs); and CM 6 (As, Bs, Cs to ground). The characteristic impedance again depends on the grounding schemes for the shields. Let us consider two scenarios. First leave the shields floating and then have all shields connected together. In both cases, the active phases are A, B, and C.

In the first case, the equivalent differential modes are A to B (A to As, As to Bs, Bs to B), and A to C (A to As, As to Cs, Cs to C), since there are no net currents in shields As, Bs, and Cs. The characteristic impedance can be obtained by considering series connections of modes (for equivalent differential inductance) or parallel connections of modes (for equivalent differential capacitance). It can be expected that these parameters will not be much different from those of unshielded cables.

In the second case, the equivalent differential modes are the same as the aforementioned. However, the characteristic impedance has to be found by connecting all shields together (to ground). Following this reasoning, the characteristic impedance is equal to that of one of the coaxial cables.

The advantages of the motor terminal RC filter include: (i) its simplicity, with only resistors and capacitors, and (ii) it is relatively independent of inverter characteristics or cable length. Its disadvantages include: (i) it needs to be placed at the motor terminal, which may be inconvenient or even impossible in some applications; (ii) it only reduces overvoltage but has no impact on dv/dt; and (iii) its relatively high loss, and therefore, only for higher power and lower switching frequency applications, as already discussed earlier.

12.2.3.3 *dv/dt* Filter

The dv/dt filter is placed at the inverter AC output terminal to slow down dv/dt. One common dv/dt filter configuration is shown in Figure 12.25, which is basically a second-order low-pass filter. Since the resonance frequency is usually selected very high to reduce the filter component (L_f and C_f) values, even higher than the switching frequency, the resistor R_f is needed to provide damping.

Note that in normal operations, each switching event of the inverter is independent of the others. The filter equivalent circuit for one switching event can be illustrated as in Figure 12.26 [19]. In this case, one inverter phase generates a voltage pulse, and the other two phases that are not switching will maintain their voltages with respect to the switching phase and therefore can be considered shorted. The connecting cables and load are assumed to be open since their impedance should be much larger than the filter impedance at high frequency. The equivalent L, R, and C can be found as in (12.66). For a two-level VSI, the inverter pulse voltage V_{in} will be either V_{DC} or $-V_{DC}$.

$$L = \frac{3}{2}L_f, R = \frac{3}{2}R_f, C = \frac{2}{3}C_f \tag{12.66}$$

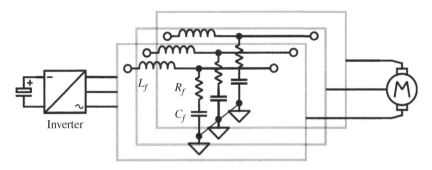

Figure 12.25 Inverter output *dv/dt* filter.

(a)

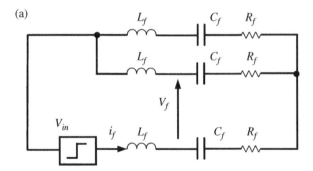

(b)

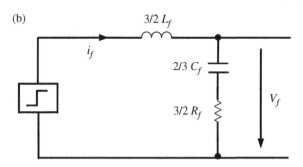

Figure 12.26 (a) Filter equivalent circuit during one switching event and (b) its simplification. *Source:* From [19]/with permission of IEEE.

Assuming an open circuit at the output side of the *dv/dt* filter, the filter output voltage can be determined either in frequency domain using Laplace transform or in time domain to directly solve the differential equation. Taking the time domain approach and conservatively assuming that the inverter terminal voltage is an ideal step voltage with the magnitude of V_{in}, the differential equation for capacitor voltage can be written as:

$$LC\frac{d^2v_C}{dt^2} + RC\frac{dv_C}{dt} + v_C = V_{in} \tag{12.67}$$

With initial condition $v_C(0) = 0, i(0) = \dfrac{dv_C}{dt}\bigg|_{t=0} = 0$, the capacitor voltage v_c can be determined as in (12.68), where p_1 and p_2 are characteristic roots or the poles of the second-order system (12.67), and can be found as in (12.69):

$$v_C = V_{in} - \frac{V_{in}}{(p_2 - p_1)}(p_2 e^{p_1 t} - p_1 e^{p_2 t}) \tag{12.68}$$

$$p_{1,2} = \frac{-R \pm \sqrt{R^2 - \frac{4L}{C}}}{2L} \tag{12.69}$$

The current through the filter can be determined as in (12.70) and the voltage drop on the filter inductor as in (12.71):

$$i(t) = C\frac{dv_C}{dt} = \frac{V_{in}}{L(p_1 - p_2)}(e^{p_1 t} - e^{p_2 t}) \tag{12.70}$$

$$v_L(t) = \frac{V_{in}}{(p_1 - p_2)}(p_1 e^{p_1 t} - p_2 e^{p_2 t}) \tag{12.71}$$

Therefore, the filter voltage

$$v_f(t) = V_{in} - v_L(t) = V_{in} - \frac{V_{in}}{(p_1 - p_2)}(p_1 e^{p_1 t} - p_2 e^{p_2 t}) \tag{12.72}$$

Depending on L, C, and R values, the filter can be overdamped $(R > 2\sqrt{\frac{L}{C}})$, critically damped $(R = 2\sqrt{\frac{L}{C}})$, or underdamped $(R < 2\sqrt{\frac{L}{C}})$. Different damping conditions result in different filter loss and voltage overshoot. In general, more damping will have better controlled voltage overshoot but shorter rise time, i.e. higher average dv/dt. Depending on the application, the filter damping can be designed accordingly.

Under different damping conditions, different relationships can be established for the filter design. For overdamped case, Eq. (12.72) can be rewritten as (12.73), where $\delta = \frac{R}{2L}$, $\omega_0 = \sqrt{\frac{1}{LC}}$, $\xi = \frac{R}{2\omega_0 L} = \frac{R}{2\sqrt{\frac{L}{C}}} = \frac{\delta}{\omega_0}$. Clearly, overdamped condition refers to $\xi > 1$, which matches the classical definition for a second-order system.

$$v_f(t) = V_{in} - \frac{V_{in}e^{-\delta t}}{2}\left[\left(1 - \frac{\xi}{\sqrt{\xi^2 - 1}}\right)e^{\left(\omega_0\sqrt{\xi^2 - 1}\right)t} + \left(1 + \frac{\xi}{\sqrt{\xi^2 - 1}}\right)e^{-\left(\omega_0\sqrt{\xi^2 - 1}\right)t}\right] \tag{12.73}$$

Using $\frac{dv_f(t)}{dt} = 0$, the peak value of v_f will occur at t_r, which can be determined with (12.74) and is the rise time of the filter voltage. The corresponding peak voltage v_{f-max} can be found using (12.73).

$$t_r = \frac{1}{2\omega_0\sqrt{\xi^2 - 1}}\ln\left(\frac{2\xi^2 - 1 + 2\xi\sqrt{\xi^2 - 1}}{2\xi^2 - 1 - 2\xi\sqrt{\xi^2 - 1}}\right) \tag{12.74}$$

For underdamped case, Eq. (12.72) can be rewritten as (12.75), where $\omega = \omega_0\sqrt{1 - \xi^2}$, $\beta = \tan^{-1}\frac{\omega}{\delta} = \sin^{-1}\frac{\omega}{\omega_0} = \sin^{-1}\sqrt{1 - \xi^2}$. The peak value of $v_f(t)$ is achieved at the rise time t_r, which can be determined again using $\frac{dv_f(t)}{dt} = 0$ as in (12.76). The corresponding peak voltage can be found as (12.77).

$$v_f(t) = V_{in} + \frac{\omega_0}{\omega}V_{in}e^{-\delta t}\sin(\omega t - \beta) \tag{12.75}$$

$$t_r = \frac{2\beta}{\omega} = \frac{2\sin^{-1}\sqrt{1 - \xi^2}}{\omega_0\sqrt{1 - \xi^2}} \tag{12.76}$$

$$v_{f-max} = V_{in} + V_{in}e^{-\frac{2\xi}{\sqrt{1 - \xi^2}}\left(\tan^{-1}\frac{\sqrt{1 - \xi^2}}{\xi}\right)} \tag{12.77}$$

For critically damped case, the general expression for $v_f(t)$ becomes (12.78), where in this case $\delta = \frac{R}{2L} = \sqrt{\frac{1}{LC}} = \omega_0$. The peak value of $v_f(t)$ is achieved at the rise time t_r, which can be determined again using $\frac{dv_f(t)}{dt} = 0$ as in (12.79). The corresponding peak voltage can be found as in (12.80), which has a fixed voltage overshoot of 13.53%.

$$v_f(t) = V_{in} - V_{in}(1 - \delta t)e^{-\delta t} = V_{in} - V_{in}(1 - \omega_0 t)e^{-\omega_0 t} \tag{12.78}$$

$$t_r = \frac{2}{\delta} = \frac{2}{\omega_0} \tag{12.79}$$

$$v_{f-max} = V_{in} + V_{in}e^{-2} = V_{in}\left(1 + e^{-2}\right) \approx 1.1353 V_{in} \tag{12.80}$$

Based on the aforementioned analysis, it can be concluded that if a greater than 13.53% voltage overshoot can be tolerated, then underdamped filter design can be used; otherwise, critically damped or overdamped filter design must be adopted. For the filter design, there are generally three constraints: (i) operating frequency voltage drop on the filter inductor, which will limit the maximum L value; (ii) dv/dt limit, which can be from the motor or cable requirement. In [20], a motor terminal dv/dt limit of 4 kV/μs is used; and generally, the limit may be tied to motor voltage rating and converter DC-link voltage. For a long cable, the dv/dt should be limited to be below the critical rise time as discussed earlier. The lower dv/dt from these two criteria should be selected as the design constraint; (iii) voltage overshoot limit. The voltage drop constraint can be expressed as in (12.81), where ω_e is the rated (or maximum) motor electrical angular frequency, I_m is the rated RMS current, V_{ph-m} is the phase RMS voltage, and ε is the allowed ratio. ε is usually selected to be less than 5%, although a higher percentage could be allowed, considering that the high current occurs at high power factor as discussed in Chapter 7, which is rare for motor drives.

$$\omega_e I_m L_f \leq \varepsilon V_{ph-m} \tag{12.81}$$

In some cases, the filter power loss may also be a constraint. Using the equivalent circuit in Figure 12.26, for a two-level VSI, each switching event (i.e. each turn-on or turnoff) will incur an energy loss of $\frac{1}{2}CV_{DC}^2 = \frac{1}{3}C_f V_{DC}^2$. For a continuous three-phase PWM, in each switching cycle, there are six switching events, and the total filter power loss can be found as in (12.82). For a discontinuous two-phase PWM, each switching cycle will have four switching events, and the total filter power loss can be found as in (12.83). The loss in each phase should be one-third of the total loss. Note that in a physical filter, the inductor and capacitor themselves can also incur loss, which must be included in the design.

$$P_f = \frac{1}{3}C_f V_{DC}^2 \times 6f_{sw} = 2C_f V_{DC}^2 f_{sw} \tag{12.82}$$

$$P_f = \frac{1}{3}C_f V_{DC}^2 \times 4f_{sw} = \frac{4}{3}C_f V_{DC}^2 f_{sw} \tag{12.83}$$

Once the design constraints are determined, the dv/dt filter design can be carried out using the relations developed earlier for peak voltage overshoot v_{f-max}, rise time t_r, and maximum inductance L_f. The intermediate design variables are L_f, C_f, and R_f values. The design variables should be variables associated with the physical inductors, resistors, and capacitors. The design procedure can follow the optimization procedure for other subsystems, and the objective can be minimum cost, size, weight, or loss.

Here, we will show an example of the dv/dt filter design for L_f, C_f, and R_f values. The motor drive system parameters are listed in Table 12.3, which follow those in [20].

The design constraints are: (i) voltage drop on inductor is <1.5%; (ii) $dv/dt < 4$ kV/μs; and (iii) voltage overshoot <10%. Assume that the design objective is to minimize the filter power loss. If the

Table 12.3 Specifications of the example motor drive system.

Rated RMS current	24 A	Motor output power	11 kW
Rated line-to-line RMS voltage	460 V	DC-link voltage	700 V
Rated motor frequency	500 Hz	PWM scheme	DPWM
Maximum motor frequency	1000 Hz	Switching frequency	8 kHz

inductor power loss is not considered, it is desirable to have the smallest capacitance and largest inductance possible. Under the voltage drop constraint, L_f can be found as:

$$L_f = \frac{0.015 \times 460 \div \sqrt{3}}{2 \times \pi \times 500 \times 24} = 52.8 \, \mu H$$

Therefore, $L = \frac{3}{2}L_f = 79.3 \, \mu H$. Given that the voltage overshoot should be less than 10%, overdamped design should be adopted. For an average dv/dt of $4 \, kV/\mu s$, the rise time can be determined via (12.84) such that $t_r \geq \frac{700 \times 1.1}{4000} = 0.1925 \, \mu s$:

$$\left(\frac{dv_f}{dt}\right)_{avg} = \frac{v_{f-Max}}{t_r} \tag{12.84}$$

If we select $C = 0.15 \, nF$, $R = 2000 \, \Omega$, then $\omega_0 = \sqrt{\frac{1}{LC}} = 9.17 \times 10^6 \, rad/s$, $\xi = \frac{R}{2\omega_0 L} = 1.376$. Using (12.73) and (12.74), it can be found that the peak line-to-line filter output voltage is approximately $1.09 V_{DC} = 1.09 \times 700 = 763 \, V$, which occurs at $t_r = 0.194 \, \mu s$. The corresponding dv/dt is $3.93 \, kV/\mu s$. The filter power loss can be estimated with (12.83) as:

$$P_f = \frac{4}{3} C_f V_{DC}^2 f_{sw} = \frac{4}{3} \times \frac{3}{2} \times 0.15 \times 10^{-9} \times 700^2 \times 8000 = 1.176 \, W$$

The estimated loss is very close to the original example in [20], even though a different dv/dt filter topology is used there (Figure 12.27). Clearly, in this example, the dv/dt filter loss due to the capacitor charge and discharge is very small, compared to the rated power of 11 kW or rated kVA of 19 kVA. Note that in a real dv/dt filter, the inductor loss can be a significant portion

Figure 12.27 An alternate dv/dt filter topology.

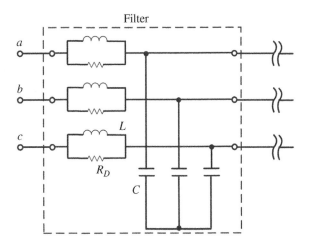

and must be considered. In the example in [20], the inductor loss is 24 W, which dominates the total filter loss. Therefore, for a drive inverter with relatively low switching frequency (e.g. 8 kHz in this example), a larger C and smaller L are preferred. If we reduce the inductance to half, double the capacitance, and at the same time reduce the resistance to half, then the filter performance will stay the same. With the new parameters, the inductor loss will be reduced to half and capacitance charge/discharge loss will double, leading to a significantly reduced total filter loss.

In addition to those in Figures 12.25 and 12.27, there are other dv/dt filter topologies available for further loss reduction and other purposes. The same design procedure as discussed here can be applied.

12.2.3.4 Sinewave Filter

The dv/dt filter also has limitations. One limitation is that for very long cable, e.g. >300 m, the critical dv/dt needed to mitigate voltage doubling effect becomes very low, such that the needed L and C become very large and difficult to design. Another limitation is that with increased switching frequency, the filter loss becomes too high. In long cable cases, even with limited dv/dt, the cable loss can be high when the switching frequency is high as discussed earlier. In these cases, sinewave filters can be adopted.

A typical sinewave filter utilizes only L and C components as shown in Figure 12.28. As will be explained later, the resonance frequency of the sinewave filter is much lower than the inverter switching frequency. Consequently, no damping resistor is practically needed. Similar to the dv/dt filter, the sinewave filter is usually placed at the inverter output terminal. The sinewave filter is a low-pass filter and, with proper design, may solve both reflected wave motor overvoltage and line-to-ground CM noise current problems for both short and long cable conditions. Low dv/dt of the fundamental frequency sinewave output will practically eliminate any overvoltage at the motor terminal even for a very long cable (e.g. 3000 m cable). Low sinewave dv/dt also ensures that the line-to-ground CM current is limited to low peak amplitudes associated with only motor frequency cable charging/discharging current.

The sinewave filter in Figure 12.28 has a resonance frequency at $f_0 = \frac{\omega_0}{2\pi} = \frac{1}{2\pi}\sqrt{\frac{1}{L_f C_f}}$. Above f_0, there is a -40 dB/dec attenuation for harmonic frequencies, and below f_0, there is no attenuation. In order to achieve sinewave quality, the inverter equivalent switching frequency f_{SW} must be much greater than f_0, such that PWM harmonic voltages of the inverter will be attenuated by -40 dB. Practically, there should be $f_{SW} > k_1 f_0$, where k_1 should be at least 4–5, or even 10 if possible. On the other hand, f_0 should be much higher than the motor operating frequency f_e to prevent system

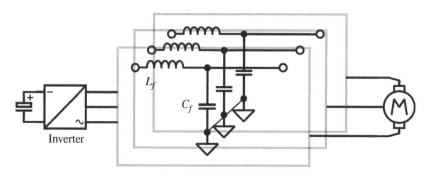

Figure 12.28 Sinewave filter.

instability, since there is no attenuation below f_0. Practically, there should be $f_0 > 10f_e$. It should be pointed out that for the dv/dt filter discussed earlier, it generally satisfies the relation $f_0 \gg f_{SW} \gg f_e$, compared to $f_{SW} \gg f_0 \gg f_e$ in the case of the sinewave filter here. Clearly, the sinewave filter is more suitable for high switching frequency application. For an inverter with a given switching frequency, it can be seen that the filter L_f and C_f need to be much larger than their dv/dt filter counterparts. Since the maximum L_f must be limited due to the voltage drop constraint (12.81), a much larger C_f must be selected. A large C_f can lead to large reactive current/power both at the startup and at motor operating frequency, which can lead to increased inverter rating need for the same motor load and potential instability.

Due to their much larger size and power loss, sinewave filters are seldom used in slow switching Si IGBT-based motor drives, except in some very long cable applications. However, for WBG-based motor drives capable of high switching frequency, sinewave filters can be more feasible and even more advantageous. Consequently, the sinewave filter design has received more attention in recent years.

The design of the sinewave filter is to select L_f and C_f values for the filter and then carry out the corresponding physical design. There are several steps for the filter design.

1) Select a resonant frequency f_0 or ω_0 between f_e and f_{SW}. It should be as high as possible but much lower than f_{SW}.
2) Select inductance value L_f. Given that C_f tends to be large for a sinewave filter, it is desirable to select an L_f value to be as large as possible. The constraint for L_f is the motor frequency voltage drop constraint (12.81).
3) Select capacitance value C_f. With L_f selected in (2), it is apparent that C_f can be determined as $C_f = \frac{1}{\omega_0^2 L_f}$. However, since for the sinewave filter, the performance of the filter at motor frequency and low-order harmonic frequencies is important, the cable and motor impedance cannot be neglected. Note that these impedances are in parallel with the filter capacitor, and the impedance under no-load conditions may be more critical than the impedance under loaded conditions. Assuming that the cable and motor inductances are L_{cable} and L_{motor}, respectively, the total equivalent inductance is as in (12.85). The filter capacitance can be determined as in (12.86).

$$L_{eq} = \frac{L_f \cdot (L_{cable} + L_{motor})}{L_f + L_{cable} + L_{motor}} \tag{12.85}$$

$$C_f = \frac{1}{\omega_0^2 L_{eq}} \tag{12.86}$$

In the example system in [20], for the sinewave filter, the switching frequency is increased to 50 kHz for the SiC-based motor drive. Given that the maximum motor frequency is 1 kHz, the resonance frequency is selected to be $f_0 = 12.5$ kHz, which is greater than 10 times of the motor top-speed frequency and is ¼ of the switching frequency. Assuming the maximum motor frequency voltage drop due to the filter inductor can be 4.5%, the filter inductance value can be determined as 0.158 mH according to (12.81). Given that the cable inductance of 5 µH and the motor inductance of 6.5 mH, the equivalent inductance is 0.155 mH according to (12.85). From (12.86), C_f is determined to be 1.047 µF.

Like the dv/dt filter, the sinewave filter will also incur losses, including the inductor loss and capacitor charging/discharging loss. In the aforementioned example, the sinewave filter loss is larger than the dv/dt filter loss, dominated by the inductor loss in both cases. However, in certain applications, sinewave filters may be more advantageous than the dv/dt filter, when the switching frequency is high and motor frequency is also high. One such application is aircraft or EV application with SiC- or GaN-based motor drives. In these cases, both the power loss and the weight of the sinewave filter can be smaller than those of the counterpart dv/dt filter.

Another advantage of the sinewave filter is the lower energy charged and discharged on other parasitic capacitances in the system, including the cable capacitances. This could lead to lower losses in the overall system. As already discussed earlier and can be seen from (12.8) and (12.9), for cables connecting the inverter and the motor, the pulsed voltage will cause the power loss due to cable capacitances to be proportional to the switching frequency. With the sinewave filter, the loss will be proportional to the fundamental frequency of the sinusoidal waveform.

12.2.4 Motor Drive Grounding

Grounding design is an essential part of the motor drive design, especially for high-power motor drives. Typically, a high-power AC-fed motor drive has a dedicated input transformer as shown in Figure 12.29, which is the configuration used in a 5–10 MW GE Si IGCT-based MV regenerative main drive. It is a back-to-back three-level NPC VSC with a three-phase input transformer. Since the transformer galvanically isolates the drive from the grid, the drive grounding must be provided and designed as part of the motor drive application design.

There could be a number of options for the grounding point selection for the motor drive in Figure 12.29, including grounding the input transformer neutral point on the secondary side (i.e. the motor drive side), the DC-link midpoint, the motor stator winding neutral point, one of the phases on the input AC side (i.e. the source side), one of the phases on the output side (i.e. the motor side), or positive or negative DC buses. For high-power, MV motor drives, symmetry is desirable. Given that motor neutral is usually not accessible, practically speaking, there are only two options, the transformer neutral point grounding or DC-link midpoint grounding. The DC-link midpoint grounding is often preferred, since it will limit the highest voltage in the drive system to be within $\pm \frac{V_{DC}}{2}$, which is beneficial when selecting components voltage and insulation capabilities. Consequently, the motor drive in Figure 12.29 adopted the DC midpoint grounding. It also makes good sense for a three-level VSC, since there is a natural DC-link midpoint anyway. For a two-level VSC, a DC midpoint must be created to realize the midpoint grounding, which is generally not too

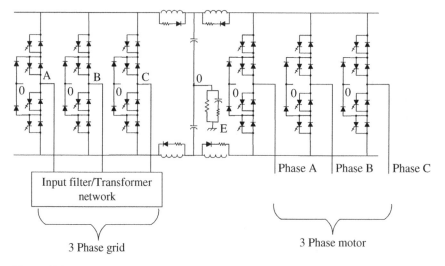

Figure 12.29 Fully regenerative IGCT three-level NPC VSC. *Source:* From [21]/with permission of IEEE.

much of an extra burden, since for high-power motor drives, multiple capacitors in parallel and in series are often needed.

It should be pointed out that a single-point grounding is generally preferred for a high-power motor drive for better ground fault tolerance and for reducing CM ground current. In fact, one of the functions of the input transformer is to provide isolation to improve ground fault tolerance. The input transformer can also provide the needed filter inductance required by the VSC and help to step down or step up the grid voltage to the needed voltage level for the motor drive. In fact, the GE drive in Figure 12.29 does not usually have a built-in AC inductor on the input side and relies on the transformer to provide the filter (or boost) inductance. The input transformer also has benefits in grid transient conditions such as during switching and lightning transients, which will be discussed in next section for grid applications.

Note that in Figure 12.29, the DC-link midpoint is not solidly grounded; instead, it is grounded through a passive component network. Here, we will discuss the design considerations for such a grounding network. The load-side ground loop configuration of the three-level NPC VSC is shown in Figure 12.30, where R_{1N}, R_{2N}, and C_{2N}, comprise the adopted converter grounding circuit, and C_M represents the equivalent parasitic capacitance between the motor neutral point and the ground. The grid-side should have a similar configuration, with the isolation transformer instead of the motor.

The design of the grounding network needs to satisfy the following requirements:

1) The grounding circuit impendence should be significantly lower than parasitic impedance between the ground and other parts of the converter, thus fulfilling the grounding purpose. This should cover wide frequency range from DC, to low AC frequency (i.e. the motor operating frequency and its harmonics), and to high frequency (i.e. the switching frequency and its harmonics).
2) The grounding impedance should be sufficiently high so as to limit the fault current during a converter single-point ground fault. In fact, in many cases, it is desirable to have the motor drive keep running during the single point to ground fault. Therefore, the steady-state fault current, including DC, low-frequency, and high-frequency current components, needs to be very low.
3) During normal operation, i.e. when there is no ground fault, the CM voltage dropped on the grounding circuit V_{OE} should be small. There are two reasons for this requirement. One is that high CM voltage will shift DC bus voltage potential with respect to ground. If the shift is too high, it will cause significant extra stess on converter components and system insulation. The other reason is that DC midpoint voltage can be used for ground fault detection. In this case, the CM voltage should be low enough under the normal condition, compared with the condition when a ground fault occurs in order to identify the ground fault.

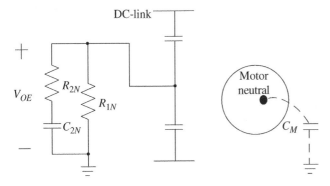

Figure 12.30 Ground loop configuration of the motor-side three-level NPC VSC.

Based on these requirements, Figure 12.30 neutral point grounding circuit is a suitable configuration. The resistor R_{1N} is chosen for DC and low-frequency impedance; and the capacitor C_{2N} is for high-frequency impedance with R_{2N} for damping purpose. Here, we will discuss the design of this grounding circuit considering its dual function as a ground fault detection circuit. The three-level NPC VSC motor drive is used as the example.

Under normal conditions, most CM voltage should drop on the grid-side transformer neutral and the load-side motor neutral such that the DC bus voltage will not significantly exceed $\pm \frac{V_{DC}}{2}$ (except for the cases where the CM voltage on motor neutral need to be reduced). Therefore, the DC midpoint voltage V_{OE} with respect to the true ground should be a small portion of the CM voltage, or in the extreme case, be zero. On the other hand, during a ground fault condition, V_{OE} will be different as a result of the ground loop change. For example, for a load-side phase A ground fault, V_{OE} will become $-V_{AO}$, which is generated by PWM modulator and is dramatically different from the normal case. Therefore, it is conceivable to detect the ground fault by monitoring the neutral point voltage V_{OE}, which is often sensed and monitored anyway for the integrity of the system.

During the ground fault, the current going into the midpoint O through grounding impedance can be formulated by (12.87), where I_{mid} is the DC midpoint current, Z is the equivalent grounding impendence, and V_{OE} can be phase-to-DC midpoint voltage during a phase-to-ground fault or half of the DC-link voltage during a DC link to ground fault:

$$I_{mid} = \frac{V_{OE}}{Z} \tag{12.87}$$

Since the capacitor C_{2N} will block DC and low-frequency current components, the value of the DC and low-frequency current should be determined by the resistor R_{1N} as in (12.88):

$$I_{mid} \approx \frac{V_{OE}}{R_{1N}} \tag{12.88}$$

The larger value of V_{OE} for aforementioned two types of ground faults should be half of the V_{DC}. Based on the power rating of the converter, the tolerable midpoint current limit I_{mid_L} can be determined. So, the lower limit of R_{1N} value is given by:

$$R_{1N} \geq \frac{1}{2} \frac{V_{DC}}{I_{mid\_L}} \tag{12.89}$$

During normal operation, the voltage V_{OE} will be the CM voltage shared by parasitic capacitor C_M and the grounding circuit impedance. For the three-level NPC converter, the CM voltage between the DC midpoint and the motor neutral varies with PWM switching and can be as high as $\pm V_{DC}/2$, with other possible values at $\pm V_{DC}/3$, $\pm V_{DC}/6$ or 0 depending on modulation schemes and operating points. For the frequency equal or beyond switching frequency, the CM voltage distribution is determined by C_{2N} and C_M. C_{2N} should be chosen large enough (significantly larger than C_M) for effective high-frequency grounding and also for midpoint ground fault detection scheme to function. For an NPC, the selection criterion should satisfy that the peak midpoint voltage should be less than 10% of the DC-link voltage during the normal operation.

Without loss of generality, in [22], design and simulation are carried out for a scaled 1-kW experimental NPC system, with a DC-link voltage of 320 V, rated AC phase RMS voltage of 115 V, and rated AC line RMS current of 2.9 A or peak current of 4.1 A. The carrier frequency is 10 kHz, and the DC-link capacitance is 135 μF. With $I_{mid\_L} = 10$ mA, which is less than 0.3% of the peak line current, R_{1N} should be greater than 16 kΩ. In the final design, $R_{1N} = 30$ kΩ is selected. With the system parasitic CM capacitance $C_M = 4$ nF, $C_{2N} = 10$ nF and $R_{2N} = 20$ Ω are selected. The design has been validated in [22].

12.3   Grid Applications

Grid applications of three-phase AC converters can generally refer to any such converters connected to power grid, including grid-tied AC-fed motor drives and three-phase rectifiers including three-phase PFC. They can also refer to those converters specifically designed for serving grid functions, such as power converters for renewable energy and energy storage systems, and converters for electric power transmission and distribution systems. In Chapters 5 and 7, we have already discussed some general requirements for rectifiers and grid-side inverters, such as harmonics and inrush conditions after restart. In this section, we will discuss other grid conditions and requirements, and their impact on the converter design.

To facilitate the discussion while without loss of generality, this section will use the converter topology shown in Figure 12.31 as the example, which is also used in Chapter 11 for a scaled 100-kW prototype. It is a modular, transformer-less (i.e. no 50 or 60 Hz transformers), three-phase four-wire DC/DC/AC power conditioning system (PCS) converter, intended for bidirectional power flow between MV distribution grid and LV DC bus. The LV DC bus is intended to interface with various distributed energy resources (DERs), such as PV panels, wind turbines, battery energy storage systems, microturbines, fuel cells, reciprocating engines, and loads. The PCS converter consists of a DC/DC stage and a DC/AC stage. The DC/DC stage boosts the LV DC to MV DC and realizes galvanic isolation, and the DC/AC stage converts the MV DC to MV AC and interfaces with the MV AC grid. The rated voltage for the LV DC bus is 850 V and the rated voltage for the MV AC grid is 13. 8 kV. The MV DC voltage (6.7 kV in Figure 12.31) is determined by the design.

Design specifications and requirements of the PCS converter are shown in Table 12.4. The converter power rating is 1 MW, and the full load efficiency needs to be above 98%. The LV DC voltage needs to consider 5% voltage variation, and the MV AC grid voltage needs to consider −12% and +10% variations based on the nominal 13.8-kV grid voltage. The PCS converter needs to achieve the four-quadrant operation at full kVA. The control bandwidths of the AC voltage and current are required to be 300 Hz and 1 kHz, respectively, for stability and power quality enhancement.

The basic converter specifications refer to performance constraints and conditions for the PCS converter when it operates at normal steady state with balanced loads. In addition to basic

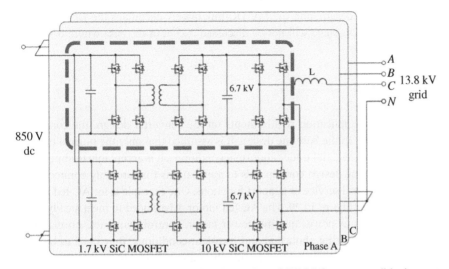

Figure 12.31   A medium-voltage grid-tied three-phase DC/DC/AC power conditioning system.

Table 12.4 The PCS converter specifications and grid requirements.

Basic converter specifications	Power rating	1 MW
	LVDC voltage rating	850 V ($\pm$5%)
	MVAC grid voltage rating	13.8 kV ($-$12 to 10%)
	MVAC grid frequency	60 Hz ($\pm$1.2 Hz)
	Efficiency	98%
	Power factor	Four-quadrant operation
	Ambient temperature	5–35 °C
	Cooling	Water cooling
	Voltage control bandwidth	>300 Hz
	Current control bandwidth	>1 kHz
Grid requirements	IEEE Std 1547	• 5% TDD in MVAC-side current • Low-/high-voltage ride-through • Low-/high-frequency ride-through
	Other system requirements	• MVAC grid faults • MVAC grid unbalance operation • 33% unbalance load support in the islanded mode • Lightning and switching transients

converter specifications, the grid requirements also need to be met. Since the PCS converter acts as the interface of DERs to the grid, it needs to comply with IEEE Std. 1547 [23]. Specifically, the PCS converter needs to meet power quality, voltage ride-through, and frequency ride-through (FRT) requirements. Moreover, the safe, stable, and reliable operation of the PCS converter under other abnormal, transient, and unbalanced conditions needs to be achieved, such as grid faults, unbalanced loads and sources, and lightning and switching transients.

The focus of the discussion is on the DC/AC stage of the PCS converter. Since grid requirements mainly impact the DC/AC stage design, and the MV DC-link largely decouples the DC/DC stage from the DC/AC stage, the DC/DC stage will not be discussed except in the case where it will be impacted.

12.3.1 Baseline Design

To illustrate the impact of grid application requirements on the converter design, the converter baseline design is explained first. In the baseline design, it is assumed that only the normal operating condition needs to be considered, and under abnormal conditions, it is acceptable to shut down or trip the converter. The baseline design corresponds to applications that are only connected to grid but do not need to provide grid service or support functions. One example is for AC-fed motor drives, such as the main drive in Figure 12.29. The AC-fed motor drives need to meet steady-state performance requirements, such as power quality, power factor, control dynamics, normal grid voltage variation, and overload. During abnormal and transient conditions, they need to be able to protect themselves but do not necessarily need to ride-through these conditions. For example, the Figure 12.29 AC-fed drive was designed with arresters to protect against the overvoltage due to lightning surges but was not designed to sustain potential overcurrent due to the same event as it can trip under such conditions.

Therefore, in the baseline design for the example system, only the steady-state performance requirements listed in Table 12.4 are considered, including basic converter requirements and grid-side current harmonic requirement (TDD<5%). The TDD is considered as it is normally applicable to motor drives. To facilitate the comparison, the design objectives/metrics are total weight and size of major components and the subsystems of the DC/AC stage.

12.3.1.1 Semiconductor Devices

From the rated line-to-line voltage of 13.8 kV and considering the −12% and +10% variations, the maximum peak phase voltage during normal operation is determined as 12.4 kV. Given that each phase has two stacked H-bridges, the DC-link voltage for each H-bridge is selected to be 6.3 kV, slightly higher than half of the maximum phase voltage peak value of 6.2 kV. Assuming a sinusoidal AC current, the peak value of the phase current is 59.2 A.

Considering the converter control bandwidth requirement, the carrier frequency or equivalent switching frequency is selected at 10 kHz. Based on the voltage rating and switching frequency requirements, it is appropriate to use 10-kV SiC MOSFET. To meet the overall converter efficiency requirement, i.e. 98%, the DC/AC stage is designed to have an efficiency of 99%. Using the third generation 10 kV/300 mΩ SiC MOSFET from Wolfspeed (see a discrete version in Figure 11.4), their forward and reverse on state resistances R_{dson-f} and R_{dson-r} can be found as a function of the junction temperature T_j through curve fitting as in (12.90) and (12.91). The turn-on and turnoff energy $E_{on\_6.3kV}$ and $E_{off\_6.3kV}$ at the DC-link voltage of 6.3 kV can be found as a function of the device current I_d through curve fitting as in (12.92) and (12.93). Note that the switching energies, which are obtained from DPT with the turn-on gate resistance of 15 Ω and turnoff gate resistance of 3 Ω, are relatively independent of T_j.

$$R_{dson-f}(T_j) = 1.09 \times 10^{-5} T_j^2 + 2.13 \times 10^{-3} T_j + 0.242 \,\Omega \tag{12.90}$$

$$R_{dson-r}(T_j) = 8.23 \times 10^{-6} T_j^2 + 2.21 \times 10^{-3} T_j + 0.236 \,\Omega \tag{12.91}$$

$$E_{on_{6.3kV}}(I_d) = 0.00601 \times I_d^2 + 0.29671 \times I_d + 2.19276 \,\text{mJ} \tag{12.92}$$

$$E_{off_{6.3kV}}(I_d) = -0.0017 \times I_d^2 + 0.2101 \times I_d + 2.0675 \,\text{mJ} \tag{12.93}$$

To achieve desired power loss and converter efficiency, three dies are paralleled at each switch position to obtain a 10 kV/100 mΩ half-bridge power module. According to Eqs. (12.90) through (12.93), with the DC-link voltage of 6.3 kV, switching frequency of 10 kHz, and peak phase current of 59.2 A, the total loss of each switch position is 370 W, of which 190 W is the switching loss and 180 W is the conduction loss. Although the packaged SiC devices can work at junction temperature of 175 °C, the loss is calculated at 105 °C with sufficient margin. The power loss calculation also considers the LV condition at 0.95 p.u. and uses a phase-shift modulation. Based on the 10 kV SiC MOSFET half-bridge in [24], the scaled 10-kV power module has a junction-to-case thermal resistance of 0.13 °C/W per switch position, a dimension of 6.5 cm × 12.5 cm × 2.4 cm, and a weight of 0.2 kg. Therefore, for each H-bridge, the device volume is 390 cm^3, and the weight is 0.4 kg.

12.3.1.2 Device Cooling

Water cooling is used in this 1 MW converter. Assume that a thin layer of thermal grease with a thickness of 100 μm is applied between the device modules and the cold plate. Given the device module baseplate area of 6.5 cm × 12.5 cm or 81.25 m^2, with a typical thermal conductivity of

1 W/(m·°C) for the thermal interface material (thermal grease), its thermal resistance for each power module can be estimated to be

$$R_{th-TIM} = \frac{100\,\mu m}{1\,\frac{W}{m\cdot°C} \times 65\,mm \times 125\,mm} = 0.0123\,°C/W$$

The ATS-CP-1004 cold plate is used to accommodate two half-bridge modules of each H-bridge. Assuming an open cooling loop, the maximum inlet temperature T_{in} is 35 °C, then the maximum outlet temperature can be calculated as:

$$T_{out} = T_{in} + \frac{4 \cdot P_{device}}{mc_p} = 35\,°C + \frac{4 \times 370\,W}{1.4\,GPM \times 4182\,J/(kg\cdot°C)}$$

$$= 35\,°C + \frac{4 \times 370\,J/s}{0.08833\,kg/s \times 4182\,J/(kg\cdot°C)} = 39.01\,°C$$

where P_{device} is the loss of one of the four devices on the cold plate, m is the flow rate in kg/s (1.4 GPM or 0.08833 kg/s for water in this case), and c_p is the specific heat capacity of the coolant (water). Then, the maximum cold plate surface temperature, considering that each H-bridge uses one cold plate, is calculated as:

$$T_{cp} = T_{out} + 4P_{device} \cdot R_{th-cp} = 39.01 + 4 \times 370 \times 0.006 = 47.89\,°C$$

where the cold plate thermal resistance R_{th-cp} is conservatively assumed to be 0.0064 °C/W for 30% glycol plus water at a flow rate of 1 GPM based on the datasheet. The maximum junction temperature for the hottest device in the H-bridge T_{jmax} can be estimated as:

$$T_{jmax} = P_{device} \cdot R_{th-jc} + 2P_{device}$$
$$\cdot R_{th-TIM} + T_{cp} = 370 \times (0.13 + 2 \times 0.0123) + 47.89 = 105.09\,°C$$

which is slightly higher than the junction temperature 105 °C assumed for the loss calculation. Considering most devices will have a lower junction temperature than T_{jmax} and also the conservative value for the R_{th-cp}, the selected cold plate and flow rate are acceptable. The dimensions of the selected cold plate are 162 mm × 172 mm × 20 mm or 557 cm^3, and its weight is 1.1 kg.

12.3.1.3 DC-Link Capacitor

Each H-bridge operates as a single-phase converter. Assume that each H-bridge in the DC/AC stage has a sinusoidal voltage and current output of

$$\begin{cases} v_o(t) = V_{pk}\sin(\omega t) \\ i_o(t) = I_{pk}\sin(\omega t + \varphi) \end{cases} \tag{12.94}$$

The output power of each H-bridge can be calculated as:

$$p_o(t) = v_o(t)i_o(t) = \frac{V_{pk}I_{pk}}{2}\cos\varphi - \frac{V_{pk}I_{pk}}{2}\cos(2\omega t + \varphi) = P_{DC} + p_{AC}(t) \tag{12.95}$$

Similar to the single-phase AC converter, in (12.95), there is a DC average power, $\frac{V_{pk}I_{pk}}{2}\cos\varphi$, and a second-order ripple power, $\frac{V_{pk}I_{pk}}{2}\cos(2\omega t + \varphi)$. The AC component will induce a second-order voltage ripple on the DC-link capacitor, which needs to be considered when sizing the DC-link capacitor.

Since the DC load or source often cannot tolerate ripple currents, the DC/DC converter in the PCS converter should block the second-order frequency ripple through control. Therefore, all the second-order ripple power needs to be provided by the DC-link capacitor. As a result, the relationship between the DC-link capacitance and the second-order voltage ripple can be derived based on the energy equation in one charging or discharging cycle, as in (12.96), where S is the three-phase converter kVA rating, N is the number of H-bridges per phase, C_{DC} is the DC-link capacitance of each H-bridge, V_{DC_max} and V_{DC_min} are the maximum and minimum DC-link voltages, and ε is the DC-link peak-to-peak voltage ripple in percentage of the rated DC-link voltage V_{DC}:

$$\left| \int_0^{\frac{\pi}{2\omega}} P_{AC}(t) \cdot dt \right| = \frac{V_{pk}I_{pk}}{\omega} = \frac{S}{3N\omega} = \frac{1}{2}C_{DC}V_{DC\_max}^2 - \frac{1}{2}C_{DC}V_{DC\_min}^2 \approx \varepsilon C_{DC}V_{DC}^2$$

(12.96)

From (12.96), the H-bridge DC-link capacitance can be determined as:

$$C_{DC} = \frac{S}{3N\omega\varepsilon V_{DC}^2}$$

(12.97)

As mentioned earlier, the rated DC-link voltage of each H-bridge is determined to be 6.3 kV, considering the MV AC grid nominal voltage range of −12% to +10% of 13. 8 kV. To limit the DC-link voltage ripple within ±5%, the DC-link capacitance is calculated to be

$$C_{DC} = \frac{S}{3N\omega\varepsilon V_{DC}^2} = \frac{1\,\text{MVA}}{3 \times 2 \times 2\pi \times 58.8\,\text{Hz} \times 0.1 \times (6.3\,\text{kV})^2} = 114\,\mu\text{F}$$

In the aforementioned calculation, the lower frequency limit 58.8 Hz allowed for DERs in IEEE Std. 1547-2018 is used. The DC-link capacitor RMS current is also calculated based on the converter operation parameters. When power factor is 0, the DC-link capacitor RMS current has the maximum value, 26 A. Also, the dominant part of the RMS current is the second-order component.

Based on the voltage, capacitance value, and RMS current, the 2 kV/440 μF film capacitor B25620B1447K983 from TDK is selected. Four capacitors are connected in series to withstand the 6.3 kV DC-link voltage and achieve a total capacitance of 110 μF. It has a current capability of 80 A, and the power loss and temperature rise of each capacitor can be calculated as:

$$\begin{cases} P_{cap} = I_{rms}^2 \times ESR = (26\,\text{A})^2 \times 2.5\,\text{m}\Omega = 1.7\,\text{W} \\ \Delta T_{cap} = P_{cap} \times R_{th-cap} = 1.7\,\text{W} \times 1.8\,°\text{C/W} = 3.1\,°\text{C} \end{cases}$$

where R_{th-cap} is the thermal resistance of the capacitor from the datasheet. Therefore, this capacitor can withstand the current. The overall capacitor size in each H-bridge module is 11 412 cm³, and the total weight is 11.4 kg.

Note that in the aforementioned analysis for the impact of single-phase power and energy ripple on DC-link power and DC-link capacitance sizing, the impact of the AC line inductors has been neglected. This is because the AC line inductance is small as a result of high switching frequency for SiC MOSFETs. Reference [25] provides a more accurate analysis that considers the impact of AC inductors, which is needed when inductance is large and/or fundamental frequency is high.

12.3.1.4 Filter Inductor

In this example system, only an inductor filter is employed and it is sized based on the grid-side current harmonic requirements for the baseline design. According to the simulation result, 4.4 mH (0.009 p.u.) inductance is needed, and the current RMS and peak values are 44 A and 66 A, respectively. Considering the core loss caused by high-frequency current ripples, an amorphous

Table 12.5 The inductor baseline design results.

Inductance	4.4 mH
Current RMS value	44 A
Current peak	66 A
Core material	Amorphous
Core size	AMCC1000 × 2
Winding turn number	60
Wire gauge	AWG #5
Air gap	4.8 mm
Winding loss	52 W
Core loss	48 W

Table 12.6 The inductor insulation requirements.

Condition	Voltage
Steady-state operation	12.6 kV
Low-frequency short-term insulation capability	31 kV
Lightning surge	95 kV

core is used. The inductor design process follows the similar procedure introduced in Chapter 4, and the results are shown in Table 12.5. As used in the 13.8-kV grid, based on IEC 60071-1 2006 [26], the inductor insulation requirement needs to consider several conditions, as shown in Table 12.6. As explained earlier, even though the baseline converter does not need to ride through abnormal or transient conditions like lightning transients, its insulation needs to survive such conditions. Since the PCS converter in Figure 12.31 is a transformer-less converter, the filter inductor is the first component that will encounter the grid-side overvoltage and therefore must be designed to withstand such conditions.

As shown in Figure 12.32, the inductor winding is wound on a bobbin, and 60 turns are divided into four layers, with 15 turns in each layer. The insulation between winding layers is Dupont

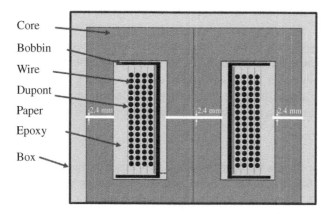

Figure 12.32 Inductor insulation design.

Table 12.7 The inductor voltage stresses.

Conditions	Voltage stress between adjacent turns in the same layer	Voltage stress between adjacent turns in different layers	Voltage stress between winding and core/container (kV)
Lightning	/	/	95
Short term	/	/	31
Steady state	110 V	2.1 kV	12.6

Table 12.8 Dielectric strength of insulation materials.

Conditions	Dupont Nomex 410 paper (kV/mm)	Epoxy (kV/mm)
Lightning	50	13.33
Short-term	30	6
Long-term	1.6	6

Nomex 410 paper. The core is set in an aluminum container, so the core and container box have the same voltage potential. The whole inductor is potted with epoxy in the container, and the container is grounded. Therefore, the insulation between the winding and the core/container is provided by epoxy.

To determine the thickness of the Nomex 410 paper and the distance between the winding and the core/container, the voltage stress and the material electrical strength have to be determined. As shown in Table 12.7, the turn-to-turn and the layer-to-layer insulation are mainly determined by the steady-state operation, while the winding-to-core insulation needs to consider all conditions.

The dielectric strength of insulation materials is shown in Table 12.8. Although the Nomex paper has a much higher dielectric strength for lightning surge, the continuous operation value is only 1.6 kV/mm without partial discharge. Epoxy has a 6 kV/mm dielectric strength for continuous operation. Given that there is a time factor of 0.45 for the dielectric strength of certain materials under one-minute stress compared to that under lightning surge [27], the dielectric strength of the epoxy for lightning surge can be conservatively estimated as $6.7 \div 0.45 = 13.33$ kV/mm. Therefore, considering the worst case, the Nomex paper thickness between winding layers needs to be 1.31 mm, and the distance between the winding and the core/container needs to be 7.13 mm.

For thermal design, the temperature distribution simulation results using Ansys/Icepak are shown in Figure 12.33. The hot spot is located in the winding, and it has a temperature of 104 °C. Since the ambient temperature in the simulation is 20 °C, the temperature rise is 84 °C.

Temperature capabilities of inductor materials are shown in Table 12.9. Clearly, the core material, which has a maximum working temperature of 155 °C, is the limiting material. Considering the maximum ambient temperature of 35 °C, the maximum allowed temperature rise could be up to 120 °C. Considering at least a 20 °C margin for uneven loss distribution and loss estimation discrepancy, the inductor temperature rise in simulation needs to be limited to 100 °C. Therefore, the

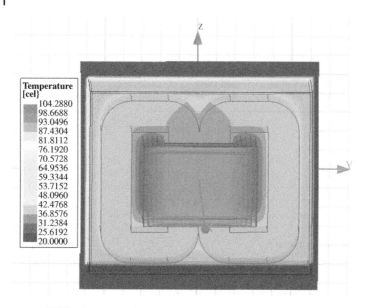

Figure 12.33 Baseline inductor thermal simulation result.

Table 12.9 Temperature capability of inductor materials.

Material	Temperature range (°C)
Core	−20 to 155
Magnetic wire	<200
Bobbin	<170
Nomex paper	<200
Epoxy	<260

designed inductor can meet the thermal requirement. Although there is still some thermal margin, further reducing the wire size or core size cannot shrink the inductor size much due to electrical insulation constraint.

Each phase inductor has a size of 6447 cm^3 and a weight of 23.2 kg.

12.3.1.5 Precharge Circuit

Since the PCS converter may need to start from the MV AC grid side, a precharge circuit is needed to avoid a large inrush current during the start-up. A resistor is connected at the MV AC terminal of the PCS converter to charge the DC-link capacitors in each H-bridge through body diodes of SiC MOSFETs. After the DC links are charged to the value corresponding to the MV AC grid voltage, the precharge resistor will be shorted by a relay. In the example system, considering the average power rating and precharge time, the precharge resistor is selected to be 10 kΩ. The precharge time is around 10 seconds, and the average power of the precharge resistor is about 281 W. The resistor TAP600K10KE which has a power rating of 600 W is selected. It has a continuous dielectric strength of 6 kV RMS, and a short-term (one-minute) dielectric strength of 12 kV RMS. The size and weight of each precharge resistor are 125 cm^3 and 0.5 kg, respectively.

The relay used to bypass the precharge resistor needs to consider both the insulation capability and the current rating. A double-pole single throw relay, RL 38-h, is selected. It has an insulation voltage of 10 kV AC RMS and a current rating of 30 A per pole. To meet the converter current rating, the two poles can be connected in parallel to achieve a 60-A total current. The size and weight of each relay are 1250 cm^3 and 2 kg, respectively.

12.3.1.6 Arrester
Arresters are installed at the MV AC-side terminals (phase to ground) to protect the converter from lightning or switching transient surges. The arrester selection is commonly based on the system voltage level as well as the energy associated with the lightning current and line length (i.e. inductance). Assuming a 10-kA lightning current and 20 km line, and following the selection guide in [28], the SIEMENS arrester 3EK7 120-3AC4 is selected. Its voltage corresponding to an 8/20 μs, 10 kA lighting impulse current is 37.3 kV. The estimated energy according to [28] is 0.5 kJ/kV$_{mcov}$, well below the capability of the selected arrester energy class of 5 kJ/kV$_{mcov}$. The arrester MCOV (maximum continuous operating voltage) is 10.2 kV, and each of the three arresters needed has a volume of 3030 cm^3 and a weight of 1.85 kg.

12.3.1.7 Baseline Design Summary
The component size and weight summary of the three-phase DC/AC converter baseline design are shown in Figures 12.34 and 12.35, respectively. Note that the size and weight of the gate driver (GD) and the gate driver power supply (GDPS) are based on the designs for the 10 kV/20 A SiC MOSFET [29, 30], since the size and weight of these components should not change much for the scaled 10-kV SiC MOSFET half-bridge power modules. Also, although the baseline design does not require a dynamic braking circuit, the item is included in Figures 12.34 and 12.35 for comparison purpose with the later designs.

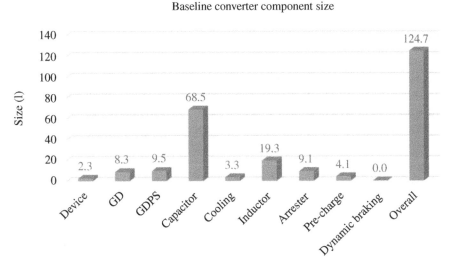

Figure 12.34 Component size summary of the DC/AC converter baseline design.

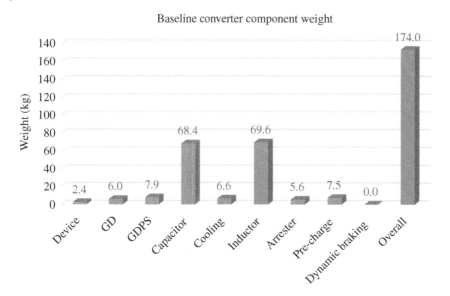

Figure 12.35 Component weight summary of the DC/AC converter baseline design.

It can be seen that the H-bridge DC-link capacitor dominates the converter size, followed by the filter inductor. The filter inductor has a higher weight than the DC-link capacitor. The total size of the considered components is around 125 l, and the total weight is 173 kg.

Note that the total liquid cooling system only accounts for a small portion of size and weight of the baseline PCS DC/AC converter design. This is because: (i) the converter efficiency is high and the power loss is quite small; (ii) the cooling is only for devices, and practically inductors also need to be cooled to maintain the surrounding temperature to be sufficiently low; and (iii) the baseline design assumes an open cooling loop and only considers the cold plate size and weight. In real high-power converters, the cooling loop is often closed, and pumps, pipes, and coolant-to-air heat exchanger should be included. However, these omissions will not impact the results of the design comparison.

12.3.2 Impact of High- and Low-Voltage Ride-Through

For converters to interface with grid and/or to provide grid support, they need to meet converter interface requirements. One of such requirements is to ride through various abnormal and transient grid conditions, rather than to simply disconnect from the grid or trip. One well-known grid-interface standard is IEEE Std. 1547 as specified in Table 12.4 for the PCS converter [23]. The standard requires the LV ride-through (LVRT) and high-voltage ride-through (HVRT) capability of DERs. There are three operation performance categories, and their LVRT and HVRT range and duration are different. Category III has the widest voltage ride-through range and longest time duration, while category I has the narrowest range and the shortest duration. In category III, the continuous operation range is 0.88–1.1 p.u. (see Figure 12.36). It requires the power converter interfaced DER to keep operating within the voltage range of 0.5–0.88 p.u. for 21 seconds and 1.1–1.2 p.u. for 13 seconds. For voltage below 0.5 p.u. or above 1.2 p.u., it is not mandatory to ride through. Therefore, the PCS converter should be designed to operate continuously in voltage range 0.88–1.1 p.u., and also to ride through the LV range 0.5–0.88 p.u. and HV range 1.1–1.2 p.u. Here, the impact of the LVRT and HVRT will be analyzed, and the design and/or control methods will be used to ensure the converter can ride through low or high voltages.

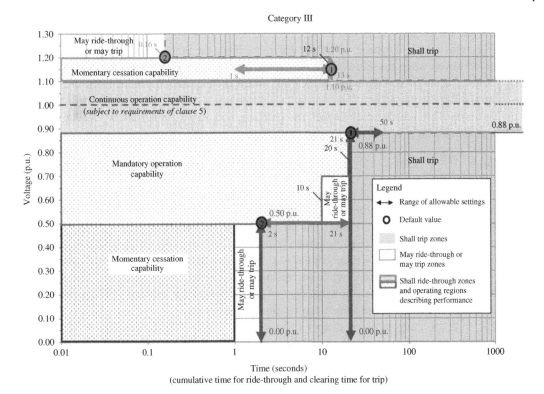

Figure 12.36 LVRT and HVRT requirements in IEEE Std. 1547 for category III [23].

12.3.2.1 Impact of Inrush Current

Figure 12.37 illustrates in one line form the grid-connected VSC and its generic control loop of sampling, calculation or control processing, and PWM generation. With only an inductor filter, the inductance voltage and current equation is

$$v_L = L\frac{di_L}{dt} = v_{PWM} - v_{grid} \tag{12.98}$$

Figure 12.37 Generic voltage source converter control loop.

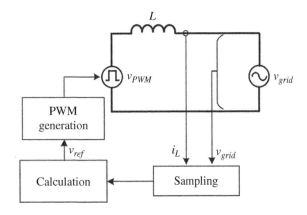

During the steady-state operation, with proper control, since the inductance value L is small in per unit sense, the averaged value of the PWM voltage (i.e. the fundamental component of the PWM voltage) is close to the grid voltage, with their difference being the fundamental voltage drop on the inductor. However, when the grid voltage, v_{grid}, has a sudden change, the converter cannot immediately update its output voltage to follow the grid voltage and maintain the inductor current due to the control and converter delays. Then, the voltage difference between the inductor terminals suddenly increases. The inductor current will also increase, and an inrush current will occur, which can be estimated as:

$$\Delta i_L \approx \frac{v_{grid0} - v_{grid1}}{L} T_d \tag{12.99}$$

where v_{grid0} and v_{grid1} are the grid voltages before and after the sudden grid voltage change, respectively, and T_d is the total delay time, which consists of all the delays in the converter control loop, including sampling, calculation, PWM modulation, and device switching.

From (12.99), it can be seen that the inrush current is approximately proportional to the total delay time, T_d, and is inversely proportional to the filter inductance. A larger filter inductance results in a smaller inrush current, while increasing the converter size, weight, and cost. Control methods, such as grid voltage feed-forward control, can speed up the response of the converter to the grid voltage disturbance, but some delays cannot be eliminated.

Assume that the grid voltage suddenly changes from 1.1 to 0.5 p.u. per Figure 12.36 and $T_d = 2.5T_s$ (T_s is switching period), which includes delays of 0.5, 1, and $1T_s$ for sampling, calculation, and switching, respectively. Using (12.99), the inrush current can be determined as:

$$\Delta i_L = \frac{\dfrac{13.8\,\text{kV}}{\sqrt{3}} \times \sqrt{2} \times (1.1 - 0.5)}{4.4\,\text{mH}} \times 250\,\mu s = 390\,\text{A}$$

which is 6.5 p.u. of the rated current of 59 A.

The components designed and selected in the baseline PCS converter design did not consider the inrush current caused by the sudden grid voltage change and therefore will not be able to tolerate the resulted inrush currents. Either those components, including devices, DC-link capacitors, and filter inductors, must be redesigned, or the inrush current must be controlled. The control-based solution is preferred without impacting hardware. One such control method is the PWM mask method [31]. When the phase current is detected larger than the preset threshold value, the PWM signals in that phase will be temporarily masked to shut down the switching. As a result, the inrush current flows through the body diodes of the 10-kV SiC MOSFETs, and the converter output voltage will be either the total positive DC-link voltage or the negative DC-link voltage depending on the current polarity. The polarity of the voltage applied to the inductor will be inverted, and the inductor current starts to decrease. When the current goes back down to be below the threshold value, the PWM signals will be released, and the converter goes back to normal control. The method has been validated both through simulation and experiments in [31], where the inrush current was reduced from 6.0 to 2.5 p.u. with the PWM mask during a sudden grid voltage drop from 1 to 0.25 p.u.

12.3.2.2 Higher DC-Link Voltage

For the LVRT, as long as the inrush current is limited with the PWM mask approach, the DC-link voltage variation is small. Also, during the steady state, the DC-link voltage is sufficient to output the AC voltage and the modulation index is lower than 1.0. On the other hand, the grid voltage during HVRT is in the range of 1.1 and 1.2 p.u. The DC-link voltage of 6.3 kV of the baseline design, which is selected based on 1.1 p.u. grid voltage, is not sufficient for the 1.2 p.u. grid voltage. Since the AC filter inductance is small (<0.01 p.u.), the DC-link voltage needed for each H-bridge can be estimated as:

$$V_{DC} = \frac{\sqrt{2} \times 13.8\,\text{kV} \times 1.2}{\sqrt{3} \times 2} = 6.76\,\text{kV}$$

In the actual implementation, the DC-link voltage is increased from 6.3 to 6.67 kV, which is slightly lower than 6.76 kV calculated earlier. Therefore, a slight modulation saturation can occur, which is acceptable considering the short period of the HVRT. Utilizing 6.67 kV rather than a higher value is to limit the device voltage stress. Also, 6.67 kV is the averaged value, with ±5% voltage ripple on the DC link, the maximum DC-link voltage will be around 7 kV, which is considered to be within the safe operating region of the device.

12.3.2.3 Power Delivery
Sudden drop of grid voltage during LVRT can cause the DC/AC converter to hit the current limit with a given power flow from the DC/DC converter, which could lead to DC-link overvoltage. Through control coordination between the DC/DC stage and the DC/AC stage and between the PCS converter and DERs/loads, the impact of the power delivery during LVRT on PCS converter design can be reduced.

12.3.3 Impact of Grid Faults

The main concern of grid faults is temporary overvoltage. Per IEEE Std. 1547, during the HVRT, the maximum overvoltage considered is 1.2 p.u. However, when a grid fault occurs, the overvoltage could be more than 1.2 p.u. The overvoltage induced by grid faults is related to the system grounding. For an ungrounded three-phase system, a single phase-to-ground fault can cause the other two phases to see the line-to-line voltages, and the temporary overvoltage for those phases can be as high as $\sqrt{3}$ p.u. under the nominal condition, and even higher under high-line condition (e.g. 1.82 p.u. for a 5% high-line condition). For a three-phase four-wire system, depending on the grounding configuration and wire configuration, the overvoltage during a ground fault could be between 1.25 and 1.5 p.u. [32]. The duration of a temporary overvoltage is related to fault clearance time. Typically, it will be at least as long as the fault detection and breaker opening time, which can be several cycles or tens of milliseconds.

The temporary overvoltage (>1.2 p.u.) ride through is not required in standards for the grid-connected converter. However, the converter needs to survive a grid fault before the protection acts. It is also desirable for the converter to quickly return to operation after the fault is cleared for resiliency. Therefore, converter design requirements considering grid faults are:

1) Ride through grid faults with overvoltage lower than 1.2 p.u. This is similar to HVRT.
2) For faults with overvoltage higher than 1.2 p.u., the converter can temporarily stop operation and survive the fault-related overvoltage and overcurrent before the protection acts, including breaking from the grid through circuit breakers. Once the grid voltage goes below 1.2 p.u., the converter needs to return to operation within a certain time, e.g. three seconds.

The temporary overvoltage can cause inrush current and DC-link overvoltage. When the grid AC voltage peak is higher than the overall DC-link voltage, even though PWM signals are masked, the inrush current can still flow through the body diode of the SiC MOSFET to charge the DC-link. The acceptable DC-link overvoltage depends on the voltage ratings of the devices, DC-link capacitors, DC busbars, and other related components. Also, for a relatively long temporary overvoltage case, the limited inrush current through SiC MOSFET body diodes (or even channels with some clever control), which is higher than the current allowed for continuous operation, can thermally damage the devices. Therefore, another method, in addition to the PWM mask method, is needed to address the overvoltage issue.

The inrush process through MOSFET body diodes is similar to the case of passive rectifiers in Chapter 5 and therefore can be analyzed similarly. As analyzed there, the peak DC-link voltage and current depend on the timing of the fault, and the maximum peak values correspond to the peak of the AC phase voltage. Assuming that the inrush process ends before the breaker opens, as shown in Eq. (5.42), the maximum DC-link voltage is determined by the initial DC-link voltage and the peak AC phase voltage. On the other hand, the maximum inrush current can be affected by the characteristics of AC filter inductors. If the inductor saturation is neglected, the peak inrush current can be determined by the AC filter inductance L and DC-link capacitance C_{DC}, as well as the initial DC-link voltage and the peak AC phase voltage. A case study is shown here to analyze the maximum overcurrent and the DC-link overvoltage as a function of fault timing (or angle) and converter power factor before the fault, assuming the grid voltage suddenly changes from 1 to 1.4 p.u. As shown in Figure 12.38, when the fault occurs near the peak of the grid phase voltage,

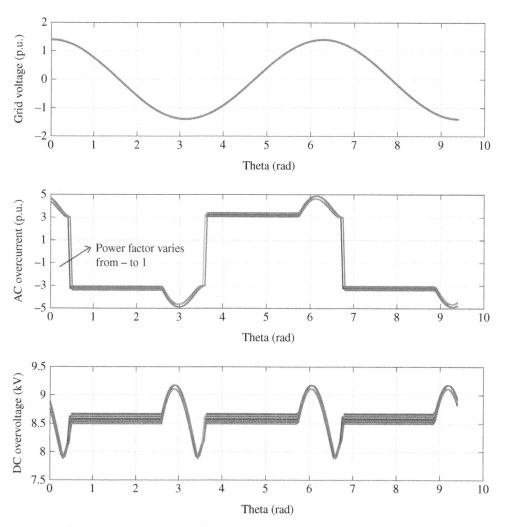

Figure 12.38 Overcurrent and DC-link overvoltage as function fault angle and converter power factor when grid voltage changes from 1 to 1.4 p.u.

the overcurrent and DC-link overvoltage reach their maximum values of 5 p.u. and 9.18 kV, respectively. The converter power factor has little impact as expected. Note that the L and C_{DC} in the baseline design in Section 12.3.1 are used to evaluate the impact.

Considering the voltage capability limitation of the 10-kV SiC MOSFETs, the DC-link capacitors, busbars, and other components, the steady-state DC-link voltage cannot be further increased beyond 6.67 kV selected for HVRT. Thus, the converter cannot ride through an overvoltage of more than 1.2 p.u. The PCS converter can temporarily stop operation during a fault if the overvoltage is higher than 1.2 p.u.; when the grid voltage is recovered after the fault is cleared, the converter needs to return to operation quickly, within three seconds in this case. Therefore, solutions must be provided through protection and control to limit the converter overvoltage during the fault and bring the converter back to normal after the overvoltage.

One common and effective solution for DC-link overvoltage control is dynamic braking circuit. As shown in Figure 12.39, it consists of a braking resistor, $R_{braking}$, and a braking switch $T_{braking}$. The braking resistor needs to be sufficiently small so that it can discharge the DC-link voltage from 9.2 to 7.1 kV within three seconds. The braking switch, which needs to withstand the DC-link voltage, can either use several LV Si or LV SiC devices connected in series, or use MV SiC devices. Since there is no need for fast switching, the series device voltage balancing can be easily realized. Also, the current rating of the braking switching can be relatively small, depending on the discharging speed need. For example, selecting the braking resistance to be 10 kΩ such that the DC link can be discharged within 0.3 seconds, then the maximum current flowing through the braking switching is less than 1 A for the 9.2-kV DC voltage. The DC-link voltage can be limited to lower value by setting the threshold of the braking voltage lower.

Figure 12.40 shows a simulation for a case design. A grid fault occurs at time $t = t_1$. As a result, the voltages of phases A, B, and C are changed from 1 to 0.5 p.u., 1.3 p.u., and 1.4 p.u., respectively. An inrush current of up to 4 p.u. occurs in phase C. To speed up the simulation, the braking resistor is selected to be 1 kΩ. Due to the filter inductor, DC-link voltages of phase C are charged to around 8.7 kV, which is higher than the grid voltage peak. The braking switch is controlled to close when the DC-link voltage is higher than 8.2 kV and to open when the DC-link voltage drops below 8 kV. This control can decrease the DC-link voltage to below 8 kV and reduce the voltage stress on the 10-kV SiC MOSFET and the DC-link capacitor. With 8 kV on the DC link of each H-bridge, the overall DC-link voltage in one phase is around 16 kV, which is higher than the grid peak voltage corresponding to 1.4 p.u. or 15.8 kV, so the DC-link will not be further charged. Instead, the DC-link voltage will gradually decrease as shown in the period of $[t_2, t_3]$ in Figure 12.40 because of the DC-link voltage balancing or bleeder resistors, which are sized at 5 MΩ. The grid fault is cleared

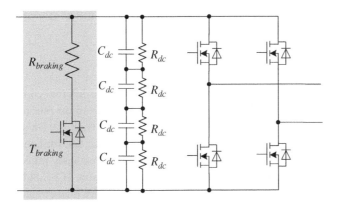

Figure 12.39 The additional braking circuit on each MV DC-link of the PCS converter.

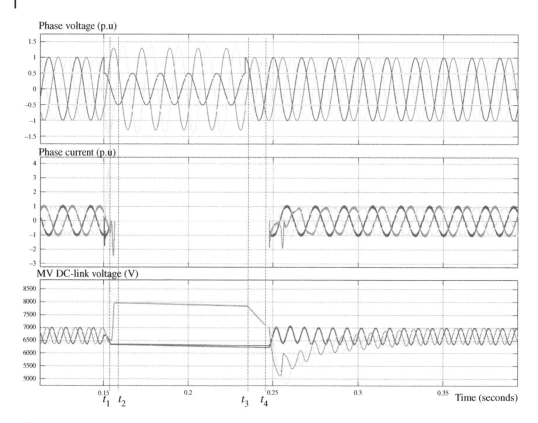

Figure 12.40 Simulation with the additional braking circuit on each MV DC link.

after five cycles, and the converter detects that the grid voltage is recovered to normal at time $t = t_3$, then the braking switch is turned on to discharge the DC link. At time $t = t_4$, DC-link voltages are all below 7.1 kV, and the converter is restarted. With the dynamic braking circuit, the converter can quickly restart after the fault is cleared, and it does not affect the steady-state operation efficiency.

Although the converter is temporarily turned off during the fault, the inrush current still flows through the filter inductor and the body diode of the SiC MOSFET. The maximum inrush current is about 5 p.u. The device transient current rating needs to be higher than the inrush current, and the filter inductor has to consider the inrush current to avoid saturation, which otherwise will result in an even larger inrush current. The large inrush current is not a big concern for diodes or thyristors in passive rectifiers in Chapter 5 but is critical for SiC MOSFETs. The impact of this inrush current on the component design and selection will be discussed later in this section.

12.3.4 Impact of Frequency Ride-Through and Grid Voltage Angle Change

The FRT is also mandated in IEEE Std. 1547-2018. The required FRT ranges for three different converter categories are the same. However, they have different requirements on maximum rate of change of frequency (ROCOF), and the values for categories I, II, and III are 0.5 Hz/s, 2.0 Hz/s, and 3.0 Hz/s, respectively. The continuous operation frequency range is 58.8–61.2 Hz, and the converter needs to ride through the low-frequency range of 50–58.8 Hz and the high-frequency range of 61.2–66 Hz with a duration of 299 seconds. In addition to the frequency variation, IEEE Std. 1547-2018 also requires DERs to ride through 20 electrical degrees of positive-sequence phase angle change within a sub-cycle-to-cycle time frame and up to 60 electrical degrees of individual phase angle change.

12.3.4.1 Frequency Ride-Through

The frequency variation in electric power systems is usually slow, and the maximum ROCOF requirement in IEEE Std. 1547-2018 is 3.0 Hz/s. As shown in [33] through the worst impact simulation study, it is found the frequency variation has little impact on the converter operation.

12.3.4.2 Angle Variation

A voltage phase angle sudden change will lead to a sudden voltage change on the AC filter inductor, which results in an inrush current and corresponding DC-link overvoltage, similar to that in LVRT and HVRT cases. Similarly, with the PWM mask method, they both can be limited.

12.3.5 Impact of Lightning Surge

The converter needs to be designed such that it will not be damaged or even trip when a lightning surge occurs.

12.3.5.1 Inrush Current

The power stage of the converter, through which a lightning surge may pass, can be simplified as shown in Figure 12.41. To further simplify the analysis, the DC/DC stage converters of the PCS converter are represented by current sources I_{dc1} and I_{dc2} and are decoupled from DC/AC stage converters. Because of the three-phase four-wire configuration, the neutral wire can be grounded through the converter or the AC grid grounding, and an inductor L_{ng}, which could also be a resistor or both, represents the neutral line to ground impedance.

When a lightning strikes, the arrester absorbs a large surge current and clamps the converter terminal voltage V_{arr} to a certain level, which is dependent on the surge current for a given arrester, for example, 37 kV with a 10-kA lightning surge. In the United States, the lightning current amplitude I_L with a unit of kA can be represented with an empirical probability P_I [34]:

$$P_I = \frac{1}{1 + (I_L/31)^{2.6}} \tag{12.100}$$

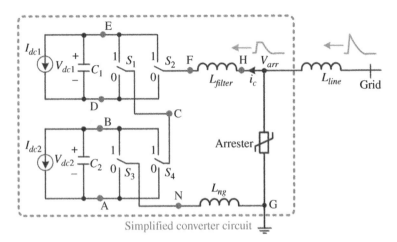

Figure 12.41 Simplified converter circuit during a lightning surge.

It can be seen from (12.100) that 50% of lightning will have a current amplitude greater than 31 kA and 10 kA is on the low end. However, with the extreme nonlinear characteristics of the modern ZnO arresters, the clamping voltage is a weak function of the current, although the current will affect the energy and the arrester selection. Obviously, 37 kV is still much higher than the grid's normal operating voltage. And the maximum AC terminal voltage of the PCS converter, V_{FN}, is the sum of two DC-link voltages in each phase, which is 13.4 kV. The filter inductor L_{filter} and the neutral-to-ground inductance L_{ng} share the remaining voltage and the phase current i_c will increase. The inductor current increment during the lightning surge can be estimated as:

$$\Delta i_L = \frac{di_c}{dt}\Delta t = \frac{V_{arr} - V_{FN}}{L_{filter} + L_{ng}} = \frac{V_{arr} - (S_2 - S_1)V_{dc1} - (S_4 - S_3)V_{dc2}}{L_{filter} + L_{ng}}\Delta t \tag{12.101}$$

where S_k ($k = 1, 2, 3, 4$) represents the state of the devices in each half-bridge, and

$$S_k = \begin{cases} 1 & \text{If current flows through the upper device} \\ 0 & \text{If current flows through the lower device} \end{cases} \tag{12.102}$$

Although a lightning transient duration is very short (tens of microseconds range), the high overvoltage can induce a large inrush current because of the large voltage and small impedance. It can be seen from (12.101) that the worst case happens when:

1) the neutral-to-ground inductance $L_{ng} = 0$, which means the PCS converter is solidly grounded at the neutral point.
2) the converter output voltage, V_{FN}, has an opposite polarity to the lightning surge voltage (i.e. $S_1 = S_3 = 1$, $S_2 = S_2 = 0$).

A lightning voltage waveform is typically represented by a time characteristic of 1.2/50 μs, that is, a rise time of 1.2 μs and a fall time of 50 μs. Therefore, it can be assumed that the arrester clamps the lightning surge at 37 kV for 50 μs, which simplifies the calculation of the inrush current. Then, with a positive lightning current, the current increment during the transient in the worst case will be

$$\Delta i_{L1} = \frac{37\,\text{kV} + 6.67\,\text{kV} + 6.67\,\text{kV}}{4.4\,\text{mH} + 0} \times 50\,\mu\text{s} = 573\,\text{A} = 9.8\,\text{p.u.}$$

If the PCS converter power rating changes, the filter inductance value will change proportionally, but the per unit current should stay the same.

A more benign situation is when the lightning surge occurs, the converter switching state is $S_2 = S_2 = 1$ and $S_1 = S_3 = 0$. The converter output voltage, V_{FN}, will have the same polarity as the lightning surge voltage. This could help to reduce the rising rate of the current. Assuming that the PCS converter output voltage has the same polarity as the lightning surge, the inrush current will be

$$\Delta i_{L2} = \frac{37\,\text{kV} - 6.67\,\text{kV} - 6.67\,\text{kV}}{4.4\,\text{mH} + 0} \times 50\,\mu\text{s} = 268\,\text{A} = 4.5\,\text{p.u.}$$

If the PWM mask function is utilized, when the inrush current exceeds the threshold value, MOSFETs will be turned off, and the inrush current will flow through the body diodes. This will change the PCS converter output voltage to be the same polarity as the lightning surge voltage and help to reduce the rising rate of the inrush current.

With a solid grounding, the PCS converter may have an inrush current within a range of 4.5–9.8 p.u. depending on the switching states of the PCS converter at the moment when the lightning surge occurs, assuming that the saturation characteristic of the filter inductor is neglected and the filter inductance is constant even with a high current. If the inductor is not designed to have high saturation current capability, it may be saturated during the lightning transient and the inrush current will go even higher than the calculated value aforementioned.

The inrush current will flow through the filter inductor, devices, busbar, etc. Also, the regular protection components, such as fuse and mechanical breakers, cannot protect the converter from this inrush current because their response time is much longer than the period of a lightning surge. Therefore, if the inrush current induced by the lightning is not considered in the converter design, it could damage the PCS converter.

From the analysis, it can also be seen that the AC grid-side filter inductance and the neutral-to-ground impedance are the main impedances to limit the inrush current from a lightning surge. A larger filter inductance or a higher grounding impedance may be needed to reduce the inrush current caused by a lightning surge.

12.3.5.2 DC-Link Overvoltage

During lightning, if MOSFETs are switching, the inrush current will flow through the device channels and can either charge or discharge the DC-link capacitor depending on the state of the switches. If the PCS converter has stopped switching as a result of overcurrent protection, the inrush current will flow through MOSFET body diodes and only charge DC-link capacitors. In both cases, DC-link overvoltage may occur.

From Figure 12.41, DC-link capacitor voltages are

$$
\begin{cases}
C_1 \dfrac{dV_{dc1}}{dt} = (S_2 - S_1)i_c - I_{dc1} \\[2mm]
C_2 \dfrac{dV_{dc2}}{dt} = (S_4 - S_3)i_c - I_{dc2}
\end{cases}
\tag{12.103}
$$

Equation (12.103) can be solved with initial conditions for DC-link voltages V_{dc1} and V_{dc2}, DC/DC converter currents I_{dc1} and I_{dc2}, as well as the switching states S_1, S_2, S_3, and S_4. Consequently, the overvoltage and overcurrent of the converter can be obtained. Since i_c is related to the lightning current and arrester characteristics, closed-form analytical solutions are difficult. Alternatively, simulation can be a good tool for estimation of peak DC-link overvoltage and converter peak overcurrent, and some detailed simulation results can be found in [33]. Since the lightning transient duration is short, increasing the DC-link capacitance can significantly reduce the voltage overshoot. Also, increasing the filter inductance can help to reduce the DC-link overvoltage because of the lower inrush current.

12.3.5.3 Converter Insulation

Insulation is another important consideration for the PCS converter when facing a lightning surge. As introduced earlier, during a lightning surge, even with the arrester, the induced voltage could be around 37 kV (referenced to ground). Therefore, high potential (referenced to ground) may also occur on the converter components, which may stress the converter insulation.

With the arrester connected between the line and ground, the potential of different components to ground in the converter need to be evaluated. From Figure 12.41, different points in the converter have different potentials during the lightning surge, and they can be expressed as:

$$
\begin{cases}
V_{NG} = \dfrac{L_{ng}}{L_{ng} + L_{filter}}\left(V_{arr} - S_{21}V_{dc1} - S_{43}V_{dc2}\right) \\[2mm]
V_{AG} = V_{NG} - S_3 V_{dc2} \\[2mm]
V_{BG} = V_{NG} + (1 - S_3)V_{dc2} \\[2mm]
V_{CG} = V_{NG} + S_{43}V_{dc2} \\[2mm]
V_{DG} = V_{NG} + S_{43}V_{dc2} - S_1 V_{dc1} \\[2mm]
V_{EG} = V_{NG} + S_{43}V_{dc2} + (1 - S_1)V_{dc1} \\[2mm]
V_{FG} = V_{NG} + S_{43}V_{dc2} + S_{21}V_{dc1} \\[2mm]
V_{HG} = V_{arr}
\end{cases}
\tag{12.104}
$$

According to (12.104), the converter neutral point potential, V_{NG}, makes a substantial difference for the components' potential during the lightning transient. V_{NG} is mainly determined by the ratio of the neutral-to-ground inductance, L_{ng}, to the filter inductance, L_{filter}. A larger ratio of L_{ng} to L_{filter} will result in a higher neutral point potential during the lightning transient, and the potentials of all the components will increase accordingly; a smaller ratio will result in a smaller neutral point potential and will help to reduce the converter component potentials. In this case, the filter inductor withstands most of the clamped voltage by the arrester, which is designed considering lightning overvoltage as explained in the baseline design earlier.

Therefore, although increasing L_{ng} can reduce the lightning inrush current, it also increases the neutral point potential and then increases the insulation requirements for components in the converter. Besides, the converter or grid grounding impedance is mainly determined by the system's temporary overvoltage design, and any changes for the lightning consideration will need to be coordinated with other requirements.

The detailed behaviors of the PCS during lightning surge can be simulated as described in [33] with proper modeling of the arrester and lightning phenomenon. For the example PCS, with the help of the PWM mask function, the inrush current cab be limited to around 5 p.u.

12.3.6 Impact of Grid Requirements on Converter Hardware Design

Based on the aforementioned analysis, the impact of various grid requirements on converter hardware can be identified, and corresponding design modifications can be implemented using the baseline design as the starting point.

12.3.6.1 Voltage Ride-Through

LVRT can lead to high inrush current. Although the inrush current can be effectively limited to a lower value by the PWM mask method, the inductor design and 10 kV SiC MOSFET selection need to consider the limited inrush current.

HVRT can lead to higher DC-link voltage, which requires a higher DC-link capacitor rating. Using (12.97), the capacitance needed to limit the second-order voltage ripple within ±5% is calculated to be 99 μF. The capacitor voltage rating needs to be higher than 7 kV, considering the voltage ripple. The capacitors used in the baseline design are still used here because they are oversized for the baseline due to the limitation of available commercial products.

12.3.6.2 Grid Faults

The impact of grid faults on converter hardware is mainly the dynamic braking circuit. Considering the discharging time and the braking switch current rating, the braking resistor is determined to be 10 kΩ. Thus, the maximum discharging current is 0.9 A, and the discharging time is around 0.3 seconds. Three IXYS 4.7 kV/2 A Si MOSFETs (IXTL2N470) are used in series as the braking switch, considering their low cost and small current rating compared to the MV SiC MOSFET. The snubber circuit is necessary to balance the voltage, but considering the low current rating, they are neglected when estimating the component size and weight. Moreover, since these MOSFETs do not need to switch fast, their loss is low and no heatsink is used. The overall braking device (three in series) size and weight are around 12 cm^3 and 24 g, respectively.

Based on the DC-link capacitor energy change, the energy dissipated on the braking resistor is

$$E_R = \frac{1}{2} \times (110\,\mu F)^2 \times \left[(9.3\,kV)^2 - (7.1\,kV)^2 \right] = 1.98\,kJ$$

Two Vishay 5 kΩ-thick film chassis mount resistors (LPS0800L5001KB) are selected to be connected in series. Each resistor has a voltage rating of 5 kV, and a short-term energy rating of 1.5 kJ (at 0.3 seconds), which can meet both the insulation and energy dissipation requirements. Also, because of the short period, no heatsink is needed for the resistor. The total resistor (two in series) size and weight are 190 cm^3 and 166 g, respectively.

12.3.6.3 Frequency and Angle Change Ride-Through

Frequency and angle change can also lead to inrush current, similar to the LVRT case. Using the PWM mask function, the inrush current and corresponding DC-link overvoltage can be limited.

The main impact of FRT is on the DC-link capacitance value related to fundamental frequency variation. Based on (12.97), to keep the same voltage ripple, a lower fundamental frequency results in a larger DC-link capacitance. Since the FRT period is not short (hundreds of seconds), it needs to be treated as steady state. Considering the minimum frequency, i.e. 50 Hz for the 60 Hz system, the DC-link capacitance needs to be increased from 99 to 119 μF. In this case, the same DC-link capacitor selected earlier for the baseline design will be kept due to the limitation of available commercial products. Since the capacitance is only 100 μF, the DC-link voltage ripple will increase from ±5% to ±5.4%, which is still acceptable.

12.3.6.4 Lightning Surge

With the filter inductor already designed considering lightning overvoltage and the installed arresters, the main impact of the lightning surge on the PCS converter design is the inrush current. The inrush current needs to be considered in the filter inductor design and 10-kV SiC MOSFETs selection. The DC-link overvoltage is still within the capabilities of the device and the DC-link capacitor, so no redesign is needed in terms of overvoltage.

The impacts of grid requirements on the converter design are summarized in Table 12.10.

12.3.6.5 DC-Link Capacitor Design

The HVRT and the FRT requirements impact the DC-link capacitor design. Starting from the baseline design, the DC-link voltage changes from 6.3 to 6.67 kV after considering the HVRT. The capacitance, which is sized based on ±5% voltage ripple, changes from 114 to 99 μF after considering the

Table 12.10 Summary of the grid requirements impact on the converter design.

Grid/design requirement	Description	Impact on converter design
Voltage ride-through	The converter needs to ride through the low voltage range of 0.5–0.88 p.u. and the high voltage range of 1.1–1.2 p.u.	• Inrush current, which can be effectively limited by the PWM mask function, e.g. to 2 p.u. • The DC-link voltage needs to increase to 6.67 kV to accommodate the 1.2 p.u. grid voltage with an acceptable PWM modulation saturation • Control coordination between the DC/DC stage and the DC/AC stage etc. to avoid large DC-link overvoltage
Grid faults	• Follow the HVRT requirements if the overvoltage is lower than 1.2 p.u. • If the fault causes overvoltage >1.2 p.u., the converter can temporarily stop working; but when the grid voltage is recovered, the converter needs to restart within three seconds	• Inrush current up to 5 p.u. • DC-link overvoltage, up to 9.3 kV • Extra braking circuit, to reduce the DC-link overvoltage to 8 kV during the fault period, and to quickly discharge the DC-link capacitor after the fault is cleared
Frequency ride-through	• The continuous operation frequency range of 58.8–61.2 Hz • Ride through the low frequency range of 50–58.8 Hz and the high frequency range of 61.2–66 Hz with a period of 299 seconds	• No inrush current and large DC-link overvoltage • Need larger DC-link capacitance to limit the DC-link voltage ripple when operating at low fundamental frequency
Angle change right through	• Ride through 20 electrical degrees of positive-sequence phase angle change within a sub-cycle-to-cycle time frame and up to 60 electrical degrees of individual phase angle change	• Inrush current exists due to the sudden voltage change but can be effectively limited by the PWM mask function • DC-link voltage variation, but no need for hardware change
Lightning surge	• Avoid damage or trip when a lightning surge occurs	• Inrush current, up to 5 p.u. with PWM mask • DC-link overvoltage not too high because of the short period • Converter component insulation needs to consider the potential of the neutral point during the lightning transient

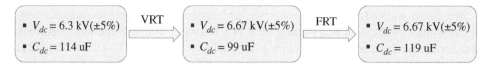

Figure 12.42 showing: $V_{dc} = 6.3$ kV(±5%), $C_{dc} = 114$ uF → **VRT** → $V_{dc} = 6.67$ kV(±5%), $C_{dc} = 99$ uF → **FRT** → $V_{dc} = 6.67$ kV(±5%), $C_{dc} = 119$ uF

Figure 12.42 DC-link capacitor parameters.

HVRT, and then increases to 119 µF after considering the FRT, as shown in Figure 12.42. Since the commercial film capacitor products have limited selections in terms of voltage ratings and capacitance values, the capacitor selected in the baseline design is still used after considering grid requirements. In this case, the voltage and capacitance margins are reduced.

12.3.6.6 AC Filter Inductor Design

The inrush current needs to be considered in the inductor design to ensure it does not saturate. In cases of LVRT, HVRT, faults, grid voltage angle change, and lightning, a larger filter inductance leads to a smaller inrush current. Therefore, the inductor design has two choices: (i) the same inductance with larger inrush current capability; or (ii) larger inductance with smaller inrush current capability. Following the second option and the area-product design approach described in Chapter 4, the inductor A_P relationship can be rewritten as:

$$A_P = L \cdot \frac{I_{rms} \cdot I_{inrush}}{K_u \cdot B_{max} \cdot J_{max}} \tag{12.105}$$

where the peak inductor current in the original formula is replaced by the peak inrush current I_{inrush} through the inductor, which is 5 p.u. in this case. The inductor electrical design parameters and results are shown in Table 12.11. Compared with the baseline design, since the current peak is significantly increased, larger cores are used, and the air gap is increased. Due to the longer winding length per turn, the winding loss is also increased. The same insulation design process as discussed in Section 12.3.1 is followed. The thermal simulation is also conducted with Ansys/Icepak, and the temperature rise is around 68 °C, which can meet the thermal requirement. The inductor's size and weight are 16 410 cm^3 and 74.5 kg, respectively. Note there are altogether three inductors with one for each phase.

12.3.6.7 Device Selection

The inrush current flows through both the inductor and the 10 kV SiC MOSFET, either the channel or the body diode, which can impact device junction temperature as well as package (e.g. wire bonds). Since there is little information about the transient current capability of the 10-kV SiC MOSFET or its body diode, we used the information and data on the more mature LV SiC devices for estimation. The maximum current ratings of LV SiC MOSFETs within a short period, e.g. between 100 μs and 1 ms, is around two to three times their current ratings. Assume the 10 kV SiC MOSFET can also achieve three times of short period current capability through thermal and package design. Then, to withstand the 5 p.u. inrush current, the device current rating needs to be 1.67 p.u., i.e. 100 A, which means each device needs five dies in parallel. This increases the scaled half-bridge power module size and weight to 390 cm^3 and 0.4 kg, respectively.

Table 12.11 Inductor electrical design results considering grid requirements.

Parameters	Baseline	Considering grid requirements
Inductance (mH)	4.4	4.4
RMS current (A)	44	44
Peak current (A)	66	300
Core material	Amorphous	Amorphous
Core size	AMCC1000 × 2	AMCC1000 × 8
Winding turn number	60	48
Wire gauge	AWG #5	AWG #5
Air gap (mm)	4.8	12
Winding loss (W)	52	90
Core loss (W)	48	104

12.3.6.8 Overall Converter Size and Weight Comparison

Figures 12.43 and 12.44 show the size and weight comparison between the baseline design and the design considering grid requirements. Due to the larger device and inductor, the overall component size increases by around 26%, and the overall component weight increases by 90%. As mentioned

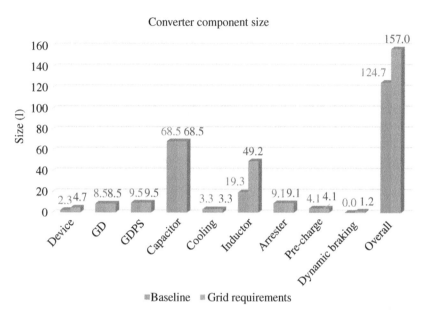

Figure 12.43 Converter size comparison between the converter baseline design and the converter designed considering grid requirements.

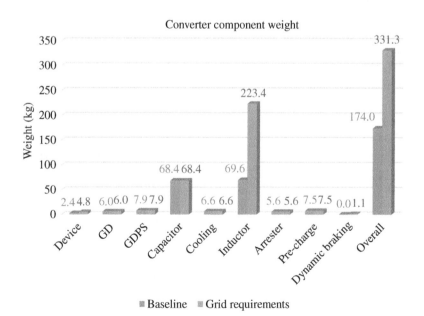

Figure 12.44 Converter weight comparison between the converter baseline design and the converter designed considering grid requirements.

earlier, the DC-link capacitor design is also impacted by the grid requirements, but the same capacitor is used in both cases due to the limitation of available commercial products.

12.3.7 Other Factors in Grid Applications

To facilitate the discussion of the grid application impact on three-phase AC converter design, the design of the specific MV PCS converter in Figure 12.31 has been. Since this converter has a transformer-less and modular configuration, with each phase unit as a module, some aspects commonly seen in other grid-tied three-phase AC converters are not reflected. This subsection will discuss these aspects.

12.3.7.1 Unbalanced Source and Load

Unlike load-side inverters for three-phase motors, grid-tied three-phase converters often need to deal with unbalanced source and loads. Since the PCS converter in Figure 12.31 basically uses three single-phase modules to realize the three-phase conversion, it can naturally support single-phase unbalanced source or loads. Other topologies, such as two- or three-level VSCs, cannot naturally support unbalance and must be designed to meet the unbalance requirements.

In order to support unbalanced three-phase loads, including single-phase loads, a three-phase, four-wire converter system will be needed, including a neutral wire. For a three-phase VSC with common DC link, Figure 12.45 shows some commonly used topologies to provide the neutral wire, including the split DC-link, the four-phase-leg, and the neutral-forming-transformer (NFT). Note that the NFT can be a shunt-connected transformer as shown in Figure 12.45, which is more suitable for high-density applications such as the aircraft system application. NFT can also be a regular transformer at the VSC AC terminal with Y/Y_0 or Δ/Y_0 connection, which can provide isolation in addition to providing neutral.

For all topologies in Figure 12.45, additional hardware will be needed to handle the current through neutral wire. For the split DC-link topology, it will require separate positive and negative DC-link capacitors and also the capacitance values need to be increased to accommodate the neutral current.

For a common DC bus VSC, e.g. two-level VSC, with spilt DC-link capacitors, the current flowing through the neutral or the DC midpoint will cause a ripple voltage. Consider the worst-case load unbalance case: (1) one phase with full load current and the other two phases open; or (2) one phase open and the other two phases with full load currents. It can be shown that both cases will have a zero-sequence current with the same magnitude and same frequency. Use case 1 as example, $i_A = I_m \sin \omega t$, and $i_B = i_C = 0$. The zero-sequence current can be found as:

$$i_0 = \frac{1}{3}(i_A + i_B + i_C) = \frac{1}{3}i_A = \frac{I_m}{3}\sin \omega t$$

Clearly, in this case, the full i_A current flows through the neutral point. It can be seen that the neutral current will be a fundamental frequency current. This current can generate a DC-link voltage ripple with a peak-to-peak value of

$$V_{ripple} = \frac{1}{2C}\int_0^{\frac{\pi}{\omega}} i_A \, dt = \frac{1}{2C}\int_0^{\frac{\pi}{\omega}} I_m \sin \omega t \, dt = \frac{I_m}{\omega C} \tag{12.106}$$

If we want V_{ripple} to be within a small fraction ε of the DC-link voltage V_{DC}, then the half DC-link capacitance C or the full DC-link capacitance C_{DC} should satisfy

$$C = 2C_{DC} \geq \frac{I_m}{\omega \varepsilon V_{DC}} \quad \text{or} \quad C_{DC} \geq \frac{I_m}{2\omega \varepsilon V_{DC}} \tag{12.107}$$

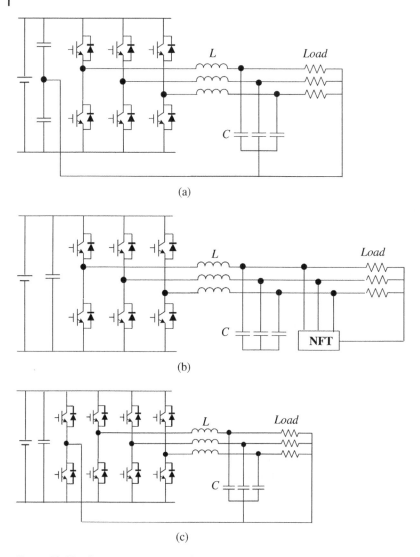

(a)

(b)

(c)

Figure 12.45 Common three-phase four-wire topologies. (a) Split DC-link topology, (b) NFT-based topology, and (c) four-phase-leg topology.

In the aforementioned unbalanced cases, there will also be negative sequence current. The negative sequence current can also generate DC-link voltage ripple. A simple way to determine the DC-link voltage ripple is through DC-link ripple power. Assume the three-phase VSC has balanced AC phase voltages:

$$
\begin{cases}
v_A = V_m \sin(\omega t) \\
v_B = V_m \sin\left(\omega t - \dfrac{2\pi}{3}\right) \\
v_C = V_m \sin\left(\omega t + \dfrac{2\pi}{3}\right)
\end{cases}
$$

For the unbalanced cases 1 and 2 aforementioned, the power on the AC terminal is as (12.108) and (12.109), respectively:

$$p = v_A i_A + v_B i_B + v_C i_C = V_m I_m \left[\sin^2(\omega t) \right] = V_m I_m = \frac{1}{2} V_m I_m - \frac{1}{2} V_m I_m \cos(2\omega t)$$

(12.108)

$$p = v_A i_A + v_B i_B + v_C i_C = V_m I_m \left[\sin^2(\omega t) + \sin^2\left(\omega t - \frac{2\pi}{3}\right) \right] = V_m I_m - \frac{1}{2} V_m I_m \cos\left(2\omega t - \frac{2\pi}{3}\right)$$

(12.109)

If we neglect the VSC power loss, the AC terminal power can be regarded as the DC-link power. From both (12.108) and (12.109), it can be seen that the unbalanced three-phase currents will result in a second-order power ripple on the DC link. The magnitudes of the ripple power are the same for the cases 1 and 2, although their ripple power have different phase angles. Clearly, the DC power or average power of the case 2 with a two-phase load is twice of the case 1 with only one-phase load. Note that in the example, we assumed the phase A voltage and current are in phase to simplify mathematical expression, which can be shown will not impact the result here. Assuming that all ripple power should be provided by the DC-link capacitance C_{DC}, and following the same method in deriving Eqs. (12.96) and (12.97), we can obtain

$$\int_{-\frac{\pi}{4\omega}}^{\frac{\pi}{4\omega}} P_{ripple} \cdot dt = \int_{-\frac{\pi}{4\omega}}^{\frac{\pi}{4\omega}} \frac{1}{2} V_m I_m \cos(2\omega t) \cdot dt = \frac{V_m I_m}{2\omega} = \frac{1}{2} C_{DC} V_{DC\_max}^2 - \frac{1}{2} C_{DC} V_{DC\_min}^2$$

$$= \frac{1}{2} C_{DC} (V_{DC\_max} + V_{DC\_min}) \cdot (V_{DC\_max} - V_{DC\_min}) \approx C_{DC} V_{DC} \cdot \Delta V_{DC\_max}$$

(12.110)

To control ΔV_{DC_max} (also the peak-to-peak ripple) to be within a small fraction ε of the DC-link voltage V_{DC}, the DC-link capacitance value should satisfy

$$C_{DC} \geq \frac{V_m I_m}{2\omega \varepsilon V_{DC}^2} = \frac{M I_m}{4\omega \varepsilon V_{DC}}$$

(12.111)

where $M = 2V_m/V_{DC}$ is the modulation index as defined in Chapter 6. In this case, the maximum modulation index is 1.0, since there cannot be any third-order harmonic injection due to the presence of the neutral wire.

If we compare the capacitance C_{DC} values in (12.107) and (12.111), clearly, the DC-link capacitance required due to zero-sequence current is more than double of the capacitance due to the negative-sequence current. In case both the zero-sequence and negative sequence currents have to be considered, the total DC-link capacitance value required needs to be higher than that required by the zero-sequence current alone (which can be shown as much as 25% higher).

12.3.7.2 Grounding

Like motor drives discussed earlier in this chapter, the grounding design is also an essential part of the grid-tied three-phase AC converter design, especially for high-power high-voltage converters. Most of the grid-tied three-phase AC converters are three-wire converters without the neutral wire; and some can be three-phase, four wire converters, such as the PCS converter in Figure 12.31. Unlike the PCS converter, many high-power three-phase converters also have their own dedicated transformers for isolation as well as for voltage step-up or step-down. In general, there are three types of grid-tied three-phase AC converter configurations as illustrated in Figure 12.46:

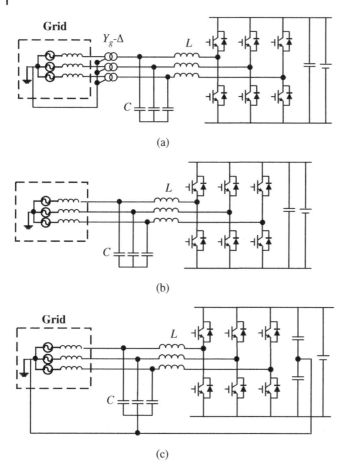

Figure 12.46 Different grid-tied three-phase AC converter configurations: (a) three-wire with isolation transformer; (b) three-wire transformer-less; (c) four-wire transformer-less.

(i) three-wires with transformer; (ii) three-wires without transformer; and (iii) four-wire without transformer. When there is a transformer, the neutral wire can be provided by the transformer if needed. Different configurations have different grounding needs. The grid-tied converter can either be connected to the grid in grid-connected mode or work alone in islanded mode. In either grid-connected or islanded mode, the converter can operate as a grid-following or grid-forming converter.

For three-phase three-wire converters with transformers, the grounding design can be similar to that of high-power motor drives. In this case, the motor is replaced with the transformer. Since the transformer insulation is generally less vulnerable that the motor insulation, and the CM voltage on transformer winding should not be much of a concern, the grounding design can favor the converter needs.

For a transformer-less three-phase three-wire converter, its grounding design needs to consider its operating modes. There are four generic operating modes, and their desired grounding design is discussed as follows:

1) Grid-Connected and Grid-Following Mode: In this mode, the converter follows a given grid voltage, e.g. through a phase-locked-loop, and provides three-phase power or current per the grid or load needs. The grid neutral point should be grounded. During the normal operation, the DC

link to ground voltages will be limited to the peak phase voltages plus the voltage drop on the AC filter. In this case, solid grounding or low impedance grounding for the converter is not desirable in order to avoid CM ground current. On the other hand, ungrounded converter is also not desirable, which can lead to overvoltage when the converter is separated from the grid during an abnormal operating condition. Therefore, a relatively high impedance grounding, e.g. a high value resistance, can be used for the DC-link midpoint grounding. As discussed earlier, high impedance grounding could lead to high-voltage neutral point (DC-link midpoint in this case) during lightning surge. In principle, a protection device, such as an arrester, can be installed at the DC-link midpoint to limit the overvoltage.

2) Grid-Connected and Grid-Forming Mode: In this mode, the converter will act as a generation source and regulate its own terminal voltage. The converter output power and current are determined by loads and other sources in the grid. With three wires, the converter will not supply any zero-sequence current, and the grid neutral point should be grounded. The converter grounding can be designed similarly as in the case of grid-connected and grid-following mode.

3) Islanded and Grid-Following Mode: In this case, some other sources and equipment in the islanded power system, e.g. a microgrid, must provide grounding. Therefore, the converter should behave similarly as in the case of the grid-connected mode, and the grounding design should also be similar.

4) Islanded and Grid-Forming Mode: In this case, the converter must reply on other equipment, such as an NFT, to provide zero-sequence current path. As a result, the grounding again can be designed similarly as the other three cases.

For a transformer-less three-phase four-wire converter, such as the PCS converter aforementioned, the neutral point is generally grounded. Its grounding design should also consider the operating modes.

1) Grid-Connected and Grid-Following Mode: Typically, no zero-sequence current is desired in this mode, which can be achieved through zero-sequence current control. In this case, the solid grounding of the converter neutral point is acceptable and preferred. With solid grounding, the converter voltage stress during normal and abnormal operation will be low. In addition, it will not contribute to zero-sequence current during external fault and therefore will not impact the grid protection settings.

2) Grid-Connected and Grid-Forming Mode: In this mode, the converter zero-sequence current can still be controlled to be zero. However, if there are unbalanced loads that need to be supplied, the zero-sequence current will be nonzero. In order not to cause overvoltage in the loads, the converter neutral point should be solidly grounded.

3) Islanded and Grid-Following Mode: In this mode, the zero-sequence current can be controlled to zero or to some other values as needed. The grounding can be solid but may or may not contribute to the system grounding.

4) Islanded and Grid-Forming Mode: In this mode, the converter should support unbalanced loads and provide neutral current path. The converter should provide the system grounding and therefore the neutral point should be solidly grounded.

Table 12.12 summarizes the desired grounding schemes for different grid-tied converter configurations in Figure 12.46 as well as corresponding capabilities. Note that the switchable NFT indicates the transformer grounding can be switched between grounded and ungrounded, or a grounding transformer can be switched in and out of the grid.

Converter and component grounding can impact parasitic capacitances to ground, which can impact switching loss [35].

Table 12.12 Grounding of the different grid-tied converter configurations.

	Three-phase three-wire w/isolation transformer	Transformer-less three-phase three-wire	Transformer-less three-phase four-wire
Grounding scheme	High impedance ground	High impedance ground	Solid ground
Maximum component to ground voltage	$\approx \pm V_{DC}/2$	$\approx \pm V_{DC}/2$	$\pm V_{DC}/2$
PWM third harmonic injection capability	Yes	Undesirable	No
Need for neutral point overvoltage protection	Maybe not (see transformer discussion below)	Maybe	No
Can operate with single line to ground fault	Yes, if fault is between converter and transformer	No	No
Suitable operating modes	All (need a switchable NFT for grid-forming mode)	All but need a switchable NFT in islanded grid-forming mode	All (need zero-sequence current or impedance control in grid-following modes)
Unbalance support	Yes, with transformer grounded on the grid side	Yes, but need an NFT	Yes

12.3.7.3 Transformer vs. Transformer-Less Design

The example PCS converter in Figure 12.31 is a transformer-less design, which gets rid of the bulky and relatively expensive 50 or 60 Hz transformer. On the other hand, many grid applications still use line frequency transformers. The converter design with a transformer provides a number of advantages: (i) galvanic isolation such that the grid-side ground and converter-side ground are separated; (ii) the isolation and insulation will also reduce the voltage stress on the converter during certain grid transients, including lightning surges; (iii) voltage conversion to allow the use of lower voltage converters and components in higher voltage grid applications; and (iv) the transformer leakage inductance to serve as filter inductance. Since the leakage inductance will not saturate, it is very valuable for overcurrent conditions. The three-phase line frequency transformer can help relieve converter overvoltage during lightning surge through its leakage inductance to limit the overcurrent. Shielding can also be added between the primary and secondary windings of the transformer to reduce the overvoltage on the converter connected to the secondary side.

As for the disadvantages of the line frequency transformers, in addition to its size, weight, cost, and loss, another one is potential saturation issue. Ideally, the transformer voltages on both the grid and converter sides should be fundamental frequency sinusoidal voltages. With PWM voltages, there could be DC voltage offset due to nonideal device characteristics, and voltage sensor and control accuracy limitations. DC voltage offset can lead to saturation. As a result, converter transformers are often designed with air gaps to avoid saturation under a given DC offset. Control can be introduced to mitigate transformer saturation, including sensing the difference between the primary and secondary currents to correct PWM voltages.

Other issues with line frequency converter transformers include harmonics and extra loss. The ripple currents need to be considered in the design. The extra losses include winding and core loss as a result of harmonic currents, as well as parasitic losses as a result of PWM voltages.

12.4 Summary

This chapter considers the impact of applications on three-phase AC converter design. For the motor drive applications:

1) The current harmonics due to PWM voltages in AC motors (induction or synchronous motors) are generally not a concern due to relatively high motor impedance at the switching frequency and its harmonics. One exception is the harmonics due to the switching dead-time in a phase-leg, which has been discussed in Chapter 10. The voltage harmonics can also cause extra motor core and cable loss.

2) The PWM voltages will generally produce CM voltages, which can appear on motor stator neutral and in turn be coupled to the motor rotor shaft. The shaft voltage and corresponding bearing currents can be detrimental to the motor and even the application system. Different schemes to mitigate the issues are introduced with trade-offs for the inverter and the motor designs and performance. The schemes include overall CM voltage reduction through topologies or control; CM voltage redistribution through grounding, filter, and parasitics design; and blocking bearing currents through motor modifications.

3) The impact of PWM voltages with high dv/dt on motors include voltage doubling at the motor terminals due to long cables, highly nonlinear initial voltage distribution in motor windings that overstress the insulation of the first several coils, and high current through parasitic capacitances. Faster switching WBGs with higher dv/dt exacerbate the problems. Common mitigation methods have been discussed and compared, including the motor terminal filter, the dv/dt filter, and the sinewave filter. The motor terminal filter is only applicable in special applications and for drives with relatively low switching frequency or low voltage due to the filter loss. The dv/dt filter is more generic, but it is not suitable for long cables (e.g. >300 m) because of the required large filter components. It may also not be able to reduce cable and other parasitic losses, which can be a concern at high switching frequency and/or high voltage. The sinewave filter is most effective but rarely used due to its bulky size. However, for motor drives with very long cables and/or high switching frequency, the sinewave filter can be competitive with the dv/dt filter.

4) The motor drive grounding design will impact component voltage stress, insulation design, CM voltage and current, and protection. For high-power motor drives, DC-link midpoint generally should be grounded through a grounding impedance network to achieve both DC/low-frequency and high-frequency grounding.

For grid applications:

1) The three-phase converters in grid applications need to comply with grid codes such as IEEE Std. 1547, as well as to survive abnormal operating conditions. The relevant grid requirements include LVRT/HVRT, grid faults, low-/high-frequency ride-through, voltage angle variation, lightning surge, and unbalanced load.

2) The LVRT, angle variation, and FRT can all cause excessive inrush current, which will overstress inductors and devices. The HVRT can cause temporary overvoltage, which will impact capacitor and device voltage ratings. The low-frequency ride-though can lead to higher DC voltage ripple, and therefore more capacitance will be needed to limit such ripple increase. The grid

faults can cause voltage potential shift and excessive DC-link voltage, and therefore extra circuits such as a dynamic braking circuit may be needed. The unbalanced load will require higher DC-link capacitance to control the second-order ripple voltage.

3) The lightning surge can be controlled first by adding arresters at the converter AC terminals. However, the clamped voltage by the arrester is still much higher than the converter normal operating voltage, possibly causing a large inrush current and the corresponding DC-link over-voltage when the converter neutral point is solidly grounded. A higher neutral-to-ground impedance can reduce the inrush current but will increase the neutral point potential, which is undesirable.

4) After considering the grid requirements, the overall component volume and weight of the example 13.8 kV, 1 MW converter are increased by 26% and 90%, respectively. The increase is a result of about 5 p.u. inrush current, 6% increased DC-link voltage, slightly increased capacitance value, and added dynamic braking circuit. The lightning arresters are needed even for AC motor drives connected to grid. Most of the size and weight increase is due to the AC filter inductor.

5) The unbalanced load, the grounding, and the transformer vs. transformer-less interface for the grid-tied three-phase converters are discussed, including the pros and cons of different approaches and their impact on the converter design.

References

1 R. Krishnan, "*Electric Motor Drives: Modeling, Analysis, and Control*," Prentice Hall, 2001.

2 R. Ueda, T. Sonoda and S. Takata, "Experimental results and their simplified analysis on instability problems in PWM inverter induction motor drives," *IEEE Transactions on Industry Applications*, vol. 25, no. 1, pp. 86–94, 1989.

3 A. Boglietti, P. Ferraris, M. Lazzari and M. Pastorelli, "Influence of the inverter characteristics on the iron losses in PWM inverter-fed induction motors," *IEEE Transactions on Industry Applications*, vol. 32, no. 5, pp. 1190–1194, 1996.

4 P. Alger and H. Samson, "Shaft currents in electric machines," *Transactions of the American Institute of Electrical Engineers*, vol. XLIII, no. 43, pp. 235–245, February 1924.

5 S. Chen, T. Lipo and D. Fitzgerald, "Source of induction motor bearing currents caused by PWM inverters," *IEEE Transactions on Energy Conversion*, vol. 11, no. 1, pp. 25–32, 1996.

6 S. Chen, T. Lipo and D. Novotny, "Circulating type motor bearing current in inverter drives," in *Conference Records of the IEEE IAS Annual Meeting*, San Diego, CA, October 6–10, 1996.

7 T. A. Hatfield, "Common Preventable Electric Motor Failures: Bearings," HECO, 4 January 2018. [Online]. Available: https://hecoinc.com/common-preventable-electric-motor-failures-bearings/. Accessed in 12 October 2022.

8 D. Busse, J. Erdman, R. Kerkman, D. Schlegel and G. Skibinski, "System electrical parameters and their effects on bearing currents," *IEEE Transactions on Industry Applications*, vol. 33, no. 2, pp. 577–584, 1997.

9 S. Chen, T. Lipo and D. Fitzgerald, "Modeling of motor bearing currents," *IEEE Transactions on Industry Applications*, vol. 32, no. 6, pp. 1365–1370, 1996.

10 D. Busse, J. Erdman, R. Kerkman, D. Schlegel and G. Skibinski, "Bearing currents and their relationship to PWM drives," *IEEE Transactions on Power Electronics*, vol. 12, no. 2, pp. 243–252, 1997.

11 F. Wang, "Motor shaft voltage and bearing currents and their reduction in multi-level medium voltage PWM voltage source inverter applications," *IEEE Transactions on Industry Applications*, vol. 36, no. 5, pp. 1336–1341, September/October 2000.

12 D. Busse, J. Erdman, R. Kerkman, D. Schlegel and G. Skibinski, "The effects of PWM voltage source inverters on the mechanical performance of rolling bearings," *IEEE Transactions on Industry Applications*, vol. 33, no. 2, pp. 567–576, 1997.

13 J. Erdman, R. J. Kerkman, D. Schlegel and G. Skibinski, "Effect of PWM inverters on AC motor bearing currents and shaft voltages," *IEEE Transsctions on Industry Applications*, vol. 32, no. 2, p. 250–259, 1996.

14 A. v. Jouanne and P. Enjeti, "Design considerations for an inverter output filter to mitigate the effects of long motor leads in ASD applications," *IEEE Transactions on Industry Applications*, vol. 33, no. 5, pp. 1138–1145, 1997.

15 IEEE. "*IEEE Guide for Testing Turn Insulation of Form-Wound Stator Coils for Alternating-Current Electric Machines, IEEE Std 522-2004 (Revision of IEEE Std 522-1992),*" IEEE, 2004.

16 IEEE Rotating Machinery Committee, "Impulse voltage strength of AC rotating machines," *IEEE Transactions on Power Apparatus and Systems*, vol. 100, no. 8, pp. 4041–4053, 1981.

17 A. Greewood, "*Electrical Trasients in Power Systems,*" New York: Wiley-Interscience, 1971.

18 R. Kerkman, D. Leggate and G. Skibinski, "Interaction of drive modulation and cable parameters on AC motor transients," *IEEE Transactions on Industry Applications*, vol. 33, pp. 722–730, 1997.

19 E. Velander, G. Bohlin, Å. Sandberg, T. Wiik, F. Botling, M. Lindahl, G. Zanuso and H.-P. N. Tra, "An ultralow loss inductorless dv/dt filter concept for medium-power voltage source motor drive converters with SiC devices," *IEEE Transactions on Power Electronics*, vol. 33, no. 7, pp. 6072–6081, 2018.

20 M. M. Swamy, J.-K. Kang and K. Shirabe, "Power loss, system efficiency, and leakage current comparison between Si IGBT VFD and SiC FET VFD with various filtering options," *IEEE Transactions on Industry Applications*, vol. 51, no. 5, pp. 3858–3866, 2015.

21 J. Lyons, V. Vlakovic, P. Espelage, F. Boettner and E. Larsen, "Innovation IGCT main drives," in *Conference Records of IEEE Industry Applications Society Annual Meeting*, Phoenix, AZ, October 3–7, 1999.

22 F. Wang, R. Lai, X. Yuan, F. Luo, R. Burgos and D. Boroyevich, "Failure-mode analysis and protection of three-level neutral-point-clamped PWM voltage source converters," *IEEE Transactions on Industry Applications*, vol. 46, no. 2, pp. 866–874, March/April 2010.

23 IEEE Std 1547-2018, "*IEEE Standard for Interconnection and Interoperability of Distributed Energy Resources with Associated Electric Power Systems Interfaces,*" IEEE, 2018.

24 D. Johannesson, M. Nawaz and K. Ilves, "Assessment of 10 kV, 100 A silicon carbide MOSFET power modules," *IEEE Transactions on Power Electronics*, vol. 33, no. 6, pp. 5215–5225, 2018.

25 R. W. F. Wang, D. Boroyevich, R. Burgos, R. Lai and P. Ning, "A high power density single phase PWM rectifier with active ripple energy storage," *IEEE Transactions on Power Electronics*, vol. 26, no. 5, pp. 1430–1443, May 2011.

26 IEC 60071-1, "*Insulation Co-ordination – Part 1: Definitions, Principles and Rules,*" IEC, 2006.

27 A. Küchler, "*High Voltage Engineering: Fundamentals-Technology-Applications,*" Springer Vieweg, 2018.

28 Siemens Energy, "*Medium-Voltage Surge Arresters,*" Raleigh, NC: Siemens Energy, 2021.

29 X. Huang, S. Ji, J. Palmer, L. Zhang, D. Li, F. Wang, L. M. Tolbert and W. Giewont, "A Robust 10 kV SiC MOSFET Gate Driver with Fast Overcurrent Protection Demonstrated in a MMC Submodule," in *IEEE Applied Power Electronics Conference and Exposition (APEC)*, March 15–19, 2020.

30 L. Zhang, S. Ji, S. Gu, X. Huang, J. E. Palmer, W. Giewont, F. Wang and L. M. Tolbert, "Design considerations for high-voltage-insulated gate drive power supply for 10-kV SiC MOSFET applied in medium-voltage converter," *IEEE Transactions on Industrial Electronics*, vol. 68, no. 7, pp. 5712–5724, July 2021.

31 H. Li, Z. Gao, S. Ji, Y. Ma and F. Wang, "An inrush current limiting method for grid-connected converters considering grid voltage disturbances," *IEEE Journal of Emerging and Selected Topics in Power Electronics*, vol. 10, no. 2, pp. 2608–2618, April 2022.

32 IEEE Std C62.92.4-1991, "*IEEE Guide for the Application of Neutral Grounding in Electrical Utility Systems, Part IV-Distribution*," IEEE, 1992.

33 H. Li, "Design, Control, and Test of Emerging Grid-connected Converters Considering Grid Requirements," Ph.D. Diss., University of Tennessee, Knoxville, 2022.

34 EPRI, "*Transmission Line Reference Book: 345 kV and Above/Second Edition*," EPRI (Electric Power Research Institute), 1982.

35 H. Li, Z. Gao and F. Wang, "A PWM Strategy for Cascaded H-bridges to Reduce the Loss Caused by Parasitic Capacitances of Medium Voltage Dual Active Bridge Transformers," in *IEEE Energy Conversion Congress and Exposition (ECCE),* Detroit, MI, 2022.

Part III

Design Optimization

.

13

Design Optimization

13.1 Introduction

Throughout this book, the design has been treated as an optimization problem at all levels for a three-phase AC converter. In previous chapters, components (e.g. a magnetic inductor) and subsystems (e.g. rectifier or inverter power stage, EMI filter) have been designed (optimally), and their models have been established for design purpose. Because an inductor or even an EMI filter is relatively simple, its design can be done manually or only use computer as an aid (e.g. through case simulations). For the design of a whole three-phase AC converter, which consists of many components and subsystems, to meet certain design objective(s) (e.g. cost, size, and/or efficiency), these components and subsystems need to be designed/selected together to achieve an overall optimal design. Given the complexity, it is generally desirable or even necessary to involve design automation (full or partial). It is therefore useful to build or use a design tool. This chapter provides basic theory and method on design optimization, also the relationship between the whole converter design and the design of its constituent components and subsystems. The design automation tool development examples will be presented.

Note that design automation is still based on the fundamental understanding and modeling of relationships between various design parameters and variables, which have been covered at component and subsystem levels in previous chapters. This chapter will also discuss how to combine and integrate them at the converter system level.

13.2 Design Optimization Concept and Procedure

13.2.1 Concept and Mathematical Formulation

The concept for design optimization has been introduced in Chapter 1 and used throughout the book. System design conditions, constraints, variables, and objective functions have been defined, explained, and applied to different three-phase AC converter subsystems. In summary, the converter design should be an optimization process, involving identifying the optimal design from all feasible designs. The feasible designs are found from available design space for all design variables under given system conditions that meet system constraints.

Optimization methodology requires engineers to design in a structured way, and a general design problem can be formulated as follows.

Design of Three-phase AC Power Electronics Converters, First Edition. Fei "Fred" Wang, Zheyu Zhang, and Ruirui Chen.
© 2024 The Institute of Electrical and Electronics Engineers, Inc. Published 2024 by John Wiley & Sons, Inc.

Minimize or maximize objective function $F(X)$:

$$\min \text{ or } \max F(X) \tag{13.1}$$

Subject to constraints, including inequality constraints, equality constraints, and integer constraints:

$$\begin{cases} g_i(X, U) < 0, & i = 1, ..., m \\ h_j(X, U) = 0, & j = 1, ..., p \\ 1 \leq x_k \leq K, & k \in (1, N) \end{cases} \tag{13.2}$$

where X and U are the design variable and design condition vectors with a dimension of N and O, respectively,

$$X = \begin{bmatrix} x_1 \\ x_2 \\ \vdots \\ x_N \end{bmatrix} \quad U = \begin{bmatrix} u_1 \\ u_2 \\ \vdots \\ u_O \end{bmatrix} \tag{13.3}$$

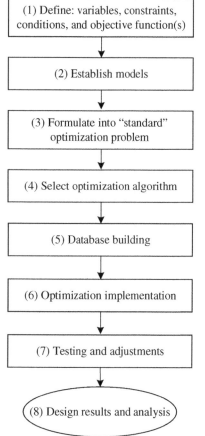

Figure 13.1 Design optimization procedure.

Usually, the objective function has only one target. If there is more than one objective, one common approach is to assign a weight for each objective and sum up to one overall objective function by using weighting coefficients.

13.2.2 Optimization-Based Design Procedure

Assuming that we will start a design optimization from the very beginning, there are generally eight steps in the procedure, which are illustrated in Figure 13.1.

Compared to the experience-based or rule-based manual design, the optimization-based design has several advantages:

1) It can lead to a true "optimal" design by exhausting almost all feasible designs. It is very difficult to cover all design choices via manual design;
2) It can generally solve a larger design problem, using mathematical formulation and design automation, without relying on certain design sequence that is often necessary in manual design procedure. This again will lead to a more "optimal" design. For example, in a converter design, it is almost inevitable to break the design tasks into the designs of several subsystems and components, and very often the interactions of these subsystems are either ignored or greatly simplified.
3) It can avoid using some experience-based design rules and design margins, which are often conservative and not very scientific. This is often related to the point 2. When breaking the design tasks of a system into the

designs of several subsystems, it is often necessary to impose some constraints at the boundaries that may not be necessary if the designs are combined together. For example, when designing a source-side inverter with *LCL* filter manually, the filter is separately designed. To simplify the design, a 10–20% ripple current limit is often assumed for the converter-side inductor, although the true limits should only be temperature for the inductor and devices through which the ripple current will flow, and to a lesser extent, the saturation flux density.

The optimization-based design can be applied to the whole converter as well as to components and subsystems. Very often, it is quite challenging to employ only one optimizer for the whole converter design, as there will be a large number of design variables involved, and the performance constraints and corresponding models are numerous and complex. In order to reduce the design complexity and to improve the design efficiency, the whole converter design optimization problem is often separated into design optimization problems of several different converter function blocks or subsystems. The interactions of these designs are handled either through design interactions or design rules. The design optimization problems for different subsystems of the three-phase converter have been covered in Part II.

13.2.3 Multi-objective Optimization and Pareto Front

For converter design, several objectives can be used for design optimization, including efficiency, cost, volumetric power density, and specific power (or weight density). Usually, there are trade-offs among these objectives, and it is difficult to optimize all of them at the same time in a design. For example, there is often a trade-off relationship between efficiency and power density (or cost). If we only try to design a high efficiency converter, certain components can be oversized, which will benefit the converter efficiency but will increase the converter size, and probably cost. There can also be trade-offs between cost and power density, as certain high-quality components that will help with better density are usually more costly.

In many cases, single-objective optimization is acceptable. In other cases, it is desirable to consider more than one objective in a converter design optimization. There are two general methods to achieve this. The first method is to still use only one objective as the design target, while setting limits for other objectives. For example, if we want to consider both power density and efficiency in the design, we can choose the power density as the design objective, while using minimum converter efficiency as a constraint. The second method to consider multiple objective functions is to use multi-objective optimization (MOO) method. The MOO method in [1] is based on the weighted sum method given as in (13.4), where w_i's are the weighting coefficients of different objectives and $f_i(x)$ is the i-th of the m objective functions. The weighting coefficients can be selected based on practical requirements and experience.

$$\min F(x) = \sum_{i=1}^{m} \omega_i \cdot f_i(x) \tag{13.4}$$

Another way to solve a MOO problem is to convert other objectives to one key objective. In PV applications, the life cycle cost (LCC) function is derived in [2], converting the converter efficiency to the life cost function. With this method, the selection of weighting coefficients for different objectives can be avoided.

Furthermore, to identify optimal design result(s) for MOO, the Pareto front concept can be employed [2, 3]. As illustrated in Figure 13.2, through performance function (objective function), design vectors (combinations of all design variables) in the feasible design space D are mapped into the feasible performance space P. The Pareto optimality is defined as a set of design variables that no

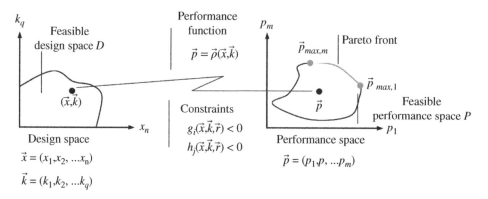

Figure 13.2 Multi-objective optimization with Pareto front curve.

other designs can further improve all performance (all objectives) simultaneously. Therefore, based on this definition, the Pareto front can be identified in the performance space, which is illustrated in Figure 13.2. With the Pareto front, an optimal design result can be selected from Pareto fronts, considering other converter performance (e.g. reliability and parametric uncertainty).

13.2.4 Mathematical Properties of Power Converter Design Optimization Problems

Optimization problems can be classified in many different ways. Some of the classifications relevant to converter designs are briefly introduced here.

13.2.4.1 Linear vs. Nonlinear Problems

In the problem defined by (13.1), (13.2), and (13.3), if the objective function $F(X)$, and all constraints $g_i(X, U)$'s and $h_j(X, U)$'s are linear functions of design variables X, then the optimization problem is a linear optimization or linear programming problem. If any one of these functions is a nonlinear function of X, then the problem is nonlinear. Most of the engineering problems, including the power converter design problems, are nonlinear problems.

13.2.4.2 Continuous, Discrete, or Hybrid (Mixed) Problems

This classification is determined by the characteristics of design variables. If all design variables in a design problem can take continuous values, then the problem is a continuous optimization problem. If all variables can only have discrete values, the problem is a discrete optimization problem. If some variables have continuous values and others have discrete values, the problem is a hybrid optimization problem. An integer constraint could still be considered continuous, e.g. in the case of inductor turns number. It should be mentioned that most methods used in nonlinear optimization or nonlinear programming are for continuous problems. To solve discrete or hybrid variable problems, special methods are often needed. Power converter design is obviously a hybrid problem with both discrete and continuous variables. The discrete design variables include inductor core, wire gauge, device candidates, heatsink and fan types, modulation schemes, and so on; the continuous variables include operating point selection, e.g. switching frequency, DC-link voltage in an AC-fed motor drive, and flow rate for thermal management system (TMS). As already mentioned earlier, some design variables are even nonnumeric, e.g. modulation scheme, filter structure, and topology.

13.2.4.3 Convex vs. Nonconvex Problems

Convex problems are attractive for optimization since their local minimum is also a global minimum. There are three rules that can be applied to determine whether an optimization problem is a convex optimization [4]. First, if the objective function is nonlinear, the problem is not a convex optimization. Second, if design variables include discrete-type variables, the problem is not a convex function. Third, if one constraint is written as "$g(x) < 0$" and $g(x)$ is not a convex function, the problem is not a convex optimization. The power converter design problems usually break at least one of the three rules in most cases. Hence, power electronics problems are treated as nonconvex optimization.

13.2.4.4 Constrained vs. Unconstrained Problems

In the problem defined by (13.1), (13.2), and (13.3), (13.2) represents the design constraints. If $m \neq 0$ or $p \neq 0$, the problem is a constraint one; if $m = p = 0$, the problem is unconstrained. Solving the unconstraint problem is to find the extrema of the objective functions in design space without any constraints. Practically, unconstrained problems do not exist. However, sometimes, a constrained problem can be converted to an unconstrained problem and then it can be more easily solved. Naturally, a power converter design is always a constrained problem, with specifications and performance requirements for the converter.

13.2.4.5 N-Dimensional vs. One-Dimensional Problem

Even though a constrained problem can be converted to an unconstrained problem, the unconstrained problem solution will still be N-dimensional problem with N-th order variable X. Many unconstrained problems can be converted step by step from N-dimensional design space to one-dimensional space. As a result, the extrema can be determined along a line through one-dimensional search. Power converter design is naturally a multidimensional problem.

Another feature of the power converter design problem is that its objective function is usually nonexplicit. The efficiency, cost, or power density of a power converter generally needs to be calculated using many different models and are based on some design procedures, which are usually not explicit functions of design variables. Therefore, for the converter system, generally, no specific and explicit objective function can be provided for an optimization algorithm. In contrast, for components design, if analytical models are employed, an explicit objective function may be obtained.

13.3 Optimization Algorithms

13.3.1 Optimization Algorithm Classification

The optimization algorithms are generally divided into two categories: linear optimizations and nonlinear optimizations [5]. In theory, a global optimal solution can be definitively obtained in a linear optimization or linear programming problem with certain algorithms such as simplex method, duality theory, and interior point method [4]. However, for nonlinear optimization problems, no global optima can be guaranteed with different nonlinear optimization methods such as gradient-based algorithms, penalty function-based algorithms, heuristic algorithms, and the linear fractional programming [5]. As many as possible different initial groups of design variables should be used for searching better local optima, while global optima searching in the nonlinear programming is an NP (nondeterministic polynomial-time) hard problem. The classification of different optimization algorithms is shown in Figure 13.3.

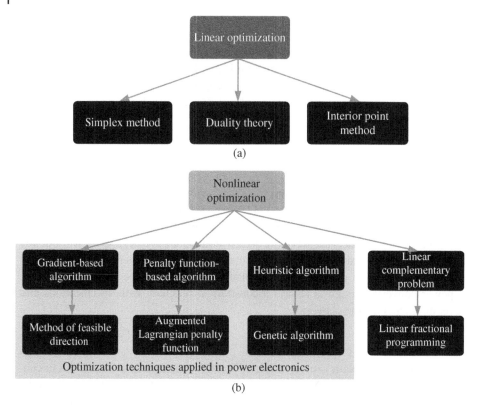

Figure 13.3 Classification of optimization algorithms in two categories. (a) Linear optimization algorithms. (b) Nonlinear optimization algorithms.

As discussed earlier, the power converter design optimization is a nonconvex optimization problem, and the global optima searching in nonlinear programming is an NP-hard problem, perhaps one of the reasons that optimization has not been widely applied. Especially for the whole converter design, there have been limited work. There have been attempts on subsystem and local optimizations for power converter designs. As shown in Figure 13.3b, a number of nonlinear programing optimization methods have been adopted in these works, including gradient-based algorithm, the penalty function-based algorithm, and the heuristic algorithm. In addition, the brute-force method has also been used. Each of these algorithms are briefly discussed here.

The simplest method is the brute-force algorithm, which enumerates all possible values for design variables in the design space. It can guarantee global optimization if sufficiently small steps are used in varying design variable values during the optimization process, but it is also the most tedious and time-consuming. The brute-force algorithm with a small step and a well-defined design space have been adopted in some power electronics system-level design [6–8].

The method of feasible direction (MFD), which falls under the gradient-based algorithm, is adopted in [9] for device loss minimization in a three-phase VSI. The basic idea of this method is, from a starting feasible point and by searching along the descent direction, to find a new feasible point that decreases the value of the objective function. The key for this algorithm is to select the searching direction and the moving step size for design variables.

The augmented Lagrangian penalty optimization algorithm, which is a penalty function-based algorithm, is adopted to minimize the weight of a half-bridge isolated DC/DC converter [10]. The algorithm constructs an augmented penalty function in an outer design iteration loop from a combination of the objective function and design constraints. An inner iteration loop is then used to solve the resulting unconstrained optimization problem.

Genetic algorithms (GAs)-based optimization techniques have been used in both the component and converter design optimizations. In the optimizer discussed in Chapter 5 for the passive rectifier, a GA-based algorithm called DARWIN is used to minimize the volume and cost of the passive components in a passive rectifier [11].

GA is one of the few optimization algorithms that works directly with discrete design variables. They are also excellent all-purpose discrete optimization algorithm because they can handle non-linear and noisy search spaces by using objective function information only. Compared to gradient-based optimizers, GA optimizers are more likely to find the globally optimal design. In addition, they are also capable of finding many near-optimal designs, providing the user with many options when selecting a final design configuration.

The GA procedure starts by selecting an initial population of randomly chosen strings (each of which represents a design). For the passive rectifier design problem, for example, each design consists of a string of integers representing the inductor core, the capacitor, and the wire, and a string of real values representing the number of wire turns and the air gap. The size of the population remains constant throughout the optimization, although the members of the population evolve over time. In order to form successive generations, parents are chosen from the current population based on their performance (designs with the best performance are given the highest probability of being selected as parents). After parents have been selected, genetic operators (crossover and muta-tion) are applied to create children. Depending on the selection procedure that is used to determine the next population of designs, selected child designs will replace their parents in the next gener-ation. One generation after another is created until some convergence criterion is met. The reader is referred to [12] for more information on evolutionary and genetic algorithms.

Particle swarm optimization (PSO) is another popular nonlinear optimization algorithm for power converter design. In [13], it was used for optimizing an EMI filter by selecting different filter configurations and parameters. PSO and GAs are both evolutional algorithms, and their differences include:

1) PSO algorithm keeps the previous best individual results and information, while GAs do not keep the previous memory and therefore the previous information may be lost with the change of the population.
2) GAs have a heavier computation burden than PSO due to coding, crossover, and mutation pro-cedure but generally have better convergence capability than PSO [14, 15].
3) PSO has a faster convergence speed than GAs. GAs share the information among the different individuals with the crossover procedure, so the whole population will move to the optimal zone evenly. In contrast, PSO does not share individual information, and it only uses the current opti-mal point in the population to update the search and the population. As a result, in most cases, PSO achieves a faster convergence speed than GAs.
4) PSO is mainly applied for continuous variable problems, while GAs can be applied for both con-tinuous and discrete variables problems.

13.3.2 Optimization Algorithm Selection

As shown in Figure 13.4, three commonly used optimization algorithm types are illustrated for comparison. They are the brute-force search scheme, the gradient-based algorithm, and the evolu-tionary algorithm. The surface plot in z-direction in each of the subfigures is the same and repre-sents the objective function. The lines in x–y plane beneath the objective function plots are also the

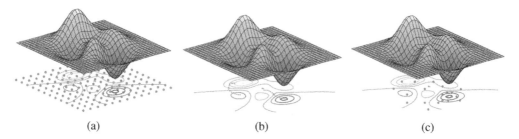

(a)   (b)   (c)

Figure 13.4 Commonly applied optimization algorithm types: (a) brute-force search, (b) gradient-based algorithm, (c) evolutionary algorithm (GA, PSO).

same for the three subfigures and represent different regions with local optima. The dots in the x–y plane are the design variable values traversed by the respective optimization algorithms. Gradient-based algorithms can be the fastest approach to search for a global optimum; however, they require an explicit and differentiable objective function and continuous design variable type, which is generally not the case for converter design problems. Also, depending on the initial point, the final search result may be trapped into a local optimum. Evolutionary algorithms have fewer limitations on the objective function and design variable types, which are suitable for the converter design problems. The main limitation is how to guarantee the design space diversity when the problem involves a large number of design variables. Hence, it is often more suitable for a subsystem design optimization problem with not too many design variables. Moreover, the searching performance of evolutionary algorithms highly depends on the population size and initial populations. The brute-force scheme is the most robust scheme to ensure the design space searching diversity while it is the slowest scheme that requires sweeping all possible design options and combinations.

13.4 Partitioned Optimizers vs. Single Optimizer for Converter Design

13.4.1 Partitioned Optimizers

The ultimate goal of the power converter design optimization is to optimize the whole converter. Sometimes, a converter can be optimized as a whole with a single-design optimization problem, especially for a simple converter such as a single-switch DC/DC converter. In many other cases, including three-phase AC converters, it is more manageable to partition the converter into several subsystems and optimize them separately. This book has followed the partitioning approach in establishing optimization problems for subsystems in three-phase AC converters. Another reason for partitioned optimizers is the optimization solution process itself, since certain optimization algorithms, e.g. GAs, work better with smaller systems with fewer design variables.

The key to the partitioning approach is to separate the design problem into several relatively independent subproblems and to search the optimal design for each individual optimizer. The system-level optimal design can be established by combining the local optimal design results and trade-offs among different optimizers.

Reference [16] adopted a design tool for the power stage of the AC-fed three-phase general-purpose industrial drive, with its configuration shown in Figure 1.22. The development of this tool will be further described in the next section. To reduce the complexity of the converter design optimization design, the design is partitioned into three relatively independent subsystems: the

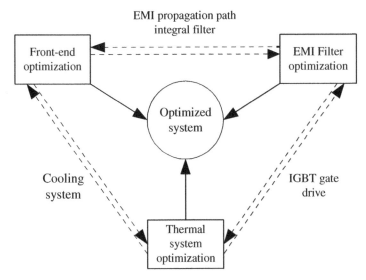

Figure 13.5 Approach for system optimization for a motor drive. *Source:* From [16]/with permission of IEEE.

front-end diode rectifier, the Si IGBT-based inverter and its TMS, and the source-side EMI filter. Figure 13.5 shows the three optimizers and their relationships. The benefit of this structure is that it can find the individual optimal design in each optimizer, and the optimizer is individually developed as a module and can be reused by other converters containing these modules. For example, the inverter power stage and EMI optimizers can be employed in many other three-phase AC inverters, including DC-fed inverters, or motor drives with active front-end. GA-based optimization techniques, i.e. DARWIN as mentioned earlier, were used for fast searching for the optimal design. The interactions and trade-offs among these partitioned optimizers are also recognized as illustrated in Figure 13.5. The front-end rectifier will act as propagation path to influence EMI filter design, and its loss will impact the TMS design. The inverter power stage design, specifically, the switching speed through the gate drive, will impact the EMI filter design.

Reference [17] describes an optimization for a two-stage converter structure of a three-phase active front-end rectifier cascaded with an isolated DC/DC converter for aircraft applications as shown in Figure 13.6. There are three levels for optimization: the system-level, the converter-level, and the component level. In this case, there is one system-level optimizer, two converter-level optimizers for the active front-end rectifier and the DC/DC converter, respectively, and a number of component number optimizers.

Figure 13.7 illustrates the corresponding three-level optimization diagram. The top-level optimization is a whole system-level optimization; the middle-level optimization is for converter-level

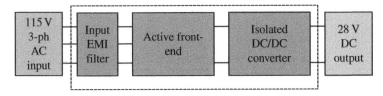

Figure 13.6 Two-stage power supply in an aircraft application. *Source:* From [17]/with permission of IEEE.

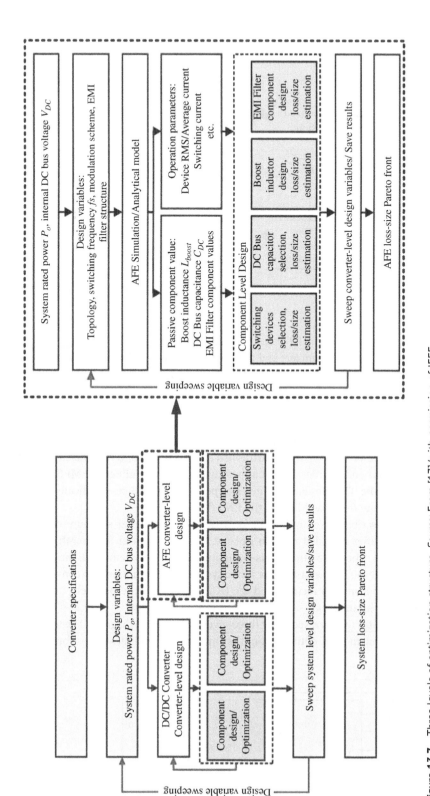

Figure 13.7 Three levels of optimization structure. *Source:* From [17]/with permission of IEEE.

optimizations, including one for the active front-end rectifier and the other for the DC/DC converter, and the bottom-level optimization is for component optimizations. The design variables for each level are defined in Figure 13.7. With predefined design variables (power and internal DC bus voltage, which can also be swept) from system-level optimization, each subsystem (converter and component) should be designed independently, calling for the well-defined design decomposition and links among different subsystems. The converter-level design variables include topology, switching frequency, modulation scheme, and EMI filter structure. From the converter-level design optimization, intermediate design results including passive component values and device ratings can be determined, which will set the requirements for component-level design optimization. From the component-level design optimization, the optimal component design and selection (devices, DC capacitor, and boost inductor) can be obtained. The power loss and size of components can be estimated, which can be the attributes of the design results associated with the design objectives, e.g. efficiency or power density. It can be seen that both simulation and analytical models are used in the design optimization.

Note that the converter design in Figure 13.7 does not explicitly consider the cooling and mechanical system, assuming those functions are part of the larger aircraft system. Also, the example did not use any optimization algorithm and effectively used the brute-force sweep. To speed up the process, the switching frequency is limited from 10 to 150 kHz and DC bus voltage is limited from 340 to 550 V with a 115-V RMS line-to-neutral input voltage, based on prior experience.

It is also noteworthy that the example in [17] utilized Pareto front in the design optimization, considering more than one objective in the design. Figure 13.8 shows relationship of higher-level

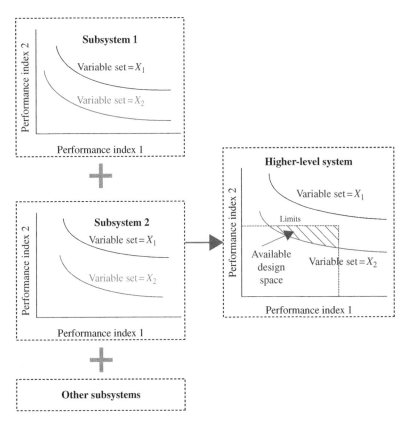

Figure 13.8 Higher-level system Pareto front determined by synthesizing subsystem Pareto fronts. *Source:* From [17]/with permission of IEEE.

system Pareto front with Pareto fronts of subsystems. In this case, Pareto curves are plotted for three different design levels: the design of the DC/DC stage, the design of the active rectifier stage, and the design of the converter components. The system-level Pareto front can be obtained by summing up subsystem Pareto fronts. The two performance indices can be loss and size.

Overall, the partitioned optimizer scheme is a suitable approach for system or whole converter design optimization. It reduces the complexity of optimization for the whole converter and separates the optimization problem into independent ones, which indirectly reduces the number of combinations of design variables and the number of design iterations. The interface conditions between different optimizers have been discussed in previous chapters for subsystem designs, which can be incorporated into the overall optimization. One challenge is how to deal with coupling effect between subsystems, which will be illustrated further through an example in the next section.

13.4.2 Single Optimizer

With a single optimizer for the system or the whole converter, it can consider all the relationships together, including couplings between different subsystems. Obviously, this approach is more likely to find the global optimal design, while the optimizer is more complex to establish and more difficult to solve.

Understandably, the single-optimizer approach has found more applications and success in some relatively simple applications, like DC/DC and single-phase AC converters. In [18], the cost of a single-phase boost power factor correction (PFC) is optimized with the switching frequency as a key variable. The optimizer is also based on analytical models to facilitate the optimization iteration. The optimization algorithm adopted belongs to the class of gradient-based methods. If the design space contains several local minima, there is a possibility that a gradient-based optimizer may be trapped by a local minimum, and the result will depend on the selection of the initial design point. In order to increase the probability of finding the point with the smallest objective function value (the global minimum), several different initial designs are used to start the optimization. It was found that even though there were local minima in the design space, in all cases studied, the local minima were better than the manual design. In this case, all design variables are continuous, including the turns number for the inductor, which is a continuous integer.

A similar optimizer is applied in a single-phase AC/DC/DC flyback converter design with the switching frequency again as a key design variable [19]. In this case, the minimum volume is the design objective.

For three-phase AC converter applications, one example is the design optimization for a three-phase three-level T-type AC/DC/AC converter with both input and output EMI filters [6]. Even though the design includes two-stage converters, the optimizer directly chooses one subset of whole design variables for two stages, and there is no interface or trade-off design variables between two stages. In addition to the single optimizer, the Pareto front is adopted, and final design results are shown in performance space of efficiency η and power density ρ as in Figure 13.9. The selected optimal design can be chosen by a multi-objective function with two weighting coefficients for the η and ρ of the whole converter.

It should be pointed out that the T-type AC/DC/AC converter in [6] is part of a UPS system. It also has a DC/DC converter, which was not included in the design optimization. Strictly speaking, the example is also a subsystem optimization, although the subsystem already consists of a two-stage AC/DC converters and filters.

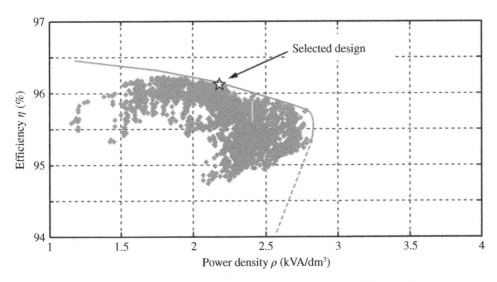

Figure 13.9 Pareto curve of the $\eta-\rho$ performance space and the selected design point. *Source:* Adapted from [6].

One issue with using a single optimizer for the whole converter is the size of the design space and the number of the design variables. One approach to overcome this issue is to fix certain variables and intermediate design results as constant based on experience and other knowledge/information. Reference [20] adopts such approach using a single optimizer to maximize power density in the design of a three-phase DC/AC inverter. In this case, the modulation scheme is preselected to minimize EMI noise based on designer's experience. Also, the AC current ripple of the converter is set at 10% of the peak inductor current. All these assumptions reduce the number of combinations and iterations, leading to a smaller design space and simpler optimization problem, with the drawback to potentially miss the global optimal design.

Overall, the single-optimizer approach less depends on the human experience and is more likely to find the global optimal design. However, since there are a great many combinations for design variables, some experience-based assumptions are often made to reduce the design space for optimization.

The main difference between the single-optimizer and the partitioned optimizer structures is the arrangement and grouping of design variables. For a three-phase application, there can be hundreds or even thousands of design variables that require huge combinations of design variables and iteration times. The two different optimizer structures can lead to large differences in the numbers of iterations. For example, one can use a system of four design variables to see the difference. Assuming the variables a, b, c, and d will each individually iterate N_a, N_b, N_c, and N_d times, respectively, the single optimizer will have a total of $(N_a \cdot N_b \cdot N_c \cdot N_d)$ times of iterations; for the partitioned optimizer structure, if design variables b, c, and d are decoupled, and variable a is assumed as the link between b, c, and d, the iteration times can be only $N_a \cdot (N_b + N_c + N_d)$, which will be dramatically reduced compared with the single optimizer. However, the partitioned optimizer structure needs clearly defined design decomposition and links among optimizers; otherwise, it could risk to lose the global optimal design.

13.5 Design Tool Development

13.5.1 Design Tool and Its Desired Features

From the aforementioned discussion, the benefit of using optimization methods in a converter design is obvious, i.e. to provide the best opportunity to achieve the "optimal design" among all feasible designs. However, the design optimization involves many iterations, especially, for complex systems like three-phase AC converters. Design automation using a design tool is desirable. Compared to the simulation tool, the design tool for power electronics converters is more complex and less mature, with no widely accepted commercial software available. Hence, it is worthwhile discussing design tool development for three-phase AC converters.

There are some desired features for converter design tools, which are summarized here.

1) *The Objective of the Tool Should Be Clear*: Conceptually, the tool should provide the "optimal" design. However, this "optimal" design is not absolute, but only for a certain objective function. For example, it is difficult, or even impossible, to achieve a converter design that will be the best in all aspects of cost, power density, and efficiency. In reality, the design optimization can be only be on one objective, or on an equivalent objective combining more than one performance index with weighting factors for each index, or equivalently using Pareto front. In addition, optimization cannot guarantee the global optimal design, especially for nonlinear programming problems like power converter designs. Moreover, mathematical optimum may not be the optimum that a real engineering solution needs. For example, once a converter design is close to the final "optimal" point, it may require many more iterations to get there, while the improvement in performance may be minimal. Given that there are errors in models, data, and manufacturing, there is very little benefit to search for the true mathematical "optimum." So, from the engineering sense, the "optimal" design is really the "most desirable" design. Therefore, the first feature of the design tool is that the design result should be a "desirable" result, or an "ideal" result in engineering sense.

2) *The Design Tool Should Be Efficient*: The main reason to adopt a design tool is to save time and effort by humans. Clearly, the design tool should be as automated as possible and allow a design to be obtained as fast as possible. One method to speed up the design tool is to use more analytical or data-driven numerical models, instead of time-domain simulations, although both approaches are often used.

3) *The Design Tool Should Be User-Friendly and Easy to Use*: This includes: (i) database establishment, models, and optimization algorithm selections. The design tool should include as many components in database as possible. Moreover, if certain components are not available in the database, the tool should allow easy addition or removal of components for a particular design. It is also desirable for the design tool to conveniently add/remove topologies, controls, and their corresponding models. There should also be options for different optimization algorithms; (ii) preprocessing interface that is easy for users to provide design conditions and constraints, and to define design objective(s); (iii) interface for postprocessing easy for users to review, compare, select, and save design results. This includes good visualization of converter performance with a particular design result. For example, it is desirable that the design tool can display EMI or harmonic spectrum of the converter for a particular design.

4) *The Design Tool Should Be Correct and Accurate*: Obviously, the design results from using the tool should represent a correct converter design. This requires that the models, the optimization software, and component database are all correct. It is generally difficult to prove the correctness

of a design tool theoretically. A tool should be based on well-established and validated models and algorithms, and should continue to be verified and improved through testing and use.

Following these requirements, two design tool development examples are presented here.

13.5.2 A GA-Based Optimization Tool for General-Purpose Industrial Motor Drive

As mentioned earlier, a GA-based optimization tool was developed for an AC-fed general-purpose industrial motor drive with the configuration of three relatively independent optimizers: the front-end passive rectifier, the two-level VSI and its TMS, and the input AC-side EMI filter, as illustrated in Figure 13.5.

13.5.2.1 Optimizer Formulation

The design tool development followed the same steps we have used in this book. The first step is to understand system conditions, variables, constraints, and design objectives. After formulating analytical (or physical) relationships and composing parts database, the design objective as minimum cost or volume can be realized by the optimizer.

13.5.2.1.1 Front-End Rectifier This optimizer focuses on the passive components of the front-end rectifier since they often account for a significant portion of cost and size of the total converter, and it has been described in Chapter 5 and will not be repeated here.

13.5.2.1.2 Inverter and Thermal Management System The second optimizer is for three-phase VSI and TMS design, with the focus on device (e.g. IGBT) selection and heatsink design. The most important analytical relationships are device loss models and system thermal models considering design constraints, such as maximum junction temperature of the IGBT module.

Device Loss Model: Similar to Chapters 2 and 6, the optimizer expresses the device losses as analytical functions of design parameters to facilitate device selection and TMS design. Multivariable behavioral models for IGBT losses are used, which are functions of current, voltage, temperature, and gate drive circuit parameters. The conduction loss and switching loss characteristics are implemented by a simple direct curve-fitting method for flexibility. With instantaneous loss model and considering the effect of modulator, the total loss can be calculated.

Thermal Model: Thermal impedance network model is needed to account for both the steady-state and transient (e.g. overload) conditions. The thermal impedance of IGBTs and antiparalleling diodes corresponding to output AC frequency f_o can be calculated by (13.5) and (13.6), where r_k and τ_k are kth (k ranges from 1 to i) thermal resistance and time constant for the ith order equivalent thermal impedance network for IGBT/diode, which are normally given in the device datasheet.

$$Z_{jc-IGBT} = \sum_{k-1}^{i} \frac{1}{\dfrac{1}{r_{k-IGBT}} + j \cdot 2 \cdot \pi \cdot n \cdot f_o \cdot \dfrac{\tau_{k-IGBT}}{r_{k-IGBT}}} \quad (r_{k-IGBT} > 0) \tag{13.5}$$

$$Z_{jc-diode} = \sum_{k=1}^{i} \frac{1}{\dfrac{1}{r_{k-diode}} + j \cdot 2 \cdot \pi \cdot n \cdot f_o \cdot \dfrac{\tau_{k-diode}}{r_{k-diode}}} \quad (r_{k-diode} > 0) \tag{13.6}$$

Thermal Analysis: The flowcharts for the loss and thermal calculations are shown in Figure 13.10. Due to a large difference between the thermal time constants of the IGBT module

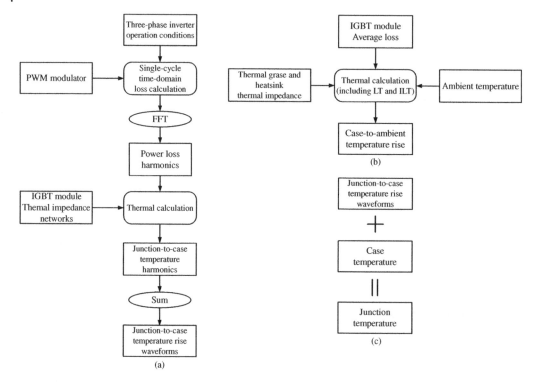

Figure 13.10 Flowcharts for the losses and thermal calculations: (a) junction-to-case temperature, (b) case temperature, and (c) junction temperature.

and the heatsink, there are two main parts in the thermal calculation: one part is the junction-to-case temperature rise calculation, as shown in Figure 13.10a; the other part is the case-to-ambient temperature rise calculation, as shown in Figure 13.10b. Note that this part is similar to the discussion in Chapter 6.

The key idea of the loss and thermal analysis is to calculate the power losses of the IGBT module just for a single cycle in the time domain and to obtain the thermal prediction results in the frequency domain. For each thermal calculation case, all the information can be obtained from just a single-cycle loss calculation according to the output frequency of the VSI. Using FFT, the average power loss and the power loss harmonics for each device are calculated. Inserting the loss harmonics into the IGBT module thermal impedance networks will result in the temperature harmonics. The sum of the average junction-to-case temperature rise and the junction-to-case temperature harmonics (i.e. waveforms) is the time-domain junction-to-case temperature rise.

Figure 13.11 shows the structure of the inverter and thermal optimizer.

13.5.2.1.3 EMI Filter The third optimizer is to minimize the AC input EMI filter cost or size. The EMI noise modeling and characterization of IGBT switching in a motor drive follow the approach in [21]. The filter topologies considered in the optimization are passive filters as shown in Figure 13.12, which are the popular and effective noise reduction schemes used in industry. The optimizer itself is not restricted to the particular filter topology, as long as the filter behavior can be modeled. One of the characteristics of the topology is that the DM inductors are realized by the leakage inductance of the CM choke.

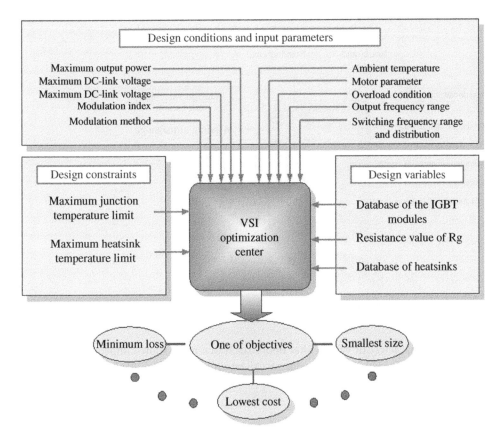

Figure 13.11 Inverter and thermal system optimization.

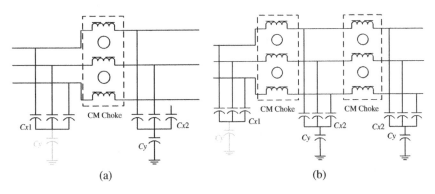

Figure 13.12 (a) One stage and (b) two-stage EMI filters.

The EMI design optimization approach is illustrated in Figure 13.13. The key analytical relations are inductor and capacitor high-frequency impedance models, CM choke thermal models, and CM choke core saturation calculations. With these models, the high-frequency performance of the CM choke can be related to the selection of specific core, wire gauge, and winding turn number, which are the design variables of the optimization program.

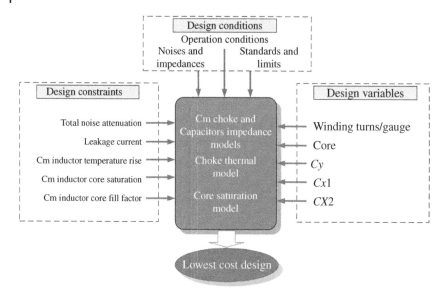

Figure 13.13 Structure of the EMI filter optimizer.

The other distinct characteristic of this filter optimizer is that noise source impedances are considered, along with the noise level. The impedances can be obtained through EMI measurements or modeling.

13.5.2.2 Optimizer Development

13.5.2.2.1 Design Tool Environment For the tool to be user-friendly and flexible to accommodate all three optimizers, it is built on a commercial software environment called ModelCenter® from Phoenix Integration. ModelCenter® enables diverse software applications (such as MATLAB) to communicate and seamlessly interact with one another. Applications are linked together and run in an intuitive visual environment. The design environment is open and extensible. The developed software environment includes prebuilt modules for the front-end rectifier, the inverter and TMS, and the EMI filter for the motor drive. Moreover, the software is extensible to other (user-defined) design problems. The modules can be modified and updated by the user and can be written in almost any languages, including MATLAB.

13.5.2.2.2 Optimization and Analysis Capabilities As mentioned before, the optimization algorithm is DARWIN, an evolution algorithm tailored specifically for engineering system design, which is able to work directly with discrete design variables and is also an excellent all-purpose discrete optimization algorithm.

Figure 13.14 shows an interface window for the example tool, which allows the user to define the objective function (e.g. total cost or volume in this case) and to specify design (e.g. EMI noise attenuation, temperature rise). The lower portion of the window is for design variables, either continuous or discrete ones. The optimization can be performed after defining evolution algorithm parameters such as population size and number of generations for convergence. The algorithm will stop when a user-specified number of populations have been generated without any improvement in the best design.

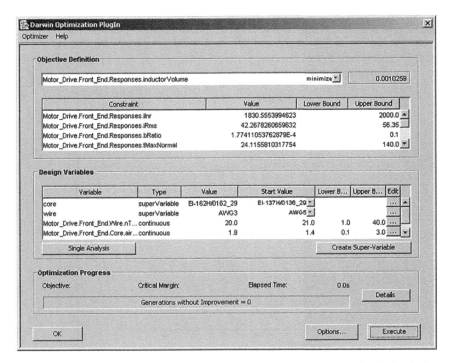

Figure 13.14 Optimization software interface (front-end rectifier example, similar for inverter/thermal and EMI filter).

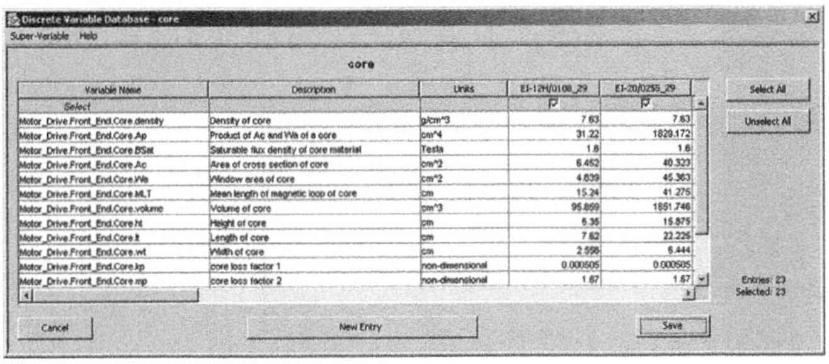

Figure 13.15 Editable component database (magnetic core example).

The optimization software environment also provides the user with other design capabilities, including

- Editing or modifying component databases (Figure 13.15)
- Viewing optimization progress (Figure 13.16)
- Examining optimization result details (Figure 13.17)
- Performing a single analysis of a selected design

The ModelCenter® environment also allows the user to connect the individual simulation modules together. For example, the RMS current output from the front-end module can be used as an input to the EMI module. Thereafter, whenever the EMI module is run, ModelCenter® will

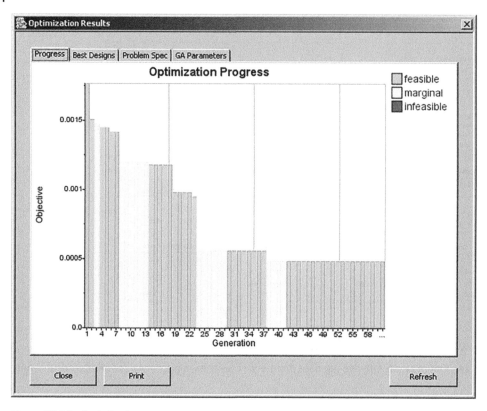

Figure 13.16 Optimization progress.

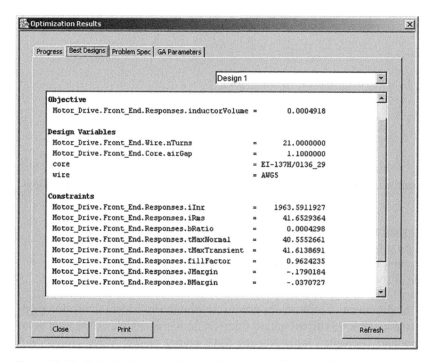

Figure 13.17 Optimization result details (front-end rectifier example).

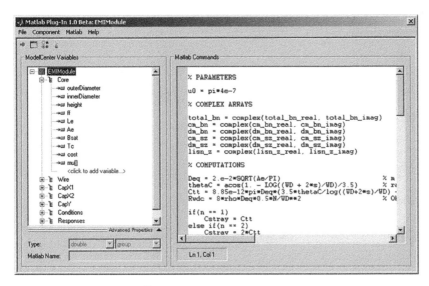

Figure 13.18 ModelCenter® MATLAB interface.

automatically run the front-end module as well (if necessary). In this way, complex systems (in this case the motor drive) can be modeled and designed.

The user may edit the modules provided with the software (front-end rectifier, IGBT inverter and thermal, and EMI filter) and may even introduce completely new modules. The modules can be in common software languages including MATLAB. MATLAB modules are easily edited via the ModelCenter® MATLAB interface (Figure 13.18).

13.5.2.3 Application Examples

13.5.2.3.1 Front-End Rectifier The front-end optimizer has been used in Chapter 5 for a 50-kW motor drive design example to minimize the passives, mainly the inductors for the rectifier. The example showed that the optimized inductor volume is significantly smaller than the manual design result, validating the benefit of the design tool. The example also showed that the DC inductor is smaller and therefore better than the counterpart three-phase AC inductor. Here, we will go through another example using the DC inductor.

Figure 13.19 shows the high-level user interface for the front-end rectifier passive components design optimizer. It can be seen that the inductor type can be selected between the three-phase AC and DC. Also, the optimization can be either for inductor only or for both the inductor and the capacitor.

The design conditions for the DC inductor design are given in Table 13.1. The design variables are given in Table 13.2. Note that the number of winding turns is restricted to integer but can be considered as continuous in the optimization algorithm. In contrast, cores and wires are selected from database and are discrete variables.

The objective of the optimization for the example is to minimize the "enclosed" volume of the front-end inductor. Note that the enclosed volume is the volume of a three-dimensional box that completely encloses the inductor. The design constraints are presented in Table 13.3. The magnetomotive force (MMF) ratio is defined as the ratio of MMFs in core and air gap. It is needed to make sure that for a design, the MMF in air gap is dominant. The current density margin is defined as $(J_{rated}/J_{max} - 1)$, where J_{rated} and J_{max} are rated and maximum current densities, respectively.

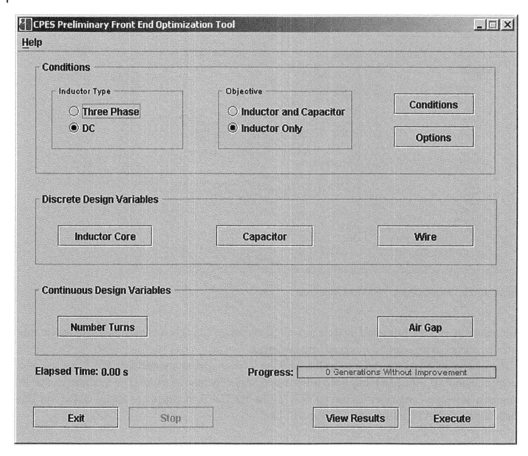

Figure 13.19 User interface of the front-end passive components design optimizer.

Table 13.1 DC inductor design conditions.

Condition	Value
Maximum power	13.5 kW
Line frequency	60 Hz
Maximum line-to-line RMS voltage	264.5 V
Rated line-to-line RMS voltage	200 V
Minimum DC-link voltage	200 V
AC line inductance	17 uH
Maximum ambient temperature	60 °C

The flux density margin is defined as $(B_{rated}/B_{sat} - 1)$, where B_{rated} and B_{sat} are rated and saturation flux densities, respectively. In this case, we choose not to have any margins for fill factor, current density, and the flux density.

DC optimization results are presented in Table 13.4. The optimized design indicates the "best" design that meets all the design constraints in Table 13.3. This design is limited by the inrush

Table 13.2 Front-end rectifier design variables (DC inductor case).

Variable	Type	Lower bound	Upper bound	Step
Core type and size	Database	—	—	—
Wire gauge	Database	—	—	—
Number of turns	Continuous	1	40	1
Air gap	Continuous	0.1 mm	3 mm	0.1 mm

Table 13.3 Front-end rectifier design constraints (DC inductor case).

Constraint	Upper bound
Inrush current	2000 A
AC RMS current	56.35 A
MMF ratio	0.1
Rated condition temperature rise	140
Overload temperature rise	150
Inductor fill factor	1
Current density margin	0
Flux density margin	0

Table 13.4 Front-end rectifier design optimization results (13.5 kW drive with DC inductor case).

Parameter	Optimized design	Alternate design
Capacitor (μF)	6600	6600
Core (Silicon steel)	EI-137H/0136_29	EI-112H/0112_29
Winding turns N	21	15
Air gap (mm)	1.4	1.0
Inductance (μH)	482.8	230.9
B_{rated} (T)	1.29	**1.619**
I_{RMS} (A)	42.2	47.0
I_{pk} (A)	68.5	85.9
I_{inrush} (A)	1971	**2718**
Wire type	AWG#5	AWG#6
J_{rated} (A/mm^2)	3.08	4.32
ΔT (°C)	41.7	57.7
$\Delta T_{transient}$ (°C)	42.7	59.4
Volume of the inductor (cm^3)	491.8	227

current constraint. Note that although the optimized design presented in the table is for an air gap of 1.4 mm, several additional designs (all identical except for the air gap) were also found by the optimizer. All of these designs have the same volume, so they are all equally good.

An alternate design is also shown in the table, which satisfy all the constraints except the saturation flux density and inrush current. This alternate design can better utilize the thermal capability of the inductor, and with a lower inductance value, also allows high harmonic current. Consequently, this design has a much smaller inductor volume. However, because it violates some constraints, it cannot be considered a feasible design. The parameters for the magnetic cores of the two inductor designs are given in Table 13.5.

The DC inductor based on the alternate design has been built and tested with the 13.5-kW drive. The rectifier input current is measured and compared with the calculation as shown in Figure 13.20. More design verification test results can be found in [11].

Table 13.5 Inductor core parameters.

Variable name	EI-137H/0136_29	EI-112H/0112_29
Inductor type	DC	DC
Weight density (g/cm^3)	7.63	7.63
Core-window area product A_p (cm^4)	111.587	50
Saturation flux density B_{sat} (T)	1.6	1.6
Cross-sectional area A_c (cm^2)	12.198	8.165
Window area W_A (cm^2)	9.148	6.124
Mean-length-per-turn (MLT) (cm)	20.955	17.145
Core volume (cm^3)	252.213	137.223
Core height h_t (cm)	8.73125	7.14375
Core length l_t (cm)	10.4775	8.5725
Core width w_t (cm)	3.528	2.88
Core loss coefficient k	5.05E-04	5.05E-04
Core loss f coefficient m	1.67	1.67
Core loss B coefficient n	1.36	1.36
Material	Silicon steel	Silicon steel

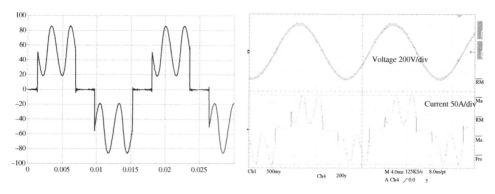

Figure 13.20 Calculated (left) and measured (right) AC input currents.

13.5.2.3.2 Inverter and Thermal System   The application example for the inverter and thermal system optimizer is for a 7.5-kW motor drive. The design conditions are given in Table 13.6. Note that we have fixed the modulation scheme and switching frequency, although in general they are design variables. Alternatively, we can repeat the design by changing the modulation and switching frequency for each design to determine the best modulator and switching frequency.

The design variables are given in Table 13.7. The gate resistance is continuous, and IGBT module and heatsink are discrete from the database.

The objective of the optimization for the example design is to minimize the total cost of the IGBT module(s) and the heatsink. The design constraints for the optimizer are listed in Table 13.8.

Table 13.6   The inverter and thermal system design conditions.

Condition	Value
Inverter rated AC frequency	50 Hz
Inverter switching frequency	12 000 Hz
Inverter output power	7500 W
DC-link voltage	560 V
Overload peak phase current	24 A
Overload time duration	60 s
Motor power factor	0.92
Rated inverter RMS current	17.6 A
Maximum ambient temperature	60 °C
Modulator type	60° DPWM

Table 13.7   The inverter and thermal system design variables.

Variable	Type	Lower bound	Upper bound	Step
IGBT module	Database	—	—	—
Heatsink	Database	—	—	—
Gate resistance R_g	Continuous	1.6 Ω	37.0 Ω	0.1 Ω

Table 13.8   The inverter and thermal system design constraints.

Constraint	Upper bound (°C)
Heatsink temperature at rated condition	90
Maximum IGBT junction under overload condition	130
Maximum diode junction temperature under overload condition	130
Maximum IGBT junction temperature at rated condition	125
Maximum diode junction temperature at rated condition	125

Results obtained using the inverter and thermal system optimizer are presented in "Optimized" column of Table 13.9. For comparison purposes, the optimizer was run (a single analysis) for a design that has the same voltage and current ratings as an existing commercial 7.5-kW motor drive. The results of this analysis are presented in the "Single analysis" column. The IGBT module used to perform the single analysis (FS35R12KE3G) has the same voltage and current ratings as the module used in the commercial motor drive. The data for heatsink "Heatsink 2" were obtained from database (see Table 13.10), which was the same as in the commercial drive. It can be seen that the cost of the optimized design is 82% of the commercial design. The cost reduction of the optimized design is primarily due to the use of a smaller IGBT module, which more fully utilizes the temperature capabilities of the IGBT module and the heatsink. With proper design, the total loss is almost the same for the two designs.

Table 13.9 The inverter and thermal system optimization results.

Parameters	Single analysis	Optimized
IGBT module	FS35R12KE3G (1200 V, 35 A, three-phase)	FS25R12KE3G (1200 V, 25 A, three-phase)
Heatsink	Heatsink 2	Heatsink 2
Gate resistance (Ω)	27.0	36.0
Maximum IGBT junction to case temperature (°C)	17.45	24.51
Maximum diode junction to case temperature (°C)	14.11	23.74
Maximum IGBT junction temperature (°C)	100.17	107.43
Maximum diode junction temperature (°C)	96.83	106.66
Overload maximum IGBT junction to case temperature (°C)	23.82	33.91
Overload maximum diode junction to case temperature (°C)	19.20	32.84
Overload maximum IGBT junction temperature (°C)	110.00	120.54
Overload maximum diode junction temperature (°C)	105.39	119.47
Heatsink temperature at rated condition (°C)	78.58	78.75
Overload heatsink temperature (°C)	80.57	80.87
Total IGBT module loss (W)	206.49	208.33
Overload total IGBT module loss (W)	280.96	287.90
Total cost of IGBT and heatsink (%)	100	82.6

Table 13.10 Heatsink database.

Variable name	Units	Heatsink 1	Heatsink 2	Heatsink 3
Cost	$	11.4	18.6	25
Weight	kg	1.4	2.1	2.42
Thermal resistance R_{th}	°C/W	0.19	0.09	0.0446

13.5.2.3.3 EMI Filter The application example for the EMI filter optimizer is also for a 7.5-kW motor drive. The design conditions are given in Table 13.11. The filter attenuation margin was selected to be zero, but there is a 6-dB built-in margin when designing DM and CM filters. Although the output power is 7.5 kW, the rectifier kVA can be as high as 18 kVA, considering motor power factor and efficiency, as well as input harmonics. It is also customary to have a motor drive to be oversized for the next power level as well as with a 60-second overload capability. As a result, the peak current can be as high as 81 A in this case.

Other design conditions include bare noise in Figure 13.21 based on the measurements from a real 7.5-kW motor drive. In the design, the EMI filter topology is given (Figure 13.22). So, in this case, the topology is also a design condition.

Table 13.11 EMI filter design conditions.

Condition	Value
Filter attenuation margin	0 dBμV
Input AC RMS current	22.2 A
Ambient temperature	60 °C
Input phase RMS voltage	288.7 V
AC line frequency	60 Hz
Peak input AC current	81 A

(a)

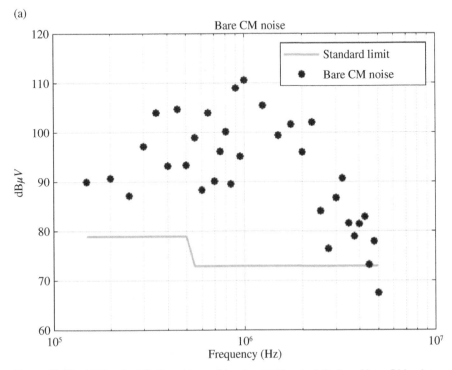

Figure 13.21 (a) Standard limit and bare CM noise; (b) Standard limit and bare DM noise.

(b)

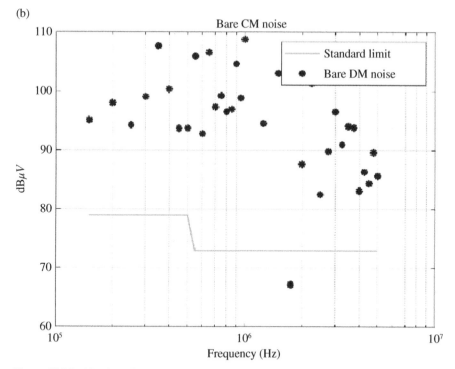

Figure 13.21 (Continued)

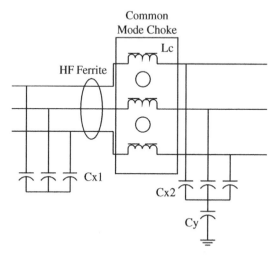

Common
Mode Choke

Figure 13.22 EMI filter topology in the design example (right side connecting to the drive and left side to LISN)).

The design variables are given in Table 13.12. The capacitors and the inductor are illustrated in Figure 13.22. All variables are discrete except for the turns number. Note with the CM choke, no air gap is needed.

The objective of the optimization procedure is to minimize the total cost of the EMI filter. The design constraints for the optimizer are presented in Table 13.13. Physical constraints including temperature and saturation flux density limits are core material-dependent. Note that the fill factor

Table 13.12 EMI filter design variables.

Variable	Type	Lower bound	Upper bound	Step
DM capacitor Cx1	Database	—	—	—
DM capacitor Cx2	Database	—	—	—
CM capacitor Cy	Database	—	—	—
CM choke core	Database	—	—	—
Inductor wire	Database	—	—	—
Number of winding turns	Continuous	1	40	1

Table 13.13 EMI filter design constraints.

Constraint	Upper bound
Total noise	EN61800 Class A
Leakage current	100 mA
Inductor core temperature	Curie temperature
Flux density	Saturation flux density
Fill factor	1
Current density	100 A/mm^2

Table 13.14 EMI filter design optimization results.

Parameters	Existing manual design	Optimized
Capacitor Cx1	1 μF MKP X2	B32924_A3225
Capacitor Cx2	1 μF MKP X2	B32924_A3105
Capacitor Cy	330 nF Y-cap	B81122-A1683
CM choke core	FT-3M F1AH0704	42915-TC-OW
CM choke wire	AWG 12	AWG 14
CM choke turns number	13	6
Maximum B field (T)	—	0.238
Core temperature (°C)	—	83.89
I_LeakageCurrent (mA)	—	7.4
Worst case CM noise margin (@ 1 MHz)	15 dB	0.36 dB
Worst case DM noise margin (@ 150 kHz)	—	0.30 dB
Worst case total noise margin (@ 150 kHz)	10 dB	0.31 dB
CM choke fill factor	—	0.82
Total cost ($)	—	2.86

for CM choke is defined as 360° over the winding coverage angle, so the limit is 1 with each phase covering 120°. Usually, each phase will cover around 100–110°, so the fill factor is around 1.15.

The EMI filter design results using the optimizer are presented in the "Optimized" column in Table 13.14. The characteristics associated with the selected capacitors and choke core are given in Tables 13.15, 13.16, and 13.17. A manually designed filter in the commercial 7.5 kW drive used for comparison is also included in Table 13.14.

The EMI filter design by the optimizer has characteristics of just sufficient attenuation as shown in Figure 13.23.

Table 13.15 EMI filter DM capacitor data.

Capacitor parameter	Units	B32924_A3225	B32924_A3105
Manufacturer	—	EPCOS	EPCOS
Cost	$	0.6	0.4
Capacitance	μF	2.20	1.00
ESL	nH	15.9	22
ESR	Ω	0.01	0.029

Table 13.16 EMI filter CM capacitor data.

Capacitor parameter	Units	B81122-A1683
Manufacturer		EPCOS
Cost	$	0.45
Capacitance	nF	68
ESL	nH	7.82
ESR	Ω	0.049

Table 13.17 EMI filter CM choke core data.

Core parameter	Units	42915-TC-0W
Manufacturer		Magnetics
Cost	$	1.047
Outer diameter	mm	29
Inner diameter	mm	19
Height	mm	15.2
Fill factor	—	1.15
Mean magnetic loop length L_e	cm	7.32
Cross-sectional area Ae	cm^2	0.74
Saturation flux density B_{sat}	Tesla	0.4
Curie temperature T_c	°C	125
Relative permeability @DC $\mu(0)$	—	10 000
Relative permeability @10 kHz $\mu(1)$	—	7840
Relative permeability @150 kHz $\mu(2)$	—	3781

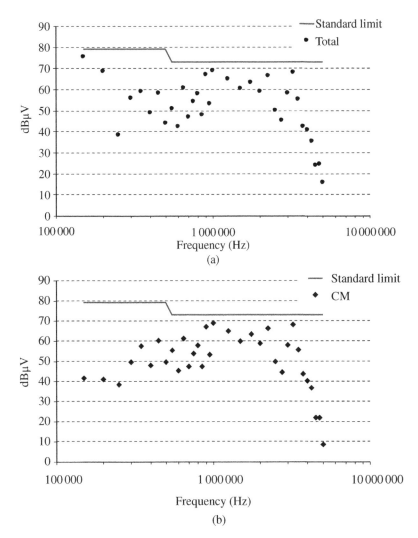

Figure 13.23 Calculated filter performance: (a) total noise and (b) CM noise.

The prototypes of the designed EMI filters using the components listed in "Optimized" column in Table 13.14 were implemented and tested with the 7.5-kW drive. The test setup includes the drive with a motor load, three LISNs, and the EMI filter. The drive inverter switching frequency was 16 kHz. The details of the measurement setup and data acquisition procedure can be found in [22, 23]. The measured EMI characteristics with the optimal filter are shown in Figure 13.24, which verify the calculated results. The noise level margin is less than 0.5 dB, which indicates the optimal filter design has been achieved. Figure 13.25 shows the measured result for the filter in Table 13.14 in the commercial drive. Clearly, the existing filter has more margins. Figure 13.26 shows the schematic layout comparison of different filter designs. The existing filter and the one-stage filter are the manual design and optimized design, respectively. Although the cost was used as the objective function in this optimization example, the optimal EMI filter footprint was also reduced to 61% compared with the manually designed existing filter.

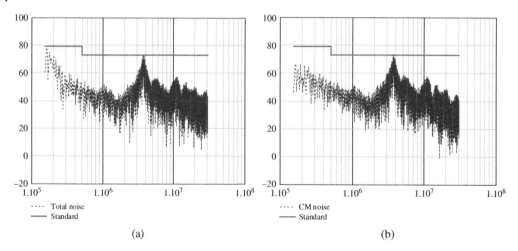

Figure 13.24 Measured noise of the optimized EMI filter: (a) total noise and (b) CM noise.

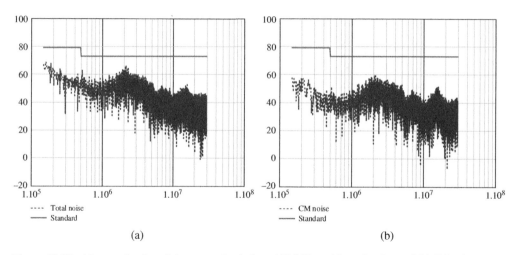

Figure 13.25 Measured noise of the manually designed EMI filter: (a) total noise and (b) CM noise.

Note that a two-stage filter design is also shown in Figure 13.26, which is in fact larger than the one-stage design. In addition, the two-stage design showed worse EMI performance due to the coupling effect between the two stages, which must be managed as discussed in Chapter 8.

13.5.2.4 Discussions on System Optimization

The GA-based optimization tool primarily focuses on optimizing separately the three subsystems of the drive power stage. The links between these subsystems can be established, and the optimizer can perform system-level optimization. As shown in Figure 13.7, physical relationships between any two of the three subsystems, the front-end passive, inverter loss and thermal, and conducted EMI noise level, have been identified.

As an example, the gate resistance effects on both EMI noise and inverter losses have been studied. For certain 1200 V/150 A IGBT half-bridge module SKM150GB123D, four different gate

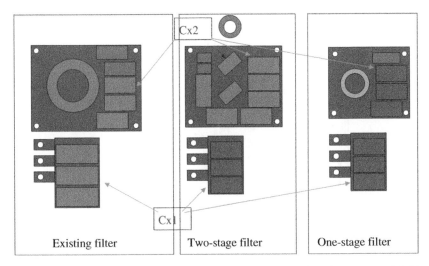

Figure 13.26 Comparison of the manually designed existing filter in a commercial drive, the designed two-stage and one-stage filter.

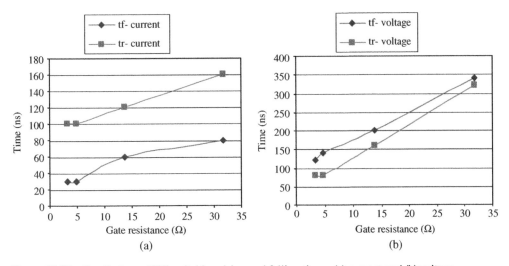

Figure 13.27 R_g effects on IGBT switching rising and falling times: (a) current and (b) voltage.

resistances, $R_g = 3.9\ \Omega$, $4.8\ \Omega$, $13.8\ \Omega$ and $31.8\ \Omega$, have been used, and the device switching rising and falling times are measured as shown in Figure 13.27, under the test conditions of load current $I_{load} = 100$ A and DC bus voltage $V_{DC} = 600$ V. Also, the conducted DM and CM EMI noises are measured for these R_g values as in Figure 13.28. Although this device test setup uses the half-bridge circuit, instead of the VSI, the hard-switching process is kept same, so the loss and EMI noise generations are similar to the VSI. It is clear that a higher gate resistance will lead to a slower IGBT switching and a larger switching loss, but smaller CM and DM noises. Therefore, R_g could be used as a common design variable in the inverter and thermal optimizer and the EMI filter optimizer.

Using the relationships, any two sets of optimization results can be combined to achieve a new optimal design considering the two together. Similarly, it can be extended to all three. The example of linking the front-end rectifier and the EMI filter is shown in Figure 13.29.

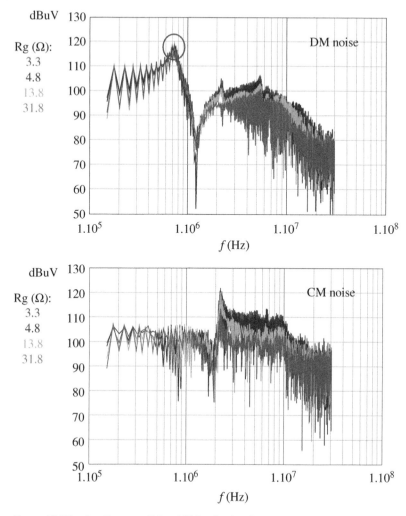

Figure 13.28 R_g effects on CM and DM noise levels.

13.5.3 A Design Tool for High-Density Inverters

13.5.3.1 Design Tool Function and Architecture

The second example of the design tool development is a tool mainly for the weight optimization of the motor drive intended for electrified aircraft applications. The basic system architecture is a DC-fed motor drive in Figure 13.30. Several commonly used topologies and PWM modulation schemes are included in the software as listed in Table 13.18. Users can select one of the topologies and PWM schemes or compare them for the optimal design.

The formulation of the design optimization follows the approach presented in this book and is shown in Figure 13.31. For EMI and THD constraints, options in the tool include DO-160E and MIL-STD-461.

Like the GA-based tool aforementioned, the database, component and subsystem models, design optimization procedure or algorithms, and simulation or calculation are four critical

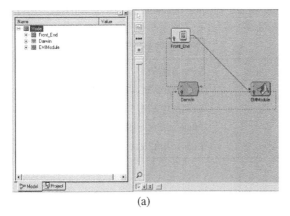

(a)

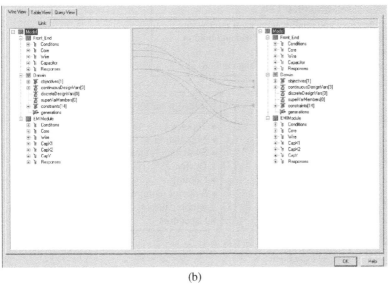

(b)

Figure 13.29 (a) Link operation window and (b) variable linkages.

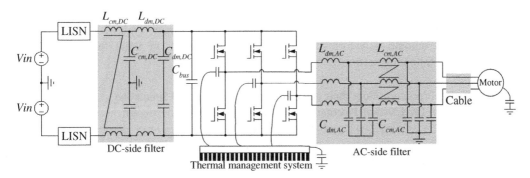

Figure 13.30 DC-fed three-phase motor drive system with the example two-level VSC.

Table 13.18 Topologies and PWM modulations schemes available for the design tool.

Topologies	Modulation schemes
2-level	SVM, DPWM
3-level ANPC	SVM, DPWM, CMR (CM reduction)
3-level T-type	SVM, DPWM, CMR
Multilevel flying capacitor clamping	Phase shift, level shift
Multilevel diode clamping	Phase shift, level shift

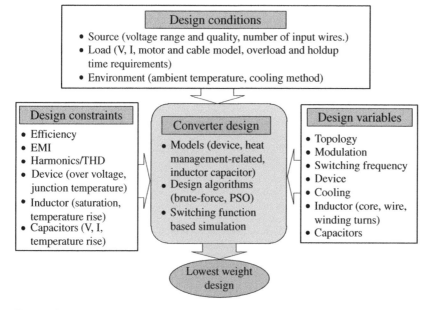

Figure 13.31 Formulation of converter design optimization.

aspects of the design tool architecture as shown in Figure 13.32. Unlike the GA-based tool, which solely relies on analytical models, simulation is also selectively used here to obtain the needed time-domain waveforms for the design. The details of these four aspects are discussed in the following sections.

13.5.3.2 Design Procedure/Algorithms and Switching Function-Based Simulation

13.5.3.2.1 System-Level Design Procedure/Algorithm In this tool, the design procedures can be separated into the system level and converter level. The system-level design refers to the configuration or function/subsystem design. For system-level optimization, the design tool enables evaluation of different topologies, PWM schemes, switching frequencies, filter configurations (single-stage or two-stage, *LC* or *CL* structure), and cooling methods (passive or active, forced-air or liquid cooling) as shown in Figure 13.33. All the design parameters aforementioned for the system-level design is related to configuration or function (thus, all discrete), with the exception of the switching frequency, which impacts multiple subsystems or functions (e.g. topology,

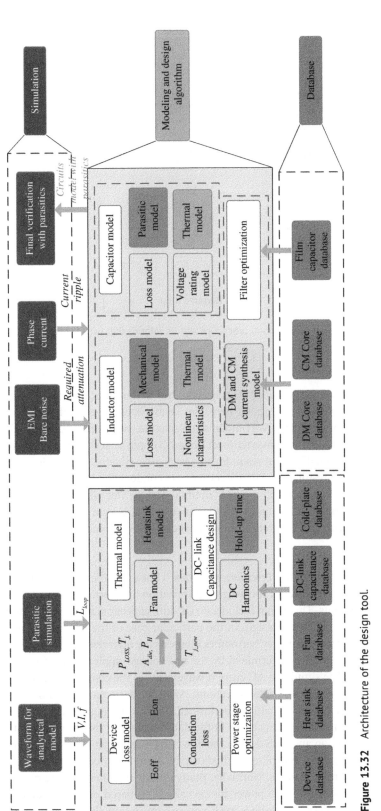

Figure 13.32 Architecture of the design tool.

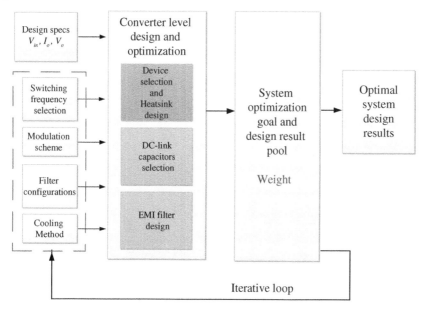

Figure 13.33 System-level optimization procedure and its relationship with converter-level design.

modulation, and filter configuration). The system-level design does not involve physical designs, which is termed converter-level design here. For each system-level configuration, which has a combined selection of topology, PWM modulation scheme, switching frequency, filter configuration, and cooling method, the converter-level design (i.e. physical design) optimization is carried out to achieve an "optimal" design result for this particular system-level configuration. The result is placed in the design result pool. Then, iterations are conducted with different configurations. In the end, the design result pool will include "optimal designs" for all selected system-level configurations. By comparing all results in the design result pool, the best configuration with the optimized design result (i.e. the minimum weight) can be determined.

13.5.3.2.2 Converter-Level Design Procedure/Algorithm
For a specific case with a selected configuration, the design is conducted for the weight optimization by selecting different devices, passives, and cooling components. The design includes (i) DC-link capacitors selection, (ii) devices selection and TMS design, and (iii) filter design. In the particular case of the DC-fed motor drive, there is generally no need for low-frequency harmonics filters, so the filter refers to EMI filter or dv/dt filter.

The DC-link capacitance is determined by the bus voltage ripple and hold-up or ride-through time requirements. The simulated AC current waveforms of the converter are used, which would not be influenced by device and filter selections. After the determination of the capacitance, the physical selection of capacitors can be performed, with also considering voltage and current requirements, by searching the capacitors in the database for the minimum weight combination.

For the device selection and TMS design, the flowchart is shown in Figure 13.34 with an example of the forced-air cooling. Other cooling methods can be similar. Devices are first selected to meet the voltage and current requirements. Normally, many device combinations are qualified. For each qualified device combination, the power loss is calculated based on the simulation waveforms. Then, based on the power loss, surface area, and maximum junction temperature requirements

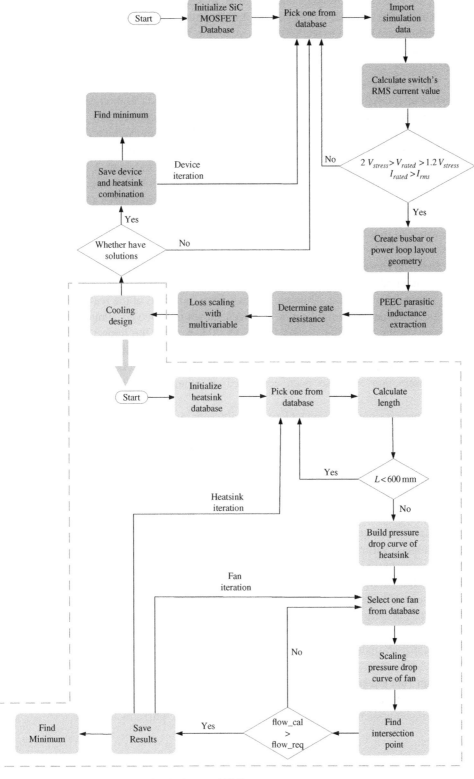

Figure 13.34 Design flowchart for devices and TMS.

of the devices, the TMS can be designed and optimized for each case. By comparing these qualified combinations of devices and TMSs, the one with the minimum weight is selected. In this example, the device selections are limited to SiC MOSFETs. Also, the design can consider the impact on loss by layout parasitics, which can be estimated by partial element equivalent circuit (PEEC) method, etc.

Filters in the DC-fed drive may include the DC-side EMI filter, and AC-side EMI filter or *dv/dt* filter depending on the system requirements. For EMI filters, DM and CM filter designs are coupled because leakage inductance of CM inductor can function as the whole or part of the DM inductance. Take the design optimization of a single-stage EMI filter as an example. With the required DM and CM corner frequencies from the bare noise attenuation analysis, two independent design variables CM capacitance C_{CM} and k (the ratio between DM and CM inductances) are iterated, with corresponding inductances and DM capacitance C_{DM} determined through corner frequencies relationship. For each combination, physical designs of inductors and capacitors are conducted with the simulated waveforms. The iteration and optimization utilizing the cores, wires, and capacitors in database will lead to the minimum weight filter design.

13.5.3.2.3 Switching-Function-Based Simulation For each aforementioned design step, simulation waveforms are required. To speed up the design process, the switching-function-based simulation described in Chapter 7 is adopted. Although the switching transients of the real switching actions would not be considered, the simplification has been shown to have very limited impact on harmonics, or even on the EMI filter design. Figure 13.28 does show the impact of gate resistance R_g (i.e. switching speed or waveform slope) on EMI noise level for a Si IGBT. However, the impact is mainly in the high-frequency range, which will not determine the filter corner frequency or filter design. The observation has also been confirmed in other reported work involving WBG devices [24].

While switching transients will impact loss and voltage or current overshoot, in the design tool, the loss is calculated using the energy loss model associated with the switching event, similar to the models in Chapter 6. The voltage and current overshoots are not considered.

13.5.3.3 Component Models

There are two types of design routines available for the design tool: fast and accurate design routines. The main difference is the models used in the design. The fast design routine uses simplified models to reduce the design time. The accurate design routine considers some nonlinear characteristics (e.g. inductor saturation) to more accurately represent components, such that the design results will be closer to the eventual hardware. Fast routine is useful to down-select several promising candidates for a design problem, and then the accurate routine can be used for further optimization. The selectable design routines benefit from the modular structure of the software.

Models used in the design tool are summarized in Table 13.19. Most of these models have been described in previous chapters, with some other references provided here. These models can be individually improved, updated, and replaced by tool developers/users.

13.5.3.4 Design Tool Integration and Features

Based on the aforementioned design procedure, algorithms, and models, the design tool has been implemented for comprehensive design optimization of the three-phase motor drive.

13.5.3.4.1 User Interface (UI) and Tool Configurations The software tool integrating all design functions and models is developed in MATLAB App Designer with a user-friendly UI. Design

Table 13.19 Main device and component models used in the design tool.

Category	Models	Fast design routine	Accurate design routine
Device models	Conduction loss	R_{dson} at maximum junction temperatures set by users	
	Switching loss	Scaled E_{on} and E_{off} with voltage and current stress	Switching transient discrete-time modeling with parasitics, device nonlinear characteristics, and R_g [25]
TMS-related models	Forced or natural air thermal resistance	Correction factors including airflow velocity, heatsink length, and temperature rise [26]	
	Fan model	Airflow velocity determined by pressure drops of the fans and heatsinks	
	Liquid cooling thermal resistance	Cold plate thermal resistance determined by flow rate based on datasheet	
Magnetics models	Inductance model	Inductance per single turn squared A_L provided by manufacturers [27]	Current-bias-dependent permeability and frequency-dependent imaginary permeability chapter
	Leakage inductance model	Leakage inductance model of common mode choke from [28, 29]	
	Core loss model	iGSE	
	Winding loss model	Sum up separated losses at different frequencies based on FFT analysis of current [30]	
	Thermal model	Level 1 thermal model [31]	
Capacitor models	Impedance model	Series LRC circuit with ESL and ESR [32]	
	Thermal model	Thermal resistance from datasheet	

conditions and constraints for a DC-fed three-phase motor drive can be configured in the UI. Figure 13.35 provides all configuration options for the converter design.

Figure 13.35a shows the basic configurations of converters provided by the developed design tool. Through this tab, five topologies and their possible modulation schemes can be selected. For multilevel topologies, the arbitrary voltage level is available for the configuration with the help of the switching-function-based simulation scheme. Also, the switching frequency range and ambient temperature can be set in this tab.

Figure 13.35b shows the configuration tab for the AC-side filter and DC-link capacitors. Based on the application requirement, the AC-side filter can be configured either as an EMI filter or as a dv/dt filter. For the dv/dt filter, the corner frequency can be determined by either the desired dv/dt value or the overvoltage level as discussed in Chapter 12. Also, design constraints for DC-link capacitors can be selected considering four factors discussed in Chapters 5 and 6: the harmonic current limit, voltage ripple limit, load change requirement, and ride-through time requirement.

Figure 13.35c shows the EMI filter configuration options, where both the stage number and filter structure can be configured. In terms of the inductor design, three cooling methods (natural, forced-air, and forced-liquid) are available for selection, and corresponding analytical models are implemented. Also, the winding wire can be chosen from the general AWG wires or Litz wires with equivalent AWG.

Figure 13.35d and e show the device selection and the device cooling configurations, respectively. The power semiconductor and heatsink candidates are loaded from the database and displayed

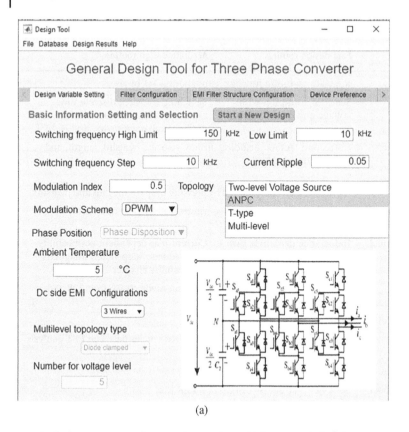

(a)

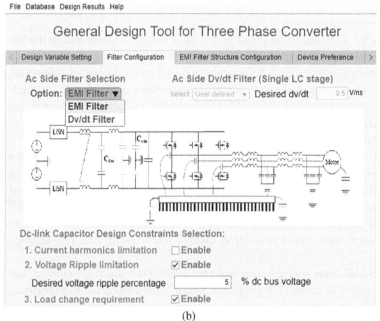

(b)

Figure 13.35 Converter design configuration options in the design tool. (a) Topology and modulation selection (b) AC-side filter configuration (c) Filter structure and design configuration (d) Device selection configuration (e) Cooling design configuration (f) System-level operating condition configuration.

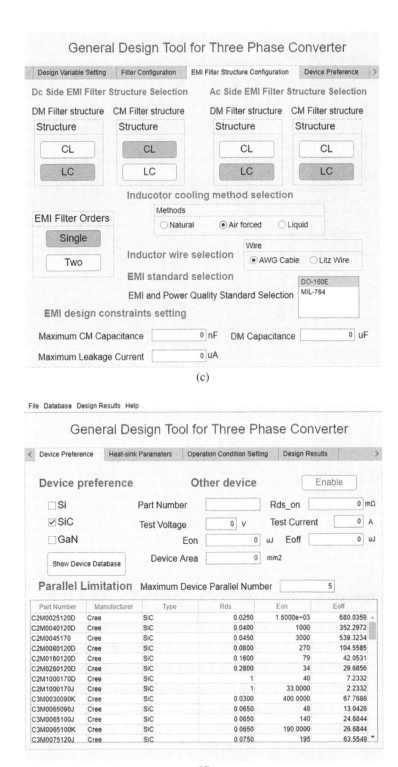

(c)

(d)

Figure 13.35 (Continued)

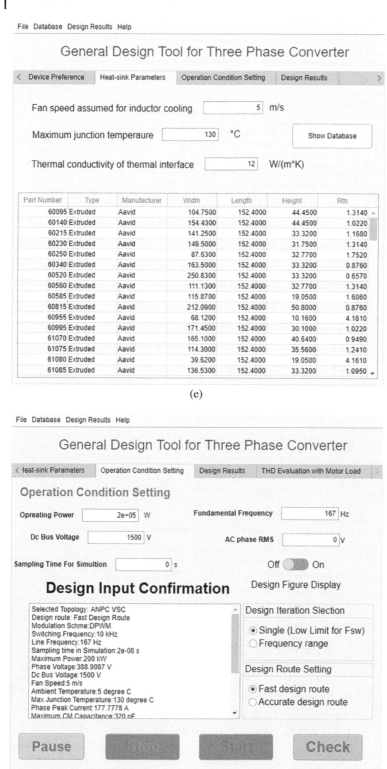

File Database Design Results Help

General Design Tool for Three Phase Converter

⟨ Device Preference | Heat-sink Parameters | Operation Condition Setting | Design Results | ⟩

Fan speed assumed for inductor cooling   5   m/s

Maximum junction temperaure   130   °C   Show Database

Thermal conductivity of thermal interface   12   W/(m*K)

Part Number	Type	Manufacturer	Width	Length	Height	Rth
60095	Extruded	Aavid	104.7500	152.4000	44.4500	1.3140
60140	Extruded	Aavid	154.4300	152.4000	44.4500	1.0220
60215	Extruded	Aavid	141.2500	152.4000	33.3200	1.1680
60230	Extruded	Aavid	149.5000	152.4000	31.7500	1.3140
60250	Extruded	Aavid	87.6300	152.4000	32.7700	1.7520
60340	Extruded	Aavid	163.5000	152.4000	33.3200	0.8760
60520	Extruded	Aavid	250.8300	152.4000	33.3200	0.6570
60560	Extruded	Aavid	111.1300	152.4000	32.7700	1.3140
60585	Extruded	Aavid	115.8700	152.4000	19.0500	1.6060
60815	Extruded	Aavid	212.0900	152.4000	50.8000	0.8760
60955	Extruded	Aavid	68.1200	152.4000	10.1600	4.1610
60995	Extruded	Aavid	171.4500	152.4000	30.1000	1.0220
61070	Extruded	Aavid	165.1000	152.4000	40.6400	0.9490
61075	Extruded	Aavid	114.3000	152.4000	35.5600	1.2410
61080	Extruded	Aavid	39.6200	152.4000	19.0500	4.1610
61085	Extruded	Aavid	136.5300	152.4000	33.3200	1.0950

(e)

File Database Design Results Help

General Design Tool for Three Phase Converter

⟨ Heat-sink Parameters | Operation Condition Setting | Design Results | THD Evaluation with Motor Load | ⟩

Operation Condition Setting

Opreating Power   2e+05   W   Fundamental Frequency   167   Hz

Dc Bus Voltage   1500   V   AC phase RMS   0   V

Sampling Time For Simultion   0   s   Off ⬤ On

Design Input Confirmation
Design Figure Display

Selected Topology: ANPC VSC
Design route: Fast Design Route
Modulation Schme:DPWM
Switching Frequency:10 kHz
Line Frequency:167 Hz
Sampling time in Simulation:2e-08 s
Maximum Power:200 kW
Phase Voltage:388.9087 V
Dc Bus Voltage:1500 V
Fan Speed:5 m/s
Ambient Temperature:5 degree C
Max Junction Temperature:130 degree C
Phase Peak Current:177.7778 A
Maximum CM Capacitance:320 nF

Design Iteration Slection

⦿ Single (Low Limit for Fsw)
○ Frequency range

Design Route Setting

⦿ Fast design route
○ Accurate design route

Pause	Stop	Start	Check

(f)

Figure 13.35 (Continued)

in two tabs. Figure 13.35f shows the system-level operation condition setting tab, including the power rating, AC RMS voltage, and DC-link voltage. This window also allows the user to change settings for different running options, single switching-frequency vs. frequency sweeping in a range, and fast routine vs. accurate routine.

Figure 13.36 shows the load configurations in the design tool. The parasitics of the AC-side cables are calculated by the tool based on the input parameters of the employed cables. In addition to the AC-side cable, three load types including the high-frequency motor model, the low-frequency motor model, and the *RLC* model are provided. The high-frequency motor model is applied for the EMI noise evaluation and filter design, while the low-frequency motor model is applied for THD evaluation. In addition, the *RLC* load is provided as the general load model for other potential applications.

Another useful function in the tool is shown in Figure 13.37. Since much effort is required to configure the details and configuration of one design case may be time-consuming, the tool can save the current design configurations as a MAT-file in MATLAB. To recover the previous design configuration quickly, the tool can load the saved configuration file to save time.

After the design configuration is fully set up, the UI then recalls *m*-file functions with the configured design conditions and constraints to execute the selected models with the design algorithm to obtain the optimal design result. Figure 13.35f shows pushbuttons for "Start," "Pause," and "Stop" for the design process. The "Check" pushbutton is to confirm the correctness of the input parameters before the design process, with the parameters shown in the display window. With the same configuration tab, the user can also select the iteration setting (single switching-frequency vs. frequency range) and the design routine setting (fast vs. accurate). There is also a time step setting for simulation, with default at 0.01 μs.

13.5.3.4.2 Database Construction and Management

The design tool uses commercial components as candidates, which should lead to more practical design results. Eight databases are built with Microsoft Excel covering devices, heatsinks, fans, DM cores, CM cores, DM capacitors, CM capacitors, and DC-link capacitors. Note that wires data are also in a database but are not user-manageable in this tool due to the simplicity of the wire data. The user interface to manage the database is shown in Figure 13.38, and the detailed information for each component is displayed in the table for convenient look-up.

To add or remove an item in the database, the software provides a function to open the file externally. Then all the items in the opened database can be edited easily with Microsoft Excel.

13.5.3.4.3 Design Results Visualization

The design tool provides visualization functions for EMI noise spectrum, AC-side current ripple, AC current harmonics and THD results, weight and loss breakdown for the single converter; and weight comparison with different switching frequencies.

Figure 13.39 shows the design results tab. In this tab, the design results with the single switching-frequency design mode and with sweeping frequency design mode can be visualized with pie and bar charts. Figure 13.40 shows the harmonics and THD design results tab. For the example of pie chart visualization, Figure 13.41 displays the weight breakdowns for the multilevel flying capacitor converter and two-level voltage source converter. Comparing Figure 13.41a and b, Figure 13.41a has one extra item in the weight of flying capacitors as a result of the topology difference.

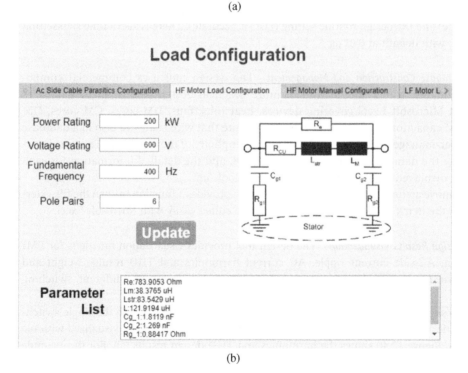

Figure 13.36 Load configurations in the design tool. (a) Cable model configuration (b) high-frequency motor model configuration.

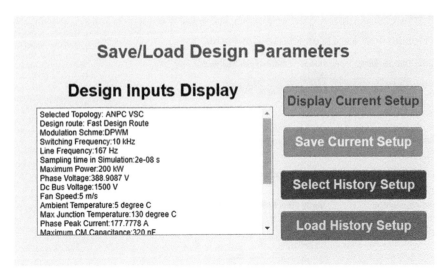

Figure 13.37 Design parameters save and load function in the developed tool.

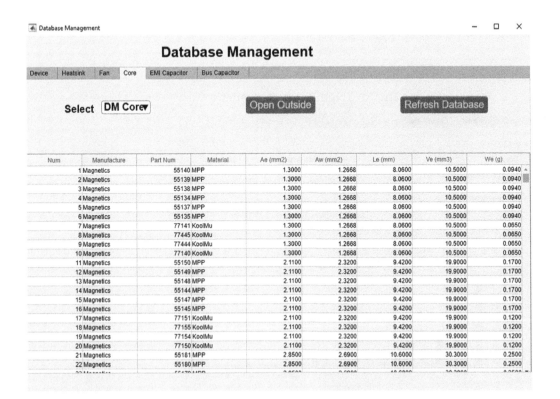

Figure 13.38 Database management of the design tool.

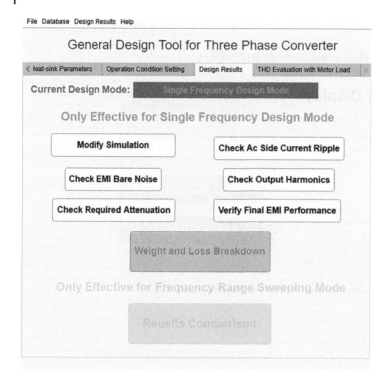

Figure 13.39 Design results display tab.

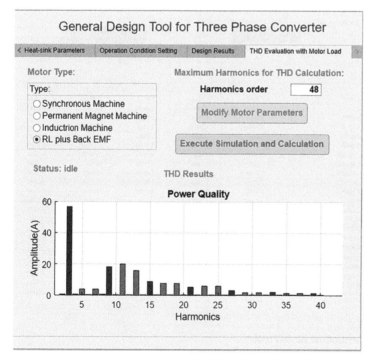

Figure 13.40 Harmonics and THD results tab.

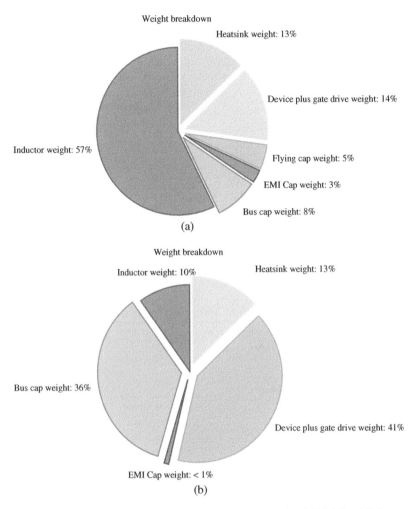

Figure 13.41 Weight breakdown displays for two topologies. (a) Multilevel flying capacitor converter weight breakdown (b) two-level VSC weight breakdown.

Similarly, the tool can display design results comparison with different switching frequencies for the selected topology and configuration. Figure 13.42 shows bar charts for two converters with different AC-side filter configurations and topologies. Figure 13.42a includes the weight of flying capacitors, while Figure 13.42b does not because of the difference in selected topologies. Figure 13.42b includes the weight of the dv/dt inductors, while Figure 13.42 does not due to the difference in AC-side filter configurations.

In addition to graphical visualization, an information query program is developed in the tool to look up the detailed information for design results. As Figure 13.43 shows, the weight and power loss values for different parts can be checked in the "Weight and Loss" tab. Also, the physical design results and component selection results can be reviewed. For example, for the inductor design, the number of cores in the stack, turns number, wire gauge, temperature rise, and total weight and loss are displayed in the "Inductor Info" tab.

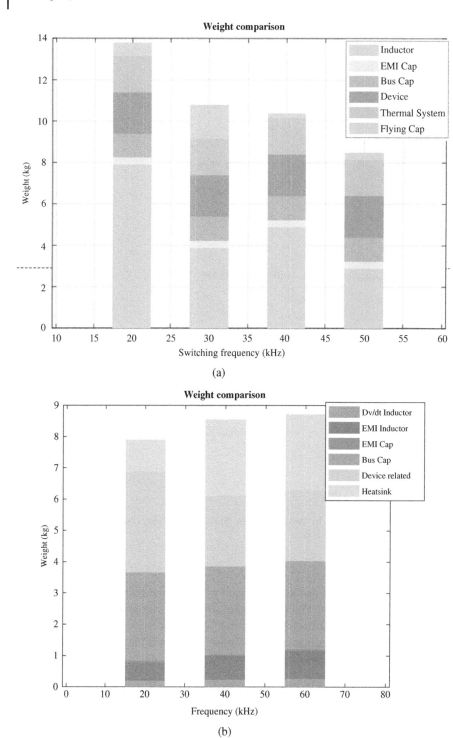

Figure 13.42 Weight comparison at different switching frequencies. (a) Multilevel flying capacitor converter with AC-side EMI filter. (b) ANPC converter with AC-side *dv/dt* filter.

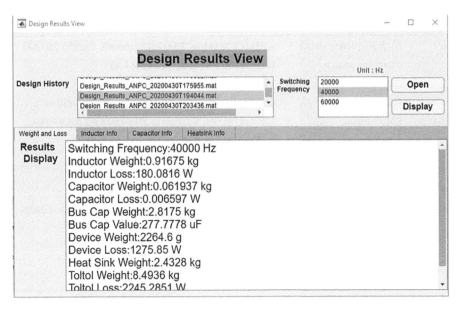

Figure 13.43 Detailed design results and the selected component information.

13.5.3.5 Design Examples

13.5.3.5.1 Design Specifications Using the developed design tool, an example design is carried out for a three-phase motor drive. The drive specifications are listed in Table 13.20, which includes design conditions, constraints, and the objective. For this specific case, the topology is selected as three-level ANPC, the modulation used is continuous SVM, and the switching frequency selected is 60 kHz. The design tool is used to optimize the device and TMS selection, and passives design. So, this is a converter-level design using the classification in Figure 13.32 architecture of the design tool Figure 13.33.

13.5.3.5.2 Design Results The weight of the optimized converter is 11.34 kg (specific power density: 17.64 kW/kg). The efficiency is 98.7% (total power loss: 2.5 kW). Detailed design results of components selection are listed in Table 13.21. The weight breakdown of the design is shown in Figure 13.44.

Table 13.20 Converter design specifications.

Design conditions	Source	± 500 V DC input with three wires
	Load	PM motor: $V_{LL\text{-}RMS} = 673.6$ V, $P = 200$ kW, $f_0 = 1$ kHz
	Environment	$T_{amb} = 40\,^\circ$C, liquid cooling for devices and forced-air cooling for inductors
Design constraints	EMI	DO-160E for both DC and AC sides
	Harmonics	DO-160E on DC side
	Device max. junction temperature	150 $^\circ$C
	Efficiency	>98%
Design objective	Minimum weight or maximum specific power	

Table 13.21 Example design results.

Device selection	High-frequency switch	APTMC120AM09CT3AG (Microsemi, 1200 V, 220 A)
	Clamping switch	HT-3292-R-VB (Cree, 900 V, 664 A)
TMS	Cold plate	180-10-24C (Wakefield-Vette)
DC-link capacitors	Film capacitor	C4AEJBW5600A3NJ (KEMET, 60 μF, 10 parallel)
DC-side EMI filter	CM inductor	176.7 μH (Core: Vitroperm 500FT60006-L2063-V110, winding: AWG 4/0, 3 turns)
	DM inductor	Leakage inductance of CM inductor: 0.93 μH
	CM capacitor	22 nF (B32022A3223M)
	DM capacitor	6.6 μF (3 parallel, BFC2 339 10225)
AC-side EMI filter	CM inductor	1.97 mH (3 series, Core: Vitroperm 500FT60006-L2102-W468, winding: AWG 2/0, 8 turns)
	DM inductor	Leakage inductance of CM inductor: 6.51 μH
	CM capacitor	23.5 nF (2 series, B32022A3473M)
	DM capacitor	2.64 μF (8 parallel, R463I33300001M)

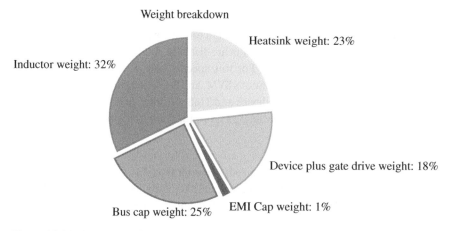

Figure 13.44 Weight breakdown of the example design.

The accurate routine is selected for the case design. It takes around three hours to get the design results using a desktop computer with an Intel(R) Core(TM) i7-7700 CPU @ 3.60 GHz processor, 16 GB of RAM, and a 64-bit operating system. For the fast routine, it takes around two hours for the same case with the same desktop computer.

One thing worth discussion about the design results in Table 13.21 is the wire size. CM inductors on DC and AC sides are designed with 4/0 and 2/0 wires, respectively, which are quite large. One reason the tool selected these wires is because the tool does not allow paralleling of multiple wires, a clear limitation. Other limitations include the database size. It is important to recognize the limitations of any design automation tools. An "optimal" design may not be really optimal due to these limitations.

13.5.4 Partition-Based Design vs. Whole Converter Design

We have discussed and promoted the subsystem partition-based approach to simplify the converter design. In many cases, the approach is adequate, especially, when subsystems are relatively independent and decoupled, and their interactions can be appropriately characterized through the interface parameters. Here, we will use another design example to illustrate the issues that may exist with the partitioned approach by comparing the design with the whole converter design approach.

The example is an AC-fed motor drive assumed for aircraft applications. The motor and inverter have the same design conditions as the example in Section 6.4 (see Tables 6.9 and 6.10). The active rectifier design conditions is in Table 13.22. The high-frequency parasitic parameters for the inverter, motor, and cable are also included in this design, with the same values as presented in Section 6.4. Since the rectifier is assumed connected to a generator with cable, the same high-frequency parasitics are assumed for the rectifier side as well. Note that due to non-unity power factor for the inverter, the DC load for the rectifier is 41 kW, lower than the inverter and motor rating of 45 kVA.

The design constraints are listed in Table 13.23. Most are the same as those in Section 6.4. The DC DM capacitance limit (defined for phase to neutral) is to avoid low power factor for the rectifier as well as to avoid potential generator control issue. The two-level VSC for both the rectifier and inverter, and 600 V DC-link voltage are preselected for comparison simplicity.

The design variable results are given in Tables 13.24 and 13.26, respectively, for the partitioned design case and the whole converter design case. Their weight and loss comparison are given in

Table 13.22 AC source and rectifier design conditions.

AC source	DC load[a]	Thermal management system	Mechanical system	Control	EMI filter
230 V RMS, 400 Hz	41 kW	Forced-air cool, maximum ambient temperature 65 °C	CM parasitics included	Recommended gate parameters	AC source side only with LISN

[a] Denotes conditions for separate rectifier design only.

Table 13.23 Design constraints.

Efficiency	EMI	DC ripple voltage	Input power factor	DC voltage
>98% for rectifier and inverter	DO-160E both input and output AC sides	<1% peak-to-peak voltage	Unity	600 V

DC DM capacitance	AC current THD	Component temperatures	Flux density	Fill factor
<1.4 μF	N/A	<Actual component limits	<80% of actual core saturation limit	<0.8

Table 13.24 Design variable results for the separately designed inverter and rectifier.

Subsystem	Component	Inverter result	Rectifier result
Inverter or rectifier power stage	Device	Same for both: Wolfspeed "C3M0065090J," 900 V, 22 A @100 °C, 4 in parallel	
	PWM scheme	Same for both: continuous SVM	
	Gate resistance	Same for both: 6 Ω per device (data sheet value with 2.5 Ω external)	
	Gate voltage	Same for both: V_{GH}: 15 V, V_{GL}: −4 V	
	Switching frequency	Same for both: 70 kHz	
	DC-link capacitor (97.6 µF for inverter, 36.2 µF rectifier)	Kemet "C4AEGBW6100A3MK," 100 µF, 450 V, 2 in series, 2 in parallel	Kemet "C4AEGBW5400A3FK," 40 µF, 450 V, 2 in series, 2 in parallel
Thermal management system	Heatsink	Aavid extruded aluminum heatsink, part no: 61325, 141 mm, 3 in parallel	Aavid extruded aluminum heatsink, part no: 61325, 100 mm, 3 in parallel
	Fan	Same for both: NMB "1608VL-04W-B40-B00"	
EMI filter	Configuration	Same for both: two-stage *LCLC*	
	DM capacitor	Kemet "R76II3150(1)30(2)," 150 nF, 180 V, 2 in series, 5 in parallel	Kemet "R76IN3820(1)30(2)," 820 nF, 180 V, 2 in series, 3 in parallel
	CM capacitor	Kemet "R76II3150(1)30(2)," 150 nF, 180 V, 2 in series, 2 in parallel	Kemet "R76II3680(1)30(2)," 680 nF, 180 V, 2 in series, 2 in parallel
	DM inductor	18.9 µH, core: "Magnetics77912" koolMu; winding: AWG 7, 29 turns	26.1 µH, core: "Magnetics55615" MPP; winding: AWG 8, 21 turns
	CM inductor	52.9 µH, core: Vitroperm 500F T60006-L2025-W380, winding: AWG 8, 2 turns	82.1 µH, core: Vitroperm 500F T60004-L2025-W622; winding: AWG 10, 4 turns

Tables 13.25 and 13.27, respectively. As expected, some components and variables are the same in both cases, including devices, PWM scheme, switching frequency, gate parameters, and fan. Note that 70-kHz switching frequency was obtained as the optimal point for the whole converter design case and fixed for the partitioned design case to facilitate comparison.

The difference of the two designs is mainly on EMI filters, especially, the CM filters. The weight of CM inductors and capacitors is significantly higher for the combined case than the partitioned case, while the weight of the DM capacitors is not very different. The weight of DM inductors increases considerably for the combined case because the DM inductors also contribute to overall CM inductance, as discussed in Chapter 6 example already.

The fact that the difference of the two design cases is mainly on CM EMI filters confirms that the CM noises of the rectifier and the inverter are coupled, and the partitioned design have not properly addressed this coupling. In this case, the partitioned design could lead to unacceptable result.

In order to verify this issue, we examine three EMI results. Figure 13.45 shows the CM noise on both the inverter and rectifier AC side with the whole AC-fed motor drive designed and simulated together. Clearly, the CM EMI noise on both sides meet the standard requirement. Figure 13.46

Table 13.25 Weight and loss breakdown for the separately designed inverter and rectifier.

Component	Inverter		Rectifier	
	Weight (g)	Loss (W)	Weight (g)	Loss (W)
DM inductor	2369	307.6	2215	219.8
CM inductor	26	7.6	22	11.8
DM capacitor	218	1.7	494	1.7
CM capacitor	87	4.0	165	8.4
DC-link capacitor	435	—	188	—
Devices	412	354 (conduction: 293, switching: 61)	412	310 (conduction: 252, switching: 58)
Heatsink and fan	562	—	432	—
Total	4109 (specific power: 10.95 kW/kg)	674.9 (efficiency: 98.36%)	3928 (specific power: 10.44 kW/kg)	551.7 (efficiency: 98.67%)
Overall	Weight 8.04 kg (specific power 5.1 kW/kg), loss 1226.6 W (efficiency 97%)			

Table 13.26 Design variable results for the inverter and rectifier designed together.

Subsystem	Component	Inverter result	Rectifier result
Inverter or rectifier power stage	Device	Same for both: Wolfspeed "C3M0065090J," 900 V, 22 A @ 100 °C, 4 in parallel	
	PWM scheme	Same for both: continuous SVM	
	Gate resistance	Same for both: 6 Ω per device (data sheet value with 2.5 Ω external)	
	Gate voltage	Same for both: V_{GH}: 15 V, V_{GL}: − 4 V	
	Switching frequency	Same for both: 70 kHz	
	DC-link capacitor (243 μF)	Kemet "C4AEGBW5700A3LK," 70 μF, 450 V, 2 in series, 7 in parallel	
Thermal management system	Heatsink	Aavid extruded aluminum heatsink, part no: 61325, 162 mm, 3 in parallel	Aavid extruded aluminum heatsink, part no: 61325, 101 mm, 3 in parallel
	Fan	Same for both: NMB "1608VL-04W-B40-B00"	
EMI filter	Configuration	Two-stage LCLC	
	DM capacitor (0.7 μF for inverter., 1 μF for rectifier)	Kemet "R76II3470(1)30(2)," 470 nF, 180 V, 2 in series, 3 in parallel	Kemet "R76II3680(1)30(2)," 680 nF, 180 V, 2 in series, 3 in parallel
	CM capacitor (0.07 μ F for inverter., 0.4 μF for rectifier)	Kemet "R76IF2680(1)30(2)," 82 nF, 180 V, 2 in series, 2 in parallel	Kemet "R76II3330(1)30(2)," 470 nF, 180 V, 2 in series, 2 in parallel
	DM inductor	12.3 μH, core: "Magnetics77102" koolMu, winding: AWG 8, 18 turns	25.5 μH, core: "Magnetics77735" koolMu, winding: AWG 7, 20 turns
	CM inductor	94.8 μH, core: Vitroperm 500F "T60004-L2194-W908," winding: AWG 7, 4 turns	98.2 μH, core: Vitroperm 500F "T60006-L2063-W985," winding: AWG 5, 6 turns

Table 13.27 Weight and loss breakdown for the inverter and rectifier designed together.

Component	Inverter		Rectifier	
	Weight (g)	Loss (W)	Weight (g)	Loss (W)
DM inductor	2670	312.7	2489	298.2
CM inductor	450	67.7	204	41.8
DM capacitor	379	29.7	494	3.7
CM capacitor	63	12.2	253	23.5
DC-link capacitor		—	1087	—
Devices	412	372 (conduction: 307, switching: 65)	412	312 (conduction: 255, switching: 57)
Heatsink and fan	627	—	438	—
Total	4601 (specific power: 8.91 kW/kg)	794.4 (efficiency: 98.36%)	5377 (specific power: 10.96 kW/kg)	679.2 (efficiency: 98.67%)
Overall	Weight 9.98 kg (specific power 5.53 kW/kg), loss 1404 W (efficiency 97%)			

shows the CM noise on both the inverter and rectifier AC side with the inverter and the rectifier in the AC-fed motor drive designed and simulated separately. The CM EMI noise for the inverter and the rectifier also meet the standard requirement on their own. Figure 13.47 shows the CM noise on both the inverter and rectifier AC side with the inverter and the rectifier in the AC-fed motor drive designed separately but then simulated together. In this case, it can be seen that the EMI noise violates the EMI standard. This confirms that the partitioned design did not consider the CM coupling well.

With this example and previous discussion, it can be concluded that, if possible, the whole converter or system design would be preferred. However, very often, it is difficult due to the complexity and scale of the problem. In fact, in the example, when expanding the design tool in [8] from the DC-fed motor drive to AC-fed motor drive, the original PSO algorithm had to be replaced by a more robust but less efficient Bayesian optimization algorithm in MATLAB. Moreover, the whole converter design in the design tool also follows a procedure to design subsystems separately as illustrated in Figure 13.33.

13.6 Virtual Prototyping

The design optimization tool presented in this chapter will help to automate the three-phase AC converter design. The results are automatically selected converter topology, filter configuration, control, and components. From these results, the performance such as power quality and efficiency, as well as design objective function, such as cost, size, or weight, can be determined. This is a big step forward compared to the manual design, both in terms of efficiency (saving times and resources) and performance (more optimal design with less unnecessary margins). However, after the "optimal" design is found, it is still necessary to build a converter prototype to validate the design. In almost all cases, the prototype will perform differently compared with the paper design. In some cases, the difference can be substantial.

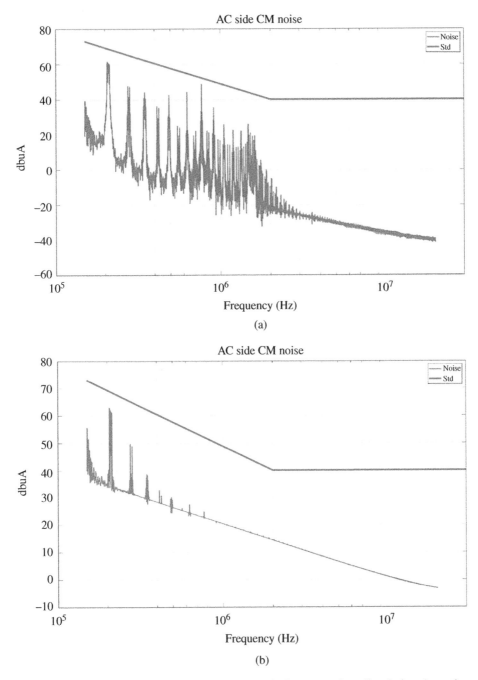

Figure 13.45 CM noise in the whole converter with the inverter and rectifier designed together. (a) Inverter AC CM noise (b) rectifier AC CM noise.

There are many reasons for the gaps between the paper design and actual prototype. One obvious reason is the accuracy of the models used in the design tool. This whole book essentially focuses on various models and modeling techniques, and has presented numerous cases of improving models of a component or a subsystem by considering various factors and refinements. The advancements

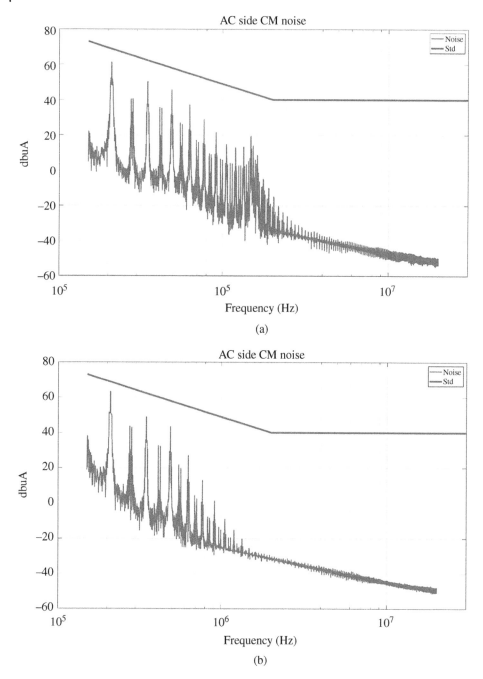

Figure 13.46 CM noise in the separately designed inverter and rectifier. (a) Inverter AC CM noise (b) rectifier AC CM noise.

on converter modeling should continue. Other reasons for the gaps include the uncertainty of the models and tolerance of the components and subsystems. Some methods to consider the uncertainty and tolerance has been developed [1, 33].

One approach to reducing the gaps between paper design and the prototype is virtual prototyping, that is, to consider prototyping as part of the design. Virtual prototyping concept is not new, and it

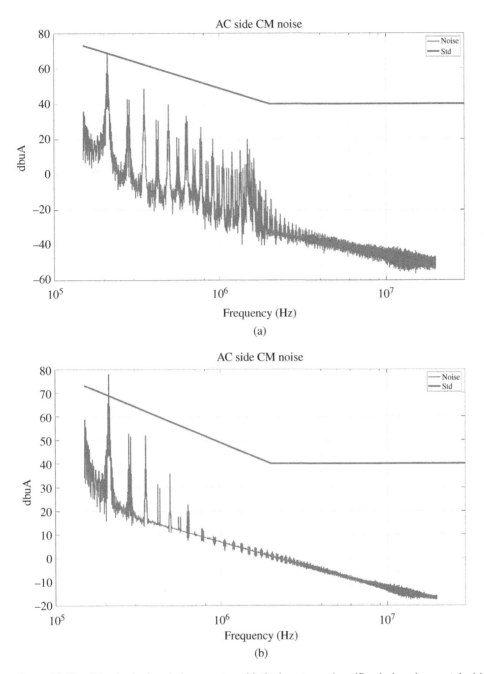

Figure 13.47 CM noise in the whole converter with the inverter and rectifier designed separately. (a) Inverter AC CM noise (b) rectifier AC CM noise.

can have different definitions and meanings in power electronics. The virtual prototyping here refers to approaches to consider physical/mechanical implementation of the converter in the design to reduce the gaps between the paper design and implementations. For example, in the models we have adopted so far, most geometry and layout-related information has not been systematically considered, although we have discussed some layout-related issues in Chapter 11, e.g. busbar layout.

The general couplings, such as the electro-thermal coupling in the loss calculation and cooling design, electromagnetic coupling in the filter design, and geometry-related design impacts in the device selection, are not covered. Hence, the purpose of the virtual prototyping is to reduce the actual prototyping and testing efforts, and trial-and-error procedures by bridging the gaps between the design and the physical prototype.

There have been some efforts on converter virtual prototyping. A rapid multi-domain virtual prototyping method in [34] to incorporate the electro-thermal power module design with 3D inductive and thermal models at typical operating and start-up conditions are evaluated. Similarly, a multi-time scale electro-thermal simulation approach [35] is developed to have a more accurate loss calculation and thermal performance prediction. Also, a virtual prototyping scheme integrated with a multi-domain simulation platform [36] enables the mechanical design evaluation and its impacts on the electro-thermal performance.

For the electromagnetic coupling, a 3D electromagnetic modeling-based design approach combined with the PEEC method and the boundary integral method [37, 38] is proposed to consider different effects of the PCB layout, self-parasitic parameters, mutual coupling, and shielding on the filter attenuation performance, allowing an optimal EMI filter design with accurate EMI filter performance prediction.

To incorporate the geometry-related design factors, a design approach for the device selection and cooling design combined with the device switching transient model and PEEC numerical model is proposed in [25]. Compared with the conventional approach, the power loop layout and gate drive resistance selection are covered, and the power loop layout can be automatically created based on the selected device package and DC-link capacitor dimensions. Then, the switching speed for the turn-on (dv/dt value) and turn-off (di/dt value) can be determined by the extracted parasitic inductance and voltage stress constraints. Considering the switching speed and power loop layout, a more accurate loss calculation and cooling design can be achieved, which are closer to the practical converter performance.

More research is needed on virtual prototyping for three-phase AC converters. It is desirable that a converter-level automatic layout algorithm can be developed, such that the design result will be more than just selected or designed components and subsystems; instead, their mechanical placement and layout will also be part of the design. With this, more accurate and realistic models can be achieved. For example, parasitics can be directly extracted for a more accurate switching waveform calculation. The coupling in EMI filters can be better identified and modeled. Moreover, the PCB or busbar weight and size not commonly considered in the design stage can be accurately estimated. The overall mechanical structure can be better modeled and visualized. All these should lead to a design closer to the real converter.

13.7 Summary

This chapter discusses methodology to achieve optimal or desired design results efficiently for three-phase AC converters through automated design optimization. Design optimization concept, general procedure, mathematical formulation, and classification for optimization problems are introduced. Optimization algorithms are discussed with focus on nonlinear, hybrid (with both continuous and discrete variables), and constrained problems, as in general, all converter design problems fall into these categories. The partitioned and centralized optimizer structures for converter design are presented. The relationship of subsystem and converter design optimization is discussed. The desired features for an automated design optimization tool is introduced.

The development of two example design tools for three-phase AC converters is presented. One is a GA-based design tool for general-purpose industrial motor drives, which relies on analytical models. It adopts the partitioned architecture covering front-end rectifier, inverter and thermal system, and EMI filter design separately. The benefit of the tool is shown through design examples, which demonstrated the design using the tool leading to smaller or lower cost designs for front-end rectifier passives, Si IGBT-based inverter power stage, and EMI filter. The results are also validated in experiments.

The other example design tool is for minimum weight design of the DC-fed high-density motor drive intended for electrified aircraft applications. It adopts a whole converter design approach but follows an iterative design procedure for subsystems. Both analytical models and switching function-based time-domain simulation are used. Brute-force search and PSO algorithm are both used. A design case showed that this tool can perform a motor drive design to find an "optimal" design result within hours on a regular desktop personal computer. This tool is also capable of handling commonly used topologies. Sample designs also showed the need to consider coupling for the partitioned optimizers.

Virtual prototyping is briefly introduced as a means to further improve the model accuracy and design for power converters.

References

1 N. Rashidi, Q. Wang, R. Burgos, C. Roy and D. Boroyevich, "Multi-objective design and optimization of power electronics converters with uncertainty quantification—part II: model-form uncertainty," *IEEE Transactions on Power Electronics*, vol. 36, no. 2, pp. 1441–1450, 2021.

2 R. M. Burkart and J. W. Kolar, "Comparative life cycle cost analysis of Si and SiC PV converter systems based on advanced η-ρ-σ multiobjective optimization techniques," *IEEE Transactions on Power Electronics*, vol. 32, no. 6, pp. 4344–4358, 2016.

3 R. M. Burkart, "Advanced modeling and multi-objective optimization of power electronic converter systems." PhD Diss., ETH Zurich, 2016.

4 S. Boyd and L. Vandenberghe, "*Convex Optimization*," Cambridge University Press, 2004.

5 M. S. Bazaraa, H. D. Sherali and C. M. Shetty, "*Nonlinear Programming: Theory and Algorithms,*" John Wiley & Sons, 2013.

6 H. Uemura, F. Krismer, Y. Okuma and J. W. Kolar, "η-ρ Pareto optimization of 3-phase 3-level T-type AC-DC-AC converter comprising Si and SiC hybrid power stage," in *2014 International Power Electronics Conference (IEEE ECCE ASIA)*, Hiroshima, Japan, 2014.

7 R. Ren, "Modeling and Optimization Algorithm for SiC-based Three-phase Motor Drive System." PhD Diss., The University of Tennessee, Knoxville, 2020.

8 Z. Dong, R. Ren, F. Wang and R. Chen, "An automated design tool for three-phase motor drives," in *IEEE Design Methodologies for Power Electronics Conference*, Bath, UK, July 2021.

9 H. Kragh, F. Blaabjerg and J. K. Pedersen, "An advanced tool for optimised design of power electronic circuits," in *Conference Record of Thirty-Third Industry Applications Society Annual Meeting*, St. Louis, MI, USA, October 12–15, 1998.

10 R. B. Ridley, C. Zhou and F. C. Lee, "Application of nonlinear design optimization for power converter components," *IEEE Transactions on Power Electronics*, vol. 5, no. 1, pp. 29–39, 1990.

11 F. Wang, G. Chen, D. Boroyevich, S. Ragon, V. Stefanovic and M. Arpilliere, "Analysis and design optimization of diode front-end rectifier passive components for voltage source inverters," *IEEE Transactions on Power Electronics*, vol. 24, no. 5, pp. 2278–2289, September 2008.

12 D. Goldberg, "*Genetic Algorithms in Search, Optimization, and Machine Learning,*" Addison-Wesley Publishing Co., Inc., 1989.

13 F. Viani, F. Robol, M. Salucci and R. Azaro, "Automatic EMI filter design through particle swarm optimization," *IEEE Transactions on Electromagnetic Compatibility*, vol. 59, no. 4, pp. 1079–1094, 2017.

14 R. C. Eberhart and Y. Shi, "Comparison between genetic algorithms and particle swarm optimization," in *International Conference on Evolutionary Programming*, San Diego, CA, USA, March 25–27, 1998.

15 R. Hassan, B. Cohanim, O. D. Weck and G. Venter, "A comparison of particle swarm optimization and the genetic algorithm," in *46th AIAA/ASME/ASCE/AHS/ASC Structures, Structural Dynamics and Materials Conference*, Austin, TX, April 18–21, 2005.

16 F. Wang, W. Shen, D. Boroyevich, S. Ragon, V. Stefanovic and M. Arpilliere, "Voltage source inverter – development of a design optimization tool," *IEEE Industry Applications Magazine*, vol. 15, no. 2, pp. 24–33, March–April 2009.

17 Q. Wang, X. Zhang, R. Burgos, D. Boroyevich, A. White and M. Kheraluwala, "Design and optimization of a high performance isolated three phase AC/DC converter," in *IEEE Energy Conversion Congress and Exposition (ECCE)*, Milwaukee, WI, September 18–22, 2016.

18 S. Busquets-Monge, J.-C. Crebier, S. Ragon, E. Hertz, D. Boroyevich, Z. Gurdal, M. Arpilliere and D. Lindner, "Design of a boost power factor correction converter using optimization techniques," *IEEE Transactions on Power Electronics*, vol. 19, no. 6, pp. 1388–1396, 2004.

19 C. Larouci, M. Boukhnifer and A. Chaibet, "Design of power converters by optimization under multiphysic constraints: application to a two-time-scale AC/DC–DC converter," *IEEE Transactions on Industrial Electronics*, vol. 57, no. 11, pp. 3746–3753, 2010.

20 I. Laird, X. Yuan, J. Scoltock and A. J. Forsyth, "A design optimization tool for maximizing the power density of 3-phase DC–AC converters using silicon carbide (SiC) devices," *IEEE Transactions on Power Electronics*, vol. 33, no. 4, pp. 2913–2932, 2018.

21 Q. Liu, F. Wang and D. Boroyevich, "Model conducted EMI emission of switching modules for converter system EMI characterization and prediction," in *Conference Record of 39th IEEE IAS Annual Meeting*, Seattle, WA, October 3–7, 2004.

22 W. Shen, F. Wang, D. Boroyevich, S. Ragon, V. Stefanovic and M. Arpilliere, "Optimizing EMI filter design for motor drives considering filter component high-frequency characteristics and noise source impedance," in *IEEE Applied Power Electronics Conference and Expostion (APEC)*, Anaheim, CA, February 22–26, 2004.

23 W. Shen, F. Wang, D. Boroyevich and Y. Liu, "Definition and acquisition of CM and DM EMI noise for general-purpose adjustable speed motor drives," in *IEEE Power Electronics Specialists Conference (PESC)*, Aachen, Germany, 2004.

24 D. Han, S. Li, Y. Wu, W. Choi and B. Sarlioglu, "Comparative analysis on conducted CM EMI emission of motor drives: WBG versus Si devices," *IEEE Transactions on Industrial Electronics*, vol. 64, no. 10, pp. 8353–8363, 2017.

25 R. Ren, Z. Dong and F. Wang, "Bridging gaps in paper design considering impacts of power-loop layout," in *IEEE Energy Conversion Congress and Exposition (ECCE)*, October 11–15, 2020.

26 AAVID, "*AAVID Product Selection Guide,*" AAVID, 2010.

27 M. K. Kazimierczuk, "*High-Frequency Magnetic Components,*" John Wiley & Sons, 2009.

28 Z. Dong, R. Ren, B. Liu and F. Wang, "Data-driven leakage inductance modeling of common mode chokes," in *IEEE Energy Conversion Congress and Exposition (ECCE)*, Baltimore, MD, USA, September 29–October 3, 2019.

29 R. Ren, Z. Dong, B. Liu and F. Wang, "Leakage inductance estimation of toroidal common-mode choke from perspective of analogy between reluctances and capacitances," in *IEEE Applied Power Electronics Conference and Exposition (APEC)*, March 15–19, 2020.

30 E. Bennett and S. C. Larson, "Effective resistance to alternating currents of multilayer windings," *Electrical Engineering*, vol. 59, no. 12, pp. 1010–1016, 1940.

31 V. C. Valchev and A. V. d. Bossche, *"Inductors and Transformers for Power Electronics,"* CRC Press, 2005.

32 L. Jun and O. Lorentz, "Modeling ceramic and tantalum capacitors by automatic SPICE parameter extractions," in *IEEE Applied Power Electronics Conference and Exposition*, Austin, TX, March 6–10, 2005.

33 N. Rashidi, Q. Wang, R. Burgos, C. Roy and D. Boroyevich, "Multi-objective design and optimization of power electronics converters with uncertainty quantification—part I: parametric uncertainty," *IEEE Transactions on Power Electronics*, vol. 36, no. 2, pp. 1463–1474, 2021.

34 P. L. Evans, A. Castellazzi and C. M. Johnson, "Design tools for rapid multidomain virtual prototyping of power electronic systems," *IEEE Transactions on Power Electronics*, vol. 31, no. 3, pp. 2443–2455, March 2016.

35 K. Li, P. Evans and M. Johnson, "Using multi time-scale electro-thermal simulation approach to evaluate SiC-MOSFET power converter in virtual prototyping design tool," in *IEEE 18th Workshop on Control and Modeling for Power Electronics (COMPEL)*, Standford, CA, 2017.

36 P. Solomalala, J. Saiz, A. Lafosse, M. Mermet-Guyennet, A. Castellazzi, X. Chauffleur and J.-P. Fredin, "Multi-domain simulation platform for virtual prototyping of integrated power systems," in *European Conference on Power Electronics and Applications*, Aalborg, Denmark, 2007.

37 I. F. Kovačević, T. Friedli, A. M. Muesing and J. W. Kolar, "3-D electromagnetic modeling of EMI input filters," *IEEE Transactions on Industrial Electronics*, vol. 61, no. 1, pp. 231–242, 2014.

38 I. Kovacevic, A. Muesing and J. W. Kolar, "PEEC modeling of toroidal magnetic inductor in frequency domain," in *IEEE Energy Conversion Congress and Exposition (ECCE) ASIA*, Sapporo, Japan, June 21–24, 2010.

Index

a

AC current distortion   296–299
AC-fed motor drives. *See also* Motor drives
  for aircraft applications   635
  baseline design   548
  components and weight contributors of   25
  front-end diode rectifier   135
  system conditions   21
  system configuration   507, 508
  system design   22
  system variables   21
  with two-level VSI   167–168
AC filter inductor design   569
AC input harmonic current   136
  commutation angle   142, 145, 164
  individual harmonic current   145–146
  RMS current   146–148
  switching functions   141–143
  voltage and current waveforms   143, 144
AC line filters   247, 252–253
AC load harmonic current   169
  average model   175
  computation times   185, 189
  load impedance   174
  reference voltage vector   180
  sine-triangle PWM   176–179
  space vector diagram, for two-level
     VSC   179–180
  SVM   181–184
  switching angle generation process
     principle   177–178
  switching function, of two-level VSI   175
  voltage harmonics comparison
    DPWM case   188–189
    SPWM case   185–186
    SVM case   186–187

AC power cycling   63
AC source equivalent impedance   248
AC source harmonic currents
  AC line filter   252–253
  LCL filter (*See LCL* filter)
  and TDD   252
AC source voltage characteristics   248
Active/hybrid filter   358–363
Active neutral-point clamped (ANPC)   12
  busbar design, for three-level converter
    capacitor selection   494
    flowchart   493
    layout design procedure   494–498
    power module selection   493–494
    switching loops in   491–493
    validation   498–502
  commutation loops   295
  simulation time comparison   283, 285
  three-phase three-level   281–283
Active rectifiers/source-side inverters
  AC-fed motor drive with   245, 246
  AC source harmonic currents
    AC line filter   252–253
    LCL filter (*See LCL* filter)
    and TDD   252
  CM noise   636, 638–641
  control architecture   264
  control performance   264–266
  DC-link stability   266–267
  description of   245
  design problem formulation   245
    conditions   248–249, 635
    constraints   247–248
    objectives   249–251
    optimization   249, 251, 280
    variables   246–247, 635–637

Design of Three-phase AC Power Electronics Converters, First Edition. Fei "Fred" Wang, Zheyu Zhang, and Ruirui Chen.
© 2024 The Institute of Electrical and Electronics Engineers, Inc. Published 2024 by John Wiley & Sons, Inc.

Active rectifiers/source-side inverters (*cont'd*)
  EMI filter   300
  interfaced subsystems   300, 301
  reliability   267
    component models   269–274
    consideration in design   269
    failure rate, MTBF, and lifetime   268
    reliability-oriented design   274–279
  topologies
    AC current distortion   296–299
    circuit modeling for different   280–286
    device voltage stress   294–297
    I-shaped three-level converters   288–290
    PWM voltage   296–299
    switching loss   290–294
    T-type three-level converter   287, 288
    Vienna-type rectifier   286–288
  weight and loss breakdown   636–638
Active zero state PWM (AZSPWM)
        354–357
ADC. *See* Analog-to-digital conversion (ADC)
AECs. *See* Aluminum electrolytic
        capacitors (AECs)
Air convection cooling   404–405
Air-core decoupling inductor   226–227
All-electric ships   5
Aluminum electrolytic capacitors (AECs)   77–78,
        98, 271, 278, 279
  failure and self-healing capability of   97
  failure modes, mechanisms, and critical
        stressors   96
American Society for Testing and Materials
        (ASTM)   268
Amorphous alloys   106
Analog-to-digital conversion (ADC)
  peripheral unit with higher resolution   415
  sensor output   438
  sigma-delta   436
ANN. *See* Artificial neural network (ANN)
ANPC. *See* Active neutral-point clamped (ANPC)
Ansys Q3D   498
A_p-based design method   114, 154
Application control layer   412–414
Application-specific integrated circuit
        (ASIC)   414
Arrester(s)   555–556, 563–566, 570
Artificial neural network (ANN)   217
ASIC. *See* Application-specific integrated
        circuit (ASIC)

ASTM. *See* American Society for Testing and
        Materials (ASTM)
ATS-CP-1004 cold plate   550
Augmented Lagrangian penalty optimization
        algorithm   588–589
Auxiliary circuits   18–19
Average value, of envelope waveform   340
AZSPWM. *See* Active zero state PWM (AZSPWM)

b
Bandwidth, of voltage sensor   434
Baseline design, of grid application
  AC-fed motor drives   548–549
  arresters (*See* Arrester(s))
  component size and weight summary   555–556
  DC-link capacitor   550–551
  device cooling   549–550
  inductor filter   551–554
  precharge circuit   554–555
  semiconductor devices   549
  total demand distortion   549
Basic isolation   416
Bearing current
  damages and failures   516–517
  evaluation   520–522
  measurement   516
  mitigation   521–523
  reduction schemes   523
  types of   516
Bonded/fabricated fins, heatsinks with   378–379
Boost inverter   11
Bootstrap gate driver   429–430
Boundary integral method   642
Breakdown voltage   43–45
Brute-force algorithm   588–590
Buck inverter   11
Busbar design
  mechanical system   19–20, 474
    component selection and busbar
        geometry   483–485
    current distribution analysis   487–488
    dimension and current density
        requirements   486–487
    electrical insulation   488–489
    flowchart   483, 484
    parasitic capacitance and resistance   490
    parasitic inductance   489–490
    problem formulation   482–483
  for three-level ANPC converter

capacitor selection   494
flowchart   493
layout design procedure   494–498
power module selection   493–494
switching loops in   491–493
validation   498–502

C

Cables
characteristic impedance   533–537
harmonics in   513
Cables models
for EMI filter   338
Capacitance   84–86
parasitic grounding   350, 351
Capacitive coupling, in T-type filter   348
capacitor winding   127–128
aluminum electrolytic   77–78, 271
bank, parallel/series connections   98
configuration and voltage balancing
98–99
parasitic inductance reduction, layout
for   99–100
CeraLink   80, 82
ceramic   76
CM and DM   337
in converter design   82–84
decoupling   168–170, 218–222
definition of   75
electrical characteristics of   78, 80
EMI filters   611–612
energy and power densities of   78, 81
equivalent circuit model (*See* Equivalent circuit
model, of capacitor)
failure rate and lifetime models   271–272
film   271, 272, 279
LCL filter   255
lifetime model   96–98, 271–272
loss and thermal model   93–96
mica   76
nanostructure multilayer   80
paper   76
PLZT-based ceramic   78–80, 82
poly-film   77
related constraints   157
relevance to converter design   100–101
scaling   101
supercapacitor   80–81
tantalum electrolytic capacitors   78

technologies   75, 78, 79, 81
voltage and current capability model (*See*
Voltage and current capability model)
Carrier-based PWM   176
third-order harmonics injection   176, 179,
191–194
Carrier frequency   177, 178, 328, 546, 549
Cascaded multilevel converter   11, 12
Case-to-ambient thermal resistance   201
of overload period end   204
Casting heatsinks   378
Cauer model. *See* Continued-fraction circuit;
Thermal impedance model
CeraLink capacitor   80, 82
Ceramic capacitors   76, 101
PLZT-based   78–80, 82
Characteristic impedance, motor terminal
filters   531–533
three-phase cable characteristics   533–536
for three-phase shielded cables   536–537
for unshielded cables   536
Chopper mode bias (CMB)   63
Circuit modeling
circuit simulation   280
switching function-based simulation
for multilevel VSC   281–286
for two-level VSC   281
CISPR standards   339, 341
Class-X capacitors   101
Class-Y capacitors   101
CM. *See* Common mode (CM)
CMB. *See* Chopper mode bias (CMB)
CMC. *See* CM choke (CMC)
CM choke (CMC)   113
CMOS. *See* Complementary metal-oxide-
semiconductor (CMOS)
CMRR. *See* Common-mode rejection
ratio (CMRR)
CMTI. *See* Common-mode transient
immunity (CMTI)
CM voltages, motor drives
bearing current   170, 516–517, 520–523
detrimental effects   513
drive inverter-motor side   515
PWM drive inverter   514
shaft voltage   170, 516–519, 521–523
stator winding capacitance coupling   515
Coefficient of thermal expansion (CTE)   380, 382
Commercial off-the-shelf heatsinks   388–390

Common mode (CM)
  current 170
  dv/dt in power converter 436
  EMI filter 612, 636, 638
  EMI noise 233–236
    active rectifier, equivalent circuit for 330, 331
    capacitors 337
    choke inductance 322, 365
    components 314, 315
    corner frequencies for 319–321, 342,
      363–365
    current of LISNs 359–362
    decoupling 310
    definition of 306
    diode rectifier paths 330, 331
    equivalent circuit, for filter
      attenuation 316–318
    filter topology 354, 355, 357, 358
    grounding impedance effect on 344
    impedance balancing circuit ratio 358
    impedance measurement 337, 338
    inductor core saturation and temperature
      rise 333–334
    inductor equivalent circuit 333
    inductor impedance 333–336
    LISN, impedance of 322
    load impedance, currents through 310
    and mixed-mode noise 351
    noise current cancellation 359–361
    noise propagation 353
    noise source impedance 331, 332
    one-and two-stage filters 323–324
    Π-type filter, couplings in 346–347
    single-phase system 306, 309
    three-phase system 310–312
    voltages 328, 354, 355, 359, 360, 362
  inductance 232, 324
  noise voltage 445
  standard limit and bare noise 609–610
  voltages (*See* CM voltages, motor drives)
Common-mode rejection ratio (CMRR) 444
Common-mode transient immunity (CMTI)
  of commercial analog iso-amp 437–438
  isolated power supply 424
  signal isolators 417, 423–424
Complementary metal-oxide-semiconductor
  (CMOS) 437, 438, 441–442
Component models, in design tool 622, 623
Composite waveform hypothesis (CWH) model
  lightning surge 118–119
Composition, of three-phase AC converters

controller and auxiliary circuits 18–19
  energy storage passives 17
  filters 17–18
  mechanical assembly 19–20
  motor drive, function block of 16, 17
  switching devices 17
  thermal management system 18
Conduction loss
  calculation 38, 196
  delta-connected current source rectifier 14, 15
  of inverter 190–195
  of SiC MOSFET 455, 456
Constrained optimization problem 587
Continued-fraction circuit 58, 94
Continuous optimization problem 586
Control and auxiliaries 24
  control architecture, for power
    electronics 411, 412
    application control layer 412–414
    converter control layer 413, 414
    hardware control layer 414
    switching control layer 413
    system control layer 411–412
  control hardware selection and
    design 414–415
  converter-level protection 450–451
  current sensors 431, 432
  deadtime setting 455–461
  device-level protection 445
    crosstalk suppression 448–450
    short-circuit protection 445–448
  gate driver (*See* Gate driver)
  high-frequency WBG converter (*See* DC bias)
  high-voltage sensors 432–434
  interfaced subsystems 466, 467
  isolation
    isolated power supply 417–418
    for low-power converter design 418
    in power converter design 415
    signal isolator 415–417
  power converter design 430–431
  printed circuit boards 452–455
  temperature sensors 432, 433
  in three-phase power converter
    system 411, 412
  voltage sensors 431
Control layers 18
Control performance 170
  active rectifiers/source-side inverters 247,
    264–266
  load-side inverters 198–200

Conventional cooling methods, for power electronics 370
  forced-air convection 371
  forced-liquid convection 371–372
  natural convection 370–371
Converter control layer 413, 414
Converter insulation 565–566
Converter-level protection 450–451
Converter main circuit 23
Converter sensing and control technology 20
Convertible voltage-source converter (C-VSC) 275–277
Convex optimization problem 587
Cooling technology 20, 369–370
  baseline design 549–550
  comparison of 375–377
  conventional cooling methods, for power electronics 370–372
  forced-liquid convection 400–403
  heat pipes 373–374
  heatsinks
    forced convention cooling 383–384
    thermal interface material 382–383
    thermal materials 380–382
    types and configurations 377–380
  heat transfer coefficient ranges for 375, 376
  jet impingement 374
  liquid immersion and pool boiling 372–373
  microchannel/mini-channel liquid convection 375
  for passive components 403
    air convection cooling 404–405
    improving internal thermal conduction 403
    liquid convection cooling 405–407
  spray cooling 374–375
  surface creation 405
  temperature rise *vs.* surface heat flux 376, 377
Core loss 114
  empirical approach 115–120
  fringing effect on 127
  loss separation approach 115
Core saturation
  CM inductor 333–334
  flux density and 113
  inrush current with/without,152
  surge voltage with/without 153
Core window area product 114
Corner frequency
  for CM and DM filters 319–321
  vs. switching frequency 342–343

Coupling, EMI filter
  capacitive coupling, in T-type filter 348
  inductive coupling, in Π-type filter 346–347
  in multistage filters 349
Crosstalk suppression 448–450
CSCs. *See* Current source converters (CSCs)
CSI. *See* Current source inverter (CSI)
CTE. *See* Coefficient of thermal expansion (CTE)
Current density 113–114
Current distribution
  EMI filters 365–366
  mechanical system 504
Current falling stage, MOSFET 209–212
Current polarity detection 465
Current rise stage, MOSFET 212–213
Current sensor technology 431, 432
Current source converters (CSCs) 10, 14, 15
Current source inverter (CSI) 11
Custom-designed heatsinks
  average channel velocity 391
  FEA simulation 392
  fin effect 389–390
  hydraulic diameter 392
  models comparison 393
  pressure drop of 391, 392
  weight density 392
Cut cores 105–107
C-VSC. *See* Convertible voltage-source converter (C-VSC)
CWH model. *See* Composite waveform hypothesis (CWH) model
Cycloconverter topology 15

d
DAB. *See* Dual active bridge (DAB)
DARWIN optimization algorithm 600
Data centers 9–10
DC bias
  dv/dt induced 436–437
  on isolated voltage sensing 434–436
  mitigation solutions 437–438
  origin of 436
  from RFI (*See* Radio-frequency interference (RFI))
DC busbar 478–480
DC current-bias-dependent permeability 112
DC-fed motor drive 231, 616, 617. *See also* Motor drives
  EMI filters for 316
  three-phase two-level VSI 350
DC harmonic impedance 138

DC impedance   239

DC-link capacitor   136, 168, 620
  baseline design   550–551, 567–568
  busbar connects   474
  configuration of   279
  current distribution among   487
  design comparison   278
  motor drive with active rectifier   330
  paralleled, on busbar   498
  in rectifier design   246
  selection problem   83–84
  in three-level ANPC converter   491, 492

DC-link RMS current   238

DC-link stability
  active rectifiers/source-side inverters   248,
      266–267
  passive rectifier   137, 149

DC-link under-and overvoltage
      trip levels   138

DC-link voltages
  decoupling capacitor   218
  harmonic impedance   141
  HVRT   558–559
  lightning surge   565
  minimum   170
  peak   170–171
  temporary overvoltage   559
  voltage source inverter   195
  under normal operating
      conditions   136–137, 148

DC power cycling   63

Dead time
  compensation   456, 461–466
  harmonics   512
  setting   455–461

Decoupling capacitor   168–170
  gate driver   428–429
  load-side inverter   218–222
  in three-level ANPC converter   494

Decoupling inductor   169, 170
  load-side inverter
    air-core inductor   226–227
    dv/dt reduction   227, 229, 230
    high-frequency impedance and circuit
        model   222–226
    on motor terminal voltage   227, 229, 230
    resonant current, during switching
        transient   224, 225
    switching waveform comparisons   227, 228
    voltage commutation time   224, 225

Delta-connected current source rectifier   14, 15
Deltamax   106
Desaturation protection scheme   446, 447
Design automation   583
Design conditions   20–21
  active rectifier and source-side
      inverter   248–249, 253
  DC inductor   603, 604
  EMI filter   325, 609
  inverter and thermal system   607
  load-side inverter   170–172
  mechanical system   477–478
  passive rectifier   138, 147
  thermal management system   386

Design constraints   21
  active rectifier and source-side
      inverter   247–248
  EMI filter   325, 610–611
  front-end rectifier   603, 605
  inverter and thermal system   607
  load-side inverter   169–170, 232
  mechanical system   476–477
  partition-based *vs.* whole converter design   635
  passive rectifier design   136–138, 147
  thermal management system   385

Design objectives
  active rectifier and source-side
      inverter   249–251
  EMI filter   326
  load-side inverter   172–173
  mechanical system   478
  passive rectifiers   139–140
  thermal management system   386–387

Design optimization   22–23
  active rectifier and source-side inverter   249,
      251, 280
  algorithms
    linear optimizations   587, 588
    nonlinear optimizations   587–589
    selection   589–590
  concept and mathematical
      formulation   583–584
  converter design tools, features for   596–597
  EMI filters   326, 327, 343–344, 611
  formulation of   616, 618
  GA-based optimization tool
    EMI filter   609–615
    front-end rectifier   603–606
    inverter and thermal system   607–608
    optimizer development   600–603

optimizer formulation   597–600
system optimization   614–617
high-density inverters (*See* High-density
    inverters tool)
load-side inverter
    DC-fed motor drive system   231
    design conditions   230–231
    design constraints   232
    design variable results   232, 233
    EMI noise   233–236
    high-frequency motor and cable
        models   231–232
    motor load conditions and parameters   230
    weight and loss breakdown   233
mathematical properties, of power converter
    design   586–587
multi-objective optimization method   585–586
Pareto front   585–586
partition-based *vs.* whole converter
    design   635–641
partitioned optimizers   590–594
passive rectifiers design   139, 140, 157–161
    DC-link capacitor volume   158
    front-end passive components   157, 158
    inductor volume   158
    magnetic core parameters   160
    observations   160–161
    three-phase inductor, with manual and
        automated approaches   159
procedural steps   584–585
single optimizer   594–595
thermal management system   386, 387, 393–397
virtual prototyping   638–642
Design problem partition   23–24
Design procedure   20, 21, 25–26
Design variables   21–22
    active rectifier and source-side
        inverter   246–247
    EMI filter   324–325, 610, 611
    front-end rectifier   603, 605
    inverter and thermal system   607
    load-side inverter   168–169
    mechanical system   475–476
    partition-based *vs.* whole converter
        design   635, 636
    passive rectifier   136
    thermal management system   385
Device behavior model   71, 72
Device cooling mechanical design   480–481
Device-level protection   445

crosstalk suppression   448–450
short-circuit protection   445–448
Device loss model   597
Device maximum junction temperature, load-side
    inverters   200–204
Device selection-related constraints, passive
    rectifiers   149–150
    air gap magnetic field intensity   152
    DC/single-phase AC inductor   153, 154
    diode rectifier, equivalent circuit of   150, 151
    maximum inductor current   151
    simulated inrush currents   152
    three-phase inductor core   150
Device under test (DUT)   55–56
DFA. *See* Double Fourier analysis (DFA)
DFIG. *See* Doubly fed induction generator (DFIG)
di/dt transients
    with controllable coil inductance   448
    in control system design   240
    hardware-based detection   450
    overcurrent protection scheme   447
    switching transient   19
    WBG-based converter   434, 443
Differential mode (DM)
    EMI filter   612
        capacitors   337
        components   314, 315
        corner frequencies for   319–321, 363–365
        decoupling   310
        definition of   306
        equivalent circuit, for filter
            attenuation   316–319
        inductor   336, 365
        LISN, impedance of   322
        load impedance, currents through   310
        one-and two-stage filters   323–324
        topologies comparison   322, 323
    EMI noise
        diode rectifier paths   330, 331
        grounding impedance effect on   344, 345
        impedance measurement   337–338
        vs. MM noise   351, 352
        noise propagation path   353
        single-phase system   306, 309
        three-phase system   310–312
        voltages, three-phase VSC for   328
    inductance   232
    noise voltage   445
    standard limit and bare noise   609–610
Digital signal processor (DSP)   414, 415

Digital-to-analog converter (DAC)   438
Diode-based rectifiers   2, 10
   front-end   135, 138
   motor drive with   330
   passive rectifiers   133–135
Diode-clamped multilevel converter   283–285
Diode neutral-point clamped (DNPC)
      phase   288–289
Diodes   10
   junction capacitances in   46
   on-state characteristics   38, 39
   output characteristics   35, 36
Discrete optimization problem   586
Discrete packages   64, 65
Distributed winding circuit model   528
DM. *See* Differential mode (DM)
DNPC phase. *See* Diode neutral-point clamped
      (DNPC) phase
Double Fourier analysis (DFA)   175–177
   EMI noise source   327–329
   voltage harmonics comparison using   185–189
Double isolation   416
Double pulse test (DPT)   55–57
   based loss calculation   293–294
   load current measured by   224, 225
   switching energy and   292
Doubly fed induction generator (DFIG)   8–9
DPT. *See* Double pulse test (DPT)
DPWM   354–356
   minimum-loss   193
   phase reference voltages and duty
      cycles   261
   phase voltage switching pattern   182, 183
   switching loss for   196, 197
   voltage harmonics comparison   188–189
DSP. *See* Digital signal processor (DSP)
Dual active bridge (DAB)   472
DUT. *See* Device under test (DUT)
dv/dt filters, motor drive
   capacitor voltage and current   538–539
   damping conditions   539
   design constraints   540
   equivalent circuit, for one switching
      event   537, 538
   filter power loss   541
   inverter output   537
   switching cycle   540
   system parameters   540, 541
   topology   541–542

dv/dt transients
   induced DC bias   436–437
   WBG-based converter   434, 443
Dynamic on-state resistance model   66–67

e

EDM. *See* Electric discharge machining (EDM)
Electrical insulation, of busbar   488–489
Electrical system
   electrified aircraft propulsion   6, 7
   more-electric aircraft   6
Electric discharge machining (EDM)
   bearing current *vs.* shaft voltage., 520–522
   definition of   516
Electric vehicles (EVs)
   powertrain power system architecture   4
Electrified aircraft propulsion (EAP)   6, 7
Electrified railway traction system, configuration
      of   6, 8
Electrified trains   5
Electrified transportation
   EAP electrical systems   6, 7
   electric vehicles   4
   electrified rail traction system, configurations
      of   6, 8
   more-electric aircraft   5, 6
   shipboard electric power system   5
Electrochemical capacitor. *See* Supercapacitor
Electrolyte vaporization   96
Electrolytic capacitors   76
   aluminum electrolytic capacitors   77–78
   tantalum electrolytic capacitors   78
Electromagnetic compatibility (EMC)
   definition of   305
   filters for   18
   standards   306–308
Electromagnetic environmental impact
      technology   20
Electromagnetic interference (EMI)
   in busbar design   19
   filter (*See* EMI filters)
EMC. *See* Electromagnetic compatibility (EMC)
EMC standard CISPR11, 306, 307
EMI. *See* Electromagnetic interference (EMI)
EMI rejection ratio (EMIRR)   442
EMI filters   138, 317
   AC source   249
   active/hybrid filter   358–363
   active rectifiers/source-side inverters   300

attenuation   316–319
bare noise   163, 236
capacitors and choke core   611–612
CM and DM noise   233–236, 609–610
   corner frequencies for   319–321
   decoupling   310
   definition of   306
   load impedance, currents through   310
   single-phase system   306, 309
   three-phase system   310–312
core saturation   113
corner frequency *vs.* interleaving angle   364
corner frequency *vs.* switching
   frequency   342–343
couplings   346–349
for DC-fed motor drive   316
definition of   305
design problem formulation
   conditions   325, 609
   constraints   325, 610–611
   objectives   326
   optimization   326, 327, 343, 611
   variables   324–325, 610, 611
grounding effect   344–345
integration techniques   365
interfaced subsystems   365–366
load-side inverter   171, 239
mechanical system   502–503
mixed-mode noise   350–352
models   326–327
mode transformation   353
modulation schemes   354–356
noise level   613, 614
noise measurement
   CM and DM components   314–316
   correction factor, for LISN
     capacitor   313, 314
   current probe transfer impedance   315, 316
   equipment under test   312–314
   experiment setup   311, 312
   LISN   312–314
   measurement setup   314–315
   noise voltage measurement   313, 314
one-stage filter   613, 615
optimizer formulation   598–600
paralleled converters   363–365
parameter and physical design/selection   322
passive rectifier   162
performance calculation   612, 613

standards   306–308
switching frequency   354
thermal management system   398
topologies   354–355, 357–359, 609, 610
   multistage filter   323–324
   selection, rule for   322–323
   two-stage filter design   614, 615
EMI noise source model   327
   DFA-based model   327–329
   impedance   331–332
   switching function simulation-based
     model   329–330
   from two converters   330–331
EMI propagation path impedance model
   cables model   338
   CM capacitors   337
   CM inductor   333–336
   DM capacitors   337
   DM inductor   336
   impedance characteristics   332
   motor model   337–338
   test receiver   338–341
EMIRR. *See* EMI rejection ratio (EMIRR)
EMI test receiver   338–339
   envelope detector modeling   340
   IF filter modeling   339
   peak, quasi-peak, and average
     detection   339–341
Empirical core loss models
   comparison of   120
   composite waveform hypothesis   118–119
   Generalized Steinmetz equation   116–117
   improved Generalized Steinmetz equation   117
   improved iGSE   118
   loss map methods   119–120
   Modified Steinmetz equation   116
   Natural Steinmetz equation   117–118
   Steinmetz equation   115, 116
   Steinmetz premagnetization graph   119
   Waveform coefficient Steinmetz equation   118
Endurance testing   63
Energy equivalent capacitance   291
Energy storage passives   17
Envelope detector modeling   339, 340
Environmental testing   63
Equipment under test (EUT)   306, 312–314
Equivalent circuit model, of capacitor
   capacitance   84–86
   equivalent series inductance   88

Equivalent circuit model, of capacitor (*cont'd*)
  equivalent series resistance 86–88
  impedance characteristics 84, 85
  nonideal capacitor 84
  simplified capacitor 84
Equivalent series inductance (ESL) 88, 99, 348,
    428–429
Equivalent series resistance (ESR) 78, 86–88,
    93, 127
Equivalent thermal capacitance 156–157
ESL. *See* Equivalent series inductance (ESL)
ESR. *See* Equivalent series resistance (ESR)
EUT. *See* Equipment under test (EUT)
Evolutionary algorithm 589, 590
EVs. *See* Electric vehicles (EVs)
Extrusion heatsinks 378

f

Failure-in-time (FIT) 269–270
Failure mode and effect analysis (FMEA) 60–62
Failure rate, of reliability 268
  capacitors 271–272
  for magnetic components 273–274
  of power device 269–271
Fast Fourier transformation (FFT) 201, 338, 339
FEA simulation. *See* Finite element analysis (FEA)
    simulation
Ferrite cores 108
  inductance curve with current bias for 112
  permeability curves for 111, 112
FFT. *See* Fast Fourier transformation (FFT)
Field-programmable gate array (FPGA) 414, 415
Fill factor 108, 113, 138, 154, 161, 336,
    604, 611
Film capacitors 101, 279
  dv/dt capability of 93
  lifetime of 271, 272
  mica 76
  overvoltage capability of 88, 89
  paper 76
  poly-film 77
  3D FEA simulation model of 94, 95
  in three-level ANPC converter 494
Filters 17–18. *See also* EMI filters
Finite element analysis (FEA)
      simulation 127, 487
  custom-designed heatsinks 392
  of film capacitors 94, 95
  stray inductance with 480
  thermal impedance 95–96

3D model-based 490, 503
First-generation thermal materials 380, 381
FIT. *See* Failure-in-time (FIT)
Flux density 113
  constraint 333
  fringing effect on 124–126
  margin 604
  peak 125–127
  saturation 106–108, 113, 125, 232
Flux waveform coefficient (FWC) 118
Flying capacitor multilevel converter 11, 12,
    283–285
FMEA. *See* Failure mode and effect
    analysis (FMEA)
Folded fins, heatsinks 379
Forced-air convection 371
Forced-air cooling system thermal
      model 383–385, 388
Forced-liquid convection cooling 371–372,
    405–407
  closed-loop 401–403
  cold plate thermal resistance/pressure
      drop 402, 403
  high-power three-phase AC converters 400
  open cooling system 401
Foster model. *See* Partial-fraction circuit;
    Thermal impedance model
Four-switch VSC (4Sw-VSC) 275–277
FPGA. *See* Field-programmable gate
    array (FPGA)
Frequency-dependent permeability 111–112
Frequency-domain impedance 138
Frequency ride-through (FRT) 562, 563
Fringing effect
  on core loss 127
  on inductance and flux density 124–126
Front-end rectifiers
  current density margin, wire 603
  DC inductor design conditions 603, 604
  design constraints 603, 605
  design optimization 604–606
  design variables 603, 605
  enclosed volume 603
  flux density margin, core 604
  high-level user interface for 603, 604
  inductor core parameters 606
  load-side inverter to 236
  optimizer formulation 597
  rectifier input current 606
FWC. *See* Flux waveform coefficient (FWC)

g

GA-based optimization tool
  EMI filter
    application examples   609–615
    optimizer formulation   598–600
  front-end rectifier
    application examples   603–606
    optimizer formulation   597
  inverter and thermal management system
    application examples   607–608
    optimizer formulation   597–599
  optimizer development
    design tool environment   600
    optimization and analysis
      capabilities   600–603
  system optimization   614–617
Gallium nitride (GaN)   17
  HEMTs   66, 446
  transistor, discrete package of   64, 65
Galvanic isolation   415, 416
GaN. *See* Gallium nitride (GaN)
Gap-filling materials   382
Gapped cores, fringing effect of   124
  on core loss   127
  on inductance and flux density   124–126
Gapped three-phase inductor core   110
Gate charge, power semiconductor
    devices   49–50
Gate control parameters   169
Gate drive board   480, 481
Gate drive IC   426–427
Gate driver (GD)   555
  bootstrap   429–430
  circuit with detailed components   422, 423
  decoupling capacitor   428–429
  definition of   418
  dynamic characteristics   420, 422, 423
  functional blocks of   419–420
  gate drive IC   426–427
  gate resistor   427–428
  isolated power supply   424–426
  motor drives   418, 419
  in phase-leg-based VSC   418–419
  signal isolator   423–424
  static characteristics   420–422
Gate driver power supply (GDPS)   555
Gate loop equivalent circuit   459
Gate resistor   427–428
Gate-to-source voltage   45, 207
Gate turn-off (GTO) thyristors   33

Gate voltage fall   51, 52
GD. *See* Gate driver (GD)
GDPS. *See* Gate driver power supply (GDPS)
Generalized Steinmetz equation (GSE)   116–117
Genetic algorithms (GAs)   589
Gradient-based algorithm   588–590
Grid applications
  baseline design (*See* Baseline design, of grid
      application)
  converter hardware design,
      requirements on   567
    AC filter inductor design   569
    DC-link capacitor design   567–568
    device selection   569
    frequency and angle change ride-
        through   567
    grid faults   567
    lightning surge   567
    size and weight comparison   570–571
    voltage ride-through   566–567
  faults impact   559–562
  frequency ride-through impact   562–563
  grid voltage angle change   562–563
  grounding design   573–576
  high and low-voltage ride-through impact
    DC-link voltage   558–559
    inrush current   557–558
    power delivery   559
    requirements   557, 558
  lightning surge
    composite waveform hypothesis (CWH)
        model   118–119
    converter insulation   565–566
    DC-link overvoltage   565
    inrush current   563–565
  power conditioning system converter   547–548
  transformer *vs.* transformer-less
      design   576–577
  unbalanced source and load   571–573
Grid faults   567
Grid-forming inverter, control structure
    of   264, 265
Grid voltage angle change   562, 563
Ground fault detection   451
Grounding effect
  EMI filter   344–345
    impedance effect   344, 345
    one-point *vs.* multipoint, for multistage
        filter   344–346
    grid-tied three-phase AC converters   573–576

Grounding effect (*cont'd*)
   motor drive   544–546
GSE. *See* Generalized Steinmetz equation (GSE)
GTO thyristors. *See* Gate turn-off (GTO) thyristors

h

Hall sensors   444
Hardware control layer   414
Harmonics, source side   247
  in cables   513
  core loss   512–513
  current   251, 252–263, 509–511
  dead-time   512
Heating, ventilation, and air
     conditioning (HVAC)   4
Heat pipe cooling system   373–374
Heatsinks   385
  commercial off-the-shelf   388–390
  custom-designed   389–393
  database   608
  dimensions   393
  fan/blower, pump, duct, pipe, and
     exchanger   383–384
  fan cooling design   394
  for forced-air cooling system   388
  geometrical configuration of   384
  geometry design   397
  with graphite fins   381–382
  for natural convection   370, 371
  temperature   400
  thermal interface material   382–383
  thermal materials   380–382
  types and configurations   377–380
HEMTs. *See* High-electron-mobility
    transistors (HEMTs)
Herbert curve   119
Heuristic algorithm   588
High-density inverters tool
  component models   622, 623
  converter design specifications   633
  database construction and
     management   627, 629
  design outcomes   634
  design procedure/algorithm
    converter-level   620–622
    system-level   618, 620
  function and architecture   616–619
  switching-function-based simulation   622
  user interface and tool configurations   622–629
  visualization function   627, 630–633
  weight breakdown, of design   633, 634

High-electron-mobility transistors (HEMTs)
  gallium nitride   66, 446
  junction capacitances in   46, 47
  on-state characteristics   38–40
  output characteristics   35, 36
  transfer characteristics of   42
High-frequency motor and cable models   231–232
High power supply   3–4
High-temperature gate bias (HTGB)   63
High-temperature gate switching (HTGS)   63
High-voltage controllable source   45
High-voltage ride-through (HVRT)   566–567
  DC-link voltage   558–559
  inrush current   557–558
  power delivery   559
  requirements   557, 558
High-voltage sensors   432–434
HTGB. *See* High-temperature gate bias (HTGB)
HTGS. *See* High-temperature gate
    switching (HTGS)
HVAC. *See* Heating, ventilation, and air
    conditioning (HVAC)
HVRT. *See* High-voltage ride-through (HVRT)
Hybrid optimization problem   586
Hysteresis loss density   115

i

I-DEAS FEA software   397
IEEE Std. 1547 standard   556, 557, 559, 562
IF filter model. *See* Intermediate-frequency (IF)
    filter model
IGBTs. *See* Insulated gate bipolar
    transistors (IGBTs)
IGCTs. *See* Integrated gate-commutated
    thyristors (IGCTs)
iGSE. *See* Improved generalized Steinmetz
    equation (iGSE)
iiGSE. *See* Improved iGSE (iiGSE)
ILT. *See* Inverse Laplace transform (ILT)
Improved generalized Steinmetz equation
    (iGSE)   117, 119
Improved iGSE (iiGSE)   118
Incremental permeability   111
Inductance   110–111
  CM and DM   232
  DC current-bias-dependent permeability   112
  frequency-dependent permeability   111–112
  fringing effect on   124–126
  LCL filter   255
Inductive coupling, in Π-type filter   346–347
Inductors   136

common mode
  core saturation and temperature
    rise   333–334
    equivalent circuit   333
    impedance in filter design   334–336
  core parameters   606
  DC/single-phase AC   153, 154
  decoupling   169, 170, 222–230
  differential mode   336, 365
  energy stored in   151–152
  fill factor   108, 113, 122, 129, 138, 154, 161, 164,
    325, 336, 604, 610–611
  filter baseline design   551–554
  for power converters   108–109
  related constraints   154–157
    A_p-based design method   154
    power loss and temperature rise
     155–157
  three-phase, with manual and automated
    approaches   159
  volume   158
  winding capacitance   127–128
  winding loss   120–121
Input capacitance   47, 48, 212, 422
Inrush current   557–558
  device related constraints   149–154
  grid faults   560–561
  lightning surge   563–565
  temporary overvoltage   559
Instrumentation amplifier (in-amp)   444–445
Insulated gate bipolar transistors (IGBTs)   10,
    16, 17
  with antiparallel diode   200–201
  capacitance for   205
  heatsink thermal resistance   376
  junction capacitances in   46, 47
  junction temperature variation of   203
  lifetime of   270
  on-state characteristics   38–40
  output characteristics   35, 36
  silicon based   33
  switches   521
  switching loss   196
  transfer characteristics of   42
  two-level VSI based on   189–190, 200
Insulation, lightning surge   565–566
Integrated gate-commutated thyristors
    (IGCTs)   33, 544
Integrating device models, for converter design   71
Interface design   24
Interleaved converters   11–13

Intermediate-frequency (IF) filter model   339
Internal thermal conduction, passive
    components   403
Interphase inductors   12
Inverse Laplace transform (ILT)   204, 532
Inverter
  application example for   607–608
  CM noise   636, 638–641
  design variable   635–637
  mechanical system   502
  optimizer formulation   597–599
  passive rectifier   162
  power characteristics   238
  voltage margin for   199
  thermal management system   398
  weight and loss breakdown   636–638
I-shaped three-level converters   288–290
Isolated power supply   417–419
  gate driver   424–426
Isolation
  amplifier (*See* DC bias)
  isolated power supply   417–418
  for low-power converter design   418
  in power converter design   415
  signal isolator   415–417
  voltage sensing, DC bias on   434–436

j

Jet impingement   374
JFET. *See* Junction-gate field-effect
    transistor (JFET)
Junction capacitance   46–48
Junction-gate field-effect transistor (JFET)
    441
Junction-to-case temperature, procedure
    of   201–204
Junction-to-case thermal resistance   58

k

Kapton films   77
Kapton sheets/epoxy   497
Karenbauer transformation   534
Koch approach   128
Kool Mμ core   107

l

Ladder network. *See* Continued-fraction circuit
Laminated cores   105–107
Laplace transform (LT)   94, 532
  for transient overload conditions   204
  voltage overshoot through   205–206

LC filter
  common mode   351
  differential mode   351, 352
  second-order   318
  single-stage/multiple-stage   320, 343
  two-stage   345
LCL filter
  CM voltage   256
  filter capacitor   255
  inductance values   255
  modulation index and PWM voltages   254
  peak-to-peak ripple current   258–262
  per-phase equivalent circuit   253, 254
  resonant frequency   262–263
  source-side three-phase voltages   255–256
  SPWM switching cycle   258–260
  VSC-based active rectifier
    SPWM reference phase voltage waveforms
      for   256–257
    switching function of   255
Lead lanthanum zirconium titanate
    (PLZT)   78–80, 82
Leakage current   43–45
Leakage inductance   123–124
L filter. *See* AC line filters
LFT. *See* Low-frequency transformer (LFT)
Life cycle cost (LCC) function   585
Lifetime model, for capacitor   96–98
Lifetime, of reliability   268
  capacitors   271–272
  for converter power stage   279
  for magnetic components   273–274
  of power device   269–271
Lightning surge, grid applications
  composite waveform hypothesis (CWH)
      model   118–119
  converter insulation   565–566
  DC-link overvoltage   565
  inrush current   563–565, 567
Linear optimization problem   586
Line impedance stabilization network
    (LISN)   208, 306, 309, 317
  CM current of   359–362
  CM/DM impedance of   322
  EMI noise measurement   312–314
  impedance characteristics   332
  time-domain voltage waveforms on   338
Line-to-ground capacitors. *See* Class-Y capacitors
Line-to-line capacitors. *See* Class-X capacitors
Liquid immersion   372–373, 405

Liquid nitrogen (LN) cooling   407
LISN. *See* Line impedance stabilization
    network (LISN)
Litz wires   120–121
Load efficiency   171
Load impedance   171
Load power   171
Load power factor   171
Load-side inverters
  AC load harmonic current (*See* AC load
      harmonic current)
  control performance   198–200
  control system   237–238, 240
  decoupling capacitor   218–222
  decoupling inductor   222–230
  design optimization   230–236
  design problem formulation
    conditions   170–172
    constraints   169–170
    DC-link capacitor   168
    motor drive with two-level VSI   167–168
    objectives   172–173
    optimization   173, 174
    variables   168–169
  device maximum junction
      temperature   200–204
  EMI filter   236, 239
  interfaced subsystems   236–238
  mechanical system   236–237, 239
  power loss   188–190
    conduction loss   190–195
    switching loss   195–198
  rectifier/DC source   236, 238–239
  switching overvoltage   204
    turn-off overvoltage   205–212
    turn-on overvoltage   212–217
  thermal management system   236
  two-level VSI, AC-fed motor drive
      with   167, 168
Loss and thermal model   93–96
Loss map methods   119
Loss separation approach   115
Low-frequency transformer (LFT)   6
Low-power converter design, isolation strategies
    for   418
Low-voltage ride-through (LVRT)   566–567
  DC-link voltage   558–559
  inrush current   557–558
  power delivery   559
  requirements   557, 558

LRC series resonant networks   222
LT. *See* Laplace transform (LT)

m

Magnetic relaxation process   118
Magnetics
  core loss   114
    empirical approach   115–120
    loss separation approach   115
  core window area product   114
  current density   113–114
  cut cores   105–107
  definition of   105
  failure rate and lifetime models for   273–274
  ferrite cores   108
  fill factor   113
  flux density and core saturation   113
  gapped cores, fringing effect of   124–127
  inductance and permeability   110–111
    DC current-bias-dependent
      permeability   112
    frequency-dependent permeability   111–112
  inductors, for power converters   108–109
  laminated core   105–107
  leakage inductance   123–124
  liquid cooled   405–407
  powder cores   107
  relevance to converter design   127–128
  tape wound core   105–107
  temperature rise   121–123
  types of   105
  winding loss   120–121
Magnetomotive force (MMF) ratio   603
Manufacturing technology   20
Massarini approach   128
Maximum junction temperatures
  devices   57, 173
  of inverter   200–204
  thermal management system   399–400
Maximum regenerative energy   171
Maximum thermal impedance
  for device junction temperature   200–204
  for thermal management system   385
Mean-time-between-failures (MTBF)   268, 279
Mechanical assembly   19–20
Mechanical characteristics, of power
    devices   63–65
Mechanical system   24
  busbar design (*See* Busbar design, mechanical
    system)

cabinet   473
connectors   475
current distribution   239
description of   471
design problem formulation   475
  conditions   477–478
  constraints   476–477
  DC busbar   478–480
  device and gate drive   480, 481
  device cooling   480–481
  mechanical layout and support   481–482
  objectives   478
  PCBs   478, 479
  variables   475–476
drawers   473
electrical/optical cables and wires   475
fasteners   475
interfaced subsystems
  current distribution and power loss   504
  EMI filter   502–503
  output from   503
  parasitic capacitances   504
  parasitic inductances   503–504
  rectifier and inverter   502
  temperature range   504
  thermal management system   503
inverter power stage design to   236–237
load-side inverter power stage design   169
medium-voltage three-phase AC/DC/DC power
    converter   471–473
passive rectifier   162
supporting boards and spacers/
    standoffs   474–475
thermal management system   398
voltage distribution   239
Medium-voltage (MV)
  DC-link voltage   561, 562
  grid-tied three-phase converters   547
  PCB in   453, 479
  three-phase AC–DC–DC power converter   472
  voltage considerations for   28
Metallized polypropylene (MPP) capacitors   98
  failure and self-healing capability of   97
  failure modes, mechanisms, and critical
    stressors   96
Metal-oxide-semiconductor field-effect transistors
    (MOSFETs)   10, 16
  bootstrap gate driver   429
  breakdown/leakage testing   45
  conduction loss   190, 192

Metal-oxide-semiconductor field-effect transistors (MOSFETs) (*cont'd*)
  grid faults impact on selection of   567
  inrush process through   560
  junction capacitances in   46, 47
  on-state characteristics   38–40
  output characteristics   35, 36
  silicon based   33
  silicon carbide   222, 224 (*See also* Silicon carbide (SiC))
  switching loss   196
  transfer characteristics of   42
  turnoff switching overvoltage   206, 207
    current falling stage   209–212
    cutoff (*See* gate voltage fall)
    flowchart for   210
    turnoff delay stage   207–208
    voltage rise stage   208–209
  turn-on switching overvoltage
    current rise stage   212–213
    flowchart for   216
    ringing stage   214–217
    turn-on delay stage   212
    voltage falling stage   213–214
  VSI based on   190
Method of feasible direction (MFD)   588
Mica capacitors   76
Microchannel cooling method   375
Miller capacitance. *See* Reverse transfer capacitance
Miller voltage   207, 213
Mini-channel liquid convection   375
Mixed-mode (MM) noise, in EMI filter   350–353
MLCC. *See* Multilayer ceramic capacitor (MLCC)
MMC. *See* Modular multilevel converter (MMC)
MMF ratio. *See* Magnetomotive force (MMF) ratio
ModelCenter®   600, 601, 603
Modified Steinmetz equation (MSE)   116
Modular multilevel converter (MMC)   11, 12
Modulation schemes, EMI filters   354–356
Molypermalloy powder (MPP) cores   107
MOO method. *See* Multi-objective optimization (MOO) method
More-electric aircraft (MEA) electrical system   5, 6
MOSFETs. *See* Metal-oxide-semiconductor field-effect transistors (MOSFETs)

Motor drives   2–3. *See also* AC-fed motor drives; DC-fed motor drive
  AC-fed general-purpose industrial (*See* GA-based optimization tool)
  with active rectifier, equivalent CM circuit for   330, 331
  CM voltages
    bearing current   516–517, 520–523
    detrimental effects   513
    drive inverter-motor side   515
    PWM drive inverter   514
    shaft voltage   516–519, 521–523
    stator winding capacitance coupling   515
  control architecture   198
  with diode rectifier   330
  dv/dt filter   537–542
  with EMI/EMC standards   137
  function block of   16, 17
  gate driver   418, 419
  grounding design   544–546
  harmonics
    in cables   513
    core loss   512–513
    current   509–511
    dead-time   512
  sinewave filters   542–544
  sinusoidal *vs.* PWM voltages   507–508
  system optimization for   590–591
  terminal filters   531–537
  terminal overvoltage (*See* Motor terminal overvoltage)
Motor model, EMI filter   337–338
Motor terminal filters
  characteristic impedance   531–533
    three-phase cable characteristics   533–536
    for three-phase shielded cables   536–537
    for unshielded cables   536
  configuration   531
  inductance and capacitance   533
  voltage and current waves   531–532
Motor terminal overvoltage
  boundary conditions   529
  coil impulse voltage envelope   527
  distributed winding circuit model   528
  equivalent winding circuit   528
  external filters   531
  normalized initial distribution   530
  reflected wave   524–525
  turn-to-turn voltage   530

voltage and current waves 523–524
voltage distribution 528–530
voltage doubling effect 524, 527
MPP capacitors. *See* Metallized polypropylene
(MPP) capacitors
MPP cores. *See* Molypermalloy powder
(MPP) cores
MSE. *See* Modified Steinmetz equation (MSE)
MTBF. *See* Mean-time-between-failures (MTBF)
Multi-domain virtual prototyping method 642
Multilayer ceramic capacitor (MLCC) 76, 78
failure and self-healing capability of 97
failure modes, mechanisms, and critical
stressors 96
ferroelectric materials used in 78–79
Multilevel converters 11, 12
Multi-objective optimization (MOO)
method 585–586
Multiphase converters 13
Multistage filters
coupling in 349
EMI 323–324
Multi-time scale electro-thermal simulation 642

n

Nanocomposites 106–107
Nanocrystalline materials/cores 106–107
inductance curve with current bias for 112
permeability curves 111, 112
NanoLam capacitors 80
Nanostructure multilayer capacitors 80
Natural convection 370–371
Natural Steinmetz equation (NSE) 117–118
Nave's method, to predict leakage
inductance 123
N-dimensional optimization problem 587
Nearest three space vector (NTSV) 342
Near state PWM (NSPWM) 354–356
Negative sequence current 169, 247, 572, 573
Negative Temperature Coefficient (NTC)
thermistor 432
Neutral busbar, surface current density
of 498, 499
Neutral-forming-transformer (NFT) 571, 572
Neutral-point clamped (NPC) 518
multilevel converter 11, 12
three-level converter 546
Neutral-point-clamped voltage-source converter
(NPC-VSC) 276, 277

NFT. *See* Neutral-forming-transformer (NFT)
Nickel/iron powder core 107
Nomex 410 paper 553
Nominal load power 138
Nonconvex optimization problem 587
Nondeterministic polynomial time (NP) hard
problem 587, 588
Nonideal capacitor equivalent circuit model 84
Nonlinear optimization
algorithms 587–589
problem 586
Nonstandard characteristics, of power
devices 66–67
Norton's Theorem 283
NPC. *See* Neutral-point clamped (NPC)
NPC-VSC. *See* Neutral-point-clamped voltage-
source converter (NPC-VSC)
NSE. *See* Natural Steinmetz equation (NSE)
NSPWM. *See* Near state PWM (NSPWM)
NTC thermistor. *See* Negative Temperature
Coefficient (NTC) thermistor
NTSV. *See* Nearest three space vector (NTSV)
Nusselt number 391

o

Objective function 22, 584
One-dimensional (1-D) calculation model 120
One-dimensional optimization problem 587
On-state characteristics 38–41
Operating junction temperature, of power
device 58, 59
Optimization software interface 600, 601
Optimizer formulation
EMI filter 598–600
front-end rectifier 597
inverter and thermal management
system 597–598
Original Steinmetz equation (OSE) 115–116
Orthonol 106
OSE. *See* Original Steinmetz equation (OSE)
Output capacitance 47, 48, 205, 296
charge-based equivalent 299
charging/discharging 459
equivalent 221
loss calculation 290–291
MOSFET 218
Output characteristics 35–38
Overcurrent protection 446–448,
450–451

Overlap energy loss   291–292
Overload   138
Overtemperature protection   451
Overvoltage capability, of film capacitor
        88, 89
Overvoltage protection   451

p
Paper capacitors   76
Paralleled capacitors layout   99, 100
Parasitic capacitances   46, 400
    busbar   490
    cable and motor   229
    common mode   231, 361
    of heatsink   398
    mechanical system   504
    of motors   517, 518, 520
    motor neutral-to-ground   282
    power stage   358
    thermal management system   400
    winding   129
Parasitic grounding capacitance   350, 351
Parasitic inductances
    capacitor bank   99–100
    busbar   489–490
    DC-link capacitor   218–220
    loop, of busbar   482, 483, 502
    mechanical system   503–504
    voltage across   447–448
Parasitic resistance
    busbar   490
    DC bus   220
    of inductors   253
Pareto front
    in design optimization   585–586
    partitioned optimizers   593–594
    single optimizer   594, 595
Partial discharge extinction voltage (PDEV)
        level   476
Partial discharge inception voltage (PDIV)
        level   476
Partial element equivalent circuit
(PEEC)   622, 642
Partial-fraction circuit   58
Particle swarm optimization (PSO)   589
Partition-based design, *vs.* whole converter
        design   635–641
Partitioned optimizers
    aircraft application, two-stage power
        supply in   591
    converter-level design variables   593

Pareto front   593–594
    system-level optimal design   590–591
    three-level optimization diagram   591–593
Passive component technology   20
    cooling system for   403
        air convection cooling   404–405
        internal thermal conduction   403
        liquid convection cooling   405–407
Passive rectifiers
    AC input harmonic current   141–148
    capacitor-related constraints   157
    classifications   161–162
    component loss   163
    DC-link stability   149
    DC-link voltage   148
    design problem formulation   589
        conditions   138
        constraints   136–138
        motor drive front-end diode
            rectifier   135–136
        objectives   139–140
        optimization   139, 140, 157–161
        variables   136
    device selection-related constraints   149–154
    diode-based   133–135
    EMI bare noise   163
    inductor-related constraints   154–157
    inrush   149–154
    interfaced subsystems   162, 163
    models   140–141
    ride-through/holdup time without input
        power   149
    topologies   133, 134
    voltage distribution   163–164
PCBs. *See* Printed circuit boards (PCBs)
PCS converter. *See* Power conditioning system
        (PCS) converter
Peak current, inductor   125
Peak flux density, inductor   125, 126
Peak overvoltage   220
Peak power, load-side inverter   171
Peak power, passive rectifier   138
Peak value, of envelope waveform   340
PEEC. *See* Partial element equivalent
        circuit (PEEC)
PEN. *See* Polyethylene naphthalate (PEN)
Penalty function-based algorithm   588
Permalloy   106
Permanent magnet (PM)
    generator   8, 9
    motor, equivalent circuit of   511

Permeability 110–111
  DC current-bias-dependent permeability 112
  frequency-dependent permeability 111–112
PFC. *See* Power factor correction (PFC)
Photovoltaic (PV) utility-scale system 9
Physical constraints 136–138
Physical design 24, 25
Physical environmental impact
      technology 20
Π-type filter, inductive coupling, in
      346–347
Π-model. *See* Thermal impedance model
PLZT-based ceramic capacitors 78–80, 82
PM. *See* Permanent magnet (PM)
Polycarbonate film capacitors 77
Polyester (PET) 89–90
  film capacitors 77
Polyethylene naphthalate (PEN) 89–90
Poly-film capacitors 77
Polyphenylene sulfide 77
Polypropylene (PP) 90
  film capacitors 77
Polystyrene 77
Pool boiling system 372–373
Powder cores 107
Power conditioning system (PCS) converter
  filter inductor 552
  insulation 565, 566
  maximum AC terminal voltage of 564
  with solid grounding 565
  specifications and grid requirements 547–548
  stop operation during fault 561
  three-phase AC-DC-DC 547
  transformer-less design 576
Power converter design optimization 586–587
Power factor 137
Power factor correction (PFC) 3, 594
Power loss
  active rectifiers/source-side inverters 247
  inductor-related constraints 155–157
  inverter 169, 188–190
    conduction loss 190–195
    switching loss 195–198
  mechanical system, busbar 504
Power module selection, in three-level ANPC
      converter 493–496
Power semiconductor devices
  active power switch, transfer characteristics
      of 41–43
  gate charge 49–50
  junction capacitance 46–48

leakage current and breakdown voltage
      43–45
  mechanical characteristics 63–65
  nonstandard characteristics 66–67
  on-state characteristics 38–41
  output characteristics 35–38
  relevance to converter design 70–72
  reliability characteristics 60–63
  safe operating area 60, 61
  scalability, parallel/series 68–70
  silicon based 33
  switching characteristics 50–57
  thermal characteristics 57–59
  wide bandgap 34
Power stage. *See* Converter main circuit
Power switching network technology 20
Power switch technology 20
Prandtl number 391
Precharge circuit 554–555
Printed circuit boards (PCBs) 99, 452–455
  based busbar 478, 479
  ground plane on 345
  mechanical system design 478, 479
Prototype busbar 500–501
PSO. *See* Particle swarm optimization (PSO)
PSpice model 296
Pulsed V-I testing 37–38
Pulse-width modulation (PWM) 168
  angular frequency of 198
  with commutations per switching cycles 513
  conduction losses 192, 193
  drive inverter 514, 515
  generations 415
  with high switching frequency 195
  inverters, closed-form harmonic solutions
      for 175–176
  mask method 558, 564, 566
  modulations schemes 616, 618
  phase voltages 281, 282
  phase voltages *vs.* sinusoidal voltages
      507–509
  reference voltage 179
  reliability-oriented design 277
  third-order harmonic injection function 194
  three-level modulator, space vector
      diagram of 514
  voltage harmonic spectrum 296–299
  voltage modulation index 512–513
  voltages 254
  VSI-generated CM voltages 516
PWM. *See* Pulse-width modulation (PWM)

q

Quasi-peak detector   340, 341

r

Radio-frequency interference (RFI)
  Hall sensors   444
  in in-amp and mitigation   444–445
  low-frequency distortion   439–442
  mitigation approaches   442–444
  sensor output   438–441
Random PWM (RPWM)   354
Rate of change of frequency (ROCOF)   562, 563
Reactive power   248
Rectifier
  active (*See* Active rectifiers/source-side
      inverters)
  mechanical system   502
  passive (*See* Passive rectifiers)
  thermal management system   398
Reference voltage vector   180
Reinforced isolation   416
Relevance to converter design   70–72
Reliability   248
  component models   269–274
  consideration in design   269
  converter design   267
  failure rate and lifetime   268
  MTBF   268
  reliability-oriented design   267
    components selection   277–278
    design comparison   278–279
    operating conditions   274–275
    switching frequency and PWM scheme   277
    VSC topology, two-level   275–277
Reliability characteristics, of power
      devices   60–63
Renewable energy systems   6, 8–9
Resistance temperature detectors (RTDs)   432
Reverse transfer capacitance   47, 448, 459
Reversible permeability   111
Reynolds number   391–392
RFI. *See* Radio-frequency interference (RFI)
Ride-through/holdup time, without input
      power   137, 149
Ringing stage, MOSFET   214–217
RMS voltage/current   89–92, 146–148, 551
RPWM. *See* Random PWM (RPWM)
RTDs. *See* Resistance temperature
      detectors (RTDs)

s

Safe operating area (SOA)   60, 61
Safety capacitors   101
Scalability, of power devices   68–70
Schottky diode   27, 34, 35, 212, 215, 293
Schwarz-Christoffel Transformation   125
Second-generation thermal materials   380, 381
Semiconductor devices, baseline design   549
Series capacitors layout   99, 100
SFA. *See* Single Fourier analysis (SFA)
Shaft voltage
  EDM current   521, 522
  evaluation   517–519
  measurement   516
  mitigation   521–523
Shielding, of high-speed signal traces   434
Shipboard electric power system   5
Short-circuit protection   445–448
SiC. *See* Silicon carbide (SiC)
Signal isolator   415–417, 419
  gate driver   423–424
Signal-to-noise ratio (SNR)   434
Silicon (Si) based semiconductor switching
      devices   33
Silicon carbide (SiC)   17
  MOSFET   222, 224, 569
    gate drive   480, 481
    structure of   34
    switching transient waveforms of   428, 498
Silicon-controlled rectifier (SCR). *See* Thyristor
Simplified capacitor equivalent circuit model   84
Sine-triangle PWM (SPWM)
  AC load harmonic current   176–179
  conduction loss   191, 192
  reference phase voltage waveforms for
      VSC   256–258
  switching cycle   258, 259
  voltage harmonics comparison   185–186
Sinewave filters   542–544
Single Fourier analysis (SFA)   177
  voltage harmonics comparison using   185–189
Single optimizer   594–595
Sinusoidal transition square (STS) voltage
      waveforms   118
Si steel   106
Small-signal analysis approach   221–222
SNR. *See* Signal-to-noise ratio (SNR)
SOA. *See* Safe operating area (SOA)
Soft magnetic alloy   105–107

Soft-switched converters 13–14

Source-side inverters. *See* Active rectifiers/source-side inverters

Space vector modulation (SVM) 176, 354, 355
  centered continuous 193
  continuous
    normalized phase reference voltage for 182
    phase reference voltages and duty
      cycles 261
    phase reference voltage waveforms for 181
    phase voltage switching pattern for 181
  discontinuous
    normalized phase reference voltage
      for 183, 184
    phase reference voltage waveforms 182, 183
    double Fourier integral limits for 328
    voltage harmonics comparison 186–187

Space vector pulse width modulation
    (SVPWM) 354, 356

SPG. *See* Steinmetz premagnetization graph (SPG)

Spray cooling 374–375

SPWM. *See* Sine-triangle PWM (SPWM)

Stamped heatsinks 378

Static characteristics, of power devices
  active power switch, transfer characteristics
    of 41–43
  equivalent device 68–69
  gate charge 49–50
  junction capacitance 46–48
  leakage current and breakdown voltage 43–45
  on-state characteristics 38–41
  output characteristics 35–38

Steinmetz equation 115, 116, 155

Steinmetz premagnetization graph (SPG) 119

Supercapacitor 80–81

Supermendur 106

Surface current density, of neutral
    busbar 498, 499

SVM. *See* Space vector modulation (SVM)

SVPWM. *See* Space vector pulse width
    modulation (SVPWM)

Switching characteristics, of power
    devices 50–57
  datasheet 54, 55
  double pulse test 55–57
  model for 50, 51
  switching trajectory 51, 53
  turn-on/turn-off 52–54
  waveform for 51, 52, 54

Switching control layer 413

Switching devices 17, 168

Switching frequency 169
  corner frequency *vs.*, 342–343
  DC bias 437
  design procedure 25
  EMI filters 354
  reliability-oriented design 277
  voltage harmonics comparison using 185–189
  weight comparison at different 632

Switching-function-based simulation 622

Switching functions
  AC input harmonic current 141–143
  simulation-based model 329–330

Switching loops
  busbar layout 3D view for 496
  equivalent circuits of 496–497
  in three-level ANPC converter 491–493

Switching loss
  DPT data 293–294
  gate voltage and drain-source voltage 290
  in inverter 195–198
  output capacitance calculation 290–291
  overlap energy calculation 291–292
  switching energy *vs.* DPT data 292

Switching overvoltage 204
  turn-off overvoltage 205–212
  turn-on overvoltage 212–217

Switching transitions, deadtime
    compensation 465

Switching waveforms
  diagram of 460
  during turn-off transient 458

4Sw-VSC. *See* Four-switch VSC (4Sw-VSC)

System conditions. *See* Design conditions

System constraints. *See* Design constraints

System control layer 411–412

System design 22–23

System-level design procedure/
    algorithm 618, 620

System-level optimization 614–617

System variables. *See* Design variables

t

Tantalum electrolytic capacitors (TECs) 78

Tape wound cores 105–107

TDD. *See* Total demand distortion (TDD)

TECs. *See* Tantalum electrolytic capacitors (TECs)

Teflon 77

Temperature rise 121–123
  capacitor 93, 94, 140, 551
  CM inductor 333–334
  inductor-related constraints 155–157
  thermal stress through 93–95
Temperature sensor technology 432, 433
THD. *See* Total harmonic distortion (THD)
Thermal characteristics, of power devices 57–59
Thermal conductivity 380, 381
Thermal impedance model 58, 94, 597
  measurement 59
  of power device 58, 59
Thermal interface material (TIM) 200, 382–383, 389
Thermal management system (TMS) 18, 24, 169
  active rectifier and source-side inverter 247
  application example for 607–608
  back-to-back VSI converter, weight distribution for 369, 370
  cooling technology (*See* Cooling technology)
  definition of 369
  design flowchart for 620–622
  design problem formulation
    conditions 386
    constraints 385
    forced-air cooling system 384, 385
    objectives 386–387
    optimization 386, 387, 393–397
    variables 385
    verification 397, 398
  device loss model 597
  EMI filter design constraints 325
  forced-air cooling system thermal model 388
  heatsink dimensions 393
  interfaced subsystems
    control system 399
    EMI filter 398
    heatsink temperature 400
    inverter and rectifier 398
    maximum junction temperatures 399–400
    mechanical system 398
    output to 399
    parasitic capacitances 400
    voltage distribution 399
  inverter to 236
  losses and thermal calculations, flowcharts for 597–598
  mechanical system 503
  passive rectifier 162

  structure of 598, 599
  thermal capacitance 204, 389
  thermal impedance
    commercial off-the-shelf heatsinks 388–390
    custom-designed heatsink 389–393
    network model 597
    thermal resistance of 201, 204
Thermal materials 380–382
Thermal resistance 58, 94–95
Thermocouple 94, 95, 432
Third-generation thermal materials 380, 381
3D-printed aluminum air-cooled heatsink 379, 380
Three-phase AC/AC converter 2, 282
Three-phase four-wire converters 13
Three-phase inverter 1, 2, 11
Three-phase rectifiers 1, 2, 10
Three-phase shielded cables, characteristic impedance for 536–537
Three-phase to three-phase matrix converter topology 15–16
Threshold voltage 42, 49, 207, 212, 448, 449, 520
Thyristors 133, 135
  on-state characteristics 38–40
  output characteristics 35, 36
TIM. *See* Thermal interface material (TIM)
T-model. *See* Continued-fraction circuit
TMS. *See* Thermal management system (TMS)
Toroidal core 129
  fill factor 325, 336, 611
Total demand distortion (TDD) 136, 549
  AC source harmonic currents and 247, 252, 253
Total harmonic distortion (THD) 136, 169, 179
  AC current 232
  voltage results comparison 184
Transconductance 42, 207, 212, 422
Transfer characteristics, of active power switch 41–43
Transient overload conditions 204
T-type filter, capacitive coupling in 348
T-type three-level converter 287, 288
T-type Vienna-type rectifier 296, 298
Turnoff delay stage, MOSFET 207–208
Turnoff switching overvoltage
  MOSFET 206, 207
    current falling stage 209–212
    flowchart for 210

turnoff delay stage   207–208
  voltage rise stage   208–209
 overshoot and resonance   205–206
 in Si IGBT-based inverters   205
 WBG devices   206
Turn-on delay stage, MOSFET   212
Turn-on switching overvoltage   212–217
 MOSFET
  current rise stage   212–213
  flowchart for   216
  ringing stage   214–217
  turn-on delay stage   212
  voltage falling stage   213–214
Turn-to-turn capacitance   128
Two-dimensional electron gas (2DEG)
     channel   66

u

UI. *See* User interface (UI)
Unconstrained optimization problem   587
Undervoltage protection   451
Uninterruptible power supplies (UPS)   4
Unshielded cables, characteristic impedance
     for   536
UPS. *See* Uninterruptible power supplies (UPS)
User interface (UI)
 converter design configuration   623–627
 design parameters   627, 629
 load configurations   627, 628
 software tool   622–623
Utility-scale PV system   9

v

Variable frequency PWM (VFPWM)   354
VFC. *See* Voltage-to-frequency converter (VFC)
VFPWM. *See* Variable frequency PWM (VFPWM)
Vienna-type multilevel converter   11, 12
Vienna-type rectifier   135, 286–288, 293,
     296, 298
Virtual prototyping   638–642
Voltage and current capability model, capacitor
 maximum AC voltage/current *vs.*
     frequency   89–92
 maximum permissible continuous DC voltage
     vs. temperature   89
 overvoltage capability   88, 89
 pulse withstanding capability   92–93
Voltage balancing method   99
Voltage distribution

EMI filters   365
 thermal management system   399
 load-side inverter   239
 passive rectifier   163–164
Voltage doubling effect   524, 527
Voltage falling stage, MOSFET   213–214
Voltage margin, for inverter   199
Voltage ride-through   566–567
Voltage rise stage, MOSFET   208–209
Voltage sensor technology   431
Voltage source converter (VSC)   10–11, 14
 AC line filter   252, 256
 busbar layout   483
 control loop   557
 converter topologies comparison   276
 convertible   275–277
 diode NPC phase   288–289
 four-switch   275–277
 galvanic isolation   416
 MOSFET-based DC-link decoupling
     capacitor in   218
 multilevel, switching function-based simulation
     for   281–286
 nine-level flying capacitor   283, 285
 NPC-VSC   275–277
 phase-leg, gate driver in   418–419
 SPWM reference phase voltage waveforms
     for   256–257
 switching commutation   195
 switching frequency   276–277
 switching function of   255
 three-level fully regenerative IGCT   544
 topology   275
 two-level   280, 330, 350
  AC-fed motor drive with   245–246
  continuous SVM for   181
  DPWM phase voltage switching
      pattern   182, 183
  phase leg   327–328
  space vector diagram for   179–180
  switching function-based simulation for   281
 Z-source converter   275–277
Voltage source inverter (VSI)   11, 136–137
 back-to-back converter, weight distribution
     for   369, 370
 hard-switching   218
 phase-leg voltage, deadtime impact
     on   461, 462
 PWM motor drives   516

Voltage source inverter (VSI) (*cont'd*)
  two-level
    continuous SVM for   182
    on IGBT   189–190, 200, 203
    MOSFET   190
    motor drive with   167–168
    switching function of   175
Voltage stress, reliability   294–297
Voltage-to-frequency converter (VFC)   433
VSC. *See* Voltage source converter (VSC)
VSI. *See* Voltage source inverter (VSI)

w

Waveform coefficient Steinmetz equation
    (WcSE)   118
WBG. *See* Wide bandgap (WBG)
WcSE. *See* Waveform coefficient Steinmetz
    equation (WcSE)
WFSM. *See* Wound field synchronous
    motor (WFSM)
Whole converter design, partition-based design *vs.*,
    635–641
Wide bandgap (WBG)
  deadtime setting   455
  device protection   445–447

gate driver design   420
high-frequency converter (*See* DC bias)
power semiconductor device   1, 17, 19–20,
    34, 54
  device package for   64
  diode characterization of   57
  fast-switching   18
  turn-off switching overvoltage   206
  turn-on switching overvoltage   206
Winding loss   120–121
Wind turbine generators   8
Wound field synchronous motor (WFSM)
  damper windings   511
  equivalent circuit of   510

z

ZCS. *See* Zero-current switching (ZCS)
ZCT. *See* Zero-current transition (ZCT)
Zero-current switching (ZCS)   13
Zero-current transition (ZCT)   13–14
Zero-sequence current   571, 573,
    575
Zero-voltage switching (ZVS)   13
Z-source converter   275–277
ZVS. *See* Zero-voltage switching (ZVS)

Printed and bound by CPI Group (UK) Ltd, Croydon, CR0 4YY

16/04/2025

14658605-0003